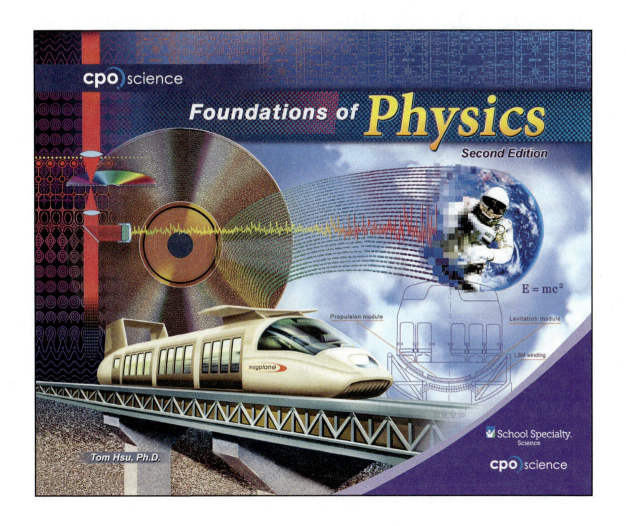

Author

Thomas C. Hsu, Ph.D., Applied Plasma Physics, Massachusetts Institute of Technology

Contributing Writers

Scott Eddleman, M.Ed., Harvard University
Patsy Eldridge, M.Ed., Tufts University
Erik Benton; Barbara Foster; Laine Ives

Editorial Team

Lynda Pennell – Senior Editor
Polly Crisman – Graphics Manager/Illustrator
Bruce Holloway – Senior Designer/Illustrator
Jesse Van Valkenburgh – Designer/Illustrator
Susan Gioia – Administrator
Sara Desharnais – Electronic Production Specialist
Lynn L'Heureux – Technical Consultant

Contributors

Stacy Kissel, Kristy Beauvais, Dr. Darren Garnier, Dr. David Guerra, Dr. Jeffrey Williams, Thomas C. Altman, Dr. Mitch Crosswait, Mary Ann Erickson, David H. Bliss, Wayne Brown. Editorial Consultant: Alan Hull

Equipment Design and Materials

Thomas Narro – Senior Vice President
Danielle Dzurik – Mechanical Engineer
Kathryn Gavin – Purchasing and Quality Control Manager

Reviewers

Dr. Jeff Schechter – Boston, MA
Dr. Tim Daponte – Houston, TX
Wanda Pagonis – Lylte, TX
Dr. Willa Ramsay – San Diego, CA
DeMarcus Wright – Grand Prairie, TX
Beverly T. Cannon – Dallas, TX
Betsy Nahas – Westford, MA
Neri Giovanni – San Antonio, TX

Scott Hanes – Hooks, TX
Bruce Ward – Boston, MA
Lee DeWitt – San Antonio, TX
Steve Heady – Houston, TX
Dr. Michael Saulnier – Boston, MA
Gigi Nevils – Houston, TX
Lebee Meehan – Houston, TX
Dr. Manos Chaniotakis – Cambridge, MA

Jay Kurima – Fort Worth, TX
James DeHart – Beaumont, TX
David Binette – Ithaca, NY
Rebecca DeLozier – Shady Shores, TX
Valerie Felger – New Braunfels, TX
Alan Hull – Atlanta, GA
David Bliss – Mohawk, NY

Foundations of Physics - Second Edition

Copyright © 2016 CPO Science

ISBN: 978-1-62571-836-5
Part Number: 1576080

Printing 2—May 2017
Printed by Webcrafters, Inc., Madison, WI

All rights reserved. No part of this work may be reproduced or transmitted in any form or by any means, electronic or mechanical, including photocopying and recording, or by any information storage or retrieval system, without permission in writing. For permission and other rights under this copyright, please contact:

CPO Science
80 Northwest Boulevard
Nashua, New Hampshire 03063
(800) 932-5227
www.cposcience.com

Foundations of Physics

Table of Contents

Introduction

Chapter 1: **The Science of Physics** 1
   1.1 The Science of Physics 2
   1.2 Scientific Inquiry and Natural Laws 16
   1.3 The Nature of Scientific Knowledge 22
Connection: Biomimicry and New Color Display Technology.. 27
Chapter 1 Assessment 29

UNIT 1 — Measurement and Motion

Chapter 2: **Measurement and Units** 31
   2.1 Space and Time 32
   2.2 Mass, Matter, and Atoms 38
   2.3 Experiments and Data 43
Connection: Nanotechnology 53
Chapter 2 Assessment 54

Chapter 3: **Position, Speed, and Velocity** 57
   3.1 Space and Position 58
   3.2 Graphs of Speed and Velocity 64
   3.3 Working with Equations 68
Connection: Slow-Motion Photography 73
Chapter 3 Assessment 75

Chapter 4: **Accelerated Motion in a Straight Line** .. 79
   4.1 Acceleration 80
   4.2 A Model for Accelerated Motion 86
   4.3 Free Fall and the Acceleration due to Gravity .. 90
Connection: Antilock Brakes 94
Chapter 4 Assessment 96

UNIT 2 — Motion and Force in One Dimension

Chapter 5: **Newton's Laws: Force and Motion** 99
   5.1 The First Law: Force and Inertia 100
   5.2 The Second Law: Force, Mass, and Acceleration 103
   5.3 The Third Law: Action and Reaction 109
Connection: Biomechanics 113
Chapter 5 Assessment 115

Chapter 6: Forces and Equilibrium 117
   6.1 Mass, Weight, and Gravity 118
   6.2 Friction 122
   6.3 Equilibrium of Forces and Hooke's Law 128
Connection: The Design of Structures 134
Chapter 6 Assessment 136

UNIT 3 Motion and Force in 2 and 3 Dimensions

Chapter 7: Using Vectors: Motion and Force 139
    7.1 Vectors and Direction 140
    7.2 Projectile Motion and the Velocity Vector 146
    7.3 Forces in Two Dimensions 154
Connection: Robot Navigation 160
Chapter 7 Assessment 162

Chapter 8: Motion in Circles 165
    8.1 Circular Motion 166
    8.2 Centripetal Force 170
    8.3 Universal Gravitation and Orbital Motion 174
Connection: Satellite Motion 177
Chapter 8 Assessment 179

Chapter 9: Torque and Rotation 181
    9.1 Torque 182
    9.2 Center of Mass 187
    9.3 Rotational Inertia 190
Connection: Bicycle Physics 194
Chapter 9 Assessment 196

UNIT 4 Energy and Momentum

Chapter 10: Work and Energy 199
    10.1 Machines and Mechanical Advantage 200
    10.2 Work 207
    10.3 Energy and Conservation of Energy 211
Connection: Hydroelectric Power 218
Chapter 10 Assessment 220

Chapter 11: Energy Flow and Power 223
    11.1 Efficiency 224
    11.2 Energy and Power 229
    11.3 Energy Flow in Systems 234
Connection: Energy from Ocean Tides 238
Chapter 11 Assessment 240

Chapter 12: Momentum 243
    12.1 Momentum 244
    12.2 Force is the Rate of Change of Momentum ... 250
    12.3 Angular Momentum 253
Connection: Jet Engines 257
Chapter 12 Assessment 259

UNIT 5 Waves and Sound

Chapter 13: Harmonic Motion 263
    13.1 Harmonic Motion 264
    13.2 Why Things Oscillate 270
    13.3 Resonance and Energy 275
Connection: Quartz Crystals 279
Chapter 13 Assessment 280

Chapter 14: Waves 283
    14.1 Waves and Wave Pulses 284
    14.2 Motion and Interaction of Waves 290
    14.3 Natural Frequency and Resonance 296
Connection: Freak Waves 302
Chapter 14 Assessment 304

Chapter 15: Sound 307
    15.1 Properties of Sound 308
    15.2 Sound Waves 313
    15.3 Sound, Perception, and Music 320
Connection: Sound from a Guitar 325
Chapter 15 Assessment 327

Table of Contents

UNIT 6 Light and Optics

Chapter 16: Light and Color 331
    16.1 Properties and Sources of Light 332
    16.2 Color and Vision 338
    16.3 Photons and Atoms 344
Connection: Color Printing 347
Chapter 16 Assessment 349

Chapter 17: Optics 351
    17.1 Reflection and Refraction 352
    17.2 Mirrors, Lenses, and Images 360
    17.3 Optical Systems 366
Connection: The Telescope 371
Chapter 17 Assessment 373

Chapter 18: Wave Properties of Light 377
    18.1 The Electromagnetic Spectrum 378
    18.2 Interference, Diffraction, and Polarization 383
    18.3 Special Relativity 388
Connection: Holography 394
Chapter 18 Assessment 396

UNIT 7 Electricity and Magnetism

Chapter 19: Electricity 399
    19.1 Electric Circuits 400
    19.2 Current and Voltage 404
    19.3 Electrical Resistance and Ohm's Law 408
Connection: Hybrid Gas/Electric Cars 414
Chapter 19 Assessment 416

Chapter 20: Electric Circuits and Power 419
    20.1 Series and Parallel Circuits 420
    20.2 Analysis of Circuits 426
    20.3 Electric Power, AC, and DC Electricity 430
Connection: Wiring in Homes and Buildings 435
Chapter 20 Assessment 437

Chapter 21: Electric Charges and Forces 439
    21.1 Electric Charge 440
    21.2 Coulomb's Law 446
    21.3 Capacitors 452
Connection: Rival Projector Technologies 456
Chapter 21 Assessment 458

Chapter 22: Magnetism 461
    22.1 Properties of Magnets 462
    22.2 Magnetic Properties of Materials 466
    22.3 Earth's Magnetic Field 469
Connection: Magnetic Resonance Imaging 473
Chapter 22 Assessment 475

Chapter 23: Electricity and Magnetism 477
    23.1 Electric Current and Magnetism 478
    23.2 Electromagnets and the Electric Motor 484
    23.3 Induction and the Electric Generator 489
Connection: Trains That Float by Magnetic Levitation 494
Chapter 23 Assessment 496

Chapter 24: Electronics 499
    24.1 Semiconductors 500
    24.2 Circuits with Diodes and Transistors 506
    24.3 Digital Electronics 510
Connection: Electronic Addition of Two Numbers 514
Chapter 24 Assessment 516

Table of Contents

UNIT 8 Matter and Energy

Chapter 25: Energy, Matter, and Atoms 519
    25.1 Matter and Atoms 520
    25.2 Temperature and the Phases of Matter 526
    25.3 Heat and Thermal Energy 534
Connection: The Refrigerator 538
Chapter 25 Assessment 540

Chapter 26: Heat Transfer 543
    26.1 Heat Conduction 544
    26.2 Convection 548
    26.3 Radiant Heat 552
Connection: Energy-Efficient Buildings 556
Chapter 26 Assessment 558

Chapter 27: The Physical Properties of Matter 561
    27.1 Properties of Solids 562
    27.2 Properties of Liquids and Fluids 569
    27.3 Properties of Gases 578
Connection: The Deep Water Submarine Alvin 582
Chapter 27 Assessment 584

UNIT 9 The Atom

Chapter 28: Inside the Atom 587
    28.1 The Nucleus and Structure of the Atom 588
    28.2 Electrons and Quantum States 596
    28.3 The Quantum Theory 602
Connection: The Laser 607
Chapter 28 Assessment 609

Chapter 29: Nuclear Reactions and Radiation 613
    29.1 Radioactivity 614
    29.2 Radiation 619
    29.3 Nuclear Reactions and Energy 625
Connection: Nuclear Power 631
Chapter 29 Assessment 633

Chapter 30: Frontiers in Physics 637
    30.1 The Origin of the Universe 638
    30.2 Gravity and General Relativity 642
    30.3 The Standard Model 646
Connection: Smash! The Large Hadron Collider 650

Glossary 653

Index 667

INTRODUCTION

CHAPTER 1

The Science of Physics

Throughout history, people have wondered: Why is the Sun hot? Why do things fall down? What is light? The natural world is full of mysteries. Over several thousand years of recorded history, wonderment became experiment, which slowly grew into knowledge. We now understand that nuclear reactions release heat in the Sun, that gravity pulls objects toward the center of Earth, and that light is a wave of electromagnetic energy. Physics is the science that describes nuclear reactions, light, and gravity. Physics also deals with sound, motion, forces, electricity, and every fundamental process in both the natural world and the world of human technology.

Objectives:

By the end of this chapter you should be able to:

- ✔ Describe what physics is about.
- ✔ Use the ideas of energy and electric current to explain how a battery lights a bulb.
- ✔ Explain how there can be many forces acting on an object that is not moving.
- ✔ Give examples of an oscillator and a wave.
- ✔ Describe the physical differences, other than color, between blue light and red light.
- ✔ Give an example of how hot matter is different from cold matter.
- ✔ Describe the process of inquiry and the relationship between inquiry and learning physics.
- ✔ Describe the difference between matter and energy.
- ✔ Explain the relationship between a theory, a hypothesis, a natural law, and an experiment.

Key Questions:

- What kinds of things can physics help us understand?
- How is scientific inquiry important to the study of physics?
- How are scientific models useful in physics?

Vocabulary

energy	inquiry	objectivity	scientific evidence
experiment	matter	process	theory
hypothesis	natural law	repeatability	

1.1 The Science of Physics

Physics is the science that seeks to understand the most fundamental workings of the universe, and everything in it. Physics includes the explanation for natural phenomena such as atoms, sunlight, and sound. Physics also includes human technology such as musical instruments, television, and space travel. In fact, nothing that occurs in the universe is outside the laws of physics. The laws of physics govern every action of the tiniest bacteria and the largest star, from throwing a ball to launching the space shuttle. In many ways, physics provides the foundation upon which rests the human understanding of other major areas in science, such as biology and chemistry (Figure 1.1).

Figure 1.1: *The universe is connected and nature draws no boundaries among biology, chemistry, physics, or any other branch of science.*

Biology Biology concerns life, and life is perhaps the ultimate expression of complexity. On the ecological scale, living things populate the planet in a constant dance of survival and adaptation. On a smaller scale, inside every living organism are thousands of chemicals such as DNA or proteins which interact in thousands of ways in complex functions such as eating, growing, and reproducing. While everything in biology follows the laws of physics, the complexity of life is far too great to be explained in detail by physics. The detailed workings of biology are, in fact, explained by chemistry.

Chemistry Chemistry is the science of matter and energy. Matter includes everything material—air and water, skin and bone, rocks and minerals, plants and animals. Chemistry describes how 92 naturally-occurring elements combine to make up the incredible diversity of matter around and within us. Chemistry also tells us how one substance can change into another substance which might be very different, such as how fire changes wood from cellulose into ashes and smoke. Chemistry describes how energy and temperature determine whether a substance such as water is solid ice, liquid, or even vaporized steam.

Physics For the explanation of why chemistry occurs, we need to dig down to the lowest level of understanding—physics! Physics provides the ground rules for how matter and energy behave. Forces and particles are the subject of physics. In many ways, physics creates the raw materials, like particles and forces, and the stage, like space and time, on which the building blocks of chemistry are constructed. Chemistry arranges the particles and forces into the ingredients and processes that combine into the complex dance of life that is the subject of biology (Figure 1.2).

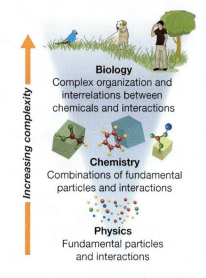

Figure 1.2: *Increasing complexity relates the sciences.*

A grand tour of physics

What is in Chapter 1? This chapter is a grand tour through the different areas that make up "physics" as a science. This chapter and investigations will give you a "big picture" view of what physics is and how we are going to approach it. Each page of this chapter will represent a whole chapter later, or perhaps two or three chapters. So, don't worry if you don't get every detail. Many details are not even there for you to get! What you *should* get is a taste of how our understanding of matter and energy creates a foundation for understanding both the natural world and the world of human technology. Let's start with a question.

Figure 1.3: *A battery, some wire, and a light bulb can be connected so the bulb lights.*

What do nerves, cell phones, and lightning have in common?

Electricity is part of physics The answer is *electricity*, and understanding electricity is part of physics. Consider a battery, wires, and a light bulb (Figure 1.3). When the bulb is connected, it lights up. Light is a form of energy, so the energy must come from somewhere, and you already know the battery is the "where." But exactly how does this occur?

Electricity comes from electrons Inside all ordinary matter are *atoms*. Inside atoms are still smaller particles, including *electrons* (Figure 1.4). The electrons are responsible for most of what we call "electricity." Every bit of matter, including you, contains electrons. In metals however, many electrons can move around. Ordinarily, these mobile electrons are randomly moving this way and that, bumping into each other. When a battery is connected, it creates an electric force that pushes the electrons in one specific direction. Moving electrons make an electric current. Electric current is what flows inside wires and carries energy from the battery to the light bulb.

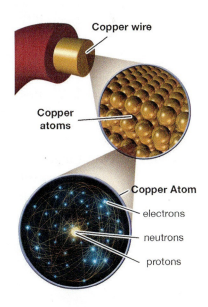

Figure 1.4: *Inside all matter are atoms, and inside atoms are smaller particles. One of these particles is called an electron. The movement of electrons is the source of ordinary "electricity."*

Conductors and insulators

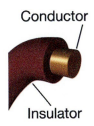

Wires are made of metal because metal is a *conductor*. Conductors are materials in which many electrons are free to move around. *Insulators* are the opposite of conductors. In an insulator, such as rubber, each electron is firmly attached to one atom. Current cannot flow though an insulator because the electrons can't move. That's why you can touch the cord of an operating appliance. The insulating plastic around the wire keeps the current from flowing through you!

1.1 THE SCIENCE OF PHYSICS

Electricity and energy

The universe contains matter and energy

Our universe is made of **matter** and **energy**. Matter is "stuff" that has mass and takes up space. You are made of matter, and so is a rock and the air around you. Energy is the ability to make things change. Energy is exchanged any time anything gets hotter, colder, faster, slower, or changes in any other observable way. A new idea in physics adds a third component to the universe: *information*. For example, information describes how the matter in this book is organized into pages and printed words.

Why electricity is useful

One reason electricity is useful is because devices like batteries can store electrical energy so it can be easily used later (Figure 1.5). A second reason is that electricity can quickly move a lot of energy a great distance. Think about using energy from a horse, the way people did only 200 years ago. You cannot use energy from the horse's muscles unless you are right there, with the horse. For example, the electrical energy you use today might be generated in a power plant 100 miles away. You don't need the noisy, smelly, and resource-hungry power plant inside your house, or even on the same street!

How electrical energy is used

A third reason electricity is so useful is that electrical energy can easily be transformed into many other kinds of useful energy. Electric motors turn electrical energy into energy of motion. Stoves turn electrical energy into heat energy. Light bulbs turn it into light energy. Computers turn it into both heat and light, and information processing.

Energy transformation

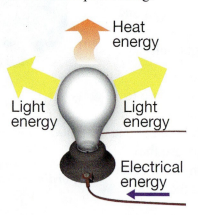

Figure 1.5: *The usefulness of electricity includes energy storage, energy transmission, and energy conversion.*

This last idea is important. Energy is active and things change by exchanging energy and transforming energy. Think about a light bulb as an *energy transformer* that turns electrical energy into heat energy and light energy.

Energy

Some examples of energy

What is the real difference between a dead battery and a fresh one? They weigh the same but one has more *energy* than the other. Energy is also an essential ingredient in all food but it adds no taste. Energy makes red different from blue, hot different from cold, and fast different from slow. Energy is the fundamental constituent of the universe. Yet, in its pure form, energy cannot be tasted, touched, seen, smelled, or heard.

Different forms of energy

Energy is the ultimate master of disguise, and appears in many forms. Food contains chemical energy, which is released by digestion. Plant life gets energy from sunlight. Light itself is a pure form of energy and different colors mean different energies. Red light has less energy than blue light and more than green light. The human eye "sees" color because molecules in your retina respond to the different amounts of energy. Heat is another form of energy. A cold cup of coffee has less heat energy than a hot one. Motion is yet another form of energy. A fast-moving car has more kinetic energy than the same car stopped at a street light.

A joule, the unit of energy

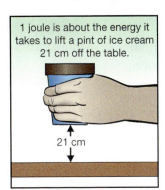

All types of energy are measured in units called *joules*. One joule is the tiny amount of energy it takes to raise a pint of ice cream about 21 centimeters. A single electric light bulb might use 100 joules of electrical energy every second and a typical car engine releases 75,000 joules of energy each second. A calorie is another unit of energy. One calorie equals 4.184 joules and 1 food Calorie (C) equals 4,184 joules. It might surprise you that eating a pint of ice cream provides your body with 1,000,000 joules of energy!

The importance of energy

Thinking about energy is not limited to learning science. A great deal of public concern today is centered around energy. The issues of global climate change, pollution, and even poverty are intrinsically connected to human access and control of energy and energy resources.

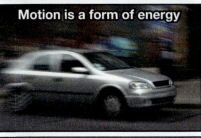

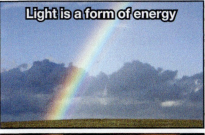

Electricity and electrical energy

Energy from a battery

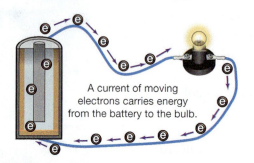

A current of moving electrons carries energy from the battery to the bulb.

Consider a battery and light bulb. Inside the battery are chemicals that react with each other to force electrons out of one end and accept them into the other end. The battery gives electrons enough energy to expel them, and the bulb converts the energy into heat and light.

Electric current An invisible *current* of tiny electrons moving between atoms is what flows and carries electrical energy. Disconnect the battery and the electrons are still in the wire, but they are no longer flowing in an organized way. Removing the battery removes the source of electrical energy that was pushing them. In fact, any break in the wire, such as opening a switch, stops the electrons from moving and cuts the flow of energy to the bulb.

The ampere is the unit of electric current Electrical energy is carried by an electric current. Electric current is measured in *amperes*, or amps. Electrons are exceedingly small. One amp of current means that more than six billion billion electrons per second flow in the wire.

Volts measure available electrical energy Electric current flows in response to differences in electrical energy. Electrical energy is measured in *volts* (Figure 1.6). A regular alkaline battery has a voltage of 1.5 volts. That means each amp of current flowing through the battery carries 1.5 joules of energy each second (Figure 1.7). The electrical outlet on the wall has an energy rating of 120 volts. That means each amp of current carries 120 joules of energy each second, which is one reason outlets are more dangerous than batteries. There is almost 100 times more energy per amp of electric current from a wall outlet compared to current from a battery.

The energy content of batteries Electric cars are difficult to make because of the huge energy requirement for a car. Under acceleration, even a small car might require 50,000 joules of energy every second. A fully-charged D battery contains about 90,000 joules of total usable energy but can release it slowly at a rate of 10 joules per second. At that rate, you would need 9,000 batteries to run a car!

Voltage	Energy
9 v	18,500 J
1.5 v	6,200 J
1.5 v	15,500 J
1.5 v	42,100 J
1.5 v	91,800 J

Figure 1.6: *Different batteries have different energy capacities.*

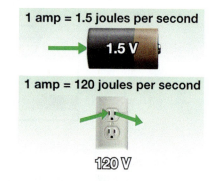

Figure 1.7: *A rating of 1.5 volts means that this battery carries 1.5 joules of energy each second.*

Mechanics

Mechanics is the science of forces and motion

To most people, *mechanics* is the skill of fixing cars. In physics, however, mechanics is the science of forces and motion. Everything that you can see or touch, including yourself, buildings, cars, and even a mosquito, is described by mechanics. Mechanics has two parts: dynamics and statics. Understanding motion is the *dynamic* part of mechanics. Understanding why things stand still or break is the *static* part of mechanics.

Force

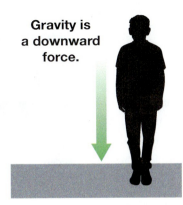

Mechanics starts with the idea of *force*. A force is an action such as a push or a pull. Your weight is a force from the gravity of Earth pulling on the mass of your body. The idea of action is important. Forces are the cause of action, even when nothing is moving. You can sit still and gravity still exerts a force on you even though the force seems to be doing nothing. If the floor was to suddenly vanish, however, that same force would immediately cause you to fall. Forces have the potential to cause motion or stop motion, even if they do not always do it.

Stability and change

Imagine a perfectly smooth, perfectly level, frictionless sheet of ice that extends forever. Now imagine rolling a ball. What happens to the ball once it leaves your hand? The answer is that the ball keeps going in a straight line, at exactly the same speed and in exactly the same direction, *forever*. If there are no forces to change its motion, no friction, and no other nudges, then the ball is stable and moves the same way forever (Figure 1.8). The point is that it takes force to start the ball moving and force to change anything about its subsequent motion, direction, or speed.

The first law

Any time motion changes, a force must be present because motion changes only through the action of forces. The key word is *changes*. Changes include starting, stopping, turning, speeding up, and slowing down. Force is required to change motion. No force means that motion cannot change. An object at rest stays at rest and an object in motion continues in motion at the same speed and in the same direction *forever*. This is known as Newton's first law, the law of inertia.

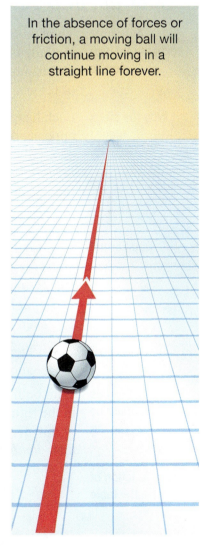

Figure 1.8: *A kicked ball on a level sheet of frictionless ice would travel forever in a straight line.*

Equilibrium and statics

A thought experiment with forces
Think about the block in Figure 1.9. Two weights are attached with strings. Nothing is moving. Are there forces acting on the block? What happens if one of the strings is cut? Why does the block suddenly start moving? Was a force added or removed?

Why the block moves
Each hanging weight makes a force. However, the forces are equal and in opposite directions. If we use the value +2 to represent the force on the right then −2 represents the force on the left. Adding +2 and −2 gives a total of zero or *zero net force*. Cutting a string removes one force. The result is that the total or net force is no longer zero, and the block moves.

Two ideas are important here.

Equilibrium
1. If nothing is moving, then the net force must be zero, and the block is stable. This is called *equilibrium*. In any normal situation there are always forces acting, one of which is usually gravity. If nothing is moving, then physics tells us at least *two* forces must be acting, with one canceling the other. This is true even when you don't know about the second force.

2. If there is any net force, no matter how small, there must be a consequent change in motion. We'll talk about this more on the next page.

Other forces that must be acting
With both strings attached, the block doesn't move right or left because the forces in the right–left directions cancel exactly to zero. Why doesn't the block move up or down? Gravity is pulling it down so there is a downward force. Of course, the table *holds* the block up. But how exactly is the table "holding the block up?"

The function of the table
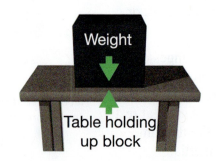

The answer is that "holding up" means the table exerts a force on the block that exactly cancels the force of gravity. If forces were visible we would see two arrows in the "up–down" direction; the "down" arrow of gravity, and the "up" arrow from the "holding up" force the table exerts on the block.

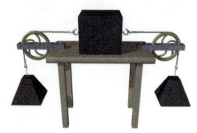
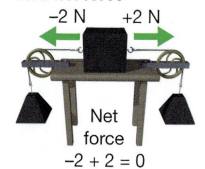
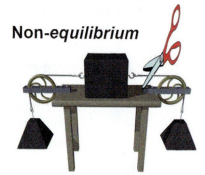

Figure 1.9: *These three situations involve force and motion.*

Dynamics

Acceleration and motion
Let's return to the block and the two strings (Figure 1.10). At the very instant the string is cut, the block is not yet moving. It immediately starts moving but *it starts from rest*. The force from the connected string *accelerates* the block by causing its speed to increase.

The word *accelerate* has so many meanings that it is often confusing. In fact, the true meaning of acceleration eluded human beings for thousands of years. These people include Galileo, who spent years trying to understand motion at the University of Pisa around 1589. Here is the meaning of acceleration in physics.

Acceleration is a change in speed that occurs over time.

Here is why acceleration is important.

Unbalanced forces cause objects to accelerate in the direction of the force.

The common-sense meaning of acceleration
This seems quite sensible on the face of it. When you push something, it speeds up in the direction you push it.

Think about pushing the accelerator pedal when a traffic light turns green. The car starts at rest, then slowly increases its speed as you hold the pedal down. The force acting between the wheels and the pavement causes the car to accelerate in the same direction as the force.

Negative acceleration or deceleration
To slow something down, push opposite to its direction of motion. The speed is reduced in the direction of your push. Positive acceleration increases speed and occurs when the force is in the same direction as the motion. Negative acceleration decreases speed when the force is in the opposite direction of the motion.

Force causes acceleration
The big idea in dynamics is that unbalanced force always causes acceleration. Acceleration always causes a change in speed, whether to speed up, slow down, or change direction.

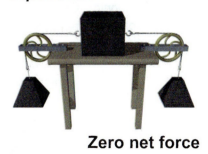

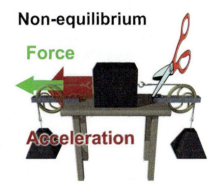

Figure 1.10: *Force and acceleration act on a block and strings.*

Harmonic motion

Harmonic motion is motion that repeats

Compare a ball rolling downhill with a swing going back and forth. The rolling ball is one kind of motion, with characteristics of speed and position. The swing is another kind of motion because it *repeats in patterns*. Repetitive motion, also called *harmonic motion*, has all the properties of ordinary motion, but includes some new properties as well.

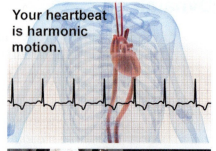
Your heartbeat is harmonic motion.

Oscillators

A swing is an example of an *oscillator*. An oscillator is an object or system that repeats over and over again in cycles. Many examples of oscillators occur in nature and also in human technology. Your own heartbeat is an important natural oscillator. In this case, the repetition is caused by rhythmic electrical impulses delivered to your heart muscle (Figure 1.11).

The vibration of a guitar string is harmonic motion.

Frequency

The *frequency* of a repeating motion is how often it occurs in a given unit of time. A frequency of once per second means the motion repeats once per second. A typical swing has a frequency of about $\frac{1}{5}$ of a cycle per second. A *cycle* is a complete back-and-forth movement of the swing. The value of $\frac{1}{5}$ cycle per second means the swing makes $\frac{1}{5}$ of a complete back-and-forth cycle each second, so the entire cycle takes 5 seconds.

The orbits and rotations of Earth and planets are harmonic motions.

Why a swing is an oscillator

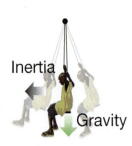

A swing and a pendulum are examples in which the repetition is caused by the natural response of the system to an initial condition. Pull a swing back and gravity pulls it back toward its center or *equilibrium* position. As it passes its lowest point, the force pulling the swing toward equilibrium becomes zero. But, inertia keeps the swing going and it overshoots past center. Now the force pulls it back so that the swing slows, reverses, and goes back through the center again.

The up, down, and rotating movements in an engine are harmonic motions.

Figure 1.11: *Some examples of repeating motion and oscillators are seen here.*

Waves

Making a wave
Consider a tank of water containing a floating object that displaces some of the water (Figure 1.12). What happens when the object is suddenly removed?

If you look right where the object was, you see the water oscillate up and down. Water rushes in to fill the space, then rises due to inertia. Some water is now higher than the rest, falls, and sinks due to its inertia, and so on.

Waves spread oscillations all around
Now look a little farther away. The water some distance away is also oscillating up and down! The disturbance traveled through the water in the form of a *wave*. A wave is a traveling form of energy that results from oscillations. Water supports waves because every bit of water is *connected*. Displacing some water in one place affects the water right next to it, and that affects the water right next to *it*, and so on all through the tank.

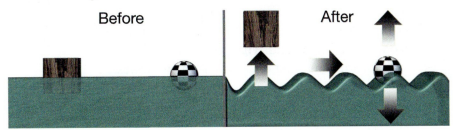

Start with block in water

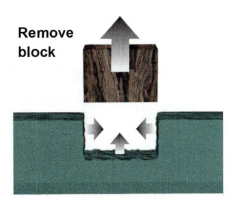

Remove block

The disturbance creates a wave

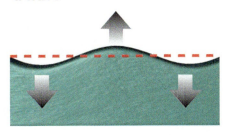

Two basic results of waves
Waves do two important things:

1. They carry information, like the wave's frequency, from one place to another. If I measure the frequency at which the water oscillates up and down over here, I will see that the water everywhere else in the tank starts to oscillate at the same frequency. A wave spreads its frequency wherever it travels.

2. Waves carry energy. The water at the far end of the tank was not moving initially but started moving later. Movement means energy and the energy arrived in the form of the wave.

What moves, the water or the wave?
Here's a curious fact. The water itself, on average, stays in the same place. A float at one end of the tank stays in the same place even as it bobs up and down. Waves move energy through a substance. The energy is the traveling oscillation, but not the substance itself.

Figure 1.12: *Making a wave in a tank of water.*

Sound

Sound and light are waves Two very important examples are the reason why the subject of waves is in this chapter. The first is sound and the second is light (Figure 1.13). Light waves and sound waves carry energy and information through traveling oscillations. Both are everywhere in human experience, in the natural world, and in human technology.

The nature of sound Sound is a wave of air pressure. When a guitar string vibrates back and forth, the soundboard of the guitar pushes the air that touches it back and forth. Like the water wave, that air pushes the air next to it, which pushes the air next to it, and so on, all the way to your ear drum. The air pushing back and forth on your eardrum is what ultimately creates the sensation of sound in your brain.

The frequency of sound waves Compared to swings and water waves, sound is a very rapid oscillation, so fast that the guitar string appears as a blur. The musical note "A" has a frequency of 440 cycles per second, or 440 hertz (Hz). One hertz is one cycle per second. That means the guitar soundboard is moving back and forth 440 times each second.

Figure 1.13: *Sound and light are waves.*

Sound waves are very small oscillations Another big difference is that sound waves are small. The actual movement of the guitar top is barely discernable, a fraction of a millimeter. This is less than the thickness of a sheet of paper. The resulting oscillation in air pressure is less than one hundredth of one percent, one part in a hundred thousand. But the sound it makes can be quite loud! The sound is loud because the human ear is a fantastically sensitive detector of sound. Your ear can detect a change as small as one part in 10 million.

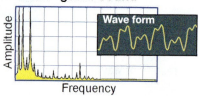

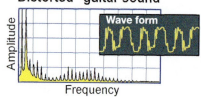

The frequency range of hearing An average human ear can detect sound waves in a frequency range from about 20 Hz to a maximum of 20,000 Hz. Some animals, such as bats, can hear much higher frequencies, up to 80,000 Hz. Other animals, such as whales, can hear much lower frequencies, below 0.1 Hz. Medical ultrasound equipment uses sound waves 100,000 Hz and higher.

Figure 1.14: *Physics is used to record and measure the sound of an electric guitar.*

Light

Look around you

What colors do you see immediately around you? Most places have an astounding variety of color. Leaves are many shades of green; the sky is shades of blue or grey. Clothing comes in innumerable colors. This variety leads to questions—what is light and what is the explanation for color?

Light is a wave of electricity and magnetism

Imagine you have a small magnet in your hand. If you move the magnet up and down, the force acts on another nearby magnet. How does the force get from one magnet to the other? Does the force travel instantaneously? In reality, your moving magnet creates a light wave that carries the force though space from one magnet to the other.

Radio waves

Light is a very rapid wave of electricity and magnetism. If you move your magnet up and down 1 million times per second, you would have a radio wave. The number 104 FM on the radio means 104 megahertz, or 104 million oscillations per second. Radio waves are actually a form of light, too. They have too low an energy to be detected by your eyes, however.

The frequency of light waves

Now imagine moving the magnet up and down 5 thousand trillion times per second (5×10^{15} Hz). You can imagine this takes more energy, and it does. You would make red light! Red light has enough energy to see. The oscillation is so fast that it cannot move the whole magnet in response. But red light has enough energy to shake the individual atoms in the opposing magnet so they give off their own reflected light and that is why you can see the other magnet. The other magnet also heats up a bit.

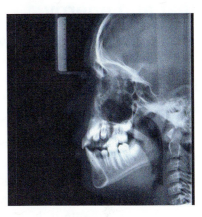

Figure 1.15: *This X-ray is of a human skull.*

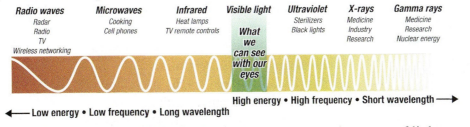

X-rays and other forms of light

The colors between red and blue in the rainbow represent the range of light energy that our eyes can detect. However, light has an infinite range of energies. The X-rays used by doctors are also a form of light, but with much higher energy than visible light (Figure 1.15).

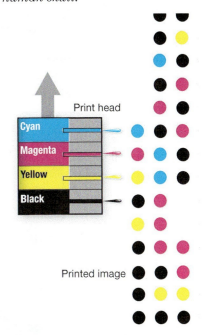

Figure 1.16: *Ink jet printers use the physics of color to reproduce full-color images.*

1.1 THE SCIENCE OF PHYSICS

Heat and energy

What is temperature? We all know the difference between hot and cold. A more difficult question is what precisely causes hot and cold? That question took 3,000 years to answer. To make the point more clear, suppose you have two identical cups of water at different temperatures (Figure 1.17). What is different about the water to make the temperatures different? Where does temperature come from?

Temperature measures heat, a form of energy The simplest answer is *energy*. The water in both cups is the same and the difference is their energy content. Temperature measures a form of energy called *heat*. All matter contains heat energy. The higher the temperature, the more heat energy there is. You can make the cold water into hot water by adding heat energy to it. If you have 1 gram of water, adding 4.18 joules of energy makes the temperature increase by exactly 1 degree Celsius (Figure 1.18). Whenever anything gets warmer or colder, energy is being exchanged. You put energy *in* to make something warmer. You take energy *out* to make something colder.

Technology, heat, and energy Many of the most useful human inventions ultimately transform heat into other forms of energy. For example, a moving car has kinetic energy, or energy of motion. The kinetic energy comes from the heat of burning gasoline in the engine. A car engine turns chemical energy in gasoline into heat energy which is converted by the engine and transmission into energy of motion of the car.

Energy flow diagrams

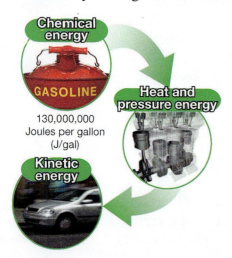

You can learn a great deal about anything in the universe by tracing the flow of energy. How much energy is there? Where does it go? A gallon of gasoline contains a fixed amount of chemical energy, and no more. The ultimate fuel-efficient gasoline-powered car would take all the energy available in gasoline and turn it into energy of motion of the car. But, physics puts a limit on how much energy can be derived from one gallon of gasoline. You can never get more energy out than the gallon intrinsically contains.

Figure 1.17: *What is different about two identical cups of water that causes them to have different temperatures?*

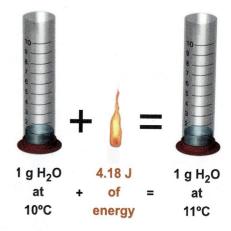

Figure 1.18: *It takes 4.18 joules of energy to raise the temperature of 1 gram of water by 1 degree Celsius.*

Heat and atoms

Brownian motion

Today we take for granted that atoms exist but it was not always that way. Who would believe in something no one could see? In 1827, Scottish botanist Robert Brown noticed that a speck of pollen floating in water moved around in a jerky, and continuously agitated way (Figure 1.19). This occurred even when the water was absolutely still. *Brownian motion*, as Brown's discovery was called, comes from the impacts of trillions of atoms. We now know that atoms are constantly moving and jostling each other. Brownian motion is related to atomic motion, which is dependent on *temperature*.

Kinetic energy

Later we will learn that the energy of objects comes in two basic forms. Energy of motion is called *kinetic energy*. Kinetic energy depends on mass and speed. The faster you go, the more kinetic energy you have. Kinetic energy also depends on mass. If two objects are going at the same speed, then the one with more mass has more kinetic energy.

Potential energy

The second type of energy is called *potential energy*. Potential energy is stored energy. Stored energy may become active, but it can also remain stored and unused. When you lift a rock off the ground you give the rock potential energy. The rock can fall back down and give up the energy. When you stretch a rubber band, you give the rubber band potential energy. That energy can become active by exerting a force on a stack of papers it is holding together.

Heat is atomic-sized kinetic energy

So what kind of energy is *heat*? Heat acts like potential energy, because heat can be stored, like a thermos bottle keeps its contents warm. However, heat is actually kinetic energy (Figure 1.20). The sub-microscopic atoms in matter are constantly in motion. Individual atoms are so small that you don't feel their individual motion. Instead you feel their average energy as temperature. When something "feels" hot, what is really happening is that the atoms of "something" are more energetic than the atoms of your skin. When these "hotter" atoms bump into your skin, they transfer some energy. The "feeling" of heat is the energy moving between more-energetic atoms to the less-energetic atoms of your skin via trillions of tiny collisions.

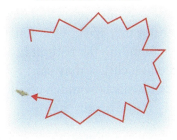

Figure 1.19: *Brownian motion is due to constant movement of atoms.*

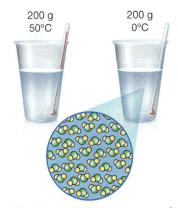

Heat is the kinetic energy of moving atoms in matter.

Figure 1.20: *Heat is the total kinetic energy associated with the thermal motion of atoms.*

1.2 Scientific Inquiry and Natural Laws

We believe all events that happen in nature obey a set of *natural laws* that do not change. **Natural laws** are rules that govern the fundamental workings of the universe. Science is about discovering and understanding the natural laws. For example, there are natural laws that govern motion. If you apply a force to a cart, it may start rolling with a certain acceleration. The natural law that tells you how much force causes how much acceleration, and in what direction, is called Newton's second law, and you will study it in Chapter 5.

Discovering the natural laws

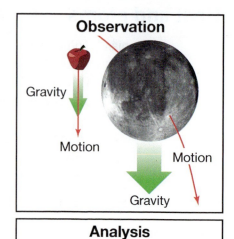

Observation

Sir Isaac Newton
(1642 – 1727)

The origin of the natural laws

Unfortunately, no one is born knowing *any* of the natural laws. By clever thinking, over thousands of years, humans have deduced many of the natural laws, often by trial and error. Fortunately, nature is reliable in that it *always* obeys the same natural laws. If you push many different carts in many different ways, eventually you understand the relationship between how hard you push and the resulting motion. This is exactly what Newton did in 1687 and why a law is named after him.

Experiments, observations, and analysis

That brings us to a key aspect of science: the **process** of deducing the natural laws. This part of physics includes *observation*, *experimentation*, and *analysis*. Observation means looking at something that happens in a careful way and recording all the important details. An **experiment** is a situation set up specifically to observe something, like the relationship between applied force and the subsequent motion of a cart. Analysis is the process of thinking about observations to determine what they mean.

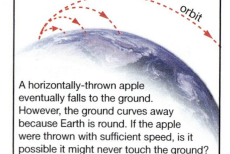

A horizontally-thrown apple eventually falls to the ground. However, the ground curves away because Earth is round. If the apple were thrown with sufficient speed, is it possible it might never touch the ground? If so, might the very same force that pulls the apple also pull the Moon into its orbit?

Natural law

$$F = G\frac{m_1 m_2}{R^2}$$

The force of gravity attracts all masses to each other with a force proportional to the product of the masses divided by the square of the distance between them.

Real life is often complex

Many times, analysis means breaking down a complex situation or problem into smaller parts that can be understood in terms of the natural laws. Real-life situations are rarely simple and must be analyzed before knowledge of physics can be applied. For example, the motion of an airplane depends on more than just the force from the engines. The wind speed and the air resistance are important factors. To apply the laws of motion to an airplane in a meaningful way, you need to analyze its motion in order to identify all the important factors.

Hypotheses and the importance of experiments

How do we know right from wrong? How do you know when you are right? In a court of law, a judge may decide right and wrong. However, judges are human. A judge must decide based on interpretation of both the law and the evidence that has been shown. Judges rarely have *absolutely conclusive* evidence, nor are their decisions without mistakes. So how can anyone be sure they are right about how *nature works*? Are the opinions of people, even judges, sufficient to tell the right explanation from a wrong one?

How science decides right from wrong The real distinction between science and other forms of human thought is how a scientist decides which explanations are right and which are wrong. In science, the right explanation is the one that correctly explains everything that *actually happens*. People's opinions do not matter, even other scientific opinions. What is written in books does not matter, not even science books. The real truth is decided *only* by actually observing nature.

Hypotheses The search for scientific knowledge begins with thinking about something that occurs. A **hypothesis** is a tentative explanation that can be tested by an experiment or observation. For example, you might have the hypothesis that the speed of falling objects increases with weight (Figure 1.21), or "heavier objects fall faster." How do you know if you are right? Do you ask someone? Do you read what has been written about falling? Do you watch a cartoon character fall on TV?

Experiments are the test of truth The only way to really decide is to actually measure the rate of falling for different objects and see whether your hypothesis is supported by your actual observations. That is the purpose and the importance of experiments. Experiments are the test of truth. Scientists constantly test what they think they know with experiments.

Hypotheses are always improving Hypotheses usually start rough or even wrong and are gradually made better over time. A mature hypothesis successfully agrees with every relevant detail of every observation or experiment. This can take years and often involves the work of many people. Most of them rarely get the correct hypothesis without doing many experiments. The validity of a hypothesis improves as more experiments are completed and analyzed. The hypothesis in Figure 1.21 seems reasonable. What changes are suggested by the actual results of observations?

Hypothesis
A tentative explanation for something that may be tested by observation or experiment

Example
The speed of falling increases with weight.

Test with experiments and observations

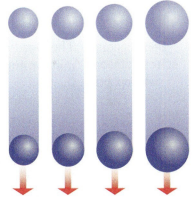

In fact they all fall at nearly the same rate. What does this mean?

Figure 1.21: *Experiments provide the test of truth for hypotheses. When an experiment contradicts a hypothesis, scientists know they do not have the right explanation.*

Scientific theories and facts

The definition of a theory
A **theory** is a comprehensive, well-tested explanation of how and why a process in nature works the way it does. Usually, one or more natural laws fit into a single comprehensive theory. For example, the classical theory of motion includes Newton's three laws as well as conservation laws for energy and momentum.

Theories grow and evolve
Like hypotheses, theories are rarely correct at first. For example, it was once believed that heat is a fluid that moves from hot to cold. Hot objects were thought to have more of this fluid than cold objects. People tested this theory and eventually found it to be incorrect. But the process of testing the theory led to a better theory of heat that we still believe today. New scientific theories are discovered gradually, by many people working and thinking together.

The nature of scientific research
The *purpose* of scientific research is to do experiments which show that existing theories do *not* give the right prediction. A theory that correctly explains 1,000 experiments but fails to explain the 1,001st cannot be wholly complete. An unexplained experiment points out where the theory is lacking and to a scientist is a pointer to something new waiting to be discovered.

Don't confuse scientific theory with opinion
The word *theory* in common language can mean "an opinion or a vague idea that may or may not be right." In science, the opposite is true. Scientific theories are the best explanations for observations and evidence of how the universe works. By the time a successful theory makes it into a book like this, it has been repeatedly tested and confirmed for years. The classical theory of motion was developed hundreds of years ago, and it is still the best explanation for motion on a large scale. With the discovery of atoms and the particles within them, scientists found that motion within an atom did not always conform to the classical theory. As they tried to figure out how atoms behaved, their work led, over time, to the development of quantum theory. With this book, you are actually learning through the experiments of all the scientists who developed those theories.

Be wary of science on the Internet
Beware of "science knowledge" you find on the Internet! Some is correct but much is not. Anyone can post their theories on the Internet, including people who do not know, are misinformed, or simply just want to advance their own, often untested, hypotheses.

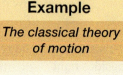

Theory
A self-consistent and comprehensive explanation that is supported by a large body of scientific evidence

Example
The classical theory of motion
- Newton's 1st law
- Newton's 2nd law
- Newton's 3rd law
- The law of conservation of energy
- The law of conservation of momentum

Beware!
Theory ≠ Opinion
The Internet contains a lot of incorrect or false information which may *appear* to be a scientific theory but which is actually a misinformed opinion.

Scientific evidence

Scientific evidence
The goal of an experiment is to produce **scientific evidence**. Scientific evidence may be in the form of measurements, data tables, graphs, observations, or any other information that describes what happens in the experiment (Figure 1.22). Like a mystery, evidence is used to sort out the right explanations from the wrong ones.

The qualities of scientific evidence
Two important characteristics of scientific evidence are that it be *objective* and *repeatable*. *Objective* means the evidence should describe only what actually happened as exactly as possible. The personal opinions of the person doing the experiment do not count as scientific evidence. Repeatable means that others who repeat the same experiment observe the same results. Good scientific evidence must pass the tests of being both **objective** and **repeatable**.

Observations may produce scientific evidence without experiments
Scientific evidence may also be produced by observing nature without actually doing an experiment. However, the same rules of objectivity and repeatability apply. For example, Galileo used his telescope to observe the Moon and recorded his observations by sketching what he saw (Figure 1.22). His sketches describe in detail what he actually saw through the telescope, therefore they pass the test of objectivity. Others who looked through his telescope saw the same thing, therefore the sketches pass the test of repeatability. The scientific evidence of Galileo's sketches convinced people that the Moon was actually a world like Earth with mountains and valleys. This was not what people believed prior to Galileo's time.

Words in science must have well-defined meanings
It is important that scientific evidence be communicated clearly, with no room for misunderstanding. This means we must attach careful meanings to concepts like "force" and "work." Many "everyday" words will be defined very precisely in physics. Usually, the physics definition is similar to the way you already use the word, but more exact. For example, a quantity of "work" in physics means using a certain strength of force to move an object a certain distance. This definition allows a quantity of work to be defined with a precise measurement that tells someone else exactly how much force and distance were involved. Words like "a lot" or "easy" leave too much room for misunderstanding. Scientific evidence requires careful definition of words and measurements so others know exactly what you mean.

Examples of scientific evidence

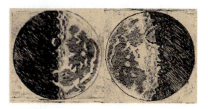

Pictures or sketches that show actual observations

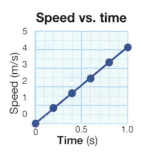

Measurements and data

Graphs and charts made from measurements

Figure 1.22: *Some examples of scientific evidence.*

Learning physics through inquiry

Scientific inquiry Think of science as having two parts. One part is the accumulated knowledge of generations of human beings. That part is well represented in books like this. The other part is the process for discovering and verifying new knowledge by adding to what is known. This second part is called *scientific inquiry*. **Inquiry** is a process of learning by asking questions and making observations that allows you to deduce answers and pose better questions (Figure 1.23). For example:

- Why does it take more force to move heavier objects?
- Does it take twice as much force if the object weighs twice as much?
- How can we measure force?
- How can we measure motion in a way that we can relate it to force?

Who uses scientific inquiry? Inquiry is a skill you use every day. You continually test your ideas against your observations, and over time, you develop *common sense*. Common sense is actually the physics of everyday life, to a large degree. For example, consider pushing a cart of groceries. You *know* more force is required to push heavier carts and less force to push lighter carts. Common sense says that heavier carts take more force to push. However, you were not born with this knowledge. The knowledge that makes up common sense came from trying it yourself and drawing conclusions from what you experienced. The common-sense understanding of force is part of a fundamental natural law of physics.

Why learning inquiry is important Physics is not memorizing what is in this book. The book may seem thick and full of knowledge, but it is barely a footnote to the many unsolved mysteries of the real universe. Someday, maybe tomorrow, you will be stumped by something you don't know. Maybe your bicycle is broken or the average temperature of Earth's atmosphere is rising, or *anything*. "Anything" is a lot of territory, even for a thick book. In fact, it is very unlikely that this book or any other book will have taught you to understand the *exact* thing that has you stumped. However, when you master the process of scientific inquiry, you have the power to *find* good answers to almost anything!

Figure 1.23: *Observation leads to inquiry and inquiry is how we learn.*

Figure 1.24: *Physics provides a way to approach and analyze any problem, easy or hard.*

Inquiry is an excellent way to learn physics

Learning by investigations

As you do the investigations that accompany this book, you will try to deduce the natural laws of physics using your own process of inquiry. For example, you will apply force to moving objects and measure how fast they move. Through this experience, you discover the natural law that connects force and motion (Figure 1.25). This is an extremely powerful law because it applies to *all* forces and *all* motions, whether the moving object is a golf ball or a planet.

Understanding how to use scientific knowledge

There are other ways to learn the natural laws. One way is to have them recited to you. Another way is to read them. Reading and listening are necessary, but often are not enough to allow you to *understand* and *use* the knowledge. For example, in soccer there is a rule that says a team is penalized if an offensive player is "off sides." This rule is difficult to understand unless you know a lot more about soccer, including the definition of "off sides" and who counts as an offensive player. Just reading the rule in a book will not help you use the rule. You need to see how the rule is applied in a game, how it affects play, and under what circumstances it is used.

Reasons why learning through inquiry is useful

Three valuable things are learned by discovering the natural laws through inquiry.

1. You will learn the basic natural laws that describe how the universe and everything in it works. This includes very practical things such as bicycles, antilock brakes, televisions, and digital recording of music.

2. You will learn how to deduce how things work and solve problems that are *not* in this book. You will face many such problems in your life, and physics gives you a powerful approach to solving many real-world problems.

3. Learning the natural laws by observing them in real systems will help you use what you learn. Like most knowledge, the value in knowing really comes from understanding how to *apply* what you know to situations *different* from the one in which you first learned the knowledge.

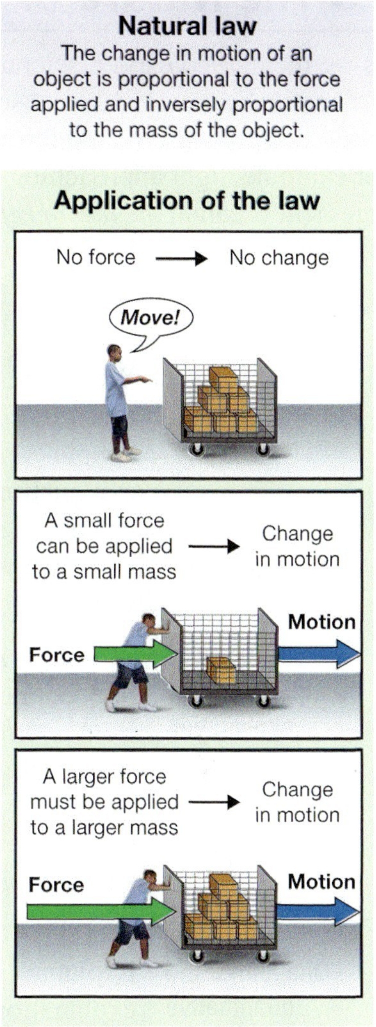

Figure 1.25: *An example of a natural law and its application.*

1.3 The Nature of Scientific Knowledge

As we mentioned earlier, science has two parts. One is a rigorous process of scientific inquiry that tests explanations against actual observations. The second part is the accumulated body of knowledge about nature that was discovered through the process of scientific inquiry which we called the natural laws. For example, light approaching a mirror at a certain angle reflects off the mirror at the same angle. This statement is called the *law of reflection*. The law of reflection correctly predicts how light reflects off of a mirror every time. This section is about the second part—what *is* scientific knowledge?

Explanations

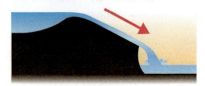

The meaning of an "explanation" When you *understand* something, you create a mental model called an *explanation* to connect what you observe with what you know. For example, a simple explanation for rivers and streams is that water flows downhill. This explanation is useful because it allows you to predict what will happen to water on any given shape of land. It also allows you to make water go where you want. If you need water to flow somewhere, you arrange the land so the place you want the water to go is downhill from the water source (Figure 1.26).

An example of a simple scientific explanation Scientific knowledge grows by systematically extending simple explanations to include more complex situations. For example, you can make water in a straw flow *uphill* by sucking on the straw. A complete explanation for why water flows must include the possibility of suction forcing water to flow uphill. To describe *suction*, we introduce the concept of pressure. Water can be sucked up a straw because water flows from higher pressure to lower pressure (Figure 1.27). This is a new explanation that both explains the old observations of water flowing downhill and the new observations of water flowing uphill in a straw.

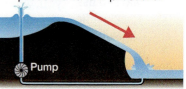

Qualitative and quantitative Explanations can to be *qualitative* and tell you *why* but not *how much*. The explanation that water flows from high pressure to low tells you that to make water flow through a hose, you need a higher pressure on one end than on the other. It does not tell you exactly how much pressure difference will make 1 gallon per minute of water flow out of a 1/2-inch hose that is 25 feet long. For this you need a more sophisticated explanation that is *quantitative*. Quantitative means "with numbers."

Figure 1.26: *Explanations can start simply and be improved to explain new observations.*

Qualitative

Explains how and why but does not provide numbers

Quantitative

Explains how and why and also provides numbers and measurements that tell how much

Figure 1.27: *Qualitative explanations provide conceptual understanding, such as "how" and "why." Quantitative explanations or models tell you a lot more, such as "how much" and "when."*

Models

Models and variables The most useful form of scientific knowledge is a *model*. A model tells you precisely how the *variables* in a *system* are related to each other. Variables are the parameters that describe things. For example, force, angle, time, and weight are variables that describe the motion of a car moving down a ramp.

Systems A system is a group that we choose to include all the variables we are interested in that affect each other. Choosing a system allows us to ignore things that do not affect what we want to know. For example, the color of the paint in the room, the amount of sunlight, and the temperature are all variables. However, we can exclude these variables in an experiment about the motion of cars on ramps because they really do not affect the outcome of the experiment.

Graphical models Simple models have two common forms. One is a graph. A graph relates two variables to each other. If you know the value of one variable, you can use the graph to predict the value of the other variable. For example, Figure 1.28 shows a graphical model for how the speed of a rolling cart changes as it goes downhill. You can see from the graph that the speed of the car is 231 cm/s at a position of 55 cm down the ramp.

Formulas The second form of simple model is a formula. Formulas are more powerful because they can include more than two variables and have a much wider range of values. For example, the speed (v) of a car is its initial speed (v_0) plus the acceleration (a) multiplied by the time (t). If a car moving at 30 miles per hour (mph) accelerates at a rate of 5 mph per second for 5 seconds, its final speed is 55 mph. The formula tells you exactly what to do with the values of each variable.

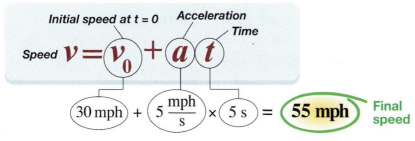

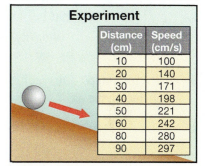

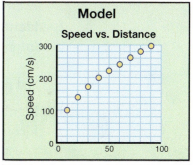

Figure 1.28: *We use experiments to test and evaluate theories. A natural law is "discovered" when a theory successfully explains the results of every single experiment over a long period of time.*

1.3 THE NATURE OF SCIENTIFIC KNOWLEDGE

Problem-solving techniques

What problem solving means Problem solving means using what you *do* know to find something you *don't* know. Analysis and problem solving go together. You cannot solve a problem until you analyze it to find out which laws or models apply and what information you know or are given. This book presents and demonstrates some useful techniques that help you apply your knowledge of physics to solving problems. Solved *example problems* appear in every section. You can recognize example problems by the icon and colored boxes around them, as shown here.

A four-step technique Our technique for problem solving has four steps.

1. **Determine what the problem asks you to find.**
   Be very specific. For example, when a problem asks how fast something is moving, the answer is the *speed*. Do not get sidetracked looking for things you do not need.

2. **Identify information you are given.**
   Information may be measurements, such as mass or length. Information may also be descriptive. For example, "at rest" means something is not moving and has a speed of zero.

3. **Identify laws or relationships.**
   Identify relationships that involve what you need to know and/or information you are given. For example, if you want to know speed and are given distance and time, you would write down the formula for speed that includes distance and time. Some problems require more than one relationship.

4. **Apply the given information and the relationships.**
   Once you have collected the information and the relationships, you will be able to see how to use what you know to get the solution. The solution may take more than one step. For example, you may need to use one relationship to find something that is needed for a second relationship.

These problem-solving steps are a useful technique for finding solutions in many subjects and situations, not just physics!

Calculating time from speed and distance

How long does it take to travel 2,000 kilometers at a speed of 100 kilometers per hour?

1. You are asked to find time.
2. You are given the distance in km and speed in km/h.
3. Time is distance ÷ speed
4. Time = 2,000 km ÷ 100 km/h
           = 20 h

A good theory that started out wrong

All the natural laws we know today started as theories that were often wrong! The discovery of oxygen and its importance to fire and life is a good example. When a candle is burned in a closed container, the flame burns for a moment then quickly goes out. If a living organism such as a mouse is placed in a closed container, it will eventually suffocate.

The phlogiston theory of combustion An early theory of fire proposed that all materials contained a substance called *phlogiston*, pronounced flow-JIS-ton. When wood burned, it released its phlogiston (Figure 1.29). The ash of the burned wood was the true material, without its phlogiston. Too much phlogiston was thought to be toxic to life. A mouse eventually died in a closed container because phlogiston was given off during breathing and suffocated the mouse.

The discovery of oxygen The phlogiston theory received wide support by most scientists throughout much of the 18th century. Eventually, however, accurate scientific evidence from experiments revealed problems. For example, when a metal burned, it was supposed to lose phlogiston. However, the metal ash left over after burning weighed *more* than the metal did before it had supposedly lost phlogiston. This implied that the removed phlogiston must have weighed less than zero. The work of Antoine Lavoisier (1743–1794) finally disproved the phlogiston theory. Through careful experiments, Lavoisier concluded that as a metal burns, it increases in weight because it gains oxygen. Lavoisier's theory of oxidation soon replaced the phlogiston theory and remains a part of modern chemistry.

The wrong answer is often the beginning of the right answer While the phlogiston theory was not correct, it led to the discovery of oxygen while people were looking for phlogiston. This book contains a collection of natural laws and explanations that have passed the test of time and are man's best knowledge about how things work. What is left out is that the laws and explanations were nearly always wrong at first! We learn new things by putting forth an idea and then testing it to see if it is right. *We cannot learn without starting from an idea, even the wrong idea.* As you learn physics, never be afraid to suggest an explanation for what you see. It does not matter if it is the right explanation at first. The evidence you gather in experiments will eventually lead you to the correct explanation.

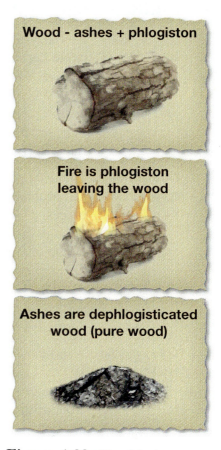

Figure 1.29: *The phlogiston theory was an early attempt to explain fire.*

Scientific knowledge and the solar system

Early observations of the sky
Each day, you see the Sun rise in the east and set in the west. The Sun seems to circle Earth as do the Moon and stars. Early theories of astronomy assumed that the Sun, Moon, and planets actually *did* move around Earth. Ptolemy, a second-century Greco-Egyptian astronomer and geographer, is credited with creating a model in which Earth is at the center of the universe and the stars, Sun, and planets are affixed to the celestial spheres which rotate around Earth's flat disk (Figure 1.30, middle).

Early civilizations thought Earth was covered by a dome on which the Sun, stars, and planets moved.

Problems with an early theory
To most people, Ptolemy's model explained what they saw. To others, his model had problems, even before Galileo turned his telescope skyward. One of the biggest problems was that the planets sometimes reverse direction as they move through the night sky! This is because the apparent motion of the planets depends on the relative positions of Earth and the planets in their respective orbits around the Sun. For example, Mars sometimes appears to move backwards in sky. Nicolaus Copernicus (1473–1543) correctly deduced that the planets revolve around the Sun (Figure 1.30, bottom) but he was not widely believed at any time during his life.

During the Middle Ages, people thought the Sun, stars, and planets circled Earth which sat in the center.

Galileo and the sun-centered universe
A more convincing argument for a Sun-centered solar system was made almost 70 years later by Galileo. With his telescope, Galileo observed that Jupiter had four moons that moved around it. He correctly reasoned that if moons could move around Jupiter, why could Earth not also move around the Sun? Convinced of the truth of Copernicus's theory, Galileo spent much of his life trying to convince people to believe the evidence of their own observations. Then, as now, people often prefer to keep on believing what they already think they know, even when the evidence clearly shows such beliefs to be wrong!

Today we know Earth and planets orbit around the Sun, and the stars are very far away.

Figure 1.30: *Three different models for the solar system were believed at different times in history.*

The start of the scientific revolution
Many historians believe the scientific revolution began with the acceptance of the Copernican model of the solar system. The importance of this event is not that Copernicus had the right answer. The importance is that, for the first time, scientific evidence was used to decide the difference between the right explanation and the wrong one. Remember, scientific evidence comes from direct observation of *what is actually measured or seen to happen*. Before the scientific revolution, people believed more in tradition than in what they observed with their senses.

Chapter 1 Connection

Biomimicry and New Color Display Technology

George de Mestral noticed many tiny objects clinging to both his clothes and his dog's fur during an outdoor excursion into the Swiss Alps. These tiny objects were burrs—small, elongated capsules that contained seeds from the aptly named burdock plant. On this particular occasion, their tenacious clinging ability impressed him and he became curious.

Inspirations from nature Inspired, he removed several burrs upon returning home and put them under a microscope. Each burr had a teardrop shape, with a tiny hook at the very tip. The tiny hook, though small, held quite strongly to fur, hairs, or any other fibrous-material-like clothing. He visualized a system of fastening objects together using two complimentary structures; one made of many tiny hooks like he saw on the burr, and another comprised of small fibers with loops to which the hooks could attach. Today his hook-and-loop fastening invention named Velcro has been used in thousands of ways as a handy, flexible, reliable, and simple fastening system. What George de Mestral had done was demonstrate a thorough, deliberate process of engineering and design based on observations from nature (Figure 1.31). That idea is now a fast-growing area of science called biomimicry.

Burdocks

Biomimicry Like George de Mestral, scientists and engineers are embracing the concept of biomimicry by taking a hard look at natural objects and organisms to carefully study their structures, behaviors, and the way they organize as groups. Scientists have found that while many organisms are colored by pigments, others use microscopic structures on the surface of their body parts, creating shimmering, iridescent coloration based on the reflection and interference of light waves. Scientists now know how to control and use this structural color-making process.

Structural color The bright, shiny feathers of the peacock, the changing metallic colors of the longhorn beetle, and the beautiful iridescent blue of the Morpho butterfly are all examples of structural color. Specialized scales, feathers, and transparent coatings can create a dazzling array of colors by causing one reflected light ray to interfere with another reflected light ray. This process is actually quite similar to the iridescent colors created by a small amount of gasoline on wet pavement, and also by the thin, spherical walls of a soap bubble. One light ray is reflected off of the surface of a transparent layer of material, and the other is transmitted through,

Soap bubble

Peacock feather   Morpho butterfly

Figure 1.31: *Natural structures help inspire designers.*

Chapter 1 Connection

only to be reflected off the bottom of the transparent material. The color we see is based on the distance between the two reflecting layers. White light is made up of all colors, and during this process most colors will experience destructive interference and won't be seen. Each color has its own particular distance that will produce constructive interference, based on its wavelength. In fact, infrared and ultraviolet light can also be reflected, but since the human eye cannot see those wavelengths, areas that reflect those wavelengths appear black.

Controlling reflected light to make colors

Scientists have devised a way to mimic this color-producing process to make a new type of display screen. To make an actual picture, a display screen is divided up into pixels, and those pixels are divided up into sub-pixels (Figure 1.32). A pixel is one unit or "dot" of a picture. Each pixel is assigned a particular color. To create any color, red, blue, and green light are used, along with black. Each sub-pixel is made up of three small cells, one red, one blue, and one green, each measuring about 10 by 10 microns. Every cell has a top reflecting surface and a bottom reflecting surface. By changing the distance between the two reflecting layers in the cells, red, green, and blue light can be created. These cells can be switched on and off with a small, quick electronic signal using very little power. In their on position, they are the particular color they were designed to be. When they are switched off, the bottom layer moves up toward the top layer. This changes the distance between the two layers, creating ultraviolet light, which is seen as black.

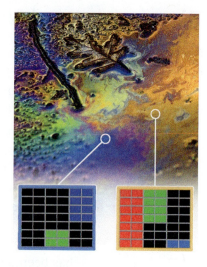

Figure 1.32: *A dark-blue pixel is made up of one particular combination of red, green, and blue subpixels, while an orange pixel is made up of a different combination of red, green, and blue subpixels.*

Power savings and efficiency

Traditional displays using liquid crystal and light-emitting diode technologies need their own internal light sources, and also must refresh each pixel up to 120 times a second, which is the main power draw in most phones and portable devices. This new technology only switches the cell on or off if the tiny part of the picture where the cell is located changes. If that part of the picture stays the same no signal is sent to the cell and no power is used. Phones using this technology get around twice the battery life of current phones because they do not need to use as much power to run the display. And since these phones use reflected light, their display can be more vivid under any light conditions. In direct sunlight, these displays look clear and crisp, while traditional displays are often washed out and hard to see. It is for these reasons of increased versatility and expanded power efficiency that this new technology stands ready to revolutionize the display of your future mobile phone and other portable electronic devices.

Chapter 1 Assessment

Vocabulary

Select the term that best completes the sentences.

natural laws	harmonic motion	analysis	potential energy
experiment	theory	energy	scientific inquiry
objective	repeatable	hypothesis	matter
observation	cycle	kinetic energy	scientific evidence
model	variable	system	joule
amps	volts	mechanics	force
equilibrium	heat	wave	qualitative
quantitative	oscillator		

1. A(n) _____ is a connected group of objects and interactions that is chosen for investigation.
2. A situation that is set up specifically to observe how something occurs is called a(n) _____.
3. Energy that is due to temperature is called _____.
4. A measurement of the color of a certain star would be an example of a(n) _____.
5. Energy that may be stored or is otherwise inactive is called _____.
6. _____ is the process of asking questions and seeking answers through objective observations and experiments.
7. _____ are units of electric current.
8. In _____ the net force is zero.
9. A(n) _____ relationship allows the computation or prediction of numerical results.
10. _____ repeats in identical cycles.
11. One _____ is one complete oscillation.
12. A scientific _____ describes how variables relate to each other and may be qualitative or quantitative.
13. A(n) _____ is a quantity that is often measured in an experiment and which typically appears in an equation.
14. A(n) _____ model provides an explanation of how and why but cannot make numerical predictions.
15. _____ are units of electrical potential energy.
16. A(n) _____ is a tentative answer to a question which may be tested by comparison with observations.
17. An object at rest has no _____ but may have other forms of energy.
18. A system that repeats a behavior in cycles is an example of a(n) _____.
19. A(n) _____ description describes only what actually happens without opinions or interpretations.
20. The part of physics that deals with forces and motion is called _____.
21. A(n) _____ is an oscillation that travels.
22. _____ is a process of thinking about scientific evidence which often seeks to deduce relationships between observed variables.
23. The _____ are the rules that humans have deduced for how we believe the universe operates.
24. A system with zero _____ has no ability to change itself or any other system.
25. A(n) _____ is the fundamental unit of energy.
26. It takes a(n) _____ to change the speed or direction of a moving object.
27. A scientific _____ is a comprehensive explanation of something in nature that is backed up by substantial observational evidence.
28. When multiple groups can obtain the same result an experiment is said to be _____.

Concept review

1. What do nerve cells, cell phones, and lightning have in common?
2. Why are wires not made of plastic?
3. Give two reasons why electricity is useful to humans.

4. What is the practical difference between a charged battery and a dead battery?
5. Describe how an animal can be an energy transformer.
6. Suppose you could throw an object in a place where there was no air and no gravity. Describe the object's motion.
7. Which has more energy, dim light or bright light? Does color matter?
8. Describe a situation in which there are forces acting but nothing is moving.
9. Give an example of positive and negative acceleration from your daily life.
10. Is the motion of a bicycle wheel an example of harmonic motion or not? Explain your answer.
11. Give an example of something with high frequency and something with low frequency.
12. What is the explanation for the different colors of light?
13. What is heat and in what unit is heat measured?
14. Provide an example of a human law and a natural law. What is the difference? (*Hint*: How are they enforced?)
15. When evaluating the truth of a hypothesis, which would you trust more: the Internet or the results of an experiment you did yourself?
16. Give an example of a hypothesis you know to be incorrect. Give another example of a hypothesis you believe to be correct. On what do you base your evaluation of correct or incorrect in both cases?
17. Suppose one person claims to have seen a flying saucer yet no one else makes a similar claim. How do you decide if there was or was not a flying saucer? Now apply the same reasoning to a particle so small you cannot see it, but someone claims it causes disease. How do you decide the truth or falsehood of the claim?
18. Which of the following would you include in a system to study the effect of nearby electric wires on the behavior of fish in a fish tank? Pick all that apply and give a reason for each choice.
    a. the water temperature
    b. the amount of light in the fish tank
    c. the color of the walls in the room
    d. the amount of time the wires are carrying electric current
    e. the distance of the wires from the fish tank

Problems

1. Which transfers more energy: 2 amps at 12 volts or 1 amp at 50 volts?
2. What is the acceleration of a car that changes its speed from 0 to 60 mph in 6 seconds?
    a. 6 mph per second
    b. 10 mph per second
    c. 60 mph per second
    d. 360 mph per second
3. What is the frequency of a rubber band that vibrates back and forth 50 times in 1 second?
4. Which of the following graphs matches the data table?

| Mass (g) | 5.0 | 3.8 | 6.8 | 2.5 | 5.6 |
| Time (h) | 2.5 | 3.5 | 1.0 | 4.5 | 2.0 |

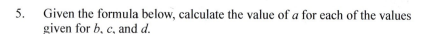

5. Given the formula below, calculate the value of a for each of the values given for b, c, and d.

$$a = \frac{b}{c} + d$$

   a. $b = 6$, $c = 2$, $d = 7$
   b. $b = 5.5$, $c = 1.5$, $d = 2.2$
   c. $b = 3.1$, $c = 12{,}000$, $d = 0.0125$
   d. $b = 4 \times 10^6$, $c = 2 \times 10^{-4}$, $d = 0$

6. A flashlight uses 3.5 joules of energy from its battery every second. The flashlight is powered by a fresh AA batteries containing 15,500 joules of energy. How long will the battery last?

UNIT 1 **MEASUREMENT AND MOTION**

CHAPTER 2

Measurement and Units

Objectives:

By the end of this chapter you should be able to:

- ✔ Express lengths in metric and English units.
- ✔ Convert measurements and calculated quantities between different units.
- ✔ Calculate the surface area and volume of simple shapes and solids.
- ✔ Work with time intervals in hours, minutes, and seconds.
- ✔ Describe two effects you feel every day that are created by mass.
- ✔ Describe the mass of objects in grams and kilograms.
- ✔ Use scientific notation to represent large and small numbers.
- ✔ Design a controlled experiment.
- ✔ Create and use a graphical model based on data.

Key Questions:

- How do you measure distance, time, and mass?
- What makes up all matter, and why is matter so varied?
- What are some important qualities of valid experiments?

Vocabulary

accuracy	data	English system	gas	inertia	plasma	solid	variable
atoms	density		graph	length	precision	speed	volume
compound	dependent variable	experimental variable	graphical model	liquid	procedure	surface area	weight
control variable	distance	exponent	independent variable	mass	scientific notation	time interval	x-axis
conversion factor	element	friction		metric system	SI system	trial	y-axis
				mixture			

2.1 Space and Time

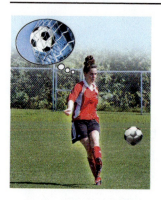

Suppose you want to kick a soccer ball through a goal. Imagine the motion of the ball as it traces a position in space every second of its flight. *Space* in physics means the three dimensions of up-down, left-right, and front-back. You can say precisely where the ball is at any moment by measuring three distances: up, forward, and to the right or left of you. The three dimensions of space are described with length units, such as meters, inches, and feet. *Time* provides another dimension for describing *when* something occurs. For example, the ball is 2 meters high, 5 meters in front, and 1 meter to the left one second after it is kicked. Together, space and time are the stage on which the universe unfolds.

Thinking about distance

Measurements   A measurement is a precise value that tells how much. How much *what*, you ask? That depends on what you are measuring. The important concept is that a measurement communicates an amount in a way that can be understood by others. For example, 2 meters is a measurement because it has a quantity, 2, and gives a unit, meters (Figure 2.1).

Units   All measurements need units. Without a unit, a measurement cannot be understood. For example, if you told someone to walk 10, she would not know what to do—10 feet, 10 meters, 10 miles, 10 kilometers are all 10, but the units are different and therefore the distances are also different. Units allow people to communicate amounts. For communication to be successful, physics uses a set of units that have been agreed upon around the world.

Distance and length   **Distance** is the amount of space between two points (Figure 2.2). You can also think of distance as how far apart two objects are. You probably have a good understanding of distance from everyday experiences, like the distance from one house to another, or the distance between two states.

Distance is measured in units of length   Distance is measured in units of **length.** You are already familiar with some units of length, like inches and miles. Others you may not have used before, like kilometers and millimeters.

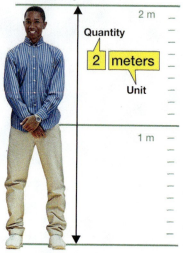

Figure 2.1: *Measurements always include both a quantity and a unit. The unit is how you know what the quantity means.*

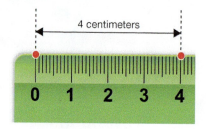

Figure 2.2: *The definition of distance.*

Two common systems of length units

There are two common systems

There are two common systems of standardized units that are used for measuring distance: the **English system** and the **metric** or **SI system**. The English system uses *inches* (in), *feet* (ft), *yards* (yd), and *miles* (mi) for length. The metric system uses *millimeters* (mm), *centimeters* (cm), *meters* (m), and *kilometers* (km). You probably have contact with both systems of units every day. For example, driving distances are often expressed in miles (Figure 2.3), but races in track and field are sometimes expressed in meters (Figure 2.4).

Scientists use metric units

Almost all fields of science use metric units because they make calculations easier. In the English system, there are 12 inches in a foot, 3 feet in a yard, and 5,280 feet in a mile. In the metric system, there are 10 millimeters in a centimeter, 100 centimeters in a meter, and 1,000 meters in a kilometer. Factors of 10 are easier to remember and use than 12, 3, and 5,280. The diagram below will help you get a sense of the metric units of distance.

Figure 2.3: *Some distances are measured using units in the English system, such as miles.*

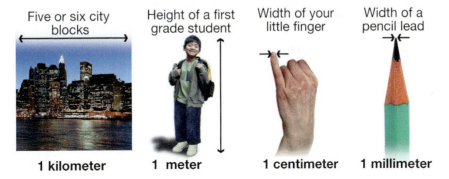

Figure 2.4: *Some distances are measured using units in the metric system, such as meters.*

You will use both systems of measurement

To solve real-world problems, you will need to know both sets of units, English and metric. For example, a doctor will measure your height and weight in English units. The same doctor will prescribe medicine in milliliters (mL) and grams (g), which are metric units. Plywood is sold in 4-foot-by-8-foot sheets, but the thickness of some types of plywood is given in millimeters. Some of the bolts on a car have English dimensions, such as 1/2 inch. Others have metric dimensions, such as 13 millimeters. Virtually all of the bolts on a bicycle are in metric units. Because both units are used, it is a good idea to know both metric and English units.

Measuring length

Scientists use the metric system
The metric system is officially named *Le Système International d'Unités* from the French name and is abbreviated SI. As you probably learned in other courses, there are four length units related to the meter (Table 2.1). We will use all four at different times because each is convenient for describing a different range of lengths.

Table 2.1: Metric length units

Length	Equivalent length
1 m	100 cm
1 m	1,000 mm
1 cm	10 mm
1 km	1,000 m

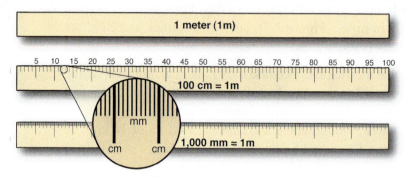

Table 2.2: English length units

Length	Equivalent length
1 inch	0.0833 feet
1 foot	12 inches
1 yard	3 feet
1 mile	5,280 feet

Inches, feet, yards and miles
The common units of measurement in the English system are inches, feet, yards, and miles. An inch is about the length of the first joint of your thumb. Things smaller than an inch are measured in fractions of inches, like one-fourth or one-eighth of an inch. Notice these fractions are multiples of one-half, not of one-tenth, like in the metric system. Table 2.2 shows the relationships between inches, feet, yards, and miles. Remembering these ratios is not as easy as remembering powers of 10.

Converting from one unit to another
It is often necessary to take a measurement in one unit and convert it into a different unit. This is done with relationships between units called **conversion factors**. To do a conversion, you arrange the conversion factors as ratios and multiply them so the units cancel out. The example shows how to convert 100 kilometers to miles.

Converting units

$$100 \text{ km} \times \frac{1{,}000 \text{ m}}{1 \text{ km}} \times \frac{1 \text{ mile}}{1{,}609 \text{ m}} = \frac{100 \times 1{,}000 \text{ miles}}{1{,}609} = \mathbf{62.1 \text{ miles}}$$

Cancel units in pairs

Converting length in yards to meters

A football field is 100 yards long. What is this distance expressed in meters?

1. You are given distance in yards.
2. You are asked to find distance in meters.
3. 1 yard = 3 feet, 1 foot = 0.3048 meters
4.

$$100 \text{ yds} \times \frac{3 \text{ ft}}{1 \text{ yd}} \times \frac{0.3048 \text{ m}}{1 \text{ ft}} = 91.4 \text{ m}$$

Time

The meaning of time

Time is a fundamental property of the universe. Time is difficult to define but intuitively you know the passage of seconds, minutes, and hours marks the steady flow of the past into the present and future. Our bodies change with time (Figure 2.5). Earth and the universe change with time. Learning how things change with time motivates much of our study of nature. Many key physics concepts are based on how certain quantities change with time, such as your location.

Two ways to think about time

In physics, just as in your everyday life, there are two ways to think about time (Figure 2.6). One way is to identify a particular moment in time. The other way is to describe a quantity of time. The same word—time—means two things that are very different in science.

What time is it?

If you ask, "What time is it?" you usually want to identify a moment in time relative to the rest of the universe and everyone in it. To answer this question, you would look at a clock or at your watch at one particular moment. For example, 3:00 p.m. Eastern time on April 21, 2014, tells the time at a certain place on Earth.

How much time?

If you ask, "How much time?" or how long did something take to occur, you are looking for a quantity of time. To answer such questions, you need to measure an interval of time with both a beginning and an end. For example, you might measure how much time has passed between the start of a race and when the first runner crossed the finish line. A quantity of time is often called a **time interval**. Whenever you see the word *time* in physics, it usually means a time interval. Time intervals in physics are almost always in seconds and are represented by a lower case letter *t*.

Figure 2.5: *The flow of time is an important part of our experience of life. In order to understand nature, we need to investigate how things change with time.*

Figure 2.6: *There are two different ways to understand time: "What time is it?" and "How much time passed?"*

How is time measured?

Units for measuring time

You probably know the common units for measuring time: seconds, hours, minutes, days, and years. In calculations, you usually need to convert time intervals to seconds and that means using the relationships between units of time. Table 2.1 gives some of the most useful relationships between units of time.

Table 2.1: Some units for time

Time unit	... in seconds...	... and in days
1 second	1	0.0001157
1 minute	60	0.00694
1 hour	3,600	0.0417
1 day	86,400	1
1 year	31,557,600	365.25
1 century	3,155,760,000	36,525

Figure 2.7: *Electronic timers have displays that show mixed units. Colons and a period separate the units.*

Choosing the right unit is important

Imagine that somebody asks you how old you are. You would probably not give your age in seconds. The number would be too big and would change too fast. A year is a better unit for giving a person's age. Different units allow useful amounts to be described with more convenient numbers, between 1 and 1,000.

Physics requires time in seconds

Minutes and hours are convenient for everyday time measurement. To make matters even more confusing, time is often given in mixed units, such as 2 minutes and 15 seconds. In physics, calculations usually require time in seconds. If you have a time interval that is in mixed units, converting it to seconds will make the calculations easier.

Convert mixed units to seconds

To convert a time into seconds, you first separate the total into the amount of time in each unit. Then you convert the amount of time in each unit to seconds. Finally, you add up the number of seconds in each amount to get the total time in seconds.

Reading a digital timer

Most timing equipment displays time in hours, minutes, and seconds with colons separating each unit. The seconds number may also have a decimal point that shows fractions of a second (Figure 2.7). To read a timer, you need to recognize and separate the different units.

Converting a mixed time to seconds

How many seconds are in 1 hour, 26 minutes, and 31.25 seconds?

1. You are asked for time in seconds.
2. You are given a time interval in mixed units.
3. 1 hour = 3,600 seconds
   1 minute = 60 seconds
4. Do the conversion:
   1 hour = 3,600 seconds
   26 minutes = 26 × 60
                 = 1,560 seconds
   Add all the seconds:
   t = 3,600 + 1,560 + 31.25
     = 5,191.25 seconds

Time scales in physics

One second — One second is about the time it takes to say "one thousand." There are 60 seconds in a minute and 3,600 seconds in an hour. There are 3,600 times 24, or 86,400 seconds in an average day of 24 hours. The second is the basic unit of time in both the English and metric systems.

Time in physics — Events in the universe happen over a huge range of time intervals. Figure 2.8 gives a few examples of time scales that are considered in physics as well as other sciences. The average life of a human being is 2.2 billion seconds. The time it takes a mosquito to flap its wings once is 0.0005 seconds. The time it takes light to get from this page to your eyes is 0.000000002 seconds.

Time in experiments — In many experiments, you will be observing how things change with time. For example, when you drop a ball, it falls to the ground. You can make a graph of the height of the ball versus the time since it was released. The *time* is the time interval measured from when the ball was released until it hits the ground. This graph shows how the height of the ball changes with time. The graph shows that it takes the ball about 0.45 seconds to fall a distance of 1 meter. Many of the experiments you will do in the lab will involve measuring times between 0.0001 seconds and a few seconds. When making graphs of results from experiments, the time almost always goes on the horizontal or *x*-axis.

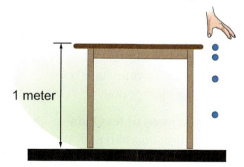
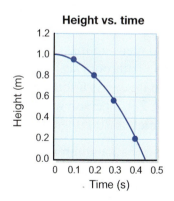
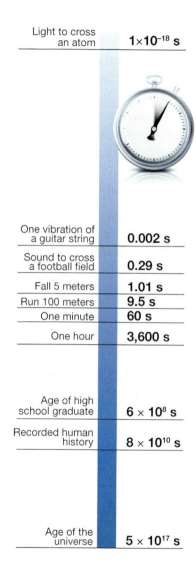

Figure 2.8: *Some time intervals in physics are shown here.*

2.2 Mass, Matter, and Atoms

Mass is the third of three basic quantities that are the foundation of physics, along with length and time. Once you can describe mass, length, and time, you have a framework for understanding the existence and motion of things. This section is about mass and also about atoms. Virtually all of the mass in things around you is in the form of atoms. A block of steel looks smooth because individual atoms of steel are so small you cannot see them directly. The air is transparent because the atoms in air are far apart and light goes right between them.

Two ways to think about mass

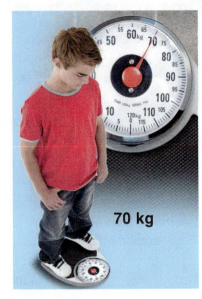

Figure 2.9: *Weight is a measure of the force of gravity pulling on mass.*

Mass is a measure of matter — Very simply, **mass** is the amount of "stuff" an object contains. Matter is anything that has mass and takes up space. *You* are matter, and so is the air you breathe, and the paper in this book. Mass is a measure of how much matter there is. A grain of salt has a very small mass relative to the mass of a planet like Earth.

The two effects of mass — The amount of mass an object contains has two important effects:

1. The more mass an object has, the more **weight** it has. Weight is the force exerted by Earth's gravity (Figure 2.9). Gravity is proportional to an object's mass, which means twice the mass creates twice the weight.

2. The more mass an object has, the harder it is to start it moving or stop it from moving. The tendency of an object to resist changes in its motion is called **inertia**. Inertia comes from mass.

Weight — The most common way to measure mass is to measure the force of gravity acting on it. When you stand on a bathroom scale, you are measuring the force of gravity acting on your mass. Gravity exerts a force of 2.2 pounds for every kilogram of mass. As shown in Figure 2.9, a person with a mass of 70 kilograms would have a weight of 154 pounds, which is 70 kilograms times 2.2 kilograms per pound.

Inertia — It is easy to pick up a tennis ball and throw it. A tennis ball has a mass of less than 0.1 kilogram. It is much more difficult to pick up a 5-kilogram bowling ball and throw it (Figure 2.10). Throwing a ball involves a large change in speed. Throwing a bowling ball is harder because a bowling ball has 50 times as much mass as a tennis ball and therefore has 50 times as much inertia to overcome.

Figure 2.10: *Objects with more mass are harder to move than objects with less mass.*

Measuring mass

The kilogram Mass is measured in kilograms. One kilogram was originally defined as the mass of 1 liter of water. A liter is a volume of 1,000 cubic centimeters (Figure 2.11). The 1-liter bottle of soda you buy in the grocery store is mostly water and has a mass of about 1 kilogram.

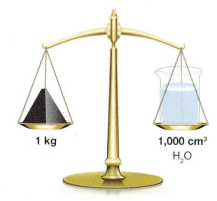

Figure 2.11: *The kilogram is defined as the mass of 1,000 cubic centimeters of water.*

Equal masses mean equal amounts of matter When we say an object has a mass of exactly one kilogram, we are saying that the object contains the same amount of matter as one kilogram of water. Even if we do not know what the object is made of, we can measure how much matter it has. Air is very light, but air still has mass. One cubic meter of air has a mass of about 1 kilogram. If you put your hand out of the window in a moving car you can feel the mass of the air.

Grams For small amounts of matter, like medicines, the kilogram is too large a unit of mass to be convenient. One gram is one thousandth of a kilogram. Equivalently, 1 kilogram is 1,000 grams. One grain of rice is about a gram.

The range of masses Like distance and time, you will encounter a very wide range of masses in physics. A single bacteria has a mass of 0.000000001 kilogram. Later in this book we will do calculations with atoms, which have masses a *thousand billion* times smaller than a bacteria. We will also analyze the motion of Earth and the Moon. The mass of Earth is about 6 trillion trillion kilograms. That is a 6 with 24 zeroes after it. Figure 2.12 shows the range of masses of some known objects in the universe.

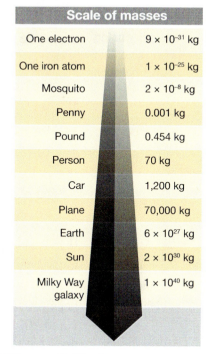

Figure 2.12: *The range of masses in the universe.*

2.2 MASS, MATTER, AND ATOMS

Very large and very small numbers

Describing large and small quantities
It is hard to imagine doing math with a number like 6 trillion trillion. Fortunately, there is a shorthand method to deal with numbers this large. The method is called **scientific notation** and it also works for extremely small numbers too, like the mass of an atom. You will need to learn this method of working with large and small numbers because physics covers such a wide range of length, time, and mass.

Scientific notation for large numbers
In scientific notation, numbers are written as a value between 1 and 10, multiplied by a power of 10 called the **exponent**. For example, the number 6 trillion trillion (6,000,000,000,000,000,000,000,000) is written as 6.0×10^{24}. Can you see why scientists like to use scientific notation to express very large or very small numbers? Below is a step-by-step example of how to write a large number in scientific notation.

Steps in writing a number using scientific notation
The average distance from Earth to the Sun is 150,000,000 kilometers. Write this value using scientific notation. The steps are shown in Figure 2.13.

1. Move the decimal until you get a value that is between 1 and 10. Count the number of times you move the decimal.
2. Write the new number without all of the zeros.
3. Write "× 10" after the number.
4. Write the number of times you moved the decimal as the exponent, or power of 10. If you moved the decimal to the *left*, the exponent will be positive. If you moved the decimal to the *right*, the exponent will be negative.

If you did all of the steps correctly, you will get 1.5×10^8 km.

Scientific notation for numbers smaller than one
Scientific notation also works for numbers less than one. For example, the number 0.0025 can be written as 2.5×10^{-3}. The exponent is negative because you had to move the decimal to the right to make a number between 1 and 10. It is important to remember that a *negative* sign on the exponent does not mean the whole number is negative. Negative exponents mean a value that is less than one.

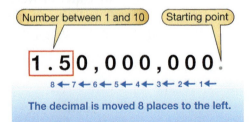

Figure 2.13: *Writing a number using scientific notation.*

Matter and atoms

Atoms Some of the most fundamental insights into how the universe works, like explaining temperature, deal with matter on its smallest scale—**atoms**. You do not experience atoms directly because they are so small. The head of a pin contains 10^{20} atoms. Aluminum foil is thin but still more than 200,000 atoms thick (Figure 2.14). A single atom is about 10^{-10} meters in diameter. That means you might lay 10,000,000,000 (10^{10}) atoms side by side in a 1-meter long space.

Whether matter is a solid, liquid, or gas depends on how the atoms are organized.

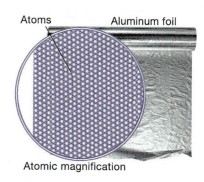

Figure 2.14: *A sheet of aluminum foil is 200,000 atoms thick.*

Solid A **solid** is "solid" because the atoms inside are connected firmly to each other and do not exchange places. Solids tend to move as a whole. They do not flow like a liquid or expand like a gas (Figure 2.15). Imagine a marching band marching in place with *everyone holding hands.* People move but they stay in the same spot relative to others. When the marching band moves, everyone moves together, like the atoms in a solid.

Liquid A **liquid** flows because atoms are more loosely connected to each other. Atoms are still close together, like in a solid, but they can easily exchange places with their neighbors (Figure 2.15). Imagine a room full of people dancing. The crowd generally stays together, *but people can move around and switch partners*, like the atoms in a liquid.

Gas A **gas** consists of atoms that are spaced far apart from each other (Figure 2.15). Gas can expand because atoms are free to move independently. Imagine many people running fast in different directions. Every person is moving independently with a lot of space between people, like the atoms in a gas.

Plasma At temperatures greater than 11,000°C, atoms in a gas start to break apart. In a **plasma**, atoms of matter separate into electrons and ions. Because the electrons are free to move independently, plasma can conduct electricity. Lightning and the Sun are good examples of plasma.

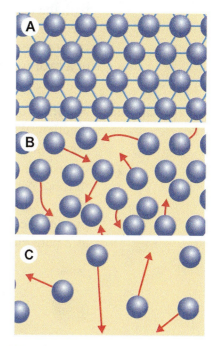

Figure 2.15: *A) Atoms in a solid stay bonded together; B) in a liquid are close together but can move and change places; and C) in a gas are far apart and move fast.*

2.2 MASS, MATTER, AND ATOMS

The diversity of matter

The incredible variety of matter
There is an incredible diversity of matter around you—air, water, sugar, skin, concrete, glass, wood, plastic, and so on. This diversity comes from combining *elements* into *compounds*, then compounds into *mixtures* of compounds. All the millions of variations in the properties of matter, such as hardness, color, and density, come from mixtures and compounds of only 92 different elements.

The elements and their atoms
There are 92 different types of atoms in ordinary matter. Each type is called an **element**. Each element has is own properties, such as mass and the ability to combine with other elements. The atoms with the smallest mass are those of the element hydrogen. The mass of a single hydrogen atom is 1.67×10^{-27} kilograms. Most of the hydrogen on Earth is found in water. The naturally-occurring atoms with the largest mass are those of the element uranium. Uranium is found in certain kinds of rock, combined with oxygen and other elements. The element iron is in between hydrogen and uranium. An atom of iron has about 56 times the mass of a hydrogen atom, at 9.3×10^{-26} kilograms. This is still an incredibly small mass, but there are a *lot* of atoms in any piece of iron big enough to see!

Compounds
A **compound** is a substance that contains two or more different elements bonded together. Water is an example of a compound. If you could look at water with a powerful atomic microscope you would find each particle of water is made from one oxygen atom and two hydrogen atoms. Another compound, glucose is a sugar in food. A single glucose molecule is made of carbon, oxygen, and hydrogen atoms. Almost all the atoms in ordinary matter are combined into compounds; pure elements are quite rare. One reason gold is prized by humans is that it exists as a pure element and not as a compound.

Figure 2.16: *Matter is everywhere you look.*

Figure 2.17: *Wood is a mixture of many compounds, with the most abundant being cellulose and water.*

Mixtures
The matter you normally experience is made of **mixtures** of compounds. Wood is a mixture that contains water and more than 100 other compounds. Air is a mixture that contains oxygen, nitrogen, water, and other elements and compounds.

2.3 Experiments and Data

An experiment is a situation you set up specifically to observe something. **Data** are the measurements and calculations that you make during the experiment. A well-designed experiment provides data that is clear and easy to interpret. An experiment that gives data that is hard to interpret can lead to the wrong conclusion. This section will provide you with guidelines for doing good experiments that tell you meaningful things, both in science class and also in real life.

Fundamental and derived quantities

Figure 2.18: *Speed limit signs in the United States have implied units of miles per hour (mph).*

Fundamental quantities   Length, mass, and time are fundamental quantities. Each has its own unit that is not made of other units. For example, a meter is a fundamental length unit and it does not depend on time, gravity, mass, or anything else.

Derived quantities   Some things you measure in experiments are fundamental quantities, but others are derived quantities. Derived quantities can be measured but are often calculated from things you measure.

Speed is a derived quantity   **Speed** is a derived quantity. Speed is calculated from distance and time. The units of speed are a combination of distance units and time units. Consider a speed limit sign that says "65 miles per hour." In this example, the units are the mile and the hour. Speed always has units of length divided by time, such as meters per second or miles per hour. Other derived quantities include energy, density, acceleration, intensity, and frequency. Each of these units can be broken down into combinations of the fundamental units of length, mass, and time.

Converting derived quantities   When converting derived quantities, you may have to convert each unit separately. For example, if you want to compare a laboratory car with a real car, you might want to convert 2 meters per second (2 m/s) into miles per hour. This means converting meters to miles and seconds to hours.

$$2 \text{ m/s} = \frac{2 \text{ m}}{1 \text{ s}} \times \boxed{\frac{3{,}600 \text{ s}}{1 \text{ h}}} \times \boxed{\frac{1 \text{ mi}}{1{,}609 \text{ m}}} = \frac{2 \times 3{,}600}{1{,}609} \frac{\text{mi}}{\text{h}} = 4.5 \text{ mph}$$

1st conversion factor   2nd conversion factor

Converting a speed from cm/s to mph

A car on a ramp is measured to go 45 cm in 1.5 s. What is the speed in miles per hour?

1. You are asked for speed in mph.
2. You are given a time interval in seconds and distance in cm.
3. You know:
   speed = distance/time
   1 hour = 3,600 s
   1 mile = 1,609 m
   1 m = 100 cm

$$\text{speed} = \frac{\text{distance}}{\text{time}} = \frac{45 \text{ cm}}{1.5 \text{ s}}$$

$$= \frac{45 \text{ cm}}{1.5 \text{ s}} \times \frac{3{,}600 \text{ s}}{1 \text{ h}} \times \frac{1 \text{ m}}{100 \text{ cm}} \times \frac{1 \text{ mi}}{1{,}609 \text{ m}}$$

$$= \frac{45 \times 3{,}600}{1.5 \times 100 \times 1{,}609} \frac{\text{mi}}{\text{h}}$$

$$= 0.67 \text{ mph}$$

Surface area and volume

Surface area Many processes in nature require you to describe how much surface or volume an object has. The units of surface and volume are derived from the unit of length. For example, a surface that measures 2 meters by 2 meters has a **surface area** of 4 square meters (m²). That means four 1-meter squares completely cover the surface. If the measurements of the sides are in inches, then the area will be in square inches. Figure 2.19 shows the surface area of three simple plane figures.

Volume The **volume** of an object is equal to the number of unit cubes that completely fill the object. For rectangular objects, volume is length times width times height. For example, the volume of a 2-meter cube is 8 cubic meters (2 m × 2 m × 2 m = 8 m³). Eight 1-meter cubes completely fill the space inside. The volume for some simple solid shapes is shown in the diagram (Figure 2.20).

Surface area of solid objects A solid object has surface area as well as volume. For example, the 2-meter cube has a surface area of 24 square meters (24 m²). There are six faces on a cube and each face is 4 square meters, making a total of 24 m².

Converting area and volume You need to apply the conversion factor *twice* when converting areas from one unit to another. This is because the units of area are length times width. For example, the 2-meter square has an area of 40,000 square centimeters (cm²). For a volume conversion you apply the conversion factor *three* times because the units of volume are length times width times height.

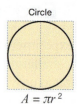

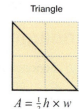

Figure 2.19: *The area of plane figures is expressed in square units.*

Surface area	
Sphere	$S = 4\pi r^2$
Cylinder	$S = 2\pi r^2 + 2\pi rh$
Rectangular solid	$S = 2hw + 2hl + 2wl$

Volume	
Sphere	$V = \frac{4}{3}\pi r^3$
Cylinder	$V = \pi r^2 h$
Rectangular solid	$V = lwh$

Figure 2.20: *The surface area and volume formulas are made up of the area formulas for plane figures.*

Calculating surface area and volume

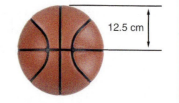

A basketball has a radius of 12.5 centimeters. Calculate the surface area and volume of the ball.

1. You are asked to find surface area and volume.
2. A ball is shaped like a sphere; you are given the radius.
3. Surface area: $A = 4\pi r^2$; volume: $V = (4/3)\pi r^3$
4. Solve:

Surface area

$A = 4(3.14)(12.5)^2 = 1{,}963 \text{ cm}^2$

Volume

$V = \frac{4}{3}(3.14)(12.5)^3 = 8{,}181 \text{ cm}^3$

Density

Is size an indication of mass? Imagine you have three solid 1-meter cubes. One is made of polyethylene plastic, one of iron, and the third of glass (Figure 2.21). Think about how they are different. Each of the three is clearly a different *type* of matter. However, all three cubes are the same size and shape and therefore have the same volume. Do they contain the same *amount* of matter?

Density is mass per unit volume. A block of plastic and a block of iron may be the same size but one has a lot more mass than the other, and therefore contains more matter. Because of the difference, plastic floats in water and iron sinks. Whether an object floats or sinks in water is related to its density. **Density** describes how much mass is in a given volume of a material. The units of density are mass divided by volume, such as kilograms per cubic meter, kg/m^3. Iron has a high density; it contains 7,800 kilograms of mass per cubic meter, $7,800 \text{ kg/m}^3$. A 1-centimeter cube of polyethylene plastic contains only 940 kilograms of matter, 940 kg/m^3.

Density is a property of matter, independent of size or shape Solids range in density from cork, with a density of 120 kg/m^3, to platinum, a precious metal with a density of $21,500 \text{ kg/m}^3$. The density of water is about $1,000 \text{ kg/m}^3$ or 1 g/cm^3. The density of air is much lower, about 1 kg/m^3.

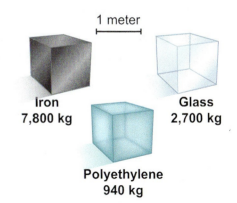

Figure 2.21: *Identically-sized cubes of iron, polyethylene, and glass contain different amounts of mass.*

Table 2.2: Densities of common substances

Material	Density (kg/m^3)	Material	Density (kg/m^3)
Platinum	21,500	Nylon Plastic	2,300
Lead	11,300	Rubber	1,200
Iron	7,800	Liquid water	1,000
Titanium	4,500	Polyethylene plastic	940
Aluminum	2,700	Ice	920
Glass	2,700	Oak (wood)	600
Granite	2,600	Pine (wood)	440
Concrete	2,300	Cork	120

Converting density measurements Like other derived units, converting between units of density involves more than one conversion factor. Water is $1,000 \text{ kg/m}^3$ or 1 g/cm^3. The conversion factor of 1 meter per 100 centimeters is applied three times because of the exponent 3 in m^3.

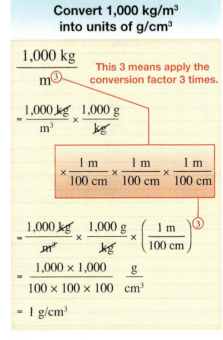

Figure 2.22: *Converting between density units.*

Accuracy and precision

Measurements are never "perfect"
Measurements are never exact. For example, you cannot determine that something takes *exactly* 10 seconds. Why not? Because all measurements of time are made with clocks, and all clocks have a limit to how small a time interval they can measure. For example, suppose you have a very good clock that can measure time to 0.01 seconds. You claim the time is exactly 10 seconds because your clock shows 10.00. However, suppose the time was in fact 10.002 seconds. Your clock rounded the measurement off to 10.00. So the best way to describe your time measurement is 10.00 $\pm$ 0.005 seconds. The actual time could have been different from 10.00 seconds by up to 0.005 seconds in either direction, and your clock would not show it.

Precision
The word **precision** means how small a difference a measurement can show. A clock that can read to 0.01 seconds has a precision of 0.01 seconds. Many of your lab experiments will be done with clocks that are precise to one ten-thousandth of a second or 0.0001 s.

Accuracy
The word **accuracy** in physics means how close a measurement is to the true value. For example, a meter stick that has been stretched can make a measurement of length that is precise to one millimeter. But the measurement will not be accurate because the meter stick is no longer a meter long! Figure 2.23 illustrates the meanings of accuracy and precision.

Why accuracy and precision are important
Accuracy and precision are important because experiments are done to see whether they agree or disagree with what you believe will happen. A measurement that is not accurate may give you the wrong conclusion. A measurement that is not precise may not help distinguish agreement from disagreement.

Comparing measurements
In physics and in life, whether two things are the same depends on how closely you need to look. For many of your lab experiments, two time measurements of 0.0233 s and 0.0234 s can be considered the same because they differ by only 0.0001 s. This time interval is the precision limit of the clock (Figure 2.24). A similar limit, 0.001 m, exists for measuring length with a meter stick. When analyzing the observations you make in the lab, you must consider both the accuracy and the precision of your measurements before making a conclusion.

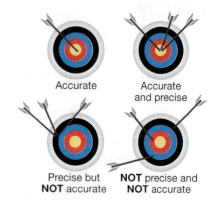

Figure 2.23: *A measurement can be accurate and precise or neither.*

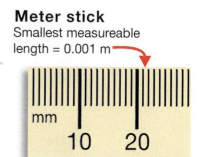

Figure 2.24: *The precision of a measuring tool is related to the smallest measurement it can make.*

Variables and relationships

Variables Factors that affect the results of an experiment are called **variables**. For example, consider a car rolling on a level track. Speed is a variable. Time is another variable. The distance the car moves is another variable. The whole science of physics can be thought of as *the search for the relationships between all the variables that describe everything.* Of course, "everything" is a lot to consider, so to learn about something in nature, scientists instead select a small set of related variables called a *system*. For example, to learn about speed you might choose to look for the relationship between time and distance.

Distance and time data for car on a ramp

Time	Distance
0.00	0.0
0.49	25.0
1.02	50.0
1.48	75.0
2.01	100.0
2.52	125.0
3.04	150.0
3.46	175.0
4.00	200.0
4.45	225.0
5.11	250.0
5.49	275.0
6.05	300.0

Data from an experiment The data in Figure 2.25 show measurements of the distance traveled by a small car rolling on a perfectly level track with no friction. The time it takes the car to reach each distance is also shown. What does this data tell us about the relationship between distance and time for this car?

$$\text{speed} = \frac{\text{distance}}{\text{time}} = \frac{50 \text{ cm}}{1.02 \text{ s}} = 49.0 \text{ cm/s}$$

Calculating speed Notice that three different pairs of data are used to calculate the speed of the car. Each time it turns out to be around 50 centimeters per second. The experiment tells us the car is moving at a constant speed of 50 cm/s. That is the relationship between the time and distance. For every second the car travels, it moves 50 cm.

$$\text{speed} = \frac{\text{distance}}{\text{time}} = \frac{125 \text{ cm}}{2.52 \text{ s}} = 49.6 \text{ cm/s}$$

$$\text{speed} = \frac{\text{distance}}{\text{time}} = \frac{225 \text{ cm}}{4.45 \text{ s}} = 50.6 \text{ cm/s}$$

Relationship

$$\frac{\text{distance}}{\text{time}} = 50 \text{ cm/s}$$

Speed is a relationship between time and distance The value of speed is the distance an object moves in one unit of time. A speed of 50 cm/s means the car moves 50 cm for each second it travels. If it travels for five seconds, then we expect it to move a distance of 50 cm/s times 5 s or 250 cm. The data do indeed show that the car moves 250 cm in about 5 seconds. This calculation shows us another powerful way to use units. If we want to find a distance and we know the speed and the time, we can arrange the variables so the units cancel out except for the one we want. This tells us distance is speed times time.

Figure 2.25: *Collected data from an experiment observing the motion of a car rolling down a ramp can be used to calculate its average speed.*

Original relationship → **Multiply by time and cancel** → **New form of relationship**

$$\text{speed} = \frac{\text{distance}}{\text{time}} \quad\quad \text{time} \times \text{speed} = \text{time} \times \frac{\text{distance}}{\text{time}} \quad\quad \text{time} \times \text{speed} = \text{distance}$$

Variables for a car on a ramp

The speed of a ball on a ramp

When a car is set on a ramp, it rolls down. The farther down the ramp the car rolls, the faster it goes. Many of the important relationships in the physics of motion were discovered by observing objects rolling down ramps. For example, both Galileo and Isaac Newton conducted experiments with balls on ramps. The laws of motion you will learn in Chapter 5 are physical models that were deduced by observing balls rolling up and down ramps. Of course, the laws apply to much more than just balls on ramps! However, a ball on a ramp is a simple system to observe accurately and therefore makes the laws clear to see.

The initial speed

Consider an experiment that measures the speed of a car at different positions as it rolls down a straight ramp. The speed depends on a number of variables. The most important ones are shown in Figure 2.26. One variable is the speed of the car, if any, when it is released. The *initial speed* is the speed an object has at the start of the experiment. The initial speed often depends on the starting point of the motion. The starting point is another variable.

The angle

The *angle* of the ramp is another important variable. Downhill motion is caused by gravity. On a level surface, gravity pulls directly against the surface and no motion results. The steeper the angle, the larger the force of gravity directed along the ramp. As we will see, larger forces create more rapid increases in speed.

Friction

Friction is another variable. All motion creates **friction**. Friction acts to reduce motion, or slow things down. A certain fraction of the force of gravity is taken up by overcoming friction. If there were no friction, a car would have even greater speed at the bottom of a ramp.

Size, mass, and shape are also important variables

The mass and size of the car are also variables that affect its motion. It may surprise you to know that mass does not have as big an effect on the speed of the car as you might think. The way the car's wheels are made also matters. A light car with heavy wheels will increase speed at a different rate compared to a heavy car with light wheels. The difference comes from *rolling* motion compared to straight-line motion in which rolling is not involved. Rolling motion can be complicated and is discussed later, in Chapters 8 and 9.

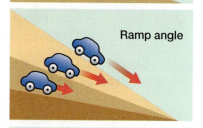

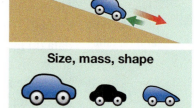

Figure 2.26: *The most important variables that affect the speed of a car rolling down a ramp are seen here.*

Experimental design

Control and experimental variables

We do experiments to find out what happens when we change a variable, such as the angle of a ramp. The variable that is changed is called the **experimental variable**. The variables that are kept the same are called the **control variables** (Figure 2.27). When you change one variable and control all of the others, we call it a *controlled experiment*. Controlled experiments are the best way to get reliable data. If you observe that something happens, such as the ball going faster, you know *why* it happened—the ramp became steeper. There is no confusion over which variable caused the change.

Experiments often have several trials

Many experiments are repeated many times. For example, you might roll a ball down a ramp 10 times. Each repetition of the experiment is called a **trial**. To be sure of your results, each trial must be as close to identical as possible to all the other trials. In an ideal experiment, the only allowed change from trial to trial is in the one variable you are testing, the experimental variable.

Experimental technique

Your *experimental technique* is how you actually do the experiment. For example, you might release the ball using one finger. If this is your technique, you want to do it the same way every time. By developing a consistent technique, you make sure your results accurately show the effects of changing your experimental variable. If your technique is not replicable, you may not be able to tell if any differences you observe are due to the technique or to changes in your experimental variable.

Procedures

The **procedure** is a collection of all the techniques you use to do an experiment. Your procedure for testing the ramp angle might have several steps. Good scientists keep careful track of their procedures so that they can come back another time and repeat their experiments. Writing the procedures down in a lab notebook is a good way to keep track of them (Figure 2.28).

Scientific results must always be repeatable

What good would a new discovery be if nobody believed you? Having good techniques and procedures is the best way to be sure your results are *repeatable*. Discoveries must always be able to be confirmed by someone other than you. If other people can follow your procedure and get the same results, then most scientists would accept your results as being true. Writing good procedures is the best way to ensure that others can repeat and verify your experiments.

Experimental question
How does the angle of a ramp change the speed of the ball?

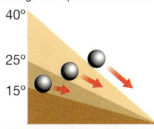

Experimental variable	Control variables
angle	initial speed
	position
	mass
	size
	shape
	starting point

Figure 2.27: *Experimental variables and control variables depend on the purpose of the experiment.*

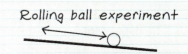

Rolling ball experiment

1. Measure from the top of the ramp.
2. Use a single photogate to measure the speed of the ball.
3. Speed is diameter divided by time through the beam.
4. Release the ball using one finger.
5. Make measurements every 10 cm.

Position (cm)	Time (s)	Speed (cm/s)
10	0.0208	91.2
20	0.0147	129
30	0.0120	158

Figure 2.28: *An example of a procedure with some data written in a lab notebook is seen here.*

Graphical data

Graphs A graph shows how two variables are related with a picture that is easy to understand. This makes graphs a good way of representing data. The example shows that the speed of a ball changes as it rolls downhill. You can see from the graph that the farther the ball goes, the greater its speed. The information in the graph is the same as the information in the table, too. We make graphs because they are easier to read than tables of numbers.

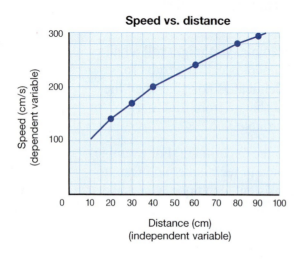

Distance (cm)	Speed (cm/s)
20	140
30	171
40	198
60	242
80	280
90	297

How to graph data accurately

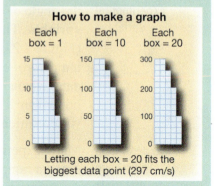

Letting each box = 20 fits the biggest data point (297 cm/s)

1. Decide what variables should be on the *x*- and *y*-axes.
2. Make a scale for each axis by counting boxes to fit your largest value. Count by multiples of 1, 2, 5, or 10 to make plot points. Make the graph big and include units on each axis. Use as much area on the graph paper as you can.
3. Plot your points by finding the *x*-value and drawing a line upward until you get to the right *y*-value. Put a dot for each point.
4. Draw a smooth curve or straight line that shows the pattern of the points. Do not simply connect the dots.
5. Create a title for your graph.

The dependent variable To a scientist, a graph is a picture that shows the relationship between two variables. By convention, graphs are drawn a certain way just like words are spelled certain ways. The **dependent variable** goes on the **y-axis** which is vertical. In the example, speed is the dependent variable because we believe the speed *depends* on how far down the ramp the ball gets.

The independent variable The independent variable goes on the horizontal or **x-axis**. In the example, distance is the **independent variable**. We say it is *independent* because we are free to make the distance anything we want by choosing where on the ramp to measure. The variable *time* is sometimes an exception to this rule. Time usually goes on the *x*-axis, even though we are not always able to control it.

Graphical models

A graph is also a model

A **graphical model** uses a graph to make predictions based on the relationship between the variables on the *x*- and *y*-axes. A **graph** is a form of a mathematical model because it shows the connection between two variables (Figure 2.29).

Using a graphical model to make a prediction

Suppose you want to find out what the speed of the ball would be 50 centimeters from the start. You did not measure the speed there, but the graph can give you a very accurate answer.

1. To predict the speed, start by finding 50 centimeters on the *x*-axis.
2. Draw a line vertically upward from 50 centimeters until it hits the curve you drew from the data.
3. Draw a line horizontally until it reaches the *y*-axis.
4. Use the scale on the *y*-axis to read the predicted speed.
5. For this example, the model graph predicts the speed to be about 220 cm/s.

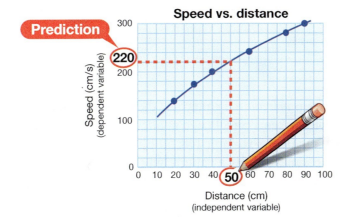

Checking the accuracy of a model

If the graph is created from accurate data, the prediction will also be accurate. You could check by doing another experiment and measuring the speed of the ball at 50 centimeters. You should find it to be very close to the prediction from your graph. Although useful, graphical models are limited because a graph does not show *why* the connection exists or *how* one variable affects the other.

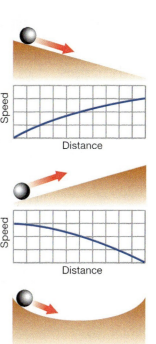
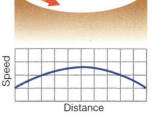
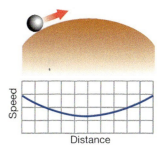

Figure 2.29: *The speed versus distance graphs show how the ball's speed changes as the shape of the ramp changes.*

Recognizing relationships in data

Cause and effect relationships In many experiments, you are looking for a cause-and-effect relationship. How does changing one variable affect another? Graphs are a simple way to see whether there is a connection between two variables or not. You cannot always tell from looking at tables of data. With a graph, the connection is easier to see.

Patterns indicate relationships When there is a relationship between the variables, a graph shows a clear pattern. For example, the speed and distance variables show a strong relationship. When there is no relationship, the graph looks like a collection of dots. No pattern appears. The number of rock bands a student can name in one minute and the last two digits of the student's phone number are two variables that are not related.

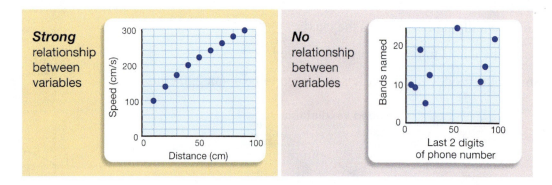

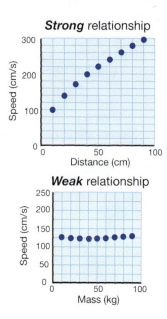

Figure 2.30: *In a strong relationship (top graph), a big change in distance creates a big change in speed. In a weak relationship, a big change in mass causes almost no change in speed.*

Strong and weak relationships You can tell how strong the relationship is from the pattern. If the relationship is strong, a small change in one variable makes a big change in another. If the relationship is weak, even a big change in one variable has little effect on the other. In weak relationships, the points may follow a pattern but there is not much change in one variable compared with big changes in the other (Figure 2.30).

Inverse relationships In inverse relationships, when one variable increases, the other decreases. If you graph how much money you spend against how much you have left, you see an inverse relationship. The more you spend, the less you have. Graphs of inverse relationships often slope down to the right (Figure 2.31).

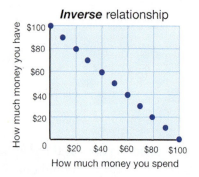

Figure 2.31: *In an inverse relationship, the dependent variable decreases as the independent variable increases.*

Chapter 2 Connection

Nanotechnology

The prefix *nano* means extremely small. A *nanometer* is 10^{-9} meters, which is a thousand times smaller than a bacterium. Nanotechnology is the technology of creating devices the size of bacteria—or smaller. Nanotechnology is a relatively new area of science and engineering. Only in the past decade have scientists been able to manipulate atoms and matter on a small enough scale to make nanotechnology possible. Future applications for nanotechnology include robots that can enter your arteries and clean out blood clots, and miniature satellites that could explore the planets. The nanotechnology of today is mostly in computers and sensors that are based on the techniques for making computer chips.

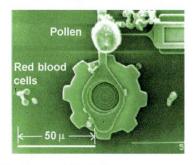

Figure 2.32: *This tiny gear is part of an experimental pump for moving cells. (Courtesy Sandia National Laboratory.)*

The size of nanotechnology   The length unit used for nanotechnology is the *micron* (μ). One micron is one-millionth of a meter (1×10^{-6} m). The width of a single strand of your hair is about 50 μ. Figure 2.32 shows a tiny blood pump with a gear that is about the width of a hair. Single cells are between 1 μ and 10 μ; several red blood cells can be seen in the microphotograph of the blood pump. With a width of 0.18 μ, the wires inside computer chips are about the smallest structures regularly produced by human technology. These wires are 300 times smaller than a hair and 10 times smaller than a cell.

Figure 2.33: *This tiny MEMS turbine has blades that are 300 microns high. It could be used to make an electric generator to power other small machines. (Courtesy MIT.)*

MEMS   One category of nanotechnology is MEMS (Micro Electronic Mechanical Systems). MEMS are tiny machines with micron-sized moving parts. There is a MEMS sensor in the air bag in your car. This sensor has a tiny arm with a small mass at the end. If the car comes to a sudden stop, the mass tries to keep moving because of its inertia. An electronic circuit detects the movement and triggers the air bag. The entire sensor is so small that a dozen would fit inside the 0 in the 2003 on a penny. An experimental MEMS turbine (Figure 2.33) might allow the creation of electrical generators smaller than the point of your pencil.

Micron-sized mirrors   Another successful MEMS device is the micro-mirror. A high-definition video projector uses an array of 1,280 by 720 of these tiny mirrors that measure 14 μ on a side (Figure 2.34). A TV display consists of thousands of points that can change color and brightness. The micro-mirror array has a single mirror for each point (or pixel) of the display. Each mirror can flip up and down to turn a pixel on or off.

Figure 2.34: *These micro-mirrors are so small they are dwarfed by the leg of a bug. (Courtesy Texas Instruments.)*

Chapter 2 Assessment

Vocabulary

Select the correct term to complete the sentences.

centimeters	conversion factor	distance	English
meter	inch	kilometer	length
surface area	metric	accuracy	millimeter
time interval	volume	gram	precision
exponent	atom	mass	kilogram
inertia	gas	weight	mixture
molecule	liquid		scientific notation
solid	plasma		

1. An English system unit of length approximately the same as the width of your thumb is the _____.
2. The space occupied by a certain number of unit cubes is known as an object's _____.
3. *Le Système International d'Unités*, or SI, is the official name for the _____ system of measurement.
4. In one inch there are 2.54 metric units of length known as _____.
5. A length of 1,000 millimeters is equivalent to one _____.
6. The system of measurement using length units of inches, feet, and miles is the _____ system.
7. The metric unit of length that is used to measure larger distances such as the distance from one city to another is the _____.
8. The total outside area of a solid is referred to as the object's _____.
9. A ratio used to change units from one measuring system to another is called a(n) _____.
10. The amount of space between two points is called _____.
11. The width of a pencil point is most nearly equivalent to the metric length unit of a(n) _____.
12. The inch, foot, centimeter, and meter are all units of _____.
13. The word in physics that means how close a measurement is to the true value is _____.
14. A quantity of time is called a(n) _____.
15. The smallest unit of measurement a ruler can measure determines the _____ of the ruler.
16. The mass of a 1-liter volume of water is 1 _____.
17. A number between 1 and 9 multiplied by some power of 10 is said to be expressed using _____.
18. The phase of matter that occurs at high temperatures and causes ionized electrons to break loose from atoms is called _____.
19. The mass of one grain of rice is close to the metric unit of a(n) _____.
20. The phase of matter that maintains a nearly constant shape and volume is _____.
21. The measurement of the force of gravity on an object is called _____.
22. The tendency of an object to resist a change in motion is called _____.
23. A group of two or more atoms that are joined together in a definite ratio is called a(n) _____.
24. The most common form of matter, made by molecules combined in no definite ratio, is known as a(n) _____.
25. In the number 1.3×10^4, the number 4 is called a(n) _____.
26. The measure of the amount of matter in an object is the _____ of the object.
27. Water normally occurs in the _____ phase of matter at temperatures between 0°C and 100°C.
28. The phase of matter that can expand or contract to fill its container is known as a(n) _____.

Concept review

1. Why are units important when measuring quantities?
2. Why is it important to understand both English and metric systems of units?
3. Give an example of a quantity that is often measured in metric units and a quantity that is often measured in English units.
4. A student expresses the volume of a fish tank in cm². Explain her mistake.
5. What are the two different meanings of the word *time*?
6. When making a graph, is time usually plotted on the *x*-axis or the *y*-axis?
7. Explain the difference between the terms *precision* and *accuracy*.
8. Heather uses a balance to measure the mass of her kitten. She repeats the measurement three times and finds the mass to be 1.25 kg each time. The actual mass of the kitten is 0.80 kg. Were Heather's measurements accurate? Were they precise?
9. Give an example of an element, a compound, and a mixture.
10. Many measurements are so common that units are not usually used. Show the units for these numerical measurements.
    a. You attend a car race called the Indianapolis 500.
    b. If you go to Canada, the speed limit is 100.
    c. A famous basketball player is 7'2".
    d. The speed limit on a road in the United States is 65.
    e. You buy a pair of men's pants that are 29/32.
11. State whether you would measure each quantity in kilometers, meters, centimeters, or millimeters.
    a. the width of a room
    b. the height of a soda bottle
    c. the length of an ant
    d. the distance from your house to school
12. List two units commonly used for measuring mass.

13. Two quantities, A and B, are measured using different dimensional units. Which arithmetic operations could be physically meaningful?
    a. A + B
    b. A/B
    c. B − A
    d. AB
14. Of the following, the units which can be used to express volume are
    a. cm.
    b. cm².
    c. cm³.
    d. cm⁴.
15. The smallest particle of water that can be identified is a(n)
    a. mixture.
    b. molecule.
    c. atom.
    d. element.
16. The diagrams below represent matter in solid, liquid, and gas phases. Identify each phase.

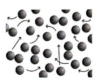 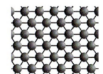 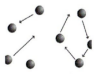

    a. _____
    b. _____
    c. _____
17. All of the following lengths are equivalent *except*
    a. micron.
    b. 1×10^{-6} m.
    c. micrometer.
    d. millimeter.
18. The mass of a person measured in *grams* is most nearly
    a. 7.
    b. 700.
    c. 7,000.
    d. 70,000.

Problems

1. Convert.
   a. 25 kilometers = _____ miles
   b. 400 centimeters = _____ meters
   c. 3 miles = _____ feet
   d. 7 inches = _____ millimeters

2. A box is 0.8 m long, 1 m wide, and 0.5 m tall.
   a. Calculate its surface area in m^2.
   b. Calculate its surface area in cm^2.
   c. Calculate its volume in m^3.

3. A cylindrical can is 15 cm tall and has a radius of 3 cm.
   a. Calculate the can's surface area in cm^2.
   b. Calculate the can's volume in cm^3.

4. How many minutes are in one year?

5. Express each number in scientific notation.
   a. 10,000
   b. 520
   c. 300,000,000
   d. 0.000001
   e. 0.000023
   f. 10.00444

6. Express each number in expanded notation.
   a. 2.33×10^6
   b. 9.9999×10^4
   c. 9.13×10^2
   d. 1.3×10^{-1}
   e. 5.2×10^{-7}
   f. 8.01×10^{-3}

7. The mean radius of Earth is 6.37×10^6 m. Calculate (a) the volume and (b) the surface area of Earth. (*Remember*: $V = 4/3\pi r^3$ and $A = 4\pi r^2$.)

8. A barrel of oil is spilled onto the ocean's surface and creates an oil slick approximately 1.0 nanometer in thickness. How large an area will be covered by the oil slick created by spilling one barrel of oil? (1 barrel = 115.6 L, 1 m^3 = 1,000 L). Compare this to the area of a soccer field that is 100 meters by 173 meters.

Applying your knowledge

1. The definition for the second, the kilogram, and the meter have changed throughout history. Describe the change that has been made in one of these fundamental units and give the reason for the change.

2. Fundamental units are established using natural phenomena. For example, Earth's daily rotation on its axis was used to describe the standard unit of time. What other natural phenomena could serve as alternative time, length, or mass standards? Choose one fundamental quantity and devise a fundamental unit. Explain why it may or may not be as useful as the standard for that quantity as used by today's scientific community.

UNIT 1 MEASUREMENT AND MOTION

CHAPTER 3

Position, Speed, and Velocity

Objectives:
By the end of this chapter you should be able to:

- ✓ Calculate time, distance, or speed when given two of the three values.
- ✓ Solve an equation for any of its variables.
- ✓ Use and interpret positive and negative values for velocity and position.
- ✓ Describe the relationship between three-dimensional and one-dimensional systems.
- ✓ Draw and interpret graphs of experimental data, including velocity versus position, and speed versus time.
- ✓ Use a graphical model to make predictions that can be tested by experiments.
- ✓ Derive an algebraic model from a graphical model and vice versa.
- ✓ Determine velocity from the slope of a position versus time graph.
- ✓ Determine distance from the area under a velocity versus time graph.

Key Questions:
- How do we describe the arrangement and movement of objects in three-dimensional space?
- How do position and velocity appear on graphs?
- Where do the equations in physics come from?

Vocabulary

average speed	coordinate system	instantaneous velocity	rate	vector
constant speed	displacement	origin	slope	velocity
coordinates	instantaneous speed	position	time	

3.1 Space and Position

This chapter is about motion, which is the movement of things from here to there—going fast or slow, or in a straight or curved path. A good starting point is to set up a system for describing precisely where "here" and "there" are relative to each other.

Position and distance

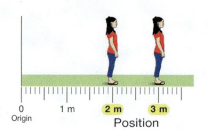

Figure 3.1: *In physics, your position describes where you are relative to the origin at one instant of time.*

Position In physics, the word **position** refers to the location of an object at one instant. For example, suppose you stretch a measuring tape and stand at the 2-meter mark. Relative to the edge of the tape, your position would be 2 meters. If you move to the 3-meter mark, your new position would be 3 meters (Figure 3.1).

The origin A position is always specified relative to an **origin**. The net change in position relative to the origin is called **displacement**. The origin is a reference point that stays fixed. In the previous example of a measuring tape, the origin would be the start of the tape. In straight-line laboratory experiments, position is usually given in meters from a starting point. When navigating on Earth's surface, position is specified in longitude and latitude (Figure 3.2). The reference for zero latitude is Earth's equator. The reference for zero longitude is the Royal Observatory in Greenwich, England.

Position and distance Distance is related to, but different from, position. You have already learned that distance is a measure of length *without* regard to direction. To understand the difference, suppose you walk 10 meters in one direction, and turn around and walk 10 meters back. The distance you walked is 20 meters. But your position is zero meters, because you walked right back to where you started from.

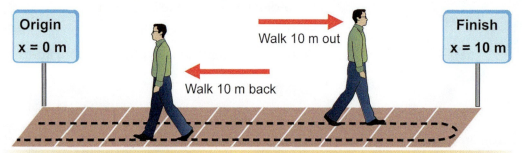

You travel a distance of 20 meters but your final position is 0 meters from the origin.

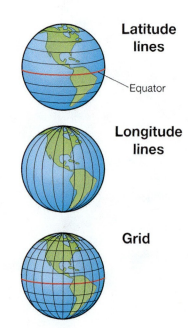

Figure 3.2: *Position on Earth's surface is specified in terms of longitude and latitude lines that make a grid around the globe.*

Position in Three Dimensions

The three directions of space
Space is three dimensional and so position must also be a three-dimensional variable. Over the years, a **coordinate system** has been devised for defining position in three dimensions. Imagine the center of mass as the origin (Figure 3.3). The three dimensions of space are assigned as follows.

- Right-left is assigned to the *x*-axis with positive being to the right.
- Up-down is assigned to the *y*-axis with positive being up.
- Forward-backward is assigned to the *z*-axis with positive being forward.

Three coordinates identify a point in space
Any position in space can be precisely specified with three numbers called **coordinates**. The term *three dimensional* refers to the fact that *only* three numbers are needed to locate any point in space. The first coordinate locates the position along the *x*-axis, such as 2 meters. The second locates a position along the *y*-axis, such as 3 meters. The third coordinate locates a position along the *z*-axis, such as 8 meters. The diagram in Figure 3.4 shows a ball that is at coordinates of $x = 2$ m, $y = 3$ m, $z = 6$ m.

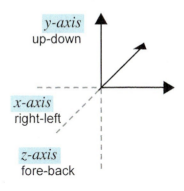

Figure 3.3: *The coordinate directions of the* x-, y-, *and* z-*axes in space.*

Positive values
If you use positive numbers, you can only describe positions in one eighth of space: to your right, in front of you, and above your center. What about the rest of space? What about behind you or to your left?

Positive and negative numbers
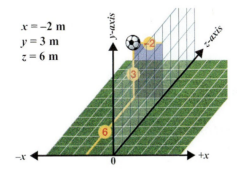
Negative numbers are used to distinguish positions to the left, behind, and below the origin. For example, the coordinates $x = -2$ m, $y = 3$ m, and $z = 6$ m put the ball the same distance above and in front, but 2 meters to the *left* instead of 2 meters to the right. Allowing *x*, *y*, and *z* to have positive and negative values allows coordinates to locate any position in all of space.

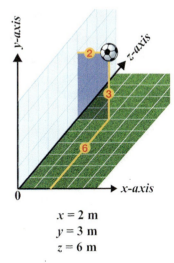

Figure 3.4: *The soccer ball is located at the coordinates* x = *2 meters,* y = *3 meters, and* z = *6 meters.*

3.1 SPACE AND POSITION

One dimensional problems

Vectors

In three-dimensional space, position is a **vector**. A vector is a variable that contains all three coordinate values. If we call the position vector $\vec{r}$ in the last example, then $\vec{r}$ can be written as an ordered triple, $\vec{r} = (2, 3, 6)$ meters. By convention, the first coordinate is the x-axis position, the second is the y-axis, and the third is the z-axis. The little arrow over the letter $\vec{r}$ indicates that this is a vector.

Choosing coordinate directions can make problems easier

There is no one "right" way to choose the origin or even the orientation of the three coordinate axes x, y, and z. The three axes must be perpendicular to each other. But apart from that, we are completely free to choose which way the axes point. Because of this freedom, it is not always necessary to use all three dimensions of space.

A car on a ramp is one dimensional

For example, suppose you are looking at the motion of a car on a straight ramp. The ramp means the car moves in a straight line. We can choose the x axis to be along the direction the car moves (Figure 3.6). This means nothing happens in the y or z directions, so these dimensions can be ignored. We say the problem is now *one dimensional* because the position of the car on the ramp can be completely described with one coordinate.

Motion in a straight line is one dimensional

Motion in a straight line is easiest to analyze because it is one dimensional. For this reason, Chapters 3, 4, 5, and 6 develop the ideas of velocity, acceleration, and force in one dimension. However, everything you learn in one dimension applies to all three dimensions, as you will see in Chapters 7, 8, and 9.

Positive and negative values

Even in one dimension there is an origin and there are positive and negative values. Positive positions are typically to the right of the origin, like on a number line. Negative positions are typically to the left of the origin. For example, the three soccer balls below are at positions of −4, −0.5, and 2.5 meters.

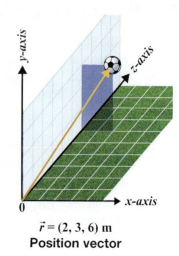

$\vec{r} = (2, 3, 6)$ m
Position vector

Figure 3.5: *The position vector for the soccer ball is $\vec{r} = (2, 3, 6)$ meters.*

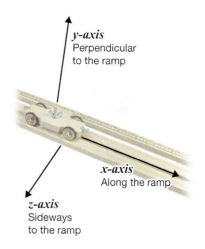

Figure 3.6: *Choosing the x-axis along the ramp means nothing happens in y or z directions, which can be ignored.*

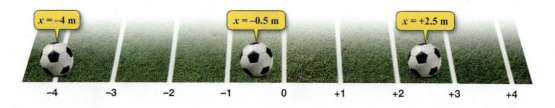

Speed and distance

Speed is the rate of change in distance
Distance gives us a new way to think about speed. Speed is the *rate* at which distance changes. In physics, the word **rate** means the ratio of how much something changes divided by how long the change takes. A high rate means a large change in a short time. A low rate means either a small change or a long time for the change to happen.

Fast and slow
In physics, just saying that something is fast is not enough description to truly understand its motion. You can easily walk faster than a turtle, yet you would not say walking speed was fast compared with the speed of a moving car.

An example of speed
The graphs below show the location of two bicycles at different times. To understand the concept of speed, think about the following two questions;

- How many meters of distance is traveled by each bicycle each second?
- Does the distance each bicycle travels change by the same number of meters every second?

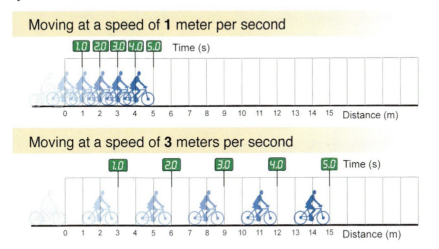

The precise meaning of speed
The speed of a bicycle is the change in distance divided by the time it takes to travel that distance. One bicycle's distance traveled is 1 meter each second so its speed is 1 m/s. The other bicycle's distance traveled is 3 meters each second so its speed is 3 m/s. Both bicycles in the diagram are moving at **constant speed**. Constant speed means the same change in distance is traveled every second.

The speed limit of the universe

Meters per second are good units to use for describing everyday motion. Consider traveling in a car at 50 miles per hour. In the car, this may not seem fast, but 50 miles per hour is a little more than 22 meters per second. Twenty-two meters is the height of a five-story building. Twenty-two meters per second means traveling this distance every second.

Light moves at 300 million meters per second (3×10^8 m/s). If you could make light travel in a circle, it would go around Earth 7-1/2 times in one second. We believe the speed of light is the ultimate speed limit in the universe. The laws of physics prevent matter from traveling faster than the speed of light.

Calculating speed

Speed is distance divided by time

The change in position is a *distance* traveled in a given amount of *time*. Therefore, to calculate the speed of an object, you need to know two things:

- the distance traveled by the object
- the **time** it took to travel the distance

Speed is calculated by dividing the distance traveled by the time taken. For example, if you drive 90 miles in 1.5 hours (Figure 3.7), then the speed of the car is 90 miles divided by 1.5 hours, which is equal to 60 miles per hour.

What does *per* mean?

The word *per* means "for every" or "for each." The speed of 60 miles per hour is short for saying 60 miles *for each* hour. You can also think of the word *per* as meaning "divided by." The quantity before the word per is divided by the quantity after it. For example, 90 miles divided by 1.5 hours equals 60 miles per hour.

Units for speed

Since speed is a ratio of distance over time, the units for speed are a ratio of distance units over time units. If distance is in miles and time in hours, then speed is expressed in miles per hour (mph). In the metric system, distance is often measured in centimeters, meters, or kilometers. Metric units for speed are centimeters per second (cm/s), meters per second (m/s), or kilometers per hour (km/h). Table 3.1 shows some of the different units commonly used for speed.

Figure 3.7: *A driving trip with an average speed of 60 miles per hour.*

Table 3.1: Speed Abbreviations

Distance	Time	Speed	Abbreviation
meters	seconds	meters per second	m/s
kilometers	hours	kilometers per hour	km/h
centimeters	seconds	centimeters per second	cm/s
miles	hours	miles per hour	mph
inches	seconds	inches per second	in/s, ips
feet	minutes	feet per minute	ft/min, fpm

Calculating speed in meters per second

A bird is observed to fly 50 meters in 7.5 seconds. Calculate the speed of the bird in meters per second.

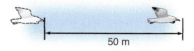

1. You are asked to find speed in m/s.
2. You are given the distance in m and time in s.
3. $v = d \div t$
4. $v = 50 \text{ m} \div 7.5 \text{ s}$
   ≈ 6.67 m/s

The velocity vector

Speed in opposite directions
Imagine two cars traveling 100 kilometers per hour in opposite directions (Figure 3.8). One is going east and the other is going west. Both are going the same speed, yet their motion is clearly different. How do we communicate this difference?

The velocity vector
We need to extend the concept of speed to include direction. This new quantity is called *velocity*. The **velocity** of an object tells you both its speed and its direction of motion. A velocity can be positive or negative. Speed is always greater than or equal to zero. Mathematically, speed is the absolute value of velocity. Practically, speed is what you read off a *speedometer*, which tells you how *fast* you are going, but nothing about what *direction* you are going. Constant velocity indicates that both the speed and the direction an object is traveling remains constant.

Positive and negative tell direction
If we choose to define east as positive and west as negative, then one car has a constant velocity of +100 km/h and the other car has a constant velocity of –100 km/h. The positive and negative signs tell you the direction of motion.

Calculating the sign of velocity
The positive or negative sign for velocity is based on the calculation of a change in position. Velocity is the change in position divided by the change in time. The diagram below shows the position and time for both cars at one hour intervals. The change in position is the final position minus the initial position (Figure 3.9). For the car going to the east, the final position is 300 km and the initial position is 200 km. The change is +100 km. For the car going to the west, the final position is 100 km and the initial position is 200 km. The change is –100 km and the velocity is therefore –100 km/h.

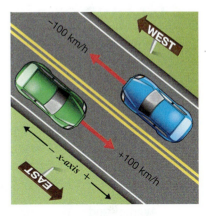

Figure 3.8: *Two cars are going opposite directions at the same speed. The speed is the same but the velocities are different—one is positive and the other is negative.*

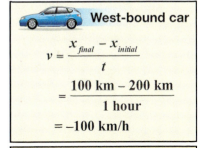

West-bound car

$$v = \frac{x_{final} - x_{initial}}{t}$$

$$= \frac{100 \text{ km} - 200 \text{ km}}{1 \text{ hour}}$$

$$= -100 \text{ km/h}$$

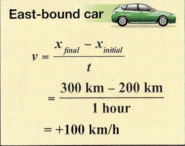

East-bound car

$$v = \frac{x_{final} - x_{initial}}{t}$$

$$= \frac{300 \text{ km} - 200 \text{ km}}{1 \text{ hour}}$$

$$= +100 \text{ km/h}$$

Figure 3.9: *You can calculate velocity from changes in position.*

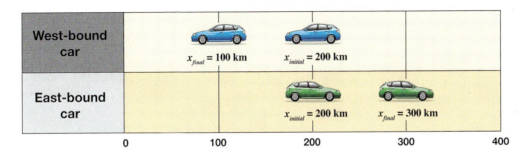

3.2 Graphs of Speed and Velocity

There are many graphs involving the terms speed, velocity, distance, position, displacement, and time. This section explores the meaning of graphs such as *speed versus distance* and *position versus time*.

The distance versus time graph

An example trip Suppose you travel between two cities that are 90 miles apart. You drive at 60 mph for 1 hour, take a half-hour rest, then drive for another 1/2 hour at the same speed. At the end, you meet a friend who calculates your speed to be 90 miles divided by 2 hours, or only 45 mph. Is your friend's calculation accurate?

Average and instantaneous speed Your friend's calculation is an accurate calculation of your **average speed**. Average speed is the total distance traveled divided by the total time taken. The average speed is lower because you took some time for a stop. The moving speed of 60 mph is higher than the average. The speedometer of a car shows **instantaneous speed**. Instantaneous speed is the speed at any moment.

The position versus time graph A position versus time graph shows the details of the actual motion during the trip. During the first hour, your position gradually increased from 0 to a point 60 miles away. The graph shows a stop between 1 hour and 1.5 hours, and that you started driving again at 1.5 hours until you reached a point 90 miles from the starting point.

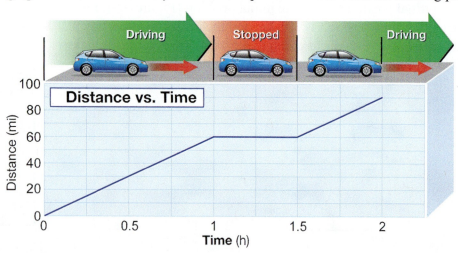

Interpreting a position versus time graph

The position versus time graph below shows a boat traveling through a long canal. The boat has to stop at locks for changes in water level.

1. How many stops does it make?
2. What is the boat's average speed for the whole trip?
3. What is the highest speed the boat reaches?

1. The boat makes three stops because there are three horizontal sections on the graph.
2. The average speed is 10 km/h (100 km ÷ 10 h).
3. The highest speed is 20 km/h. You can tell because of the position changes by 20 km in one hour for the first, third, and fifth hours of the trip.

Speed is the slope of the distance versus time graph

Position versus time
The graph below shows the distance versus time for a ball rolling along a level floor. The ball rolled a distance of 10 meters in 10 seconds. The average speed is 1 m/s. If the graph is a complete description of the motion, you should be able to calculate the speed of the ball from the graph, and you can.

The definition of slope
The **slope** of a line is the ratio of the line's *rise*, or vertical change, to its *run*, or horizontal change. The rise is determined by finding the height of the triangle shown. The run is determined by finding the length along the base of the triangle. Here, the x values represent time and the y values represent position.

The slope is the speed
Remember that speed is defined as the distance traveled divided by the time taken. As long as you travel in a straight line, distance is just the difference in position between where you finished and where you started. This is equal to the rise on the graph. The run on the graph is the time taken for the trip. The slope is rise divided by run, which is the distance traveled divided by the time taken, which is the speed. *This is an important result!* The slope of a distance versus time graph is the speed.

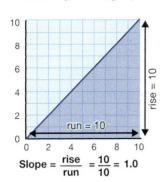

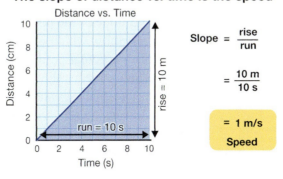

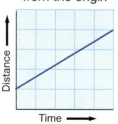

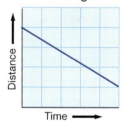

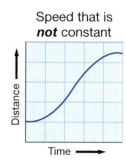

Figure 3.10: *These examples of distance versus time graphs show three different types of motion.*

Straight lines mean constant speed
A straight line has a constant slope. If the slope changed, the line would curve, like the example in Figure 3.10. The straight line in the graph of distance versus time for the rolling ball shows that the speed is constant during the motion. *This is a corollary to the important result above:* A straight line on a distance versus time graph tells you that the motion is at a constant speed.

3.2 GRAPHS OF SPEED AND VELOCITY   65

Positive and negative velocities on a position versus time graph

A trip with forward and backward motion

When the direction of motion is part of the calculation, changes in position are referred to as displacement. On the graph below, maximum displacement reaches 90 miles at a time of 2 hours. It starts decreasing at 3 hours and reaches zero when the round trip is completed at 4.5 hours. **Average velocity** uses the values of displacement and elapsed time from the position versus time graph. At point C the displacement is 45 miles after 3.75 hours. The average velocity at C is 12 mph.

Three ways to write the equation for velocity

$$v = \frac{\text{change in position}}{\text{change in time}}$$

$$= \frac{\Delta x}{\Delta t}$$

$$= \frac{x_{final} - x_{initial}}{t}$$

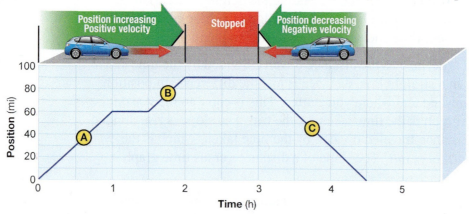

Figure 3.11: *Velocity is calculated by dividing the change in position by the change in time.*

Slope and speed

Steeper slope indicates greater velocity. The slope of the position versus time graph at any one time is called **instantaneous velocity**, which is usually synonymous with the term velocity when referring to one particular point in time. At A and B on the graph, velocity is positive. Positive velocity means the change in position is increasing with time. At C, the velocity has a negative value, –60 mph. Negative velocity means the change in position is decreasing relative to the starting point at that moment. Notice on the graph that positive slope corresponds to positive velocity and negative slope corresponds to negative velocity.

Reading positive and negative velocities

Figure 3.12 shows velocities from 0 to 100 m/s in two directions drawn on a graph. Notice two things about Figure 3.12 and the graph in the middle of the page:

1. A flat, horizontal line means position is not changing, so the velocity is zero.
2. When velocity is in a straight line in one direction, its absolute value is speed. Velocities of +100 m/s and –100 m/s both have a speed of 100 m/s.

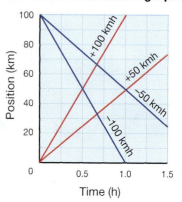

Figure 3.12: *The graphs illustrate the difference between positive and negative velocities.*

The velocity versus time graph

The velocity versus time graph

The velocity versus time graph has velocity on the *y*-axis and time on the *x*-axis. This graph tells a very accurate history of how the velocity of a moving object changes with time. It is useful even if the velocity is not changing. For example, the graph in Figure 3.13 shows velocity versus time for a ball rolling at constant velocity on a level floor. On this graph, a constant velocity is a straight horizontal line. If you follow the line over to where it intersects the *y*-axis, you can see that the velocity of the ball is 1 m/s. The graph also shows that the velocity stays constant at 1 m/s for 10 seconds.

Distance equals area on the velocity versus time graph

Information about an object's position is also present in the velocity versus time graph, but you need to know how to find it. The distance traveled is equal to the velocity of the ball multiplied by the time it moves. Suppose we draw a rectangle on the velocity versus time graph between the line that shows the velocity and the *x*-axis. The area of a rectangle is equal to its length times its height. On this graph, length is equal to time and height is equal to velocity. Therefore, area on the graph is velocity multiplied by time, which is the distance traveled. *This is an important result.* The area on a velocity versus time graph is equal to the distance traveled. This rule is true as long as the *x*-axis represents a velocity of zero.

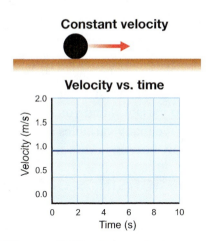

Figure 3.13: *The graph shows a constant velocity of 1 m/s.*

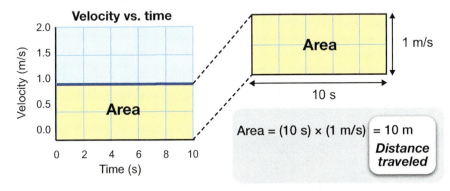

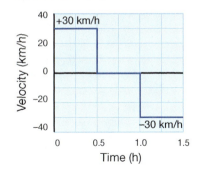

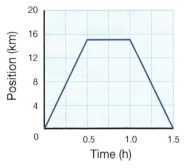

Figure 3.14: *The graphs compare position and velocity for the same event.*

Relating *x* versus *t* and *v* versus *t* graphs

A velocity versus time graph can show positive and negative velocities. The velocity versus time graph in Figure 3.14 shows a velocity of +30km/h going out and a velocity of –30km/h on the return trip. Notice that the lower graph, a position versus time graph, can be interpreted to yield the same information using the slope to calculate velocity at corresponding time intervals.

3.3 Working with Equations

A scientific model is a description of the relationship between variables. A graph is a model that describes the relationship between two variables. An equation is a much more powerful form of model than a graph. Most of the natural laws are most accurately described with equations. Graphs are limited to two variables and you cannot easily read a graph outside the limits on its *x*- and *y*-axes. Equations can have many variables and can be used over a wide range of values. Equations can also be rearranged to show how any one variable depends on all the others.

Relationships between distance, speed, and time

Mixing up distance, time, and speed How far do you go if you drive for 2 hours at a speed of 100 kilometers per hour? This is a common form of question in physics. You know how to get speed from time and distance. How do you get distance from speed and time? The answer is the reason mathematics is the language of physics.

Let the letter *v* stands for "speed," the letter *d* stands for "distance traveled," and the letter *t* stands for "time taken." If we remember that the letters stand for those words, we can now write an equation for speed.

SPEED

$$\text{Speed (m/s)} \quad v = \frac{d \quad \text{Distance traveled (m)}}{t \quad \text{Time taken (s)}}$$

Three forms of the speed formula There are three ways to arrange the three variables that relate distance, time, and speed. You should be able to work out how to get any of the three variables if you know the other two.

The equation	gives you	if you know
$v = d \div t$	speed	time and distance
$d = vt$	distance	speed and time
$t = d \div v$	time	distance and speed

Using formulas Remember that the words or letters are placeholders for the values that the variables have. For example, the letter *t* will be replaced by the actual time when we plug in numbers for the letters. You can think about each letter as a box that will eventually hold a value.

Calculating time from speed and distance

How far do you go if you drive for 2 h at a speed of 100 km/h?
1. You are asked for distance.
2. You are given time in h and speed in km/h.
3. $d = vt$
4. $d = 2 \text{ h} \times 100 \text{ km/h} = 200 \text{ km}$

Solving an equation

Solving equations
Many physics problems require you to "solve" an equation. In this context, "solve" means to get a desired variable *by itself* on one side of an equals sign. The other side must not contain the variable. For example, the equation $d = vt$ is solved for the distance because d is alone on one side of the equals sign and does not appear on the other side. This is useful because you can now calculate d from values of t and v.

The rules of algebra
Solving an equation is like solving a puzzle. There are rules that you must follow and you test different things, mostly by trial and error, until you get the result you want. The first rule is that *whatever you do to the left of the equals sign you must do exactly the same on the right*. This preserves the equality. For example, consider the statement $5 = 5$. If you add 10 to the left you must add 10 to the right, which seems obvious: $10 + 5 = 5 + 10$ or $15 = 15$.

Algebra with variables
What is less obvious is the same rule applies to variables and other operations like multiplying and dividing. Multiplying by t on the left means you must also multiply by t on the right. It does not matter what specific value t has; the equality is maintained. Notice that multiplying by t on both sides of the equation for speed allows you to cancel a t from the numerator and the denominator on the right (Figure 3.15).

Isolating a variable
This gives you a new formula, $d = vt$, which says that distance is speed multiplied by the time. Mathematically, you have solved the speed equation for distance and isolated the variable d on one side of the equals sign.

Substitute numbers after solving the equation
Get in the habit of solving an equation before you plug in numbers. This will seem slow and unnecessary at first, but it is absolutely necessary later. That is because more complex problems require you to substitute whole equations for single variables. This is how you incorporate many different factors, such as friction, into the basic equations of motion.

> True statement
> $5 = 5$

> True statement
> $10 + 5 = 10 + 5$

> True statement for any value of t
> $5 + t = 5 + t$

> If this is true...
> $v = \dfrac{d}{t}$
> Then this is also true for any value of t.
> $vt = \dfrac{d}{t} \times t$

Some useful algebra examples

Given the equation
$$v = \frac{d}{t}$$
how do you get the variable t in the numerator?

You multiply both sides by t and cancel the t from the right side.

$$t \times \left(v = \frac{d}{t} \right) \rightarrow vt = \frac{d\cancel{t}}{\cancel{t}}$$

$$\boxed{vt = d}$$

Given the equation
$$v = \frac{x_f - x_i}{t}$$
solve for x_f the final position.

First, bring the t out of the denominator by multiplying both sides by t.

$$v \times t = \frac{x_f - x_i}{\cancel{t}} \times \cancel{t}$$

$$vt = x_f - x_i$$

Next, add x_i to both sides so it cancels with the x_i on the right.

$$\boxed{vt + x_i = x_f}$$

Figure 3.15: *Some of the rules of algebra are seen here, and examples of when to use them.*

The position versus time equation

Starting at a position other than the origin
Consider a car that starts 100 kilometers east of the origin then travels at a constant velocity of 100 km/h in a straight line east for three hours. Where is the car at the end of the three hours?

You know that traveling at 100 km/h for three hours results in a change of position, or displacement, of 300 km. If you start 100 km away then travel an additional 300 km, you finish 400 km away.

The position versus time equation
Now, repeat the calculation with variables. Let x_0 be the place you start. The subscript 0 is often used to indicate the position at time $t = 0$. Your displacement is given by $d = vt$, where v is the velocity and t is the time. We can now put the equation together. In plain words, the equation says your position, x, is equal to the position you started at, x_0, plus the additional amount you traveled, vt.

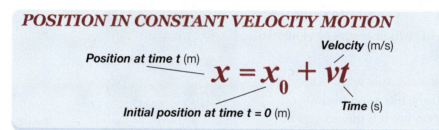

POSITION IN CONSTANT VELOCITY MOTION

$$x = x_0 + vt$$

Position at time t (m) — Velocity (m/s) — Initial position at time $t = 0$ (m) — Time (s)

Using the equation
Once you have the equation, you can use it to calculate the position for *any* value of initial position, velocity, and time. For example, suppose the car started at 100 km east and traveled west for three hours at 100 km/h. West is the opposite direction so the velocity is –100 km/h. The equation tells you the car ends up at a position of –200 km, which is 200 km west of where it started.

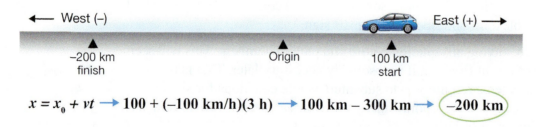

$x = x_0 + vt$ → $100 + (-100\text{ km/h})(3\text{ h})$ → $100\text{ km} - 300\text{ km}$ → $\boxed{-200\text{ km}}$

Solving the position versus time equation

A car moving in a straight line at constant velocity starts at a position of 10 meters and finishes at 30 meters in five seconds. What is the velocity of the car?

1. You are asked for velocity.
2. You are given that the motion is at constant velocity, two positions, and the time.
3. $x = x_0 + vt$
4. First, solve for v:

$$x - x_0 = vt$$

$$\frac{x - x_0}{t} = v$$

Then, substitute numbers:

$$v = \frac{30\text{ m} - 10\text{ m}}{5\text{ s}} = 4\text{ m/s}$$

Relating equations and graphs

Equations often come from graphs
An equation contains all the information in a graph and more. In many real problems, scientists and engineers use graphs as the starting point for developing an equation. The graphs come from actual experiment data. Statistical software can be used to determine an equation that fits any number of combinations of variables from an experiment. Figure 3.16 shows the data from such an experiment and its graph.

The equation of a straight line
The most recognizable relationship is a straight line. In math classes, you learned that one equation for a straight line is the slope-intercept form, where x and y are the variables corresponding the x and y axes. This is known as the slope-intercept form of an equation.

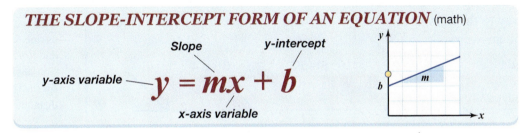

The position equation is a straight line
In science and engineering, *any* two variables can be used in the equation for a line, not just x and y. The slope and the y-intercept may not be m and b but may have other meanings, such as velocity and initial position. The equation on the previous page is actually the equation for a straight line on a graph of position versus time. In this case, the y corresponds to x, the position at any time; the x corresponds to t, time; the slope, m, corresponds to the velocity, v; and the y-intercept, b, corresponds to the initial position, x_0. The equation is not a theory of physics, but the equation of a straight line on the x versus t graph.

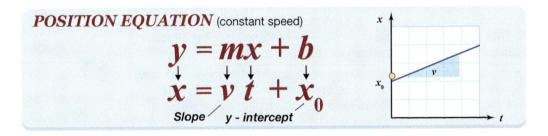

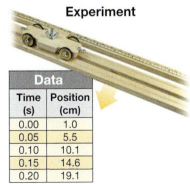

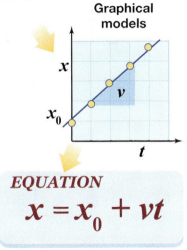

Figure 3.16: *The process of developing a model or theory in science starts with actual experiments and data, and produces a validated model in the form of an equation.*

3.3 WORKING WITH EQUATIONS

CHAPTER 3 — POSITION, SPEED, AND VELOCITY

How to solve physics problems

Physics problems Physics problems usually give you some information and ask you to find something else that is related to the information you are given. Solving physics problems is excellent practice because it teaches you to analyze information and think logically about how to get an answer. This skill is important in all careers. For example, financial analysts are expected to look at information about businesses and determine which companies are succeeding. Doctors collect information about patients and must diagnose what is causing pain or an illness. Mechanics gather information about a car and have to troubleshoot what is causing a malfunction and how to fix it. All these examples use problem-solving skills.

A four-step technique Our technique for solving problems has four steps. Follow these steps and you will be able to see a way to the answer most of the time and will at least put you on the path to finding the answer.

Step 1 — What do you want to find?
Step 2 — What do you know?
Step 3 — Identify useful relationships
Step 4 — Solve the problem ✓

Step	What to do
1	Identify clearly what the problem is asking. If you can, figure out exactly what variables or quantities need to be in the answer.
2	Identify the information you are given. Sometimes this includes numbers or values. Other times it includes descriptive information you must interpret. Look for words like *constant* or *at rest*. In a physics problem, saying something is constant means it does not change. That is useful information. The words *at rest* in physics mean the speed is zero. You may need conversion factors to change units.
3	Identify any relationships that involve the information you are asked to find and also the information you are given. For example, suppose you are given a speed and time and asked to find a distance. The relationship $v = d \div t$ relates what you are asked for to what you are given.
4	Combine the relationships with what you know to find what you want to know. Once you complete Steps 1–3, you will be able to see how to solve most problems. If not, start working with the relationships you have and see where they lead.

Calculating distance from time and speed

A space shuttle is traveling at a speed of 7,700 meters per second. How far in kilometers does the shuttle travel in one hour? At an altitude of 300 kilometers, the circumference of the shuttle's orbit is 42 million meters. How long does it take the shuttle to go around Earth one time?

1. This is a two-part problem asking for distance in kilometers and time in hours.
2. You are given a speed and time for the first part, and a speed and distance for the second.
3. $d = vt$, and $t = d \div v$
   1 h = 3,600 s
   1 km = 1,000 m
4. Part 1:
   $d = (7,700 \text{ m/s})(3,600 \text{ s})$
   $\quad = 27,720,000 \text{ m}$
   Convert to kilometers:
   $\quad = 27,720,000 \text{ m} \div 1,000 \text{ km/m}$
   $\quad = 27,720 \text{ km}$
   Part 2:
   $t = 42 \times 10^6 \text{ m} \div 7,700 \text{ m/s}$
   $\quad = 5,455 \text{ s}$
   Convert to minutes:
   $\quad = 5,455 \text{ s} \div 60 \text{ s/min}$
   $\quad = 90.9 \text{ minutes}$

Chapter 3 Connection

Slow-Motion Photography

You have probably seen slow-motion photography if you watch sports on television. It looks as if everything takes much longer than it does in real time. Slow-motion photography is also very useful in science. Many things in nature move so quickly it is hard to see enough detail to determine what is happening. For example, a slow-motion camera can allow you to see the motion of a hummingbird's wings or the turning of an engine. At their normal speeds, these motions are far too rapid to see clearly.

Creating the illusion of motion A video camera does not photograph moving images. It takes a sequence of still images called *frames* and changes them fast enough that your brain perceives a moving image. The standard for video is to change frames 30 times per second. At this rate, a sequence of still images is perceived as smooth motion.

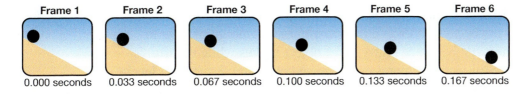

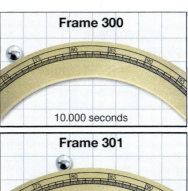

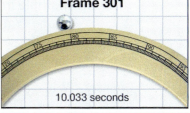

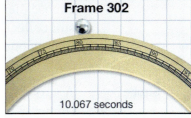

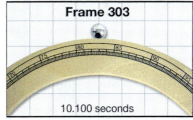

Making a slow-motion movie To take a slow-motion movie, the camera captures more frames per second. Instead of taking 30 per second, it may take 120 frames per second. The slow-motion effect comes from playing the movie back at the standard rate of 30 frames per second. The motion appears four times slower because the playback rate is four times slower than the rate at which the pictures were actually taken.

Using video to analyze motion You can use an ordinary video camera to analyze motion in laboratory experiments. Many professional quality video cameras allow you to advance a video one frame, at a time. Each frame represents 1/30 of a second. The speed of an object can be determined by comparing the object's position in two consecutive frames. The speed of the object is equal to the distance it moved between frames divided by 1/30 of a second. Figure 3.17 shows a sequence of frames taken from video of a ball rolling over the top of a hill. The ball slows down near the top of the hill. You can tell because the distance the ball moves between frames gets less and less as the ball gets near the top of the hill.

Figure 3.17: *A sequence of video images showing the motion of a ball rolling over the top of a hill. You can see that the ball slows down as it goes over the hill.*

Chapter 3 Connection

Strobe photography

Long-exposure photography with strobe lights

There is a second way to do stop-motion photography. In this method, a special light called a *strobe* light is used. A strobe light repeatedly flashes very bright pulses of light. Suppose the flashes are one-tenth of a second apart. The camera sees the moving object only every tenth of a second, when the strobe flashes. If the camera is left with a long exposure time, all of the flashes from the strobe appear on the same image. As a result, the image looks like Figure 3.18. There are multiple images of the moving object separated by the time between flashes of the strobe. You can see the basketball slowing down as it reaches its high point, then speeding up as it returns to the ground. You can tell because the distance between flashes decreases on the way up, then increases as it goes back down.

Figure 3.18: *This strobe photograph shows a bouncing basketball.*

Ordinary cameras

When a normal still-frame camera photographs a moving object, the image can be blurry. The blur occurs because it takes a certain minimum amount of light to expose the film. An image becomes blurry if the object moves while the film is still collecting light. Still-frame cameras can open and close the shutter that lets in light in one-thousandth of a second. This seems fast, but it takes only a few millimeters of motion to create a blurry image.

Strobe lights

A strobe light can be used to take a sharp image of a rapidly moving object (Figure 3.19). To do this, the strobe is flashed only once with a very bright but very short pulse of light. The flash is so short that the object does not move during the light pulse. A strobe light can flash a pulse many times faster than the mechanical shutter on a camera can move. The single rapid flash provides enough light to capture the image without blurring. The strobe is able to overcome the limitations of the camera shutter to capture motion on an image.

Figure 3.19: *A strobe photograph showing a drop of milk splashing on a hard surface.*

Chapter 3 Assessment

Vocabulary

Select the correct term to complete the sentences.

position	origin	coordinate system
coordinates	vector	rate
constant speed	distance	time
velocity	average speed	instantaneous speed
slope	initial	model
y-axis	x-axis	at rest
graph		

1. The _____ at which something changes is calculated by dividing the amount of change by the time over which the change occurs.
2. The term _____ means a speed that is unchanging.
3. The speed an object has at the beginning of its motion is known as its _____ speed.
4. A representation of a system used to make predictions is called a(n) _____.
5. The vertical axis on a graph is also called the _____.
6. A picture model scientists use to show the mathematical relationship between two variables is called a(n) _____.
7. The horizontal axis of a graph is the _____.
8. The location of an object at any moment is its _____.
9. An object that is not moving is said to be _____.
10. A fixed point of reference from which all changes are measured is called the _____.
11. On a graph, the ratio of the vertical change to the horizontal change is known as the _____ of the graph.
12. A(n) _____ is a spatial framework in which each position can be specified with a set of unique values.
13. A set of values that specify a position, typically values for *x*, *y*, and *z*.
14. A(n) _____ is a type of variable that contains information about both quantity and direction.
15. _____ is the separation between two points in space.
16. Calculate speed by dividing the distance traveled divided by the _____ taken to travel that distance.
17. The _____ of an object tells you both its speed and its direction of motion.
18. The _____ for a trip would be the total distance traveled divided by the total elapsed time.
19. A speedometer shows _____.

Concept review

1. What is the difference between the terms *position* and *distance*?
2. In a standard right-handed coordinate system, assign the labels *x*, *y*, and *z* to the three of the perpendicular directions—up, down, left, right, forward, and backward.
3. Give an example of a motion that can be accurately described with one dimension and another motion that requires two or three dimensions to describe.
4. Which two quantities are needed to determine the speed of an object?
5. Fill in the missing information in the table showing common units for speed.

Distance	Time	Speed	Abbreviation
meters	seconds		
			km/h
		centimeters per second	
miles			mph
		inches per second	
			ft/min

CHAPTER 3 ASSESSMENT   75

6. Write the three meaningful formula arrangements of the variables speed, distance, and time. Let v = speed, t = time, and d = distance.

7. Average speed and instantaneous speed are generally different quantities. Describe a situation in which the two quantities could be the same.

8. A student at the top of a building throws Ball A upward with a speed of v_i and throws a second, identical Ball B downward with the same initial speed. Compare the speed of Ball A and Ball B when they strike the ground.

9. How is the slope of a straight, best-fit line determined?

10. What does the slope of a position versus time graph represent?

11. Sam rolls down his driveway on a skateboard while Beth keeps track of his position every second for 15 seconds. When they make a graph of the data, the best-fit line for the position versus time graph is a curved line. What does this indicate about Sam's speed?

12. What is the physical meaning of the area under a speed versus time graph?

13. If the position versus time graph is horizontal, what value must the speed versus time graph have at the corresponding time?

14. Sketch the shape of the position versus time graph for each situation.
    a. an automobile stopped at a traffic light
    b. a cyclist traveling at a constant speed on a highway
    c. an airplane gradually rolling to a stop on its runway

15. Sketch the shape of the speed versus time graph for each situation in question 14.

Problems

1. The sound from an underwater explosion is recorded by sailors on a submarine 4.80×10^3 meters away 3.2 seconds after the explosion occurs. According to their data, what is the speed of sound in water?

2. If it takes 500 seconds for the light from the Sun to reach Earth, what is the distance to the Sun in meters? (*Hint*: The speed of light is 3.00×10^8 m/s.)

3. If Lexi bikes at an average speed of 22 mph, how many hours will it take for her to cover a 110-mile course?

4. You travel on the highway at a rate of 65 mph for 1 hour, at 55 mph for 1.5 hours, and 47 mph for 3 hours.
    a. What is the total distance you have traveled?
    b. What is your average speed during the trip?

5. On July 2, 1988, Steve Cram of England ran a mile in 3.81 minutes. Calculate his speed in miles per hour.

6. The uniform motion of a cart is represented on the graph below. What is the speed of the cart?

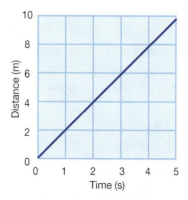

7. The graph below represents the relationship between the distance and time for the motion of an object.
   a. During which time period is the object's speed changing?
   b. What is the speed of the object at time, $t = 5$ seconds?

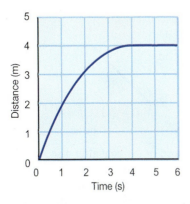

8. A data recorder on a ship records the following data while the ship travels out and back.

Time (h)	0–2	2–4	4–6	6–8
Velocity (km/h)	25	20	−15	−10

   a. Construct the graph of speed versus time.
   b. Assume the origin is where the ship starts at $t = 0$. Construct the position versus time graph by calculating the area under each section of the graph from part a.
   c. What is the ship's position after eight hours?

9. A plane flying in a straight line travels 700 kilometers in 2 hours at a constant speed. The plane then runs into a storm and slows down. The plane only covers 500 kilometers in the next 2 hours. Which of the following is best match for the speed versus time graph?

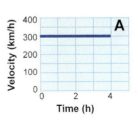 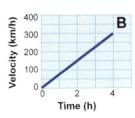 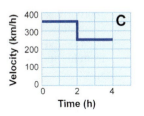

10. The graph below represents the relationship between the position and time for a car.
    a. During which time interval is the car at rest?
    b. During which time interval is the car moving at a constant speed?

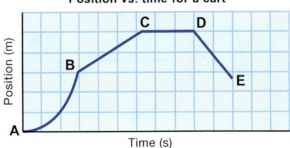

11. A Global Positioning Satellite (GPS) receiver uses communications with orbiting satellites to determine position anywhere on the surface of Earth. Many people have GPS receivers in their cars and phones to help them with directions.

    John drives his car to a friend's house which is 3.2 kilometers away on the same straight road as John's house. John uses a GPS receiver to measure his position every 0.40 kilometers. He records his position and also the speed from his speedometer in the data table shown below. Construct a graph of speed versus position to describe John's trip.

Speed (m/s)	0	10	15	20	23	23	21	17	0
Position (km)	0	0.4	0.8	1.2	1.6	2.0	2.4	2.8	3.2

12. Examine the graphs below and write a brief description of the motion represented by each.

    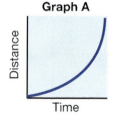
    Graph A

    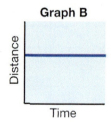
    Graph B

    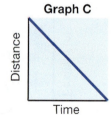
    Graph C

13. The graph represents the speed of a boat for the first 10 seconds as it moves away from a dock. Answer the following questions based upon the graph.

    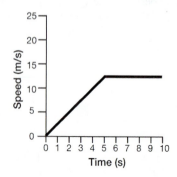

    a. What is the speed of the boat after 4 seconds?
    b. How far has the boat traveled after 5 seconds?
    c. Describe the motion of the boat from the 5th second to the 10th second.
    d. How far has the boat traveled after 10 seconds?

14. A tortoise, "racing" along at 8.00 cm/s is able to beat a hare whose speed is 15 times greater than the tortoise's. Because the hare stops in the middle of the race for a 2.00 minute rest, the tortoise finishes 15 cm ahead of the hare.
    a. How much time does the race take?
    b. How long is the race?

Applying your knowledge

1. "The speedometer in your car shows instantaneous speed." Is this statement totally accurate? Why or why not? Suggest an alternative method by which you can find instantaneous speed.

2. Dr. Harold Edgerton created interesting photographs using strobe photography. Were there other achievements that can be credited to his inventive genius?

3. Strobe lights provide interesting demonstrations and are useful in timing certain events. Are there any associated dangers or precautions that must be taken when using strobe lights?

UNIT 1 **MEASUREMENT AND MOTION**

CHAPTER 4

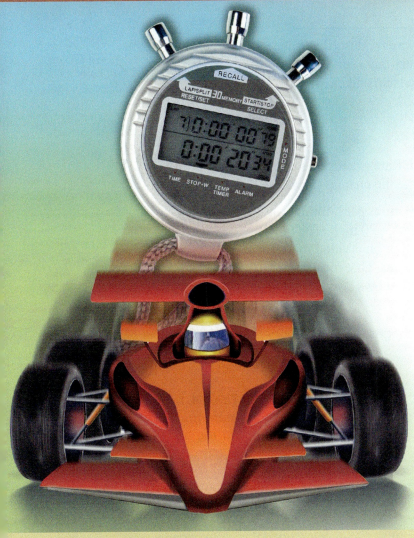

Accelerated Motion in a Straight Line

Objectives:

By the end of this chapter you should be able to:

- ✔ Calculate acceleration from the change in speed and the change in time.
- ✔ Give an example of motion with constant acceleration.
- ✔ Determine acceleration from the slope of the speed versus time graph.
- ✔ Calculate time, distance, acceleration, or speed when given three of the four values.
- ✔ Solve two-step accelerated motion problems.
- ✔ Calculate height, speed, or time of flight in free fall problems.
- ✔ Explain how air resistance makes objects of different masses fall with different accelerations.

Key Questions:

- How do you know if a moving object is accelerating?
- How do you predict the motion of an accelerating object?
- How does gravity affect the motion of a falling object?

Vocabulary

acceleration	constant acceleration	initial speed	terminal velocity
acceleration due to gravity (g)	delta (Δ)	m/s^2	time of flight
air resistance	free fall	term	uniform acceleration

4.1 Acceleration

The speed of moving objects rarely stays the same for long. Acceleration is the way we describe change in speed, and this chapter is about acceleration. You experience acceleration every day. You speed up and slow down as you walk, and you probably ride in a car or bus that also speeds up and slows down. Any time your speed changes, you experience acceleration.

Acceleration of a car

Acceleration

Acceleration is the rate of change in the speed of an object. Rate of change means the ratio of the amount of change divided by how much time the change takes.

An example of acceleration

Suppose you are driving and your speed goes from 20 to 60 miles per hour in four seconds (Figure 4.1). The change is 40 miles per hour, 60 mph – 20 mph. The time it takes to change speeds is 4 seconds. The acceleration is 40 mph divided by 4 s, or 10 mph/s. Your car accelerated 10 miles per hour per second. That means your speed increased by 10 miles per hour each second. Table 4.1 shows how your speed changed during those four seconds of acceleration.

Acceleration is the rate at which speed changes.

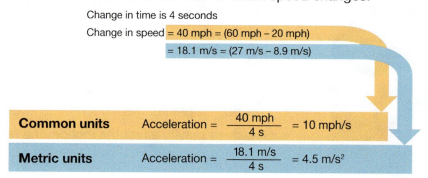

Change in time is 4 seconds
Change in speed = 40 mph = (60 mph – 20 mph)
= 18.1 m/s = (27 m/s – 8.9 m/s)

Common units Acceleration = $\frac{40 \text{ mph}}{4 \text{ s}}$ = 10 mph/s

Metric units Acceleration = $\frac{18.1 \text{ m/s}}{4 \text{ s}}$ = 4.5 m/s^2

Acceleration in metric units

In metric units, the car's speed increases from 8.9 m/s to 27 m/s. The acceleration in metric units is 18.1 m/s divided by 4 seconds, or 4.5 meters per second per second. The interpretation is that the speed increases by 4.5 meters per second every second. The unit of meters per second per second is usually written as meters per second squared (**m/s^2**).

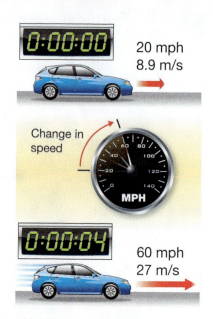

Figure 4.1: *A car is shown accelerating from 20 miles per hour to 60 miles per hour in four seconds.*

Table 4.1: Speedometer readings every second during acceleration

Time	Speed
0 (start)	20 mph
1 second	30 mph
2 seconds	40 mph
3 seconds	50 mph
4 seconds	60 mph

The difference between velocity and acceleration

Comparing velocity and acceleration
Speed and velocity are fundamentally different from acceleration. Because direction is important the discussion will use both velocity and speed. Speed is the absolute value of velocity. Velocity can be positive or negative and is the rate at which an object's position changes. The units of velocity are meters per second. Acceleration is the rate at which *velocity* changes. Acceleration is measured in meters per second *per second*, or meters per second squared (m/s^2). If an object has an acceleration of 1 m/s^2, its velocity increases by 1 m/s every second.

$$\text{Acceleration} = \frac{\text{Change in velocity}}{\text{Change in time}} = \frac{\frac{\text{meters}}{\text{second}}}{\text{second}} = \frac{\text{meters}}{\text{second} \times \text{second}} = \frac{m}{s^2}$$

Acceleration and velocity in the same direction
The acceleration of an object can be in the same direction as its velocity or in the opposite direction. Velocity increases when acceleration is in the same direction. For example, a ball rolling down a ramp has an acceleration in the same direction as its velocity. The data table in Figure 4.2 shows a *speed* that increases by 1 m/s every second.

Acceleration and velocity in opposite directions
A ball rolling *up* a ramp has an acceleration in the opposite direction to its velocity. As the ball goes up, its speed *decreases*. When the acceleration is $-1\ m/s^2$, the speed decreases by 1 m/s every second. The data in Figure 4.3 show the decreasing speed of a ball as it rolls up a ramp.

Using positive and negative signs
Because velocity and acceleration can have opposite directions, it is useful to assign positive and negative signs. A common choice is to make positive to the right, and negative to the left. A positive velocity means the object is moving to the right. A negative velocity describes an object moving to the left.

Positive and negative acceleration
When velocity and acceleration are the *same* sign, the speed increases. When acceleration and velocity have the *opposite* sign the speed decreases. If *both* velocity and acceleration are negative the speed increases but the motion is still in the negative direction. For example, suppose a ball is rolling down a ramp sloped downhill to the left. Motion to the left is defined to be negative so the velocity and acceleration are both negative. The velocity of the ball gets larger in the negative direction, which means the ball moves faster to the left. The speed *increases* since speed is the absolute value of velocity.

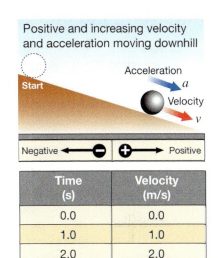

Figure 4.2: *Acceleration of a ball rolling down a ramp.*

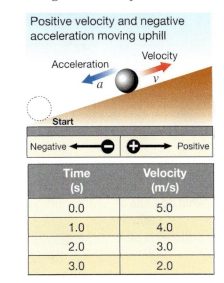

Figure 4.3: *Acceleration of a ball rolling up a ramp.*

4.1 ACCELERATION

Calculating acceleration

You accelerate coasting downhill

Acceleration is the change in velocity divided by the change in time. The Greek letter **delta (Δ)** means "the change in." When you see the Δ symbol, replace it in your mind with the phrase "the change in." The acceleration can then be written as $\Delta v/\Delta t$, which translates to "the change in velocity divided by the change in time."

ACCELERATION (definition)

$$a = \frac{\Delta v}{\Delta t}$$

Acceleration (m/s²) — Change in velocity (m/s) / Change in time (s)

Acceleration from experiments

The formula for acceleration can also be written in a form that is convenient for experiments. In experiments, you typically measure a sequence of velocities at different times. For example, v_1 is the velocity at one time, t_1, and v_2 is the velocity at a later time, t_2. The change in velocity is $v_2 - v_1$. The corresponding change in time is $t_2 - t_1$. The acceleration can be calculated using the formula below. This is actually the same formula as the previous one except the quantity Δv has been replaced by $v_2 - v_1$ and Δt has been replaced by $t_2 - t_1$.

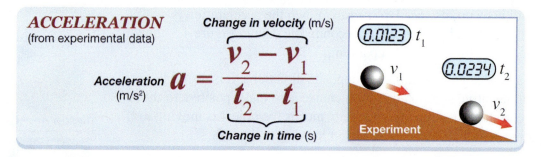

ACCELERATION (from experimental data)

$$a = \frac{v_2 - v_1}{t_2 - t_1}$$

What do units of seconds squared mean?

Many physics problems will use acceleration in m/s². Writing acceleration in units of meters per *second squared* is really just a mathematical shorthand. The units of square seconds do not have physical meaning in the same way that square inches mean surface area. It is better to think about acceleration in units of speed change per second. If you encounter an acceleration of 10 m/s², this number means the speed is increasing by 10 m/s every second.

Calculating acceleration

A student conducts an acceleration experiment by coasting a bicycle down a steep hill. A partner records the speed of the bicycle every second for five seconds. Calculate the acceleration of the bicycle.

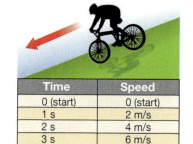

Time	Speed
0 (start)	0 (start)
1 s	2 m/s
2 s	4 m/s
3 s	6 m/s
4 s	8 m/s
5 s	10 m/s

1. You are asked for acceleration.
2. You are given a table of times and speeds from an experiment.
3. $a = (v_2 - v_1) \div (t_2 - t_1)$
4. Choose any two pairs of speed and time data.

$$a = (6 \text{ m/s} - 4 \text{ m/s})$$
$$\div (3 \text{ s} - 4 \text{ s})$$
$$= 2 \text{ m/s}^2$$

For this experiment, the acceleration should be the same for any two points.

Constant speed and constant acceleration

Zero acceleration An object has zero acceleration if it is traveling at constant speed in one direction. You might think of zero acceleration as "cruise control." If the speed of your car stays the same at 60 miles per hour, your acceleration is zero. A ball rolling along a level floor with no friction also has zero acceleration and moves at constant speed. Motion with zero acceleration appears as a straight horizontal line on a speed versus time graph.

Constant acceleration An object moving with **constant acceleration** has a velocity that changes by the same amount every second. In one-dimensional motion, constant acceleration appears as a sloped line on a velocity versus time graph. Constant acceleration is sometimes called **uniform acceleration** in physics problems. A ball rolling down a ramp and a dropped object in free fall have constant acceleration.

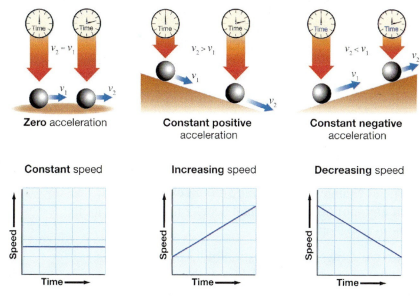

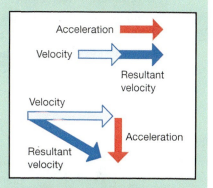

Acceleration from changing direction

In general, acceleration is a vector, like velocity. That means the acceleration can be directed sideways from velocity. Sideways accelerations can cause a change in speed, a change in direction, or both.

In Chapter 7, you will see that it is even possible for an object to be constantly accelerating and have its speed stay constant. An object moving at constant speed in a circle is an example of this kind of motion. The speed stays the same but the direction constantly changes.

In fact, *any* change in an object's state of motion—either direction or speed—is due to acceleration. In the next chapter, you will see that any acceleration also implies the presence of forces.

Acceleration with zero speed An object can have acceleration but no speed. Consider a ball rolling *up* a ramp. As the ball slows down, eventually its speed becomes zero and at that moment the ball is at rest. However, the ball is still accelerating because its velocity continues to change. The moment after it stops, the ball reverses direction and starts moving down. Its speed increases in the negative direction.

The speed versus time graph for accelerated motion

Analyzed data with speed versus time graph
The graph below shows data collected from an experiment with a ball rolling down a ramp. The time is measured between when the ball is released and when its speed is measured somewhere along the ramp. You can see the speed of the ball increases the longer it rolls down. In this experiment, velocity and acceleration are in the same direction. No negative quantities appear, and the analysis may simply use the speed instead of the velocity.

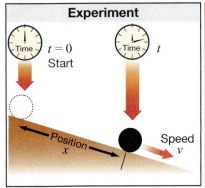

Time (s)	Speed (m/s)	Position (m)
0.0	0.00	0.00
0.2	0.83	0.08
0.4	1.66	0.33
0.6	2.50	0.75
0.8	3.33	1.33
1.0	4.16	2.08

The graph shows a straight line
The straight line on the graph means the speed of the ball increases by the same amount every second. The graph and data show that the speed of the ball increases by 0.83 m/s every two-tenths of a second. This graph shows *constant acceleration* because the speed changes by the same amount every second. The speed-versus-time graph of any motion with constant acceleration always looks like a straight line with a non-zero slope.

Acceleration
The graph in the diagram above shows an acceleration of 4.16 m/s². This is calculated by dividing the total change in speed, 4.16 m/s, by the total change in time, 1 second. You could do the calculation using any two points on the graph and find the same acceleration. For example, if you looked between 0.4 and 0.6 seconds, you would also calculate an acceleration of 4.16 m/s².

The acceleration of cars

The advertisement for a powerful sports car claims the car can go from 0 to 60 miles per hour in 4 seconds. This claim is all about acceleration. A speed of 60 mph is equal to 26.8 m/s. The car's average acceleration is 6.7 m/s² (26.8 m/s ÷ 4 s).

An average car accelerates from rest to 60 miles per hour in 12 to 20 seconds. At 12 seconds, the acceleration is 2.23 m/s². At 20 seconds, the acceleration is 1.34 m/s².

The practical limit for cars is 9.8 m/s². Greater acceleration than this would take a force greater than the force of gravity holding the car on the road.

Acceleration from the speed vs. time graph

Slope From the last section, you know that the slope of a graph is equal to the ratio of *rise* to *run*. On the speed versus time graph, the rise and run have special meanings, as they did for the distance versus time graph. The *rise* is the amount the speed changes. The *run* is the amount the time changes.

Acceleration and slope Remember, acceleration is the change in speed over the change in time. This is exactly the same as the ratio of rise over run on a speed versus time graph. *This is an important result!* The slope of the speed versus time graph is the acceleration.

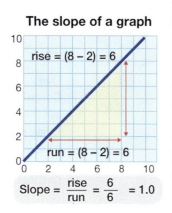
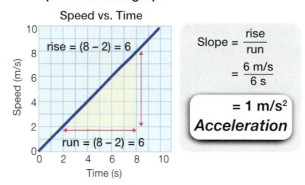

Make a triangle to get the slope To determine the slope of the speed versus time graph, take the rise, or change in speed, and divide by the run, or change in time. It is helpful to draw the right triangle shown to help figure out the rise and run. The rise is the height of the triangle. The run is the length of the base of the triangle.

Complex speed versus time graphs You can use slope to recognize when there is acceleration and when there is not on complicated speed versus time graphs. The greatest acceleration occurs when the slope is the steepest, at (B) on the graph. Level sections on the graph (A) show an acceleration of zero. Sections that slope down (C) show negative acceleration or slowing down.

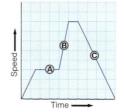

Calculating acceleration from a speed versus time graph

The following graph shows the speed of a bicyclist going over a hill. Calculate the maximum acceleration of the cyclist and calculate when in the trip it occurred.

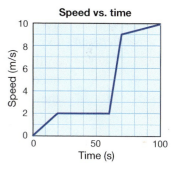

1. You are asked for acceleration.
2. You are given a graph of speed versus time.
3. a = slope of graph
4. The steepest slope is between 60 and 70 seconds, when the speed goes from 2 to 9 m/s.
   $a = (9 \text{ m/s} - 2 \text{ m/s}) \div (10 \text{ s})$
   $= 0.7 \text{ m/s}^2$

4.2 A Model for Accelerated Motion

The speed, distance, and time for a moving object are related by a formula: $v = d \div t$. This formula is a model that tells you the speed if you know the distance traveled and time taken. This section introduces a similar model for accelerated motion. The model includes the variables of distance, time, speed, and acceleration. Because there are more variables, the model includes two formulas. One formula relates speed, acceleration, and time. A second formula gives the distance traveled.

The speed of an object that is accelerating

Begin with the acceleration formula — To get a formula for the speed of an accelerating object, we can rearrange the experimental formula we had for acceleration. Consider a ball that starts with an initial speed, v_0 (Figure 4.4). At time, $t = 0$, the ball encounters a ramp and starts to accelerate. A time t later, the speed of the ball is v. The acceleration a of the ball is the change in speed divided by the change in time, or $a = (v - v_0) \div t$.

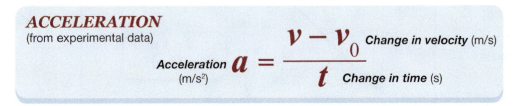

ACCELERATION (from experimental data)

$$\underset{(m/s^2)}{\text{Acceleration }} a = \frac{v - v_0 \;\; \text{Change in velocity (m/s)}}{t \;\; \text{Change in time (s)}}$$

Solve for the speed — If we rearrange this formula, we can get the speed, v, of the ball at any time, t, after it starts moving.

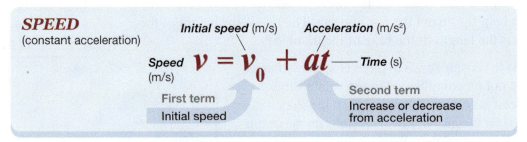

SPEED (constant acceleration)

$$\underset{(m/s)}{\text{Speed }} v = v_0 + at$$

First term — Initial speed
Second term — Increase or decrease from acceleration

How to interpret the formula — You can think of this formula in two pieces. In physics, a piece of an equation is called a **term**. The first term of the formula is the object's starting speed, or its **initial speed**. The second term is the amount the speed changes due to acceleration.

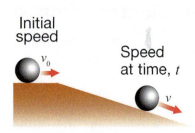

Figure 4.4: *A ball begins with an initial speed, and its speed increases as it accelerates down the ramp.*

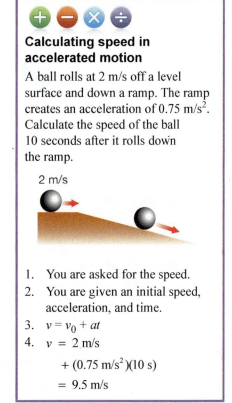

Calculating speed in accelerated motion

A ball rolls at 2 m/s off a level surface and down a ramp. The ramp creates an acceleration of 0.75 m/s². Calculate the speed of the ball 10 seconds after it rolls down the ramp.

1. You are asked for the speed.
2. You are given an initial speed, acceleration, and time.
3. $v = v_0 + at$
4. $v = 2$ m/s
   $+ (0.75 \text{ m/s}^2)(10 \text{ s})$
   $= 9.5$ m/s

Distance traveled in accelerated motion

An example experiment The distance traveled by an accelerating object can be found by looking at the speed versus time graph. The graph shows a ball that started with an initial speed of 1 m/s (Figure 4.5). The ball starts accelerating and after one second its speed has increased to 5.16 m/s. How far does the ball move down the ramp?

Distance from the speed versus time graph The distance traveled is equal to the area under the line representing the motion. This area has two pieces. The first piece is a rectangle that represents the distance the ball would have gone had its speed stayed constant. The second piece is a triangle. The triangle represents the additional distance the ball moves because its speed is increasing.

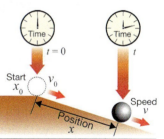

Time (s)	Speed (m/s)	Position (m)
0.0	1.00	0.00
0.2	1.83	0.28
0.4	2.66	0.73
0.6	3.50	1.35
0.8	4.33	2.13
1.0	5.16	3.08

Figure 4.5: *Data from an accelerated-motion experiment using a ball rolling down a ramp. The ball starts down the ramp with an initial speed of 1 m/s.*

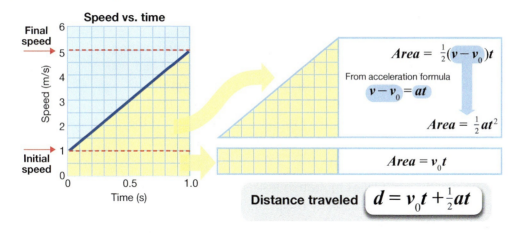

Calculating areas on the graph The area of the shaded rectangle is the initial speed v_0 multiplied by the time t, or $v_0 t$. This is the first term in the equation. The second term is the area of the shaded triangle. The height of the triangle is the change in speed, or $v - v_0$. This quantity is also the acceleration multiplied by the time, since $v - v_0 = at$. The base of the triangle is t, so the area of the triangle is one-half the base times the height, or $\frac{1}{2} at^2$. The total distance, d, the ball moves is the sum of the areas of the triangle and the rectangle, giving the equation $d = v_0 t + \frac{1}{2} at^2$.

Calculating the distance traveled This formula allows us to calculate how far the ball moves. The calculation is shown in Figure 4.6. You can see that the calculated distance of 3.08 meters agrees with the position of the ball one second after being released (Figure 4.5).

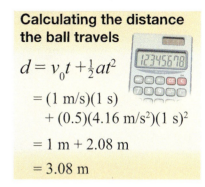

Figure 4.6: *Calculating the distance.*

A model for accelerated motion

Including initial position in the model

We need to add one more detail to complete a model for motion with constant acceleration in a straight line. It is possible that a moving object may not start at the origin. Let x_0 be the starting position. The distance an object moves is equal to its change in position ($x - x_0$). We can replace the distance traveled, d, with the change in position. The final formula describes the position of an object moving with constant acceleration.

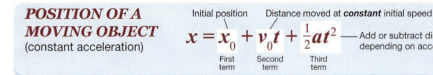

POSITION OF A MOVING OBJECT (constant acceleration)

$$x = x_0 + v_0 t + \frac{1}{2}at^2$$

- First term: Initial position
- Second term: Distance moved at *constant* initial speed
- Third term: Add or subtract distance depending on acceleration

What the formula means

The formula for the position has three terms. The first term is the starting position. The second term is the distance the object would have moved if its speed had stayed constant. The third term adds or subtracts distance depending on the acceleration. If the acceleration is negative, this term will *decrease* the total distance traveled because the object moves slower for part of its motion. If the acceleration is positive, this term will *increase* the total distance traveled because the object moves faster than its initial speed for part of its motion.

The model has two formulas

We now have both formulas for a model of motion in a straight line with constant acceleration. The first formula describes velocity or speed. The second formula describes position. In many physics problems both formulas will be used. When using the formulas outside of textbook physics problems, be aware that constant acceleration is usually only an approximation.

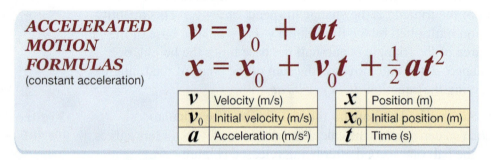

ACCELERATED MOTION FORMULAS (constant acceleration)

$$v = v_0 + at$$
$$x = x_0 + v_0 t + \frac{1}{2}at^2$$

v	Velocity (m/s)		x	Position (m)
v_0	Initial velocity (m/s)		x_0	Initial position (m)
a	Acceleration (m/s²)		t	Time (s)

Calculating position from speed and acceleration

A ball traveling at 2 m/s rolls up a ramp. The angle of the ramp creates an acceleration of –0.5 m/s². What distance up the ramp does the ball travel before it turns around and rolls back? (*Hint*: The ball keeps rolling upward until its speed is zero.)

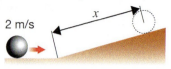

1. You are asked for distance.
2. You are given an initial speed and acceleration. You may assume an initial position of 0.
3. $v = v_0 + at$
   $x = x_0 + v_0 t + \frac{1}{2}at^2$
4. At the highest point the speed of the ball must be zero.

$$0 = 2 \text{ m/s} - 0.5t$$
$$t = 4 \text{ seconds}$$

Now use the time to calculate how far the ball went.

$$x = 0 + (2\text{m/s}^2)(4\text{s}) + \frac{1}{2}(-0.5\text{m/s}^2)(4\text{s})^2$$
$$= 8\text{m} - 4\text{m}$$
$$= 4\text{m}$$

At its highest point, the ball moves 4 meters up the ramp.

Solving motion problems with acceleration

Many practical problems involving accelerated motion have more than one step. That means you cannot use just one formula and plug in the numbers to find the answer. You need a strategy for working a way to the answer. The following questions will help you solve multiple-step problems.

List variables 1. Make a list of all the variables that might appear and assign values to those you know. This list usually includes: x, x_0, v, v_0, a, and t.

Cancel terms that are zero 2. Are any of the variables equal to zero? If a variable is zero, any terms including that variable will be zero and will cancel out. In many problems, the initial position, x_0, and initial speed, v_0, are zero. If no initial position is given, you may assume it is zero.

When is speed zero? 3. If an object is moving upward, its speed becomes zero when it reaches its highest point.

Use both formulas 4. When you look at any one of the formulas, could you get an answer if only you knew one more piece of information? This "extra information" is often the acceleration or the time. See if you can use the other formula to find the missing piece of information.

Calculating position from time and speed

A ball starts to roll down a ramp with zero initial speed. After one second, the speed of the ball is 2 m/s. How long does the ramp need to be so that the ball can roll for 3 seconds before reaching the end?

1. You are asked to find the length of the ramp.
2. You are given $v_0 = 0$, $v = 2$ m/s at $t = 1$ s, $t = 3$ s at the bottom of the ramp, and you may assume $x_0 = 0$.
3. After canceling terms with zeros, $v = at$ and $x = \frac{1}{2}at^2$
4. This is a two-step problem. First, you need the acceleration, then you can use the position formula to find the length of the ramp.

$$a = v \div t = (2 \text{ m/s}) \div (1 \text{ s})$$
$$= 2 \text{ m/s}^2$$

$$x = \frac{1}{2}at^2 = (0.5)(2 \text{ m/s}^2)(3 \text{ s})^2$$
$$= 9 \text{ meters}$$

Calculating time from distance and acceleration

A car at rest accelerates at 6 m/s². How long does it take to travel 440 meters, or about a quarter-mile, and how fast is the car going at the end?

1. You are asked for the time and the speed.
2. You are given $v_0 = 0$, $x = 440$ m, and $a = 6$ m/s²; assume $x_0 = 0$.
3. $v = v_0 + at$
   $x = x_0 + v_0 t + \frac{1}{2}at^2$
4. Since $x_0 = v_0 = 0$, the position equation reduces to

$$x = \frac{1}{2}at^2$$
$$440 \text{ m} = (\frac{1}{2})(6 \text{ m/s}^2)t^2$$
$$t^2 = 440 \div 3 = 146.7$$
$$t = 12.1 \text{ s}$$

Now use the time to calculate the speed.

$$v = (6 \text{ m/s}^2)(12.1 \text{ s})$$
$$= 72.6 \text{ m/s}$$

This is 162 miles per hour.

4.3 Free Fall and the Acceleration due to Gravity

From experience, we know that objects tend to fall to the ground. Gravity causes objects to accelerate as they fall down. In fact, a definition of *down* is "the direction objects fall." Whether released from rest, tossed up into the air, or thrown down, gravity causes a downward acceleration on free-falling objects. This section is about motion under the influence of gravity.

The acceleration due to gravity

Free fall   An object is in **free fall** if it is moving under the sole influence of gravity *only*. For example, if you drop a ball, it is in free fall from the instant it leaves your hand until it hits the ground. A ball thrown upward is also in free fall, because once it leaves your hand, its motion is determined by the influence of gravity.

Gravity accelerates objects at 9.8 m/s²   You know from experience that free-falling objects speed up, or accelerate, as they fall. When objects accelerate due to gravity, they always accelerate at the *same* rate. Gravity causes all free falling objects to accelerate at 9.8 m/s² toward Earth's center (Figure 4.7). The acceleration of 9.8 m/s² is so important, it is given its own name and symbol—**acceleration due to gravity (*g*)**. When you see a lowercase italic *g* in a physics equation, it usually stands for an acceleration of 9.8 m/s².

The sign of *g*   Whether the acceleration is positive or negative depends on how you choose to set up a problem. For some problems, it is convenient to assign the direction away from Earth's surface, or "up," to be the positive direction. This choice makes "down" the negative direction. For this choice, the acceleration is negative, or –9.8 m/s². For problems that involve only downward motion, it is often more convenient to make *down* the positive direction. If down is positive, the acceleration due to gravity is also positive, or +9.8 m/s².

***g* decreases slowly with altitude**   The value of *g* depends on distance from the center of the planet. The acceleration of gravity is equal to 9.8 m/s² only at Earth's surface. At high altitudes, the acceleration is smaller but the effect is not very important even on the highest mountains. The radius of Earth is 6,380 kilometers. The top of Mt. Everest is barely 8 kilometers above sea level. The acceleration due to gravity becomes appreciably less than 9.8 m/s² only when the altitude becomes significant compared to Earth's radius.

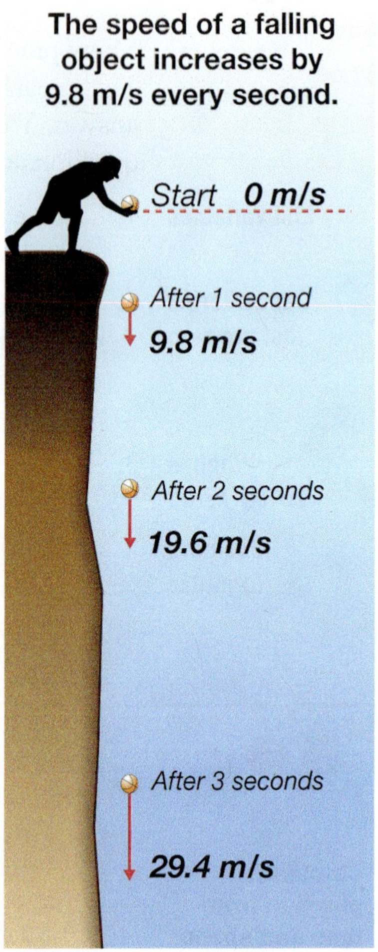

Figure 4.7: *The speed of a ball dropped off a cliff increases by 9.8 m/s every second when air friction is neglected.*

Free fall with initial velocity

Motion formulas for free fall — The motion of an object in free fall is described by the equations for speed and position with constant acceleration. The acceleration, a, is replaced by the acceleration due to gravity, g. The variable, x, has also been replaced with y since height is usually shown on the y-axis of a graph. Care must be taken when setting up a problem to determine the correct sign of the acceleration. The formulas assume up is positive, therefore the acceleration due to gravity is $-g$. If down is positive, then the negative signs in both formulas must be changed.

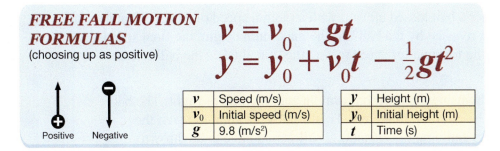

FREE FALL MOTION FORMULAS
(choosing up as positive)

$$v = v_0 - gt$$
$$y = y_0 + v_0 t - \tfrac{1}{2}gt^2$$

v	Speed (m/s)		y	Height (m)
v_0	Initial speed (m/s)		y_0	Initial height (m)
g	9.8 (m/s²)		t	Time (s)

Upward and downward motion — When an object's initial speed is downward, the acceleration due to gravity increases the speed until the object hits the ground. When the initial speed is upward, at first the acceleration due to gravity causes the speed to *decrease*. For example, a ball thrown straight up slows down as it rises. At the highest point, the speed is zero. After reaching the highest point, an object in free fall starts back down and its speed increases exactly as if it were dropped from the highest point with zero initial speed.

An example of free fall with upward motion — Figure 4.8 shows the motion of a ball thrown upward with an initial velocity of 19.6 m/s. The acceleration due to gravity reduces the speed of the ball by 9.8 m/s every second. After one second, the ball is still traveling upward with a velocity of 9.8 m/s. After two seconds, the ball has a speed of zero and has reached its maximum height. After three seconds, the velocity of the ball is –9.8 m/s. The negative sign indicates the ball is moving down. After four seconds, the ball has a velocity of –19.6 m/s and has returned to the height at which it was first thrown upward. (*Note:* Air resistance has been ignored in this example.)

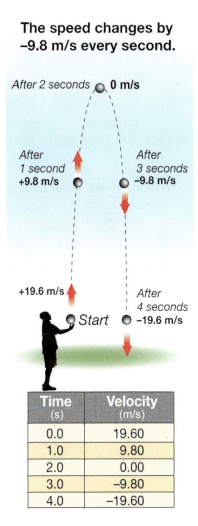

Figure 4.8: *The motion of a ball launched upward with the speed of 19.6 m/s.*

Time (s)	Velocity (m/s)
0.0	19.60
1.0	9.80
2.0	0.00
3.0	–9.80
4.0	–19.60

4.3 FREE FALL AND THE ACCELERATION DUE TO GRAVITY

Solving problems with free fall

Acceleration is 9.8 m/s² in free fall
Free fall problems are like other problems with constant acceleration. The chief difference is that you know the acceleration is 9.8 m/s² downward, even if it is not given in the problem.

Types of free fall problems
Most free-fall problems ask you to find either the height or the speed. Height problems often make use of the knowledge that the speed becomes zero at the highest point of an object's motion. In many situations, this knowledge allows you to calculate the time of flight and use that to find the height.

A problem with upward initial speed
For example, consider a ball tossed upward with an initial speed of 5 m/s. What is the maximum height reached by the ball? At the maximum height, the final speed is zero. We can use the formula for speed to find the time it takes the ball to reach its maximum height.

Finding the height
You find the height by using the position formula and the time of flight. The ball rises to a maximum height of 1.28 meters. This is not very high; to reach the height of a three-story building (12 meters), a ball would have to be thrown upward with a speed of 15 m/s (34 miles per hour).

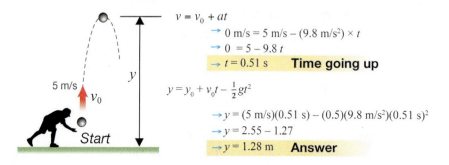

$v = v_0 + at$
→ $0 \text{ m/s} = 5 \text{ m/s} - (9.8 \text{ m/s}^2) \times t$
→ $0 = 5 - 9.8\,t$
→ $t = 0.51$ s **Time going up**

$y = y_0 + v_0 t - \frac{1}{2}gt^2$
→ $y = (5 \text{ m/s})(0.51 \text{ s}) - (0.5)(9.8 \text{ m/s}^2)(0.51 \text{ s})^2$
→ $y = 2.55 - 1.27$
→ $y = 1.28$ m **Answer**

Finding the time
If a problem asks for the **time of flight**, remember that an object takes the same time going up as it takes coming down. For example, the ball that was tossed upward took 0.51 seconds to reach its highest point. It takes another 0.51 seconds for the ball to fall back down to the height at which it started, so the total time of flight is 1.02 seconds.

Calculating height from the time of falling

A stone is dropped down a well and it takes 1.6 seconds to reach the bottom. How deep is the well? You may assume the initial speed of the stone is zero.

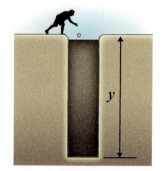

1. You are asked for distance.
2. You are given an initial speed and time of flight.
3.
$$v = v_0 - gt$$
$$y = y_0 + v_0 t - \frac{1}{2}gt^2$$

4. Since $y_0 = 0$ and $v_0 = 0$, the height equation reduces to

$$y = -\frac{1}{2}gt^2$$
$$y = -(0.5)(9.8 \text{ m/s}^2) \times (1.6 \text{ s})^2$$
$$y = -12.5 \text{ meters}$$

The negative sign indicates the height is lower than the initial height by 12.5 m.

Air resistance and mass

The acceleration of gravity does not depend on mass

The acceleration due to gravity does not depend on the mass of a falling object. This statement seems like it cannot possibly be true because a feather and a brick do not fall at the same rate. However, the reason they do not fall at the same rate is because gravity is *not* the only force acting. Objects that are dropped in air are *not* truly in free fall because air creates *friction*. Friction is a force that resists the motion of objects. Air friction can be substantial. If you put your hand out the window of a moving car, you feel a significant force pushing back against your hand. The feather falls slower because the resistance of the air is proportionately greater on the feather than on the brick. If one were to repeat the experiment in a chamber from which the air had been pumped out, the brick and feather would fall at exactly the same rate (see sidebar).

When air friction can be ignored

All of the formulas and examples discussed in this section are exact only in a vacuum, meaning no air. Fortunately, for small, dense objects at low speeds, friction from **air resistance** is so small it may be neglected. You may safely assume that $a = g = 9.8$ m/s^2 for speeds of up to several meters per second.

Air resistance and surface area

Air resistance may not be ignored for objects that have a large surface area and a low weight. A flat piece of paper is a good example of this sort of object. The more surface area an object has, the more air it has to push aside at it moves. The more air it has to push, the greater will be the force of air friction. Crumpling a piece of paper into a small ball greatly reduces its surface area. The air friction on a crumpled ball of paper is so small that the ball falls at the same rate as a brick.

Terminal velocity

Air friction increases rapidly with speed. That is because the faster something falls, the more air it pushes out of the way each second. Eventually, the speed of a falling object increases so much that the force from air friction becomes as strong as the force from gravity. At this point the net force becomes zero, acceleration also becomes zero, and the object falls with constant speed. Unless the object changes shape, it can never travel any faster. The maximum speed at which an object falls when limited by air friction is called its **terminal velocity**. The terminal velocity depends on shape, size, and weight. Thin, streamlined objects such as an arrow have a higher terminal velocity than light, wide objects such as a sheet of paper or a feather.

Parachutes and air resistance

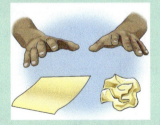

Take two identical sheets of paper. Crumple one of the pieces into a tight ball. Drop both at the same time from the same height. Which hits the ground first? Try to explain your observations using the concept of air resistance.

Parachutes use air resistance in order to reduce the terminal velocity of a skydiver. Alone, the skydiver has a small amount of surface area and a terminal velocity over 100 mph. The parachute increases the surface area dramatically, creating much greater air resistance. The skydiver reaches a slower terminal velocity—one that allows for a safe landing.

Chapter 4 Connection

Antilock Brakes

Brakes are often considered the most important devices on a vehicle. Antilock braking systems (ABS) are standard on all new vehicles since 2012. With the help of constant computer monitoring, these systems give the driver more control when stopping quickly. ABS prevents a car from skidding by ensuring that the wheels of the vehicle do not lock up during heavy braking. If the wheels keep turning, the driver is better able to steer and keep the vehicle under control.

Friction and traction

Figure 4.9: *Traction comes from an action-reaction pair of forces acting between the tire and the road.*

Traction — Friction between the tires and the road, called *traction*, is what allows the road to exert forces on the tires of a car. These forces are necessary for steering, acceleration, and braking (Figure 4.9). Without traction, a moving vehicle tends to continue moving in the direction it is going because of its inertia. Think of a moving car on ice. The car cannot turn, stop, or speed up because the wheels have no traction. Anything that gets between the tires and the road can cause dangerous skids. Ice, sand, and water are common hazards that can lead to loss of traction.

Braking — Friction is also used inside the braking system to slow or stop a vehicle. When you step on the brake pedal, a force is transferred to brake pads that squeeze against a disk attached to the wheel (Figure 4.10). The brake pads create friction that resists the motion of the car, and the car decelerates.

Locking the brakes — If brakes are applied too hard or too fast, a rolling wheel *locks up*, which means it stops turning and the car skids. This usually results in the tires making a screeching sound, and the car eventually skids to a stop. Wheel lockup happens because the car has more inertia than its wheel. Steering requires that the wheel be rolling on the road. *Once a wheel starts skidding it is no longer possible to steer.*

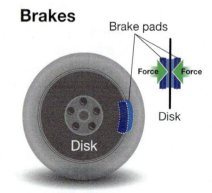

Figure 4.10: *When brakes are applied, the brake pads push against the disk, creating friction that slows the rotation of the wheel.*

Before ABS, drivers were taught to "pump" the brakes repeatedly in emergency braking situations. Pumping the brakes means to repeatedly brake hard and then take your foot off the brakes to let the wheels roll briefly, providing some minimal steering ability. This technique is difficult and requires practice to be used effectively. It takes a skilled driver to sense the changing balance between friction from the brakes and traction with the road, allowing for maximum braking and steering at the same time.

Chapter 4 Connection

Antilock braking systems

Wheel-speed sensor

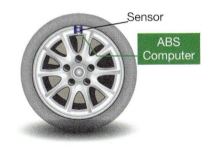

Using what is there
ABS essentially pumps the brakes automatically, much faster than any human could. ABS prevents wheel lockup under a wide range of traction conditions including water, ice, snow, and sand. ABS uses the same basic braking system but adds a control computer that constantly monitors the vehicle's wheel speed. To an ABS engineer, the wheel speed is the rate at which the vehicle's wheels are turning.

Wheel-speed sensors
With four-wheel ABS, the computer monitors input from four *wheel-speed sensors*, one on each wheel (Figure 4.11). As the wheel spins, each sensor constantly monitors the speed of the wheel and reports the information to the control computer. The computer compares the rate at which the speed of each wheel is changing, its acceleration, with the maximum rate at which a wheel can decelerate without losing traction. The computer bases its decision on average traction measurements taken from different tires in many driving conditions.

Figure 4.11: *The wheel-speed sensor sends information about each wheel's rotation to the ABS computer.*

What the control computer does
When the control computer decides the deceleration in a wheel is too rapid, the computer activates a small control called an actuator valve (Figure 4.12). The actuator valve releases pressure in the brake line which decreases the force applied to the brakes. This happens very quickly, before a wheel has come to a complete stop and begun to skid. Once the brakes have been released, traction with the road quickly spins the wheel back up to normal speed. When the wheel reaches normal speed, the ABS computer starts the cycle over by closing the actuator valve and allowing braking to resume. The automatic pumping of the brakes continues until the computer no longer senses a skid is about to occur.

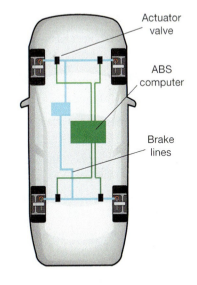

How to know when ABS is working
When the ABS computer is actively pumping the brakes, the driver hears a "gr-r-r-r" sound that comes from the rapid on-off cycling of the braking system. Some systems can apply and release the brakes up to 15 times a second. It may sound bad but this is a normal noise and the sign of a correctly functioning ABS. Some cars employ a similar system to prevent wheels from spinning while accelerating or turning corners. This system is often called traction control. Traction control can be very useful on snow-covered roads.

Figure 4.12: *The ABS computer can open and close the actuator valves on brakes for all four wheels.*

4.3 FREE FALL AND THE ACCELERATION DUE TO GRAVITY

Chapter 4 Assessment

Vocabulary

Select the correct term to complete the sentences.

acceleration	delta (Δ)	m/s²
initial	constant	term
slope	free fall	time of flight
acceleration due to gravity (g)	air resistance	
friction	terminal velocity	

1. The phrase "the change in" can be represented by the Greek letter _____.

2. An object whose speed changes by the same amount each unit of time has _____ acceleration.

3. A term for the rate of change of speed is _____.

4. One unit for acceleration of an object can be abbreviated as m/s/s, or _____.

5. The speed an object has when its motion is first being measured is called its _____ speed.

6. The ratio of rise to run on a graph is known as the _____ of the graph.

7. The amount of time during which an object is in free fall is its _____.

8. The maximum speed attained by a falling object when limited by the friction due to air resistance is _____.

9. An object falling near the surface of Earth gains speed at a rate of 9.8 m/s², commonly referred to as _____.

10. A force that opposes the motion of one surface over another or of objects through air is _____.

11. When the only force acting on an object is gravity it is described as being in a condition of _____.

12. _____ is the restriction created by the friction due to air on a falling object.

13. In physics, a piece of an equation may be called a(n) _____.

Concept review

1. Distinguish between average speed and instantaneous speed.

2. How can the instantaneous speed of an object be determined from a graph of position versus time?

3. Name the three things to know about an object to calculate its acceleration. Arrange them in an equation for calculating acceleration.

4. A bicycle racer may experience positive, negative, or zero acceleration at various times during a race.
   a. Identify when each of these accelerations would occur.
   b. Explain what happens to the biker's speed when she is moving in a straight line with positive, negative, and then zero acceleration.

5. A toy car starts from rest at the top of a hill and rolls down the hill with a constant acceleration.
   a. Sketch the shape of the speed versus time graph that describes its motion.
   b. Explain how to use the graph to determine the car's acceleration.

6. A graph is made of the speed and time of a plane as it flies from San Francisco to the Kahului Airport on Maui. How could the distance traveled by the plane be determined from the graph?

7. Sketch the shape of the speed versus time graph for
   a. a car coming to a stop at a red light.
   b. a cue ball on a pool table rolled at constant speed from one end to the other.

8. A coin is tossed into the air. Compare the direction of its velocity to the direction of its acceleration while it is moving upward, at its peak height, and moving downward.

9. What is the meaning of a negative sign when discussing motion relative to Earth?

10. A mouse races a rhinoceros from the roof of a tall building to the ground. Both jump from the roof at the same time.
    a. Who hits the ground first? Explain your answer. (*Note*: Assume that air resistance has no effect on either one.)
    b. On which animal is the force applied by the air greater?

11. While watching a skydiver from the ground, you see that when the parachute opens, he slows down but continues to fall toward the ground. Later you are watching a movie about skydivers. The jumpers appear to move upward after opening their parachutes. Explain the difference between your observations in the two cases.

12. Which has more acceleration when moving in a straight line: a car increasing its speed from 90 km/h to 100 km/h in one second or a cyclist accelerating from 0 km/h to 10 km/h in one second?

Problems

1. A plane flies north 600 kilometers in 2 hours and then turns west and flies 800 kilometers in 5 hours.
    a. Calculate the average speed of the plane during the first 2 hours of the trip.
    b. Calculate the average speed of the plane during the last 5 hours of the trip.
    c. Calculate the average speed of the plane for the entire trip.

2. Base your answers to the following questions on the graph below which represents the motion of two cars (A and B) on a straight road. Car A is initially at rest some distance down the road. Car A starts moving at the same instant that Car B passes. The time is measured from the moment Car A starts moving, when both cars are at the same position along the road.
    a. What is the instantaneous speed of each car at $t = 50$ s?
    b. Compare the distance traveled by Car A and Car B from $t = 0$ to $t = 60$ s.
    c. Over what time interval are the speeds of Car A and Car B constant?
    d. Calculate the acceleration of Car A and Car B from $t = 0$ to $t = 60$ s.
    e. During the time intervals given below, which car traveled the greatest distance?

        1. Car A from $t = 0$ to $t = 30$ s
        2. Car A from $t = 30$ to $t = 60$ s
        3. Car B from $t = 0$ to $t = 30$ s
        4. Car B from $t = 30$ to $t = 60$ s

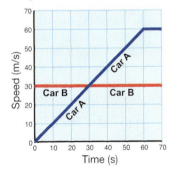

3. Seth starts from rest at the end of his driveway and accelerates down the road on his skateboard at 0.40 m/s² for 12 seconds. What is his speed at the end of the 12 seconds?

4. A roller coaster launches its cars out of the loading area. The cars accelerate for 2 seconds over a distance of 50 meters. Assuming the cars accelerate uniformly, use this information to answer the following questions.
    a. What is the average speed of a car while traveling the 50 meters?
    b. What is the instantaneous speed of a car once it has traveled the 50 meters?
    c. What is the acceleration?

5. If you toss an object upward from Earth's surface, it rises to a certain height and then falls back down. The faster you toss it, the farther it goes before reversing direction. If you throw it fast enough, it will not reverse direction and come back. The escape speed is the minimum speed an object needs to completely break free from the gravity of a planet. Earth's escape speed is approximately 40,270 km/h or 11,190 m/s. If an object starts from rest and accelerates at 20 m/s², how many seconds will it take to reach escape speed? What distance will it travel during this time?

6. A rocket is launched from the ground. Its speed is measured each second as it climbs. The graph shown is prepared from this data. Use the graph to answer the questions below.

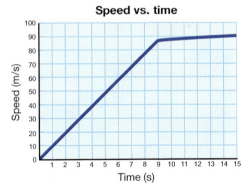

    a. What is the rocket's acceleration from $t = 0$ s to $t = 6$ s?
    b. How far does the rocket travel between the fourth and eighth second?
    c. What is the acceleration of the rocket from $t = 9$ s to $t = 15$ s?

7. Tom rides his bicycle from home to his friend Tony's house in 25 minutes. He spends 35 minutes there and returns home in 30 minutes. Use the graph to find the distance from Tom's house to Tony's house.

8. The graph to the right represents the motion of Amy's radio-controlled go-kart as she operates the kart in her driveway. The kart begins at the garage door and moves in a straight line. Briefly describe the motion of the kart for each interval on the graph (A–B, B–C, C–D, D–E, E–F, F–G, and G–H).

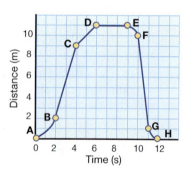

9. A ball is dropped out of a window. Which graph of speed versus time shown below best represents the motion of the ball as it moves toward the ground?

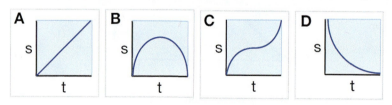

10. The graph below represents the relationship between speed and time for a kart that travels in a straight line.

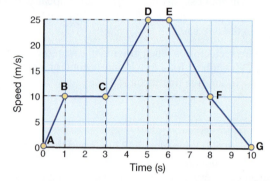

   a. What is the total distance traveled by the kart during the first three seconds?
   b. During which intervals does the kart have a positive acceleration?
   c. What is the acceleration of the kart during interval C–D?

11. A ball is thrown straight up in the air at a speed of 114 m/s. What is the speed and direction of the ball after 8 seconds?

12. A ball is thrown straight up at a speed of 58.8 m/s. Calculate its position after 7.2 seconds.

13. A cannon ball is shot straight up several meters and falls straight back into the cannon seconds later. Assuming "up" to be positive and that there is no friction due to air, sketch the following graphs for the ball from the moment it leaves until it falls back into the cannon again.
   a. velocity versus time
   b. position versus time

Applying your knowledge

1. How would a complete absence of air resistance affect people on Earth?
2. Some animals are able to accelerate to surprisingly high speeds. Research the world's fastest animals. Select one of the top 10 and explain the adaptations that make your choice a fast mover.
3. Basketball players who jump in the air seem to "hang" there for several seconds. Research the greatest height to which NBA players jump. Show by calculation that no NBA player has ever "hung" in the air for several seconds.

UNIT 2 MOTION AND FORCE IN ONE DIMENSION

CHAPTER 5

Newton's Laws: Force and Motion

Objectives:

By the end of this chapter you should be able to:

✔ Describe how the law of inertia affects the motion of an object.
✔ Give an example of a system or invention designed to overcome inertia.
✔ Measure and describe force in newtons (N) and pounds (lb).
✔ Calculate the net force for two or more forces acting along the same line.
✔ Calculate the acceleration of an object from the net force acting on it.
✔ Determine whether an object is in equilibrium by analyzing the forces acting on it.
✔ Draw a diagram showing an action-reaction pair of forces.
✔ Determine the reaction force when given an action force.

Key Questions:

- When do you encounter Newton's laws of motion in daily life?
- How are force, mass, and acceleration related?
- What are some common action-reaction force pairs?

Vocabulary

action	force	net force	Newton's second law	reaction
dynamics	law of inertia	newton (N)	Newton's third law	statics
equilibrium	locomotion	Newton's first law		

CHAPTER 5 — NEWTON'S LAWS: FORCE AND MOTION

5.1 The First Law: Force and Inertia

Sir Isaac Newton (1642–1727), an English physicist and mathematician, is one of the most famous scientists who have ever lived. Before the age of 30, he made several important discoveries in physics and invented a whole new kind of mathematics—calculus. The three laws of motion discovered by Newton are probably the most widely-used natural laws in all of science. Together, Newton's laws are the model which connects the forces acting on an object, its mass, and its resulting motion. This chapter is about Newton's laws, and the first section is about the first law, the law of inertia.

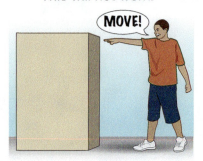

Only *force* has the ability to change motion.

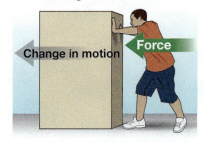

Figure 5.1: *Force is the action which has the ability to change motion. Without force, the motion of an object cannot be started or changed.*

Force

Changing an object's motion — Suppose you want to move a box from one side of the room to the other. What would you do? Would you yell at it until it moved? "Hey, box, get going! Move to the other side of the room!" Of course not! You would *push* or *pull* it across the room. In physics terms, you would apply a *force* to the box.

Force is an action that can change motion — A **force** is what we call a *push or a pull*, or *any action that has the ability to change an object's motion*. Forces can be used to increase the speed of an object, decrease the speed of an object, or change the direction in which an object is moving. For something to be considered a force, it does not necessarily have to change the motion, but it must have the ability to do so. For example, if you push down on a table, it will probably not move. But if the legs were to break, the table *could* move. Therefore, your push qualifies as a force.

Creating force — Forces can be created by many different processes. For example, gravity creates force. Muscles can create force. The movement of air, water, sand, or other matter can create force. Electricity and magnetism can create force. Even light can create force. No matter how force is created, its effect on motion is always described by Newton's three laws.

Changes in motion only occur through force — Forces create changes in motion, and *there can be no change in motion without also having a force* (Figure 5.1). Anytime there is a change in motion, a force must exist, even if you cannot immediately recognize the force. For example, when a rolling ball stops by hitting a wall, its motion changes rapidly. That change in motion is caused by the wall exerting a force that stops the ball.

Inertia

Objects tend to keep doing what they are doing

Consider that box you wish to move across the room. What if the box had been moving and you wanted to stop it? Again, yelling a command will not make it stop. The only way to stop the box is to apply enough force in a direction opposite to its motion. In general, objects tend to continue doing what they are already doing. If they are moving, they tend to keep moving, in the same direction, at the same speed. If they are at rest, they tend to stay at rest. This idea is known as Newton's first law of motion.

Newton's first law

Newton's first law states that an object will continue *indefinitely* in its current state of motion, speed, and direction, unless acted upon by a net force. Intuitively, you know that the motion of a massive object is harder to change than the motion of a lighter object. *Inertia* is a term used to measure the ability of an object to resist a change in its state of motion. An object with a lot of inertia takes a lot of force to start or stop; an object with a small amount of inertia requires a small amount of force to start or stop. Because inertia is a key idea in Newton's first law, the first law is sometimes referred to as the **law of inertia**.

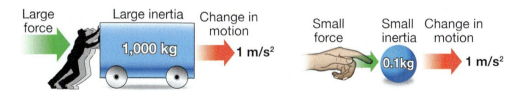

Inertia is a property of mass

The amount of inertia an object has depends on its mass. More massive objects have more inertia than less massive objects. Recall that mass is a measure of the amount of matter in an object. Big trucks are made of more matter than small cars; thus, they have greater mass and a greater amount of inertia. It takes more force to stop a moving truck because it has more inertia than a small car. This is a common-sense application of the first law.

Origin of the word *inertia*

The word *inertia* comes from the Latin word *inertus*, which can be translated to mean "lazy." It can be helpful to think of things that have a lot of inertia as being very lazy when it comes to change. In other words, they want to maintain the status quo and keep doing whatever they are currently doing.

Which systems in a car overcome the law of inertia?

The engine supplies force that allows you to change motion by pressing the gas pedal.

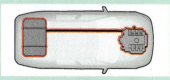

The brake system is designed to help you change your motion by slowing down.

The steering wheel and steering system is designed to help you change your motion by changing your direction.

Can you think of three parts of a bicycle that are designed to overcome the law of inertia?

Applications of Newton's first law

Seat belts and air bags Two very important safety features of automobiles are designed with Newton's first law in mind: seat belts and air bags. Suppose you are driving down the highway in your car at 55 miles per hour when the driver in front of you slams on the brakes. You also slam on your brakes to avoid an accident. Your car slows down but the inertia of your body resists the change in motion. Your body tries to continue doing what it was doing—traveling at 55 miles per hour. Luckily, your seat belt or air bag or both supplies a restraining force to counteract your inertia and slow your body down with the car (Figure 5.2).

Cup holders A cup holder does almost the same thing for a cup. Consider what happens if you have a can of soda on the dashboard. What happens to the soda can when you turn sharply to the left? Remember, the soda can was not at rest to begin with. It was moving at the same speed as the car. When your car goes left, the soda can's inertia causes it to keep moving forward (Figure 5.3). The result is quite a mess. Automobile cup holders are designed to keep the first law from making messes.

The tablecloth trick Have you ever wondered how a magician is able to pull a tablecloth out from underneath dishes set on a table? It's not a trick of magic at all, but just physics. The dishes have inertia and therefore tend to resist changes in motion. Before the magician pulls on the cloth, the dishes are at rest. So when the tablecloth is whisked away, the inertia of the dishes keeps them at rest. This trick works best when the tablecloth is pulled very rapidly and the table is small. It would be quite difficult to perform this trick with the long table in the diagram. Can you think why the long table would make the trick hard to do?

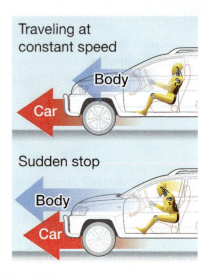

Figure 5.2: *Because of its inertia, your body tends to keep moving when your car stops suddenly. This can cause serious injury if you are not wearing a seat belt.*

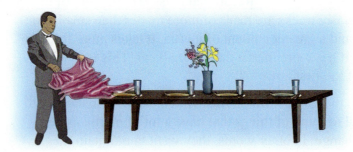

Figure 5.3: *Because of its inertia, a soda can on the dashboard will tend to keep moving forward when the car turns left.*

5.2 The Second Law: Force, Mass, and Acceleration

Newton's discovery of the connection among force, mass, and acceleration was a milestone in our understanding of science. The second law is the most widely used equation in physics because it is so practical. This section shows you how to apply Newton's second law to practical situations.

Newton's second law of motion

Force is related to acceleration — The acceleration of an object is equal to the force you apply divided by the mass of the object. This is **Newton's second law**, and it states precisely what you already know intuitively. If you apply more force to an object, it accelerates at a higher rate. If the same force is applied to an object with greater mass, the object accelerates at a lower rate because mass adds inertia. The rate of acceleration is the ratio of force divided by mass.

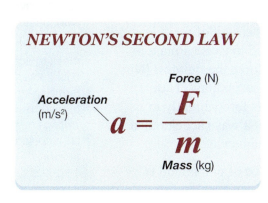

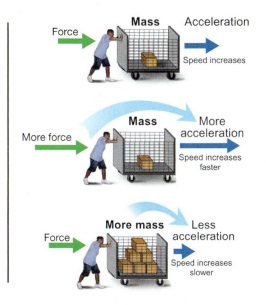

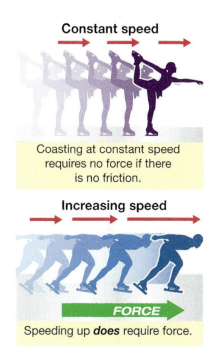

Figure 5.4: *An ice-skater can coast for quite a long time because motion at constant speed does not require force. If there was no friction, a skater could coast at constant speed forever. Force is required only to speed up, turn, or stop.*

Motion at constant speed — Force is not necessary to keep an object in motion at constant speed. An ice-skater will coast for a long time without any outside force (Figure 5.4). However, the ice-skater does need force to speed up, slow down, turn, or stop. Recall that changes in speed or direction always involve acceleration. Force *causes* acceleration, and mass *resists* acceleration.

The definition of force

Pounds   In the English system, the unit of force, the *pound*, was originally defined by gravity. One pound is the force of gravity pulling on a mass of 0.454 kilograms. When you measure your weight in pounds on a bathroom scale, you are measuring the force of gravity acting on your mass.

What is force?

The simplest concept of force is a push or a pull.

On a deeper level, force is the *action* that has the ability to create or change motion. Pushes or pulls do not always change motion. But they *could*.

The unit of force is derived from fundamental quantities of length, mass, and time. Using the second law, the units of force work out to be kg·m/s².

A force of 1 N causes a 1 kg mass to accelerate at 1 m/s². We could always write forces in terms of kg·m/s². This would remind us what force is. But, writing kg·m/s² everywhere would be a nuisance. Instead we use newtons. One newton (1 N) is 1 kg·m/s².

One pound (lb) is the force exerted by gravity on a mass of 0.454 kg.

One newton (N) is the force it takes to accelerate 1 kg at 1 m/s².

Newtons   The metric definition of force depends on the acceleration per unit of mass. A force of one newton is exactly the amount of force needed to cause a mass of one kilogram to accelerate at one m/s². We call the unit of force the **newton (N)** because force in the metric system is defined by Newton's second law. The newton is a useful way to measure force because it connects force directly to its effect on matter and motion. A net force of one newton will always accelerate a 1-kilogram mass at 1 m/s², no matter where you are in the universe.

Converting newtons and pounds   The newton is a smaller unit of force than the pound. A force of one pound is equal to about 4.448 newtons. This means a pound of force can accelerate a 1-kilogram mass at 4.448 m/s². Pounds are fine for everyday use here on Earth but inconvenient for physics because of the conversion factor of 4.448.

1 pound ≈ 4.448 newtons

Using the second law of motion

Net force The force (*F*) that appears in the second law is the **net force**. There are often many forces acting on the same object. Acceleration results from the combined action of all the forces that act on an object. When used this way, the word *net* means "total." To solve problems with multiple forces, you have to add up all the forces to get a single net force before you can calculate any resulting acceleration.

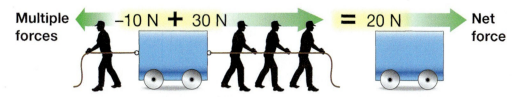

Three forms of the second law The second law can be rearranged three ways. Choose the form that is most convenient for calculating what you want to know. The three ways to write the law are summarized below.

Table 5.1: Three forms of the second law

Use...	if you want to find...	and you know...
$a = \dfrac{F}{M}$	The acceleration (*a*)	The net force (*F*) and the mass (*m*)
$F = ma$	The net force (*F*)	The acceleration (*a*) and the mass (*m*)
$M = \dfrac{F}{a}$	The mass (*m*)	The acceleration (*a*) and the net force (*F*)

Units for the second law To use Newton's second law in physics calculations, you must be sure to have units of m/s² for acceleration, newtons for force, and kilograms for mass. Many problems will require you to convert forces from pounds to newtons. Other problems may require you to convert weight in pounds to mass in kilograms. Remember also that *m* stands for *mass* in the formula for the second law. Do not confuse the variable *m* with the abbreviation *m* that stands for *meters*.

Calculating the acceleration of a cart on a ramp

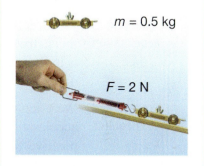

A cart rolls down a ramp. Using a spring scale, you measure a net force of 2 newtons pulling the car down. The cart has a mass of 500 grams (0.5 kg). Calculate the acceleration of the cart.

1. You are asked for the acceleration (*a*).
2. You are given mass (*m*) and force (*F*).
3. Newton's second law applies. $a = F \div m$
4. Plug in numbers. Remember that 1 N = 1 kg·m/s².
   $a = (2\text{ N}) / (0.5\text{ kg})$
   $= (2\text{ kg·m/s}^2) / (0.5\text{ kg})$
   $= 4\text{ m/s}^2$

5.2 THE SECOND LAW: FORCE, MASS, AND ACCELERATION

Finding the acceleration of moving objects

Dynamics — The word **dynamics** refers to problems involving motion. In dynamics problems, the second law is often used to calculate the acceleration of an object when you know the force and mass. For example, the second law is used to calculate the acceleration of a rocket from the force of the engines and the mass of the rocket.

Direction of acceleration — The acceleration is in the same direction as the net force. Common sense tells you this is true, and so does Newton's second law. Speed *increases* when the net force is in the same direction as the motion. Speed *decreases* when the net force is in the opposite direction as the motion.

Positive and negative — We often use positive and negative numbers to show the direction of force and acceleration. A common choice is to make velocity, force, and acceleration positive when they point to the right. Velocity, force, and acceleration are negative when they point to the left. You can choose which direction is to be positive, but once you choose, be consistent in assigning values to forces and accelerations.

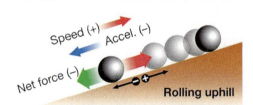

Speed **decreases** when the net force is opposite to the direction of motion.

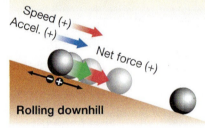

Speed **increases** when the net force is in the same direction as the motion.

The sign of acceleration — When solving problems, the acceleration always has the same sign as the net force. If the net force is negative, the acceleration is also negative. When both velocity and acceleration have the same sign, the speed increases with time. When velocity and acceleration have opposite signs, speed decreases with time. Careful use of positive and negative values helps keep track of the direction of forces and accelerations.

Acceleration from multiple forces

Three people are pulling on a wagon applying forces of 100 N, 150 N, and 200 N as shown. Determine the acceleration and the direction the wagon moves. The wagon has a mass of 25 kilograms.

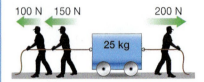

1. You are asked for the acceleration (a) and direction.
2. You are given the forces (F) and mass (m).
3. The second law relates acceleration to force and mass ($a = F \div m$).
4. Assign positive and negative directions. Calculate the net force then use the second law to determine the acceleration from the net force and the mass.

$F = -100\text{N} - 150\text{ N} + 200\text{N}$
$\quad = -50\text{N}$
$a = (-50\text{ N}) \div (25\text{ kg})$
$\quad = -2\text{ m/s}^2$

The wagon accelerates 2 m/s² to the left.

Finding force from acceleration

Amount of force needed Newton's second law allows us to determine how much force is needed to cause a given acceleration. Engineers apply the second law to match the force developed by different engines to the acceleration required for different vehicles. For example, an airplane taking off from a runway needs to reach a certain minimum speed to be able to fly. If you know the mass of the plane, Newton's second law can be used to calculate how much force the engine must supply to accelerate the plane to take-off speed.

Forces that must have been The second law also allows us to determine how much force *must have been* present to cause an observed acceleration. Wherever there is acceleration there must also be force. Any change in the motion of an object results from acceleration. Therefore, any change in motion must be caused by force. When a tennis ball hits a racquet, it experiences high acceleration because its speed goes rapidly to zero then reverses direction. The high acceleration is evidence of tremendous forces between the racquet and the ball, causing the ball to flatten and the racquet strings to stretch. Newton's second law can be used to determine the forces acting on the ball from observations of its acceleration.

Force to accelerate a plane taking off

$m = 5{,}000$ kg
$a = 5$ m/s^2

An airplane needs to accelerate at 5 m/s^2 to reach take-off speed before reaching the end of the runway. The mass of the airplane is 5,000 kg. How much force is needed from the engine?

1. You asked for the force (F).
2. You are given the mass (m) and acceleration (a).
3. The second law applies.
   $a = F \div m$
4. Plug in the numbers. Remember that
   1 N = 1 kg·m/s^2.
   $F = (5{,}000 \text{ kg}) \times (5 \text{ m/s}^2)$
   $= 25{,}000$ N

Force on a tennis ball striking a racquet

A tennis ball contacts the racquet for much less than one second. High-speed photographs show that the speed of the ball changes from −30 m/s to +30 m/s in 0.006 seconds. If the mass of the ball is 0.2 kg, how much force is applied by the racquet?

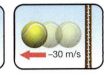

1. You are asked for force (F).
2. You are given the mass (m), the change in speed ($v_2 - v_1$), and the time interval (t).
3. Newton's second law ($a = F \div m$) relates force to acceleration. Acceleration is the change in speed divided by the time interval over which the speed changed or $a = (v_2 - v_1) \div t$.
4. Use the change in speed to calculate the acceleration. Use the acceleration and mass to calculate the force.

$a = (60 \text{ m/s}) \div (0.006 \text{ s}) = 10{,}000$ m/s^2

$F = (0.2 \text{ kg}) \times (10{,}000 \text{ m/s}^2) = 2{,}000$ N. This force is equal to three times the weight of the tennis player and 1,000 times the weight of the tennis ball!

Finding forces when acceleration is zero

Zero acceleration means zero net force
When acceleration is zero, the second law allows us to calculate unknown forces in order to balance other forces we know. Think about a gymnast hanging motionless from two rings (Figure 5.5). The force of gravity pulls down on the gymnast. The acceleration must be zero if he is not moving. The net force must also be zero because of the second law. The only way the net force can be zero is if the ropes pull upward with a force exactly equal and opposite the force of gravity pulling downward. If the weight of the gymnast is −700 newtons, then each rope exerts an upward force of +350 newtons.

Equilibrium
The condition of zero acceleration is called **equilibrium**. In equilibrium, all forces cancel out, leaving zero net force. Objects that are standing still are in equilibrium because their acceleration is zero. Objects that are moving at a constant speed and direction are also in equilibrium.

Statics problem
A **statics** problem usually means there is no motion. Most statics problems involve using the requirement of zero net force, or equilibrium, to determine unknown forces. Engineers who design bridges and buildings solve statics problems to calculate how much force must be carried by cables and beams. The cables and beams can then be designed so that they safely carry the forces that are required. The net force is also zero for motion at constant speed. Constant speed problems are treated like statics problems as far as forces are concerned.

Figure 5.5: *This gymnast is not moving so the net force must be zero. If the weight of the gymnast is 700 N, then each rope must pull upward with a force of 350 N in order to make the net force zero.*

A static force problem

A woman is walking two dogs on a leash. If each dog pulls with a force of 80 newtons, how much force does the woman have to exert to keep the dogs from moving?

1. You are asked for force (F).
2. You are given two 80 N forces and the fact that the dogs are not moving ($a = 0$).
3. Newton's second law says the net force must be zero if the acceleration is zero.
4. The woman must exert a force equal and opposite to the sum of the forces from the two dogs. Two times 80 N is 160 N, so the woman must hold the leash with an equal and opposite force of 160 N.

5.3 The Third Law: Action and Reaction

This section is about the often-repeated phrase "For every action there is an equal and opposite reaction." This statement is known as Newton's third law of motion. Newton's first and second laws of motion discuss single objects and the forces that act on them. Newton's third law discusses pairs of objects and the interactions between them. This is because forces in nature always occur in pairs, like the top and bottom of a sheet of paper. You cannot have one without the other.

Figure 5.6: *An astronaut can move in space by throwing an object in the direction opposite where the astronaut wants to go.*

Forces always occur in action-reaction pairs

Moving in space is a problem — The astronauts working on the space station have a serious problem when they need to move around in space: There is nothing to push on. How do you move around if you have nothing to push against?

Forces always come in pairs — The solution is to throw something opposite the direction you want to move. This works because *all forces always come in pairs*. If this seems like a strange idea, think through the following example. Suppose an astronaut throws a wrench. A force must be applied to the wrench to accelerate it into motion. The inertia of the wrench resists its acceleration. Because of its inertia, the wrench pushes back against the gloved hand of the astronaut. The wrench pushing on the astronaut provides a force that moves the astronaut in the opposite direction (Figure 5.6).

Forces on objects at rest — Forces also come in pairs when objects are not moving. For example, consider this book. It is probably lying open on a table. The weight of the book exerts a force on the table, the same as it would exert on your hands if the book was resting on your hands. The table pushes back upward on the book with a force equal and opposite the book's weight. A chain of force pairs keeps going because the table pushes down on the floor and the floor pushes back up on the table (Figure 5.7). The floor pushes down on the walls and Earth pushes back up on the walls to hold up the floor.

Action-reaction pairs — The two forces in a pair are called **action** and **reaction**. Anytime you have one, you also have the other. If you know the strength of one you also know the strength of the other since both forces are always equal. The two forces in an action-reaction pair always point in exactly opposite directions. They do not cancel each other because they act on *different objects*.

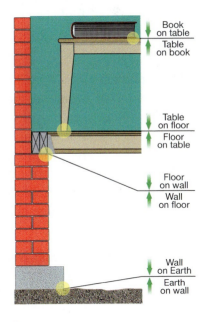

Figure 5.7: *Forces always come in action-reaction pairs. The two forces in a pair are equal in strength and opposite in direction.*

Newton's third law of motion

The first and second laws
The first and second laws apply to single objects. The first law states that an object will remain at rest or in motion at constant speed and direction until acted upon by an external force. The second law states that net force causes acceleration and mass resists acceleration.

The third law operates on pairs of objects
In contrast to the first two laws, the third law of motion applies to pairs of objects because *forces always come in pairs*. **Newton's third law** states that for every action force there has to be a reaction force that is equal in strength and opposite in direction. For example, to move on a skateboard you push your foot against the ground (Figure 5.8). The reaction force is the ground pushing back against your foot. The reaction force is what pushes you forward, because it is the force that acts on *you*. Your force against the ground pushes against Earth; however, the planet is so large that there is no perceptible motion resulting from your force.

Action-reaction forces act on different objects
Action and reaction forces act on different objects, *not* on the same object. For example, the action-reaction pair that is required to move a skateboard in the traditional way includes your foot and Earth. Your foot pushing against the ground is the action force. The ground pushing back on your foot is the reaction force. The reaction force makes you move because it acts on *you* (Figure 5.8). Why doesn't your foot make the ground move? Simply because the force is too small to accelerate Earth's huge mass. Even though the reaction force that acts on you is the same size, you are much less massive than Earth. The same size reaction force *is* big enough to accelerate you.

Stopping action and reaction confusion
It is easy to get confused about action and reaction forces. People often ask, "Why don't they cancel each other out?" The reason is that the action and reaction forces act on *different* objects. The action force of your foot acts on Earth and Earth's reaction force acts on you. The forces cannot cancel because they act on different objects.

Action and reaction
It does not matter which is the action force and which is the reaction. Whichever force you call the action makes its counterpart the reaction. The important thing is to recognize which force acts on which object (Figure 5.9). To apply the second law properly, you need to identify the forces acting on the object for which you are trying to find the acceleration.

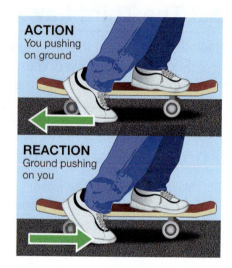

Figure 5.8: *All forces come in pairs. When you push on the ground (action), the reaction of the ground pushes back on your foot.*

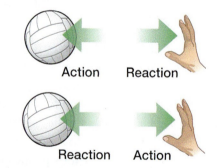

Figure 5.9: *It does not matter which force you call the action or the reaction. The action and reaction forces are interchangeable.*

Solving problems with action-reaction forces

Thinking about which force is acting on which object

In many physics problems, you are asked to determine the acceleration of a moving object from the forces acting on it. In the last section, you learned that the net force is the total of all forces acting *on* an object. Very often one of the forces will be a reaction force to a force created *by* the object. For example, consider a small cart attached to a spring (Figure 5.10). When you push the spring against a wall, a force is created. When you let the cart go, the force from the spring accelerates the cart away from the wall. But the force from the spring is pushing on the wall, so what force accelerates the cart? The answer is the *reaction force* of the wall pushing back on the spring. A force created by an object cannot accelerate the object itself, but the reaction force can.

A car attached to a spring

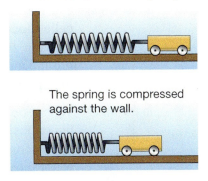

The spring is compressed against the wall.

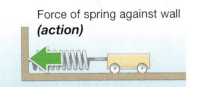

Force of spring against wall (action)

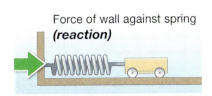

Force of wall against spring (reaction)

Figure 5.10: *Analyzing the action and reaction forces for a cart launched off a wall by a spring.*

Determining the reaction forces from people pushing a cart

Three people are each applying 250 newtons of force to try to move a heavy cart. The people are standing on a rug. Someone nearby notices that the rug is slipping. How much force must be applied to the rug to keep it from slipping? Sketch the action and reaction forces acting between the people and the cart and between the people and the rug.

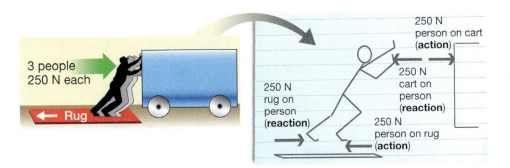

1. You are asked for how much force (F) it takes to keep the rug from slipping.
2. You are given that three forces of 250 N each are being applied.
3. The third law says that each of the forces applied creates a reaction force.
4. Each person applies a force to the cart and the cart applies an equal and opposite force to the person. The force on the rug is the sum of the reaction forces acting on each person. The total force that must be applied to the rug is 750 N in order to equal the reaction forces from all three people.

Locomotion

Locomotion The act of moving or the ability to move from one place to another is called **locomotion**. Any animal or machine that moves depends on Newton's third law to get around. When we walk, we push off the ground and move forward because of the ground pushing back on us in the opposite direction.

In the water When something swims, it pushes on water and the water pushes back in the opposite direction. As a result, the animal, submarine, or even microscopic organism moves one way, and a corresponding amount of water moves in the opposite direction. The movement of a boat through water results from a similar application of Newton's third law. When a lone paddler in a kayak exerts an action force pushing the water backward, the reaction force acts on the paddle, pushing the paddle and the kayak forward (Figure 5.11).

In the air Whether insect, bat, bird, or machine, any object that flies under its own power moves by pushing the air. Living creatures flap their wings to push air, and the air pushes back, propelling them in the opposite direction. Jets, planes, and helicopters push air, too. In the specific example of a helicopter, the blades of the propeller are angled such that when they spin, they push the air molecules down (Figure 5.12). According to Newton's third law, the air molecules push back up on the spinning blades and lift the helicopter.

The natural jet engine in a squid Squid use jet propulsion to move quickly. A squid fills a large chamber in its body with water. The chamber has a valve the squid can open and close. To move quickly, the squid squeezes the water inside its body with powerful muscles, then opens the chamber valve and shoots out a jet of water. The squid moves with a force equal and opposite in direction to the water jet that leaves its body.

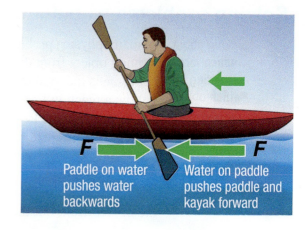

Figure 5.11: *Action and reaction forces for a kayak moving through the water.*

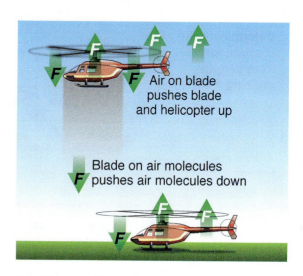

Figure 5.12: *Action and reaction forces on a helicopter.*

Chapter 5 Connection

Biomechanics

Biomechanics is the science of how physics is applied to muscles and motion. Many athletes use principles of biomechanics to improve their performance. People who design sports equipment use biomechanics to achieve the best performance by matching the equipment design to the athlete's body. Physicians, carpenters, people who build furniture, and many others also use biomechanics in their work. Any machine that relies on forces from the human body also relies on biomechanics.

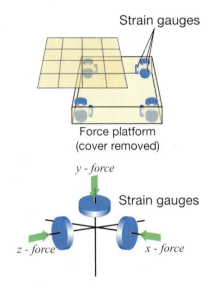

Figure 5.13: *A force platform has 12 strain gauges arranged to measure forces in the x, y, and z directions at each of the corners.*

The force platform A force platform is a very sophisticated scale that can record how forces change over time. Instead of containing springs, as your bathroom scale might, a force platform contains *strain gauges* (Figure 5.13). When a person steps or jumps on the platform, each strain gauge produces a reaction force and also a signal proportional to the strength of the reaction force. The force readings given off by the platform are referred to as ground reaction forces or GRFs.

Measuring force in three directions There are usually 12 strain gauges, three in each corner of a force platform oriented along the x-, y-, and z-axes. When force is applied to the platform, electrical signals from all 12 strain gauges are sent to a computer. The computer converts the signals to 12 separate force readings. From these readings, data is generated regarding the magnitude, direction, and sequence of GRFs being produced. Based on the relative magnitude of forces on each gauge, the center of pressure, or location of the force, can also be calculated.

Who uses force platforms? Force platforms are used in many different fields including medicine and athletics. Physicians, technicians, and therapists use force platforms in clinical settings to help in the diagnosis and rehabilitation of walking disorders. *Biomechanists*, including athletic trainers, use force platforms for research and to help athletes improve their technique. Equipment designers and manufacturers use information from force platforms in the design of sports equipment such as running shoes.

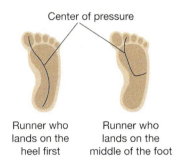

Figure 5.14: *The center of pressure for two runners with different running styles.*

The center of pressure The center of pressure is the place on your foot at which the average force is exerted against the ground. The center of pressure moves as your foot changes its contact point with the ground during walking or running. Force platform analysis is often used to evaluate the differences caused by various types of shoes, different track surfaces, walking versus running, and changes in gait patterns before and after surgery (Figure 5.14).

5.3 THE THIRD LAW: ACTION AND REACTION

Chapter 5 Connection

Force from a vertical jump

Measuring the forces from a vertical jump

The vertical jump is a common sport skill. Vertical jumps are seen in many different sports including basketball, volleyball, soccer, football, baseball, and tennis. A force platform makes an excellent tool to analyze the forces between the jumper's foot and the floor.

Newton's second law

To start the experiment, the athlete stands motionless in the middle of the platform (Figure 5.15). The "standing still" data is used to measure the weight of the athlete. That weight is converted to mass, using the second law ($m = F \div g$). The mass data is stored for later use. When given the command by the researcher, the athlete bends and jumps as high as possible. The force platform measures the force from each strain gauge at a rate of 250 to 1,000 measurements per second.

The force versus time curve

The biomechanist uses the data to generate a force versus time graph. A typical force versus time curve for a vertical jump is shown at the bottom of Figure 5.15. The total force recorded is the combination of the athlete's weight and the force produced during the jump. In this case, the athlete weighs 550 newtons. The peak GRF recorded is approximately 1,340 newtons. The time from the start of the jump until the athlete leaves the platform is just about one second. Once the athlete takes off and no longer touches the force table, the force readings drop to zero until the athlete lands back on the platform. The total time that the force is zero corresponds to the time in the air, a piece of information that allows other calculations to be made later.

Other characteristics of jumping motion

The force table data can be used to calculate many characteristics of the jumping motion. The total energy used can be calculated, as well as the maximum height reached. The force generated by the athlete's legs can also be determined along with maximum acceleration, and the balance of force between right and left legs.

Other biomechanical techniques

The technique of *electromyography* monitors the nerve signals to muscles and can determine the relative strength and sequence of contractions in the muscles being used in jumping. When combined with video equipment, the force, position, and time data can give a complete analysis of the motion that an athlete can watch to improve or evaluate her technique.

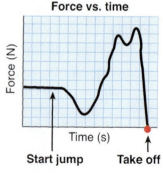

Figure 5.15: *A force platform can be used to measure the vertical force exerted during a vertical jump.*

Chapter 5 Assessment

Vocabulary

Select the correct term to complete the sentences.

force	inertia	law of inertia
Newton's first law	net force	dynamics
equilibrium	statics	Newton's second law
locomotion	newton (N)	action
reaction	Newton's third law	

1. The measure of the ability of an object to resist a change in its state of motion is called _____.
2. Any action which has the ability to change an object's motion must be a(n) _____.
3. Newton's first law is sometimes referred to as the _____.
4. The law of motion which states that all objects tend to resist changes in motion is known as _____.
5. A problem in which there is no motion is described as _____.
6. The acceleration of an object is proportional to the net force applied to the object according to _____.
7. The total force applied to an object as the result of many applied forces is called the _____.
8. A(n) _____ problem is one involving motion.
9. The force needed to accelerate a mass of one kilogram at a rate of 1 m/s² is one _____.
10. When the net force acting on an object is zero, the condition of zero acceleration is called _____.
11. Forces always occur in pairs. One force is called the _____ force, the other is called the _____ force.
12. "For every action force there is an equal and opposite reaction force," is a statement of _____.
13. The act of moving can be called _____.

Concept review

1. A glass of milk sits motionless on the kitchen table.
   a. Describe the forces acting on the glass of milk. Include their direction in the description.
   b. What word describes the state of motion of the glass of milk?
2. Name two units commonly used to measure force. How are they related?
3. Are the following statements true or false? Explain your answers using an example.
   a. Applying a force to an object will make it move.
   b. To keep an object moving, a force must be applied.
   c. A force must be applied to change the direction of a moving object.
4. To tighten the head of a hammer on its handle, it is banged against a surface as shown to the right. Explain how Newton's first law is involved.
5. How can rolling a bowling ball help you to determine the amount of matter in the ball?
6. List at least three parts of an automobile that are designed to overcome the effects of Newton's first law. Briefly explain the function of each.

7. State Newton's second law in words. Write an equation expressing the law.
8. Explain how the unit of force used by scientists, the newton, is defined.
9. In a space shuttle orbiting Earth, where objects are said to be weightless, an equal arm balance could not be used to measure the mass of an object. How could you measure the mass of an object in this situation?
10. Explain the difference between mass and weight. State common units for each.
11. What is the difference between the terms *force* and *net force*?
12. In physics problems, velocities, accelerations, and forces often appear with positive (+) or negative (−) signs. What do those signs indicate?
13. How does the sign of the force applied to an object compare with the sign of the acceleration?
14. What do motionless objects have in common with objects that are moving in a straight line with constant speed?

15. What is the difference between dynamics problems and statics problems? Give an example of each.

16. You and your little 6-year-old cousin are wearing ice skates. You push off each other and move in opposite directions. How does the force you feel during the push compare to the force your cousin feels? How do your accelerations compare? Explain.

17. You jump up. Earth does not move a measurable amount. Explain this scenario using all three of Newton's laws of motion.

Problems

1. The box pictured is being pulled to the right at constant speed along a level surface.
    a. Draw a diagram with arrows to represent the size and direction of all the forces acting on the box.
    b. Draw a diagram to represent the size and direction of the *net force* acting on the box.

2. Calculate
    a. the weight in pounds of a 16-newton object.
    b. the weight in newtons of a 7-pound object.
    c. the weight in newtons of a 3-kilogram object on Earth.
    d. the mass in kilograms of an object that weighs 12 newtons on Earth.

3. How does the inertia of a 200-kg object compare to the inertia of a 400-kg object?

4. A constant force is applied to a cart, causing it to accelerate. If the mass of the cart is tripled, what change occurs in the acceleration of the cart?

5. If the net force acting on an object is tripled, what happens to its acceleration?

6. On Venus, the acceleration due to gravity is 8.86 m/s². What is the mass of a man weighing 800 N on the surface of that planet?

7. A 60-kilogram boy on a 12-kilogram bicycle rolls downhill. What net force is acting on the boy and his bicycle if he accelerates at a rate of 3.25 m/s²?

8. A young girl whose mass is 30 kilograms is standing motionless on a 2-kg skateboard holding a 7-kg bowling ball. She throws the ball with an average force of 75 N.
    a. What is the magnitude of her acceleration?
    b. What is the magnitude of the acceleration of the bowling ball?

9. On Mercury, a person with a mass of 75 kg weighs 280 N. What is the acceleration due to gravity on Mercury?

10. A baseball player strikes the ball with his 1-kg bat. The bat applies an average force of 500 N on the 0.15-kg baseball for 0.20 seconds.
    a. What is the force applied by the baseball on the bat?
    b. What is the acceleration of the baseball?
    c. What is the speed of the baseball at the end of the 0.20 seconds?

11. The graph represents the motion of a 1,500-kg car over a 20-second interval.
    a. During which interval(s) is the net force on the car zero?
    b. What force is being applied to the car during interval C–D?

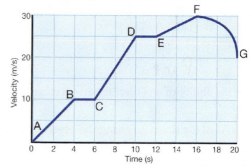

12. Two forces are applied to a 2-kilogram block on a frictionless horizontal surface as shown in the diagram. Calculate the acceleration of the block.

Applying your knowledge

1. When a sled pulled by a horse accelerates, Newton's third law says that the sled and horse exert equal and opposite forces on each other. If the horse and sled apply equal but opposite forces on one another, explain how the sled can be accelerated under these circumstances.

2. A bowling ball is positioned near the front of a stationary wagon. If the wagon is suddenly pulled forward, the bowling ball appears to move backwards in the wagon. Use each of Newton's three laws to explain what is actually happening to the wagon-ball system.

3. A 0.1-kilogram ball held at waist height is dropped and bounces back up toward the student's hand. Include all of the words from the chapter vocabulary list in describing the motion of the ball.

UNIT 2 MOTION AND FORCE IN ONE DIMENSION CHAPTER 6

Forces and Equilibrium

Objectives:

By the end of this chapter you should be able to:

- ✓ Calculate the weight of an object using the strength of gravity (*g*) and mass.
- ✓ Describe the difference between mass and weight.
- ✓ Describe at least three processes that cause friction.
- ✓ Calculate the force of friction on an object when given the coefficient of friction and normal force.
- ✓ Calculate the acceleration of an object including the effect of friction.
- ✓ Draw a free-body diagram and solve one-dimensional equilibrium force problems.
- ✓ Calculate the force or deformation of a spring when given the spring constant and either of the other two variables.

Key Questions:

- How are mass and weight related?
- What is friction, and how is it described and modeled?
- Does equilibrium mean that all forces acting on an object are zero?
- How can you tell how much force a spring will exert?

Vocabulary

ball bearings	deformation	free-body diagram	prototype	spring constant
coefficient of friction	dimensions	g-forces	restoring force	static friction
coefficient of static friction	engineering	Hooke's law	rolling friction	subscript
	engineering cycle	lubricant	sliding friction	viscous friction
compressed	extended	normal force	spring	weightless

117

6.1 Mass, Weight, and Gravity

People often use the words *mass* and *weight* interchangeably. However, in physics they are not the same. Mass is not weight. Mass is not volume either. Mass is a fundamental property describing the amount of matter in an object. Weight is *not* a fundamental property of an object. Weight is a *force* created by gravity. Since the strength of gravity is different in other places in the universe, such as on the Moon, the weight of the same object may also be different. However, the mass stays the same. In this section, you will learn about how mass, weight, and gravity are related.

Mass and weight

Mass is a measure of matter   In Chapter 1, you learned that the term *mass* is used in science to describe the amount of matter an object contains. You are made out of much more matter than a paper clip, so your mass is correspondingly greater than that of a paper clip.

Mass is a constant   The mass of an object is the same anywhere in the universe. Consider a suitcase full of clothes. The mass of the packed suitcase is a measure of the amount of matter it contains. The amount of matter does not change with location, so a 10-kilogram suitcase is always 10 kilograms, even on Mars (Figure 6.1). The only way to change the mass would be to physically remove something, such as some clothes. Actually, this is not quite true *exactly*. Mass *does* change when objects move at speeds close to the speed of light, as you will learn in Chapter 18. However, the speed of light is so great that the difference is far too small to worry about, even at the very highest speeds you are ever likely to encounter.

Weight is a force   The word *weight* is used to describe the force of gravity acting on an object. If somebody asks you, "How much do you weigh?" what they are technically asking is, "How much *force* does gravity exert on your mass?"

Weight is not constant   The strength of gravity is not constant throughout the universe, as you will see in Chapter 8. For example, the strength of gravity at Earth's surface is 2.6 times stronger than on the surface of Mars. Since weight depends on the strength of gravity, a kilogram on Mars weighs less than a kilogram on Earth. The 10-kilogram suitcase that weighs 98 newtons on Earth weighs only 38 newtons on Mars. Gravity is even weaker on the Moon, where the same suitcase weighs only 16 newtons.

Figure 6.1: *The weight of an object depends on the strength of gravity wherever the object is. The mass always stays the same.*

Calculating weight with mass and gravity

Weight varies since strength of gravity varies
The weight of an object depends on its mass and the strength of gravity. The formula gives the weight (F_w) in terms of the mass of an object, m, and the strength of gravity, g.

WEIGHT (g at Earth's surface)

Weight force (N) — $F_W = mg$ — Strength of gravity (9.8 N/kg)

Mass (kg)

F_w is the symbol for weight
We use the letter F to represent weight in this book because weight is a *force*. The little W next to the F stands for *weight* and reminds us which force is being considered. The W, or any letter or number written below another character, is called a **subscript**; we would say "F sub W" if we were telling someone what we were writing. The F and W always stay together since they function as one symbol for the force of gravity or the weight of the object.

The strength of gravity on Earth
In the equation for weight, the strength of gravity near Earth's surface is represented by the letter g. Within ±60 km of sea level, the value of g is approximately 9.8 N/kg. That means that gravity pulls on each kilogram of mass with a force of 9.8 newtons. A 1-kilogram object weighs 9.8 newtons. A 2-kilogram object weighs 19.6 N. A 10-kilogram object weighs 98 N.

The strength of gravity on Jupiter
What if you wanted to know the weight of an object somewhere else in the solar system? The strength of gravity is different on or near different planets. For example, the strength of gravity on Jupiter is much stronger than it is on Earth. At the top of Jupiter's atmosphere, the value of g is approximately 23 N/kg. That means the weight of a 1-kilogram object on Jupiter would be 23 newtons, about $2\frac{1}{3}$ times what it would weigh on Earth. The strength of gravity also decreases with altitude above a planet's surface. You will learn more about this in Chapter 8.

Two meanings for g

g	g
Strength of gravity (9.8 N/kg)	Acceleration of gravity (9.8 m/s²)

In Chapter 4, you learned that the symbol g stands for the acceleration of gravity in free fall, which is 9.8 m/s². In this chapter, you are being introduced to another meaning for g, and that is the strength of gravity, which is 9.8 N/kg.

In Chapter 4, we discussed objects that were in free fall being accelerated by the force of gravity. Here, we are discussing the force of gravity acting on all objects, whether they are accelerating or not.

When calculating weight, it is more natural to discuss gravity in N/kg instead of m/s². This is because objects may not be in motion but they still have weight. The two meanings for g are equivalent since a force of 9.8 N acting on a mass of 1 kg produces an acceleration of 9.8 m/s².

Gravity, acceleration, and weightlessness

Weightlessness An object is **weightless** when it experiences no net force from gravity. There are two ways to become weightless. One way is to get away from any source of gravity, such as planets, stars, or similar objects of large mass. The space between the planets has so little gravity that an object is essentially weightless.

Weight and acceleration A second way to become weightless is to be in *free fall*. To understand how falling can make you weightless, think about being in an elevator. Riding in an elevator that is accelerating upward makes you feel *heavier*. When the elevator is accelerating upward, the floor pushes against your feet to accelerate your body along upward with the rest of the elevator. If you are standing on a scale in the elevator, the scale reads your weight *plus* the additional force applied to accelerate you. If the elevator accelerates upward at 9.8 m/s^2, the scale reads twice your normal weight. You feel twice as heavy because the elevator pushes on your feet with an additional force, ma, equal to your weight, mg.

The human body in zero gravity

You feel weight because every bit of matter from your head to your toes feels the force of gravity. Your muscles and bones are used to supporting the weight of your body, and need to act against gravity to remain strong. Astronauts who live in weightless conditions begin to lose muscle tone and bone mass within a few days. For example, astronauts on the space station must keep a vigorous exercise schedule to minimize loss of strength from being in a low-gravity environment.

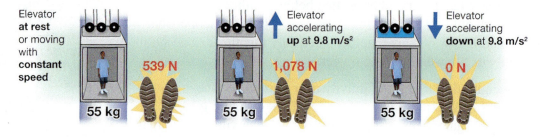

Free fall and weightlessness Now consider an elevator in free fall. The elevator is accelerating downward at 9.8 m/s^2. The scale feels *no force* because it is falling away from your feet at the same rate you are falling. As a result, you are weightless. A body in free fall is weightless because every particle of mass is accelerating downward.

g-forces Airplane pilots and race car drivers often describe forces they feel from acceleration as **g-forces** (Figure 6.2). These g-forces are not really forces at all, but are created by inertia. Remember, inertia is resistance to being accelerated. A body that is forced to accelerate upward at 1 *g* "feels" 9.8 extra newtons on every kilogram. A very fit human can withstand 6 or 7 *g*'s of acceleration before the g-forces prevent blood from flowing properly and cause loss of consciousness.

Figure 6.2: *An aircraft in a tight turn subjects the pilot to high g-forces.*

Using weight in physics problems

Do not substitute mass for force

Many problems in physics give you the mass of an object, yet ask for forces to be used or calculated. In these kinds of problems, you must often calculate the weight of the object. Like other forces, weight is measured in newtons or pounds. Mass is *not* a force. Mass is measured in grams or kilograms. If you use a quantity in kilograms in a formula that asks for a force, you will get the wrong answer.

Weight in equilibrium problems

Very often, weight problems involve equilibrium where forces are balanced. For example, how much force does it take to lift a 2-kilogram book from a table? If the book is lifted very slowly, there is almost no acceleration. Therefore, the minimum force it takes is just equal to the weight of the book. If you apply a force equal to the book's weight, the book is in equilibrium and there is no force holding it down to the table. Theoretically, the tiniest additional force causes the book to rise. Although the problem asked how much force would lift the book, you may assume that the required answer is the force needed to counteract the book's weight, which is 2 kg times 9.8 N/kg or 19.6 newtons. You may assume problems like this are always looking for the minimum force required, with no acceleration.

Weight on other planets or high altitudes

The other common type of weight problem involves other planets, or high altitudes, where the strength of gravity (g) is not the same as on Earth's surface. For these kinds of problems use the value for g that is given in the problem.

Calculating weight on Jupiter

How much would a person who weighs 490 N (110 lbs) on Earth weigh on Jupiter? Since Jupiter may not have a surface, *on* means at the top of the atmosphere. The value of g at the top of Jupiter's atmosphere is 23 N/kg.

1. You are asked for the weight.
2. You are given the weight on Earth and the strength of gravity on Jupiter.
3. $F_w = mg$.
   First, find the person's mass from the weight and strength of gravity on Earth:
   $m = (490 \text{ N}) \div (9.8 \text{ N/kg})$
   $= 50 \text{ kg}$
4. Next, find the weight on Jupiter:
   $F_w = (50 \text{ kg}) \times (23 \text{ N/kg})$
   $= 1,150 \text{ N} (259 \text{ lbs})$

Calculating force required to hold up an object

A 10-kilogram ball is supported at the end of a rope. How much force, or tension, is in the rope?

1. You are asked to find force.
2. You are given a mass of 10 kilograms.
3. The force of the weight is $F_w = mg$ and $g = 9.8$ N/kg.
4. The word "supported" means the ball is hanging motionless at the end of the rope. That means the tension force in the rope is equal and opposite to the weight of the ball.
   $F_w = (10 \text{ kg}) \times (9.8 \text{ N/kg}) = 98 \text{ N}$
   The tension force in the rope is 98 newtons.

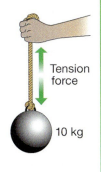

6.2 Friction

Everyone has experienced frictional forces. Frictional forces are forces that resist or oppose motion. Friction is what slows you down when coasting in a car or on a bicycle. You may not realize that friction is also necessary to speed up. Friction between the tires and the road is what allows forces to be transmitted from the road to the tires, accelerating the car or bicycle. This section is about frictional forces—what causes them, and what different types of frictional forces exist. The section also describes estimating frictional forces, and how to predict their effect on motion.

The force of friction

What causes friction? Friction results from relative motion between objects, such as the bottom of a cardboard box and the floor it is sliding across. If you looked through a powerful microscope, you would see that all surfaces—even those that appear smooth and shiny—have microscopic hills and valleys. As the surfaces slide across each other, the hills and valleys interfere with each other. This interference is one of the causes of friction. A good example of high friction would be two pieces of sandpaper rubbed together. Another cause of friction is that atoms of one surface stick to atoms of another surface when they get very close. A liquid such as oil in between two surfaces reduces this cause of friction because the liquid keeps the surfaces from contacting each other. This explains why an ice cube sliding across a wet floor has very little friction.

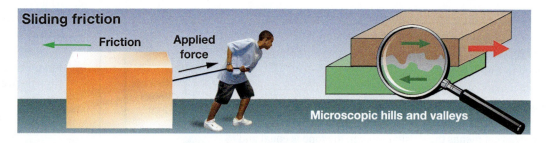

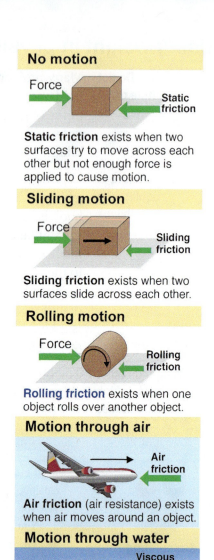

Friction resists motion Friction is a resistive force. Describing friction as resistive means that it *always works against the motion that produces it*. For example, as you push a box across a floor, the forward motion creates frictional forces that essentially push back on the box. Frictional forces make the box harder to move because they act opposite the direction of motion.

A model for friction

Friction depends on many factors The amount of friction that exists when a box is pushed across a smooth floor is very different from the amount of friction that exists when the same box is pushed across a carpeted floor. Different combinations of surfaces produce different amounts of friction. Some of the factors that affect friction include the type of material, the degree of roughness, and the presence of dirt or oil. Even the friction between two identical surfaces changes as the surfaces are polished by sliding across each other. No single model or formula can accurately describe the many processes that create friction. Even so, some simple approximations are useful.

Friction is proportional to the force holding surfaces together For dry, sliding surfaces, the force of friction is approximately proportional to the force squeezing two sliding surfaces together. For example, consider pulling a piece of paper across the table. If you pull the paper at constant speed, the force you apply is equal to the force of friction. It is easy to pull the paper alone, because the force holding it to the table is only its own weight. The force of friction is less than the weight of the paper. However, the weight of a brick placed on top of the paper creates much more force holding the paper against the surface of the table. The increased force between the paper and table creates more friction. It takes more force to pull the paper because the force of friction is larger.

The coefficient of friction The **coefficient of friction** is a ratio of the strength of **sliding friction** between two surfaces compared to the force holding the surfaces together, called the *normal force*. The coefficient of friction is most often a number between zero and one. A coefficient of *one* means the force of friction is equal to the normal force. A coefficient of *zero* means there is no friction no matter how much force is applied to squeeze the surfaces together. The symbol used for coefficient of friction is the Greek letter μ, which is pronounced "myou." The normal force, F_n, holding the surfaces together acts perpendicular to the plane of motion. This model is called dry sliding friction because it is most accurate when the surfaces are dry.

Calculating the force of friction

A 10-N force pushes down on a box that weighs 100 N. As the box is pushed horizontally, the coefficient of sliding friction is 0.25. Determine the force of friction resisting the motion.

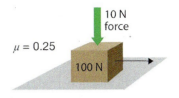

1. You are asked for the force of friction F_f.
2. You are given the weight F_w, the applied force F, and the coefficient of sliding friction μ.
3. The normal force is the sum of forces pushing down on the floor or $F_f = \mu F_n$.
4. First, find the normal force:
   $F_n = 100\text{ N} + 10\text{ N}$
   $ = 110\text{ N}$
   Use $F_f = \mu F_n$ to find the force of friction:
   $F_f = (0.25) \times (110\text{ N})$
   $ = 27.5\text{ N}$

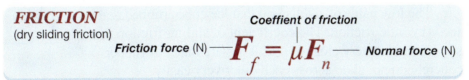

FRICTION (dry sliding friction)   Friction force (N) — $F_f = \mu F_n$ — Normal force (N), Coeffient of friction

Calculating the force of friction

The normal force
The **normal force** is the force perpendicular to two surfaces which are moving relative to each other. In many problems, the normal force is the *reaction* in an action-reaction pair. The action force of the pair is the force created by an object's weight pressing down on the supporting surface. The reaction force is the surface pushing back up to support the object. If additional forces are applied perpendicular to a sliding surface, the normal force is equal to the weight of the object *plus* the extra applied force.

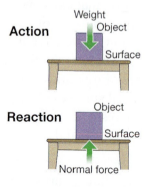

Static and sliding friction
It takes a certain minimum amount of force to make an object start sliding. The maximum net force that can be applied *before* an object starts sliding is called the force of **static friction**. Static friction results from the same interlocking and atomic attraction that creates sliding friction. Static friction can also be described by a similar model to sliding friction. The **coefficient of static friction** (μ_s) relates the *maximum* force of static friction to the normal force.

Table 6.1: Some representative values for the coefficient of friction

Surfaces	μ_{static}	$\mu_{sliding}$
rubber on concrete	0.80	0.65
wood on wood	0.50	0.20
ice on ice	0.10	0.03
glass on glass	0.94	0.40
steel on steel	0.74	0.57

STATIC FRICTION

Friction force (N) — $F_f = \mu_s F_n$ — Normal force (N)

Coefficient of static friction

Calculating the force of static friction

A steel pot with a weight of 50 N sits on a steel countertop. How much force does it take to start to slide the pot?

1. You are asked for the force to overcome static friction F_f.
2. You are given the weight F_w. Both surfaces are steel.
3. $F_f \leq \mu_s F_n$
4. $F_f \leq (0.74) \times (50\ N)$
   $\leq 37\ N$

Static friction is greater than sliding friction
The coefficient of static friction is nearly always greater than the coefficient of sliding friction. This is because it takes more force to break two surfaces loose than it does to keep them sliding once they are already moving. Table 6.1 gives some representative values for static and sliding friction.

Low μ values indicate slippery surfaces
There are coefficients of friction for each type of friction—static, sliding, rolling, air, and viscous. Notice that the sliding friction coefficient for ice on ice is 0.03, a value very close to zero. The low value of μ tells you that ice cubes rubbed across one another experience very little friction. The coefficient of sliding friction for rubber on concrete is 0.80, a value very close to one. Tires are made of rubber because rubber and concrete do not slide across each other very easily.

Friction and motion

How does friction affect acceleration?

When calculating the acceleration of an object, the F that appears in Newton's second law represents the net force. Since the net force includes all of the forces acting on an object, it also *includes the force of friction*. In many cases, the net force on an object is the applied force minus the force of friction. As a result, friction tends to reduce the acceleration produced by any forces that are applied to real objects. Physics problems are easier to solve in a frictionless, ideal world. However, the real world is never friction-free, so any *useful* physics must incorporate friction into practical models of motion.

The force of static friction

An applied force *less* than the force of static friction will produce *no* acceleration because it is not enough to start an object moving. That is because the force of static friction is equal and opposite to *any* applied force, up to a certain maximum value ($F_f = \mu_s F_n$). For example, suppose you calculate a force of static friction of 10 newtons for a box on a floor. The box does not move if a 5-N force is applied because 5 N is less than the maximum force of static friction. In this case, *the actual force of static friction is only 5 N*. Friction cannot *cause* an object to move. Unlike sliding friction, static friction is a variable force that increases itself to make the net force zero, up to its limit.

Friction reduces acceleration

Since frictional forces act against motion, any real acceleration will always be less than it would have been if there were no friction. If you know the applied force acting on an object, the weight of the object, and the coefficient of friction between the interacting surfaces, you can determine the acceleration of the object from the net force, which is the applied forces less the friction forces.

Calculating the acceleration of a car including friction

The engine applies a forward force of 1,000 newtons to a 500-kilogram car. Find the acceleration of the car if the coefficient of rolling friction is 0.07.

1. You are asked for the acceleration a.
2. You are given the applied force F, the mass m, and the coefficient of rolling friction μ.
3. Relationships that apply: $a = F \div m$, $F_f = \mu F_n$, $F_w = mg$ and $g = 9.8$ N/kg.
4. The normal force equals the weight of the car: $F_n = mg = (500 \text{ kg})(9.8 \text{ N/kg}) = 4{,}900$ N.
   The friction force is: $F_f = (0.07)(4{,}900 \text{ N}) = 343$ N.
   The acceleration is the net force divided by the mass:
   $a = (1{,}000 \text{ N} - 343 \text{ N}) \div 500 \text{ kg} = 657 \text{ N} \div 500 \text{ kg} = 1.31 \text{ m/s}^2$

Perpetual motion

Throughout history, many people have claimed to have invented a machine that will run forever with no outside force. We call these fanciful inventions perpetual motion machines, and none have ever worked.

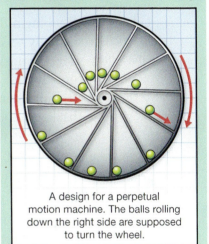

A design for a perpetual motion machine. The balls rolling down the right side are supposed to turn the wheel.

Perpetual motion machines cannot ever work because there is always some friction in any machine with moving parts. Friction always opposes motion, and every motion must inevitably slow down without an external source of energy.

If someone shows you a device that seems to go without stopping, be suspicious. There is no escape from friction. Somewhere you will find an external energy source to overcome friction.

Reducing the force of friction

All surfaces experience some friction
Frictional forces are unavoidable. Any motion where surfaces move across each other or through air or water always creates some friction. Unless a force is continually applied, friction eventually slows all motion to a stop. For example, bicycles have very low friction, but even the best bicycle slows down if you coast on a level road. Friction cannot be completely eliminated but it can be reduced. Many clever inventions have been created to reduce the force of friction.

Lubricants reduce friction in machines
Keeping a fluid such as oil between sliding services keeps the surfaces from actually touching each other. Instead of wearing away each other's bumps and depressions, surfaces separated by oil stir up the oil instead. The force of friction is greatly reduced, and surfaces do not wear out as fast. A fluid used to reduce friction is called a **lubricant**. You add oil to a car engine so the pistons can slide back and forth with less friction. Even water can be used as a lubricant under conditions where there is not too much heat. Powdered graphite, another lubricant, can be sprayed into locks so that a key slides more easily.

Ball bearings
In systems where there are axles, pulleys, and rotating objects, **ball bearings** are used to reduce friction. Ball bearings change sliding motion into rolling motion. Rolling motion creates much less friction than sliding motion. For example, a metal shaft rotating in a hole rubs and generates a great amount of friction. Ball bearings are small balls of steel that go between the shaft and the inside surface of the hole. The shaft rolls on the bearings instead of rubbing against the walls of the hole. Well-oiled bearings rotate easily and greatly reduce friction (Figure 6.3).

Magnetic levitation
Another method of reducing friction is to separate two surfaces with a cushion of air. A hovercraft floats on a cushion of air created by a large fan. Electromagnetic forces can also be used to separate surfaces. Working prototypes of a magnetically-levitated (or maglev) train have been built from several designs. A maglev train floats on a cushion of force created by strong electromagnets (Figure 6.4). Once it gets going, the train does not touch the rails. Because there is no contact, there is far less friction than with a train on tracks. The ride is smoother, allowing for much faster speeds. Maglev trains are discussed in greater detail in the Chapter 23 Connection.

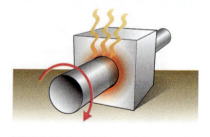
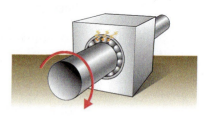

Figure 6.3: *The friction between a shaft, the long pole in the picture, and an outer part of a machine produces a lot of heat. Friction can be reduced by placing ball bearings between the shaft and the outer part.*

Figure 6.4: *In a maglev train, there is no contact between the moving train and the rail. This means that there is very little friction.*

Using friction

Friction is useful for brakes and tires There are many applications where friction is both useful and necessary. For example, the brakes on a bicycle create friction between two rubber *brake pads* and the rim of the wheel. Friction between the brake pads and the rim slows down the bicycle. Friction is also necessary to make a bicycle go. Designing tires that *maximize* friction is a multibillion-dollar industry.

Weather condition tires Rain and snow can act like lubricants to separate tires from the road. A smooth rubber tire slides easily on a wet road because water cannot drain away fast enough as the tire rolls. Practical all-weather tires have *treads,* which are patterns of deep grooves that allow space for water to be channeled away from the road-tire contact point (Figure 6.5). Special irregular groove patterns, with tiny slits on the contact surface are used on "snow tires" to provide increased traction in snow. These tires increase friction in two ways. They keep snow from getting packed into the treads, allowing the contact surface of the tire to change shape in order to grip the uneven surface of a snow-covered road.

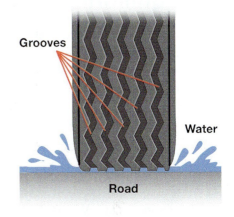

Figure 6.5: *Grooved tire treads allow space for water to be channeled away from the road-tire contact point, allowing for more friction in wet conditions.*

Nails Friction is the force that keeps nails and screws in place (Figure 6.6). A nail is actually a form of a wedge. The material into which a nail is driven, such as wood, pushes against the nail from all sides with compression forces. Each blow from the hammer drives the nail deeper into the material, increasing the contact surface and thereby increasing the total force of static friction keeping the nail in place. Over time, chemical reactions can actually cement the nail to the surrounding wood, creating a bond that results in an extraordinarily high coefficient of static friction. Removing a nail is more difficult than hammering it in.

Cleated shoes In footwear, it is desirable to increase the friction between the soles of shoes and the ground. For this reason, football players and soccer players wear special shoes with *cleats*. Cleats are teeth that stick out from the bottom of the shoe and dig into the ground. Cleats greatly increase the friction between the shoe and the ground. The increased friction means players can exert much greater forces against the ground to accelerate. Cleats also help players keep from slipping on the field.

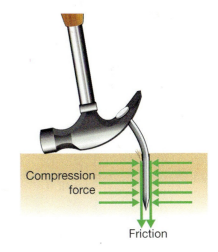

Figure 6.6: *Friction is what makes nails hard to pull out and gives them the strength to hold wood together.*

6.3 Equilibrium of Forces and Hooke's Law

In Chapter 5, you learned that the motion of an object depends on the net force acting on it. You also learned that when the net force acting on an object is zero, the object experiences no acceleration. This situation is called equilibrium. This section is about equilibrium and the important role equilibrium plays in the world outside the physics classroom. Hooke's Law describes the forces created by springs, rubber bands, and other objects that stretch. It is the natural spring-like behavior of all materials that allows normal forces between inanimate objects to maintain themselves precisely equal and opposite without need of any calculator or physics equation!

Equilibrium

Definition of equilibrium When the net force acting on an object is zero, the forces on the object are balanced. We call this condition equilibrium. Because of Newton's second law, $a = F \div m$, objects in equilibrium experience no acceleration. In equilibrium, an object at rest stays at rest, and a moving object continues to move with the same speed and direction.

Equilibrium does not mean no force Physics students sometimes do not understand, and think that if an object is in equilibrium, it must mean that there are no forces acting on the object. This is not the case. Newton's second law simply requires that for an object to be in equilibrium, the net force, or the sum of the forces, has to be zero. This *could* mean that there are no forces acting on it. But it is more probable that there *are* forces acting, and that the forces balance each other out by adding up to zero.

Reviewing Newton's 2nd law

Recall that the *F* in Newton's second law, $a = F \div m$, refers to the net force acting on an object.

Acceleration results from a net force that is *not* equal to zero.

Equilibrium results from a net force that *is* equal to zero. This does not mean there are not any forces acting, just that they cancel each other out.

Calculating the net force from four forces

Four people are pulling on the same 200-kg box with the forces shown. Calculate the acceleration.

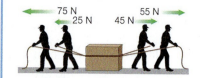

1. You are asked for acceleration.
2. You are given mass and force.
3. $a = F \div m$
4. $F = -75\text{N} - 25\text{N} + 45\text{N} + 55\text{N}$
   $= 0$ N, so $a = 0$

Equilibrium			Net Force	Non-equilibrium	
No motion	50 N → ☐ ← 50 N		+50 N −50 N 0	**Acceleration from net forces** 50 N → ☐ ← 35 N Acceleration → **Net Force** +50 N − 35 N = +15 N	**Free fall** ○ ○ ↓ ○ 9.8 m/s² **Net Force** = weight
Motion at constant speed and direction	50 N Force → ☐ ← 50 N Friction 5 m/s →		+50 N −50 N 0		

128 UNIT 2 MOTION AND FORCE IN ONE DIMENSION

Free-body diagrams

Forces on a free-body diagram
Many problems have more than one force applied to an object in more than one place. To keep track of the number and direction of all the forces, it is useful to draw a **free-body diagram**. A free-body diagram contains only the object under consideration. All connections or supports are removed and replaced by the forces they exert on the object.

An accurate free-body diagram includes every force acting on an object. All forces that are included must *act on the object*, not on other objects. This means many forces may be reaction forces.

An example situation
As an example of a free-body diagram, consider a 30-newton book resting on a table that weighs 200 newtons. The book is on one corner of the table so that its entire weight is supported by one leg. Figure 6.7 shows a free-body diagram of the forces acting on the table.

The free-body diagram for the table
Because the table is in equilibrium, the net force acting on it must be zero. The weight of the book acts *on the table*. The weight of the table acts *on the floor*. The force acting *on the table* is the reaction to its weight acting on the floor. The correct free-body diagram shows six forces. Equilibrium requires that the upward reaction at each leg be one quarter of the weight of the table (50 newtons). The leg beneath the book also supports the weight of the book (80 N = 50 N + 30 N).

The purpose of a free-body diagram
By separating an object from its physical connections, a free-body diagram makes it possible to focus on all forces and where they act. Forces due to weight or acceleration may be assumed to act directly on an object, often at its center. As a general rule, a reaction force is usually present at any point an object is in contact with another object or the floor.

Representing positive and negative forces
There are two ways to handle positive and negative directions in a free-body diagram. One way is to make all upward forces positive and all downward forces negative. The second way is to draw all the forces in the direction you believe they act on the object. When you solve the problem, if you have chosen correctly, all the values for each force are positive. If one comes out negative, it means the force points in the opposite direction from what you calculated.

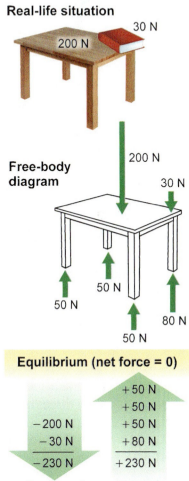

Figure 6.7: *A free-body diagram shows forces acting on a table. The table is in equilibrium, so the total upward force is equal to the total downward force.*

Applications of equilibrium

Equilibrium helps identify forces
Equilibrium and the second law are also used to prove the existence of forces that are otherwise difficult to see. Consider a book at rest on a table. If we see that an object is at rest, we know its acceleration is zero. That means the net force must also be zero. If we know one force, such as weight, we know there is another force, the normal force, in the opposite direction to make the net force zero.

Finding forces by using equilibrium
One common application of equilibrium is to find an unknown force. If an object is not moving, then you know it is in equilibrium and the net force must be zero. The condition of zero net force allows you to find an unknown force, if you know the other forces acting on the object, such as its weight. For example, suppose two cables are used to support a sign. You know the total upward force from the cables must equal the downward force of the sign's weight because the sign is in equilibrium. Therefore, the force in each cable is half the weight (Figure 6.8).

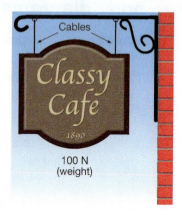

Figure 6.8: *Two cables must each exert an upward force of 50 N to support a sign with a weight of 100 N.*

Equilibrium in three directions
Real objects can move in three directions: up-down, right-left, and front-back. The three directions are called three **dimensions** and usually given the names x, y, and z (Figure 6.9). When an object is in equilibrium, forces must balance *separately* in each of the x, y, and z dimensions. For example, the block in the diagram in Figure 6.9 will move in the +z direction. The net force is zero in the x and y directions, so there will be no acceleration in the x or y direction. However, forces are not balanced in the up-down direction. Chapter 7 will discuss equilibrium of forces in three dimensions in more detail.

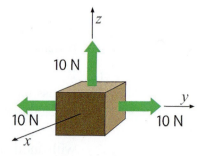

Figure 6.9: *Objects are free to move in three dimensions. Equilibrium implies balance of forces in all three dimensions. This box is not in equilibrium because there is a net force in the z-direction.*

Using equilibrium to find an unknown force

Two chains are used to lift a small boat. One of the chains has a force of 600 newtons. Find the force on the other chain if the mass of the boat is 150 kilograms.

1. You are asked for the force.
2. You are given one of two forces and the mass.
3. Relationships that apply: net force = zero, $F_w = mg$ and $g = 9.8$ N/kg.
4. The weight of the boat is $F_w = mg = (150\ \text{kg})(9.8\ \text{N/kg}) = 1{,}470$ N.

Let F be the force in the other chain, The condition of equilibrium requires that:
$F + (600\ \text{N}) = 1{,}470$ N, therefore $F = 870$ N

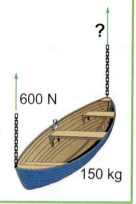

The force from a spring

Springs exert forces
A **spring** is a device designed to expand or contract, and thereby make forces in a controlled way. A jack-in-the-box is a children's toy that uses a spring. When the top comes off, the clown pops out of the box and into the air (Figure 6.10). Since the weight of the clown pulls down, the force from the spring is bigger than the weight. The difference in force accelerates the clown out of the box. Springs are used in many devices to create force. There are springs holding up the wheels in a car, springs to close doors, and a spring in a toaster that pops up the toast.

Characteristics of springs
Some common types of springs are coils of metal or plastic that create a force when they are **extended**, meaning stretched, or **compressed**, meaning squeezed. The force from a spring has two important characteristics.

1. The force always acts in a direction that tries to return the spring to its undeformed shape. For example, when you extend a spring, it pulls back. If you compress a spring, it pushes back against the applied force.

2. The strength of the force is proportional to the amount of extension or compression in the spring. For example, suppose you extend a certain spring by 10 centimeters and it makes a force of 5 newtons (Figure 6.11). If you extend the same spring twice as much, it will make a force twice as strong.

Figure 6.10: *A jack-in-the-box is a children's toy that relies on a spring.*

Forces created by a spring

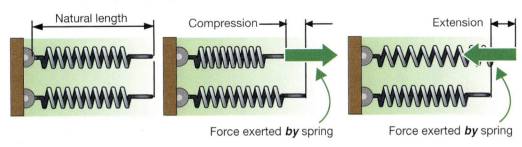

Force is proportional to change in length

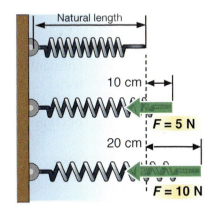

Figure 6.11: *The force exerted by a spring is proportional to the amount the spring is extended or compressed.*

Restoring force
When a spring is extended or compressed, the spring tries to "restore" itself to its original length. The force created by an extended or compressed spring is called a "restoring force" because it always acts in a direction to restore the spring to its natural length.

6.3 EQUILIBRIUM OF FORCES AND HOOKE'S LAW

Restoring force and Hooke's law

Deformation of a spring

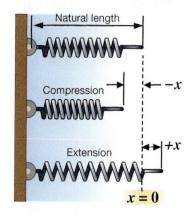

Deformation The change a natural, unstretched length from extension or compression is called **deformation**. Every spring has a natural (unstretched) length and the deformation of a spring is measured relative to its natural length (Figure 6.12). For example, suppose a spring with a natural length of 10 centimeters is compressed to 8 centimeters. The spring has a deformation of –2 centimeters because 2 centimeters is the difference between the natural and compressed lengths. Extension creates positive deformation (+x) and compression creates negative deformation (–x).

Strength of the restoring force For a given amount of deformation, the strength of the **restoring force** depends on the design of the spring and the material it is made from. Springs made from thick metal wire exert strong forces, even from small deformations. Springs made from thin wire or plastic exert only small forces, even when deformed a large amount.

The spring constant, k The relationship between the restoring force and deformation of a spring is given by the **spring constant** (k). A high value of k means the spring deforms very little even under relatively large forces. The springs in automobile shock absorbers are very *stiff* because they have a large spring constant. A loose spring, such as a coiled spring toy, has a low value of k.

Figure 6.12: *The deformation of a spring is measured relative to its natural unstretched length.*

The unit for k is N/m The spring constant has units of newtons per meter, abbreviated N/m. Looking at the units of k gives insight into what the spring constant means. The spring constant represents how many newtons of restoring force the spring exerts per meter that it is extended or compressed. For example, a spring with a spring constant of 10 N/m exerts 10 newtons of force for every meter that it is extended or compressed from its natural length.

Calculating the force from a spring

A spring with k = 250 N/m is extended by one centimeter. How much force does the spring exert?

HOOKE'S LAW (springs)

Force (N) — $F = -kx$ — Deformation (m)

Spring constant (N/m)

1. You are asked for force.
2. You are given k and x.
3. $F = -kx$
4. $F = -(250 \text{ N/m})(0.01 \text{ m})$
   $= -2.5$ N

Hooke's law The relationship between force, spring constant, and deformation is called **Hooke's law**. The negative sign indicates that positive deformation, or extension, creates a restoring force in the opposite direction. Negative deformation, or compression, creates a restoring force in the positive direction.

More about action-reaction and normal forces

Reaction forces Hooke's law provides the answer to the riddle of how an inanimate brick wall can always exert just the right amount of normal force to resist you without accelerating you. A wall cannot think for itself or calculate a physics formula.

A force applied to a wall creates an equal and opposite reaction force.

Action force (acts on wall)

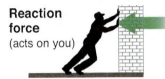
Reaction force (acts on you)

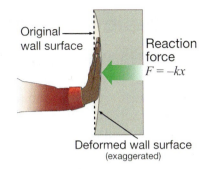

Figure 6.13: *The reaction force is created by deformation of the wall, acting in accordance with Hooke's law.*

Solid materials act like springs To better understand what creates a normal force, we must look more closely at the point of contact between your hand and the wall. If we could look at it very closely, on a microscopic level, we would see that the wall is deformed a very small amount by the force from your hand. All solid materials act like springs because they deform when forces are applied. Soft materials like rubber deform more than hard materials like wood and steel. However, even hard brick and steel deform *some* in response to an applied force.

Solid materials also exert restoring forces When you push against the wall, it deforms proportionately to the strength of your push. Like a spring, the wall exerts a restoring force back on your hand proportional to the amount it has deformed (Figure 6.13). If you push with a larger force, the wall deforms a larger amount. The larger deformation results in a larger restoring force. The restoring force from the wall is always exactly equal and opposite to the force you apply, because it is *caused by the deformation resulting from the force you apply.*

The what, the how, and the why The conditions of equilibrium allow us to verify that reaction forces exist. These reaction forces are the source of the normal force in certain situations, such as a table supporting a book. Equilibrium and Newton's second law allow us to calculate the strength of the reaction force or normal force. Finally, Hooke's law allows us to explain how the normal force is created, why its strength is always equal to the applied force, and why its direction is opposite to the applied force. A wall exerts the correct force by deforming in response to the applied force, based on Hooke's law.

Calculating the restoring force from a solid

The spring constant for a piece of solid wood is 1×10^8 N/m. Use Hooke's law to calculate the deformation when a force of 500 N (112 lbs) is applied.

1. You are asked for the deformation (x).
2. You are given the force and spring constant (F and k).
3. $F = -kx$; therefore $x = -F \div k$
4. $x = -(500 \text{ N}) \div (1 \times 10^8 \text{ N/m})$
   $= -5 \times 10^{-6}$ m,

or five-millionths of a meter, much smaller than the thickness of a hair. This is why you do not notice deflections of solid materials except under very large forces.

6.3 Equilibrium of Forces and Hooke's Law

Chapter 6 Connection

The Design of Structures

A structure is anything designed to withstand forces without breaking and without excessive deformation. We are surrounded by structures. A house is a structure, and so is a bridge, a building, a car, and almost every other object around you, even a tree. To design a structure, you first need to know what forces act and how, and where the forces are applied. The forces on a structure are found by applying the conditions of equilibrium. Most structures are designed to act in equilibrium, which means they are not accelerating. At least, they are not supposed to!

A bridge is a simple structure Consider the very simple structure of a bridge with two supports. To work, the bridge must bear weight—people, vehicles, animals—without breaking. The supports for the bridge must in turn be able to bear the weight of the bridge, to hold it up. Suppose there are 40 people on a bridge. Each person weighs 600 N and the bridge itself weighs 6,000 N. How strong must the bridge supports be?

Analysis of the forces By drawing a free-body diagram, we can identify the forces which must be acting if the bridge is not moving (Figure 6.14). The downward force is 30,000 newtons, the weight of the people and bridge (40×600 N + 6,000 N). The upward force must be equal and opposite to the downward force in order to make the net force on the bridge zero. Each support must therefore be able to withstand a force of 15,000 newtons (3,370 pounds).

Engineering and technology The analysis of forces in order to design structures is part of *mechanical engineering*. In a broad sense, **engineering** is the application of science to solving real-life problems, such as designing a bridge. A bridge and other inventions created by engineers are part of *technology*. From the invention of the plow to the computer, all technologies arise from someone's application of science to the solution of a practical problem.

Bridge

Free-body diagram

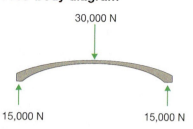

Figure 6.14: *A basic force analysis of a bridge. A more accurate analysis would allow for distributing the weight of the people unevenly all along the bridge instead of in the center.*

Chapter 6 Connection

The engineering design cycle

Design, prototype, test, then evaluate

The design of a bridge has many steps. The design starts with a concept that is an idea for how the bridge will be made. A **prototype** is constructed to test the bridge design. The prototype is tested to see if it works as designed. The test results are evaluated and used to correct any problems or improve the performance. The process of design-prototype-test-evaluate is called the **engineering cycle** (Figure 6.15). The most reliable technology goes through the cycle many times, and each cycle leads to more improvements.

Figure 6.15: *The engineering design cycle is how an invention gets from concept to reality.*

Solving an engineering design problem

A small-scale project is a good illustration of the design cycle. Suppose you are given a box of toothpicks and some glue, and are assigned to build a bridge that will hold a 25-newton brick. After doing research, you come up with an idea for how to make the bridge. Your idea is to make the bridge from four structures connected together. Your structure is a truss because you have seen bridges that use trusses. Your idea is called a *conceptual design*.

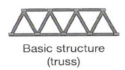
Basic structure (truss)

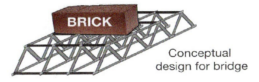
Conceptual design for bridge

The importance of a prototype

Your next step is to build a prototype and test it. If you can determine how much force it takes to break *one* structure, you would know if four structures will hold the brick. Your prototype should be close enough to the real-life bridge so that what you learn from testing can be applied to the final structure. For example, if your final bridge is to be made with round toothpicks, your prototype also has to be made with round toothpicks.

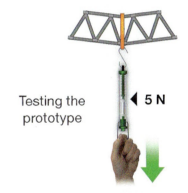

Testing the prototype

Revised design with 7 trusses

Figure 6.16: *Testing the prototype tells you if it is strong enough. Testing often leads to a revised design, such as using more trusses.*

Testing the prototype

You test the prototype truss by applying more and more force until it breaks. You learn that your truss breaks at a force of 5 newtons. Since the brick weighs 25 newtons, four trusses are not enough. Based on the test results, your bridge design may be revised to use seven trusses instead of four (Figure 6.16). The *evaluation* of test results is a necessary part of any successful design. Testing identifies potential problems and unexpected effects in the design in time to correct them.

6.3 EQUILIBRIUM OF FORCES AND HOOKE'S LAW

Chapter 6 Assessment

Vocabulary

Select the correct term to complete the sentences.

mass	weight	weightless
g-force	friction	static friction
rolling friction	viscous friction	air friction
normal force	extension	net force
free-body diagram	lubricant	equilibrium
ball bearings	dimensions	spring
Hooke's law	compression	spring constant
deformation	restoring force	coefficient of friction
engineering	design cycle	subscript
prototype	coefficient of static friction	sliding friction

1. A force measurement that depends upon the mass of an object and the strength of gravity is _____.

2. In free fall, a bathroom scale with a person standing on it would read 0 pounds. The person might be referred to as _____.

3. The amount of matter a body contains is its _____.

4. The inertial forces felt by race car drivers or jet pilots when accelerating are called _____.

5. A small number or character written below another character is called a(n) _____.

6. A measure of the strength of sliding friction between two surfaces is expressed as a number known as the _____.

7. _____ is the sum of all the forces acting on an object that could cause acceleration.

8. The minimum amount of force required to start an object moving equals the _____.

9. A fluid, such as oil, used to reduce friction between surfaces is called a(n) _____.

10. The ratio of the maximum force of static friction to the normal force is the _____.

11. A force that is perpendicular to two surfaces moving relative to each other is a(n) _____.

12. The friction that exists between plane, dry surfaces moving over one another is described as _____.

13. Ball bearings can be used to change sliding friction to _____.

14. The magnetic levitation of trains replaces the rolling friction of the wheels of conventional trains with _____.

15. _____ exists when objects move through water or other fluids.

16. Small balls of steel known as _____ surround a turning shaft to reduce friction.

17. A drawing representing all forces acting on an object is called a(n) _____.

18. Changing the length of a spring by extending or compressing it causes _____ of the spring.

19. The law which summarizes the relationship among force, spring constant, and deformation of a spring is called _____.

20. When all the forces acting on a body are balanced, a condition of _____ exists.

21. Left-right, up-down and front-back are the three _____ of space represented by the letters x, y and z.

22. A coil spring will store force when it undergoes _____ or _____.

23. The relationship between the strength and deformation of a spring is described by its _____.

24. The force which acts to return a stretched spring to its normal length is referred to as a(n) _____.

25. The application of science to solving real-life problems such as the proper construction of a bridge is called _____.

26. An object such as an elastic coil wire which tends to return to its natural length may be called a(n) _____.

27. A model that tests the design of an engineering concept can be called a(n) _____.

28. The process of design-prototype-test-evaluate is called the _____.

Concept review

1. Explain how an object's weight is related to its mass.
2. If you were to travel to the Moon, where gravity is weaker than on Earth, would your weight be the same, more, or less? Would your mass be the same, more, or less?
3. What are the two ways a person could feel weightless?
4. A backpack has a weight of 85 N and is sitting on the floor. Describe what happens to the backpack in each case.
    a. You pull up on the backpack with a force of 50 N.
    b. You pull up on the backpack with a force of 85 N.
    c. You pull up on the backpack with a force of 86 N.
5. What is the cause of sliding friction?
6. If you are pulling a sled carrying your little brother to the right, in which direction is the force of friction on him? Which type of friction is involved?
7. How is static friction different from the other types of friction? How is it similar?
8. List the type of friction involved in each of the following cases.
    a. A soccer ball slows as it rolls on the grass.
    b. You drag a heavy backpack across a desk.
    c. A skydiver's parachute slows her descent.
    d. A piano stays at rest as you attempt to push it.
    e. A diver hits the surface of the water and comes to a stop before reaching the bottom of the pool.
9. The coefficient of friction is usually a number between ____ and ____.
10. Which is usually greater for a pair of surfaces, the coefficient of static friction or the coefficient of sliding friction? Why?
11. You push a heavy box with a force of 250 N, and it does not move. What is the force of static friction between the box and the floor?
12. List three ways to reduce friction between objects.
13. Give some examples of how friction is useful in our daily lives.
14. Read each statement and decide whether it is true or false.
    a. If an object is in equilibrium, then there can't be any forces acting on it.
    b. If an object is in equilibrium, then it must be at rest.
    c. If the net force on an object is zero, then it is in equilibrium.
15. You stretch a spring to the right. What is the direction of the force the spring exerts on your hand?
16. How is a spring's force related to the distance it is extended or compressed?
17. What is the meaning of the negative sign in Hooke's law?
18. Which spring would be easier to stretch, one with a spring constant of 10 N/m or one with a spring constant of 30 N/m?
19. The driver of a speeding truck slams on its brakes and skids to a stop. If the mass of the truck was twice as great, how would the length of the skid compare if the initial speed was the same?

Problems

1. Calculate the weight of each of the following.
    a. a 60-kg person on Earth
    b. a 4-kg cat on Earth
    c. a 30-kg dog on Jupiter
2. Calculate the mass of each of the following.
    a. a car that weighs 15,000 N on Earth
    b. a frog that weighs 12 N on Jupiter
3. Chris has a mass of 75 kg. He stands on a scale in an elevator that is accelerating downward at 4.9 m/s^2. What force does the scale display in newtons?
4. Calculate the static friction between your rubber sneakers and the concrete sidewalk if your mass is 55 kg.
5. While vacuuming, you have to move the sofa. The sofa weighs 500 N, and the force of sliding friction between the sofa and the carpet is 200 N. Calculate the coefficient of sliding friction between the sofa and the carpet.
6. While ice skating, you push off of a wall with a force of 100 N. Calculate your acceleration if you have a mass of 50 kg and the coefficient of sliding friction between your skates and the ice is 0.10.
7. Draw the free-body diagram for each object.
    a. a 1-kg rock sitting on a table
    b. a 500-N box at rest that you are pushing with a force of 100 N
    c. a 20-kg monkey hanging from a tree limb by both arms
8. Two children are engaged in a game of tug-of-war. While holding onto the rope, Toni applies a force of 15 N to the right. Marie applies a force of 20 N to the left. The mass of the rope is 2.5 kg. Calculate the acceleration of rope.

9. Calculate the size of the restoring force in each scenario.
   a. a spring with a spring constant of 20 N/m is stretched 0.5 m
   b. a spring with a spring constant of 3 N/m is compressed 0.1 m
   c. a wall with a spring constant of 5×10^7 N/m is compressed 1×10^{-5} m

10. The diagram below represents a hockey puck that has been struck and is sliding across the ice to the right.
    a. Draw a free-body diagram to represent the forces on the sliding puck.
    b. If the hockey puck has a mass of 0.15 kg, calculate the normal force on the puck.
    c. Determine the coefficient of static friction between the hockey puck and the ice if the friction force is 0.147 N.

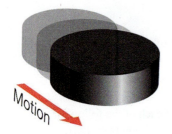

Applying your knowledge

1. If gold were sold by *weight* would you rather buy it in Denver or Death Valley? If gold were sold by *mass* would you rather buy it in Denver or Death Valley? Why?

2. ABS brakes rapidly actuate and release the brakes in a panic stop. Before the invention of ABS, drivers were cautioned to pump their brakes instead of "locking" the brakes by stepping hard on the brake pedal. Explain why "locking" the brakes is less effective than pumping the brakes.

3. A 2,000-kg car is traveling due north at a constant speed of 30 m/s. What is the net force acting on the car? Explain your answer.

4. The angle of inclination can be related to the coefficient of friction for a body resting on the incline. Investigate this relationship. Discuss any practical applications.

5. Two student drivers were moving slowly up a snowy hill. The student in the second car tried to accelerate in order to pass the first car. Instead of passing, the second student slid slowly down the hill. Explain why this might have happened.

UNIT 3 MOTION AND FORCE IN 2 AND 3 DIMENSIONS

CHAPTER 7

Using Vectors: Motion and Force

Objectives:

By the end of this chapter you should be able to:

- Add and subtract displacement vectors to describe changes in position.
- Calculate the *x* and *y* components of a displacement, velocity, and force vector.
- Write a velocity vector in polar and *x-y* coordinates.
- Calculate the range of a projectile given the initial velocity vector.
- Use force vectors to solve two-dimensional equilibrium problems with up to three forces.
- Calculate the acceleration on an inclined plane when given the angle of incline.

Key Questions:

- Why are vectors important, and how do you work with them?
- How do you use vectors to analyze projectile motion?
- How can you use force vectors to explain motion on a ramp?

Vocabulary

Cartesian coordinates	magnitude	range	scalar	trajectory
component	parabola	resolution	scale	velocity vector
cosine	polar coordinates	resultant	sine	*x*-component
displacement	projectile	right triangle	tangent	*y*-component
inclined plane	Pythagorean theorem			

7.1 Vectors and Direction

Direction is always important when describing motion. A plane, boat, or car cannot get to its destination unless the driver has information about direction as well as distance. A vector is a quantity that includes information about direction as well as quantity. This section is about how to work with vectors. In the real world, vectors are necessary to solve most problems involving motion and forces.

Scalars and Vectors

Scalars have magnitude A **scalar** is a quantity that can be completely described by one value: its **magnitude**. You can think of magnitude as size or amount, including units. Temperature is a good example of a scalar quantity (Figure 7.1). If you are sick and use a thermometer to measure your temperature, it might show 101°F. The magnitude of your temperature is 101, and degrees Fahrenheit is the unit of measurement. The value of 101°F is a complete description of the temperature because you do not need any more information.

Examples of scalars Many other measurements are expressed as scalar quantities. The length of a running race is a scalar because it is completely described with one value, such as 100 meters. Time is a scalar quantity. A time of 11 seconds has a magnitude of 11 measured in the units of seconds. Speed is also a scalar quantity and you might use time and distance to calculate the average speed of a runner in a race.

Vectors have direction The location of a place is an example where a single value, such as distance, is not enough to completely describe where the place is. For example, knowing a new pizza place is 1 kilometer away is not enough information to locate the place. The place could be anywhere on a circle with a radius of 1 kilometer.

Vectors are necessary in two and three dimensions The surface of the land is approximately a two-dimensional shape with east-west, and north-south directions. To uniquely locate any point on a two dimensional surface you need direction as well as distance. You *would* have enough information to locate the pizza place if you were told the place was 1 kilometer in a direction 40 degrees east of north (Figure 7.2). The information "1 kilometer, 40 degrees east of north" is an example of a vector. A vector is a quantity that includes both magnitude and direction. Vectors require more than one number, which is one reason they are different from scalars.

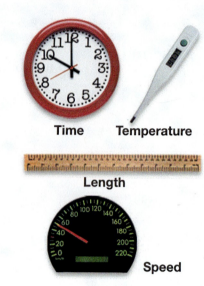

Figure 7.1: *Some examples of scalar quantities.*

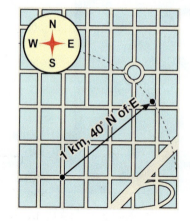

Figure 7.2: *A vector includes information about magnitude, or distance, and direction.*

Drawing the displacement vector

Using a scale A vector can be represented with an arrow. The length of the arrow shows the magnitude of the vector, and the arrow points in the direction of the vector. When drawing a vector as an arrow you must choose a **scale**. For a walk in a field, an appropriate scale would be to let 1 centimeter represent 1 meter.

The displacement vector If you walk five meters east, your *displacement* can be represented by a 5-centimeter arrow pointing to the east (Figure 7.3). The **displacement** vector describes a change in position. Displacement vectors have units of distance.

Measuring displacement Suppose you walk 5 meters east, turn, go 8 meters north, then turn and go 3 meters west. Where are you relative to your starting point? You can represent each leg of the walk by a displacement vector. One vector starts at the end of the previous one, just like each leg of the walk starts at the end of the previous leg. The diagram below shows the trip as a sequence of three displacement vectors.

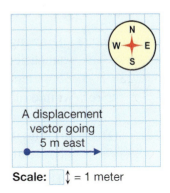

A displacement vector going 5 m east

Scale: ☐ = 1 meter

Figure 7.3: *A displacement vector that goes 5 meters east.*

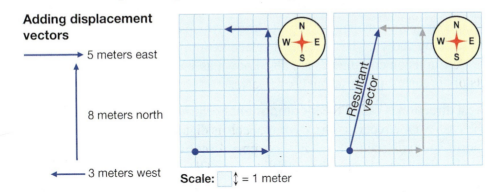

Drawing a resultant At the end of the trip, your position is 8 meters north and 2 meters east of where you started. In physics, the position is where you are. The diagonal vector that connects the starting position with the final position is called the **resultant**. The resultant is the sum of two or more vectors added as shown in the diagram above. You could walk a shorter distance by going 2 meters east and 8 meters north, and still end up at the same position. The resultant shows the most direct line between the starting position and the final position.

> **Be careful adding vectors**
>
> You cannot usually add vectors like regular numbers. For example, if you simply added 5 m + 8 m + 3 m, the result is 16 m. This *is* the distance you walked, but it is *not* your final position. When adding vectors, you must be careful to understand what you are trying to calculate. If the answer is position, you need to add displacement vectors, using a graph or by components. If the answer is the distance traveled, you can add the magnitudes of the displacement vectors and not worry about direction.

7.1 VECTORS AND DIRECTION   **141**

Representing vectors with components

Adding vectors Drawing carefully-scaled arrows to add vectors can be time-consuming. There is an easier way to add vectors mathematically. For example, suppose you walk 5 meters northeast, as shown in Figure 7.4. Notice that you end up in the exact same place as if you had walked 4 meters east then 3 meters north. In fact, every displacement vector in two dimensions can be represented by two vectors: a north-south vector, and an east-west vector. The process of describing a vector in terms of two perpendicular directions is called **resolution**.

Components of a vector The displacement vector can be written (4, 3) m. The first number in the parentheses is a vector in the east-west direction and the second number is a vector in the north-south direction. The two perpendicular vectors are called a **component** of the original vector. You actually walked the diagonal line, but your displacement can be written *as if* you did the walk in two perpendicular segments, one after the other.

Cartesian, or x-y coordinates To understand what component vectors mean, you have to know which *coordinate system* the components are being shown. The example displacement of (4, 3) m is in **Cartesian coordinates**. Cartesian coordinates are also known as *x-y* coordinates. Adding vectors is easiest in Cartesian coordinates.

x and y components The vector in the east-west direction is called the **x-component** because the *x*-axis on a graph represents the east-west direction. The *x*-component in Figure 7.4 is 4 m. The vector in the north-south direction is called the **y-component** because the *y*-axis on a graph usually represents the north-south direction. The *y*-component of the example is 3 m. The *x* and *y* components are also sometimes called horizontal and vertical components.

Polar coordinates The degrees on a compass are an example of a **polar coordinates** system. Polar coordinates use a length and an angle; for example, the same vector (4, 3) m can be expressed as (5 m, 37°) which is 5 m at 37 degrees north of east (Figure 7.5). The angles in standard polar coordinates are given relative to the positive *x*-axis. Displacements on a map used for navigation are usually given in polar coordinates because they are the natural coordinate system for use with a compass. Vectors in polar coordinates are difficult to add and subtract, and are usually converted first to Cartesian coordinates.

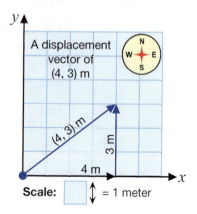

Figure 7.4: *The 5 m northeast vector has an east-west component of 4 m east and a north-south component of 3 m north.*

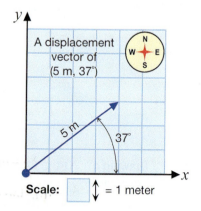

Figure 7.5: *A polar coordinate version of the same vector as above, in Figure 7.4.*

Adding and subtracting vectors

Symbols for vectors
Vectors are indicated with arrows over the symbol for the variable. For example, the displacement vector is written as $\vec{x}$. The variable x without an arrow is the distance, and is *not* a vector. In this chapter, we will introduce vectors for several familiar variables, including displacement ($\vec{x}$), velocity ($\vec{v}$), and force ($\vec{F}$).

Adding vectors
Writing vectors in components makes it much easier to add them. Suppose we take the example from the previous page. The first vector is 5 meters east. This can be written $\vec{x}_1 = (5, 0)$ m. The 5 meters is the distance east and the 0 means there was no distance north. The next vector is $\vec{x}_2 = (0, 8)$ m representing a distance of 8 meters north. The third vector is $\vec{x}_3 = (-3, 0)$ m and the -3 meters indicates that the direction is west. To get the resultant, you just add the x components and y components separately: $\vec{x}_1 + \vec{x}_2 + \vec{x}_3 = (2, 8)$ m. Adding numbers is easier and more accurate than drawing arrows on graph paper.

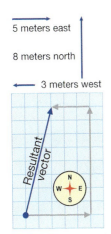

Adding component vectors $\vec{x} = \vec{x}_1 + \vec{x}_2 + \vec{x}_3$

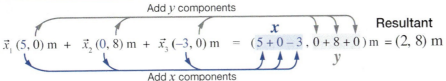

Subtracting vectors
To subtract one vector from another vector, you subtract the components. To see how this works, consider the two vectors, $\vec{x}_1 = (4, 3)$ m, and $\vec{x}_2 = (0, 1)$ m. When the vectors are subtracted, $\vec{x}_1 - \vec{x}_2 = (4, 2)$ m. Subtracting vectors is equivalent to going backward for the vector being subtracted.

Subtracting component vectors $\vec{x} = \vec{x}_1 - \vec{x}_2$

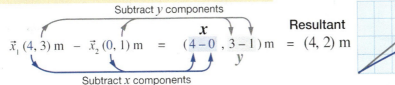

Calculating a resultant vector by adding components

An ant walks 2 meters west, 3 meters north, and 6 meters east. What is the displacement of the ant?

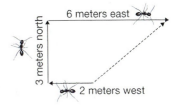

1. You are asked for the resultant vector.
2. You are given three displacement vectors.
3. Make a sketch of the ant's path, and add the displacement vectors by components.
4. Solve:

$$\vec{x}_1 = (-2, 0) \text{ m}$$
$$\vec{x}_2 = (0, 3) \text{ m}$$
$$\vec{x}_3 = (6, 0) \text{ m}$$

$$\vec{x}_1 + \vec{x}_2 + \vec{x}_3 = (-2+6, 3) \text{ m}$$
$$= (4, 3) \text{ m}$$

The final displacement is 4 meters east and 3 meters north from where the ant started.

7.1 VECTORS AND DIRECTION

Calculating vector components

Adding vectors Finding the components of a vector is easy when the vector points in one of the four compass directions along the x- or y-axis. Finding the components of a vector at an angle requires using the properties of triangles. Any displacement vector can be represented on a graph by a triangle with two sides parallel to the x- and y-axes, as shown in Figure 7.6.

Using graphs to find components To find components graphically, draw a displacement vector as an arrow of appropriate length at the specified angle. For example, Figure 7.6 shows how to draw a displacement vector $\vec{x} = (5\text{ m}, 37°)$. A protractor is used to mark the angle and a ruler to draw the arrow. The x-component of the vector is the *projection* of the arrow along the x-axis. The y-component is the projection along the y-axis.

Using mathematical representation to find components Finding components using trigonometry is quicker and more accurate than the graphical method. The variable r is used to represent the length of the vector. The angle is represented with the symbol θ. The triangle is a **right triangle** since the sides are parallel to the x- and y-axes. The ratios of the sides of a right triangle are determined by the angle. These ratios are called **sine** and **cosine**. The sine of the angle is the ratio y/r where y is the y-component of the vector. The cosine of the angle is the ratio x/r where x is the x-component of the vector. Both the sine and cosine are between 0 and 1 and are built into scientific calculators.

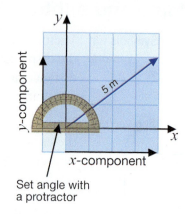

Figure 7.6: *The triangle formed by the displacement vector (4, 3) m.*

Finding the components of a vector

Vector	Triangle	Relationships	Components
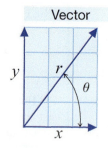	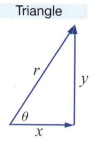	$\sin\theta = \dfrac{y}{r}$	$y = r\sin\theta$
		$\cos\theta = \dfrac{x}{r}$	$x = r\cos\theta$
		$\tan\theta = \dfrac{y}{x}$	

Sine and cosine on a calculator

Scientific calculators have buttons that calculate the sine and cosine of angles. For example, to calculate the sine of 37 degrees on most calculators, you use the following sequence:

The sine of 37 degrees is 0.6018. On most calculators you do not have to press the equals button to display the trigonometric functions.

Finding the magnitude of a vector

$a^2 + b^2 = c^2$ When you know the *x*- and *y*-components of a vector, you can find the magnitude using the **Pythagorean theorem**. This useful theorem states that $a^2 + b^2 = c^2$, where *a*, *b*, and *c* are the lengths of the sides of any right triangle. For example, suppose you need to know the distance represented by the displacement vector (4, 3). If you walked east 4 meters then north 3 meters, you would walk a total of 7 meters. This is a distance, but it is not the distance specified by the vector, or the shortest way to go. The vector (4, 3) m describes a single straight line. The length of the line is 5 meters because $4^2 + 3^2 = 5^2$.

THE PYTHAGOREAN THEOREM

$$a^2 + b^2 = c^2$$

a and *b* are the lengths of the short sides of the right triangle.

c is the length of the side opposite the right angle.

Finding the angle of a vector

In some problems, you know the sides of the triangle but want to find the angle. The inverse sine is a function that gives you the angle if you know the ratio *y/r*. The inverse tangent gives the angle if you know the ratio *y/x*.

A calculator can also do the inverse of the sine, cosine, or tangent. For example, suppose you have a vector where $y = 3$ and $x = 4$. That makes the tangent 0.75 (3/4).

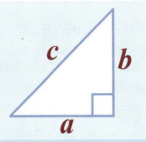

What is the angle?

Calculators either have an INV (inverse) key or a TAN⁻¹ key. Depending on your calculator, one of the following sequences of buttons should work.

Finding two vectors that have a certain resultant

Robots are programmed to move with vectors. A robot must be told exactly how far to go and in which direction for every step of a trip. A trip of many steps is communicated to the robot as a series of vectors. A mail-delivery robot needs to get from where it is to the mail bin on the map. Find a sequence of two displacement vectors that will allow the robot to avoid hitting the desk in the middle.

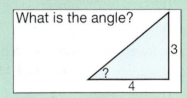

Distance in meters

1. You are asked to find two displacement vectors.
2. You are given the starting and final positions.
3. The resultant vector must go from the start to the final position.
4. Solve:
   The robot starts at (1, 1) m and the mail bin is at (5, 5) m. The displacement required is (5, 5) m − (1, 1) m = (4, 4) m. First go up 4 meters, then over 4 meters.
   $\vec{x}_1 = (0, 4)$ m, $\vec{x}_2 = (4, 0)$ m

Check the resultant: (4, 0) m + (0, 4) m = (4, 4) m

7.2 Projectile Motion and the Velocity Vector

Imagine that you are playing basketball. You must decide how fast and at what angle to throw the ball in order to hit the basket. The ball starts moving upward at an angle. Then it curves downward at an angle toward the basket. As the ball moves, its velocity changes direction, first pointing upward, then downward. This section is about the velocity vector, which describes both the speed of the ball and its direction of motion. The velocity vector of a thrown ball follows a very predictable curve under the influence of gravity. This section will explain why a ball follows a curved path and how you can determine exactly where it will land.

Projectiles and trajectories

Definition of projectile Any object that is moving through the air affected only by gravity is called a **projectile**. Examples of projectiles include a basketball thrown toward the basket, a car driven off a cliff by a stunt person, and a skier going off a jump. Flying objects such as airplanes and birds are not projectiles because they are affected by forces from their own power, and not just the force of gravity.

Trajectories The path a projectile follows is called its **trajectory**. The trajectory of a thrown basketball follows a special type of arch-shaped curve called a **parabola** (Figure 7.7). A projectile launched at a steep angle will result in a tall and narrow parabola. A wide and low parabola results from a launch at an angle close to the horizontal. The distance a projectile travels horizontally is called its **range**. The range of a projectile depends on its initial angle, speed, and height off the ground.

The velocity vector The **velocity vector** describes the speed and direction an object is moving at any point along its trajectory (Figure 7.8). The velocity vector of a basketball changes as the ball moves. Both the speed and direction are different at different places along the trajectory. To calculate the motion of a basketball, or any other projectile, we first must develop a way to work with the velocity vector.

Friction Before we start, it is worth noting that a real object's trajectory is not always a perfect parabola. Air resistance and other forms of friction add additional forces. For now, we will assume air resistance and other forms of friction are minimal and can be ignored. Once we develop a good model for curved motion, we can add the effects of friction back in to make the model more realistic.

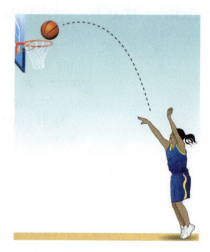

Figure 7.7: *When you throw a ball, it follows a curved path called a parabola.*

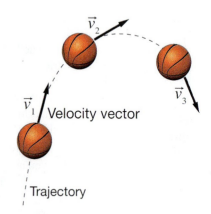

Figure 7.8: *The velocity vector usually changes its direction and its magnitude all along the trajectory of a projectile.*

The velocity vector

Describing the velocity vector
The velocity vector ($\vec{v}$) is a way to precisely describe the speed and direction of motion. For example, suppose a ball is launched at 5 meters per second at an angle of 37 degrees (Figure 7.9). At the moment after launch, the velocity vector for the ball in polar coordinates is written $\vec{v} = $ (5 m/s, 37°). In *x-y* components, the same velocity vector is written as $\vec{v} = $ (4, 3) m/s. Both representations tell you exactly how fast and in what direction the ball is moving at that moment.

Speed is the magnitude of the velocity vector
The *magnitude* of the velocity vector is the *speed* of the object, which is 5 m/s in the example. The speed is represented by a lower case *v* *without* the arrow. When a velocity vector is represented graphically, the lengths are proportional to speed, not distance. For example, the graph in Figure 7.10 shows the velocity vector $\vec{v} = $ (4, 3) m/s as an arrow on a graph.

Interpreting the *x-y* components of velocity
If there was no gravity, the ball would continue with the same initial velocity. For every 5 meters the ball moves along the 37 degree direction, it also moves 4 meters along *x* and 3 meters along *y*. The ball's speed follows the 5 m/s arrow in Figure 7.10, but it is useful to *think* about the motion as being separated into a speed along *x* and a different speed along *y*. When written as $\vec{v} = $ (4, 3) m/s, the components are really just a mathematical way to separate the velocity into individual speeds in the *x* and *y* directions.

Two ways to write a velocity vector

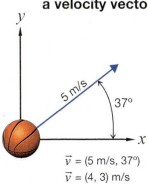

$\vec{v} = $ (5 m/s, 37°)
$\vec{v} = $ (4, 3) m/s

Figure 7.9: *Different ways to write a velocity vector.*

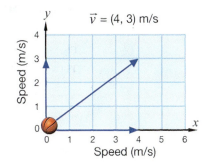

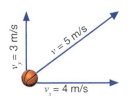

Figure 7.10: *Drawing and interpreting a velocity vector.*

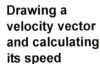

Drawing a velocity vector and calculating its speed

Draw the velocity vector $\vec{v} = $ (5, 5) m/s and calculate the magnitude of the velocity, or the speed, using the Pythagorean theorem.

1. You are asked to sketch a velocity vector and calculate its speed.
2. You are given the *x-y* component form of the velocity.
3. Set a scale of 1 cm = 1 m/s to draw the sketch.
4. The Pythagorean theorem is $a^2 + b^2 = c^2$. Solve:

$v^2 = (5 \text{ m/s})^2 + (5 \text{ m/s})^2 = 50 \text{ m}^2/\text{s}^2$

$v = \sqrt{50 \text{ m}^2/\text{s}^2} \approx 7.07 \text{ m/s}$

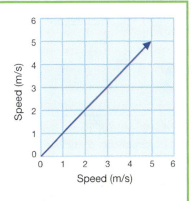

7.2 PROJECTILE MOTION AND THE VELOCITY VECTOR

The components of the velocity vector

Writing a velocity vector Velocity vectors are broken into components just like displacement vectors. For example, suppose a car is driving 20 meters per second in a direction as shown on the map in Figure 7.11. The direction of the vector is 127 degrees. The polar representation of the velocity is $\vec{v} = (20 \text{ m/s}, 127°)$.

Calculating x and y components In x-y form, a velocity vector is written $\vec{v} = (v_x, v_y)$ where v_x and v_y are the x and y components. The magnitude of the vector is the speed, v, without the arrow. You can take any velocity vector and make a right triangle for which the legs of the triangle are the components (v_x and v_y) and the hypotenuse of the right triangle is the speed (v). For drawing the triangle, the angle is $180° - 127°$, or $53°$. The sine, cosine, and **tangent** of the angle are ratios of v_x, v_y, and v as shown below. The ratios can be rearranged to solve for the components.

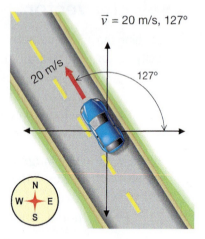

Figure 7.11: *A car driving with a velocity of 20 m/s at 127 degrees.*

Finding the components of a velocity vector

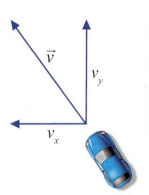

Analysis

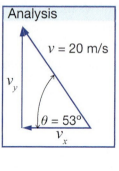

Relationships

$\sin\theta = \dfrac{v_y}{v}$ → $v_y = v \sin\theta$

$\cos\theta = \dfrac{v_x}{v}$ → $v_x = v \cos\theta$

$\tan\theta = \dfrac{v_y}{v_x}$

Application

$v_y = (20 \text{ m/s}) \sin 53° = 16 \text{ m/s}$
$v_x = -(20 \text{ m/s}) \cos 53° = -12 \text{ m/s}$

$\vec{v} = (-12, 16)$ m/s

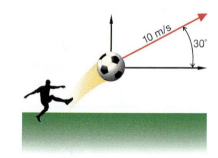

Figure 7.12: *The velocity of a soccer ball kicked at an angle.*

Calculating the components of a velocity vector

A soccer ball is kicked at a speed of 10 m/s and an angle of 30 degrees (Figure 7.12). Find the horizontal and vertical components of the ball's initial velocity.

1. You are asked to calculate the components of the velocity vector.
2. You are given the initial speed and angle.
3. Draw a diagram and use $v_x = v \cos\theta$ and $v_y = v \sin\theta$.
4. Solve: $v_x = (10 \text{ m/s})(\cos 30°) = (10 \text{ m/s})(0.87) = 8.7$ m/s
   $v_y = (10 \text{ m/s})(\sin 30°) = (10 \text{ m/s})(0.5) = 5$ m/s

Adding velocity vectors

Why you might add velocity vectors
There are circumstances where the total velocity of an object is a combination of velocities. One example is the motion of a boat on a river. The boat moves with a certain velocity relative to the water. The water is also moving with another velocity relative to the land. The velocity of the boat *relative to the land* is the sum of the boat's velocity relative to the water plus the water's velocity relative to the land. A similar situation applies to aircraft flying in a wind.

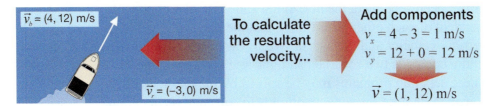

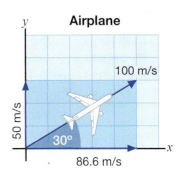

Figure 7.13: *Finding components of the airplane's velocity*

An example of adding velocity vectors
Velocity vectors are added by components, just like displacement vectors. To calculate a resultant velocity, add the *x* components and the *y* components separately. For example, suppose a boat is moving with a velocity of $\vec{v}_b = (4, 12)$ m/s. The river is moving with a velocity of $\vec{v}_r = (-3, 0)$ m/s. The resultant velocity of the boat is $\vec{v} = (1, 12)$ m/s. Any boat traveling across a current must steer slightly upstream to compensate for the velocity of the water.

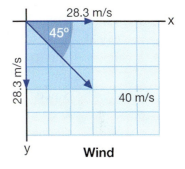

Figure 7.14: *Finding components of the wind's velocity*

Calculating the components of a velocity vector

An airplane is moving at a velocity of 100 m/s in a direction 30 degrees northeast relative to the air. The wind is blowing 40 m/s in a direction 45 degrees southeast relative to the ground. Find the resultant velocity of the airplane relative to the ground.

1. You are asked to calculate the resultant velocity vector.
2. You are given the plane's velocity and the wind velocity.
3. Draw diagrams and add the components to get the resultant velocity.
4. Figure 7.13 shows the plane velocity vector.
   $v_x = 100 \cos 30° = 86.6$ m/s, $v_y = 100 \sin 30° = 50$ m/s
   Figure 7.14 shows the wind velocity vector.
   $v_x = 40 \cos 45° = 28.3$ m/s, $v_y = -40 \sin 45° = -28.3$ m/s
   The resultant $\vec{v} = (86.6 + 28.3, 50 - 28.3) = (114.9, 21.7)$ m/s.

7.2 PROJECTILE MOTION AND THE VELOCITY VECTOR

Projectile motion

Gravity only accelerates vertical motion At the start of the chapter, we set out to understand projectile motion. For a projectile, the force of gravity makes one direction of motion different from another. Motion that is up or down is changed by the acceleration of gravity. Motion that is sideways is not similarly affected. Because gravity acts differently on vertical and horizontal motion, it is useful to separate motion into components that are vertical and horizontal.

Independence of horizontal and vertical motion Once a velocity vector has been separated, the horizontal and vertical components can be analyzed independent of each other. This means that the horizontal velocity has no effect on the vertical velocity and vice versa. What was a complicated, curved problem in x and y becomes two separate, straight-line problems, one in x and the other in y. These separate problems may be solved by the methods of Chapters 3 and 4. *The way to analyze projectile motion is to consider vertical and horizontal directions separately.*

Horizontal motion Consider a ball that rolls off a table with a velocity of 5 meters per second. Once it leaves the table, the ball is a projectile because it experiences only the influence of gravity. The horizontal velocity of a projectile remains constant during the entire time it is in the air because no horizontal force acts on it, ignoring air friction. Since there is no force, the horizontal acceleration is zero. That means the ball will keep moving to the right at 5 meters per second (Figure 7.15). The horizontal distance a projectile moves can be calculated according to the formula:

HORIZONTAL DISTANCE
projectile motion

$$x = v_{ox} t$$

Distance in x (m) = x component of initial velocity (m/s) × Time in flight (s)

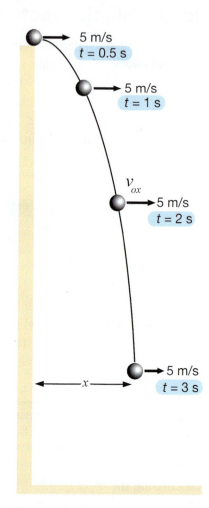

Figure 7.15: *The ball's horizontal velocity remains constant while it falls because gravity does not exert any horizontal force.*

Keeping track of variable names To keep track of things, the velocity components are labeled with subscripts for x and y, and also for initial values. For example, the variable v_{ox} is the x component of the initial velocity vector $\vec{v}_o = (v_{ox}, v_{oy})$. When solving projectile motion problems, there get to be many letters to keep track of. The key is to be very organized about writing down what each variable means, and *always write the subscripts* or you will lose track of what is what and quickly become confused.

Vertical motion

Vertical motion Analyzing the vertical motion of a projectile is more complicated because gravity accelerates objects in the vertical direction. This motion is the same as that of an object in free fall. The ball's vertical speed changes by 9.8 m/s each second.

VERTICAL VELOCITY
projectile motion

$$v_y = v_{oy} - gt$$

VERTICAL DISTANCE
projectile motion

$$y = v_{oy}t - \tfrac{1}{2}gt^2$$

y	Distance y in meters
v_y	y component of velocity (m/s)
v_{oy}	y component of **initial** velocity (m/s)
g	Acceleration of gravity (9.8 m/s²)
t	Time in flight (s)

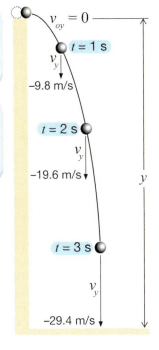

These are the same as the free-fall equations If you look back to Chapter 4, you will recognize that these are the exact same equations we derived for free fall. We can use them again for the vertical motion of a projectile because the horizontal and vertical components of the motion can be separated.

When the initial vertical velocity is zero The simplest type of projectile motion involves an object launched horizontally, like a stunt car driving off a cliff. When the car leaves the edge of the cliff, its initial velocity is entirely horizontal. Therefore, the initial vertical velocity component is zero.

Analyzing a horizontally-launched projectile

A stunt driver steers a car off a cliff at a speed of 20 meters per second. He lands in the lake below 2 seconds later. Find the height of the cliff and the horizontal distance the car travels.

1. You are asked for the vertical and horizontal distances.
2. You know the initial speed and the time.
3. Relationships that apply:
   $y = v_{oy}t - \tfrac{1}{2}gt^2$, $x = v_{ox}t$
4. The car goes off the cliff horizontally, so $v_{oy} = 0$
   $y = -(1/2)(9.8 \text{ m/s}^2)(2 \text{ s})^2$
   $= -19.6$ meters
   The negative sign shows the car is below its starting height.
   Use $x = v_{ox}t$, to find the horizontal distance.
   $x = (20 \text{ m/s})(2 \text{ s})$
   $= 40$ meters

Projectiles launched at an angle

Vertical velocity depends on launch speed and angle
A soccer ball kicked off the ground is also a projectile, but it starts with an initial velocity that has vertical and horizontal components. The launch angle determines how the initial velocity divides between vertical (*y*) and horizontal (*x*) directions. A projectile launched at a steep angle will have a large vertical velocity component and a small horizontal velocity. One launched at a low angle will have a large horizontal velocity component and a small vertical one (Figure 7.16).

Calculating velocity components
The initial velocity components of an object launched at a velocity $\vec{v}_o$ and angle θ are found by breaking the velocity into *x* and *y* components.

Components of initial velocity

$$v_{ox} = v_o \cos \theta \qquad v_{oy} = v_o \sin \theta$$

v_o	Magnitude of initial velocity (m/s)
v_{ox}	*x* component of initial velocity (m/s)
v_{oy}	*y* component of initial velocity (m/s)
θ	Launch angle (degrees)

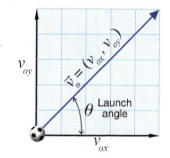

Components of initial velocity

Vertical velocity is zero at the top
The vertical velocity of an upwardly-launched projectile decreases by 9.8 meters per second each second as the object moves upward. Eventually, the vertical velocity reaches zero, and the object starts to move downward. At the top of the trajectory, the vertical velocity is zero. This does not mean the projectile has stopped moving at this point. It is still moving horizontally at the same speed it was initially.

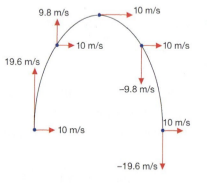

Steep angle
v_{oy} much *larger* than v_{ox} makes a high trajectory

Shallow angle
v_{oy} much *smaller* than v_{ox} makes a low trajectory

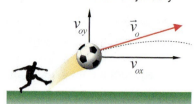

Vertical velocity along the trajectory
As the object begins to fall, its vertical velocity increases at the same rate with which it decreased. The trajectory is symmetric, and the time the projectile takes to move upward is the same as the time it takes to move downward. When it falls to its launch height, the projectile is moving at the same speed as when it was launched, but it is moving in a different direction.

Figure 7.16: *The trajectory depends on the balance of vertical and horizontal velocity components. This balance is determined by the launch angle.*

Range of projectiles

Range increases as velocity squared
The range, or horizontal distance, traveled by a projectile depends on the launch speed and the launch angle. The greater the launch speed, the greater the range a projectile will have at a specific angle. The launch range is proportional to the square of the velocity. Doubling the launch speed quadruples the range of the projectile. This means a football kicked at 10 meters per second will travel four times as far as one kicked at the same angle at 5 meters per second.

Calculating the range
The range of a projectile is calculated from the horizontal velocity and the time of flight. The vertical velocity is responsible for giving the projectile its air time. The time of flight is twice the time it takes the projectile to reach the top of its trajectory, where $v_y = 0$. A projectile travels farthest when launched at 45 degrees. At this angle, its velocity is evenly divided between horizontal and vertical.

Range of a projectile

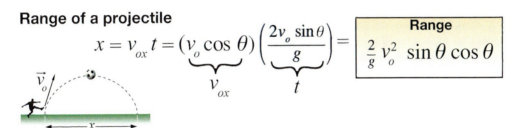

$$x = v_{ox} t = \underbrace{(v_o \cos \theta)}_{v_{ox}} \underbrace{\left(\frac{2v_o \sin \theta}{g}\right)}_{t} = \boxed{\frac{2}{g} v_o^2 \sin \theta \cos \theta} \quad \text{Range}$$

Different launch angles can have the same range
The more the launch angle varies from 45 degrees, the smaller the range. A projectile launched at 30 degrees will have the same range as one launched at 60 degrees, because both angles are 15 degrees from 45. A projectile launched at 30 degrees has a fast horizontal velocity but a short air time. The projectile launched at 60 degrees has a long air time but a slow horizontal velocity. As a result, they have the same range. This holds true for any pair of angles which add up to 90.

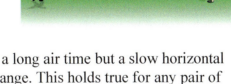

Hang time

If you have ever watched a skilled basketball player take a leap toward the basket, you have probably thought the athlete seemed almost to float through the air for a period of time. That time spent in the air is called *hang time*.

You can easily calculate your own hang time. Run toward a doorway and jump as high as you can, touching the wall or door frame. Have someone watch to see exactly how high you reach. Measure this distance with a meter stick.

The vertical distance formula can be rearranged to solve for time:

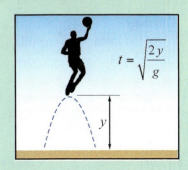

$$t = \sqrt{\frac{2y}{g}}$$

Plug in your jump height in meters for *y* and solve for the time. This represents the time you move upward; double it to find your hang time.

7.3 Forces in Two Dimensions

A block at rest on a level, flat table will remain motionless. A block on a frictionless slope will begin to slide immediately when you let it go. In both cases, gravity is pulling on the block with a force equal to its weight. But the *effect* of gravity is different in the two scenarios. The effect is different because force is a vector, and the relationship between force and motion is a relationship between vectors. This section first explores the force vector and then describes an example of how the force vector is used to solve the problem of motion down a ramp.

The force vector

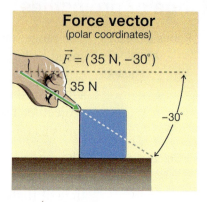

Figure 7.17: *Like other vectors, force can be represented in polar coordinates.*

What the force vector means
The force vector is a way to precisely describe the strength and direction of a force. For example, suppose you push against a block with a force of 35 newtons at an angle of 30 degrees from the horizontal (Figure 7.17). Some of your force accelerates the block, and some of your force pushes the block into the table. The force is most accurately described as a vector. Like other vectors, $\vec{F}$ can be written in polar coordinates: $\vec{F} = (35\ \text{N}, -30°)$. The force can also be written in x-y components: $\vec{F} = (30.3, -17.5)\ \text{N}$. Both representations tell you exactly how strong the force is and in what direction it is pushing.

The magnitude of the force vector
The *magnitude* of the force vector is the *strength* of the force. In the example, the magnitude is 35 newtons. Like other vectors, the magnitude is related to the x and y components by the Pythagorean theorem: $a^2 + b^2 = c^2$. For the example force in Figure 7.18, the calculation is: $35^2 = 30.3^2 + 17.5^2$. The magnitude of the force is represented by a capital letter F, but *without* the arrow. When a force vector is represented graphically, the length is proportional to the magnitude of the force.

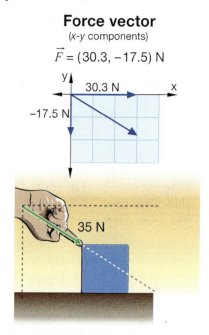

Figure 7.18: *Force can also be represented in x-y components.*

Interpreting the x-y components of force
The x and y components of a force vector can be thought of as actual forces. In fact, if the x and y component forces were applied along their appropriate axes, their effect would be *exactly the same as the single force* (Figure 7.18). You can think of the single force as the *resultant* of adding the component forces together. In many problems, the solution can be found by breaking forces up into components and analyzing each direction separately, just as with projectile motion.

Equilibrium of forces

Balanced forces If an object is in equilibrium, all of the forces acting on it are balanced and the net force is zero. If the forces act in two dimensions, then all of the forces in the *x*-direction and *y*-direction balance *separately*. As you may have guessed, the word *separately* means the forces in the *x* direction must total zero *and* the forces in the *y* direction must total zero.

Example of equilibrium To do a force balance, all forces must be represented in *x-y* components. For example, imagine a gymnast with a weight of 700 newtons who is supporting himself on two rings with his arms straight below his shoulders. If he is at rest, and therefore in equilibrium, the net force on his body is zero. Gravity pulls down with a force of 700 N, so the ropes must pull up on his arms with a total force of 700 N. Each arm holds half of his weight, or 350 N (Figure 7.19).

Forces applied at an angle It is much more difficult for a gymnast to hold his arms out at a 45-degree angle. To see why, consider that each arm must still support 350 newtons vertically to balance the force of gravity. When the gymnast's arms are at an angle, only part of the force from each arm is vertical. The total force must be larger because the vertical *component* of force in each arm must still equal half his weight.

Calculating the total force The total force in each arm is the magnitude (F) of the force vector ($\vec{F}$). The vertical component (F_y) at 45 degrees must be 350 N. The diagram below shows how to use the *y*-component to find the total force in the gymnast's left arm.

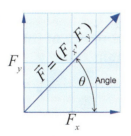

Components of a force vector

$F_x = F \cos \theta$
$F_y = F \sin \theta$

Equilibrium in *y*-direction (vertical)

$$350 \text{ N} = F_y$$
$$= F \sin \theta$$

$$F = \frac{350 \text{ N}}{\sin 45°} = 495 \text{ N}$$

The force in the right arm must also be 495 newtons because it also has a vertical component of 350 N (Figure 7.19). Although it is not shown, the horizontal components of force from right and left arms cancel each other because they have equal magnitude and are in opposite directions.

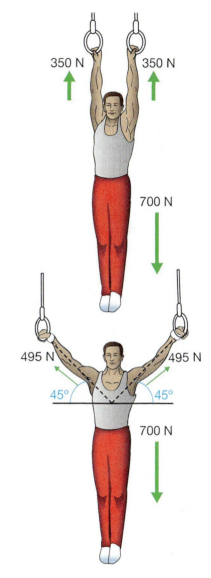

Figure 7.19: *A gymnast holding himself in equilibrium must use greater force to keep his arms at 45 degrees than is needed in keeping them vertical.*

The inclined plane

Definition of inclined plane
An **inclined plane** is a straight surface, usually with a slope. A wood ramp is a good example of an inclined plane. The angle of the incline is the angle relative to the horizontal direction (Figure 7.20). When objects move along an incline, they move parallel to the surface.

Forces on an inclined plane
Consider a block sliding down a ramp. There are three forces that act on the block. These three forces are always present with any inclined plane. The three forces are: gravity, the reaction force from the surface acting on the block, and friction. Motion along the ramp depends on the sum of these three forces. Because the ramp is usually at an angle, the three forces must be treated as vectors.

Choosing coordinates along the ramp
The best coordinates to use for an inclined plane are aligned with the surface, and not with gravity (Figure 7.21). This is because any motion must occur parallel to the surface. By lining up the coordinates with the incline, motion in the x direction is along the surface. There is usually no motion in the y direction, because it would mean the object lifting off or going through the ramp.

Resolving the weight vector in ramp coordinates
The force of gravity on an object always acts in a direction toward the center of Earth. When the surface is a ramp, the direction of gravity is still straight toward the ground but is *not* perpendicular to the surface of the ramp. To treat the force of gravity, it must be resolved into components parallel (x) and perpendicular (y) to the ramp. If the angle of incline is θ, then the weight of the block is represented by the vector $\vec{F}_w = (mg \sin\theta, mg \cos\theta)$.

An inclined plane

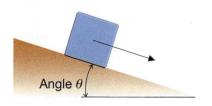

Figure 7.20: *A ramp is an example of an inclined plane.*

Coordinates for an inclined plane

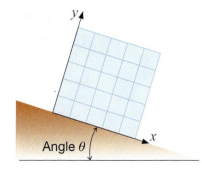

Figure 7.21: *Choosing coordinates along the incline.*

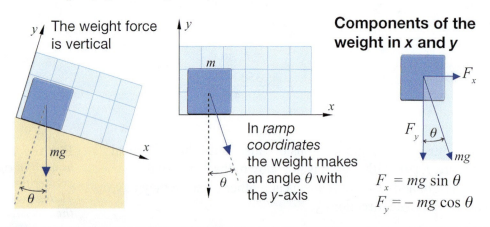

$F_x = mg \sin\theta$
$F_y = -mg \cos\theta$

Resolution of forces on an inclined plane

Force along the incline A block accelerates down a ramp because its weight creates a force parallel to the incline. If the ramp is analyzed as shown in Figure 7.22, the force parallel to the surface is given by $F_x = mg \sin \theta$.

Normal force When discussing forces, the word *normal* means "perpendicular to." For a block on a ramp, equilibrium of forces perpendicular to the ramp surface is what prevents the block from going through the ramp or lifting off of it. The normal force acting *on* the block is the reaction force from the weight of the block pressing against the ramp. To make equilibrium in the y-direction, the normal force on the block is equal and opposite to the component of the block's weight perpendicular to the ramp (F_y). The diagram below shows that the normal force acting on the block (F_N) is equal to $+mg \cos \theta$.

Force parallel to ramp

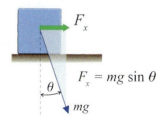

Figure 7.22: *The force parallel to the ramp (F_x).*

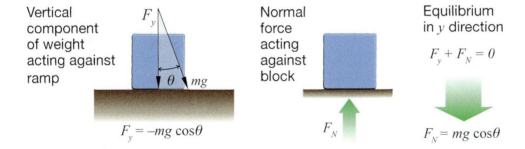

Friction If an object is moving, the force of friction acts opposite the direction of motion. On a ramp for which the coordinates are like Figure 7.23, the friction force is in the negative x direction, acting opposite to the motion. If the block is at rest on the ramp, then the direction of the friction force is opposite to the direction the block would move if there were no friction.

Magnitude of the friction force on a ramp The magnitude of the friction force between two sliding surfaces is roughly proportional to the force holding the surfaces together. For a ramp, that means the friction force is proportional to the normal force. In Chapter 6, you learned that the friction force for sliding friction is μF_N where μ is the coefficient of friction. For a ramp, the friction force is therefore $-\mu mg \cos \theta$ (Figure 7.23).

Friction force

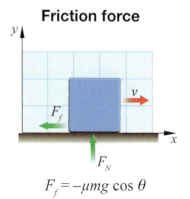

Figure 7.23: *The friction force is opposite to the velocity.*

CHAPTER 7 USING VECTORS: MOTION AND FORCE

Motion on an inclined plane

Slope and acceleration A block on a ramp accelerates downward if the force acting parallel to the ramp exceeds the force of static friction. The greater the angle of the ramp, the greater the downward acceleration. The acceleration increases because greater angles direct more of the block's weight in the direction parallel to the ramp rather than perpendicular to it.

Calculating the acceleration Newton's second law can be used to calculate the acceleration once you know the components of all the forces on an incline. According to the second law, $a = F \div m$, where a is the acceleration and m is the mass. Since the block can only accelerate along the ramp, the force that matters is the net force in the x direction, parallel to the ramp. With no friction, the net force is $F_x = mg \sin\theta$. The acceleration is therefore $a = g \sin\theta$. The net force in the y direction is always zero. This must be true because an object on an inclined plane does not accelerate off of the surface or sink through it.

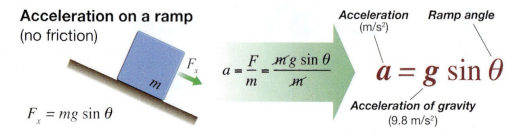

Calculating the acceleration of a skier on a slope

A skier with a mass of 50 kg is on a hill with a 20-degree slope. The friction force is 30 N. What is the skier's acceleration?

1. You are asked to find the acceleration.
2. You know the mass, friction force, and angle.
3. The relationships that apply are $a = F \div m$ and $F_x = mg \sin\theta$.
4. Calculate the x component of the skier's weight:
$F_x = (50\ \text{kg})(9.8\ \text{m/s}^2) \times (\sin 20°) = 167.6\ \text{N}$

Calculate the force:
$F = 167.6\ \text{N} - 30\ \text{N} = 137.6\ \text{N}$

Calculate the acceleration:
$a = 137.6\ \text{N} \div 50\ \text{kg}$
$= 2.75\ \text{m/s}^2$

Accounting for friction If friction is included, the acceleration is reduced from $g \sin\theta$. Including friction, the net force acting along a ramp is $F_x = mg\sin\theta - \mu mg\cos\theta$. The resulting acceleration is $a = g(\sin\theta - \mu\cos\theta)$. For a smooth surface, the coefficient of friction (μ) is usually in the range of 0.1–0.3.

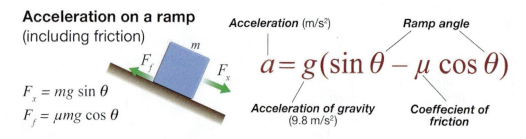

The vector form of Newton's second law

The acceleration vector When we introduced Newton's second law in Chapter 5, we said the acceleration caused by a force was also in the direction of the force. Although we did not say it at the time, this statement implies that acceleration is a vector since it has direction. In fact, an object moving in three dimensions can be accelerated in the x, y, and z directions. In component form, the acceleration vector can be written in a similar way to the velocity vector: $\vec{a} = (a_x, a_y, a_z)$ m/s².

The vector equation is three component equations The most general form of Newton's second law is a relationship between two vectors, force and acceleration, and a scalar, mass.

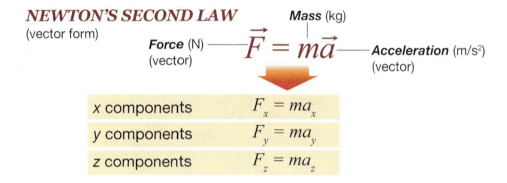

Now that we have worked with vectors and components, it is more useful to think of the second law as three separate equations. There is one equation for each of the coordinate directions. Forces in the x direction cause acceleration in the x direction. Forces in the y direction cause acceleration in the y direction, and likewise for z.

Dynamics If you know the forces acting on an object, you can predict its motion in three dimensions. For example, this is how the computers that control space missions determine when and for how long to run the rocket engines. The computers determine the magnitude and direction of the required acceleration and use the engines to get exactly the right force. The process of calculating three-dimensional motion from forces and accelerations is called *dynamics*.

Calculating the acceleration from 3-D forces

A 100-kg satellite has many small rocket engines pointed in different directions that allow it to maneuver in three dimensions. If the engines make the following forces, what is the acceleration of the satellite?

$\vec{F}_1 = (0, 0, 50)$ N
$\vec{F}_2 = (25, 0, -50)$ N
$\vec{F}_3 = (25, 0, 0)$ N

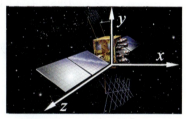

1. You are asked to find the acceleration.
2. You know the mass and forces.
3. The relationships that apply are $\vec{a} = \vec{F} \div m$ and $\vec{F}$ = net force.
4. Calculate the net force by adding components.
   $\vec{F} = (50, 0, 0)$ N

Calculate the acceleration:

$a_y = a_z = 0$
$a_x = 50$ N $\div$ 100 kg
$ = 0.5$ m/s²
$\vec{a} = (0.5, 0, 0)$ m/s²

7.3 FORCES IN TWO DIMENSIONS

Chapter 7 Connection

Robot Navigation

Imagine you wanted to make a map of the sea floor around a shipwreck deep in the ocean. Instead of swimming down to the bottom yourself, you could send a robot in your place. Robots use vectors to keep track of where they have gone and where they want to go. Another example of robots using vectors is a self-driving car. In the future, you may be able to buy a car that does not require a human driver. Imagine getting into your car and saying, "To the museum, please!" Prototype cars that drive themselves have been built and tested.

Controlling robots Robots are machines, usually controlled by a computer, intended to perform one or more tasks. These tasks are usually ones that humans cannot do, like making a map of the ocean floor, or that humans do not want to do, like driving to the same place over and over. A driving robot uses maps stored in its memory and signals from satellites to plot a course from its current position to its destination.

1. The robot's computer looks up the coordinates of the destination.

Destination	Latitude	Longitude
Office	76.30	76.00
Home	75.27	75.30
School	65.11	65.15
Museum	72.43	72.61
Beach	85.20	82.15
Cinema	73.03	73.01
Store	73.20	73.15

2. Using the GPS to find its current position, the robot determines the vector displacement to the destination.

3. With information stored about streets, the robot creates a path made of many vectors. The sum of these is the displacement.

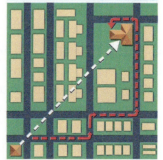

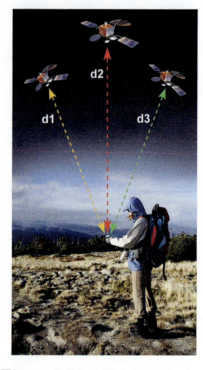

Figure 7.24: *A Global Positioning System (GPS) receiver determines its position to within a few meters anywhere on Earth's surface. The receiver works by comparing signals from three different GPS satellites.*

Global Positioning System (GPS) Twenty-four satellites orbit Earth and transmit radio signals as part of a positioning or navigation system. At any one time, up to 12 of those satellites are in the sky, all transmitting their own unique code and location. A GPS receiver reads the codes, and uses vectors to find its position relative to the satellites (Figure 7.24).

Chapter 7 Connection

Knowing your position

Triangulation Using the location of other objects to find your position is called *triangulation*. On the planet's surface, knowing your distance to one object reduces your possible locations to a circle, two objects eliminates everything but two points, and with three objects, you can determine your location exactly. GPS receivers use triangulation to calculate a position from the signals from three satellites.

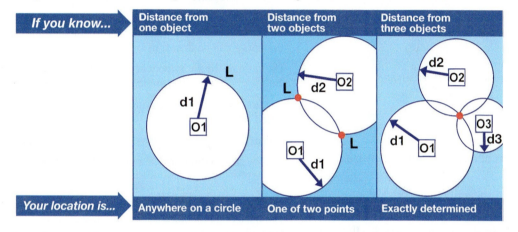

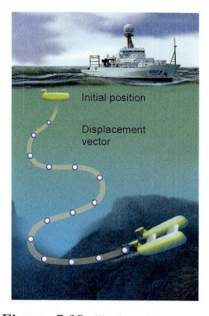

Figure 7.25: *The inertial navigation system (INS) on a robot submarine is used to calculate the submarine's motion in many small displacement vectors by sensing its acceleration.*

Inertial navigation system (INS) Sometimes, robots cannot use the GPS. Radio waves do not travel through water, so underwater robots use a different navigation system. If you have ever ridden on a bus with your eyes closed, you can sense which direction the bus is turning because your body can sense acceleration. Electronics and gyroscopes give a robot a very accurate sense of acceleration in all three dimensions. The inertial navigation system (INS) used by robot submarines uses acceleration to constantly update its displacement vector (Figure 7.25). For each small amount of time, the INS "feels" which direction it is traveling, and represents the distance it has traveled in that time as a vector. By recording its starting position and adding all the vectors together, the INS provides the control computer with a current position vector.

If the robot is underwater for too long, it can start to get confused. Small errors add up over time to produce significant errors in the position calculated by the INS. In order for a robot submarine to navigate long distances, it must come to the surface every so often to reorient itself with GPS information.

7.3 FORCES IN TWO DIMENSIONS

CHAPTER 7 USING VECTORS: MOTION AND FORCE

Chapter 7 Assessment

Vocabulary

Select the correct term to complete the sentences.

vector	scalar	magnitude
x-component	y-component	cosine
parabola	Pythagorean	displacement
resultant	position	resolution
triangle	sine	dynamics
tangent	normal force	projectile
trajectory	Cartesian	range
velocity	equilibrium	inclined plane
polar	scale	component

1. The size or amount of a value, including units of measurement, is known as the _____ of the quantity.
2. The proportion that is a representation of some quantity to the original may be called a(n) _____.
3. A vector quantity representing the change in the position of an object is the _____ of the object.
4. A quantity that can be described completely by its magnitude is a(n) _____ quantity.
5. The process of describing a vector in terms of two perpendicular directions is referred to as _____.
6. An x-y coordinate system may also be called a(n) _____ coordinate system.
7. The sum of two or more vectors is called the _____.
8. A measured quantity that includes both magnitude and direction is called a(n) _____.
9. The location of an object is called its _____.
10. Length and an angle are used to represent position in a(n) _____ coordinate system.
11. In a Cartesian coordinate system, the vector in the up-down or north-south direction is the _____.
12. The ratio of the y component of a vector and the vector quantity is known as the _____ of the angle between them.
13. The mathematical theorem which states that the sum of the squares of the length of the sides of a right triangle equals the square of its hypotenuse is called the _____ theorem.
14. The graphical representation of the vector sum of the x and y components of a quantity can be drawn using a(n) _____ triangle.
15. The ratio of the x component of a vector and the vector quantity is known as the _____ of the angle between them.
16. The horizontal component in a Cartesian coordinate system may also be called the _____.
17. One of the coordinate values of a vector is known as a(n) _____.
18. The speed and direction of a projectile at a point along its path is its _____ vector.
19. The path a projectile follows is called its _____.
20. The horizontal distance a projectile travels is known as the _____ of the projectile.
21. An object moving through the air affected only by gravity is called a(n) _____.
22. The path a projectile follows is an arch-shaped curve called a(n) _____.
23. The ratio of the value of the x and y values of a vector is the _____ of the angle between the vector and its x component.
24. A force acting perpendicular to its supporting surface is a(n) _____.
25. The process of calculating three-dimensional motion from forces and accelerations is called _____.
26. The term that refers to a straight, sloped surface is _____.
27. When all forces acting on an object are balanced and the net force is zero the object is in _____.

Concept review

1. Explain the difference between scalar and vector quantities.
2. Classify each of the following as a scalar or vector.
   a. height
   b. displacement
   c. velocity
   d. area
3. Draw a vector to show a displacement of 100 meters at an angle of 30 degrees. Use a scale for your drawing.
4. You walk 1 kilometer north to the store, turn around, and return home. What is your displacement?
5. What is the maximum resultant of a 1-centimeter vector and a 4-centimeter vector? What is the minimum resultant?
6. Vectors can be expressed using Cartesian coordinates and polar coordinates. What type of quantity is each coordinate when a vector is expressed using Cartesian coordinates? What type of quantity is each coordinate when a vector is expressed using polar coordinates?
7. If a vector is at 45 degrees, what do you know about the magnitude of its components?
8. List three advantages of using components rather than a scale drawing to add a set of vectors.
9. What is the only force that affects the motion of a projectile?
10. Which component of a projectile's velocity changes as it moves through the air, the horizontal component or the vertical component?
11. A ball rolls off the edge of a horizontal table. What is the initial vertical velocity component?
12. A soccer ball is kicked off the ground at an angle of 45 degrees.
    a. At the top of its path, is its velocity entirely horizontal, entirely vertical, or a combination of both? Explain your reasoning.
    b. At the top of its path, is its acceleration entirely horizontal, entirely vertical, or a combination of both? Explain your answer.
13. At what angle should a ball be thrown for it to travel a maximum distance?
14. A ball kicked at an angle of 25 degrees will have the same range as a ball kicked at the same speed at ___ degrees.
15. What does it mean for an object to be in equilibrium?
16. Give three examples of objects moving along an inclined plane.
17. A sled is sliding down an icy hill.
    a. List the three forces that act on the sled as it moves down the hill.
    b. Draw a diagram to show the direction of each of the forces on the sled.
18. Which force acting on an object on an inclined plane causes it to accelerate?
19. Which force acting on an object on an inclined plane decreases its acceleration?
20. If an object is in equilibrium, what is the net force in the x-direction? In the y-direction?
21. As a projectile moves through its parabolic trajectory, which of the following quantities, if any, remain constant?
    a. acceleration
    b. horizontal component of velocity
    c. vertical component of velocity

Problems

1. Add the following sets of vectors (N = north, W = west, and S = south).
   a. 2 cm N + 7 cm W
   b. 5 m S + 8 cm N
   c. 30 m/s W + 50 m/s S
   d. 5 cm N + 7 cm W + 9 cm S
2. Resolve the vector (6 cm, 25°) into x-y components.
3. Calculate the components of the vector representing a velocity of 40 meters per second at an angle of 55 degrees.
4. A pilot wants to fly directly to the west. The engine pushes the plane at 100 m/s and there is a crosswind blowing to the south at 30 m/s. Determine the exact angle at which the pilot should head.
5. You and a friend are rowing a boat. You aim it toward a dock directly across the 100-meter-wide river and paddle at a speed of 1 m/s. You both are concentrating on rowing, so you do not notice that there is a 2 m/s current pushing you downstream. How far from the dock will you be when you reach the shore? Will the time it takes to cross the river be affected by the current?
6. You take a running jump off the end of a diving platform at a speed of 7 m/s and splash into the water 1.5 seconds later.
   a. How far horizontally do you land from your takeoff point?
   b. How high is the diving platform?

CHAPTER 7: USING VECTORS: MOTION AND FORCE

7. A model rocket is launched into the air so that its initial horizontal speed is 20 m/s and its initial vertical speed is 39.2 m/s. Complete the chart by finding the horizontal and vertical components of the velocity each second.

Time (s)	Horizontal speed (m/s)	Vertical speed (m/s)
0	20.0	39.2
1		
2		
3		
4		
5		
6		
7		
8		

8. A circus performer wants to land in a net 5 meters to the right of where she will let go of the trapeze. If she is 10 meters above the net, how fast must she be moving horizontally when she lets go?

9. You hit a baseball at a speed of 35 m/s and an angle of 40 degrees. A player catches the ball at the same height off the ground as it is hit.
   a. Find the horizontal and vertical components of the ball's initial velocity.
   b. How many seconds did the ball take to get to the top of its path?
   c. How much time did the ball spend in the air?
   d. How far off the ground was the ball at its highest point?
   e. How far horizontally did the ball travel?

10. A swing is designed so the ropes hang at an angle of 10 degrees from the vertical. A child with a weight of 200 newtons sits on the swing.
    a. How much tension is in each rope?
    b. How does the tension compare with the tension in a swing with ropes that are completely vertical?

11. Chris rides a sled down a hill with a slope of 22 degrees. The combined weight of Chris and his sled is 500 newtons.
    a. What is the normal force of the ground on the sled?
    b. Calculate the component of the weight that is parallel to the ground.
    c. Assuming the ground is frictionless, what is Chris's acceleration?
    d. If Chris's little sister also rides on the sled, their combined weight is 700 newtons. What is the new acceleration?
    e. Compare your answers to part c. Can you explain this?

12. A 2-kilogram object slides down a frictionless slope.
    a. Calculate its acceleration if the slope is angled at 30 degrees.
    b. Calculate its acceleration if the slope is angled at 60 degrees.
    c. Compare your answers to part a. Can you explain this?

13. A heavy cardboard box full of books slides down a wooden inclined plane with a certain acceleration. Identify whether each suggested change would result in an increased acceleration, decreased acceleration, or no change in the acceleration.
    a. greasing the inclined plane to lower the coefficient of sliding friction
    b. adding more books to the box
    c. decreasing the angle of the inclined plane

14. A 20-kilogram object slides at constant speed down an incline plane that makes a 30 degree angle with the horizontal. Using a scale of 1 cm = 20 N, make a scale drawing representing the force of gravity (F_g), the normal force (F_N), the perpendicular component of weight (F_y), the frictional force (F_f), and the parallel component of weight (F_x).

Applying your knowledge

1. What is the equation for a parabola shaped like a projectile's trajectory? How is the equation for a parabola similar to the equations for projectile motion?

2. While GPS is very good at locating a position, errors can be made. Research GPS and name a few error sources. How does DGPS help in correcting these errors?

3. A banana is thrown directly at a monkey hanging from the branch in a tree. At the instant the banana is thrown, the monkey releases the branch and falls straight down. Does the banana hit the monkey? Explain your answer.

4. The maximum range of a projectile occurs at a launch angle of 45 degrees to the horizontal if air resistance is neglected. If air resistance is *not* neglected, will the optimum angle be greater or less than 45 degrees? Explain your answer.

UNIT 3 MOTION AND FORCE IN 2 AND 3 DIMENSIONS

CHAPTER 8

Motion in Circles

Objectives:

By the end of this chapter you should be able to:

- ✔ Calculate angular speed in radians per second.
- ✔ Calculate linear speed from angular speed and vice versa.
- ✔ Describe and calculate centripetal forces and accelerations.
- ✔ Describe the relationship between the force of gravity and the masses and distance between objects.
- ✔ Calculate the force of gravity when given masses and distance between two objects.
- ✔ Describe why satellites remain in orbit around a planet.

Key Questions:

- How do you describe and measure circular motion?
- What is centripetal force, and how does it affect circular motion?
- What force exists between all objects that have mass, and how does it relate to orbital motion?

Vocabulary

angular displacement	centripetal acceleration	ellipse	linear speed	revolve
angular speed		gravitational constant	orbit	rotate
axis	centripetal force		radian	satellite
centrifugal force	circumference	law of universal gravitation		

165

CHAPTER 8 MOTION IN CIRCLES

8.1 Circular Motion

Linear motion is the kind of motion that goes from one place to another, like a car moving down a ramp. The concepts of position, velocity, and acceleration are useful for describing linear motion. Planets orbiting the Sun, a child on a merry-go-round, and a basketball spinning on a fingertip are examples of *circular motion*. Circular motion repeats itself by rotating or revolving. This chapter introduces some new ideas, such as angular velocity, that are similar to their linear motion counterparts, but also different and better suited for describing circular motion.

Describing circular motion

Rotating and revolving — A basketball spinning on your fingertip and a child on a merry-go-round both move around an imaginary line called an **axis**. The basketball's axis runs from your finger up through the center of the ball (Figure 8.1). We say the ball **rotates** about its axis because the axis is *inside* the ball. The child moves around an axis that is in the center of the merry-go-round (Figure 8.2). We say the child **revolves** around the center because the axis is *outside* the child. We say an object rotates when the axis passes through the object. An objects revolves when the axis does not pass through the object itself. Objects can rotate and revolve at the same time. Earth revolves around the Sun once each year while it rotates around its north-south axis once each day (Figure 8.3).

Angular speed — The rate at which an object rotates or revolves is described by its *angular speed*. An object's **angular speed** tells you how many turns it makes per unit of time. To calculate angular speed, you divide the number of rotations or revolutions by the time it takes. For example, if a basketball rotates 15 times in three seconds, its angular speed is 15 rotations divided by 3 seconds, or 5 rotations per second. Angular speed is sometimes called rotational speed.

Units of angular speed — There are two different ways in which angular speed is usually measured. One is by the number of complete turns per unit of time. The rpm—rotations per minute—is a unit commonly used to measure angular speed. A motor turning at 500 rpm turns 500 times each minute. The second way to measure angular speed is by the change in *angle* per unit of time. For example the second hand on a clock turns at a rate of 6 degrees per second. A full rotation is 360 degrees and the second hand takes 60 seconds at 6 degrees per second to make one full rotation.

Figure 8.1: *A basketball rotates around an internal axis.*

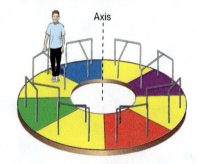

Figure 8.2: *The child revolves around an external axis.*

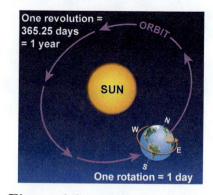

Figure 8.3: *Earth both rotates on its own axis and revolves around the Sun.*

MOTION IN CIRCLES CHAPTER 8

The relationship between linear and angular speed

Circumference of a circle How fast is a point on the edge of a turning wheel moving? As you might recall from geometry, the **circumference** is the distance around a circle. The circumference depends on the radius of the circle. A point on the edge moves a distance of one circumference in each turn.

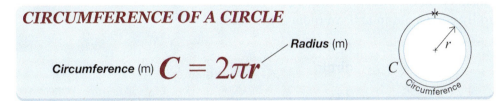

CIRCUMFERENCE OF A CIRCLE

Circumference (m) $C = 2\pi r$ — Radius (m)

Calculating linear speed The **linear speed** (v) of a point at the edge of a turning circle is the circumference divided by the time it takes to make one full turn. Automobile speedometers use this formula to calculate the linear speed of a car based on the angular speed and radius of the tires. The factor $2\pi \div t$ is defined as the angular speed, ω. The angular speed of a wheel is the same for all points on the wheel. However, the linear speed of a point on a wheel depends on the radius, r, which is the distance from the center of rotation.

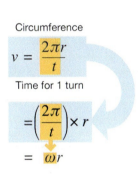

Circumference

$$v = \frac{2\pi r}{t}$$

Time for 1 turn

$$= \left(\frac{2\pi}{t}\right) \times r$$

$$= \omega r$$

LINEAR AND ANGULAR SPEED

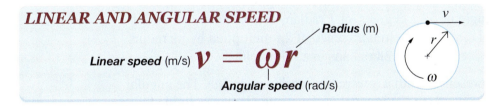

Linear speed (m/s) $v = \omega r$ — Radius (m)

Angular speed (rad/s)

Linear speed varies with radius Points on the outer edge of a rotating wheel move faster than points nearer to the center. The point at the very center of the wheel where $r = 0$ may have a linear speed of zero. This is a problem for a DVD player which must read a spinning disc. The rate at which the data pass under the laser depends on the radius. Data near the edge of the disc pass the reading head five times faster than data near the center.

Calculating linear speed from angular speed

Two children are spinning around on a merry-go-round. Siv is standing 4 meters from the axis of rotation and Holly is standing 2 meters from the axis. Calculate each child's linear speed when the angular speed of the merry-go-round is 1 rad/s.

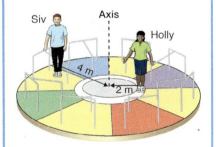

1. You are asked for the children's linear speeds.
2. You are given the angular speed of the merry-go-round and radius to each child.
3. The relationship that applies is $v = \omega r$.
4. Solve:
   For Siv:
   $v = (1 \text{ rad/s})(4 \text{ m})$
   $= 4$ m/s
   For Holly:
   $v = (1 \text{ rad/s})(2 \text{ m})$
   $= 2$ m/s

8.1 CIRCULAR MOTION

The units of radians per second

Angles in radians What are the units for angular speed? On the last page we defined the angular speed as $2\pi/t$ where t has units of seconds. The number 2π is a pure number with no units at all. The number π, approximately 3.14, is the ratio of the circumference of a circle to the diameter. One **radian** is the angle you get when you rotate the radius of a circle a distance on the circumference equal to the length of the radius. Angular speed naturally comes out in units of radians per second.

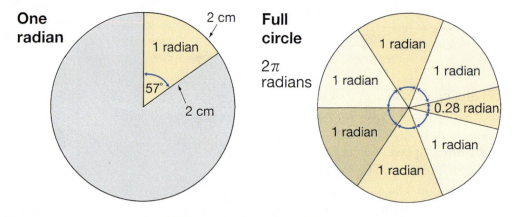

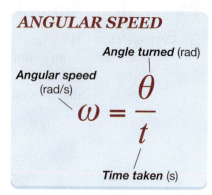

ANGULAR SPEED

Angular speed (rad/s)

$$\omega = \frac{\theta}{t}$$

Angle turned (rad)

Time taken (s)

Radians, degrees, and arc length There are approximately 57.3 degrees in one radian, so a radian is a larger unit of angle measure than a degree. An arc length is a distance measured along the curved circumference of a circle. The circle in the diagram has a radius of 2 centimeters. An angle of 57.3 degrees, or 1 radian, defines an arc length of 2 centimeters. A full circle has a circumference of 2π multiplied by its radius. Therefore, there are 2π or about 6.28 radians in a full circle.

Angular speed in radians per second Angular speed is represented with a lowercase Greek omega (ω). The angular speed in radians per second (rad/s) is equal to the change in angle divided by the change in time. Be careful when the angular speed is in full turns! To use the formula $v = \omega r$ the angular speed *must be in rad/s*. A full turn is 2π, approximately 6.28 radians. Therefore, a rotation speed of 5 turns per second is 5 times 6.28, or 31.4 rad/s. The units of angular speed are *per second* because the radian is a *dimensionless* unit. When we write *rad/s* the "rad" is there to remind us that the angle is in radians. When canceling units in a problem, *you may ignore radians*.

Calculating angular speed in rad/s

A bicycle wheel makes 6 turns in 2 seconds. What is its angular speed in radians per second?

6 turns in 2 seconds

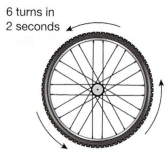

1. You are asked for the angular speed.
2. You are given turns and time.
3. There are 2π radians in one full turn. $\omega = \theta \div t$
4. Solve:
   $\omega = (6 \times 2\pi) \div (2\text{ s}) = 18.8$ rad/s

Relating angular speed, linear speed, and displacement

The speed of a rolling wheel
For a rolling object like a wheel, the forward speed is equal to the linear speed of a point at the edge of the wheel. As a wheel rotates, the point touching the ground passes around its circumference. When the wheel has turned one full rotation, it has moved forward a distance equal to its circumference. Therefore, the linear speed of a wheel is its angular speed multiplied by its radius, or $v = \omega r$.

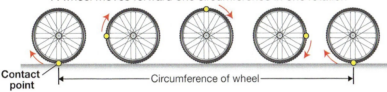

A wheel moves forward one circumference in one rotation

Speedometers and odometers
The speedometer and odometer on an automobile measure the speed and distance traveled by translating the circular motion of the wheels into linear motion. They take into account the size of tires used on cars for which they are designed. If tires of the wrong radius are used, the speed and distance will be inaccurate. If the tires are replaced with larger ones, the speedometer will display a slower speed than what the car is actually traveling. As a result, the driver might exceed the speed limit without knowing it.

Angular displacement
Linear displacement is measured in units of distance. **Angular displacement** is an angle represented by the Greek letter theta (θ) and is measured in units of angles, such as radians. The total angular displacement can be many revolutions. For example, a displacement of 12π radians is equal to six rotations, because one rotation is 2π radians. An angular displacement of 540 degrees is one-and-one-half turns, since 360 degrees is a full turn.

Calculating angular speed from linear speed

A bicycle has wheels that are 70 cm in diameter (35 cm radius). The bicycle is moving forward with a linear speed of 11 m/s. Assume the bicycle wheels are not slipping and calculate the angular speed of the wheels in rpm.

1. You are asked for the angular speed in rpm.
2. You are given the linear speed and radius of the wheel.
3. $v = \omega r$, 1 rotation = 2π radians
4. Solve:
   $\omega = v \div r$
   $= (11 \text{ m/s}) \div (0.35 \text{ m})$
   $= 31.4$ rad/s

Convert to rpm:
$$\omega = \frac{31.4 \text{ rad}}{1 \text{ s}} \times \frac{60 \text{ s}}{1 \text{ min}} \times \frac{1 \text{ rotation}}{2\pi \text{ rad}}$$
$= 300$ rpm

8.2 Centripetal Force

Imagine whirling a potato around your head on a string (Figure 8.4). According to the Newton's first law, the potato naturally tries to move in a straight line. The string forces the motion to bend into a circle. The instant the string breaks, the potato flies off in a straight line and no longer moves in a circle. To keep anything moving in a circle, there must be something that exerts a force that always points toward the center, like the string. A planet in orbit, a race car going around a track, and a child riding on a merry-go-round all must have a center-pointing force to keep them in circular motion.

Centripetal force

Acceleration — We usually think of acceleration as a change in speed. But because velocity includes both speed and direction, acceleration can also be a change in the direction of motion. In circular motion, the acceleration is radially inward, toward the center of the circle, and perpendicular to the velocity.

Figure 8.4: *An object on a string is kept moving in a circle by the inward force of the string.*

Centripetal force causes circular motion

Direction of centripetal force — Whether a force causes a change in speed, direction, or both, depends on the direction of the force. A force in the same direction as the velocity causes an object to speed up or slow down. A force applied *perpendicular* to an object's velocity vector causes an object to change its path from a line to a circle, without changing its speed.

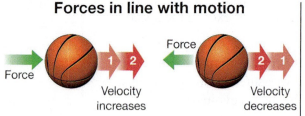
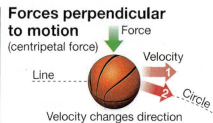

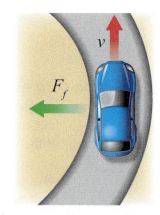

Figure 8.5: *Friction between the road and tires allows the car to turn the bend. The centripetal force of friction is toward the center of the circle, perpendicular to the velocity.*

Calculating centripetal force

Mass, speed, and radius The magnitude of the centripetal force needed to move an object in a circle depends on the object's mass and speed, and on the radius of the circle.

1. Newton's second law says force is proportional to mass. The greater the mass of an object, the greater the centripetal force needed to change its motion.
2. The strength of centripetal force is proportional to the square of the speed. As speed increases, a greater centripetal force is required to keep bending an object's path into a circle.
3. Centripetal force is inversely related to radius. The larger the radius, the smaller the required centripetal force to bend the motion into a circle. This makes sense because a larger circle means more gradual turning, which requires proportionally less force.

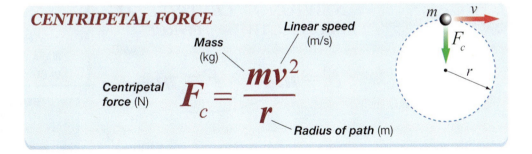

Centripetal force at the amusement park

A popular amusement park ride consists of a spinning cylindrical room in which riders stand with their backs against the wall. The room spins faster and faster, and then the floor suddenly drops out from beneath the riders' feet. The passengers seem to magically stick to the wall, suspended above the floor.

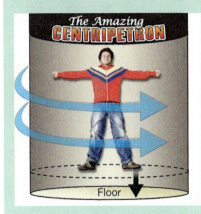

What seems to be magic can be explained easily with physics. The force of the wall pushing against a rider's back provides the centripetal force to move the rider in a circle. Friction between the rider's back and the wall prevents the person from sliding down to the floor.

Calculating centripetal force

A 50-kilogram passenger on an amusement-park ride stands with his back against the wall of a cylindrical room with radius of 3 meters. What is the centripetal force of the wall pressing into his back when the room spins and he is moving at 6 meters per second?

1. You are asked to find the centripetal force.
2. You are given the radius, mass, and linear speed.
3. The formula that applies is $F_c = mv^2 \div r$.
4. Solve: $F_c = (50 \text{ kg})(6 \text{ m/s})^2 \div (3 \text{ m}) = 600$ N

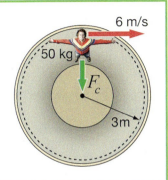

Centripetal acceleration

Centripetal force causes centripetal acceleration
Acceleration is the rate at which an object's velocity changes as the result of a force. **Centripetal acceleration** is the acceleration of an object moving in a circle due to the centripetal force. A race car driving down a straight track has no centripetal acceleration. A race car making a sharp turn has a large centripetal acceleration. Think about centripetal acceleration as the rate at which the *direction* of an object's velocity changes.

The formula for centripetal acceleration
The easiest way to find a formula for centripetal acceleration is by comparing Newton's second law and the formula for centripetal force. According to the second law, force equals mass times acceleration. Centripetal force must also equal mass times centripetal acceleration. You can see by comparing the two formulas that centripetal acceleration must be the square of the object's speed divided by the radius of the motion.

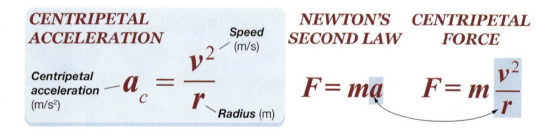

Direction of centripetal acceleration
The direction of an object's acceleration is always in the direction of the net force. Therefore, the centripetal acceleration is in the same direction as the centripetal force—toward the center of the circle. A major difference between linear and centripetal acceleration is that the direction of the centripetal acceleration constantly changes, just like the direction of the centripetal force.

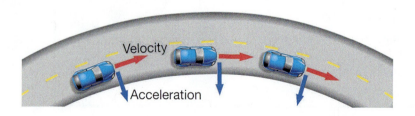

Calculating centripetal acceleration

A motorcycle goes around a bend with a 50-meter radius at 10 m/s. Find the motorcycle's centripetal acceleration and compare it with g, the acceleration of gravity.

1. You are asked for centripetal acceleration and a comparison with g (9.8 m/s^2).
2. You are given the linear speed and radius of the motion.
3. $a_c = v^2 \div r$
4. Solve:
   $a_c = (10 \text{ m/s})^2 \div (50 \text{ m})$
   $= 2 \text{ m/s}^2$

The centripetal acceleration is about 20 percent, or 1/5 that of gravity.

Figure 8.6: *The rotation of a wheel-shaped space station creates centripetal acceleration that simulates the feeling of gravity.*

Centrifugal force

Inertia As Newton's first law states, an object's inertia tends to keep the object moving in the same direction at a constant speed. An object moving in a circle is constantly changing its direction of motion. A constant centripetal force is required to cause this change in direction. Circular motion is *constantly accelerated*.

Riding in a turning car Imagine yourself sitting in the center of a smooth back seat in a moving car that suddenly turns to the left. Your body slides to the right until it is stopped by the seat belt or the door of the car. Why does this occur? Your body, like any body, tends to keep moving in a straight line unless an opposing force prevents the motion. When forces between the car's tires and the ground cause the car to turn, you continue to move straight ahead because the smooth seats do not provide a frictional force large enough to oppose your motion. While it may feel as though you are thrown to the right, you are not. *The car turns to the left below you*, making it seem like your body moves to the right. The force applied by the seat belt or the door of the car then provides the centripetal force to move you around the bend.

Centrifugal force Although the centripetal force pushes you toward the center of the circular path, it seems as if there is also a force pushing you to the outside. This apparent outward force is called **centrifugal force**. While it feels like an actual force acting on you, centrifugal force is *not a true force* exerted on your body. It is simply your tendency to move in a straight line due to inertia. This is easy to observe by twirling a small object at the end of a string. When the string is released, the object flies off in a straight line, tangent to the circle. No outward force pushes it away from the center of the circle, but the lack of a centripetal force keeps it from continuing to move in its circular path.

Banked turns

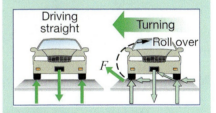

In a turn, the road exerts a sideways centripetal force on the wheels of a car, as well as the upward force reaction to the car's weight. At high speed, the resultant force on the inside wheel can lift the car and flip it to the outside of the curve.

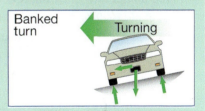

To allow safe travel at highway speeds, most roads bank the curves. In a banked curve the road slopes up, away from the center of the curve. The road acts like a ramp and the centripetal force is partly supplied by the parallel component of the car's own weight.

8.3 Universal Gravitation and Orbital Motion

One of the most important sources of centripetal force is gravity. The Sun's gravity keeps the planets moving in nearly-circular orbits. Gravity is a true centripetal force that changes direction as an object moves. This section is about gravity, orbits, and the motion of planets and satellites.

Newton's law of universal gravitation

The law of universal gravitation
Sir Isaac Newton first deduced that the force responsible for making objects fall on Earth is the same force that keeps the Moon in orbit. This idea is known as the **law of universal gravitation**. Gravitational force exists between all objects that have mass. The strength of the gravitational force depends on the mass of the objects and the distance between them. This law describes the relationship between the attractive gravitational force between two objects and their masses and distance apart.

You do not feel the attraction between ordinary masses
All objects that have mass attract each other, even small ones. You do not notice the attractive force between ordinary objects because gravity is a relatively weak force. For example, the attractive force between a person and an apple is 2×10^{-10} N (Figure 8.7). The force is small because the masses are small. Both the person and the apple *do* feel a much larger force of gravity, their *weight*. Weight is a significant force because Earth has a huge mass. Gravitational forces tend to be important only when one of the objects has a mass comparable to a planet.

Direction of the gravity force
The force of gravity between two objects always lies along the line connecting their centers. As objects move, the force of gravity changes its direction to stay pointed along the line of those centers. For example, the force between Earth and your body points from your center of mass to the center of mass of the planet itself. The direction of Earth's gravity is what we use to define "down." Down is the direction of the gravitational force between any object and Earth.

Figure 8.7: *The gravitational force between you and Earth is much stronger than the force between you and the apple because Earth is much more massive.*

The force of gravity between two masses points along a line joining their centers.

Calculating the gravitational force between objects

Mass and gravity — The attractive force of gravity between two objects is proportional to the masses of both objects multiplied together. If one object doubles in mass, then the force between the objects doubles. If both objects double in mass, then the force of gravity between them is multiplied by four.

Distance and gravity — Distance is also important when calculating gravitational force. The closer objects are to each other, the greater the force between them; the farther apart, the weaker the force. The gravitational force is inversely related to the square of the distance, so doubling the distance reduces the force to one-fourth its original value. Tripling the distance reduces the force to one-ninth its original value.

Determining distance — The distance is measured from the center of one object to the center of the other. To find the force between you and Earth, you must use the distance to the planet's center. If you climb a hill, this distance increases, so the force decreases. However, this change in distance is insignificant when compared with Earth's radius, so the difference in the force is not noticeable.

The law of universal gravitation — The law of universal gravitation allows you to calculate the gravitational force between two objects from their masses and the distance between them. The law includes a value called the **gravitational constant**, or G. This value is the same everywhere in the universe. The universal law of gravitation can be used to describe and predict the force between small objects like apples, and also between huge objects like planets, moons, and stars.

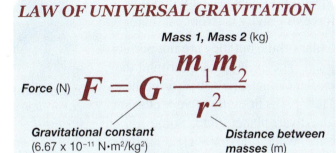

LAW OF UNIVERSAL GRAVITATION

Force (N) $\quad F = G \dfrac{m_1 m_2}{r^2} \quad$ Mass 1, Mass 2 (kg)

Gravitational constant (6.67×10^{-11} N·m²/kg²)

Distance between masses (m)

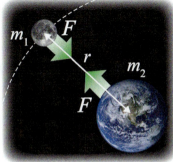

Calculating the weight of a person on the Moon

The mass of the Moon is 7.36×10^{22} kg. The radius of the moon is 1.74×10^6 m. Use the equation of universal gravitation to calculate the weight of a 90-kg astronaut on the Moon's surface.

90 kg

1. You are asked to find a person's weight on the Moon.
2. You are given the radius and the masses.
3. The formula that applies is $F_g = G m_1 m_2 \div r^2$.
4. Solve:

$$F_g = \left(6.67 \times 10^{-11} \, \dfrac{\text{N} \cdot \text{m}^2}{\text{kg}^2} \right)$$

$$\times \dfrac{(90 \text{ kg})(7.36 \times 10^{22} \text{ kg})}{(1.74 \times 10^6 \text{ m})^2}$$

$$= 146 \text{ N}$$

By comparison, on Earth the astronaut's weight would be 90 kg × 9.8 m/s² or 882 N. The force of gravity on the Moon is approximately one-sixth what it is on Earth.

Orbital motion

Satellites A **satellite** is an object that is bound by gravity to another object such as a planet or star. The Moon, Earth, and the other planets are examples of natural satellites. Artificial satellites that move around Earth include the Hubble Space Telescope, the International Space Station, the space shuttles, and satellites used for communication. The first artificial satellite, Sputnik I, which translates as "traveling companion," was launched by the former Soviet Union on October 4, 1957. Hundreds of satellites have been sent into space since then.

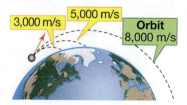

Figure 8.8: *The faster an object is launched, the farther it travels before landing. If an object is launched fast enough, it falls into orbit around the Earth.*

Orbits and gravitational force An **orbit** is the path followed by a satellite. The orbits of many natural and man-made satellites are circular, or nearly circular. The centripetal force that bends the motion of a satellite into an orbit comes from the gravitational attraction between the satellite and the planet it orbits.

Launching a satellite The motion of a satellite is closely related to projectile motion. If an object is launched above Earth's surface at a slow speed, it will follow a parabolic path and fall back to Earth (Figure 8.8). The faster it is launched, the farther it travels before reaching the ground. At a launch speed of about 8 kilometers per second, the curve of a projectile's path matches the curvature of the planet. At this speed, an object goes into orbit instead of falling back to Earth. You can think about the motion of a satellite in orbit as *falling around Earth*. A satellite in orbit is actually in free fall but as it falls, Earth curves away beneath it.

Elliptical orbits Not all orbits are perfectly circular. If an object is launched above Earth at more than 8 kilometers per second, the orbit will be a noncircular **ellipse**. An object in an elliptical orbit does not move at a constant speed. It moves fastest when it is closest to the object it is orbiting because the force of gravity is strongest there.

The orbits of planets and comets The planets in our solar system have nearly circular orbits, but they are not perfect circles. All the planet's orbits are slightly elliptical. The orbit of Earth around the Sun is only 2 percent different from a perfect circle. Comets, however, orbit the Sun in very long elliptical paths (Figure 8.9). Their paths bring them very close to the Sun and then out into space, often beyond the dwarf planet Pluto. Some comets take only a few years to orbit the Sun once, while others travel so far out that a solar orbit takes thousands of years.

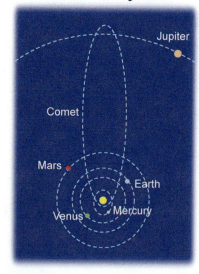

Figure 8.9: *The orbits of the inner planets, except for Mercury's, are nearly circular. Comets have highly-elliptical orbits.*

Chapter 8 Connection

Satellite Motion

In 1957, the first artificial Earth satellite was launched by the Soviets. Called Sputnik, it created a sensation and sparked the space race between the United States and the Soviet Union. This competition culminated in 1969, when U.S. astronauts Neil Armstrong and Buzz Aldrin landed on the surface of the Moon. Firsthand human exploration of the moon ended in the 1970s, though development of artificial satellites continued. Today, the satellite industry is a multibillion-dollar business involving hundreds of companies in dozens of countries. Nearly every person on Earth is affected day-to-day in some way by satellites orbiting far overhead, in the vacuum of space.

The orbit equation For a satellite in a circular orbit, the force of Earth's gravity pulling on the satellite equals the centripetal force required to keep it in its orbit. If this were not so, the satellite would either get closer to or farther away from Earth and the orbit would not remain circular. We can apply the law of universal gravitation and the equation for centripetal acceleration to determine the relationship between a satellite's orbital radius, r, and its orbital velocity, v (Figure 8.10).

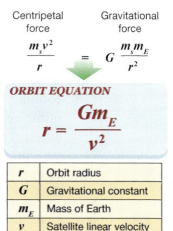

Figure 8.10: *The relationship between orbital speed (v) and the radius (r) of a satellite's orbit.*

Geostationary satellites Most people in the United States—and many millions more around the world—are familiar with weather satellites. These satellites stay above the same point on Earth so that their pictures are of that place 24 hours a day. This "hovering" over the same spot is possible because the satellites are in a special kind of orbit called *geostationary*. A satellite in geostationary orbit completes one orbit in exactly one day, so that its motion follows the motion of the ground underneath it (Figure 8.11). Many communications and TV broadcast satellites are also in geostationary orbits. This is why a satellite dish antenna can receive a signal 24 hours a day but stay in a fixed position. If the TV satellite were not in a geostationary orbit, the dish antenna would have to move around to track the satellite.

The altitude of geostationary orbit To keep up with Earth's rotation, a geostationary satellite must travel the entire circumference of its orbit ($2\pi r$) in 24 hours, or 86,400 seconds. To stay in orbit, the satellite's radius and velocity must also satisfy the orbit equation in Figure 8.10. The combination of these two conditions determines the radius of geostationary orbit, which is 42,300 kilometers from Earth's center (Figure 8.11). The altitude of the orbit is 35,920 kilometers above the planet's surface, after subtracting 6,380 kilometers for the radius of Earth itself.

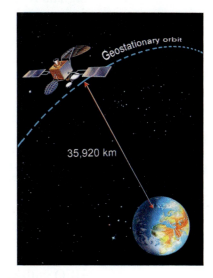

Figure 8.11: *A satellite in geostationary orbit stays above the same place on Earth as it revolves in its orbit.*

8.3 UNIVERSAL GRAVITATION AND ORBITAL MOTION

Chapter 8 Connection

Use of HEO and geostationary orbits

There are many satellites in geostationary orbit

All geostationary satellites must orbit directly above the equator. This means that the geostationary "belt" is the prime real estate of the satellite world. Currently, more than 300 satellites occupy "slots" in geostationary orbit, and their number continues to grow. There have been international disputes over the right to the prime geostationary slots, and there have even been cases where satellites in adjacent slots have interfered with each other. To avert cluttering up the geostationary belt with dead satellites, all geostationary satellites reaching the end of their lifetimes are now designed to conduct a "graveyard burn" which propels them into a higher, non-geostationary orbit, away from the congestion.

Polar regions cannot "see" geostationary orbit

Geostationary satellites provide excellent images and TV coverage for most regions of Earth, but they do not cover the polar regions. To a ground observer at latitudes above 70 degrees, a geostationary satellite appears less than 10 degrees above the horizon. At this low angle, even small hills interfere with reception of signals from the satellite.

Highly elliptical orbit (HEO)

There is no way to place a satellite directly over Earth's north or south pole permanently. However, there is a way to maximize the time a satellite spends over a polar region. Such satellites are placed in a highly-elliptical orbit (or HEO), which means the orbit is highly elongated (Figure 8.12). The orbital ellipse is arranged as such that one side of the ellipse is far away from Earth, while the other side is quite close.

Perigee and apogee

The closest approach of an elliptical orbit is called the *perigee*. The farthest point in the orbit is called the *apogee*. If you solve the equations for the trajectory of an elliptical orbit, the velocity at apogee is slower than the velocity at perigee by the ratio of r_p/r_a, the orbital radii at perigee and apogee. This means a satellite will slow down as it approaches apogee, and speed back up at it moves toward perigee. If apogee is above the north or south pole, then the satellite will spend most of its time above the pole where it is moving slowest. A communications satellite in a HEO with its apogee above the north pole is in line-of-sight of the northern hemisphere of Earth for most of its orbital period. A pair of satellites in north polar HEO can provide continuous coverage for northern regions.

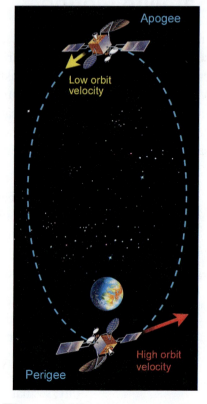

Figure 8.12: *A north polar HEO is an exaggerated elliptical orbit that allows a satellite to spend most of its time being visible from the northernmost latitudes.*

Chapter 8 Assessment

Vocabulary

Select the correct term to complete the sentences.

rotates	revolve	axis
law of universal gravitation	circumference	linear speed
angular speed	centrifugal force	radian
orbit	centripetal force	centripetal acceleration
ellipse	satellite	angular displacement
gravitational constant		

1. An angle that can be represented by an arc length equal to one radius equals one _____.
2. The distance around a circle is its _____.
3. The displacement measured around a circle in units such as radians is _____.
4. If a body moves around an internal axis it _____.
5. The rate at which an object rotates about its axis is measured as _____.
6. An object that moves about an external axis is said to _____.
7. The product of the angular speed and the radius of circular motion equals the _____ of a revolving object.
8. An imaginary line around which an object may revolve or rotate is called a(n) _____.
9. The apparent outward force exerted on an object moving in a circular path is called _____.
10. The acceleration that results when a force causes an object to move in a circle is _____.
11. The name of the force that causes an object to follow a circular path is _____.
12. The tiny quantity, $6.67 \times 10^{-11} \, N \cdot m^2/kg^2$, that appears in Newton's universal law of gravity is called the _____.
13. The path followed by a revolving satellite is called its _____.
14. The shape of orbits for all satellites, including circular orbits, is actually a(n) _____.
15. The scientific law stating "the force pulling two objects together is directly proportional to their masses and inversely proportional to the square of the distance between them" is _____.
16. A revolving object bound in orbit by the gravitational attraction of another object may be called a(n) _____.

Concept review

1. State whether each of these objects rotates or revolves.
    a. a globe
    b. a satellite orbiting Earth
    c. a toy train traveling around a circular track
    d. a fan blade
2. State whether each of the following units is appropriate to express angular speed.
    a. rotations per second
    b. meters per second
    c. rpm
3. How many radians are in a full circle?
4. Two ants are sitting on a spinning record. The ant at point A is closer to the center of the record than the ant at point B. How do their angular speeds compare? How do their linear speeds compare?

5. Give three examples of forces that can act as the centripetal force keeping an object in circular motion.
6. If centrifugal force is not a true force, explain why it is called a force at all.
7. Explain what happens to the gravitational force between two objects as they are moved closer together.
8. What happens to the gravitational force between two objects if the mass of one is doubled? What if the mass of both objects is doubled?
9. Why do you weigh slightly less in a high-flying airplane?
10. How is projectile motion similar to orbital motion?
11. Name an object that rotates and revolves at the same time.

12. A cyclist rides clockwise around a circular track at a constant speed. What is the direction of her velocity at points A, B, and C? What is the direction of the centripetal force on her at these points? What is the direction of the centripetal acceleration at each point?

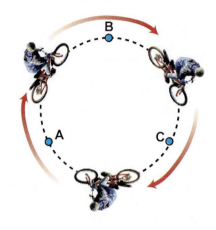

13. If the cyclist in the preceding question rides her bike at a faster speed, but in the same path, explain what happens to
    a. the centripetal force?
    b. the centripetal acceleration?

14. In some amusement parks, there is a ride that consists of a spinning cylindrical room. Describe where you would have to stand as the room turns to
    a. revolve.
    b. rotate.

15. What is represented by each of the following variables or symbols?
    a. π b. ω c. C d. v e. r

Problems

1. Marion rides her racing bicycle at a speed of 8 m/s. The bicycle wheels have a radius of 34 cm.
    a. What is the angular speed of the wheels?
    b. How many times does each wheel go around during a 10-minute ride?

2. Chris runs around a circular track with a radius of 30 meters. He makes five trips around the track in 220 seconds.
    a. What is the circumference of the track?
    b. How many meters does he run?
    c. What is his angular speed?
    d. What is his linear speed?

3. Convert the following.
    a. 30 degrees to radians
    b. 220 degrees to radians
    c. 2 radians to degrees
    d. 4.25 radians to degrees
    e. 2 revolutions to radians

4. The wheel of an in-line skate has a radius of 3 cm. What is the linear speed of the skater if the wheel turns at 25 rad/s?

5. A car tire has a radius of 33 cm. The tire turns a total of 10,250 radians during a trip to the store. How many meters did the car travel?

6. A 50-kg ice skater turns a bend at 7 m/s. If the radius of the curve is 5 m, what is the centripetal force provided by the friction between the blade of the skate and the ice?

7. A piece of clay with a mass of 0.30 kg is tied to the end of a string 6.5 meters in length, and then is whirled around in a horizontal circle at a speed of 6 m/s.
    a. What is the centripetal acceleration of the clay?
    b. What is the tension in the string?

8. Calculate the gravitational force of attraction between two 55-kilogram people standing 0.25 m apart.

9. Pluto has a mass of 1.5×10^{22} kg and a radius of 1.15×10^6 m.
    a. What is the acceleration due to gravity on Pluto?
    b. How much would a 70-kilogram person weigh on Pluto? Compare the value of that weight on Pluto with the person's weight on Earth.

10. A merry-go-round has a radius of 8 meters makes 2 revolutions every 2.5 minutes.
    a. Express the angular speed of the merry-go-round in radians per second.
    b. Express the angular speed of the merry-go-round in rpms.
    c. What is the linear speed of a child seated 7 meters from the center?

Applying your knowledge

1. As a cyclist pedals a bicycle along a bicycle path, circular motion takes place at several points on the bicycle. Name two objects associated with pedaling a bicycle that rotate and two that revolve as the cyclist moves along the path. Describe the center of turning for each object.

2. Research the speeds of CDs, DVDs, and LP vinyl records to answer the following questions.
    a. How do the angular speeds of CDs and DVDs compare?
    b. While a CD is playing, what happens to its linear speed and its angular speed? Why?
    c. Compare the angular and linear speeds of an LP vinyl record.

3. What is the difference between a geosynchronous satellite and a geostationary satellite? Describe at least one advantage and one disadvantage of using a geostationary satellite.

UNIT 3 MOTION AND FORCE IN 2 AND 3 DIMENSIONS
CHAPTER 9

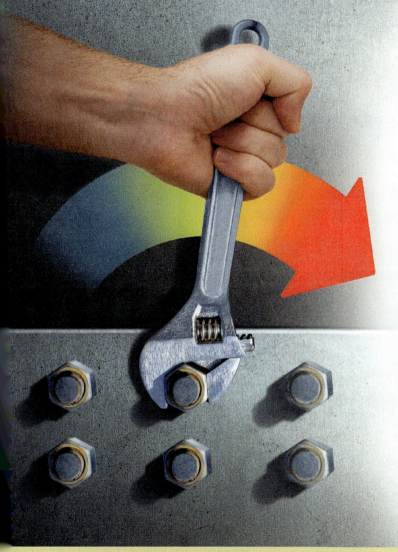

Torque and Rotation

Objectives:

By the end of this chapter you should be able to:

✔ Calculate the torque created by a force.
✔ Solve problems by balancing two torques in rotational equilibrium.
✔ Define the center of mass of an object.
✔ Describe a technique for finding the center of mass of an irregularly-shaped object.
✔ Calculate the moment of inertia for a mass rotating on the end of a rod.
✔ Describe the relationship between torque, angular acceleration, and rotational inertia.

Key Questions:

- What is the relationship between torque and rotational motion?
- Why do some objects appear stable, but others topple over easily?
- How does Newton's first law apply to rotating objects?

Vocabulary

angular acceleration	center of rotation	moment of inertia	rotational inertia
center of gravity	lever arm	rotational equilibrium	torque
center of mass	line of action		translation

9.1 Torque

Forces may cause an object to move and/or spin (Figure 9.1). Motion in which an entire object moves is called **translation**. Motion in which an object spins is called rotation. Both kinds of motion can be present at the same time, as they are in a rolling ball, for example. We understand force and its relationship to translational motion. However, in *rotational* motion, the same force can cause very different results. For example, think about trying to open a door by pushing at the hinged side of the door. The door won't swing open no matter how strong your force (Figure 9.2). Pushing the door at its handle causes it to swing open easily. For rotational motion, *torque* is what is most directly related to motion, not force. Whether you are creating rotational motion by opening a jar, screwing in a light bulb, using a wrench, or opening a door, you are using torque. This section is about torque and the relationship between torque and rotational motion.

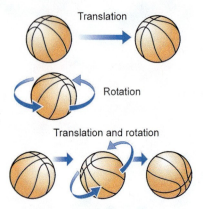

Figure 9.1: *Translation occurs when a ball moves in a line. Rotation occurs when it spins. Both can also occur at the same time.*

What is torque?

Torque and force A **torque** is an action resulting from a force that causes objects to rotate. A torque is required to rotate an object, just as a force is required to move an object in a line (Figure 9.1). Torque is the rotational equivalent of force: if force is a *push* or *pull*, then torque is a *twist*.

How torque and force differ Torque is *not* the same thing as force. Torque is created by force, but torque also depends on the point of application and the direction of the force. For example, a door pushed at its handle will easily turn and open, but a door pushed near its hinges will not move as easily (Figure 9.2). The force may be the same but the torque is quite different.

The center of rotation The point or line about which an object turns is its **center of rotation**. For example, a door's center of rotation is at its hinges. A force applied far from the center of rotation produces a greater torque than a force applied close to the center of rotation. Doorknobs are positioned far from the hinges to provide the greatest amount of torque for a given force. If you have ever accidentally tried to open a door by pushing on the hinged side, you know that even a large force did not cause the door to open. Forces applied near the hinges create very little torque *because* they are applied close to the center of rotation.

Figure 9.2: *A door rotates around its hinges and a force creates the greatest torque when applied far from the hinges.*

The torque created by a force

The line of action Torque is created when the **line of action** of a force does not pass through the center of rotation. The line of action is an imaginary line that follows the direction of a force and passes though its point of application. For example, when you pull on a wrench, the line of action of your force passes at a distance from the center of rotation, the bolt. The force creates torque because it acts to rotate the wrench (Figure 9.3). Forces applied in this way act to cause rotation.

A force that makes no torque No torque is created when the line of action goes *through* the center of rotation. For example, pushing on a wrench in its long direction is useless for tightening a bolt. Pushing in the long direction creates no torque because the line of action of the force passes through the center of rotation. The application of the force does *not* act in a way to cause the wrench to rotate.

The lever arm The direction in which the force is applied is also important. To open a door, you apply a force to the knob perpendicular to the door. If you push parallel to the door, toward the hinges, it will not open. To get the maximum torque, the force should be applied in a direction that creates the greatest *lever arm*. The **lever arm** is the perpendicular distance between the line of action of the force and the center of rotation (Figure 9.4).

Calculating torque The torque created by a force is equal to the lever arm length times the magnitude of the force. Torque is usually represented by the lower case Greek letter "tau," τ. In sketches, a torque is represented by an arc with an arrow indicating the direction of rotation the torque would cause. When calculating torque, be careful determining the length of the lever arm. If the line of action passes through the center of rotation, the lever arm is zero, and so is the torque, no matter how large a force is applied.

The line of action of a force

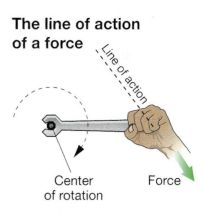

Figure 9.3: *The line of action is an imaginary line that follows the direction of a force and passes though its point of application.*

The lever arm of a force

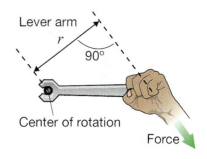

Figure 9.4: *The lever arm is the perpendicular distance between the line of action of the force and the center of rotation.*

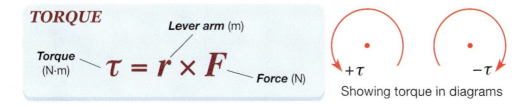

Showing torque in diagrams

9.1 TORQUE **183**

CHAPTER 9 TORQUE AND ROTATION

Calculating torque

Torques can be added and subtracted If more than one torque acts on an object, the torques are combined to determine the net torque. If the torques tend to make an object spin in the same direction, clockwise or counterclockwise, they are added together (Figure 9.5). If the torques tend to make the object spin in opposite directions, they are subtracted. Positive and negative values are used to keep the addition and subtraction from getting confused. Torques that tend to cause *counterclockwise* rotation are usually assigned *positive* values. Torques that tend to cause *clockwise* rotation are therefore *negative*. The total torque is calculated by adding up each individual torque and keeping track of the positive and negative signs.

Units of torque The units of torque are force times distance, or newton-meters. A torque of 1 N·m is created by a force of 1 newton acting with a lever arm of 1 meter. Because torque is a product of two variables, it is possible to create the same torque with different forces. For example, a 10-newton force applied with a lever arm of 0.1 meter also produces a torque of 1 N·m.

Torque about different centers Torque is always calculated around a *particular center of rotation*. For this reason, the same force may cause different torques when an object is allowed to rotate around different points. For example, a force of 1 newton applied to a 1-meter board creates a torque of 0.5 N·m when the board rotates about its center (Figure 9.6). The same force creates a torque of 1 N·m, or twice as great, when the board rotates about its end.

Positive and negative torque

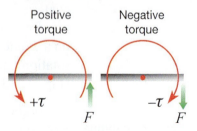

Adding torques

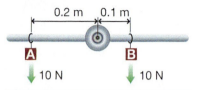

Figure 9.5: *The terms* clockwise *and* counterclockwise *are used to describe the direction of torques.*

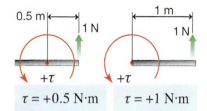

Figure 9.6: *The torque created by a force depends on the location of the center of rotation.*

Calculating a torque

A force of 50 newtons is applied to a wrench that is 30 centimeters long. Calculate the torque if the force is applied perpendicular to the wrench so the lever arm is 30 centimeters.

1. You are asked to find the torque.
2. You are given the force and lever arm.
3. The formula that applies is $\tau = rF$.
4. Solve: $\tau = (-50 \text{ N})(0.3 \text{ m}) = -15 \text{ N·m}$

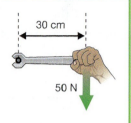

Rotational equilibrium

Net torque is zero An object is in **rotational equilibrium** when the net torque applied to it is zero. For example, if an object such as a see-saw is not rotating, you know the torque on each side is balanced. An object in rotational equilibrium can also be rotating at constant speed, like a wheel.

Using rotational equilibrium Rotational equilibrium is often used to determine unknown forces. Any object that is not moving is in rotational equilibrium *and* in translational equilibrium. Both kinds of equilibrium are often needed to find unknown forces. For example, consider a 10-meter bridge that weighs 500 newtons supported at both ends. A person who weighs 750 newtons is standing 2 meters from one end of the bridge. What are the forces F_A and F_B holding the bridge up at either end?

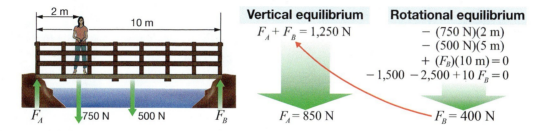

Vertical equilibrium For the bridge not to move up or down, the total upward force must equal the total downward force. This means $F_A + F_B = 1{,}250$ N. Unfortunately, the condition of zero net force in the vertical direction does not tell you how the force is divided between the two ends, F_A and F_B.

Solving for the unknown forces For the bridge to be in rotational equilibrium, the total torque *around any point* must also be zero. If we choose the left end of the bridge, the torque created by force F_A is zero because its line of action passes through the center of rotation. By setting the total of the remaining torques to zero, the force on the right support (F_B) is calculated to be 400 newtons. Since the total of both forces must be 1,250 N, that means the force F_A on the left must be 850 N. This kind of analysis is used to solve many problems in physics and engineering, including how strong to make bridges, floors, ladders, and other structures that must support forces.

Solving a rotational equilibrium problem

A boy and his cat sit on a seesaw. The cat has a mass of 4 kg and sits 2 m from the center of rotation. If the boy has a mass of 50 kg, where should he sit so that the seesaw will balance?

1. You are asked to find the boy's lever arm.
2. You are given the two masses and the cat's lever arm.
   $\tau = rF$, torque
   $F = mg$, weight
   In equilibrium the net torque must be zero.
3. Solve for the cat:
   $\tau = (2\text{ m})(4\text{ kg})(9.8\text{ N/kg})$
   $= +78.4$ N·m
4. Solve for the boy:
   $\tau = (d)(50\text{ kg})(9.8\text{ N/kg})$
   $= -490\, d$

For rotational equilibrium, the net torque must be zero.
$78.4 - 490d = 0$
$d = 0.16$ m

The boy must sit quite close (16 cm) to the center.

CHAPTER 9 TORQUE AND ROTATION

When the force and lever arm are not perpendicular

Force and lever arm are not always perpendicular
Torque is easiest to calculate when the lever arm and force are at right angles to each other. In this situation, the lever arm is the same as the distance between the point where the force is applied and the center of rotation. When the force and lever arm are *not* perpendicular, an extra step is required to calculate the length of the lever arm.

Torque in English units
In the English system of units, torque is measured in inch-pounds or foot-pounds. A torque of one foot-pound is created by a force of one pound applied with a lever arm of one foot. One foot-pound is equal to 12 inch-pounds.

Many machines are held together with nuts and bolts, or other threaded fasteners. The proper tightness of nuts and bolts is often specified in foot-pounds of torque. If a nut is too tight, it may strip the threads or snap the bolt. If a nut is too loose, it may vibrate off the bolt. Each size nut has a proper tightening torque.

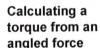

Calculating a torque from an angled force

A 20-centimeter wrench is used to loosen a bolt. The force is applied 0.20 meters from the bolt. It takes 50 newtons to loosen the bolt when the force is applied perpendicular to the wrench. How much force would it take if the force was applied at a 30-degree angle from perpendicular?

1. You are asked to find the force.
2. You are given the force and lever arm for one condition.
3. The formula that applies is $\tau = rF$.
4. Solve: The torque required to loosen the bolt
   $\tau = (50\ \text{N})(0.2\ \text{m}) = 10\ \text{N} \cdot \text{m}$

Calculate the torque with a force applied at 30 degrees:
$10\ \text{N} \cdot \text{m} = F \times (0.2\ \text{m}) \cos 30° = 0.173 F$
$F = 10\ \text{N} \cdot \text{m} \div 0.173 = 58\ \text{N}$. It takes a larger force.

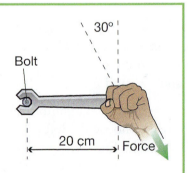

9.2 Center of Mass

The shape of an object and the way its mass is distributed affect the way the object moves and balances. For example, it is much easier to tip over a tall, thin vase than a short, wide tea kettle. This section talks about the center of mass, which is the point around which an object naturally balances or rotates. The location of the center of mass explains why some objects are stable and others easily topple over.

Finding the center of mass

The motion of a tossed object
In Chapter 7, you learned that a ball thrown into the air at an angle moves in a parabola. But what if an irregularly shaped object is thrown? If you throw a soda bottle so it spins during its flight, the general shape of its path will also be a parabola. If you watch the bottle closely, one specific point moves in a perfect parabola. This point is located on the axis about which the bottle spins.

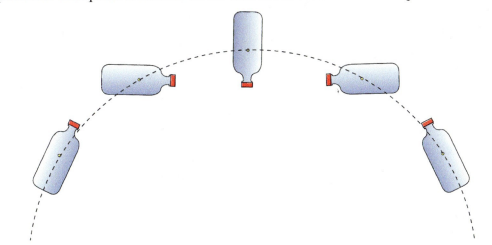

Defining center of mass
There are three different axes about which an object will naturally spin. The point at which the three axes intersect is called the **center of mass** (Figure 9.7). The center of mass is defined as the average position of all the particles that make up the object's mass. It is easy to find the center of mass for a symmetrically-shaped object made of a uniform material such as a solid-rubber ball or a wooden cube. The center of mass is located at the geometric center of the object. If an object is irregularly shaped, the center of mass can be found by spinning the object and finding the intersection of the three spin axes.

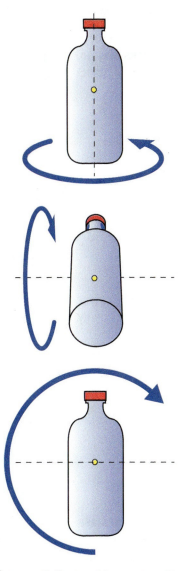

Figure 9.7: *An object naturally spins about three different axes. Their intersection is the center of mass.*

Finding the center of mass

The center of mass may not be "in" an object

There is not always material at an object's center of mass. The center of mass of a donut is at its very center, where there is only empty space. The same is true for a coffee mug, a boomerang, and an empty box.

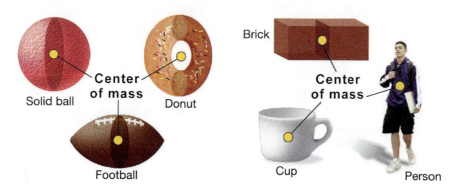

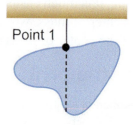

The center of gravity

Closely related to the center of mass is the **center of gravity**. The center of gravity is the average position of an object's weight. It is also the point at which we consider the force of gravity to act on an object. If the acceleration due to gravity is the same at every point in an object, the center of mass and center of gravity are at the same location. This is the case for most everyday objects, so the two terms are often used interchangeably.

Finding the center of gravity

An object's center of gravity can easily be found experimentally. If an object is suspended from a point at its edge, the center of gravity will always fall in the line directly below the point of suspension. If the object is suspended from two or more points, the center of gravity can be found by tracing the line below each point and finding the intersection of the lines (Figure 9.8).

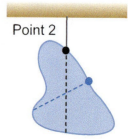

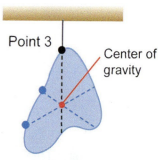

Figure 9.8: *The center of gravity of an irregularly-shaped object can be found by suspending it from two or more points.*

Centers of mass and gravity may differ

For very tall objects, such as skyscrapers, the acceleration due to gravity may be slightly different at points throughout the object. Gravity is stronger closer to the surface of Earth, so the pull of gravity at the bottom of a tall building is slightly stronger than the pull at the top. The top half therefore weighs less than the bottom half, even when both halves have the same mass. While the center of mass will be halfway up the building, the center of gravity will be slightly lower.

Balance and the center of mass

Balancing an object To balance an object such as a book or a pencil on your finger, you must place your finger directly under the object's center of gravity. The object balances because the torque caused by the force of the object's weight is equal on each side.

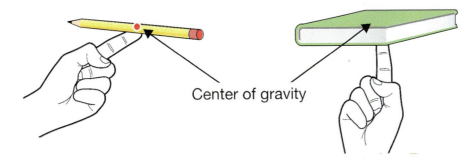

Center of gravity

The area of support For an object to remain upright, its center of gravity must be above its area of support. The area of support includes the entire region surrounded by the actual supports. For example, a table's area of support is the region bounded by its four legs. Your body's support area is not only where your feet touch the ground, but also the region between your feet. The larger the area of support, the less likely an object is to topple over.

When an object will topple over An object will topple over if its center of mass is not above its area of support. For example, a block will topple over if it is tipped far enough. Imagine a straight line drawn from the center of gravity toward Earth's center. If this line passes through the area of support, the object will not topple. If the line passes outside the area of support, the object will topple over if it is not held upright.

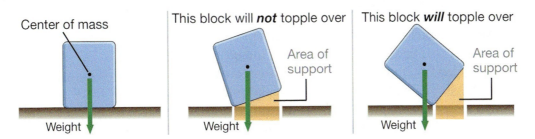

SUV rollovers

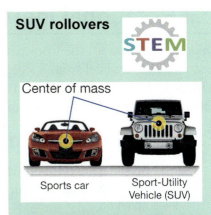

Many injuries and fatalities from auto accidents involve rollovers. Sports cars are designed with a wide wheelbase and a low center of mass to reduce the risk of rolling over at high speed. Sport utility vehicles (SUVs), however, have a relatively high center of mass compared with their wheelbase. Because of that high center of mass, an SUV is more likely to roll over in an accident.

According to the National Highway Traffic Safety Administration (NHSTA), SUVs are 75 percent more likely to experience rollovers than regular cars. And even though there have been many recent improvements in their design, the NHSTA reports that the safest SUVs are still more likely to rollover than the least-safe cars.

9.3 Rotational Inertia

Imagine that you lift the front wheel of your bicycle off the ground. A friend applies a force to the tire and makes the wheel spin. It keeps spinning until the torque provided by friction eventually slows it to a stop. Just as objects moving in a line have inertia, spinning objects have *rotational inertia*. Objects that are rotating tend to keep rotating at the same speed due to their rotational inertia.

Rotation and Newton's first law

Spinning objects tend to keep spinning — Recall Newton's first law of motion: An object at rest will remain at rest and an object in motion will remain in motion at a constant velocity unless acted on by a net force. Newton's first law also applies to objects in rotational motion. A rotating object will keep rotating at a constant angular speed *unless acted on by a net torque*. Just as a force is needed to change the motion of an object moving in a line, a torque is needed to change either the angular speed or the axis of rotation of a rotating object.

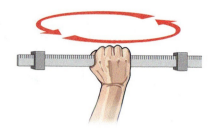

Figure 9.9: *The meter stick is easier to spin when the masses are closer to the center.*

Rotational inertia is different — **Rotational inertia** describes an object's resistance to a change in its rotational motion. With rotation, however, the *distribution* of mass matters as much as how much mass there is. This is very different from translational motion. In translational motion, mass equals inertia and the shape of the object does not matter. Two objects that have a mass of one kilogram both have the same inertia, even if one is a ball and the other is a hoop.

Mass distribution affects rotational inertia — An object's rotational inertia depends not only on the total mass, but also on the way mass is distributed. Consider rotating a meter stick with two heavy masses attached on either side of the center. When the masses are near the center, the meter stick is easy to spin (Figure 9.9). When the masses are far away from the center, the meter stick is more difficult to spin.

The reason why — Mass that is farther from the center of rotation contributes more rotational inertia than mass near the center. This is because mass at a larger radius actually *moves faster* for a given angular speed. Remember the relation between linear speed and angular speed, $v = \omega r$? Mass located farther from the axis of rotation has a larger radius, r, and therefore a larger linear velocity, v. The larger linear velocity is why mass at the end of the meter stick contributes more rotational inertia than mass near the center.

Rotational inertia

The meaning of rotational inertia Inertia is a measure of resistance to acceleration. To understand rotational inertia, we need to find a form of Newton's second law that applies to rotating motion. According to Newton's second law, $a = F \div m$, the linear acceleration is equal to force divided by the mass. Because inertia is created by mass, in most circumstances an object's *linear* inertia is equal to its mass. For rotating motion, we need an equation that relates torque, and *angular acceleration* in a similar way as force and linear acceleration are related by $a = F \div m$.

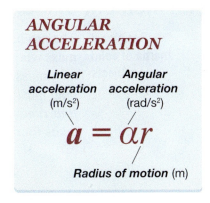

Angular acceleration The rate at which angular speed changes is called **angular acceleration.** The angular acceleration is the change in angular speed divided by the change in time. For example, suppose a wheel starts at rest. Five seconds later the wheel is rotating with an angular speed of 100 rad/s. The angular acceleration of the wheel is 20 rad/s², meaning the angular speed increases by 20 rad/s each second.

Figure 9.10: *The relationship between linear and angular acceleration.*

Linear and angular acceleration In Chapter 8, you learned that the linear speed (v) is equal to the angular speed (ω) times the radius of the motion (r). A similar relationship exists between the linear acceleration (a) and angular acceleration (α) (Figure 9.10). The units of angular acceleration are rad/s², although the radian is not a true unit (Chapter 8).

The form of rotational inertia To find rotational inertia, consider the force it takes to start moving a mass fixed to the end of rod that is free to rotate (Figure 9.11). The acceleration of the mass is given by $a = F \div m$. To put the equation into rotational motion variables, the force is replaced by the torque about the center of rotation ($F = \tau \div r$). The linear acceleration is replaced by the angular acceleration ($a = \alpha r$). The resulting formula has a quantity mr^2 that connects the torque and angular acceleration in the exact same way that mass connects force and linear acceleration. This quantity (mr^2) is therefore the rotational inertia.

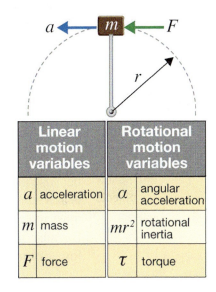

Figure 9.11: *A rotating mass on a rod can be described with variables from linear or rotational motion.*

CHAPTER 9 TORQUE AND ROTATION

The moment of inertia

Rotational inertia of solid objects

The product of mass and radius squared (mr^2) is the rotational inertia for a point mass where r is measured from the axis of rotation. Because of the power of 2 in mr^2, mass that is farther from the axis of rotation has *much* more rotational inertia than mass close to the axis. For example, a 1-kilogram mass on a 2-meter rod has *four times* the rotational inertia of the same mass on a 1-meter rod. That means it takes four times as much torque to produce the same angular acceleration.

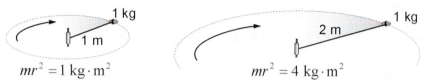

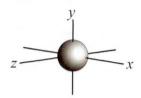

Sphere of mass, m, radius, r

Hoop of mass, m, radius, r

Bar of mass, m, length, l

The moment of inertia

A solid object contains mass distributed at different distances from the center of rotation. Because rotational inertia depends on the *square* of the radius, the distribution of mass makes a big difference for solid objects, such as steel balls. The sum of mr^2 for all the particles of mass in a solid is called the **moment of inertia** (I). Figure 9.12 shows the moment of inertia for some simple solid shapes when rotated about an axis passing through the center. Notice the moment of inertia of the same shape is different depending on which rotation axis is chosen. In Chapters 11 and 12, you will see the moment of inertia used instead of mass in many calculations involving rotational motion, including kinetic energy and angular momentum.

Comparing solid and hollow objects

A hoop has a greater moment of inertia than a solid disk of the same mass. This is because all the mass of a hoop is at a large radius. Some of the mass of a solid disk is at a smaller radius and therefore contributes less rotational inertia. In fact, the mass at the center contributes zero rotational inertia since its radius is zero. Because some mass must be near the center of rotation, a solid object always has a lower moment of inertia than a hollow object *of the same mass*.

Moment of inertia			
	Axis of rotation		
	x	y	z
Sphere	$\frac{2}{5}mr^2$	$\frac{2}{5}mr^2$	$\frac{2}{5}mr^2$
Hoop	$\frac{1}{2}mr^2$	mr^2	$\frac{1}{2}mr^2$
Bar		$\frac{1}{12}ml^2$	$\frac{1}{12}ml^2$

Figure 9.12: *The moment of inertia of some simple shapes rotated around axes that pass through their centers.*

Rotational inertia depends on the axis

The axis about which an object rotates affects its moment of inertia. If an object rotates about an axis that does not pass through its center of mass, it usually has a greater moment of inertia. You can feel this for yourself using the meter stick with the masses on it. No matter where you place the masses, it will always be easier to spin the meter stick when your hand is *between* the masses.

Rotation and Newton's second law

Angular acceleration of a wheel When a torque is applied to an object, it spins in the direction of the applied torque (Figure 9.13). Its angular speed increases at a rate directly proportional to the net torque and inversely proportional to the object's moment of inertia. For example, the moment of inertia of a bicycle wheel is about 0.1 kg·m². When the brakes are applied, a force of 100 newtons is applied at the rim of the wheel, at a radius of 0.35 m. The torque produced by this force is 35 N·m. The angular acceleration is the torque divided by the moment of inertia, or 350 rad/s² (35 N·m ÷ 0.1 kg·m²). The angular speed of the wheel increases by 350 radians per second each second the torque is applied.

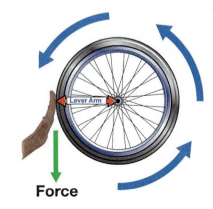

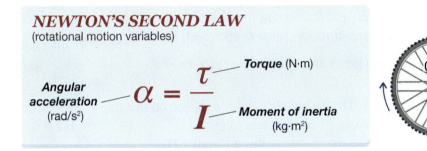

Figure 9.13: *If you apply a torque to a wheel, it will spin in the direction of the torque. The greater the torque, the greater the angular acceleration.*

New variables in familiar relationships Although rotational motion has a completely new set of variables, the relationships are similar. Force causes linear acceleration, torque causes angular acceleration. Mass is a measure of linear inertia. The moment of inertia (*I*) is a measure of rotational inertia.

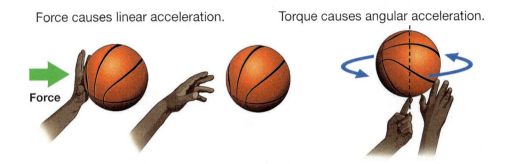

9.3 ROTATIONAL INERTIA   193

Chapter 9 Connection

Bicycle Physics

A modern bicycle is the most efficient machine ever invented for turning human muscle power into motion. Bicycles work by a series of transformations from forces to torques and back. By changing the radius of a bicycle's gears, the rider can choose different ratios between the force applied to the pedals and the force applied to the road.

Structures and functions of the drive system

The drive system of a multispeed bicycle has four major parts (Figure 9.14).

1. The crank and pedals are where force is applied.
2. The chain transmits the force to the rear wheel.
3. A set of gears called a freewheel transfers the force from the chain to the back wheel of the bicycle. The back wheel transmits the force to the road.
4. A *derailleur* allows the chain to switch gears and change the force ratio between the pedals and the road.

Figure 9.14: *The components of a bicycle drive system.*

Force and torque on the crank

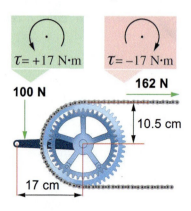

A force applied to the pedals creates a torque on the crank. A crank is 17 centimeters long and a 100-N force on the pedal creates a torque of 17 N·m. The crank on a road bicycle usually has two gears called chain wheels. The larger chain wheel has 52 teeth and a diameter of 10.5 cm. When the crank is in rotational equilibrium, the force on the chain is 162 N. The force is multiplied because the chain wheel has a smaller radius and it takes a larger force to balance the torque from the 100 N force applied to the pedals.

Force and torque on the rear wheel

The 162-newton force creates a torque on the rear wheel that depends on the gear the chain is on. The small gear on a freewheel has 14 teeth and a radius of 2.8 cm. The torque applied to the rear wheel is then 4.54 N·m, which is 162 N × 0.028 m. The torque applied by the chain is balanced by the torque created by the force of the road pushing back on the bicycle. The radius of the wheel is 35 centimeters. The force applied by the road is therefore 4.5 N·m ÷ 0.35 m = 13 N. This is the force that moves the bicycle forward (Figure 9.15).

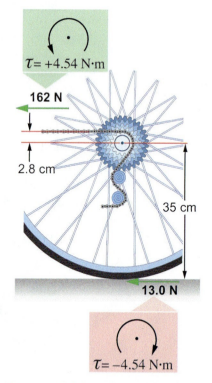

Figure 9.15: *Torques and forces on the rear wheel.*

Chapter 9 Connection

Changing gears

Force and torque with the small chain wheel

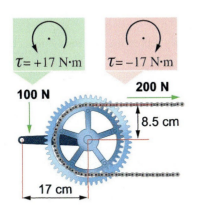

The front and rear derailleurs make it easy to move the chain onto different gears. The different gears change the ratio of forces and torques on the crank and the rear wheel. For example, the rider can select the smaller chain wheel, which has a radius of 8.5 cm. When 100 N is applied to the pedals, the force on the chain is now 200 N, compared with 162 N with the chain on the larger chain wheel. The force is greater because the torque applied by the chain on the crank must still balance the torque from the pedals (17 N·m), but at a smaller radius.

Force and torque on the rear wheel

Suppose the rider uses the derailleur to select the largest gear in the freewheel. This gear typically has 32 teeth and a radius of 6.5 cm. The 200-N force from the chain creates a torque of 13 N·m on the rear wheel (Figure 9.16). This results in a force on the ground of 37.1 N, almost three times larger than the force of 13 N obtained with the other combination of gears.

High gear

With the combination of the large chain wheel and small freewheel gear, a 100-N force applied to the pedals creates a 13-N force on the road. This is a relatively small force compared with the input force of 100 N. The advantage is that the rear wheel turns four times (52 teeth ÷ 13 teeth) for every turn of the pedals. This combination is called high gear. High gear is used for high speeds on level ground or going downhill because each turn of the pedals moves the bicycle forward four times the circumference of the wheel.

Low gear

The combination of the small (42-tooth) chain wheel and large (32-tooth) freewheel gear is called *low gear*. In low gear, the 100-N force on the pedals creates a 37.1-N force on the ground. The trade-off is in speed. One turn of the pedals in low gear turns the rear wheel only 1.3 times (42 teeth ÷ 32 teeth). Low gear is used for high-force, low-speed riding, such as climbing hills or starting from a stop.

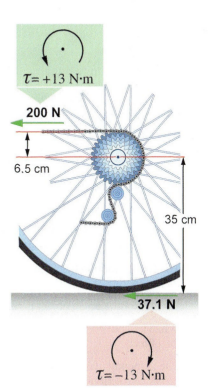

Figure 9.16: *Forces and torques on the rear wheel when the chain is on the 32-tooth gear of the freewheel.*

Chapter 9 Assessment

Vocabulary

Select the correct term to complete the sentences.

torque	center of mass	angular acceleration
rotational inertia	rotation	translation
center of rotation	rotational equilibrium	lever arm
center of gravity	moment of inertia	line of action

1. The point or line around which an object turns is called its _____.
2. An action that causes an object to rotate is known as _____.
3. The imaginary line that follows the direction of a force and passes through its center of application is called the _____.
4. The type of motion that results when all the particles in a body move parallel to one another is called _____.
5. The net torque on an object is zero when the object is in _____.
6. When an applied force causes an object to spin it creates _____ of the object.
7. The perpendicular distance between the line of action of the force and the center of rotation is called the _____.
8. The point at which the three axes of rotation intersect is referred to as the _____.
9. The average position of a body's weight is its _____.
10. The rate at which the angular speed changes is the _____.
11. The rotational inertia of an object can also be called its _____.
12. The resistance to change in the rotational motion of an object is the _____ of the object.

Concept review

1. Identify the quantity represented by the symbols below.
   a. α
   b. τ
   c. r
   d. mr^2
   e. F
   f. a

2. Supply the missing character in the blank of the following equations.
   a. $\alpha = \tau / $ _____
   b. $\tau = $ _____ $\times F$
   c. $a = \alpha$ _____

3. How are torque and force similar? How are they different?

4. What is the difference between translation and rotation? What determines which type of motion an object will experience?

5. Why does a long-handled wrench make it easier to loosen a bolt?

6. You use torque to move when you pedal a bike. How does the torque provided by your feet when they are in a vertical line compare to the torque when one foot is in front of the other? Explain.

7. Which force will create the greatest amount of torque on the shovel?

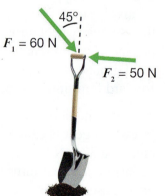

8. If the see-saw is balanced, which person has the greater weight? How do you know this?

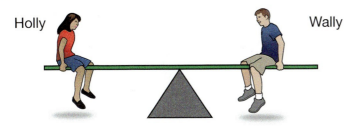

Holly   Wally

9. What does it mean to say an object is in rotational equilibrium?
10. When is it okay to use the terms center of gravity and center of mass interchangeably?
11. Choose the letter of the point that is the center of mass of each object.

12. Tightrope walkers often use long poles to help them balance. Explain why this makes sense.
13. If you spin an egg in its shell, stop it for an instant, and then release it, it will start spinning again. Explain why this happens. Will the same thing happen if the egg is hard boiled? Why or why not?
14. Explain how you can experimentally locate an object's center of gravity.
15. Which two factors determine an object's rotational inertia?
16. Which would be easier to spin, a 3-kg bowling ball or a 3-kg barbell? Why?
17. How are torque, rotational inertia, and angular acceleration related?
18. Each object's center of gravity is marked. Which object(s) will topple?

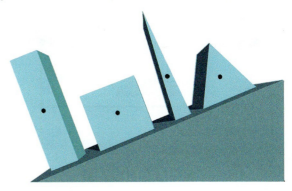

Problems

1. By holding on to his fishing pole at 1.8 meters from the tip, Toby lifts a 20-N bass over the side of his boat while holding the pole at an angle of 37° to the horizontal. How much torque must be applied at the end of the pole to support the fish as it is lifted into the boat?

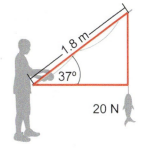

2. You use a wrench to loosen a bolt. It finally turns when you apply 250 N of force perpendicular to the wrench at a distance of 0.3 m from the bolt. Calculate the torque.

3. A carousel at an amusement park has a total moment of inertia of 60,000 kg·m^2 when children are sitting on all of its horses. When it starts spinning, the motor supplies a torque of 15,000 N·m. What is the resulting angular acceleration?

4. A torque of 60 N·m can be created by a 30-N force acting perpendicular to a 2-m lever arm. How can a 15-N force be used to create the same torque?

5. How far from the center of the see-saw should Helena sit if she wants to balance Lindsey's torque?

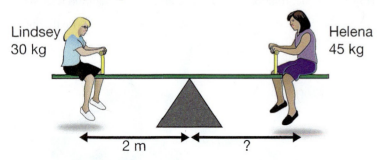

6. What is the unknown mass if the rod is balanced? Assume the rod's mass is negligible when compared with the mass of the blocks.

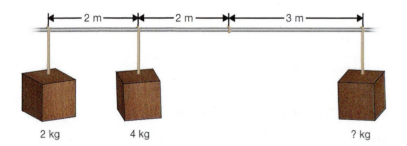

7. Calculate the moment of inertia of each of the following objects rotated about an axis through its center of mass.
   a. a solid rubber ball with a radius of 0.10 m and a mass of 2 kg
   b. a hollow rubber ball with a radius of 0.15 m and a mass of 1 kg
   c. a meter stick with a mass of 0.3 kg
   d. a ring with a radius of 0.25 m and a mass of 4 kg

8. What torque is required to spin a wheel with a moment of inertia of 2.5 kg·m² at an angular acceleration of 5 rad/s²?

9. Ed wants to determine the moment of inertia of the propeller on his helicopter. He applies a torque of 3,000 N·m and measures the propeller's angular acceleration to be 10 rad/s². What is the propeller's moment of inertia?

10. Calculate the net torque applied to the wheel by the three forces represented by the diagram.

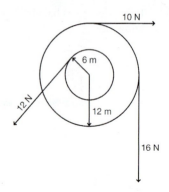

Applying your knowledge

1. A firecracker is thrown into the air and explodes at the highest point in its path. Compare the path of the firecracker before and after the explosion.

2. Name four commonly-used devices that increase torque.

3. Why do long-legged animals such as giraffes run with a slower gait than short-legged animals such as mice?

4. If you stand with your back and heels against a wall you fall forward when you bend over in an attempt to touch your toes. Explain why.

5. A solid sphere and a hoop of the same mass and diameter are rolled down an incline. Which one accelerates at a higher rate down the incline? Why?

6. Research the "Fosbury Flop." How did this technique revolutionize the high jump?

7. Experiment with cans filled with solid material and cans filled with liquid material. Which rolls down an incline faster? Explain your results.

8. A string is wrapped around a spool as pictured. It rests on a surface at point 0.
   a. If the string is pulled at Position 1, which way will the spool roll? Why?
   b. If the string is pulled from vertical Position 2, which way will the spool roll? Why?
   c. Is there a direction in which the string can be pulled that will not cause any rolling?

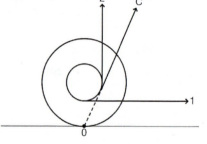

UNIT 4 ENERGY AND MOMENTUM

CHAPTER 10

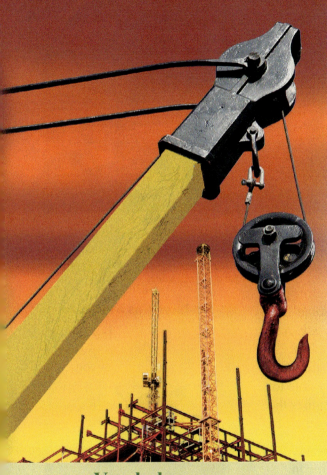

Work and Energy

Objectives:

By the end of this chapter you should be able to:

- ✔ Analyze a simple machine in terms of input force, output force, and mechanical advantage.
- ✔ Calculate the mechanical advantage for a lever or rope and pulleys.
- ✔ Calculate the work done in joules for situations involving force and distance.
- ✔ Give examples of energy and transformation of energy from one form to another.
- ✔ Calculate potential and kinetic energy.
- ✔ Apply the law of energy conservation to systems involving potential and kinetic energy.

Key Questions:

- What are simple machines, and how do they manipulate forces?
- What is the specific meaning of "work" in physics?
- What is energy, and what law governs energy transformations?

Vocabulary

chemical energy	input	machine	output	ramp
closed system	input arm	mechanical advantage	output arm	rope and pulley
law of conservation of energy	input force	mechanical energy	output force	screw
electrical energy	joule	mechanical system	potential energy	simple machine
fulcrum	kinetic energy	nuclear energy	pressure energy	thermal energy
gears	lever		radiant energy	work

CHAPTER 10 WORK AND ENERGY

10.1 Machines and Mechanical Advantage

The human body is capable of exerting forces up to a few times its own weight. How did ancient people move huge stones to build monuments like the Great Pyramid of Giza long before the invention of trucks and engines? The answer is that people are ingenious, and the ancient builders developed simple machines that allowed them to multiply by many times the force from their muscles. In this section, you will learn how simple machines manipulate forces to accomplish useful tasks.

Machines

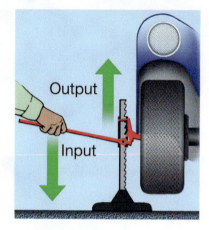

Figure 10.1: *A jack is a simple machine that allows a single person to lift a car with only the force of muscles.*

Machines and mechanical systems
The ability of humans to build buildings and move mountains began with our invention of **machines**. A machine is a device that is created by humans to do a task using force and motion. A screwdriver is a very simple example of a machine created to turn screws. A bicycle is a more complex machine. All the parts of a bicycle work together to transform forces from your muscles into speed and motion. The structure of any machine is designed for a specific function.

A simple machine you should know how to use
Suppose you must lift a 2,000-kilogram automobile so you can change a tire. The road is deserted and there is no one to help you. Fortunately, you know some practical physics, and you remember that inside the trunk of the car is a *jack*. A jack is an example of a **simple machine**. In physics, the term *simple machine* means a machine that uses only the forces directly applied and accomplishes its task with a single motion. You attach the jack under the car and are easily able to lift the car yourself, and with only one hand (Figure 10.1). The jack multiplies the force from your arm by many times. The force from your arm is transformed into a much larger force, capable of lifting a car.

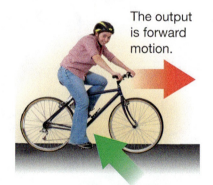

Figure 10.2: *The bicycle applies the ideas of input and output of force. The bicycle is a very efficient machine.*

The concepts of input and output
The best way to analyze what a machine does is to think about the machine in terms of **input** and **output**. The *input* includes everything you do to make a machine work. For a bicycle, one input is force applied to the pedals. The pedals also move and their motion is another input to the bicycle machine. Another input pair of force and motion are applied to the handlebar to steer. The *output* is what the machine does for you. The output of a bicycle is force applied to the road that makes the bicycle accelerate or overcome friction (Figure 10.2). The bicycle also moves forward and its motion is another output. All machines can be described in terms of input and output.

200 UNIT 4 ENERGY AND MOMENTUM

Mechanical advantage

Input and output forces A simple machine is analyzed in terms of forces applied *to* the machine and forces applied *by* the machine. A lever is a good example of a simple machine. The jack described on the preceding page is an example of a lever. In order to lift the car you apply a force to the arm of the jack (Figure 10.3). The force you apply is called the **input force**. The arm of the jack rotates around a pin and pushes a hook that lifts the car. The **output force** of the lever is the force that pushes on the hook.

Mechanical advantage **Mechanical advantage** is the ratio of output force to input force. If the mechanical advantage is bigger than one, the output force is bigger than the input force (Figure 10.3). A mechanical advantage smaller than one means the output force is smaller than the input force. For a typical automotive jack, the mechanical advantage is 30 or more. For a mechanical advantage of 30, a force of 100 newtons (22.5 pounds) applied to the input arm of the jack produces an output force of 3,000 newtons (675 pounds)—enough to lift one corner of an automobile.

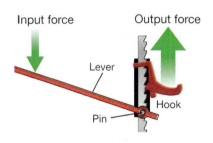

Figure 10.3: *A jack is an example of a lever. The input force is applied to the lever. The output force is applied by the lever to lift the car.*

MECHANICAL ADVANTAGE

Mechanical advantage $\quad MA = \dfrac{F_o}{F_i} \quad$ — Output force (N)
— Input force (N)

How mechanical advantage is created If you use a jack to lift a car, you will notice that you have to move the arm of the jack a lot to raise the car only a little. Machines create mechanical advantage by trading off between force and distance. On the input of the jack, a small force has to move a large distance. On the output of the jack, a much larger force moves only a small distance. Later in the chapter you will see that the inverse relationship between force and distance is characteristic of all simple machines, and is due to a powerful natural law in physics, the law of conservation of energy.

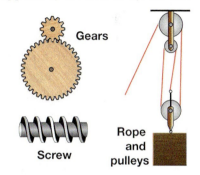

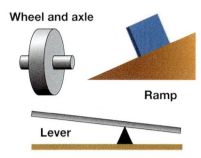

Figure 10.4: *The basic types of simple machines.*

Types of simple machines There are a few basic kinds of simple machines that create mechanical advantage. The lever, wheel and axle, rope and pulleys, screw, ramp, and gears are the most common types (Figure 10.4). Complex machines, like a bicycle, combine many simple machines into mechanical systems. A **mechanical system** is an assembly of simple machines that work together to accomplish a task.

The mechanical advantage of a lever

Example of a lever
A simple lever is a board balanced on a log (Figure 10.5). The board can rotate around the log. Pushing down on one end of the board creates an upward force on the other end. If the board is used to lift a rock, the force you apply is the input force. The force that lifts the rock is the output force.

Analyzing a lever
The essential features of a **lever** are the input arm, output arm, and fulcrum. The **fulcrum** is the point about which the lever rotates. The **input arm** is the distance between the input force and the fulcrum. The **output arm** is the distance between the output force and the fulcrum. The lower diagram in Figure 10.5 shows how to find the input and output arms and the fulcrum for the board-and-log lever.

The mechanical advantage of a lever
When the input arm is longer than the output arm, the output force is greater than the input force. If the input arm is 10 times longer than the output arm, then the output force will be 10 times greater than the input force. The mechanical advantage of a lever is therefore a ratio, the length of the input arm divided by the length of the output arm.

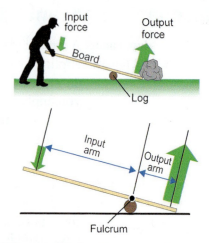

Figure 10.5: *A board and log used as a lever to move a rock.*

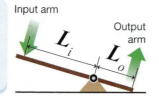

MECHANICAL ADVANTAGE OF A LEVER

Mechanical advantage $MA_{lever} = \dfrac{L_i}{L_o}$ — Length of input arm (m) / Length of output arm (m)

The three types of levers
There are three basic types of levers, as shown in Figure 10.6. They are classified by the location of the input and output forces relative to the fulcrum. All three types are used in many machines and follow the same basic rules.

The output force can be *less* than the input force
You can make a lever where the output force is *less* than the input force. The input arm is shorter than the output arm on this kind of lever. The human arm is a good example. Compare where the bicep connects to the arm (input arm) with the location of the weight (output arm) in Figure 10.6 on third-class levers. If it seems inefficient that the biceps muscle has to make such a large force, consider that the biceps contracts a few centimeters while your hand can move almost a meter. Evolution has shaped the human arm to provide a large range of motion, at a cost of requiring strong muscles.

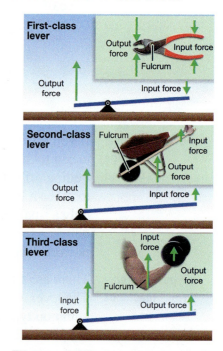

Figure 10.6: *The three classes of levers.*

How a lever works

Trading force for distance When the mechanical advantage is larger than one, the input arm of a lever moves a much larger distance than the output arm. In fact, if the mechanical advantage is three, the input arm moves three times more than the output arm (Figure 10.7). In return, the output force is three times greater than the input force. The ratio of output motion to input motion is the *inverse* of the ratio of output force to input force. This principle is true for all simple machines. Force is multiplied by trading larger motions for smaller motions.

Torque and the mechanical advantage of a lever A lever works by rotating about its fulcrum. The mechanical advantage can be deduced by calculating the *torques* created by the input and output forces. The input force creates a counterclockwise (positive) torque. The output force on the rock creates a reaction force on the lever. The torque created by the reaction force is clockwise (negative). When the lever is in equilibrium the net torque must be zero. The mechanical advantage of a lever follows directly from setting the net torque to zero (Figure 10.8).

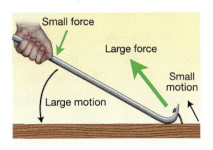

Figure 10.7: *A crowbar pulling a nail is an example of a lever. A small input force acts over a large motion to create a large output force acting over a small motion.*

Calculating the position of the fulcrum for a lever

Where should the fulcrum of a lever be placed so one person weighing 700 newtons can lift the edge of a stone block with a mass of 500 kilograms? The lever is a steel bar three meters long. Assume a person can produce an input force equal to their own weight. Assume that the output force of the lever must equal half the weight of the block to lift one edge.

1. You are asked to figure out the location of the fulcrum.
2. You are given the input force, length of the lever, and the mass to be lifted.
3. The weight of an object is equal to its mass times 9.8 N/kg. The mechanical advantage of a lever is the ratio of length of the input arm divided by length of the output arm.
4. Solve the problem:

   Output force = $F_o = \left(\frac{1}{2}\right)(500 \text{ kg})(9.8 \text{ N/kg}) = 2{,}450 \text{ N}$

   The required mechanical advantage is MA = $\frac{F_o}{F_i}$ = (2,450 N)/(700 N) = 3.5

   The mechanical advantage of a lever is the ratio of lengths: MA = $\frac{L_i}{L_o}$ = 3.5.
   From this equation you know that $L_i = 3.5\, L_o$.
   The total length of the lever is 3 meters; that means $L_i + L_o = 3$ m.
   Substitute for the length of the input arm to solve for the length of the output arm.
   $(3.5\, L_o) + L_o = 3$ m; $4.5\, L_o = 3$ m
   $L_o = 0.67$ m, the fulcrum should be placed 0.67 meters from the edge of the block.

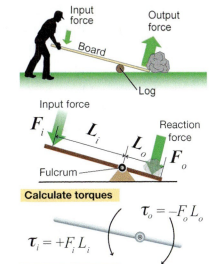

Figure 10.8: *Balancing the torques acting around the fulcrum of a lever.*

Mechanical advantage of ropes and pulleys

The forces in ropes and strings — A tension force is a pulling force acting along the direction of a rope or string. Ropes and strings carry tension forces throughout their length. If friction is small, that means the tension force in a rope is the same everywhere. If you were to cut a rope in tension and insert a force scale, the scale would measure the same force at any point.

The forces in ropes and pulleys — Figure 10.9 shows three different arrangements of **rope and pulley** machines. Each arrangement uses a different number of pulleys and has a different number of strands of rope supporting the load. Any input force applied to the rope is felt in *every part of the rope*. As a result, in case A the load feels two upward forces equal to the input force. In case B the load feels three times the input force, and in case C the load feels four times the input force.

Mechanical advantage of ropes and pulleys — If there are four strands of rope directly supporting the load, each newton of force you apply produces 4 newtons of output force. Therefore, arrangement C has a mechanical advantage of 4. The output force is four times greater than the input force. Because the mechanical advantage is 4, the input force for machine C is one-fourth the output force. If you need an output force of 20 N, you only need an input force of 5 N.

Trading force for motion — If a rope and pulley machine has a mechanical advantage of four, then the input rope must be pulled 4 meters for every meter the load is lifted. Four strands of rope support the load to get a mechanical advantage of four. When the load rises by 1 meter, each strand must be shortened by a meter, resulting in a total pull of 4 meters. Like all simple machines, ropes and pulleys obey the rule by trading off advantage in force for reduction in motion.

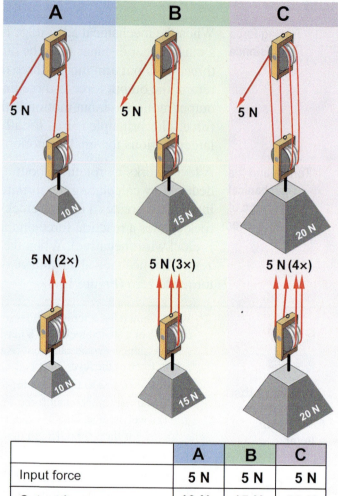

	A	B	C
Input force	5 N	5 N	5 N
Output force	10 N	15 N	20 N
Mechanical advantage	2	3	4

Figure 10.9: *The block-and-tackle machine is a simple machine using one rope and multiple pulleys. The rope and pulleys can be arranged to create different amounts of mechanical advantage depending on how many strands of rope support the load being lifted.*

Wheels, gears, and rotating machines

The wheel and axle To understand how a wheel works, consider dragging a sled that holds a heavy weight. You must overcome the entire force of friction to pull the sled. If the sled is on wheels, the friction force occurs between the wheel and axle (Figure 10.10). The radius of the axle is much smaller than the radius of the wheel. As a result, the force required to pull the load is reduced by the ratio of the axle radius to the wheel radius. If the axle is 10 times smaller than the wheel, this results in a reduction of 10 times in force.

Other advantages of the wheel and axle There are other ways a wheel and axle provide advantages. Friction occurs mostly where the wheel and axle touch. This area can be sealed against contamination from dirt and lubricated with grease to reduce friction. A second advantage is that rolling friction creates less resistance than sliding friction. Rolling motion creates less wearing away of material compared with two surfaces sliding over each other.

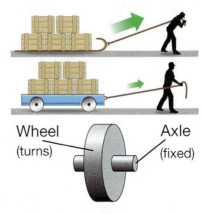

Figure 10.10: *It takes more force to drag a sled than to pull a cart with wheels and axles.*

Gears change torque and rotating speed Some machinery, such as small drills, requires low torque at high rotating speeds. Other machinery, such as mill wheels, requires high torque at low rotating speeds. Since they act like rotating levers, **gears** allow the torque carried by different axles to be changed along with the rotational speed. Gears are better than wheels because they have teeth and don't slip as they turn together. Two gears with their teeth engaged act like two touching wheels (Figure 10.11).

The gear ratio The *input gear* is the one you turn, or apply forces to. The *output gear* is connected to the output of the machine. The *gear ratio* is the ratio of output turns to input turns. Because gear teeth don't slip, moving 36 teeth on one gear means that 36 teeth have to move on any connected gear. If one gear has 36 teeth, it turns once to move 36 teeth. If the connected gear has only 12 teeth, it has to turn three times to move 36 teeth (3 × 12 = 36). Smaller gears turn faster than larger gears because they have fewer teeth.

Mechanical advantage of gears For a rotating machine, the mechanical advantage is the ratio of output torque to input torque. With gears, the trade-off is made between torque and rotation speed. An output gear will turn with *more* torque when it rotates slower than the input gear. The mechanical advantage of a pair of gears is therefore the *inverse* of the gear ratio (Figure 10.11).

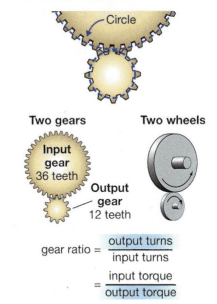

Figure 10.11: *Gears act like touching wheels, but with teeth to keep them from slipping as they turn together.*

10.1 MACHINES AND MECHANICAL ADVANTAGE

Ramps and screws

Ramps It is easier to push a heavy cart up a ramp than it is to lift the same load straight up. **Ramps** reduce input force by increasing the distance over which the input force needs to act. For example, suppose a 10-meter ramp is used to elevate a cart one meter. If the weight of the cart is 1,000 newtons, then the force required to pull it up the ramp is one-tenth that weight, or 100 newtons (Figure 10.12).

Mechanical advantage of a ramp When friction is negligible, the mechanical advantage of a ramp is equal to the distance along the ramp divided by the height of the ramp. In theory, the ramp in Figure 10.12 has a mechanical advantage of 10. Of course, with practical ramps there is always friction. And because of friction, the practical mechanical advantage is always less that the theoretical value. Friction also places limits on how a mechanical advantage can be achieved. A ramp with a mechanical advantage of 100 would probably not work as expected because the force expended to overcome friction would be greater than the force saved by making a ramp with such a small angle. The design of a ramp is a balance between mechanical advantage and friction.

Screws A **screw** is a simple machine that turns rotating motion into linear motion. A thread wraps around a screw at an angle, like the angle of a ramp (Figure 10.13). In fact, the analysis of how a screw works treats a screw thread just like a rotating ramp. Imagine unwrapping one turn of a thread to make a ramp. Each turn of the screw advances the nut the same distance it would have gone sliding up a ramp. The lead of a screw is the distance it advances in one turn. A screw with a lead of one millimeter would advance one millimeter each turn.

Mechanical advantage of a screw Screws are used to hold things together because the combination of a screw and a lever (wrench) have tremendous mechanical advantage. By itself, the mechanical advantage of a screw is similar to the mechanical advantage of a ramp. The vertical distance is the lead of the screw. The distance along the ramp is measured along the average circumference of the thread. A quarter-inch screw you find in a hardware store has a lead of 1.2 millimeters and a circumference of 17 millimeters along the thread. In theory, the mechanical advantage is 14. In real machines, the mechanical advantage is less because of friction.

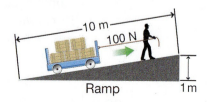
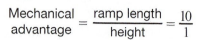

Figure 10.12: *The mechanical advantage of a ramp.*

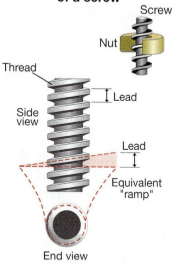

Figure 10.13: *A screw works like a rotating ramp.*

10.2 Work

All simple machines obey a rule that says any advantage in force must be compensated by applying the force over a proportionally longer distance. This rule is an example of one of the most powerful laws in all of physics. The law involves the physics meaning of *work*, which is the subject of this section.

Work in physics

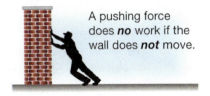

A pushing force does **no** work if the wall does **not** move.

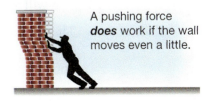

A pushing force **does** work if the wall moves even a little.

The meaning of *work* — The word *work* is used in many different ways.

- You *work* on science problems.
- You go to *work*.
- Your toaster doesn't *work*.
- Taking out the trash is too much *work*.

What *work* means in physics — In physics, **work** has a very specific meaning that is *different* from any of the meanings listed above. In physics, work is done by forces. The amount of work done is equal to the force times the distance over which the force acts. When you see the word *work* in a physics problem, it means the quantity you get by multiplying force and distance.

Figure 10.14: *Work is done (in the physics sense) only if a force causes movement.*

Work involves change — In physics, work represents a measurable change in a system, caused by a force. If you push on a wall and it does not move, when you are done pushing, the system is exactly the same as it was. Your force created no change, and therefore did no work (Figure 10.14). In reality, the wall moves a little as you push. But, when you release your push, the wall moves back, doing work on you! Over the whole process, the *net* (total) work done is zero. If a system finishes in exactly the same state it started, then the work exchanged is zero.

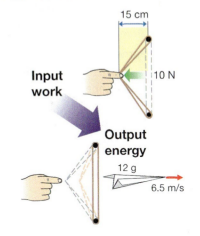

Figure 10.15: *The relationship between work and energy.*

Work and energy — The definition of work will make more sense when you see how work fits into the big picture of energy and systems. Work is one way systems change the amount of energy they have. When work is *done on* a system, its energy increases. For example, if you stretch a rubber band by pulling on it (doing work) you *increase* the energy of the rubber band. When a system *does* work, the energy of the system decreases. For example, if the stretched rubber band is used to launch a paper airplane, its energy is spent by doing work on the plane (Figure 10.15).

10.2 WORK **207**

CHAPTER 10   WORK AND ENERGY

The work done by a force

Work is done by forces on objects

In physics, work is done *by* forces. When thinking about work, you should always be clear about which force is doing the work. Work is done *on* objects. If you push a block 1 meter with a force of 1 newton, you have done 1 joule of work *on the block*. It is necessary to keep careful track of where the work goes because later you will see that it may be possible to get some or all of the work back.

Units of work

The unit of measurement for work is the **joule**. You may recognize this unit from Chapter 1 as the unit of *energy*. Work is a form of energy. One joule of work is done by a force of 1 newton acting over a distance of 1 meter. If you push a box with a force of 1 newton for a distance of 1 meter, you have done exactly 1 joule of work (Figure 10.16).

Force parallel to the distance

In general, the work done by a force is equal to the force times the distance moved *in the direction of the force*. When the force and distance are in the same direction (parallel), you can calculate the work in joules by multiplying the force in newtons by the distance in meters.

Force at an angle to the distance

When the force and distance are at an angle, only part of the force does work. For example, suppose a force was directed along an angle making a triangle with sides of lengths 3, 4, and 5. The fraction of the force that does work is four-fifths of the total force, because the distance is aligned with the 4 side of the triangle and the force is aligned with the 5 side. The work done by the force is only four-fifths of what it would have been if the force and distance were parallel.

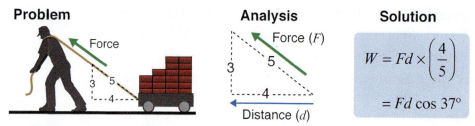

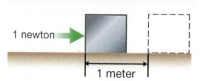

1 joule is the amount of work done by a force of 1 newton acting over a distance of 1 meter.

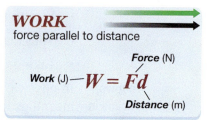

WORK force parallel to distance

Work (J) — $W = Fd$ — Force (N), Distance (m)

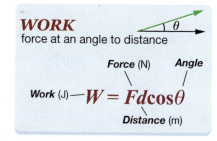

WORK force at an angle to distance

Work (J) — $W = Fd\cos\theta$ — Force (N), Angle, Distance (m)

Figure 10.16: *The definition of work and how to calculate work if you know the force and distance.*

Work and vectors

From Chapter 7, you might recognize the ratio $\frac{4}{5}$ is the same as the cosine of the angle (37°) between the force and the distance. In general, both force and distance are vectors. To calculate the work done, you multiply the magnitudes of the force and distance vectors, then multiply again by the cosine of the angle between them.

Work done against gravity

Work against gravity depends only on height

Many situations involve work done by or against the force of gravity. When you lift something off the floor, you are doing work against gravity. Because gravity always pulls straight down, the work done is easy to calculate because *it does not matter what path you take*. The work done by or against gravity is equal to the weight of the object (force) times the change in height (distance). Whether you take a zig-zag stairway or an elevator, the work done is the same.

Why the path does not matter

This remarkable fact results from another way of looking at work for motion at an angle to a force. The work done is the force times the *distance moved in the direction of the force*. This is mathematically the same as our previous definition. To understand what it means, consider climbing up a zig-zag stair. The distance moved is always along the stair. The force of gravity is always down. The distance moved *along the force of gravity* is just the height of each leg of the stair. The work done against gravity on each section of the stair is the height of that section times your weight. The total for the zig-zag path adds up to the same amount of work you would have done by jumping straight up—if you could!

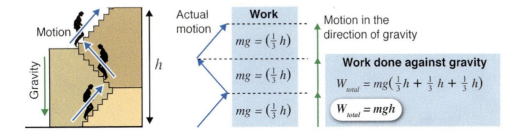

The difference between force and work

Going up a stair or ramp is certainly easier than climbing straight up. Stairs and ramps are easier because you need less force. But you do have to apply the force for a much longer distance. In the end, the amount of work done is the same. Of course if there is friction, *you* might have to do more work on a stair or ramp because you have to overcome friction. Work done *against* friction, however, is not the same as work done on *you*. Remember what we said on the last page about keeping track of work done by which force on which object.

Calculating work done against gravity

A crane lifts a steel beam with a mass of 1,500 kg. Calculate how much work is done against gravity if the beam is lifted 50 meters in the air. How much time does it take to lift the beam if the motor of the crane can do 10,000 joules of work per second?

1. You are asked for the work and the time it takes to do the work.
2. You are given mass, height, and the work done per second.
3. Use the formula for work done against gravity, $W = mgh$.
4. Solve:
   $W = (1{,}500 \text{ kg})(9.8 \text{ N/kg})(50 \text{ m})$
   $= 735{,}000 \text{ J}$

At a rate of 10,000 J/s, it takes $735{,}000 \div 10{,}000 = 73.5$ seconds to lift the beam.

CHAPTER 10 WORK AND ENERGY

Work done by a machine

Work and machines
Work is usually done when a force is applied to a simple machine. For example, when a rope and pulley machine, also known as a block-and-tackle machine, lifts a heavy load, force is applied by pulling on the rope. As a result of the force, the load moves a distance upward. Work has been done by the machine because force was exerted over a distance.

Input work and output work
All machines can be described in terms of input work and output work. As an example, consider using the block-and-tackle machine to lift a load weighing 10 newtons. The load moves a distance of one-half meter. The machine has done 5 joules of work on the load (Figure 10.17), so the work output is 5 joules.

Calculating input work
What about the work input? The force on the rope is only 5 newtons because the machine has a mechanical advantage of two. But the rope must be pulled 1 meter in order to raise the load one-half meter. The input work is the force applied to the rope times the distance the rope was moved. This is 5 newtons times 1 meter, or 5 joules. The work input is the same as the work output.

Output work can never exceed input work
The example illustrates a rule that is *true for all machines*. You can *never* get more work out of a machine than you put into it. Nature does not give something for nothing. When you design a machine that multiplies force, you pay by having to apply the force over a greater distance. The force and distance are related by the amount of work. In a perfect (theoretical) machine, the output work is exactly equal to the input work.

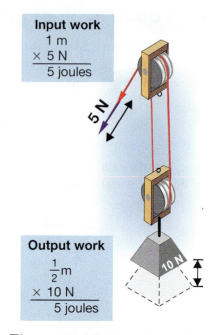

Figure 10.17: *For a frictionless rope and pulley machine, the work output equals the work input even though the output and input forces are different.*

Friction reduces output work
In a practical machine, there is always friction. When friction comes from motion and creates force, friction removes energy from the machine. In any machine, some of the input work goes to overcoming friction. The output work is always less than the input work because of the energy lost to friction.

Perpetual motion

A perpetual motion machine is a machine for which the work output equals or exceeds the work input. Many inventors have claimed to make one, and none has ever worked because the laws of physics make it impossible. The U.S. Patent and Trademark Office is always looking out for inventions claiming to be perpetual motion machines.

10.3 Energy and Conservation of Energy

Our universe is made of *matter* and *energy*. Matter is something that has mass and takes up space; you might call it "stuff." Energy describes the ability of a physical system to make things change. Energy appears in different forms, such as motion and heat. Energy can travel in different ways, such as light, sound, or electricity. The workings of the universe—including all of our technology—can be viewed as energy flowing from one place to another and changing back and forth from one form to another.

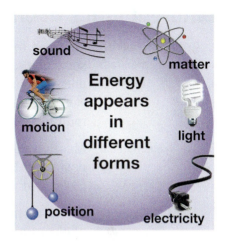

Figure 10.18: *Energy appears in many different forms.*

What is energy?

The definition of energy   Energy describes a system's ability to cause change. A system that has energy has the ability to do work. That means anything with energy can produce a force that is capable of acting over a distance. The force can be any force, and it can come from any source, such as your hand, the wind, or a spring (Figure 10.18).

1. A moving ball has energy because it can create force on whatever tries to stop it or slow it down.
2. A sled at the top of a hill has energy because it can move a distance down the hill and produces force as it goes.
3. The wind has energy because it can create forces on any object in its path.
4. Electricity has energy because it can turn a motor to make forces.
5. Gasoline has energy because it can be burned to make force in an engine.
6. You have energy because you can create forces.

Work and energy   Energy is measured in joules, the same units as work because work is the transfer of energy. Energy is the *ability* to make things change. Work is the *action* of making things change. Energy moves through the action of work. When you push a cart up a ramp, some of the energy is transferred to the cart by doing work (Figure 10.19). Your energy decreases by the amount of work done. The energy of the cart increases by the same amount of work (if there is no friction). Whenever work is done, energy moves from the system *doing* the work to the system on which the work is *being done*.

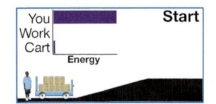

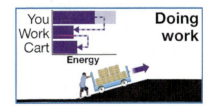

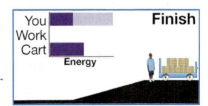

Figure 10.19: *Energy moves when work is done.*

Different forms of energy

How does energy manifest itself? One way to understand energy is to think of it as nature's money. Energy can be spent and saved in different ways any time you want to do something. You can use energy to buy speed, height, temperature, mass, and more. But you need energy to start with, and what you spend diminishes what you have left. Energy is found in multiple forms, and can be transformed into other types of energy by doing work.

Mechanical energy **Mechanical energy** is the energy possessed by an object due to its motion or its position. Mechanical energy can be either *kinetic* (energy of motion) or *potential* (energy of position).

Light energy **Radiant energy** includes light, microwaves, radio waves, X-rays, and other forms of electromagnetic waves (Chapters 18 and 26).

Nuclear energy **Nuclear energy** is energy contained in matter itself. Nuclear energy can be released when heavy atoms in matter are split up or light atoms are put together. Radioactivity also releases nuclear energy (Chapter 30).

Electrical energy **Electrical energy** is something we take for granted whenever we plug an appliance into an outlet (Chapter 20). The electrical energy we use is derived from other sources of energy. For example, the energy may start as chemical energy in gas. The gas is burned, releasing heat energy. The heat energy makes hot steam. The steam turns a turbine, making mechanical energy. Finally, the turbine turns an electric generator, producing electrical energy (Figure 10.20).

Chemical energy **Chemical energy** is energy stored in molecules. The chemical energy stored in batteries changes to electrical energy when you connect wires and a light bulb. Your body also uses chemical energy when it converts food into energy so that you can walk or think. Chemical reactions release chemical energy (Chapter 29).

Thermal energy Heat is an example of **thermal energy**. Thermal energy is energy that can be measured by differences in temperature. Hot objects have more thermal energy than cold objects (Chapter 25).

Pressure energy You may not have heard the term **pressure energy**. Pressure in gases and liquids is a form of energy (Chapter 27). It takes work to blow up a balloon. Some of the work is stored as energy in the form of higher-pressure air inside the balloon.

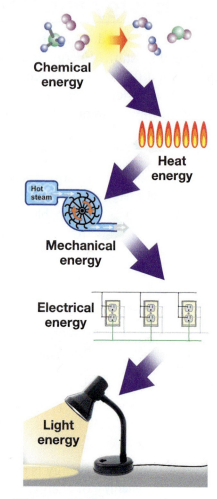

Figure 10.20: *Some of the forms energy takes on its way to your house or apartment.*

Potential energy

Potential energy When something is lifted off the ground, it can fall back down and exert force as it falls. Exerting force while falling means doing work. When you lift an object higher, you increase its **potential energy** because the higher an object is, the more ability it has to do work when it falls. Objects that have potential energy do not use the energy until they move. That is why it is called *potential* energy. Potential means that something is capable of becoming active. Any object that can move to a lower place has potential energy.

Potential energy comes from gravity An object's potential energy comes from the gravity of Earth. Consider a marble that is lifted off the table. Since Earth's gravity pulls the marble down, you must apply a force to lift it up. Applying a force over a distance requires doing work, which gets stored as the potential energy of the marble. Technically, energy from height is called *gravitational potential energy*. Other forms of potential energy also exist, such as energy stored in springs.

Calculating potential energy The potential energy an object has represents the amount of work the object can do by changing its height. In our discussion of work and machines, you may remember that the work you get out of a machine can never exceed the work you put in. The same is true of potential energy. In fact, the potential energy an object can release coming down is exactly the same as the work you must put in to move the object upward in the first place. For an object of mass (m) raised a height (h), the work done against gravity equals mgh. This is also the formula for calculating the potential energy!

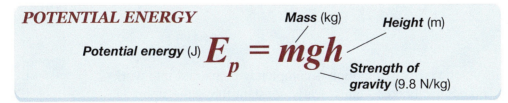

POTENTIAL ENERGY

Potential energy (J) E_p = mgh Mass (kg), Height (m), Strength of gravity (9.8 N/kg)

Energy from work There is a symmetry between work and energy that appears throughout physics. You can only do as much work as you have energy. And the energy you have is equal to the work done to create the energy in the first place.

Calculating the potential energy of a cart

A cart with a mass of 102 kg is pushed up a ramp. The top of the ramp is 4 meters higher than the bottom. How much potential energy is gained by the cart? If an average student can do 50 joules of work each second, how much time does it take to get up the ramp?

 102 kg

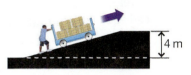

5. You are asked for the potential energy and time.
6. You are given mass, height, and the work done per second.
7. Use the formula for potential energy $E_p = mgh$.
8. Solve:
   E_p = (102 kg)(9.8 N/kg)(4 m)
   = 3,998 J

At a rate of 50 J/s, it takes 3,998 ÷ 50 = 80 seconds to push the cart up the ramp.

Kinetic energy

Kinetic energy is energy of motion
An object in motion has energy because it is moving. A moving mass can exert forces, as you would quickly observe if someone ran into you in the hall. Energy of motion is called **kinetic energy**. The kinetic energy of a moving object depends on two things: mass and speed.

Kinetic energy and mass
The kinetic energy of a moving object is proportional to the object's mass. If you double the mass, you also double the kinetic energy. For example, consider a two-kilogram rabbit moving at a speed of one meter per second; it has one joule of kinetic energy (Figure 10.21). A larger rabbit moving at the same speed has more energy. This follows from the fact that it takes more work to *stop* a heavier rabbit. A 4-kilogram rabbit moving at 1 m/s has 2 joules of kinetic energy.

Figure 10.21: *Kinetic energy is proportional to mass.*

Kinetic energy increases as speed squared
The kinetic energy of a moving object also depends on the speed of the object. Consider the 2-kilogram rabbit moving at 1 m/s. It has 1 joule of kinetic energy. If the rabbit ran at 2 m/s, it would have 4 joules of kinetic energy. If the speed of an object doubles, its kinetic energy increases four times (Figure 10.22). Mathematically, kinetic energy increases as the square of speed. If the speed increases by three, the kinetic energy increases by nine because $3^2 = 9$.

The formula for kinetic energy
The kinetic energy of a moving object is equal to one-half the object's mass times the square of its speed.

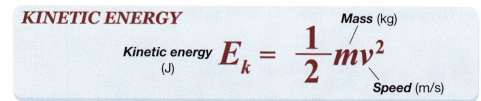

KINETIC ENERGY

$$E_k = \frac{1}{2}mv^2$$

Kinetic energy (J), Mass (kg), Speed (m/s)

Figure 10.22: *Doubling the speed multiplies the kinetic energy by four.*

Kinetic energy and driving
When a car stops, its kinetic energy of motion is completely converted into work done by the brakes (Fd). Since brakes supply nearly constant force, the stopping distance (d) is proportional to the initial kinetic energy. Going 60 mph, a car has four times as much kinetic energy as it does at 30 mph. That means it takes four times the distance to stop at 60 mph compared with 30 mph (Figure 10.23). At 90 mph, a car has nine times as much energy as it does at 30 mph, and it requires *nine times* the distance to stop.

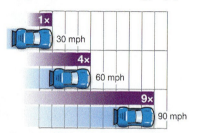

Figure 10.23: *Kinetic energy and braking distances.*

The formula for kinetic energy

Deriving the equation for kinetic energy
The kinetic energy of a moving object is exactly equal to the amount of work required to get the object from at rest to its final speed. The calculation is similar to the analysis used for potential energy, except the math has more steps. To start, suppose a ball of mass (m) is at rest. A force (F) is applied and creates acceleration (a). After a distance (d), the ball has reached speed (v).

Step 1 Work is force times distance, but force is mass times acceleration. The work done on the ball is therefore its mass times acceleration times distance.

$$W = Fd = (ma) \times d = mad$$

Step 2 The kinetic energy formula involves only mass and speed. Is there a way to get speed from acceleration and distance? In Chapter 4, you found a relationship between distance traveled, acceleration, and time.

$$d = \frac{1}{2}at^2$$

Step 3 Replace distance in the equation for work and combine similar terms:

$$W = ma(\tfrac{1}{2}at^2) = \tfrac{1}{2}ma^2t^2$$

Step 4 When an object starts from rest with constant acceleration, its speed is equal to its acceleration multiplied by the time it has been accelerating. Mathematically, $v = at$, therefore $v^2 = a^2t^2$. This is the result that is needed. Replace the a^2t^2 with v^2, and the resulting work (W) is exactly the formula for kinetic energy.

Why it works Remember, this calculation is the work done on the ball to bring it from rest up to a final speed (v). As with potential energy, the kinetic energy of a moving object is equal to the work done to create the energy.

Calculating the kinetic energy of a moving car

A car with a mass of 1,300 kg is going straight ahead at a speed of 30 m/s (67 mph). The brakes can supply a force of 9,500 N. Calculate
a) the kinetic energy of the car.
b) the distance it takes to stop.

1. You are asked for the kinetic energy and stopping distance.
2. You are given mass, speed, and the force from the brakes.
3. Kinetic energy $E_k = (\tfrac{1}{2})mv^2$
   Work, $W = Fd$
4. Solve:
   $E_k = (\tfrac{1}{2})(1{,}300 \text{ kg})(30 \text{ m/s})^2$
   $= 585{,}000 \text{ J}$

To stop the car, the kinetic energy must be reduced to zero by work done by the brakes.
$585{,}000 \text{ J} = (9{,}500 \text{ N}) \times d$
$d = 62 \text{ meters}$

Conservation of energy

Energy transformations

Energy is always moving and changing. If you skate uphill, you do work to get to the top. Going downhill, your speed increases because potential energy is converted into kinetic energy. When you apply the brakes, they get very hot and wear away. The kinetic energy partly becomes heat energy and partly goes to wearing away particles from the brakes. From the energy perspective, first, work was done to gain potential energy (top of hill), which was then converted to kinetic energy (bottom), and then heat and friction (Figure 10.24).

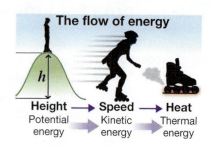

Figure 10.24: *Some energy transformations on a skating trip.*

The law of conservation of energy

The concept of energy is important because of the following fact: *The total energy in the universe remains constant.* As energy takes different forms and changes things by doing work, nature keeps perfect track of the total. No new energy is created and no existing energy is destroyed. This concept is called the **law of conservation of energy**. The rule concerning the input and output work of a machine is an example of the law of conservation of energy.

Law of conservation of energy

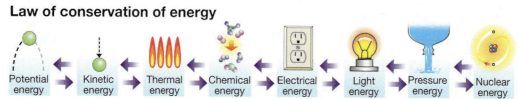

Energy can never be created or destroyed only changed from one form to another.

An example of energy conservation

What happens when you throw a ball straight up in the air? The ball leaves your hand with kinetic energy due to the speed you give it as you let go. As the ball goes higher, it gains potential energy (Figure 10.25). The potential energy gained equals the kinetic energy lost and the ball slows down as it goes higher.

Eventually, all initial kinetic energy is gone. The ball is as high as it will go and its upward speed is zero. The original kinetic energy has been completely exchanged for an equal amount of potential energy. The ball falls back down again and accelerates as it falls. The gain in speed comes from potential energy being converted back to kinetic energy. If there was no friction, the ball would return to your hand with the same speed it started, except in the opposite direction.

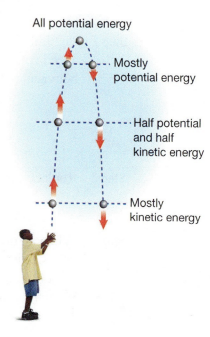

Figure 10.25: *When you throw a ball up in the air, its energy is transformed from kinetic, to potential, and back to kinetic.*

Energy in a closed system

Energy and the boundaries of a closed system

The conservation of energy is most useful when it is applied to a closed system. Remember, you choose a system to include the things you are interested in. A **closed system** means you do not allow any matter or energy to cross the boundaries of the system you choose. Because of the conservation of energy, the total amount of matter and energy in your system stays the same forever.

Conservation of energy can be used to predict and describe a system

For example, suppose your system is a ball rolling along a track with a hill and a valley (Figure 10.26). The ball is released from rest at the highest point (h_0). The total energy in the system is the potential energy of the ball at the start. Later, the ball is at a lower height (h) moving with speed (v). At this time, the ball has both potential and kinetic energy.

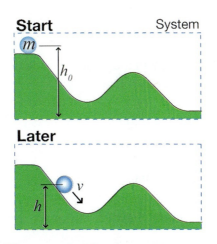

Figure 10.26: *A ball rolling on a hilly track is a system for investigating the conservation of energy.*

Energy at start

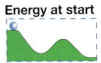 $E = mgh_0$

Energy later

 $E = mgh + \frac{1}{2}mv^2$

Accounting for all the energy in the system

Because of conservation of energy, we can write an equation that describes the total energy of the ball anywhere on the track.

Energy at start = energy later ⮕ $mgh_0 = mgh + \frac{1}{2}mv^2$

Since mass (m) appears in every term, we can cancel it out. When the equation is rearranged some interesting things can be deduced about the motion of the ball.

$\cancel{m}gh_0 = \cancel{m}gh + \frac{1}{2}\cancel{m}v^2$ ⮕ $v = \sqrt{2g(h_0 - h)}$

1. The mass of the ball does not matter since mass does not appear.
2. The speed of the ball depends only on the change in height ($h_0 - h$).

Friction can divert some energy

The law of conservation of energy holds true even when there is friction. Some of the energy is converted to heat or to the wearing away of material. The energy converted to heat or wear is no longer available to be potential energy or kinetic energy, but it was not destroyed. All the energy of the system can be accounted for.

10.3 ENERGY AND CONSERVATION OF ENERGY

Chapter 10 Connection

Hydroelectric Power

Every day in the United States the average person uses over 100 *million* joules of electrical energy. This energy comes from many sources, including burning coal, gas, and oil; nuclear power; and hydroelectric power. In hydroelectric power, the potential energy of falling water is converted to electricity. No air pollution is produced, nor hazardous wastes created. If you are lucky enough to live near a source of falling water, hydroelectric power is, in many ways, an ideal method for producing energy. Approximately 7 percent of the electricity used in this country comes from hydroelectric power.

What is a hydroelectric power system? A typical hydroelectric power system starts with a dam placed across a river. The water builds up behind the dam and creates a large difference in potential energy from the top to the bottom (Figure 10.27). Water flows down through giant tubes cast in the concrete of the dam.

Figure 10.27: *The working parts of a hydroelectric power plant.*

Energy transformation At the bottom of the dam, the water turns a turbine. A turbine is a spinning wheel specially designed to extract as much kinetic energy as possible from moving water. The turbine turns an electric generator, which produces electricity. In a hydroelectric power plant, the energy is transformed from potential energy of water, to kinetic energy of water, then to kinetic energy of the turbine, and finally to electrical energy.

Hoover Dam and Lake Mead Hoover Dam near Las Vegas, Nevada, is a famous hydroelectric power plant. It was built in 1935 to control flooding by the Colorado River and to make hydroelectric power. The dam is 221 meters high and 379 meters wide. To withstand the enormous pressure of the water, Hoover Dam is 203 meters thick at the base, narrowing to 13 meters at the top, which is not much wider than your classroom. Figure 10.28 shows an aerial view of Hoover Dam and Lake Mead, the reservoir behind it.

Energy produced depends on height difference and flow rate The energy available depends on two factors: the height difference between the inlet and outlet of the dam, and the flow rate of water. At Hoover Dam, the peak flow rate is 700 cubic meters of water per second. The water drops about 200 meters from the inlet to the turbine. The water from the dam is divided among 17 turbines and generators.

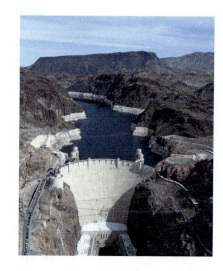

Figure 10.28: *Hoover Dam, near Las Vegas, Nevada, and its reservoir, Lake Mead.*

Chapter 10 Connection

Applying energy conservation to Hoover Dam

Conservation of energy
You can use the conservation of energy to calculate just how much energy the Hoover Dam extracts from the waters of the Colorado River. Start by calculating the potential energy of one kilogram of water 200 meters above the turbine. The potential energy works out to 1,960 joules for each kilogram of water passing through the dam.

The flow rate of Hoover Dam
The flow rate of the Colorado is 700 m³/s—a lot of water (Figure 10.29). The average classroom has a volume of 150 m³. The flow through the Hoover Dam is equivalent to almost 4-2/3 classrooms full of water every second. Another way of putting it is that 1 cubic meter of water has a mass of 1,000 kilograms. The water flow through Hoover Dam is 700,000 kilograms per second.

Energy efficiency
The turbine and generator are about 80 percent efficient. That means 80 percent of the 1,960 joules of potential energy in each kilogram of falling water is converted into electrical energy. At a water flow rate of 700,000 kg/s, Hoover Dam produces 1.1 billion joules of electrical energy every second.

Hoover Dam supplies energy for a large city
There are 86,400 seconds in a day, which makes the daily energy output of Hoover Dam approximately 95×10^{12} (trillion) joules. It takes 100 million joules per day to support an average person, so Hoover Dam can supply all the electricity for about 1 million people, the population of a large city.

Examples of other hydroelectric power plants
Ninety-five trillion seems like—and is—a huge number, but people use a lot of energy. Hoover Dam is only the ninth largest hydroelectric power plant in the United States. The Grand Coulee Dam on the Columbia River in Washington state produces five times as much electricity as Hoover Dam. When it is completed, the James Bay hydropower project in northern Quebec, Canada, will supply more than 20 times as much electricity as Hoover Dam.

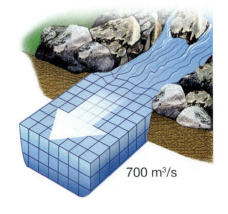

Figure 10.29: *The flow rate of the Colorado River is 700 m³/s.*

Chapter 10 Assessment

Vocabulary

Select the correct term to complete the sentences.

output force	thermal energy	ramp
gear	screw	ropes and pulleys
closed	work	lever
friction	mechanical system	simple
potential	kinetic	radiant
nuclear	mechanical	chemical
joule	pressure	conservation of energy
electrical	input	output
mechanical advantage	energy	input force
machine		

1. The ratio of the output force to the input force is equal to the _____.
2. What the machine does for you is called the _____.
3. A machine with no source of energy is known as a(n) _____ machine.
4. A device created by humans to do something useful could be called a(n) _____.
5. Everything you do to make a machine work is considered to be _____.
6. The type of simple machine that turns rotating motion into linear motion is a(n) _____.
7. An incline that decreases the input force by increasing the distance over which the force is applied can be called a(n) _____.
8. A rotating lever with teeth that allows torque to be transferred from one point to another is known as a(n) _____.
9. A simple machine that works as input and output arms rotate about a fulcrum is the _____.
10. The force applied by a machine on an object is called the _____.
11. A block and tackle system is a simple machine using _____ to create a mechanical advantage.
12. The force applied to a machine is called _____.
13. An assembly of simple machines that work together to accomplish a task can be called a(n) _____.
14. The unit of measurement for work is the _____.
15. When a force is applied and motion occurs in the direction of the force _____ is done.
16. The output work of a machine is always less than the input work because of the energy lost to _____.
17. The ability to "make things change" is a one way to describe _____.
18. The amount of work an object can do is measured as _____ energy.
19. The mechanical energy an object possesses because it is moving is called _____ energy.
20. The energy responsible for making most household appliances work is _____ energy.
21. The energy responsible for causing a balloon to expand is _____ energy.
22. Energy an object has due to its motion or potential is called _____ energy.
23. Energy released when the nuclei of light atoms are combined or nuclei of heavy atoms are split is known as _____ energy.
24. Hot objects contain more _____ energy than cold objects.
25. Energy is stored in molecules in the form of _____ energy.
26. Several forms of electromagnetic energy are commonly referred to as _____ energy.
27. A system in which no matter or energy is allowed to cross its borders is called a _____ system.
28. The scientific law based on the concept that the total energy in the universe is constant is called the law of _____.

Concept review

1. Describe the measurements you would need to take, if any, and the calculations you must do to find the mechanical advantage of each of the following simple machines.
   a. lever
   b. rope and pulley
   c. wheel and axle
   d. ramp
2. What is the unit used to represent mechanical advantage?
3. What is the major difference between first-, second-, and third-class levers?
4. "Force is multiplied by trading larger motions for smaller motions." This principle is true for all simple machines. Explain how this principle is true for a rope and pulley system.
5. The word *work* has a very specific meaning in physics. Why is it that no work is done, in the physics sense, if a force of 1,000 N is applied to a brick wall that does not move?
6. Why is the joule a unit of energy *and* a unit of work?
7. Many inventors have attempted to make a perpetual motion machine, but none have succeeded. Will a perpetual motion machine ever be invented? Why or why not?
8. Stacy eats a bowl of cornflakes for breakfast and then rides her bicycle to school. When Stacy pedals the bicycle, a mechanical device transforms the mechanical energy from the turning wheel into electrical energy to run a small flashing taillight for highway safety. Describe each energy transformation that must occur to eventually allow the taillight to work. Use the following terms in your description: mechanical energy, light energy, nuclear energy, electrical energy, chemical energy, and thermal energy. (*Hint*: Start your description all the way back with the Sun and follow the transformations to the end.)
9. A car going twice as fast requires four times as much stopping distance. What is it about the kinetic energy formula that accounts for this fact?
10. Harold bounces on a trampoline. At what point in his motion does he have the highest potential gravitational energy, and at what point does he have the highest kinetic energy? Explain.
11. In general, the work done by a force is equal to the force times the distance moved in the direction of the force. Thus, when the force is applied parallel to the distance moved, the work is equal to force times distance. What if the force and distance are at an angle? Explain how to find the work done by a force when it is exerted at an angle to the motion of an object. Give a daily life example to support your explanation.
12. Two identical twins, Tim and Tom, start from the same point at the bottom of a hill. Tim runs straight up the hill to the top. Tom walks back and forth across the face of the hill, reaching a slightly greater height with each traverse until he finally reaches the top. Neglecting friction, who has expended more energy in getting to the top of the hill? Explain your answer.
13. Most of the electrical energy in the United States is generated using coal. While coal generation is becoming cleaner, alternative methods may be "greener." Name at least three alternative methods.

Problems

1. The diagram below illustrates a cart and load being pulled by a rope that makes an angle of 12° with the horizontal surface on which it moves. How much work must be done to move the load 10 meters at constant speed?

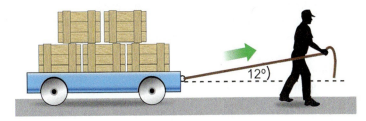

2. A weight of 200 newtons is placed 4 meters from the fulcrum of a first-class lever. An input force of 80 newtons is used to lift the weight.
   a. Draw a diagram of this lever, with forces and distances labeled.
   b. How far from the fulcrum must the input force be applied to lift the weight?
3. Michelle pulls on the input rope of a block and tackle system to raise a 400-N canoe. There are four supporting ropes in the block and tackle system.
   a. What is the tension in each supporting rope?
   b. How much input rope must Michelle pull to lift the canoe 3 meters off the ground?
   c. What is the mechanical advantage of the block and tackle system?

4. Shawn wants to set up a lever to lift the edge of his refrigerator so he can clean the floor underneath. Shawn weighs 450 N, and the refrigerator weighs 1,000 N. Shawn can produce an input force equal to his weight, and the output force of the lever must equal half the weight of the refrigerator to lift one edge. Shawn has a 2.5 m steel bar to use for the lever. Where should he place the fulcrum of the lever? Show all work and draw a labeled diagram of the lever, showing input and output forces and distances.

5. Martha must carry a 50-N package up three flights of stairs. Each flight of stairs has a height of 2 m, and the actual distance of the diagonal path she walks up the stairs is 10 m. What is the total work done on the package?

6. In one eight-hour workday, a forklift operator at a particular distribution warehouse lifts a total of 100,000 N of boxes. Each box must be lifted to a height of 1.5 m and carried an average of 10 m to the shipping dock. What is the net work done against gravity?

7. An interesting and potentially dangerous phenomenon that can occur in a large city is an exploding manhole. When underground cables become frayed, the cable insulation can smolder and catch fire, and the build-up of gases released can explode, sending the manhole cover as much as 50 feet in the air! Manhole covers can weigh as much as 300 lbs, (a mass of 136 kg). Suppose a 136-kg manhole cover is launched 50 feet in the air. Find the following.
   a. What is the potential energy of the manhole cover when it reaches 50 feet above the ground? (Don't forget to convert feet to meters!)
   b. What is the speed of the manhole cover when it hits the ground on its descent in meters per second and miles per hour?

8. The diagram below represents a 5-kg mass placed on a frictionless track at Point A and released from rest. Assume gravitational potential energy of the system to be zero at Point E. Use the diagram and your knowledge of work and energy to answer parts a through e.

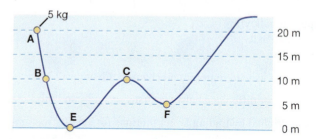

   a. Calculate the gravitational potential energy at Point A.
   b. Compare the kinetic energy of the object at Point B to the gravitational potential energy at Point C.
   c. Compare the kinetic energy of the object at Point B to the kinetic energy at Point E.
   d. If the mass is released from rest at Point B calculate its kinetic energy at Point F.
   e. If the mass is released from rest at Point A, calculate its speed at Point E.

Applying your knowledge

1. What is a perpetual motion machine? Explain why perpetual motion machines cannot work and which scientific law(s) they violate. Give one example of a perpetual motion machine and explain how it violates scientific laws.

2. A typical car is about 13 percent efficient at converting energy from gasoline to energy of motion. The average car today gets about 25 miles per gallon.
   a. Name at least four energy transformations that take place in a car.
   b. Name three things that contribute to lost energy and prevent a car from ever being 100 percent efficient.

3. Steve lifts a 5-kg toolbox 0.5 meters off the ground. Compare the amount of work that would be done on Earth to the amount that would be done on the Moon.

4. The United States produces more electricity than any other country in the world. What percentage of the world's electrical energy is produced by the United States? How do our sources for electrical energy compare to other countries? Pick another country and compare the amounts of electricity produced by various methods.

UNIT 4 ENERGY AND MOMENTUM

CHAPTER 11

Energy Flow and Power

Objectives:

By the end of this chapter you should be able to:

- ✔ Give an example of a process and the efficiency of a process.
- ✔ Calculate the efficiency of a mechanical system from energy and work.
- ✔ Give examples applying the concept of efficiency to technological, natural, and biological systems.
- ✔ Calculate power in technological, natural, and biological systems.
- ✔ Evaluate power requirements from considerations of force, mass, speed, and energy.
- ✔ Sketch an energy-flow diagram of a technological, natural, or biological system.

Key Questions:

- How do you determine a system's efficiency?
- What is power, how do you calculate it, and how does it compare to energy?
- How does energy flow in systems?

Vocabulary

carnivore	efficiency	food chain	irreversible	reversible
cycle	energy conversions	food web	power	steady state
decomposer	energy flow	herbivore	power transmission	watt
ecosystem	food calorie	horsepower	producer	

11.1 Efficiency

It is often said that "nothing is perfect," and this section is about a fundamental way in which any machine or macroscopic process *cannot* be perfect. The limitation on perfection applies to the *efficiency* with which energy changes or moves. Since nearly everything that happens in the universe involves energy, the concept of efficiency applies to more than just machines made by humans. Scientists believe that less-than-perfect efficiency is a characteristic of *all* natural processes larger than the scale of atoms. For a subtle reason, efficiency is also linked to why time only goes forward.

What *efficiency* means

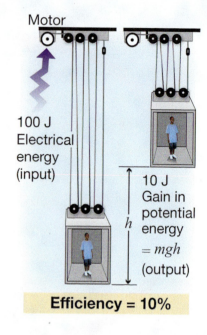

Figure 11.1: *An elevator can be viewed as a process that converts electrical energy to potential energy with an efficiency of 10 percent.*

The definition of efficiency — The **efficiency** of a process is the ratio of useful output to total input. A process that is 100 percent efficient means that 100 percent of what you start with ends up being what you get at the end of the process. For example, suppose you use an elevator to move people (Figure 11.1). The output of the elevator is potential energy; people are moved up. The input is electrical energy to a motor. A typical elevator has an efficiency of about 10 percent. That means 100 joules of electrical energy are used for every 10 joules of potential energy gained by the people inside.

Processes — Efficiency is defined for a process. A process is any activity that changes things and can be described in terms of *input* and *output*. The elevator in Figure 11.1 can be seen as a process for converting electrical energy (input) to potential energy (output). The growth of a tree is also a process. For a tree, the input is energy from the Sun, carbon from the air, and nutrients from the soil. The output is growth and reproduction of more trees. All processes can be characterized by an efficiency, including an elevator, a tree growing, and any other process you can think of.

Energy efficiency — In this chapter, efficiency means the ratio of *energy* output divided by *energy* input (Figure 11.2), and also you may assume that "efficiency" means energy efficiency. Energy efficiency is the measurable amount of useful work done per quantity of energy used.

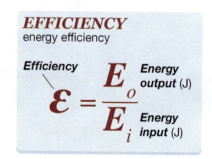

EFFICIENCY
energy efficiency

$$\varepsilon = \frac{E_o}{E_i}$$

Efficiency
E_o Energy output (J)
E_i Energy input (J)

Figure 11.2: *The definition of energy efficiency.*

Efficiency in mechanical systems

The ideal machine Efficiency is usually expressed in percent. An ideal machine would be 100 percent efficient. However, real machines are never 100 percent efficient. Some work is always done against friction. For example, a machine that is 75 percent efficient produces 3 joules of output work for every 4 joules of input work. One joule out of every four, or 25 percent, is lost to friction.

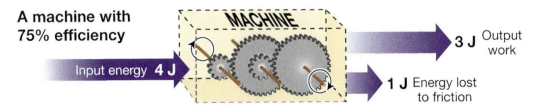

Calculating the efficiency of a rubber band

A 12-gram paper airplane is launched at a speed of 6.5 m/s with a rubber band. The rubber band is stretched with a force of 10 N for a distance of 15 cm. Calculate the efficiency of the process of launching the plane.

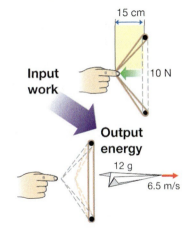

How friction affects machines Because of friction, work output is always less than work input. For example, a wheel turning on an axle gets hot. When the wheel gets hot, it means some of the input work is being converted to heat by the action of friction. The work output is reduced by the work that is converted to heat, resulting in lower efficiency.

Accounting for the energy of a system According to the law of conservation of energy, energy cannot ever be lost, so the *total* efficiency of any process is 100 percent. When we say a machine is 75 percent efficient, we mean 75 percent of the energy ends up being used in the way we *want*. For example, a car's useful output energy is kinetic energy of motion and potential energy for climbing hills. All other forms of energy are considered "losses," such as heat from the radiator and exhaust gases. When calculating the car's efficiency, only the *usable* energy is counted as output.

Automobiles In general, cars are not very efficient users of energy—13 percent is typical. This means only 13 percent of the energy released by burning gasoline is converted to work done moving the car. The rest of the energy becomes heat, wears away engine parts, moves air around the car, and is spent in other ways that do not result in work done by the wheels.

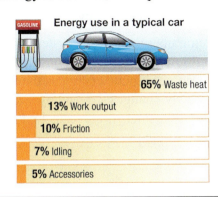

1. You are asked for the efficiency.
2. You are given the input force and distance, and the output mass and speed.
3. Efficiency is output energy divided by input energy. The input energy is work = $F \times d$. The output energy $E_k = \frac{1}{2}mv^2$.
4. Solve:

$$\varepsilon = \frac{(0.5)(0.012 \text{ kg})(6.5 \text{ m/s})^2}{(10 \text{ N})(0.15 \text{ m})}$$

$$= 0.26 \text{ or } 26\%$$

CHAPTER 11 ENERGY FLOW AND POWER

Efficiency in natural systems

The meaning of efficiency

Energy drives every process in nature—from the wind in the atmosphere to the nuclear reactions in the core of a star. In the environment, efficiency is interpreted as the fraction of energy that goes into a *particular* process. For example, Earth receives energy from the Sun. Some of the energy is absorbed and some is reflected back into space. Earth absorbs sunlight with an average efficiency of 73 percent. Out of every 100 joules that falls on Earth from the Sun, 73 joules are absorbed and the remaining 27 joules are scattered or reflected back into space.

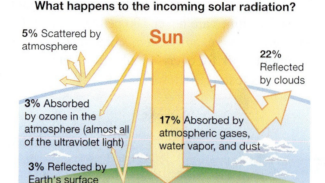

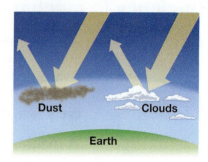

Figure 11.3: *Dust and clouds reflect light back into space, decreasing the efficiency with which Earth absorbs energy from the Sun.*

The importance of solar efficiency

The efficiency of Earth at absorbing solar energy is critical to living things. If the efficiency *decreased* by a few percent, Earth's surface would become too cold for life. Dust reflects solar energy (Figure 11.3). Some scientists believe that many volcanic eruptions or a nuclear war would decrease the absorption efficiency by spreading dust in the atmosphere. On the other hand, if the efficiency *increased* by a few percent, it would get too hot to sustain life. Carbon dioxide and methane in the atmosphere increase the absorption efficiency (Figure 11.4). Scientists are concerned that Earth has already warmed a few degrees as a result of carbon dioxide released by human technology.

Efficiencies always add up to 100 percent

It is important to remember that, in any system, all of the energy goes somewhere. For example, rivers flow downhill. Most of the potential energy lost by water moving downhill becomes kinetic energy as the water moves. Erosion takes some of the energy and slowly changes the land by wearing away rocks and dirt. Friction takes some of the energy and heats up the water. If you could add up the efficiencies for every single process, that total would always be 100 percent.

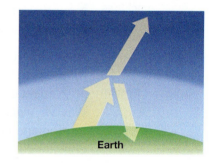

Figure 11.4: *Carbon dioxide and other greenhouse gases in the atmosphere absorb some energy that otherwise would have been radiated back into space. This increases the efficiency with which Earth absorbs energy from the Sun.*

Efficiency in biological systems

Calories in food Energy in food is measured in **food calories**. A single food calorie is technically a kilocalorie and is equal to 4,187 joules. On food labels, this energy is listed as Calories with a capital *C*. Living requires a lot of energy. A pint of ice cream contains about 1,000 kilocalories, which is a little more than 4 million joules of energy (Figure 11.5). By comparison, 1 joule is the work equivalent of lifting 1 pint of ice cream 21 centimeters.

Efficiency is low for living things In terms of output work, the energy efficiency of living things is typically very low. Almost all of the energy in the food you eat becomes heat and waste products; very little becomes physical work. Of course, living creatures do much more than physical work. For example, you are reading this book.

Estimating the efficiency of a human To estimate the efficiency of a person doing physical work, consider climbing a 1,000 meter mountain. For the average person with a mass of 70 kilograms, the increase in potential energy is 686,000 joules. The potential energy comes from work done by muscles. A human body doing strenuous exercise uses about 660 food calories per hour. It takes about three hours to climb the mountain, during which time the body uses 1,980 food calories, or 8.3 million joules. The energy efficiency is 686,000 J divided by 8.3 million J, or about 8 percent (Figure 11.6).

Baseline metabolic rate The overall energy efficiency for a person is actually lower than 8 percent. An average person uses 55–75 kilocalories per hour when sitting still. The rate at which your body uses energy while at rest is called your baseline metabolic rate (BMR). During a 24-hour period, a person with a BMR of 65 kcal/hr uses 1,536 kilocalories, or 6.43 million joules. Even if you did the equivalent work of climbing a 1,000-meter mountain every day, your average daily efficiency is only 4.6 percent.

Efficiency of plants The efficiency of plants is similar. Photosynthesis in plants takes input energy from sunlight and creates sugar, a form of chemical energy. To an animal, the output of a plant is the energy stored in sugar, which can be eaten. The efficiency of pure photosynthesis is 26 percent, meaning 26 percent of the sunlight absorbed by a leaf is stored as chemical energy. As a whole system however, plants are only 1–3 percent efficient. The system efficiency is lower than 26 percent because leaves absorb only a third of the energy in sunlight—some energy goes into reproducing, and some energy goes into growth and other plant functions.

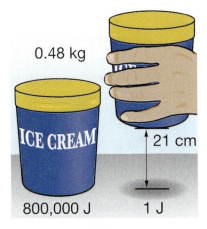

Figure 11.5: *Food contains a huge amount of energy compared with typical work output.*

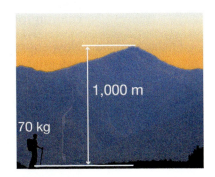

$E_p = mgh$

$= (70 \text{ kg})(9.8 \text{ N/kg})(1,000 \text{ m})$

$= 686,000 \text{ J}$

Figure 11.6: *A 70-kilogram hiker gains 686,000 joules of potential energy climbing a 1,000-meter mountain.*

Efficiency and the arrow of time

A connection between efficiency sand time

The efficiency is less than 100 percent for virtually all processes that convert energy from one form to any other form except heat. After 2,000 years of thinking about it, we believe that the inevitable "loss" of energy into heat is connected to why time flows forward and not backward. The connection between efficiency and time is not at all obvious, but read along and see if it makes sense to you.

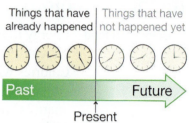

The arrow of time

Think of time as an arrow pointing from the past into the future. All processes move in the direction of the arrow, and never go backward (Figure 11.7).

Figure 11.7: *Time can be thought of as an arrow pointing from the past into the future.*

Reversible processes

Suppose a process were 100 percent efficient. As an example, think about connecting two marbles of equal mass by a string passing over an ideal pulley that has no mass and no friction (Figure 11.8). One marble can go down, transferring its potential energy to the other marble, which goes up. This ideally efficient process can go forward and backward as many times as you want. In fact, if you watched a movie of the marbles moving, you could not tell if the movie were playing forward or backward. To a physicist, this process is **reversible**, meaning it can run forward or backward in time.

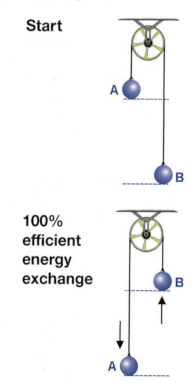

Friction and the arrow of time

Now suppose there is a tiny amount of friction. The efficiency of transferring potential energy is reduced to 99 percent. Because some potential energy is lost to friction, one marble does not go all the way down, and the other does not go all the way up. Every time the marbles exchange energy, some is lost and the marbles don't rise quite as high as they did the last time. And if you made a movie of the motion, you could tell whether the movie was running forward or backward. Because of the energy lost to friction, any process with an efficiency less than 100 percent runs only one way, *forward with the arrow of time*.

Irreversible processes

Friction turns energy of motion into heat. Once energy is transformed into heat, the energy cannot ever completely get back into its original form. Because energy that becomes heat cannot get back to potential or kinetic energy, any process for which the efficiency is less than 100 percent is **irreversible**. Irreversible processes can only go forward in time. Since processes in the universe almost always lose a little energy to friction, time cannot run backward. As you study physics, you will find that this idea connecting energy and time has many other implications.

Figure 11.8: *Exchanging energy with a frictionless, massless pulley.*

11.2 Energy and Power

In science, the words *energy* and *power* have specific meanings, and they do *not* mean the same thing. Energy is the ability to cause change and is measured in joules. But change can happen slowly or quickly. Whether you run up the stairs or walk up the stairs, your increase in potential energy is the same. What is different is the *rate* at which your energy changes. The change in energy happens quickly when you run up the stairs. The change is slower when you walk. The rate at which energy flows or changes is called *power*. Power is measured in joules per second, or *watts*, and is the subject of this section.

Doing work fast or doing it slowly

How fast work is done It makes a difference how fast you do work. Suppose you drag a box with a force of 100 newtons for 10 meters in 10 seconds. You do 1,000 joules of work. Your friend drags a similar box the same distance in 60 seconds. You both do the same amount of work because the force and distance are the same. But something is different. You did the work in 10 seconds and your friend took six times longer.

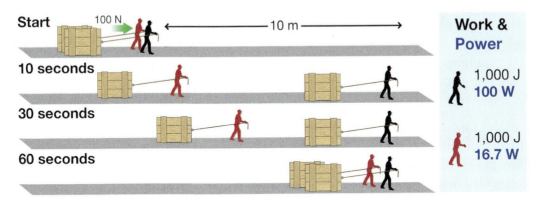

Power is the rate of doing work or using energy **Power** is the *rate* at which work is done, which is the amount of work done divided by the time it takes to do the work. In physics, when you see the word *power* you should think "energy used divided by time taken." This is similar to thinking about speed as "distance traveled divided by time taken." For example, doing 1,000 joules of work in 10 seconds equals a power of 1,000 J divided by 10 seconds, or 100 joules per second. Doing the same amount of work in 60 seconds requires a power of 1,000 J divided by 60 seconds, or 16.7 joules per second.

Calculating power in climbing stairs

A 70-kg person goes up stairs 5 m high in 30 seconds.

a. How much power does the person need to use?
b. Compare the power used with a 100-watt light bulb.

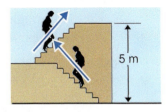

1. You are asked for power.
2. You are given mass, distance, and time.
3. Relationships that apply:
   $E_p = mgh$; $P = E \div t$
4. Solve:
   $E_p = (70\ kg)(9.8\ N/kg)(5\ m)$
       $= 3{,}430\ J$
   $P = (3{,}430\ J)/(30\ s) = 114$ watts
   a. 114 W
   b. This is a little more power than a 100-watt light bulb. Many human activities use power comparable to a light bulb.

Power

Units of power The unit of power in physics is the **watt**, named after James Watt (1736–1819), the Scottish engineer and inventor of the steam engine. One watt (1 W) is equal to 1 joule of work done in 1 second. Another commonly-used unit of power is the **horsepower**. One horsepower is equal to 746 watts. As you may have guessed, 1 horsepower was originally the average power output of a horse (Figure 11.9).

Two interpretations of power Power is used to describe two similar situations. The first situation is work being done by a force. Power is the rate at which the work is done. The second situation is energy flowing from one place to another, such as electrical energy flowing through wires. The power is the amount of energy that flows divided by the time it takes. In both situations, the units of power are joules per second, or watts.

Calculating power To calculate power, you take the quantity of work or energy and divide by the time it takes for the work to be done or the energy to move.

POWER

Power (W) $\quad P = \dfrac{E}{t} \quad$ Change in work or energy (J)

Change in time (s)

As an example, 100 kilograms of water per second fall 10 meters to turn a turbine. The change in energy is *mgh* or (100 kg) × (9.8 N/kg) × (10 m) = 9,800 J. If the energy change happens in one second, the power is 9,800 watts (Figure 11.10).

A second way to calculate power There is a second useful formula for power. Work is force times distance. Power is work divided by time. Combining these two relationships gives another way to calculate power: Power is force times speed. If you apply a constant force of 100 newtons to drag a sled at a constant speed of 2 meters per second, you use 200 watts of power. Since both force and velocity are vectors, you may multiply them to calculate power if they are in the same direction. When the force and velocity are at an angle to each other, you must also multiply by the cosine of the angle.

POWER (alternate formula) $\quad$ Force (N)

Power (W) $\quad P = \vec{F} \cdot \vec{v} \quad$ Velocity (m/s)

1 horsepower (1 hp)

The power output of a farm horse

1 watt (1 W)

The power it takes to raise your arm 3 cm in 1 second

1 watt = 1 joule per second
1 horsepower = 746 watts

Figure 11.9: *Units of power.*

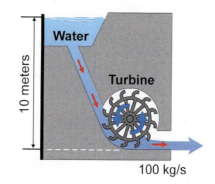

Figure 11.10: *A flow of 100 kilograms of water per second dropping 10 meters represents a power of 9,800 watts.*

Power in human technology

Ranges of power
You probably use technology with a wide range of power every day. On the high end of the power scale are cars and trucks. A typical car engine makes 150 horsepower (hp), which is 112,000 watts (W). This power is delivered in the form of work done by the wheels. Moderate power devices include appliances such as washing machines, fans, and blenders. These household machines have electric motors that do work by converting electrical energy to mechanical energy. Common electric motors found around the house range from 1 hp, or 746 W, down to 1/20th hp (37 W). Many appliances have "power ratings" that indicate their power. For example, a blender might say it uses 1/3 hp, indicating a power of about 250 W. Table 11.1 lists the power used by some everyday machines.

Table 11.1: Power used by some common devices

Machine	Power used (W)	Machine	Power used (W)
Lawn mower	2,500	Electric drill	200
Refrigerator	700	Television	100
Washing machine	400	Desk lamp	100
Computer	200	Small fan	50

Estimating power requirements
Machines are designed to use the appropriate amount of power to create enough force to do the work they are designed to do. You can calculate the power required if you know the force you need and the rate at which things must move.

For example, suppose your job is to choose a motor for an elevator. The elevator must lift 10 people, each with a mass of 70 kilograms. The specification for the elevator says it must move three meters between each floor in three seconds. The energy required is given by $E_p = mgh$. Substituting the numbers results in a value of (10 × 70 kg) × (9.8 N/kg) × (3 m) = 20,580 J. This amount of energy is used in three seconds, so the power required is 20,580 J divided by three seconds, or 6,860 W. Motors are usually sold by horsepower, so divide again by 746 W/hp to get 9.2 hp. A more accurate calculation would add the weight of the elevator car as well as the people and some extra power for rapid acceleration.

Estimating power required by a fan

A fan uses a rotating blade to move air. How much power is used by a fan that moves 2 cubic meters of air each second at a speed of 3 m/s. Assume the air is initially at rest and has a density of 1 kg/m^3. Fans are inefficient; assume an efficiency of 10 percent.

1. You are asked for power.
2. You are given volume, density, speed, and time.
3. Relationships that apply:
   $\rho = m/V$, $E_k = \frac{1}{2}mv^2$, $P = E/t$
4. Solve:
   $m = \rho V = (1 \text{ kg/m}^3)(2 \text{ m}^3) = 2 \text{ kg}$
   $E_k = (0.5)(2 \text{ kg})(3 \text{ m/s})^2 = 9 \text{ J}$

At an efficiency of 10 percent, it takes 90 J of input energy to make 9 J of output energy.

$P = (90 \text{ J})/(1 \text{ s})$
$= 90 \text{ watts}$

Power in natural systems

Stars and supernovae

Natural systems exhibit a much greater range of power than human technology. At the top of the power scale are stars. The Sun has a total power output of 3.8×10^{26} W. This is a tremendous amount of power, especially considering the Sun has been shining continuously for more than 4 billion years. Even more powerful is a supernova, the explosion of an old star at the end of its normal life. Supernova explosions are the most powerful events in the known universe, releasing 10 billion times the power of the Sun. Fortunately, supernovae are rare, occurring about once every 75 years in the Milky Way galaxy (Figure 11.11).

Energy from the Sun

Almost all of the Sun's power comes to Earth as radiant energy, including light. The top of Earth's atmosphere receives an average of 1,373 watts of solar power per square meter. In the summer at northern latitudes in the United States, about half that power, or 660 W/m², makes it to the surface of Earth. The rest is absorbed by the atmosphere or reflected back into space. In the winter, the solar power reaching the surface drops to 350 W/m². About half of the power reaching Earth's surface is in the form of infrared light. The remaining power is mostly visible and ultraviolet light.

Estimating the power in wind

The power received from the Sun is what drives the weather on Earth. To get an idea of the power involved in weather, suppose we estimate the power in a gust of wind. A moderate wind pattern covers 1 square kilometer and involves air up to 200 meters high. This represents a volume of 200 million cubic meters (2×10^8 m³). The density of air is close to 1 kg/m³, so the mass of this volume of air is 200 million kilograms.

Assume the wind is moving at 10 m/s (22 mph) and it takes three minutes to get going. The power required to start the wind blowing is the kinetic energy of the moving air divided by 180 seconds (three minutes). The result is 56 million watts, nearly the power to light all the lights in a town of 60,000 people. Compared with what people use, 56 million watts is a lot of power! But 1 square kilometer receives 1.3 *billion* watts of solar power. A 10 m/s wind gust represents only 4 percent of the available solar power. An ordinary storm might deliver 1,000 times more power than 56 million watts because much more air is moving (Figure 11.12).

Photo courtesy of NASA/ESA/JPL/Arizona State Univ.

Figure 11.11: *The Crab Nebula is the remains of a supernova explosion that was seen from Earth in 1054 CE. The supernova was so bright it could be seen during the day, according to the records of Chinese astronomers of the time.*

Image courtesy of NOAA

Figure 11.12: *A powerful storm system moves a great amount of air.*

Power in biological systems

Power output of people and animals

A physically-fit human can sustain a peak power of about 300 watts for a short time and a steady 100 watts for hours. A horse can have an average work output of 1 horsepower, or 746 watts. Only 200 years ago, a person's muscles and those of their animals were all anyone had for power. Compare that to what is available today. The average lawn mower has a power of 2,500 watts—the equivalent power of three horses *plus* three people.

Range of animal power

The power output of animals varies with the size of the animal. Big animals need more power to get up and move. For example, a blue whale can sustain a power output of 500 hp (373,000 W), about the same as a large truck. Insects use very little power. For example, the power output of a flying insect is 0.0001 watts. However, since insects outnumber people on Earth by more than 100,000 to 1, the total power output of insects is greater than the power output of people.

Most of the power output of animals ends up as heat. An average person produces 100 watts of heat continuously. A crowd of people can give off so much heat that buildings need air conditioning to remove the heat, even in the winter.

Power used by plants

The solar power used by plants is typically one-third of the power in visible light falling on their leaves. By this estimate, a large tree uses 6,000 watts of solar power in full sunlight, about the same power as a motor scooter. The power used by plants goes mostly into growth of new plant material and moving water from the roots out to the leaves. The output power from plants is input power for animals. Even though less than 1 percent of the input power from the Sun becomes food for animals, it is enough to support the entire food chain on Earth.

Estimating the average input power of a person

An average diet includes 2,500 food calories per day. Calculate the average power this represents in watts over a 24-hour period. One food calorie equals 4,187 joules.

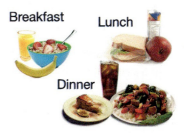

1. You are asked for power.
2. You are given the energy input in food calories and the time.
3. Relationships that apply:
   1 food calorie = 4,187 J
   $P = E \div t$
4. Solve:
   $E = (2{,}500 \text{ cal})(4{,}187 \text{ J/cal})$
   $= 10{,}467{,}500 \text{ J}$

There are $60 \times 60 \times 24 = 86{,}400$ seconds in a day.

$P = (10{,}467{,}500 \text{ J}) \div (86{,}400 \text{ s})$
$= 121 \text{ watts}$

This is a little more than the power used by a 100-watt light bulb.

CHAPTER 11 — ENERGY FLOW AND POWER

11.3 Energy Flow in Systems

Looking at the big picture, our universe is matter and energy organized in *systems*. There are large systems, like our solar system composed of the Sun, planets, asteroids, comets, smaller bits of matter, and lots of energy. There are smaller systems within the solar system, such as the ecology of Earth. In fact, there are systems within systems, ranging in scale from the solar system, to Earth, to a single animal on Earth, to a single cell in the animal, right down to the scale of a single atom. In every single system, energy flows, creating change. This section presents a few brief examples of how energy flows in systems.

Following an energy flow

The energy flow in a pendulum A pendulum is a system in which a mass swings back and forth on a string. At its highest point, a pendulum has only potential energy, because it is not moving. At its lowest point, a pendulum has kinetic energy. Kinetic energy and potential energy are the two chief forms the energy of a pendulum can take. As the pendulum swings back and forth, the energy flows back and forth between potential and kinetic, with a little lost to friction. An energy diagram might look like Figure 11.13.

Energy conversion Energy flows almost always involve **energy conversions**. In a pendulum, the major conversion is between potential and kinetic energy. A smaller conversion is between kinetic energy and other forms of energy created by friction, such as heat, air motion, and the wearing away of the string.

Model energy flow with a diagram One of the first steps to understanding an **energy flow** is to write down the forms that energy takes. If you choose the system to be a pendulum, there are three chief forms of energy: potential energy, kinetic energy, and losses from friction.

The next step is to diagram the flow of energy from start to finish for all the important processes that take place in the system.

The last step is to try to estimate how much energy is involved and what are the efficiencies of each energy conversion. Almost every conversion will involve some loss of energy to heat, wear, or another source of friction.

Pendulum

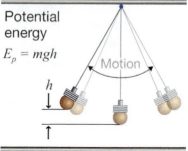

Potential energy
$E_p = mgh$

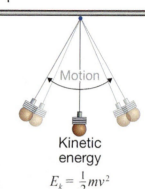

Kinetic energy
$E_k = \tfrac{1}{2}mv^2$

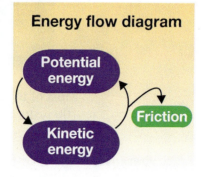

Figure 11.13: *In a pendulum, the energy mostly flows back and forth between potential energy and kinetic energy. Some energy is lost to friction on every swing.*

UNIT 4 ENERGY AND MOMENTUM

Energy flows in technology

Processes The energy flow in technology can usually be broken down into four types of processes. Complex machines often include two, three, or even all four processes. A very complex machine such as a car includes multiple types of each process.

Storage 1. **Energy storage:** Examples of energy storage technologies are batteries (chemical energy), springs (elastic potential energy), pressure (fluid energy), height (gravitational potential energy), gasoline (chemical energy), and motion (kinetic energy).

Conversion 2. **Energy conversion:** Many devices convert one type of energy to another. For example, an electric motor converts electrical energy to mechanical energy. A pump converts mechanical energy to fluid energy.

Transmission 3. **Power transmission:** Power is the rate of energy flow through a system. Some examples of different methods of power transmission are through wires (electrical); through tubes (fluid power); through mechanisms such as cables, gears, or levers (mechanical power); or through light (radiant power).

Output 4. **Output use:** This is the form in which energy is needed to accomplish a task. Mechanical work is the output of a bulldozer. Heat is the output of an electric stove. A light bulb makes light.

Energy flow in a rechargeable drill A rechargeable electric drill is a good example of a device that uses all four processes. Energy is stored in a battery (storage). Power from the battery gets to the motor by wires (electrical transmission). The motor converts electrical energy to mechanical energy (energy conversion). The rotation of the motor is transferred to the drill bit by gears (mechanical transmission). The spinning drill bit cuts wood (output work). An energy flow diagram for the drill is shown in Figure 11.14.

Efficiency Every process in an energy flow has an efficiency. For example, batteries are typically 45 percent efficient. If you put 100 joules into a battery, you only get 45 joules back out. The rest mostly becomes heat. Mechanical transmissions can be 95 percent efficient. Electric motors are moderately efficient; 65 percent is a good average. The overall efficiency of the drill is calculated by multiplying the efficiencies for each process. For the rechargeable drill, the overall efficiency is only 28 percent ($0.45 \times 0.65 \times 0.95$).

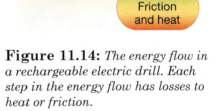

Figure 11.14: *The energy flow in a rechargeable electric drill. Each step in the energy flow has losses to heat or friction.*

Energy flows in natural systems

Steady state energy balance
The energy flow in technology tends to start and stop. For example, you turn your car's motor on, drive somewhere, and turn it off. The energy flows in natural systems tend to be **steady state**. Steady state means there is a balance between energy in and energy out, so that the total energy remains the same. A good example of a system in steady state is Earth as a planet. Energy from the Sun represents energy input. But Earth is warm compared with the chill of empty space. Consequently, Earth radiates thermal energy back into space. Earth's average energy stays about the same because the energy input from the Sun is balanced by the power radiated back into space (Figure 11.15).

Natural systems work in cycles
Many of the energy flows in nature occur in **cycles**. The water cycle is a good example. Radiant energy from the Sun is absorbed by water, mostly the oceans, and also lakes. Some water evaporates into the air, carrying energy from the warm water into the atmosphere. The water vapor goes up into the atmosphere and cools, releasing its energy to the air. The cooled water condenses into droplets as rain, which falls back to the ground. Eventually, the rainwater makes its way back to the ocean through rivers and groundwater, and the cycle begins again. The water cycle moves energy from the oceans into the atmosphere (Figure 11.16).

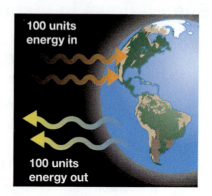

Figure 11.15: *Earth's total energy stays relatively steady because the energy input from the Sun equals the energy radiated back into space.*

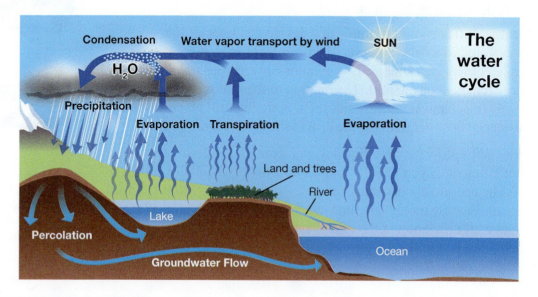

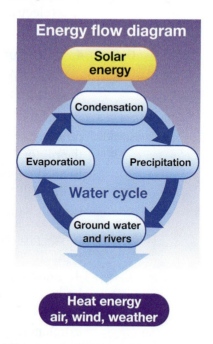

Figure 11.16: *An energy flow diagram for the water cycle.*

Energy flows in biological systems

Producers and food chains

A **food chain** is a series of processes through which energy and nutrients are transferred between living things. At the bottom of the food chain are **producers**. Producers are plants and one-celled organisms that use energy from the Sun. Producers create biological molecules such as carbohydrates, fats, and proteins, which store energy from the Sun in forms that can be passed on to animals higher in the food chain. Most producers are very small. The most numerous producers are phytoplankton, small organisms that live near the surface of the oceans.

Herbivores

The next step up the food chain are the **herbivores**. An herbivore is an organism that eats plants. Herbivores include rabbits, snails, most insects, some fish, deer, and many other land and sea animals. Herbivores concentrate the energy output from plants into proteins, fats, and animal tissue. It takes many producers to support one herbivore. Think of how many blades of grass a rabbit can eat.

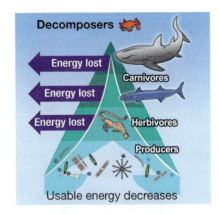

Figure 11.17: *The energy pyramid is a good way to show how energy moves through an ecosystem.*

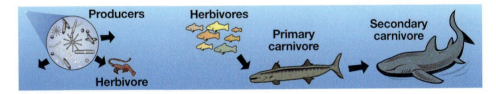

Carnivores and decomposers

Carnivores eat concentrated energy and proteins, fats, and carbohydrates in the bodies of herbivores. A *primary* carnivore eats herbivores. A hawk is an example of a primary carnivore. Hawks eat mice and other small animals that eat plants. *Secondary* carnivores eat other carnivores as well as herbivores. A shark is an example of a secondary carnivore. Sharks eat fish that eat other fish, as well as fish that eat plants. Another important group in the food chain are **decomposers**. Decomposers break down waste and bodies of other animals into nutrients that can be used by plants. Earthworms and many bacteria are examples of decomposers.

Food webs and ecosystems

An ecosystem is an interdependent collection of plants and animals that support each other. A food chain is like one strand in a **food web**. A food web connects all the producers and consumers of energy in an **ecosystem**. A good way to look at the energy flow in a food web is in the form of a pyramid (Figure 11.17). Figure 11.18 shows an energy flow diagram for an ecosystem.

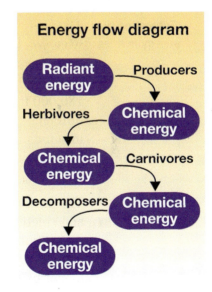

Figure 11.18: *Energy flow in an ecosystem.*

Chapter 11 Connection

Energy from Ocean Tides

Tides are enormous flows of water created by gravity in the Earth-Moon system. Tides occur for two primary reasons. First, the gravity of the Moon pulls the water directly beneath it as it passes overhead. Second, the Earth-Moon system actually rotates about its common center of mass, which is not the center of Earth, but almost three-fourths of the way toward the "Moonward" surface. If Earth could change its shape, lunar gravity would make the planet slightly egg-shaped. The long axis of the "egg" would be along the line between Earth and the Moon. Earth, however, cannot change its shape quickly enough to respond to the passing of the Moon. But the oceans are liquid and they *can* change their shape. Figure 11.19 is an exaggerated diagram of how the surface of the ocean responds to lunar gravity. Because Earth rotates once a day, a person standing in one location sees the oceans go up and down twice per day.

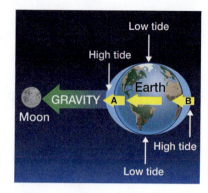

Figure 11.19: *The gravity of the moon causes the tides on Earth.*

Power in tides

Estimating the energy in tides The energy and power in tides is enormous. A simple estimate can be made using what you know about energy. The surface area of Earth is 511 billion square kilometers, 71 percent of which is covered by water. Suppose half the water in the oceans is lifted 1/2 meter higher than average (Figure 11.20). That means lifting 180,000 trillion (1.8×10^{17}) kilograms of water, and creating a potential energy difference of 1.8 million trillion joules (1.8×10^{18} J). Since tides go up and down in some places twice per day, this flow of energy occurs over 12 hours, representing a power of 41 trillion watts (4.1×10^{13} W). This simple estimate is five times the total power used by the 6 billion people living on the planet today.

The source of tidal energy The power that moves the oceans and creates tides comes from the total potential and kinetic energy of the Earth-Moon system. Tides represent a frictional force on the motion of Earth and the Moon. Every day the tides take a bit of energy away from the system. Friction from tides slows the rotation of Earth, making the day longer by 0.0016 seconds every 100 years. The Moon also takes longer to make one revolution of its orbit by a tiny fraction of a second. Fifty billion years from now, the slow energy transfer of tides will cause the rotation of Earth to become synchronized with the orbit of the moon, making a day and a month for both equal to 47 hours. Fortunately, 50 billion years is so far into the future that we need not worry much about days and months getting longer.

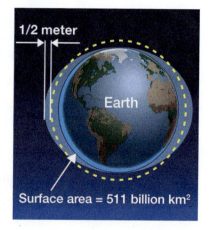

Figure 11.20: *Estimating the energy and power available in the tides.*

Chapter 11 Connection

Transforming tidal energy into electrical energy

Advantages of tidal power Many experimental projects have been built to harness the power of tides. Like hydroelectric power, energy from tides creates no pollution, nor does it use up fossil fuels such as petroleum or coal. Three promising techniques are being evaluated and several power plants of each design have been built.

Tidal barrage power The simplest approach is to use specialized dams, called barrages, to create a basin that fills up at high tide, when the ocean is at its highest level. At low tide, the basin empties out through a turbine (Figure 11.21). The potential energy of the water is converted to electricity by a turbine and generator, just like other forms of hydroelectric power (Chapter 10). The tidal power stations in the Bay of Fundy in eastern Canada produce 20 million watts of electric power when the tide goes out, twice a day. The Rance power station in France generates a peak power of 240 million watts.

The underwater turbine approach The second approach uses underwater propellers that act like wind turbines (Figure 11.22). The propeller blades can swivel so that they generate power when the tide is going out or when it is coming in. This is an advantage over the basin approach, which can only generate power when the tide goes out.

A new approach involves constructing an artificial lagoon in offshore tidal flats. The lagoon is created with rubble wall breakers. Turbines and generators convert the water's potential energy into usable electrical energy. The tidal lagoon method is similar to the tidal barrage, except that the lagoon is located offshore and does not impact the local environment, as barrages located on the shoreline can.

Tide power is active research Developing tidal power is an active area of engineering research around the world. The idea is simple but a practical tidal power plant is not simple at all. Because equipment is out in the ocean, a practical tidal power plant must be extremely rugged to withstand hurricanes, ice storms, and other violent weather. Salt water is very corrosive and moving parts must be well protected from rust and corrosion. Service underwater or in the open ocean are both difficult and expensive, so tidal power machines must be very reliable. Many clever ideas are being tried, and there is plenty of room for invention. Maybe one day you will think of a new and clever way to use the energy from tides!

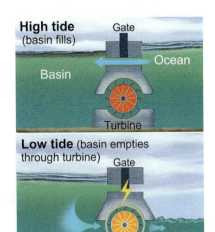

Figure 11.21: *How a tidal-basin power plant works. The turbine only makes power when the tide is low. The basin fills when the tide is high.*

Figure 11.22: *The underwater-turbine approach to tide power.*

Chapter 11 Assessment

Vocabulary

Select the correct term to complete the sentences.

efficiency	process	input
output	calories	reversible
irreversible	power	horsepower
producers	energy flow	watt
cycle	food chain	power transmission
herbivore	carnivores	decomposers
food web	energy conversions	steady state
ecosystem		

1. A process that can run forward or backward in time would be described as _____ by a physicist.
2. A ratio of energy input to energy output is the _____ of a process.
3. A process that is less than 100 percent efficient can only go forward in time and is _____.
4. Any activity that changes things can be described as a(n) _____.
5. Energy used by a living being or a machine is considered _____ energy.
6. Energy contained in the food we eat is often measured in units called food _____.
7. The work done by a machine is _____ energy.
8. The unit of power equivalent to 746 watts is the _____.
9. The rate at which energy flows or changes is called _____.
10. A light bulb consuming 100 joules of energy per second would be labeled as a 100 _____ light bulb.
11. If the energy flow into a natural system equals the energy flow out, the system is in a condition described as _____.
12. An animal that eats only plants is called a(n) _____.
13. An interdependent collection of plants and animals that support each other is known as a(n) _____.
14. Energy changes from one form to another are referred to as _____.
15. Organisms that break down waste and the bodies of other organisms are called _____.
16. A(n) _____ is a series of processes through which energy and nutrients are transferred between living things.
17. The movement of water from the oceans into the atmosphere, over land, and back to the oceans is a process of nature referred to as the water _____.
18. A food chain is one part of a(n) _____.
19. Animals that consume concentrated energy in the proteins, fats, and carbohydrates in the bodies of other animals are called _____.
20. The movement of energy through a system, often diagramed to illustrate the efficiency of conversions in the system, is referred to as _____.
21. Plants and one-celled organisms that absorb energy from the Sun to create biological molecules are _____ in a food chain.
22. The rate at which energy moves through a system would be a measure of the _____.

Concept review

1. What is the difference between an ideal machine and a real machine, in terms of efficiency?
2. A consumer foundation states that buying energy efficient products is one of the smartest ways you can reduce energy usage and help prevent air pollution.
   a. What, in general terms, is an "energy efficient" product?
   b. How will using an energy-efficient product reduce energy usage and prevent air pollution?
3. If the efficiency of pure photosynthesis is 26 percent, why is the whole system of a plant only 1 to 3 percent efficient?
4. Bicycles can have efficiencies as high as 85 percent. Is the remaining 15 percent of the input energy lost? Does this contradict the law of conservation of energy?

5. Would the efficiency of a motorcycle be higher or lower than the efficiency of a bicycle? Explain your reasoning.
6. Why is the energy efficiency of biological systems typically very low?
7. In everyday language, the words *energy* and *power* are often used interchangeably, but they actually do *not* mean the same thing. What is the difference between energy and power in the physics sense?
8. Describe the two interpretations of power and the two ways to calculate power.
9. Why do you suppose that we still use the unit called horsepower to describe the power delivered by a machine?
10. Two students are working out in the weight room. Erik lifts a 50-pound barbell over his head 10 times in 1 minute. Patsy lifts a 50-pound barbell over her head 10 times in 10 seconds. Who does the most work? Who delivers the most power?
11. Two mountain lions run up a steep hillside. One animal is twice as massive as the other, yet the smaller animal got to the top in half the time. Which animal did the most work? Which delivered the most power?
12. List two examples of technology you use each day that have a high power rating, and two examples of technology that have a relatively low power rating. Explain why these particular examples have high and low power ratings.
13. Steve lifts a toolbox 0.5 meters off the ground in one second. If he does the same thing on the Moon, does he have to use more power, less power, or the same amount of power? Explain.
14. At each level of the food web pyramid, about 90 percent of the usable energy is lost in the form of heat.
    a. Which level requires the most overall input of energy to meet its energy needs?
    b. Use the diagram information to explain why a pound of beefsteak costs more than a pound of corn.

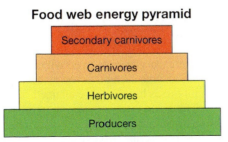

15. Use the diagram to answer questions a–d.

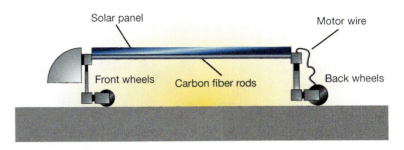

    a. What component of the model solar car represents the energy storage process?
    b. What specific energy conversion(s) take(s) place?
    c. What sort of power transmission takes place?
    d. What is the output use?

Problems

1. What is the efficiency of an escalator that uses 200 joules of electrical energy for every 20 joules of energy gained by the riders?
2. Michelle, a basketball star, takes her 75-kg body up a 3-meter staircase in 3 seconds.
    a. What is her power rating in watts?
    b. What is her power rating in horsepower?
    c. How many joules of work does Michelle do?
    d. If Michelle uses 10 food calories to do the work, what is her energy efficiency?
3. Robert's car has a 40-horsepower engine that can accelerate from 0 to 60 mph in 16 seconds. Matt's car has a 375-horsepower engine. Assuming that the cars have the same mass, both have uniform acceleration, and ignoring friction, how many seconds will it take Matt to go from 0 to 60 mph?
4. A 2-horsepower engine runs a water pump for 24 hours.
    a. Calculate the work done by the engine.
    b. What would happen if a 1-horsepower engine was used to pump the same amount of water?

5. Carmine uses 800 joules of energy on a jack that is 85 percent efficient to raise her car to change a flat tire.
   a. How much energy is available to raise the car?
   b. If the car weighs 13,600 newtons, how high off the ground can she raise the car?

6. Suppose you exert 200 newtons of force to push a heavy box across the floor at a constant speed of 2 m/s.
   a. What is your power in watts?
   b. What would happen to your power rating if you used the same force to push the box at 1 m/s?

7. Suppose your job is to choose a motor for an escalator. The escalator must be able to lift 20 people at a time, each with a mass of 70 kg. The escalator must move between the two floors, which are 5 meters apart, in 5 seconds.
   a. What energy is required to do this?
   b. What is the power rating of the motor required, in horsepower?

8. Leroy takes in about 3,000 food calories per day. Calculate the average power this represents in watts over a 24-hour period.

9. Fill in the blanks in the energy-flow diagram. Compute the output work and the total wasted energy. What is the overall efficiency of the model solar car?

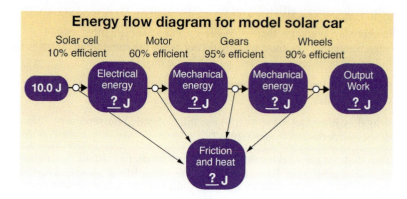

Applying your knowledge

1. Solar cells, also called photovoltaic cells, are used to power satellites in outer space, yet they are not used as commonly to power households on Earth. Use the Internet to research the use of solar cells to generate electricity for homes. Prepare a short report to answer the following questions.
   a. Why is it more difficult to use solar cells on the surface of Earth as opposed to outer space?
   b. What are the advantages and disadvantages of using solar cells to power a home?
   c. What current percent of homes in the United States use solar cells to generate at least some of their power? Is the number increasing or decreasing?
   d. Which state currently leads the nation in using solar cells to power homes? Why?

2. Nature's cycles have to do with how Earth renews itself. Living things interact with each other and also with their non-living environment to form a unit that is largely self-contained. The water cycle was explained in the this chapter and there are many more such cycles, including the life of plants and animals, energy cycles, disease cycles, and chemical cycles. Research and briefly describe two cycles and their importance in nature.

UNIT 4 ENERGY AND MOMENTUM

CHAPTER 12

Momentum

Objectives:

By the end of this chapter you should be able to:

- ✔ Calculate the linear momentum of a moving object given its mass and velocity.
- ✔ Describe the relationship between linear momentum and force.
- ✔ Solve a one-dimensional elastic-collision problem using momentum conservation.
- ✔ Describe the properties of angular momentum in a system, such as a bicycle.
- ✔ Calculate the angular momentum of a rotating object with a simple shape.

Key Questions:

- What is the law of conservation of momentum, and how is it used?
- How is momentum related to force?
- What is angular momentum, and how does it explain the motion of rotating objects?

Vocabulary

angular momentum   law of conservation of momentum   gyroscope   linear momentum

collision   elastic collision   impulse   inelastic collision   momentum

243

12.1 Momentum

According to Newton's first law, objects tend to continue in the motion they already have, with the same speed and direction. You already know it takes force to change the speed of a moving object, because you must do work to change the object's kinetic energy. But it also takes force to change an object's direction of motion, even if the speed remains the same. The more mass an object has, the more force it takes to deflect its motion. Why is this true? This section is about *momentum*, which is a *vector* property of moving matter that depends on both mass and velocity. Momentum describes the tendency of objects to keep going in the same direction with the same speed. Another way to look at force is that force is the action that changes momentum. Conversely, any change in momentum must create force.

Momentum comes from mass in motion

An example Consider two balls of the same mass moving in the same direction with different velocities (Figure 12.1). Ball A is slower than Ball B. Suppose a "sideways" 1-N force is applied to deflect the motion of each ball. What happens? Does the same force deflect each ball equally?

Momentum You probably guessed that the slower ball is deflected more than the faster ball (Figure 12.2). The difference is due to the difference in **momentum**. Momentum is a property of moving mass that resists changes in a moving object's velocity vector in either speed *or direction, or both*. The faster ball has more momentum, therefore its direction requires more force to change compared to the slower ball.

Momentum and inertia Inertia is another property of mass that resists changes in velocity. However, inertia depends *only* on mass and is a scalar, not a vector. The momentum vector depends on both mass and velocity. The momentum of a moving object increases as its mass or its velocity increases.

Kinetic energy and momentum are different Kinetic energy and momentum are different quantities, even though both depend on mass and speed. Kinetic energy is a scalar, and therefore it does not depend on direction. Two objects with the same mass and speed will always have the same kinetic energy. Kinetic energy is always positive or zero, but momentum can be positive or negative. Because momentum is a vector, it *always* depends on direction. Two objects with the same mass and speed have opposite momenta if they are moving in opposite directions (Figure 12.3).

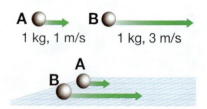

Figure 12.1: *The momentum of each ball depends on its mass and velocity. Ball B has more momentum than Ball A.*

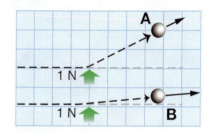

Figure 12.2: *Ball B deflects much less than Ball A when the same force is applied because Ball B has greater momentum.*

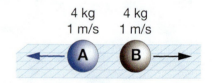

	Kinetic energy	Momentum
A	2 J	−4 kg·m/s
B	2 J	+4 kg·m/s

Figure 12.3: *Two balls with the same mass and speed can have the same kinetic energy but opposite momenta.*

MOMENTUM   CHAPTER 12

Calculating momentum

Momentum is mass times velocity

The momentum of a moving object is its mass multiplied by its velocity. For example, if a car and a truck are moving at the same velocity, the truck will have more momentum because it has greater mass. If two trucks of equal mass are moving, the one with the greater velocity has more momentum. Momentum is measured in units of kilogram-meters per second, or kg·m/s.

MOMENTUM

Momentum (kg·m/s) — $\vec{p} = m\vec{v}$ — Mass (kg), Velocity (m/s)

Momentum is a vector

When working with momentum, it is always necessary to consider the direction of motion. For many problems, it is convenient to choose positive and negative signs to indicate direction. Generally, momentum to the right is positive, and momentum to the left is negative (Figure 12.4). Because momentum has both magnitude and direction, momentum is always a vector. The symbol for the momentum vector is a lower case p with an arrow above it, $\vec{p}$.

4 kg
1 m/s

$\vec{p} = m\vec{v}$
$= (4 \text{ kg})(-1 \text{ m/s})$
$= -4 \text{ kg·m/s}$

4 kg
1 m/s

$\vec{p} = m\vec{v}$
$= (4 \text{ kg})(+1 \text{ m/s})$
$= +4 \text{ kg·m/s}$

Figure 12.4: *Velocity is usually defined to be positive to the right and negative to the left. In this coordinate system, momentum is also positive to the right and negative to the left.*

Comparing the momentum of a moving car and a motorcycle

A car is traveling at a velocity of 13.5 m/s (30 mph) north on a straight road. The mass of the car is 1,300 kg. A motorcycle passes the car at a speed of 30 m/s (67 mph). The motorcycle and rider have a combined mass of 350 kg. Calculate and compare the momentum of the car and motorcycle.

1. You are asked to calculate momentum.
2. You are given the masses and velocities.
3. Momentum is $\vec{p} = m\vec{v}$
4. Solve:
   For the car,

   $\vec{p} = (1{,}300 \text{ kg})(13.5 \text{ m/s}) = 17{,}550 \text{ kg·m/s}$

   Also, for the motorcycle and rider,

   $\vec{p} = (350 \text{ kg})(30 \text{ m/s}) = 10{,}500 \text{ kg·m/s}$

   The car has more momentum even though it is going much slower.

Momentum without mass

We have defined momentum for objects with mass. But momentum is a fundamental property of matter and energy. Light also carries momentum even though it is pure energy with no mass. The momentum of light depends on the energy of the light.

12.1 MOMENTUM   **245**

CHAPTER 12 MOMENTUM

Conservation of momentum and Newton's third law

The law of conservation of momentum
If you are on a skateboard and throw a heavy rock forward, you will move backward (Figure 12.5). The faster you throw, the faster you move backward. This effect is due to the **law of conservation of momentum**. When a system of interacting objects is not influenced by outside forces, like friction, this law says the *total momentum of the system remains constant*. When you throw a rock forward, the rock gets forward, positive momentum. Because the total momentum cannot change, you move backward with an equal amount of negative momentum. The positive momentum gained by the rock is exactly canceled by your gain of negative momentum. As a result, the total momentum of the system—you and the rock—is the same before and after the throw (Figure 12.6).

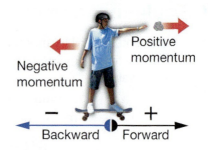

Figure 12.5: *If you throw a rock forward from a skateboard, you will move backward in response.*

Momentum and the third law
The law of conservation of momentum is a consequence of Newton's third law of action and reaction. To see the relationship, consider two balls of unequal mass connected by a spring. The balls are motionless and therefore have no momentum. When you compress the spring, the third law says the balls exert equal forces in opposite directions on one another, $-F_1 = F_2$.

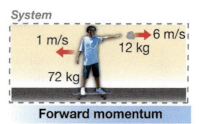

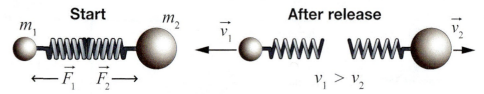

Proving the law of conservation of momentum
From Newton's second law, opposite forces create opposite accelerations, which create opposite velocities. The acceleration of each ball is inversely proportional to its mass, so the change in velocity is also inversely proportional to mass. The heavier ball has less change in velocity and the lighter ball has more change in velocity. The result is that equal and opposite forces create equal and opposite changes in momentum. The sum of equal and opposite changes is *zero*, therefore the total momentum of the system stays the same.

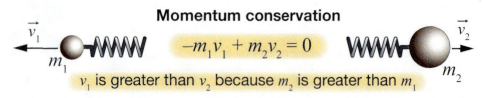

Figure 12.6: *The forward (positive) momentum of the rock exactly cancels the backward (negative) momentum from your own motion. The total change in momentum for the system (that being you and the rock) is zero.*

UNIT 4 ENERGY AND MOMENTUM

Collisions in one dimension

Momentum transfer in collisions A **collision** occurs when two or more objects interact with each other through forces. The objects may actually hit each other during the collision or they may not. During a collision, momentum is transferred from one object to another. According to the law of momentum conservation, the total momentum before the collision is equal to the total momentum after the collision.

Elastic and inelastic collisions In an **elastic collision**, the objects bounce off each other with no loss in the total kinetic energy. The collision of two steel balls is very close to a perfectly elastic collision (Figure 12.7). In an **inelastic collision**, objects change shape or stick together, and some kinetic energy is converted to other forms of energy such as heat, deformation, or friction. A clay ball hitting another clay ball is an example of an inelastic collision. So are two vehicles colliding (Figure 12.8). In both cases, some of the kinetic energy is used to permanently change an object's shape.

Momentum is conserved in all collisions Momentum is conserved in both elastic and inelastic collisions, even when kinetic energy is not conserved. Conservation of momentum allows you to determine the speeds and directions of objects *before* a collision, even if you only know the speeds and velocities *after* the collision. Accident investigators often use this approach to determine if anyone was speeding before an automobile accident.

Elastic collisions

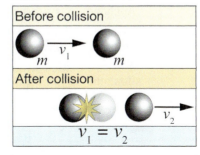

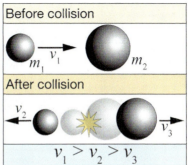

Figure 12.7: *Examples of elastic collisions.*

Inelastic collision

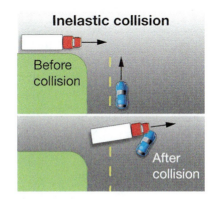

Figure 12.8: *Example of an inelastic collision.*

Collision type	Before collision	After collision
Elastic: Objects bounce off each other in the collision.	m_1 $\vec{v}_1$ → ← $\vec{v}_2$ m_2	← $\vec{v}_3$ m_1 m_2 $\vec{v}_4$ →
	$m_1\vec{v}_1 + m_2\vec{v}_2 = m_1\vec{v}_3 + m_2\vec{v}_4$	
Inelastic: Objects stick to each other after the collision.	m_1 $\vec{v}_1$ → ← $\vec{v}_2$ m_2	$(m_1 + m_2)$ $\vec{v}_3$ →
	$m_1\vec{v}_1 + m_2\vec{v}_2 = (m_1 + m_2)\vec{v}_3$	

Remember that velocity can be positive or negative. Also, the diagram only shows how to apply conservation of momentum to collisions in a straight line.

CHAPTER 12 MOMENTUM

Solving momentum problems

Applying conservation of momentum to physical situations

Using momentum to analyze problems takes practice. The first step is usually to draw a diagram and label positive and negative directions. Decide whether the collision is elastic or inelastic. The collision is *inelastic* if objects change shape or stick together. You will need to assign unique variable names to masses and any velocities that may be different before and after the collision.

Elastic collision of billiard balls

Two 0.165-kg billiard balls roll toward each other and collide head-on (Figure 12.9). Initially, the 10 ball has a velocity of 0.5 m/s. The 5 ball has an initial velocity of −0.7 m/s. The collision is elastic, and the 5 ball rebounds with a velocity of 0.4 m/s, reversing its direction. What is the velocity of the 10 ball after the collision?

1. You are asked to find the 10 ball's velocity after the collision.
2. You are given mass, initial velocities, and the 5 ball's final velocity. You can treat the collision as elastic since it involves billiard balls.
3. Make a diagram (Figure 12.9). Apply conservation of momentum to find the 10 ball's final velocity. $m_1v_1 + m_2v_2 = m_1v_3 + m_2v_4$
4. Solve:
   $(0.165 \text{ kg})(0.5 \text{ m/s}) + (0.165 \text{ kg})(-0.7 \text{ m/s}) = (0.165 \text{ kg})v_3 + (0.165 \text{ kg})(0.4 \text{ m/s})$
   $0.033 \text{ kg} \cdot \text{m/s} = (0.165 \text{ kg})v_3 + 0.066 \text{ kg} \cdot \text{m/s}$
   $v_3 = -0.6 \text{ m/s}$

The 10 ball travels at −0.6 m/s, the negative value indicating its movement in the opposite direction as the arrow in the diagram.

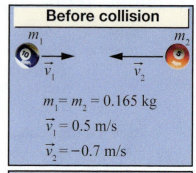

Figure 12.9: *Analyzing a collision between two billiard balls.*

Inelastic collision of train cars

A train car moving to the right at 10 m/s collides with a parked train car (Figure 12.10). They stick together and roll along the track. If the moving car has a mass of 8,000 kg and the parked car has a mass of 2,000 kg, what is their combined velocity after the collision?

1. You are asked for the velocity.
2. You are given the masses and the initial velocity of the moving car. You know the collision is inelastic because the cars stick together.
3. Apply the law of conservation of momentum. Because the cars stick together, consider the two cars to be one object after the collision.
   $m_1v_1 + m_2v_2 = (m_1 + m_2)v_3$ (from the diagram on the preceding page)
4. Solve:
   $(8,000 \text{ kg})(10 \text{ m/s}) + (2,000 \text{ kg})(0 \text{ m/s}) = (8,000 + 2,000 \text{ kg})v_3$
   $v_3 = 8$ m/s. The train cars move together to the right at 8 m/s.

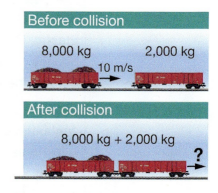

Figure 12.10: *What is the velocity of the two train cars together after one has collided with the other?*

Collisions in two and three dimensions

Two-dimensional collisions Most real-life collisions do not occur in one dimension. In a two- or three-dimensional collision, objects move at angles to each other before or after they collide. For example, two billiard balls might have a two-dimensional collision as shown in the diagram. Momentum conservation still applies and can be used to determine the directions and velocities before or after the collision. Because angles are involved, the math is more complicated. There are also more variables, so both energy and momentum equations need to be used.

Using momentum conservation In order to analyze two-dimensional collisions you need to look at each coordinate direction separately. In two dimensions, momentum is conserved *separately* in the x and y directions. Two- and three-dimensional problems are easiest to solve when the coordinate axes are chosen wisely. Usually you will want to align the x-axis with one of the initial velocities. Next, each velocity vector is resolved into its x and y components. Momentum is calculated separately in the x and y directions both before and after the collision. The total x momentum must be the same before and after the collision. The same goes for the total y momentum.

Accident reconstruction

STEM

Police forensics specialists use conservation of momentum and other physics knowledge to analyze traffic accidents. Skid marks, debris such as broken glass, and other clues allow investigators to reconstruct the events of an accident scene with surprising accuracy.

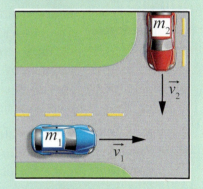

Skid marks are used to determine the directions of the vehicles before and after the crash. Skid marks can also be used to estimate velocities. With information on friction, skid distances, and directions, the forensics specialists use momentum conservation to determine the vehicles' velocities before and after the crash.

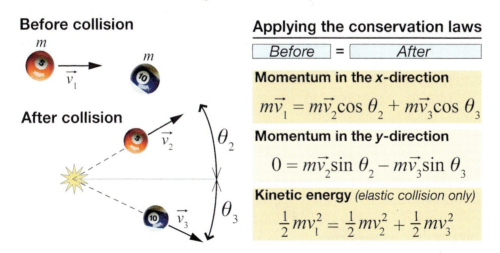

Before collision

After collision

Applying the conservation laws

| Before | = | After |

Momentum in the x-direction

$$m\vec{v}_1 = m\vec{v}_2 \cos\theta_2 + m\vec{v}_3 \cos\theta_3$$

Momentum in the y-direction

$$0 = m\vec{v}_2 \sin\theta_2 - m\vec{v}_3 \sin\theta_3$$

Kinetic energy *(elastic collision only)*

$$\tfrac{1}{2}mv_1^2 = \tfrac{1}{2}mv_2^2 + \tfrac{1}{2}mv_3^2$$

Energy conservation For elastic collisions, kinetic energy is also conserved. That means the total kinetic energy before the collision equals the total kinetic energy after the collision. Many two-dimensional problems require a separate equation for energy conservation.

12.2 Force is the Rate of Change of Momentum

Momentum changes when a net force is applied. The converse is also true—if momentum changes, forces are created. If momentum changes quickly, large forces are involved. It is sometimes useful to interpret Newton's second law as a prescription for relating force to the change in *momentum*, not acceleration. As a practical application, all cars sold today are equipped with safety devices designed to slow down any change in momentum your body might experience in a crash. The real purpose of seat belts and air bags is to reduce the force on your body by spreading out the change in momentum from a crash over a longer period of time (Figure 12.11). This section discusses the relationship between force and momentum, and provides a new interpretation of Newton's second law.

Figure 12.11: *Seat belts and air bags work together to stop passengers safely in the event of a crash.*

Car crash safety

Minimizing the forces on car passengers When a car crashes to a stop, the momentum of the car drops to zero as the body of the car crumples. Car bodies are designed to absorb the momentum of a crash by crumpling as slowly as possible. The same is not true for human bodies! Due to inertia, a passenger without a seat belt will fly forward and change momentum very fast by hitting the windshield, steering wheel, or dashboard (Figure 12.12). This rapid change in momentum is what produces the forces that cause injuries, or even fatalities, when they act on a body.

Minimizing forces in a collision Rapid acceleration creates large forces. The opposite is also true—gradual acceleration creates smaller forces. The key to remaining uninjured in a crash is to make any changes in the velocity of your body as slow as possible, and thereby reduce the magnitude of forces acting on your body. Seat belts are made of very strong fabric that stretches slightly when a force is applied. By holding you in the seat, a seat belt keeps *your* momentum from changing any faster than the momentum of the whole car.

Figure 12.12: *Crash-test dummies are used in car safety tests. Sensors placed at different locations on the dummy record the forces experienced during a crash.*

Air bags *Air bags* work together with seat belts to bring passengers to a stop as gradually as possible. An air bag inflates when the force applied to the front of a car reaches a dangerous level. The air bag deflates slowly as the passenger's body applies a force to the bag upon impact. The force of impact pushes the air out of small holes in the air bag, and the force of rapid momentum change is dissipated over time. Many cars now contain both front and side air bags.

Force and momentum change

Force is rate of change of momentum

The relationship between force and motion follows directly from the second law. Acceleration is the change in speed divided by the change in time. In this type of logic it is convenient to use as a symbol the Greek letter *delta* (Δ) which translates to "the change in." When you see the Δ symbol, replace it in your mind with the phrase "the change in." The acceleration can then be written as Δv/Δt, which translates to "the change in speed divided by the change in time."

Newton's second law $$\vec{F} = m\frac{\Delta \vec{v}}{\Delta t}$$ Translation ⟹ Force = mass × (change in velocity)/(change in time)

Figure 12.13: *Large forces produce a large change in momentum, such as when you hit a golf ball.*

Momentum form of Newton's second law

Mass multiplied by the change in velocity ($m \times \Delta \vec{v}$) is the same as the change in momentum ($\Delta \vec{p}$). We can rewrite the second law in a form that shows force is equal to the rate of change in momentum. Large forces produce a proportionally large change in momentum (Figure 12.13).

NEWTON'S SECOND LAW momentum form

Force (N) — $$\vec{F} = \frac{\Delta \vec{p}}{\Delta t}$$ — Change in momentum (kg·m/s) / Change in time (s)

The momentum form of Newton's second law is sometimes more useful than the acceleration form. For example, pure energy can have momentum, but not mass. The momentum form of the second law is used to do calculations with light and atoms because light is pure energy that has momentum without mass.

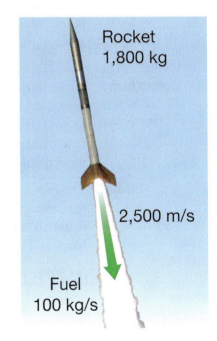

Calculating force on a rocket from the change in momentum

Starting at rest, an 1,800-kg rocket takes off, ejecting 100 kg of fuel per second out of its nozzle at a speed of 2,500 m/s. Figure 12.14 shows a diagram of the motion. Calculate the force on the rocket from the change in momentum of the fuel.

1. You are asked for the force exerted on the rocket.
2. You are given that the rocket ejects fuel at 100 kg/s at a speed of 2,500 m/s.
3. Use the equation $F = \Delta p/\Delta t$.
4. Solve: $\Delta p = (100 \text{ kg})(-2{,}500 \text{ m/s}) = -25{,}000$ kg·m/s
   $F = \Delta p / \Delta t = (-25{,}000 \text{ kg·m/s}) \div (1 \text{ s}) = -25{,}000$ N
   The rocket exerts the −25,000 N force on the fuel. The fuel exerts an equal and opposite force on the rocket of +25,000 N which makes the rocket go forward.

Figure 12.14: *Momentum change from the fuel creates the force that drives a rocket forward.*

CHAPTER 12   MOMENTUM

Impulse

Force from elastic and inelastic collisions

A rubber ball and a clay ball are dropped on the hard floor of a gym (Figure 12.15). The rubber ball has an elastic collision and bounces back upward. The clay ball hits the floor with a thud and stays there (inelastic collision). Both balls have the same mass and are dropped from the same height. They have the same speed as they hit the floor. Which ball exerts a greater force on the floor?

Bounces have greater momentum change

The best way to determine the greater force is to compare the change in momentum. For example, suppose both balls have a mass of 1 kg, and are moving at 2 m/s when they hit the floor (Figure 12.16). The momentum of the rubber ball has a larger change, −2 kg·m/s to +2 kg·m/s, a net change of +4 kg·m/s. The momentum of the clay ball goes from −2 kg·m/s to zero, for a net change of +2 kg·m/s. The rubber ball (elastic collision) has *twice* the change in momentum and can exert *twice as much force* on the floor compared to the clay ball. The momentum change is always greater when objects bounce compared with when they do not bounce.

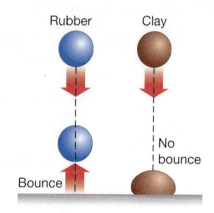

Figure 12.15: *A rubber ball bounces because it has an elastic collision with the floor. A clay ball has an inelastic collision and does not bounce.*

What can be learned from a change in momentum

This is approximately right; however, there is a detail left to solve. A change in momentum tells you how the *product* of force and time changes, but not force or time individually. A change of 4 kg·m/s could result from a force of 4 N for one second or a force of 1 N for 4 seconds. In fact, the change in momentum could result from any combination of force and time that has a product of 4 N·s when multiplied together.

Impulse

The product of force and time is called the **impulse**. To find the impulse, you rearrange the momentum form of Newton's second law. This form of the equation is useful because collisions happen fast. It is *not* always possible to calculate the force and time individually, but you can determine their product.

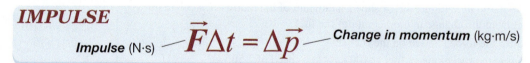

IMPULSE   Impulse (N·s) — $\vec{F}\Delta t = \Delta \vec{p}$ — Change in momentum (kg·m/s)

Units of impulse

Notice that the force side of the equation has units of N·s, while the momentum side has units of momentum, kg·m/s. However different they look, these are the same units, since 1 N is 1 kg·m/s². Impulse can be expressed in kg·m/s or in N·s.

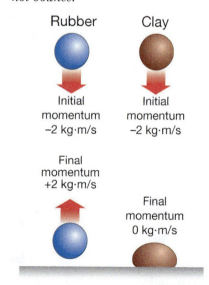

Figure 12.16: *Calculating the momentum change for the rubber ball versus the clay ball.*

12.3 Angular Momentum

An object that is spinning tends to keep spinning because of *angular momentum*. Angular momentum is a lot like linear momentum except that it applies to rotating motion. The conservation of angular momentum is why an ice skater spins faster when she pulls her arms in and why a cat can twist in midair to land on its feet. Angular momentum is also why the Moon orbits Earth instead of being drawn straight in by Earth's gravity. This section introduces angular momentum and describes a few examples of systems where angular momentum plays an important role.

What is angular momentum?

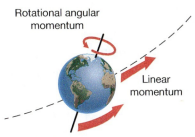

Figure 12.17: *The Earth has both linear and angular momentum from its rotation about its axis and its revolution around the Sun.*

Two types of momentum Momentum resulting from an object moving in linear motion is called **linear momentum**. Linear momentum was the subject of the first two sections of this chapter. Momentum resulting from the rotation of an object is called **angular momentum**. Like rotational inertia, an object's angular momentum depends on the mass and shape of the object, as well as its rotational speed.

Objects can have both types of momentum An object can have both types of momentum. For example, Earth has linear momentum and angular momentum. Earth has linear momentum because the entire planet moves as it travels around the Sun. Earth has two kinds of angular momentum. Earth's *rotational* angular momentum comes from its daily *rotation* on its axis. Earth's *orbital* angular momentum comes from its yearly *revolution* around the Sun (Figure 12.17). The motion of a planet is determined by the conservation of both linear and angular momentum.

Understanding angular momentum To get a sense for why objects have angular momentum, think about a wheel that is spinning (Figure 12.18). The wheel has mass and every part of the wheel is rotating except the exact center. That means each particle of mass in the wheel has linear momentum. Particles far from the center are moving faster and have more linear momentum. The angular momentum of the wheel comes from the organized motion of the *entire wheel* around a center of rotation. Angular momentum is a separate quantity from linear momentum and is always defined about a specific center of rotation.

Figure 12.18: *Mass that is close to the center of rotation has less momentum because it is moving more slowly.*

The importance of angular momentum Angular momentum is important because it obeys a separate conservation law from linear momentum. Angular momentum is also a fundamental property of elementary particles in the atom, like electrons and protons (Chapter 28).

Conservation of angular momentum

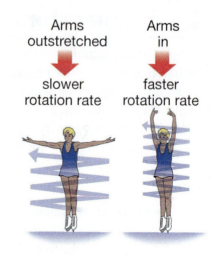

Why angular momentum is important
Angular momentum is important because it obeys a conservation law, similar to linear momentum. If the net torque acting on a closed system is zero, the total angular momentum of the system stays constant. The only way to change a system's angular momentum is to apply a torque.

Mass and speed
Like linear momentum, angular momentum is proportional to mass and velocity. The faster an object rotates, the more angular momentum it has. *Unlike* linear momentum however, the shape and distribution of an object's mass makes a big difference to its angular momentum.

The distribution of mass
Angular momentum does not depend on mass directly, but instead depends on the *moment of inertia* around the center of rotation. Mass that is far from the axis of a rotating object has a larger moment of inertia and contributes more angular momentum for a given angular velocity. Mass close to the axis has a smaller moment of inertia and contributes less angular momentum. Mass on the exact axis of rotation has *zero* moment of inertia and *zero* angular momentum.

Figure 12.19: *The skater's angular velocity increases when she decreases her moment of inertia.*

The spinning skater and the diver
A spinning figure skater provides a good example of how angular momentum affects motion (Figure 12.19). When a skater is spinning with her arms out, the mass in her arms is relatively far from her axis of rotation. She has a large moment of inertia and is spinning relatively slowly. When she pulls her arms in, her moment of inertia decreases because the mass of her arms is now closer to the axis of rotation. As her arms come in, she starts spinning faster! She spins faster because her total angular momentum must stay the same. Her moment of inertia has *decreased* so her angular velocity increases to compensate. The opposite effect happens when a spinning diver comes out of a tuck. Straightening out puts more of the diver's body mass farther from the center of rotation. This *increases* the diver's moment of inertia and slows the diver's rotational speed.

Angular momentum
moment of inertia (I)

angular velocity (ω)

Conservation of angular momentum

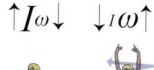

Angular momentum changes through torque
Angular momentum can change through the application of torque. The net torque on a rotating object equals the rate of change of its angular momentum, similar to linear momentum. Since friction can exert torque, a spinning figure skater eventually slows down and loses her angular momentum. To get a spin going, the skater pushes against the ice with her skates, creating a reaction torque on her body that increases her angular momentum.

Calculating angular momentum

Formula for angular momentum
Angular momentum is calculated in a similar way to linear momentum, except the mass and velocity are replaced by the moment of inertia and angular velocity. The angular momentum is the moment of inertia of an object times its angular velocity. The capital letter L is used for angular momentum, with a vector arrow above it. (For review, check Sections 9.3 for moment of inertia and 8.1 for angular velocity.) The SI units of angular momentum are $kg \cdot m^2/s$.

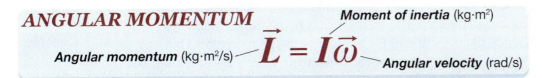

ANGULAR MOMENTUM

$$\vec{L} = I\vec{\omega}$$

Angular momentum ($kg \cdot m^2/s$) — Moment of inertia ($kg \cdot m^2$) — Angular velocity (rad/s)

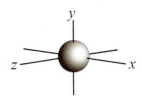

Sphere of mass, m, radius, r

Hoop of mass, m, radius, r

Bar of mass, m, length, l

Calculating angular momentum
To calculate the angular momentum, you need to know the angular velocity and the moment of inertia. Angular velocity should be in radians per second (rad/s). The moment of inertia is in units of $kg \cdot m^2$.

Moment of inertia
Remember from Chapter 9 that the moment of inertia of an object is the average of mass times radius squared for the whole object. Since the radius is measured from the axis of rotation, the moment of inertia depends on the axis of rotation. The table in Figure 12.20 gives the moment of inertia for some simple geometric shapes rotated around axes that pass through their centers.

Moment of inertia			
	Axis of rotation		
	x	y	z
Sphere	$\frac{2}{5}mr^2$	$\frac{2}{5}mr^2$	$\frac{2}{5}mr^2$
Hoop	$\frac{1}{2}mr^2$	mr^2	$\frac{1}{2}mr^2$
Bar		$\frac{1}{12}ml^2$	$\frac{1}{12}ml^2$

Figure 12.20: *Moment of inertia for some simple shapes.*

Calculating angular momentum for two objects of equal mass and different shapes

An artist is making a moving metal sculpture. She takes two identical 1-kg metal bars and bends one into a hoop with a radius of 0.16 m. The hoop spins like a wheel. The other bar is left straight with a length of 1 meter. The straight bar spins around its center, like the y-axis in Figure 12.20. Both have an angular velocity of 1 rad/s. Calculate the angular momentum of each and decide which would be harder to stop.

1. You are asked for the angular momentum.
2. You are given the angular velocity, shape, and mass.
3. Use the equation $L = I\omega$.
4. Hoop: $I = (1\ kg)(0.16\ m)^2 = 0.026\ kg \cdot m^2$: $L = (1\ rad/s)(0.026\ kg \cdot m^2) = 0.026\ kg \cdot m^2/s$
   Bar: $I = (1/12)(1\ kg)(1\ m)^2 = 0.083\ kg \cdot m^2$: $L = (1\ rad/s)(0.083\ kg \cdot m^2) = 0.083\ kg \cdot m^2/s$
   The bar has more than three times as much angular momentum as the hoop, and is therefore harder to stop.

Gyroscopes and angular momentum

The angular momentum vector

Angular momentum has a direction like any other vector. However, the direction of the angular momentum vector is *not* the direction of motion. Instead, the direction of the angular momentum vector points *along the axis of rotation* (Figure 12.21). Counterclockwise rotation is usually chosen to have positive angular momentum. Clockwise rotation therefore has negative angular momentum.

Torque resists change in angular momentum

As with linear momentum, changing the direction of the angular momentum vector creates a torque that resists the change. This is often demonstrated by holding a spinning bicycle wheel. The wheel resists being turned sideways because the turn shifts the direction of the angular momentum vector. This principle explains why bicycles stay vertical when their wheels are spinning, but fall over easily when standing still.

Gyroscopes

A **gyroscope** is a device that contains a spinning mass with a lot of angular momentum. Gyroscopes can do amazing tricks because they conserve angular momentum. For example, a *spinning* gyroscope can easily balance on a pencil point. The gyroscope falls off when it is not spinning.

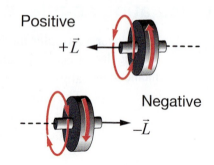

Figure 12.21: *The direction of the angular momentum vector.*

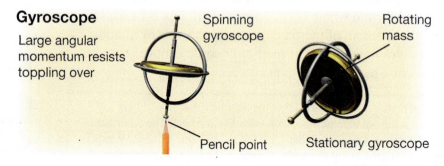

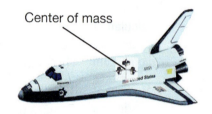

Turning the space shuttle

The space shuttle, and other spacecraft, use gyroscopes for control. The shuttle's gyroscope contains three spinning wheels aligned along each of the coordinate axes (Figure 12.22). When the shuttle makes a small rotation, the movement is resisted by the angular momentum of the gyroscope wheels. Sensors on each wheel detect the small torques created in the x, y, and z directions. By analyzing the torques from the gyroscope, an on-board computer is able to accurately measure the rotation of the shuttle and maintain its orientation in space.

Figure 12.22: *The gyroscope on the space shuttle is mounted at the center of mass, allowing a computer to measure rotation of the spacecraft in three dimensions.*

Chapter 12 Connection

Jet Engines

Nearly all modern airplanes use jet propulsion to fly. Jet engines and rockets work because of conservation of linear momentum (Figure 12.23). Air is accelerated to a very high speed as it passes through the jet engine. The change in speed of the mass of air passing through the engine is equal to the change in the air's momentum. The increase in momentum of the air results in the forward thrust force on the plane.

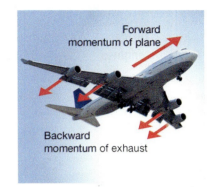

Figure 12.23: *Momentum transfer in a jet airplane.*

The turbojet engine The diagram shows a cutaway view of a turbojet engine (Figure 12.24). The main component of the engine is a high-speed rotating shaft with a series of fans. Air entering the engine passes first through the fans of the compressors. Fuel is injected into the hot air as it moves through the combustion chamber. The air leaves the engine through the nozzle as exhaust jet.

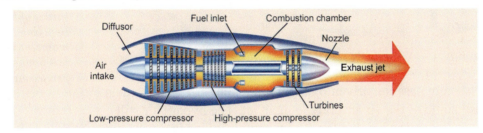

How a turbojet works Air enters the front of the jet engine. The air then passes through a series of fans that make up the compressor, where it is squeezed, causing its pressure to increase. As the pressure of the air increases, its temperature also rises. The hot compressed air then passes into the combustion chamber, where it is injected with fuel. The fuel burns, heating the air and causing it to expand to many times its original volume. The expanded hot air is expelled out the nozzle in the engine's exhaust in a very fast stream, called a *jet*, which is where the jet engine gets its name.

The "turbo" in turbojet The exhaust jet passes through a second set of fans called the *turbine*. The motion of the exhaust jet spins the turbine. The turbine extracts a small fraction of the power of the exhaust jet in order to turn the compressor fans.

Why are jet engines so loud? The exhaust speed of modern passenger jet engines is greater than 300 m/s (670 mph). The tremendous noise from an operating jet engine comes from the hot, high-velocity exhaust jet colliding with the cooler air outside the plane.

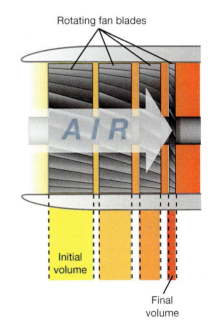

Figure 12.24: *The compressor in a turbojet engine.*

12.3 ANGULAR MOMENTUM

Chapter 12 Connection

Turbofan engines

Turbofan engines
The diagram below shows a cutaway view of a turbofan jet engine. The Boeing 757-200 passenger jet uses two of these engines and can fly 3,500 miles carrying about 185 people at a cruising speed of 270 m/s (600 mph).

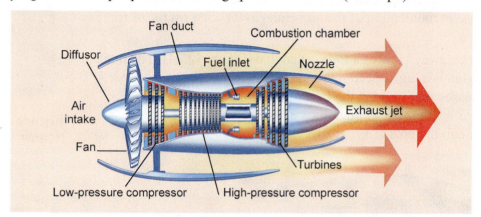

Figure 12.25: *Calculating the thrust of the turbofan jet engine by applying the conservation of linear momentum. This large jet engine has a mass flow rate of 550 kg of air per second.*

Turbofan engines are more fuel efficient
The turbofan engine has a bypass fan driven by the main jet engine. The fan acts like a propeller to push some of the air around the main jet. The bypass air still generates thrust because it is pushed by the fan. The turbofan engine is much more fuel efficient than a plain turbojet. For example, the Pratt and Whitney PW-2037 turbofan engine used on Boeing 757s generates a takeoff thrust of 170,000 N (38,000 lbs) and uses 48 percent less fuel than a turbojet engine of equal size.

Momentum conservation
The engine thrust force is equal to the rate of change in momentum. At takeoff, the PW-2037 engine moves 550 kg of air and fuel per second. By analyzing the change in momentum (Figure 12.25), you can calculate the speed of the exhaust jet. To create a thrust force of 170,000 N, the average exhaust speed of the jet is 310 m/s.

Rocket engines
A rocket engine uses the same principles as a jet, except that in space, there is no oxygen. A jet engine on Earth burns fuel using oxygen from the air. A rocket has to carry the oxygen along with the fuel. Most rockets have to carry so much oxygen and fuel that the payload of people or satellites is usually less than 5 percent of the total mass of the rocket at launch. The other 95 percent is fuel.

Chapter 12 Assessment

Vocabulary

Select the correct term to complete the sentences.

momentum	linear momentum	angular momentum
conservation of momentum	elastic	inelastic
collision	gyroscope	moment of inertia
impulse		

1. The type of collision during which no kinetic energy is lost is a(n) _____ collision.
2. The product of the mass of a moving object and its velocity is equivalent to its _____.
3. A collision during which some kinetic energy of the objects is transformed into heat, sound, or friction can be characterized as _____.
4. The total momentum of a system of interacting objects cannot change unless an external force is applied according to the law of _____.
5. A(n) _____ occurs when two or more objects hit each other.
6. The product of the force acting on an object and the time during which it acts on the object is called the _____.
7. The angular momentum of an object is equal to the product of its angular velocity and its _____.
8. Momentum resulting from the motion of an object in a straight line is known as _____.
9. A device that contains a spinning object with a lot of angular momentum is a(n) _____.
10. Momentum resulting from rotation of an object is called _____.

Concept review

1. Compare and contrast kinetic energy and momentum. How are they similar? How are they different?
2. Why is momentum considered a vector quantity?
3. Decide which of the following statements are correct about momentum. If a statement is correct, simply answer "correct." If a statement is incorrect, rewrite the statement to make it correct.
    a. The momentum of any object is conserved and remains constant.
    b. Objects with mass have momentum.
    c. All moving objects have momentum.
    d. Momentum is measured in joules.
    e. An object moving at a constant speed has no momentum.
    f. The momentum of two objects with equal mass will be equal.
    g. If two objects of different mass move at the same speed, they will have the same momentum.
    h. When an object's velocity changes, its momentum changes.
    i. The momentum of an accelerating object increases.
    j. The momentum of a faster object will always be greater than the momentum of a slower object.
    k. If an object has zero momentum, then it has zero kinetic energy.
4. Consider a bulldozer at rest and a hockey player moving across the ice. Explain each answer.
    a. Which has greater mass?
    b. Which has greater velocity?
    c. Which has greater momentum?
5. A bicycle rider doubles her speed. What happens to her momentum?
6. The law of conservation of momentum is a consequence of which one of Newton's laws of motion?

7. Howard tries to jump from a rowboat to a nearby dock. Why is this a bad idea, even if the distance from the rowboat to the dock seems short enough to jump?

8. How does the momentum of a jet engine's backward exhaust compare to the momentum of the jet going forward? Are the velocities of the backward exhaust and the forward-moving jet the same? Explain.

9. When playing a game of billiards, you use a long stick to hit a white ball (the cue ball) that then collides with numbered balls to send them into pockets or holes around the edge of the table. Sometimes, you want the cue ball to come to a sudden stop after it knocks into a numbered ball so that the cue ball doesn't follow the numbered ball into a pocket. Suppose you manage to collide the cue ball head-on with a numbered ball and the cue ball stops.

   a. What is true about the total momentum of the colliding billiard balls before and after the collision?
   b. Is this collision more like an elastic or an inelastic collision? Explain.
   c. What is true about the kinetic energy of the cue ball before and after the collision?
   d. How does the velocity of the cue ball before the collision compare to the velocity of the numbered ball after the collision? (*Note*: Assume a perfectly elastic collision.)
   e. What do you predict would happen to the balls if the cue ball were to hit the numbered ball at an angle rather than head-on?

10. What is the difference between impact and impulse?

11. Is the unit used to represent impulse equivalent to the momentum unit? Explain.

12. Is it true that in a collision, the net impulse experienced by an object is equal to its momentum change? Explain.

13. Padding is used in baseball gloves, goalie mitts, motorcycle helmets, and gymnastic mats. Explain the physics behind the use of padding in these cases.

14. What is the secret to catching a water balloon without breaking it? Explain using physics.

15. Contrary to popular belief, cars that crumple in a collision are safer than cars that rebound when they collide. Explain why this is so.

16. What is the difference between linear and angular momentum? Can an object have both types of momentum at the same time? Explain.

17. Suppose a solid cylinder and a hollow cylinder of the same mass and radius are released on an incline. Which rolls with greater acceleration? Explain.

18. Suppose two solid cylinders of the same radius but very different masses are released on an incline. How do their accelerations compare? Explain.

19. A physics teacher stands on a small rotating platform and holds a 10-pound weight in each hand with her arms outstretched. A student helps by starting the teacher in a slow rotation. The teacher then pulls her arms in and holds the weights close to her body.
    a. How does the teacher's angular momentum compare before and after pulling her arms in?
    b. How does the teacher's moment of inertia compare before and after pulling her arms in?
    c. How does the teacher's angular velocity compare before and after pulling her arms in?

20. Identify the quantity represented by each symbol or combination of symbols below.
    a. ω b. I c. L
    d. Δp e. $I\omega$ f. $F\Delta t$

Problems

1. Could a 1,000-kg elephant moving very slowly (0.5 m/s) ever have less momentum than a 0.15-kg baseball pitched by a major-league baseball pitcher? Justify your answer.

2. Which has more momentum: a boat with a mass of 1,200 kg moving at 50 m/s, or a truck with a mass of 6,000 kg moving at 10 m/s? Show your work.

3. A 0.040-kg golf ball leaves the head of a golf club with a speed of 50 mph.

   a. What is the momentum of the golf ball in kg·m/s?
   b. How fast would a 0.15-kg baseball have to be thrown to have the same momentum? Find your answer in m/s and then convert it to mph.

4. Suppose a 1,200-kg car moving at 12 m/s crashes head-on and locks on to a 1,000-kg car moving at 16 m/s in the opposite direction.
   a. Is this collision elastic or inelastic? Why?
   b. What is the velocity of the cars after the collision?

5. Two baseball players accidentally run into each other when trying to catch a fly ball. Player 1 is 75 kg and is moving east at 3 m/s. Player 2 is 70 kg and is moving west at 2 m/s. How fast are they moving together after the collision? Find the speed and specify the direction of motion.

6. Jessie rolls a 4.5-kg bowling ball down a long hallway at 5 m/s. Sue rolls a 5-kg bowling ball down the same hallway in the opposite direction at a velocity of −7 m/s. The balls collide, and the 4.5-kg ball moves away at −8.5 m/s after the collision. Find the final velocity of the 5-kg ball.

7. An average force of 20 N is exerted on an object for 2 s.
   a. What is the impulse?
   b. What is the change in momentum?

8. Tracy has a mass of 65 kg and is driving her sports car at 30 m/s when she must suddenly slam on her brakes to avoid hitting a moose in the road. The air bag brings her body to rest in 0.40 s.
   a. What average force does the airbag exert on Tracy?
   b. If Tracy's car had no air bag, the windshield and structure of the car would have stopped her body in 0.001 s. What average force would the windshield have exerted?

9. A solid ball has a mass of 0.5 kg and a radius of 0.1 m. It rolls along the ground at 3 m/s.
   a. Calculate its moment of inertia.
   b. Calculate its angular momentum.
   c. Calculate its linear momentum.

10. An ice skater spins at an angular speed of 10 rad/s with her arms down at her sides. Her moment of inertia is 1.5 kg·m².
    a. What is her angular momentum?
    b. If she lifts her arms out, her moment of inertia increases to 3 kg·m². What is her new angular speed?

11. The diagram below illustrates a 0.20 kg ball, B_1, rolling at 6 m/s toward a second identical stationary ball, B_2. Upon colliding, ball B_1 moves off at an angle of 37° to the original path while ball B_2 moves off at an angle of 53°. Calculate the speed of each ball after the collision.

    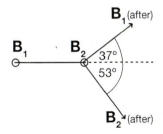

Applying your knowledge

1. During the California Gold Rush, a study of waterwheels was made by a man named Lester A. Pelton. It resulted in what has become known as a Pelton wheel. What did Pelton observe? How did he use the physics of impulse and momentum to improve the waterwheel?

2. Research how railroad cars can be connected in combinations stretching for more than one mile in length. Research the term *slack* as it applies to railroad terminology. Use this and your knowledge of momentum and impulse to explain how it is possible for one locomotive to initiate motion of an entire one-mile train of coupled cars.

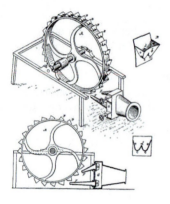

Original patent sketch of the Pelton Wheel, 1880.

A Pelton turbine

UNIT 5 WAVES AND SOUND

CHAPTER 13

Harmonic Motion

Objectives:

By the end of this chapter you should be able to:

✓ Identify characteristics of harmonic motion, such as cycles, frequency, and amplitude.
✓ Determine period, frequency, and amplitude from a graph of harmonic motion.
✓ Use the concept of phase to compare the motion of two oscillators.
✓ Describe the characteristics of a system that lead to harmonic motion.
✓ Describe the meaning of natural frequency.
✓ Identify ways to change the natural frequency of a system.
✓ Explain harmonic motion in terms of potential and kinetic energy.
✓ Describe the meaning of periodic force.
✓ Explain the concept of resonance and give examples of resonance.

Key Questions:

- What are some examples of harmonic motion and oscillators?
- Is it possible to change the natural frequency of a system?
- What is resonance, and why is it important?

Vocabulary

amplitude	hertz (Hz)	periodic force	piezoelectric effect
damping	natural frequency	periodic motion	resonance
frequency	oscillator	phase	stable equilibrium
harmonic motion	period	phase difference	unstable equilibrium

CHAPTER 13 HARMONIC MOTION

13.1 Harmonic Motion

As you watch moving things around you, there are two different categories of motion you will observe. One category includes motion that goes from place to place without repeating. This category is *linear motion*. The concepts of distance, time, speed, and acceleration come from thinking about linear motion. The second category includes motion that repeats over and over. This category is **harmonic motion**, and is the subject of this section. The word *harmonic* comes from *harmony*, which means "multiples of." Harmonic motion repeats in identical or nearly identical cycles. A pendulum swinging back and forth is a good example of harmonic motion (Figure 13.1).

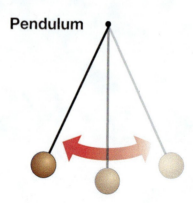

Figure 13.1: *A pendulum swings back and forth in harmonic motion.*

Cycles, systems, and oscillators

What is a cycle? The cycle is the building block of harmonic motion. A cycle is a unit of motion that repeats. The defining characteristic of harmonic motion is a repeated sequence of cycles. The diagram shows the cycles of a simple pendulum.

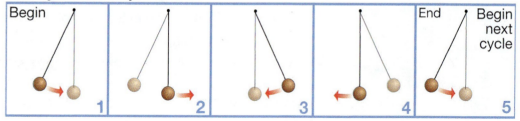

Finding the cycle The first step to investigating harmonic motion is often to identify the basic cycle. A single cycle has a beginning and an end. Between the beginning and end, the cycle has to include all the motion that repeats. The cycle of the pendulum is defined by where we choose the beginning. If we start the cycle when the pendulum is all the way to the left, the cycle ends when the pendulum has returned all the way to the left. The motion of the pendulum is one cycle after another with no gaps between cycles.

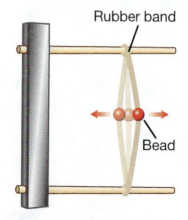

Figure 13.2: *A bead going back and forth on a stretched rubber band is a good example of an oscillator in harmonic motion.*

Definition of an oscillator A system in harmonic motion is called an **oscillator**. A pendulum is one example of an oscillator; a vibrating bead on a rubber band is another (Figure 13.2).

UNIT 5 WAVES AND SOUND

Harmonic motion is very common

Music comes from oscillations
Sound is an oscillation of the air. Musical instruments and speakers are oscillators that we design to create sounds. If you gently touch a speaker making sound, you can feel the rapid vibration that comes from the oscillation of the speaker's diaphragm. The oscillation of the speaker diaphragm causes the air to oscillate. The oscillation of the air travels to your eardrum where it causes tiny bones in your ear to oscillate in response. There is harmonic motion at every step, from the original musical instrument, to the speaker, to the detection of sound in your ear.

Oscillators are used in communications
Almost all modern communication technology relies on fast electronic oscillators. Cellphones use oscillators that make more than 100 million cycles each second. In the United States, FM radio uses oscillators between 88 million and 108 million cycles per second. When you tune a radio, you are selecting the frequency of the oscillator you want to listen to. Each station sets up an oscillator at a different frequency. Sometimes you can receive two stations at once when you are traveling between two radio towers with nearly the same frequency.

Oscillators are used to measure time
Each cycle of a perfect oscillator takes the same amount of time. This makes harmonic motion a good way to keep time. If you have a pendulum that has a cycle 1 second long, you can count time in seconds by counting cycles of the pendulum. Pendulum clocks and mechanical watches count cycles of oscillators to keep time (Figure 13.3). Even today, the world's most accurate clocks keep time by counting cycles of light from a cesium atom oscillator. Modern atomic clocks are accurate to within 1 second in 300,000,000 years!

Natural cycles involving Earth
An *orbit* is a type of cycle because it is repeating motion. Earth is a part of several oscillating systems. The Earth-Sun system has an orbital cycle of 1 year, which means Earth completes one orbit around the Sun in a year. The Earth-Moon system has a orbital cycle of approximately 1 month. Earth itself has several different cycles (Figure 13.4). Earth rotates on its axis once a day, creating the 24-hour cycle of day and night. There is also a wobble of Earth's rotational axis that completes a full cycle about every 22,000 years, moving the location of the north and south poles around by hundreds of miles. There are cycles in weather, such as El Niño and La Niña oscillations in ocean currents, that produce fierce storms every decade or so. Much of our planet's ecology depends on cycles.

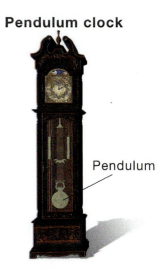

Figure 13.3: *A pendulum clock uses the period of a pendulum to count time.*

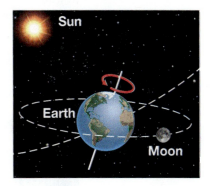

Figure 13.4: *The Earth-Moon-Sun system has many cycles.*

CHAPTER 13   HARMONIC MOTION

Describing harmonic motion

Period is the time for one cycle

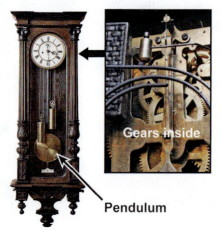

Pendulum • Gears inside

The hands on a pendulum clock are counters. They count cycles of the clock's pendulum and counting cycles is how the clock keeps time. The time it takes for the pendulum to complete one full cycle is called a **period**. The period is one of the important characteristics of all harmonic motion (Figure 13.5). A large, floor-standing clock might have a pendulum with a period of 2 seconds. A full minute would be 30 cycles of this pendulum. The clock's gears cause the minute hand to move one-sixtieth of a turn for every 30 swings of the pendulum.

The **period** is the time to complete one cycle.

Figure 13.5: *The period is the time it takes to complete one cycle.*

Frequency is the number of cycles per second

Frequency is closely related to period. The **frequency** of an oscillator is the number of cycles it makes per second. Every day we experience a wide range of frequencies. Your heartbeat probably has a frequency between one-half and two cycles per second. A plucked rubber band might have a frequency of 100 cycles per second (Figure 13.6). The sound of the musical note A has a frequency of 440 cycles per second. The human voice contains frequencies mainly between 100 and 2,000 cycles per second. Frequency and period are inversely related. The period is the time per cycle. The frequency is the number of cycles per unit of time.

Frequency is measured in hertz

The unit of one cycle per second is called a **hertz (Hz)**. A frequency of 440 cycles per second is usually written as 440 hertz, or abbreviated 440 Hz. The hertz is a unit that is the same in English and metric measurement systems. When you tune into a station at 100.6 on the FM dial, you are setting the oscillator in your radio to a frequency of 100.6 megahertz (MHz) or 100,600,000 Hz. You hear music when the oscillator in your radio is exactly matched to the frequency of the oscillator in the transmission tower connected to the radio station.

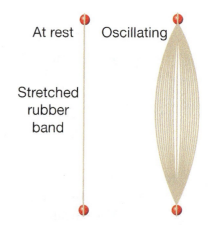

A **frequency** of 100 Hz means the oscillating rubber band completes 100 cycles each second.

Figure 13.6: *The definition of frequency.*

Amplitude

Amplitude describes the size of a cycle
The **amplitude** describes the "size" of a cycle. In this context, *size* means how far the oscillation moves the system way from its resting or average state. With mechanical systems, such as a pendulum, the amplitude is often a distance or an angle. A pendulum's resting state is hanging straight down. A large amplitude means each cycle swings relatively far to either side. A small amplitude means the pendulum only moves a little to each side. With other kinds of oscillators, the amplitude might be voltage or pressure. The amplitude is measured in units that match the oscillation you are describing.

Measuring amplitude
The value of the amplitude is the maximum amount the system moves away from equilibrium. For a pendulum, equilibrium is at the center. The amplitude is the amount the pendulum swings away from center (Figure 13.7). The amplitude is also half the total motion from one extreme to the other. For some oscillators, the most convenient way to find the amplitude is to measure the total side-to-side motion and divide by two.

Amplitude and energy
Harmonic motion has energy. Sometimes the energy is kinetic. The pendulum has its highest kinetic energy when it swings through the lowest point in its cycle. At other times the energy is potential, such as when the pendulum reaches the farthest point on either side and is momentarily stopped. As you might suspect, the energy of an oscillator is proportional to the amplitude of the motion. Large-amplitude motions have higher energy than small-amplitude motions (Figure 13.8).

Damping
Friction drains energy away from motion and slows a pendulum down. Just as with linear motion, harmonic motion is reduced by friction. The effect of friction is to slowly reduce the amplitude of the system. If you start a pendulum swinging, you will observe that the amplitude slowly decreases until the pendulum is hanging motionless. We use the word **damping** to describe the gradual loss of amplitude of an oscillator. Damping is due to friction acting to reduce motion.

Damping is a gradual loss of amplitude due to friction.

Large amplitude

Small amplitude

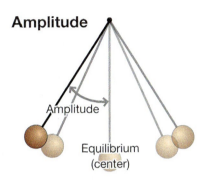

The **amplitude** is the maximum amount the system moves away from equilibrium.

Figure 13.7: *The definition of amplitude is related to equilibrium.*

Large amplitude → high energy

Small amplitude → low energy

Figure 13.8: *The energy of an oscillator is proportional to its amplitude.*

CHAPTER 13 HARMONIC MOTION

Harmonic motion graphs

Cycles and time Harmonic motion is easiest to recognize on a graph that shows how things change over time. Figure 13.9 shows two graphs that are not harmonic motion and two that are. Can you see the difference? Two graphs show repetition and two do not. The diagram below shows position versus time for a pendulum. The graph shows repeating cycles just like the motion. A repeating pattern of cycles is a clear signature of harmonic motion.

Using positive and negative numbers Harmonic motion graphs often use positive and negative values to represent motion on either side of center. We usually choose zero to be at the equilibrium point of the motion. The graph below shows a pendulum swinging from +20 centimeters to −20 centimeters and back. The amplitude is the maximum distance away from center, or 20 centimeters.

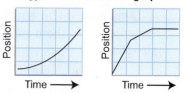

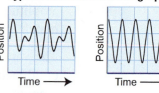

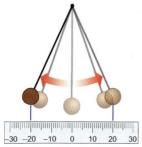 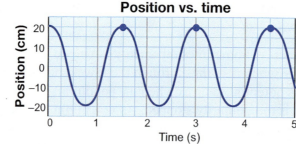

Figure 13.9: *Graphs of linear motion do not show cycles. Harmonic motion graphs show oscillation and cycles.*

Harmonic graphs repeat every period Notice that the graph above returns to the same place every 1.5 seconds. No matter where you start, you come back to the same value 1.5 seconds later. Graphs of harmonic motion repeat every period, just as the motion repeats every cycle. Harmonic motion is sometimes called **periodic motion** for this reason.

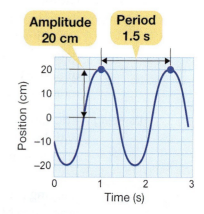

Figure 13.10: *Amplitude and period can be read from a harmonic motion graph.*

Amplitude and period on a graph To get the period from a graph, start by identifying one complete cycle. The cycle must begin and end in the same place on the graph (Figure 13.10). Once you have identified a cycle, you use the time axis of the graph to determine the period. The period is the time difference between the beginning of the cycle and the end. The amplitude is half the distance between the highest and lowest points on the graph. In Figure 13.10, the amplitude is 20 centimeters and the period is 1.5 seconds.

268   UNIT 5 WAVES AND SOUND

Circles and the phase of harmonic motion

Circular motion Circular motion is very similar to harmonic motion. A rotating wheel returns to the same position every 360 degrees. Rotation is a type of cycle. One unique aspect about rotation is that cycles due to rotating motion *always* have a length of 360 degrees (Figure 13.11).

The phase of an oscillator Degrees are also convenient to describe where an oscillator is in its cycle. For example, how would you identify the moment when a pendulum was one-tenth of the way through its cycle? If we let one cycle be 360 degrees, then one-tenth of that cycle is 36 degrees. Thirty-six degrees is a measure of the **phase** of the oscillator. The word *phase* means where the oscillator is in its cycle.

What we mean by "in phase" The concept of phase is important when comparing one oscillator with another. Suppose you observe two identical pendulums, with exactly the same period. If you start them together, their graphs would look like the diagram on the left. You would describe the two pendulums as being *in phase* because their cycles are aligned and each one is always at the same place at the same time.

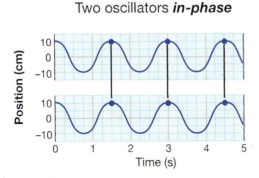
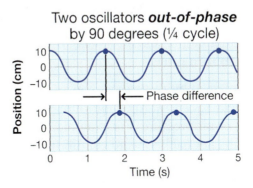

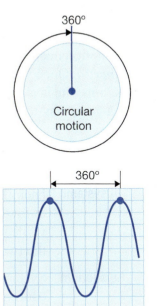

Figure 13.11: *A full turn of a circle is 360 degrees. One full cycle of circular harmonic motion is also 360 degrees.*

Out of phase If you start the first pendulum swinging a little before the second one, the graphs might look like the diagram above, to the right. Although, they have the same cycle, the first pendulum is always a little ahead of the second. The graph shows the lead of the first pendulum as a **phase difference**. Notice that the top graph reaches its maximum 90 degrees *before* the bottom graph. We say the two pendulums are *out of phase* by 90 degrees, or one fourth of a cycle.

13.2 Why Things Oscillate

This section explains why some systems tend to have harmonic motion and other systems do not. Harmonic motion occurs when a system is *stable*, has *restoring forces*, and has some property that provides *inertia*. Once you know what to look for, you can predict when harmonic motion is likely to occur. The opposite is also true, and you can predict which systems are *not* likely to oscillate. Harmonic motion is a good way to introduce the important idea of *stability*.

Restoring force and equilibrium

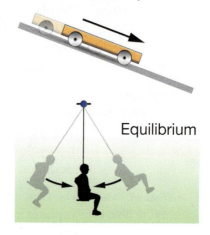

Figure 13.12: *Examples of equilibrium and non-equilibrium situations.*

Different kinds of systems If you set a wagon on a hill and let it go, the wagon rolls down and does not come back (Figure 13.12). If you push a child on a swing, the child goes away from you at first, but then comes back. The child on the swing shows harmonic motion while the wagon on the hill does not. What is the fundamental difference between the two situations? Are there ways to predict when harmonic motion will occur?

Equilibrium Systems that have harmonic motion move back and forth around a central or equilibrium position. You can think of equilibrium as the system at rest, undisturbed, and with zero net force acting on it. A wagon on a hill is *not* in equilibrium, because the force of gravity pulls the wagon down the moment you let it go. A child sitting motionless on a swing *is* in equilibrium because the swing stays put until you apply a force.

Restoring forces Equilibrium is maintained by *restoring forces*. A restoring force is any force that acts to pull the system back toward equilibrium. Gravity can be a restoring force, and so can a spring or air pressure. For example, if a child on the swing is moved forward, gravity pulls her back, toward equilibrium. If she moves backward, gravity pulls her forward, back toward equilibrium again (Figure 13.13). No matter which way the child is pushed, gravity always acts to push the swing back toward equilibrium.

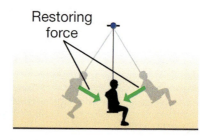

Figure 13.13: *Equilibrium is maintained by restoring forces in a system.*

When harmonic motion happens Any system with both equilibrium and restoring forces is a good candidate for harmonic motion. A disturbance in the system can start harmonic motion. For example, a pendulum at rest hangs straight down, in equilibrium. Pulling back the weight disturbs the equilibrium. The pendulum oscillates back and forth around its equilibrium position under the action of the restoring forces. Eventually, friction slows it down and the system returns to equilibrium.

Inertia

Inertia causes an oscillator to go past equilibrium

Newton's first law explains why harmonic motion happens for moving objects. According to the first law, an object in motion stays in motion unless acted upon by a force. Think about a pendulum swinging. At the bottom of the swing, the net force on the pendulum is zero because that is the equilibrium position. But the pendulum has inertia that carries it through the bottom of the swing. Inertia causes the pendulum to overshoot its equilibrium position every time. The result is harmonic motion. You can analyze the motion in five steps.

The cycle of the pendulum

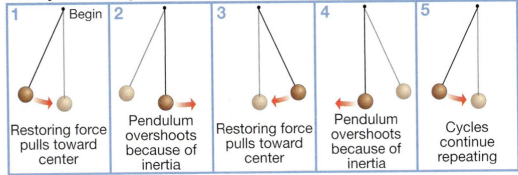

1. The restoring force pulls the pendulum toward the center.
2. The pendulum overshoots the center because of its inertia.
3. The restoring force pulls the pendulum back toward the center, slowing and reversing its direction.
4. The pendulum overshoots the center again, because of inertia.
5. The cycle repeats, creating harmonic motion.

Inertia is common to all oscillators

All systems that oscillate on their own have some property that acts like inertia, and some type of restoring force. Oscillation results from the interaction of the two effects: inertia and restoring force.

Harmonic motion in machines

So far, we have been discussing systems that move with harmonic motion on their own. Many machines create harmonic motion by actively pushing or rotating parts. In these machines, the motion is caused by mechanical constraints, not the interaction of restoring forces and inertia.

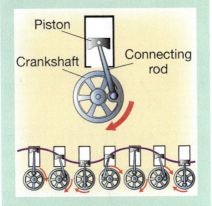

For example, the piston of a car engine goes up and down as the crankshaft turns. The piston is in harmonic motion, but the motion is caused by the rotation of the crankshaft and the attachment of the connecting rod. The up-and-down motion of the needle on a sewing machine is another example of harmonic motion created by machinery.

CHAPTER 13 HARMONIC MOTION

Stable and unstable

Unstable equilibrium Not all systems in equilibrium show harmonic motion when disturbed. A marble at the top of a hill is an example. The marble may be perfectly balanced and in equilibrium. However, if we disturb it with a little push, it rolls away—and it does *not* come back to the top of the hill. The marble on a hill is a good example of **unstable equilibrium**. In unstable systems, there are forces that act to pull the system *away* from equilibrium when disturbed. Unstable systems do not usually exhibit harmonic motion.

Unstable equilibrium
If the system moves away from equilibrium, forces tend to move it farther away.

Stable equilibrium
If the system moves away from equilibrium, forces tend to pull it back toward equilibrium again.

A valley is a stable equilibrium The bottom of a valley is an example of a place where **stable equilibrium** occurs. Stable is the opposite of unstable. If you move a marble to either side, it tends to roll back toward the bottom again. If we push it to the left, gravity provides a restoring force that pulls the marble back to the right. If we push it to the right, the gravity pulls back to the left.

Stability and energy A valley provides stability because any way you move the marble causes it to have *higher energy* than it has at the bottom. At the bottom, the energy is minimized. An equilibrium that has a minimum of energy is almost always stable. The top of a hill is where the marble is unstable because there is no minimum of energy. The energy of the marble is reduced when it rolls farther away.

Stable systems tend to exhibit harmonic motion Stable systems almost always exhibit harmonic motion when they are slightly disturbed. A marble in a valley rolls back and forth in harmonic motion when displaced to one side and released. The pendulum is another example of an oscillator that has stable equilibrium. Gravity always acts to pull the pendulum back toward its center.

Why airplanes have tails

Airplanes have tails to make them stable while flying. In Chapter 9, you learned that objects tend to rotate about the center of mass. Consider the rotation of an airplane around its center of mass. No matter which way the airplane rotates, nose up or nose down, wind blowing against the tail fins creates a force that tends to straighten the plane out again.

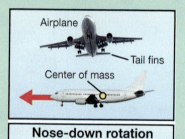

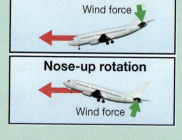

The tail is located as far as possible from the center of mass, usually at the very end of the plane. The farther back the tail is, the more effective is the restoring force at resisting the rotation of the plane.

The natural frequency

Systems tend to have a preferred frequency

If you watch a pendulum, a swing, or any other oscillator, you will observe a curious fact. In the absence of friction, once it starts moving, the system tends to oscillate with a particular frequency. For example, a certain pendulum swings with a frequency of two cycles per second (2 Hz). Every time you set the pendulum swinging, it always swings with the same frequency and no other.

Natural frequency

The **natural frequency** is the frequency at which a system tends to oscillate when disturbed. Everything that can oscillate has a natural frequency, and most systems have more than one. The natural frequency is useful to know because many inventions are designed to work at a specific frequency. For example, a pendulum clock might have a pendulum with a natural frequency of exactly one cycle per second (1 Hz). The second hand moves one-sixtieth of a turn for each swing of the pendulum. The accuracy of the clock depends on the natural frequency of the pendulum. Watches, computers, and many devices rely on the precise natural frequency of quartz crystals.

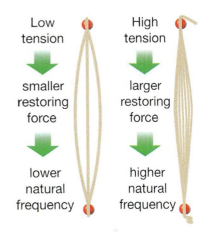

Figure 13.14: *Stretching a rubber band tighter (right) increases the restoring force and raises the natural frequency.*

Stronger restoring forces raise the natural frequency

The natural frequency comes from the interaction of force and inertia. If the restoring forces are very strong, a system tends to respond quickly and have a high natural frequency. If you pluck a stretched rubber band, it oscillates at its natural frequency. If you stretch the rubber band (Figure 13.14) and make it tighter, it oscillates at a higher natural frequency. The natural frequency goes up because the restoring forces are larger in the tight rubber band. If you pluck the steel string of a guitar, it oscillates much faster than a rubber band, generally at a frequency of more than 200 hertz. The natural frequency is higher because steel is very stiff and creates larger restoring forces than a rubber band.

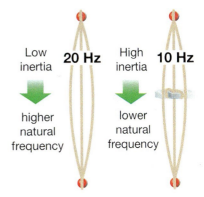

Figure 13.15: *Adding a steel nut greatly increases the inertia of a stretched rubber band. As a result, the natural frequency decreases.*

Adding inertia decreases the natural frequency

Adding inertia, by increasing mass, has the opposite effect on the natural frequency. A system with more inertia is harder to accelerate and responds slowly, which lowers its natural frequency. The bass strings on a guitar are wound with extra wire to make them heavier. The increased mass creates more inertia and the strings oscillate slower because the natural frequency is lower. Tying a steel nut on the rubber band will demonstrate the same effect. The added inertia of the steel nut greatly lowers the natural frequency of the system (Figure 13.15).

CHAPTER 13 — HARMONIC MOTION

Changing the natural frequency

Acceleration and natural frequency The natural frequency is proportional to the acceleration of a system. Systems with high acceleration have a higher natural frequency. Systems with low acceleration have lower natural frequencies.

Newton's second law Newton's second law can be applied to see the relationship between acceleration and natural frequency. The acceleration is proportional to the force divided by the mass. The force is the restoring force. The mass is the inertia. If the ratio of force divided by mass *increases*, the natural frequency *increases*. If the ratio of force divided by mass *decreases*, the natural frequency *decreases*. Figure 13.16 shows how to change the natural frequency of several common oscillators.

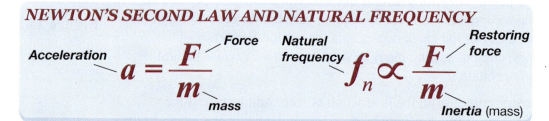

NEWTON'S SECOND LAW AND NATURAL FREQUENCY

Acceleration $a = \dfrac{F \text{ (Force)}}{m \text{ (mass)}}$ Natural frequency $f_n \propto \dfrac{F \text{ (Restoring force)}}{m \text{ (Inertia (mass))}}$

Natural frequency and the response time of a system A system with a high natural frequency can respond to changes faster than a system with a low natural frequency. This is because of the relationship between acceleration and natural frequency. This principle is applied directly to the design of cars. Cars have springs in the shock absorbers on each wheel that allow the wheel to independently follow bumps in the road. The body of the car stays relatively level as the wheels move up and down. The system of a car and shock absorbers is a mass on a spring, an oscillator with a natural frequency. The natural frequency is proportional to the ratio of the spring constant to the mass of the car. Sports cars have low mass and stiff springs, creating a high natural frequency. The high natural frequency means the car can respond quickly to changes in direction. It also means the car responds to bumps in the road which is why sports cars tend to have a stiff, bumpy ride. Other cars have a larger mass, softer springs, and a lower natural frequency. The low natural frequency means the car does not respond to high-frequency bumps in the road, and therefore rides smoother. However, a big touring car does not steer quickly.

To decrease the natural frequency

Increase the string length of a pendulum

Add more mass to a mass on a spring

To increase the natural frequency

Tighten a rubber band or a string

Use a stronger spring with a mass on a spring

Figure 13.16: *Changing variables changes the natural frequencies of some common oscillators.*

13.3 Resonance and Energy

It takes force to get a system moving, and continued application of force increases the energy of the system. With linear motion, we learned how to relate applied force to motion and energy (Chapters 5–10). With harmonic motion, the connection between force, motion, and energy is more complex. Newton's laws still apply, but the *frequency* of the force also matters. Even a small force applied with a rhythm matching the natural frequency of a system can produce surprisingly large motion and subsequent high energy. Resonance, the subject of this section, relates the frequency of an oscillating force to motion and energy.

Energy in harmonic motion

Potential energy Harmonic motion involves both potential energy and kinetic energy. As an example, consider the motion of a pendulum. At the highest point of the cycle, the pendulum is momentarily stopped (Figure 13.17). It has no kinetic energy because its speed is zero. The pendulum *does* have potential energy, because it is raised above its equilibrium position.

Kinetic energy At the low point of the cycle, all the potential energy has been converted to kinetic energy. The pendulum has its highest speed at the lowest point in the cycle (Figure 13.18). As the pendulum swings through the low point, it climbs up again, converting its kinetic energy back into potential energy.

Energy "sloshes" back and forth Oscillators like a pendulum, or a mass on a spring, continually exchange energy back and forth between potential and kinetic. The total energy is a combination of potential and kinetic energy. A graph of energy versus time shows the exchange. The potential energy is low when the kinetic energy is high and vice versa.

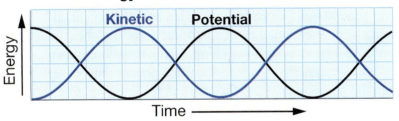

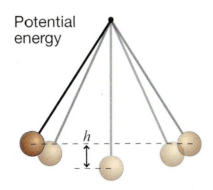

Figure 13.17: *A pendulum has potential energy at the top of its swing because it has risen to height* h *above its lowest point.*

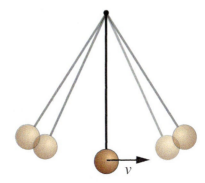

Figure 13.18: *At the bottom of the swing, a pendulum has kinetic energy because it is moving with speed* v.

CHAPTER 13 HARMONIC MOTION

Periodic forces and the natural frequency

Definition of periodic force
In linear motion, the application of steady force creates constant acceleration. Speed and position both increase. In harmonic motion, we are mostly concerned with forces that go back and forth because these types tend to be associated with back-and-forth motion. A force that oscillates in strength or direction is called a **periodic force**. Periodic forces create harmonic motion. Harmonic motion also creates periodic forces. The vibration of a motorcycle engine is caused by periodic forces created by the harmonic up-and-down motion of the pistons in the engine.

Periodic force and natural frequency
The effect of a periodic force on a system depends strongly on the match between the frequency of the force and the natural frequency of the system. Consider a child being pushed on a swing. The swing is like a pendulum and it has a natural frequency of back-and-forth motion. To get a big swinging motion going, you push every time the swing reaches the end of its motion (Figure 13.19). Each push is given at the same time in each cycle of motion. In physics language, you apply a *periodic force* at the *natural frequency* of the swing. In time, your repetition of small pushes builds up a large amplitude of motion.

Forces that are not at the natural frequency
Consider what would happen if you pushed a child on a swing at random times. Sometimes your pushes *add* to the motion and sometimes they act *against* the motion. Random pushes do not smoothly increase the amplitude of the motion. In fact, even a periodic force applied at the wrong frequency does not work. Figure 13.20 shows a periodic force applied at twice the natural frequency of the swing. One push helps the motion and the next is against the motion.

Resonance
The amplitude of harmonic motion increases *dramatically* when periodic forces are applied to a system *at its natural frequency*. When the frequencies of the force and the system are matched, each push adds in phase to the last. Even small repetitive pushes add up over time to build large motions. This behavior is called **resonance**. Resonance occurs when the frequency of a periodic force matches the natural frequency of a system. As a result of this matching, even small forces can build very large amplitudes because systems in resonance often accumulate large amounts of energy. Although the physical laws are the same as for linear motion, resonance is a unique property of harmonic motion.

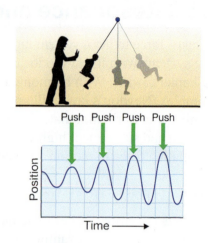

Figure 13.19: *A pushing applied once per cycle is a periodic force. The amplitude of the swing grows with each push.*

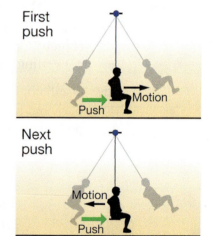

Figure 13.20: *Pushing in the same direction twice per cycle means every other push is in the wrong direction.*

Resonance

A system view of resonance A good way to understand resonance is to think about three distinct parts of any interaction between a system and a force. The first part is the periodic force itself that is applied to the system. The second part is the system itself. The third part is the response of the system to the periodic force. If the force matches the natural frequency of the system, the response is to build up a large amplitude motion—resonance occurs. If the force is *not* matched to the natural frequency of the system, the response is much smaller and there is no resonance.

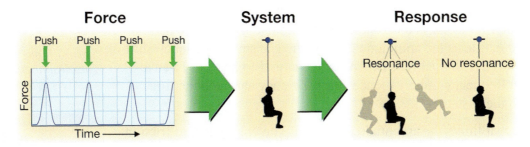

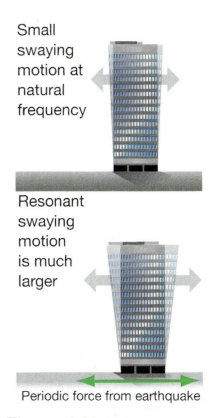

When resonance occurs Resonance occurs when the frequency of a periodic force matches the natural frequency of a system in harmonic motion. Every system that can oscillate has a natural frequency, and most systems have more than one. That means resonance occurs all around you, because every system with a natural frequency can have resonance, too. For example, buildings can sway back and forth, have a natural frequency, and have resonance. In areas where there are earthquakes, it is very important that the natural frequency of a building not be in the range of frequencies at which the ground shakes during a quake. If the natural frequency of a building matches the frequency of an earthquake, resonance occurs and the building is destroyed by its own shaking (Figure 13.21).

Figure 13.21: *Tall buildings sway when the wind blows and when the ground shakes during an earthquake. If the natural frequency of the building matches that of the earthquake, resonance occurs. In resonance, Earth's motion "pumps" the motion of the building and the amplitude of swaying grows until the building is destroyed.*

Energy flow is very efficient at resonance Energy flows through the applied force into the system causing motion. The energy flow is very efficient when the system and the applied force are in resonance. You can think of the system in resonance as storing up each small input of energy from every cycle of the applied force. Many technological systems make use of resonance to create energy at a particular frequency. For example, a cellphone transmitter is based on resonance at 2.4×10^9 Hz (2.4 GHz).

CHAPTER 13 | HARMONIC MOTION

Energy, Resonance, and Damping

Resonant systems accumulate energy

A system in resonance acts like an accumulator of energy. The energy comes from the applied force. The energy is stored as potential and kinetic energy of the oscillation. Resonant systems are found in many technologies because they are so effective at building up energy from the repetition of small forces at the right frequency (Figure 13.22).

Limits to amplitude at resonance

Suppose you apply a small periodic force at the natural frequency of a pendulum. You might think the amplitude of motion would just keep growing and growing. In reality, that is not what happens. Two factors tend to limit how large a resonant motion can get. The first factor is the system itself. In most systems, the natural frequency changes a little as the amplitude gets larger. The small frequency shift means a periodic force no longer matches the exact natural frequency. Even a small mismatch in frequency greatly reduces the efficiency at which the force pumps energy into the system.

Friction and steady state

The second factor that limits the amplitude is friction. The energy lost to friction goes up as the amplitude increases. As the amplitude increases, at some level the energy lost to friction balances the energy input from the applied force. Once the balance point is reached, the amplitude stops increasing and the system is in *steady state*. In steady state, the amplitude remains constant.

The balance between damping and energy input

Steady state is a balance between damping from friction and the strength of the applied force. If you have ever dribbled a basketball in place, then you have experienced a resonant system in steady state (Figure 13.23). The ball bounces in a steady rhythm as long as you keep supplying a constant push with your hand on every bounce. Remove your hand and damping takes over. The bounces get shorter and shorter until the ball is at rest on the floor. Start to push harder and the amplitude of the bounces increases. The steady-state amplitude is a balance between energy lost by damping and energy supplied by the applied force.

Resonance

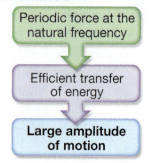

Figure 13.22: *Resonance allows small input forces to accumulate large amounts of energy in a system.*

Figure 13.23: *Dribbling a basketball on a floor is a good example of resonance with a steady-state balance between energy loss from damping and energy input from your hand.*

Chapter 13 Connection

Quartz Crystals

You may not realize how much of modern technology relies on precise timekeeping. For example, computers do not work unless each circuit in the computer is synchronized with the other circuits to within a few billionths of a second. When you see a computer advertised at "3 GHz," what the manufacturer is telling you is that the clock inside the computer ticks 3 billion times per second. Each tick of the clock means the computer can do something like add part of a number or write a character to the screen. Faster clocks mean computers can do more things each second.

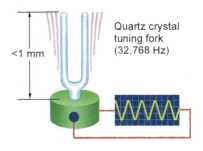

Figure 13.24: *Inside a quartz crystal watch is a tiny tuning fork made from a piezoelectric quartz crystal. The natural frequency of the tuning fork is 32,768 hertz.*

Quartz crystals are used in electronics The precise heartbeat of nearly all modern electronics is a tiny quartz crystal oscillating at its natural frequency. The resonance of a quartz crystal is so accurate that even the least expensive varieties vary by less than one in a million seconds. A $5 watch and a $10,000 computer both use the same quartz crystal technology.

The piezoelectric effect In 1880, Pierre Curie and his brother Jacques discovered that crystals could be made to oscillate by applying electricity to them. This is known as the **piezoelectric effect**. A piezoelectric crystal works two ways. If the crystal oscillates mechanically, it produces an electric oscillation of the same frequency. The reverse is also true: If an oscillating electrical signal is applied, the crystal will vibrate mechanically.

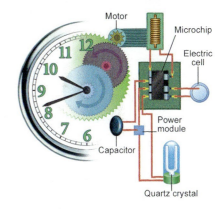

Figure 13.25: *The inside of this watch uses a quartz crystal to keep time. The crystal regulates a tiny electric motor that turns gears to move the hands of the watch.*

Quartz crystals in watches Quartz is an ideal piezoelectric crystal because it is abundant, hard but not brittle, and stable over a range of temperatures. By cutting crystals of quartz in different ways, they can be made to vibrate at almost any frequency. The quartz crystal in most watches is cut like a tiny tuning fork (Figure 13.24) and has a natural frequency of 32,768 Hz. An electric circuit applies a small electrical signal at the quartz crystal's natural frequency, causing it to resonate and produce a very stable time signal. The high-quality pocket watch in Figure 13.25 keeps time using a quartz crystal of this type.

CHAPTER 13 HARMONIC MOTION

Chapter 13 Assessment

Vocabulary
Select the correct term to complete the sentences.

harmonic motion	cycle	period
frequency	amplitude	hertz (Hz)
damping	periodic motion	periodic force
resonance	phase	phase difference
equilibrium	restoring	stable
unstable	oscillator	natural
steady state	piezoelectric effect	oscillation

1. A system that has harmonic motion is called a(n) _____.
2. Another name for harmonic motion is _____.
3. Back-and-forth movement of air molecules and rubber bands represent motion called _____.
4. The angular distance a pendulum swings from equilibrium is a measure of the _____.
5. The number of cycles an oscillator makes per second is the _____.
6. Motion that repeats itself over and over again is called _____.
7. The time for one cycle to occur is the _____.
8. Two oscillators reaching their maximum amplitudes at different times are displaying a(n) _____.
9. _____ is the gradual loss of amplitude by an oscillator.
10. The frequency unit of one cycle per second is called a(n) _____.
11. The term describing the position of an oscillator in its cycle is _____.
12. A unit of motion that repeats over and over again is known as a(n) _____.
13. A marble balanced at the top of a hill is in a state of _____ equilibrium.
14. The frequency at which a system tends to oscillate when disturbed is its _____ frequency.
15. Any force that always acts to pull a system back toward equilibrium can be called a(n) _____ force.
16. When a force always acts to return an object to its equilibrium position, the object is in _____ equilibrium.
17. A condition of resonance that occurs when the damping between friction matches the strength of applied force to the system is referred to as _____.
18. When the frequency of a periodic force matches the natural frequency of a system, _____ occurs.
19. A force that oscillates in strength or direction is called a(n) _____.
20. The oscillation of a crystal caused by the application of a periodic electric signal is described as the _____.

Concept review

1. Identify the following as examples of harmonic or linear motion.
    a. a child moving down a playground slide
    b. the vibration of a tuning fork
    c. the spinning of Earth on its axis
    d. a driver making a right-hand turn at an intersection
    e. a bouncing ball
2. Describe the difference in appearance of a graph of linear motion and harmonic motion.
3. Two players are dribbling basketballs at the same time. How would the motion of the basketballs compare if they were in phase?
4. A young child asks to be pushed on a swing. Explain how the term *resonance* is related to the motion of the child on the swing.
5. Using a swing as an example of harmonic motion, identify the
    a. period.
    b. frequency.
    c. cycle.
    d. amplitude.

6. A long spring is stretched between two friends and vibrated. The vibration causes harmonic motion so that, at one instant in time, the spring looks like the diagram at the right. Sketch the diagram on your paper and label the following.
   a. equilibrium position
   b. amplitude
   c. length of one cycle

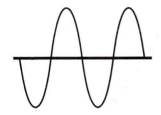

7. Your favorite radio station is 106.7. What are the units on this number and what do they mean in terms of harmonic motion?

8. How can the natural frequency of a vibrating string be changed? Give two examples of changes that will create a different natural frequency.

9. A system with harmonic motion is called an oscillator. Name at least two common oscillators.

10. Describe the difference between a damping and a restoring force.

11. What causes oscillation of a system?

12. What is the difference between stable and unstable equilibrium? State an example of each type of equilibrium.

13. Astronauts are unable to use a conventional equal-arm balance to measure mass while in orbit because of the "weightless" condition of all objects in a space station. How might mass be determined using an oscillator?

14. A mass attached to a spring is oscillating with harmonic motion. Describe the changes in its acceleration in one cycle of oscillation.

15. Describe the transformations of energy that take place with an oscillating pendulum.

16. If you rub a moistened finger on the top of a thin glass goblet it can be made to "ring." Explain this using terms such as *natural frequency*, *periodic force*, and *resonance*.

17. Resonant systems accumulate energy which increases the amplitude of oscillation. There is a limit to the magnitude of the amplitude. What is the name for the point at which this limit is reached? Under what condition is this limit reached?

Problems

1. The diagram shows the position versus time for a harmonic oscillator.
   a. Name four pairs of points that are in phase.
   b. Name four pairs of points that are 180° out of phase.

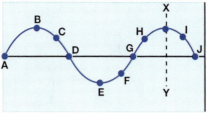

2. The diagram shows the position versus time graphs for four swinging pendulums.

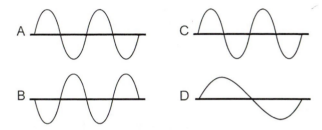

   a. Which pendulums are in phase?
   b. Which pendulums are 180° out of phase?

3. The wings of a ruby-throated hummingbird move at a frequency of 70 hertz. How long does it take to complete a wing-beat cycle?

4. Referring to the diagram below, rank the numbered positions in order from the highest to the lowest amount of kinetic energy.

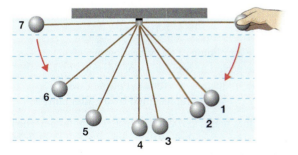

5. Referring to the diagram above, rank the numbered positions in order from the highest to the lowest amount of potential energy.

6. The diagram below represents a graph of velocity versus time for the harmonic motion of the pendulum shown. Match the lettered points on the graph to the numbered positions of the pendulum.

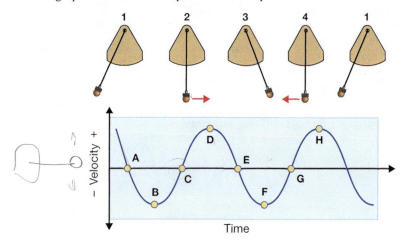

7. Use the graph of the position versus time of a vibrating mass on a spring to answer the questions.

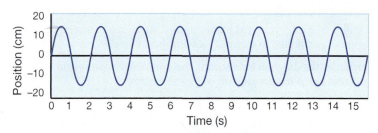

   a. What is the amplitude of the vibrating mass?
   b. What is the period of oscillation of the mass?
   c. What is the frequency of vibration of the mass?
   d. How many cycles are represented by the diagram?

8. A simple pendulum hangs vertically at rest. The bottom of the bob is at a height of 1.0 m. The 1.5-kg bob is then moved away from its equilibrium position so that it is at a height of 1.25 m. Find the speed of the bob as it passes through the equilibrium position once released. (*Note*: Assume no energy loss.)

9. Penny, a 392-newton girl, "pumps" her 5-kg swing until she is traveling at a speed of 7.5 m/s as she passes closest to the ground at a height of 0.42 m. How high above the ground will the swing travel if there is no damping?

10. Objects of known mass are attached to a spring. When the spring is stretched and released, the period of the spring with each mass is recorded. A graph of the results is plotted. Using the graph, determine the mass of an object that oscillates with a frequency of 0.286 Hz.

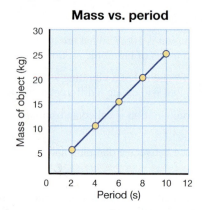

11. A spring oscillates with a frequency of 20 Hz when a 1-kg mass hangs from it. Describe two ways that you could create an oscillation with a lower frequency.

Applying your knowledge

1. While the devastating effects of earthquakes on buildings and landforms is well documented, other sources of vibration can create problems. Research other vibration sources and the problems they cause.

2. The terms *amplitude* and *frequency* are important in the oscillations that produce a radio signal. Investigate the terms *AM* and *FM*. Briefly describe the difference between the oscillations associated with AM and FM radio broadcasting.

3. When driving at certain speeds on sections of a "washboard" road, automobiles vibrate and tend to lose traction, sliding dangerously on dry pavement. At other speeds, traction is maintained and the automobile is more manageable. Explain this. Where and why is this pattern sometimes included in the design of roads?

UNIT 5 WAVES AND SOUND

CHAPTER 14

Waves

Objectives:

By the end of this chapter you should be able to:

✔ Recognize a wave in nature or technology.
✔ Measure or calculate the wavelength, frequency, amplitude, and speed of a wave.
✔ Give examples of transverse and longitudinal waves.
✔ Sketch and describe how to create plane waves and circular waves.
✔ Give at least one example of reflection, refraction, absorption, interference, and diffraction.
✔ Describe how boundaries create resonance in waves.
✔ Describe the relationship between the natural frequency, fundamental mode, and harmonics.

Key Questions:

- What is a wave, and what are some properties of waves?
- What happens when a wave hits a boundary or an opening?
- How are frequency, wavelength, and speed related?

Vocabulary

absorption	crest	longitudinal wave	reflection	trough
antinode	destructive interference	mode	refracted wave	wave
boundary	diffraction	node	refraction	wave front
boundary condition	fixed boundary	open boundary	standing wave	wave pulse
circular wave	fundamental	plane wave	superposition principle	wavelength
constructive interference	harmonic	propagation	transverse wave	
continuous	incident wave	reflected wave		

283

CHAPTER 14 WAVES

14.1 Waves and Wave Pulses

A ball floating on water can oscillate up and down in harmonic motion. But something else happens to the water as the ball oscillates. The surface of the water oscillates in response, and the oscillation spreads outward from where it started. An oscillation that travels is a **wave**.

A wave is an oscillation that travels.

All waves—including water waves, sound waves, and light waves—share common properties. When you hear a musician or theater actor on a stage, waves carry the sound through the air to your ears. When you call a friend on a cell phone, a microwave travels from the phone's antenna to the cell phone tower via a series of electromagnetic waves which carry your conversation. Astronomers believe gravity waves are created when black holes collide with each other.

Why learn about waves?

Waves carry information and energy Waves are important because we use waves to carry information and energy over great distances. In music, the sound wave that travels through the air carries information about the vibration of a guitar string from the instrument to your ear. Your ear receives the vibration of the sound wave and your brain perceives music. In a similar way, a radio wave carries sounds from a transmitter to your receiver. Another kind of radio wave carries television signals. A microwave carries cell phone conversations. Waves carry energy and information from one place to another. The information could be sound, color, pictures, instructions for a computer or other electronic device to perform in a certain way, or many other useful things.

Waves are all around us Waves are part of everyday experience. Waves occur in both natural systems, such as oceans or living creatures, and in human technology, such as cell phones and musical instruments. You might not recognize all the waves around you, but they are there. Consider standing on almost any road and looking around. Figure 14.1 gives some examples of waves that are so common that you have certainly experienced all of them.

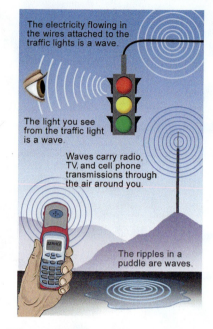

Figure 14.1: *Four examples of everyday waves are shown.*

→ *The light from the stoplight that you see with your eyes is a wave.*

→ *The ripples in a puddle of water are waves.*

→ *The electricity flowing in the wires attached to the street lights is a wave.*

→ *Waves carry radio, television, and cell phone transmissions through the air all around you.*

UNIT 5 WAVES AND SOUND

Recognizing waves around you

There are two types of waves
All waves can be categorized as one of two types: mechanical waves or electromagnetic waves.

Mechanical waves require a medium in which to travel. When mechanical waves travel, vibrations or movement can be observed.

- Any time you see or hear a vibration, there is a mechanical wave.
- Anything that makes or responds to sound uses mechanical waves.

Electromagnetic waves can travel in a vacuum, so they require no medium as they transfer energy.

- Anything that makes or responds to light uses electromagnetic waves.
- Anything that transmits information through the air or space without wires uses electromagnetic waves. This includes cell phones, radio, and television.
- Anything that allows you to "see through" objects uses electromagnetic waves. This includes ultrasound, computerized axial tomography (CAT) scans, magnetic resonance imaging (MRI) scans, and X-rays.

Electromagnetic waves carry information
You will find waves whenever information, energy, or motion is transmitted over a distance without anything obviously moving. The remote control for a TV is an example. To change the channel, you can use the remote or get up and push the buttons with your finger. Both actions carry information, the channel selected, to the TV. One uses physical motion and the other uses an infrared light wave that goes from the remote control to the television (Figure 14.2). Physics tells you that a wave must have come from that remote because information traveled from one place to another, and nothing appeared to move. You cannot see the wave from the remote control because infrared light is invisible to your eye.

Waves are a form of traveling energy
A wave is a traveling form of energy. Watch a floating stick as a wave passes and you will see the stick bob up and down in about the same place as the wave moves under it (Figure 14.3). The water under the stick does *not* move like water flowing in a stream. As the wave moves by, individual particles of water oscillate up and down, but on average, each particle remains in the same place. What *does* move is the *energy* of the wave. The energy moves through the water, causing the surface to oscillate up and down as the wave energy passes through.

Using a wave

Figure 14.2: *We change channels by sending an electromagnetic wave from a remote control.*

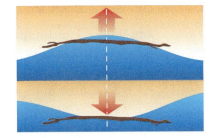

Figure 14.3: *A stick bobs up and down as a mechanical wave passes, but the stick and the water beneath it stay, on average, in the same place.*

Characteristics of waves

Basic properties Waves have cycles, frequency, and amplitude, just like oscillations. Because waves spread out and move, they have new properties of wavelength and speed. Because waves are spread out, frequency and amplitude have to be more carefully defined than they are for a stationary oscillation.

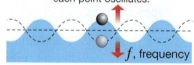

Figure 14.4: *The frequency of a wave.*

Frequency The frequency of a wave measures how often it oscillates at any given point. (Figure 14.4). To measure the frequency, you look at one place as the wave passes by. The frequency of the oscillating motion of one point is the frequency of the wave. The wave also causes distant points to oscillate up and down *with the same frequency*. A wave carries its frequency to every area it reaches.

Frequency is measured in Hz Wave frequency is measured in hertz (Hz), just like any oscillation. A frequency of 1 hertz (1 Hz) describes a wave that makes everything it touches go through a complete cycle once every second. Laboratory-size water waves typically have low frequencies, between 0.1 and 10 hertz. Sound waves have higher frequencies, from 20 hertz to 20,000 hertz. Light waves have even higher frequencies, in the range of 10^{12} Hz for visible light and higher for X-rays.

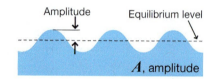

Figure 14.5: *The amplitude of a wave.*

Amplitude The amplitude of a wave is the largest amount that the wave oscillates above or below its zero-energy, or equilibrium level (Figure 14.5). The equilibrium level for a water wave is the surface of the water when it is completely still, like the dotted line in Figure 14.5. You can often measure the amplitude as one-half of the distance between the highest and lowest levels the wave reaches.

Wavelength **Wavelength** is the length of one complete cycle of a wave (Figure 14.6). For a water wave, this would be the distance from a point on one wave to the same point on the next cycle of the wave. You can measure the wavelength from high point to high point, low point to low point, or as the length of a cycle relative to the equilibrium level. Physicists use the Greek letter lambda (λ) to represent wavelength. You write a lambda like an upside down *y*.

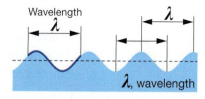

Figure 14.6: *The wavelength is the length of one complete cycle. The wavelength can be measured from any point on a cycle to the same point on the next cycle.*

Wave pulses and the speed of a wave

Speed The speed of a wave describes how fast the wave can transmit an oscillation from one place to another. Waves can have a wide range of speeds. Most water waves are slow; a few miles per hour is typical. Light waves are extremely fast—186,000 miles per *second*. Sound waves travel at about 660 miles per hour, faster than water waves but much slower than light waves.

The speed of a wave pulse A **wave pulse** is a short length of wave, often just a single oscillation. Imagine stretching an elastic string over the back of a chair, as in the diagram below. To make a wave pulse, pull down a short length of the string behind the chair and let go. This creates a "bump" in the string that races away from the chair. The moving "bump" is a wave pulse. The wave pulse moves *on* the string, but each section of string returns to the same place after the wave moves past. The speed of the wave pulse is what we mean by the speed of a wave.

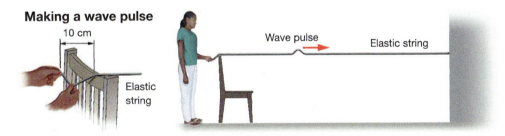

What is the speed of a wave? The speed of a wave is different from the speed of whatever the wave is causing to move. In the wave pulse example, the string moves up and down as the pulse passes. But the up-down speed of the string is *not* the speed of the wave. The speed of the wave describes how quickly a movement of one part of the string is transmitted to another place on the string. To measure the speed of the wave, you would start a pulse in one place and measure how long it takes the pulse to affect a place some distance away. A similar technique works for measuring the speed of water waves. A stone dropped in a pond starts a ripple, which is a small wave. The speed of the wave is the speed at which the ripple spreads (Figure 14.7).

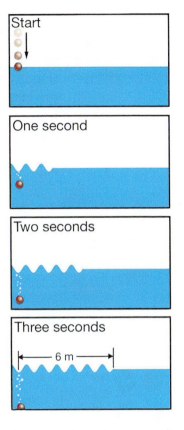

Calculating the speed of the wave

$$v = \frac{6 \text{ m}}{3 \text{ s}} = 2 \text{ m/s}$$

Figure 14.7: *The speed of a wave is the speed at which oscillations spread outward from where the wave started.*

CHAPTER 14 WAVES

The relationship between speed, frequency, and wavelength

Speed is frequency times wavelength

The speed of the wave is the speed at which a cycle moves from one place to another. In one complete cycle, a wave moves forward one wavelength (Figure 14.8). Speed is the distance traveled divided by time. A wave travels one wavelength in one period. The speed of a wave is therefore its wavelength divided by its period. Since frequency (f) is the inverse of period (T), it is usually easier to calculate the speed of a wave by multiplying wavelength and frequency. The result is true for sound waves, light waves, and even gravity waves. Frequency multiplied by wavelength is the speed of the wave.

SPEED OF A WAVE

Speed (m/s) —— $v = f\lambda$ —— Frequency (Hz), Wavelength (m)

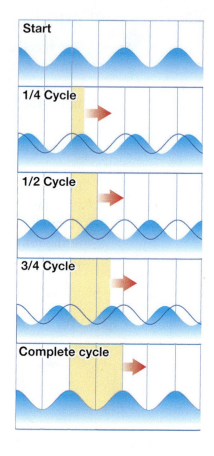

Figure 14.8: *The speed of a wave equals the frequency times the wavelength.*

$$v = \frac{\text{distance}}{\text{time}} = \frac{\text{wavelength}}{\text{period}}$$

$$= \frac{\lambda}{T} = \left(\frac{1}{T}\right)\lambda = f\lambda$$

Calculating how long it takes a wave to move from one place to another

A student does an experiment with waves in water. The student measures the wavelength of a wave to be 5 centimeters. By using a stopwatch and observing the oscillations of a floating ball, the student measures a frequency of 4 Hz. If the student starts a wave in one part of a tank of water, how long will it take the wave to reach the opposite side of the tank 2 meters away?

1. You are asked for the time it takes to move a distance of 2 meters.
2. You are given the frequency, wavelength, and distance.
3. The relationship between frequency, wavelength, and speed is $v = f\lambda$. The relationship between time, speed, and distance is $v = d \div t$.
4. Rearrange the speed formula to solve for the time: $t = d \div v$.
   The speed of the wave is the frequency times the wavelength.
   $v = f\lambda = (4 \text{ Hz})(5 \text{ cm}) = 20 \text{ cm/s} = 0.2 \text{ m/s}$
   Use this value to calculate the time:
   $t = (2 \text{ m}) \div (0.2 \text{ m/s}) = 10 \text{ seconds}$

Transverse and longitudinal waves

Waves spread through connections A wave moves along a string because the string is **continuous**. By continuous, we mean it is connected to itself. Waves spread through connections. If we were to break the string in the middle, the wave would not spread across the break. Any extended body that is continuous can support waves. A lake is a good example. Waves can travel all the way across a lake because the water is continuous from one shore to another.

Transverse waves A **transverse wave** has its oscillations perpendicular to the direction the wave moves. In Figure 14.9, the wave moves from left to right. The oscillation is up and down. Water waves are also transverse waves because the up-and-down oscillation is perpendicular to the motion of the wave.

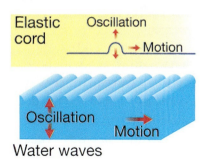

Figure 14.9: *A wave pulse on an elastic cord and ripples on water are two examples of transverse waves.*

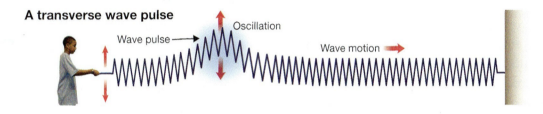

Longitudinal waves A **longitudinal wave** has oscillations in the same direction as the wave moves. Stretch a coiled spring toy with one end fastened to the wall. Give the free end a sharp push toward the wall and pull it back again. You see a compression wave of the spring move toward the wall. A compression wave on a spring is a longitudinal wave because the compression is in the direction the wave moves (Figure 14.10).

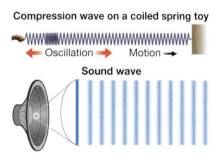

Figure 14.10: *A compression wave on a coiled spring toy and a sound wave are two examples of longitudinal waves.*

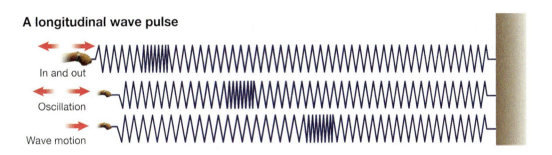

14.1 WAVES AND WAVE PULSES

14.2 Motion and Interaction of Waves

What happens when two waves meet each other or a barrier? This section discusses how waves interact with each other and with barriers. Water waves provide a familiar example to introduce concepts that apply to all waves. Water waves make a good example because a water wave is relatively easy to see and slow enough to see the details of what happens. Light and sound waves are harder to study directly because light waves are small and fast, and sound waves are invisible. Almost every process we observe with water waves also occurs with sound and light waves.

Wave shapes

Crests and troughs Think of a wave as a moving pattern of high points and low points. The **crest** of a wave represents all the high points of one cycle. A **trough** represents all the low points (Figure 14.11). The entire wave is an alternation of crests and troughs.

Wave fronts The shape and movement of a wave crest is a convenient way to show the shape and motion of the entire wave. As the wave moves, the crests move with it. The crests are sometimes called **wave fronts**. Figure 14.12 shows the wave fronts of two different shapes of waves.

One- and two-dimensional waves Waves on a string are one-dimensional because they can move along only one axis. Waves on the surface of water are two-dimensional since they can move along the x and y axes. The shape of the wave fronts of two-dimensional waves affects how the waves spread out and their direction of motion. The shape of the wave front is determined by how the wave is created and what the wave encounters as it moves. The shape of the wave front can change, as you will see later in the chapter. Waves also can be three-dimensional, moving along the x, y, and z axes. Light and sound waves are three-dimensional waves.

Plane waves and circular waves You can make wave fronts in all shapes but **plane waves** and **circular waves** are easiest to create and study (Figure 14.12). The crests of a plane wave form straight lines so the wave fronts are also straight lines. A plane wave can be created by disturbing the surface of water with a straight-edged object. The crests of a circular wave form circles. The wave fronts of circular waves are expanding circles. A circular wave is started by disturbing the water at a single point. A fingertip touching the water's surface will start a circular wave.

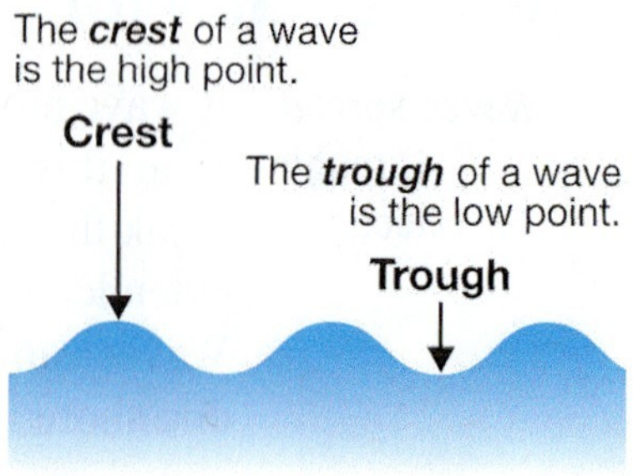

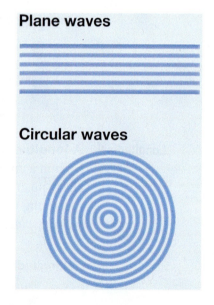

Figure 14.11: *Crests and troughs are the maximum and minimum points in a wave.*

Figure 14.12: *Plane waves and circular waves are types of two-dimensional waves.*

Propagation of waves

Why waves move

The word **propagation** means "to spread out and grow." Propagation is a good word for describing the motion of waves because it describes what happens. To see why a wave propagates, consider a water wave. When you drop a stone into water, some of the water is pushed aside and raised up by the stone (A). The higher water pushes the water next to it out of the way as it tries to get back down to equilibrium (B). The water that has been pushed then pushes on the water in front of *it*, and so on. The wave spreads through the interaction of each bit of water with the water immediately next to it (C).

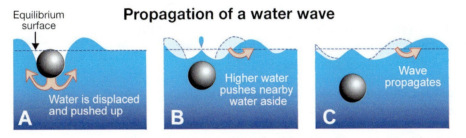

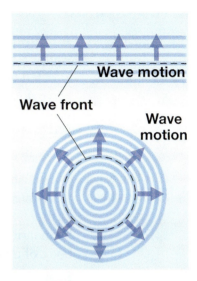

Figure 14.13: *The shape of a wave front determines the direction the wave moves. Waves move perpendicular to their wave fronts.*

Determining the direction the wave moves

The shape of its wave front determines the direction a wave moves. Each point on a wave front moves perpendicular to the wave front. Circular waves have circular wave fronts that move outward from the center (Figure 14.13). Plane waves have straight wave fronts that move in a line perpendicular to the wave fronts.

Changing the direction of a wave

To change the direction of a wave, you have to change the shape of the wave front. This occurs naturally when a wave encounters any change, such as a wall or a boundary between deeper and shallower areas. In later chapters, we will see that this is exactly how lenses work to bend light waves.

Waves propagate through continuous materials

Water waves propagate along surfaces that are continuous. If the continuity is broken, becoming *discontinuous*, a wave may not be able to propagate across the break. For example, a single pan of water has a continuous surface and waves can spread to every part of the surface (Figure 14.14). If you put a ruler across the pan the surface becomes discontinuous because the surface on one side of the wall cannot be influenced by the surface on the other side. A wave on one side will not propagate across the wall.

Figure 14.14: *A pan of water has a continuous surface. A wall across the pan makes the surface discontinuous. A wave does not spread across a discontinuous surface.*

14.2 MOTION AND INTERACTION OF WAVES

Waves and boundaries

Boundaries

A **boundary** is a place where conditions change. For example, the surface of a wall is a boundary. The end of a string is also a boundary for the string. Waves are affected by boundaries, so we use boundaries to shape and control waves. What a wave does at a boundary depends on the **boundary conditions**. The surface of a wall is a **fixed boundary** for water waves because the boundary does not move. A free end of a string is an **open boundary** for string waves because the end can move in response to the wave.

The four wave interactions

Waves can interact with boundaries in four different ways (Figure 14.15). The first three interactions—reflection, refraction, diffraction—occur at the boundary. Diffraction occurs when waves go around boundaries or through openings in boundaries. Absorption can occur at a boundary, but usually happens within the body of a material after a wave crosses a boundary.

Reflection   The wave can bounce off the boundary and go in a new direction.

Refraction   The wave can pass straight into and through the boundary.

Diffraction   The wave can bend around corners or through holes in the boundary.

Absorption   The wave can lose amplitude and/or disappear after crossing the boundary.

Combinations of the four interactions

Sometimes, the wave can do all four things at once, partly bouncing off, partly passing through, partly being absorbed, and partly going around. You may have noticed the radio in a car sometimes loses a radio station signal as you enter a tunnel. Part of the wave that carries the signal bends around the entrance to the tunnel and follows you in. Part is absorbed by the ground. The deeper you go in the tunnel, the weaker the wave gets until the radio cannot pick up the signal at all and you hear static. Simple things like mirrors and complex things like ultrasound or X-rays all depend on how waves act when they encounter boundaries.

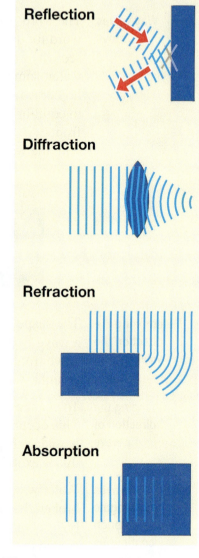

Figure 14.15: *The four basic interactions between waves and boundaries.*

Reflection and refraction

Reflection When a wave bounces off a boundary, we call it **reflection**. Any boundary where there is a sudden change in material almost always causes some reflection. For example, if water waves travel toward a wall, they will be reflected. If the boundary is straight, the wave that is reflected is like the original wave but moves in a new direction. A plane wave reflecting from a straight boundary will also be a plane wave. A circular wave reflecting from a straight boundary will be a circular wave (Figure 14.16). The wavelength and frequency of a wave are usually unchanged by reflection.

Refraction Waves can cross boundaries and pass into or through some materials. For example, placing a thin plate on the bottom of a shallow tray of water creates a boundary where the depth of the water changes. If you look carefully, you see that waves are bent as they cross the boundary where the depth of water changes (Figure 14.17). The wave starts in one direction and changes direction as it crosses. **Refraction** occurs when a wave changes direction as it crosses a boundary and the wave is *refracted* in the process of changing direction.

Curved boundaries Boundaries that are not straight can be used to change the shape of the wave fronts and therefore change the direction of a wave. For example, a curved boundary can turn a plane wave into a circular wave. Curved boundaries are used with both reflecting and refracting surfaces.

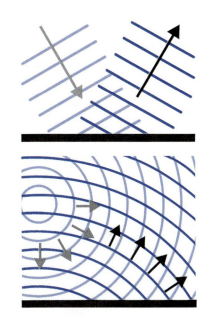

Figure 14.16: *Reflection of plane waves and circular waves from a boundary.*

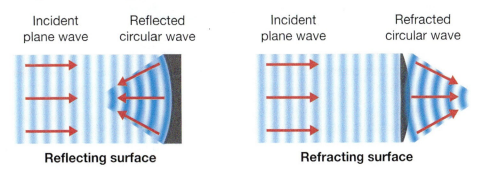

Incident, reflected, and refracted waves The wave approaching a boundary is called the **incident wave**. A wave reflected from a boundary is a **reflected wave**. A wave that is bent passing across or through a boundary is called a **refracted wave**.

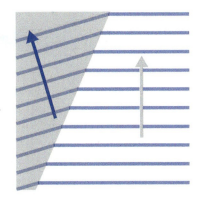

Figure 14.17: *Refraction of water waves crossing a boundary where the water's depth changes.*

Diffraction and absorption

Absorption Waves can be absorbed as they pass through objects. **Absorption** is what happens when the amplitude of a wave gets smaller and smaller as it passes through a material. Some objects and materials have properties that absorb certain kinds of waves (Figure 14.18). A sponge can absorb a water wave while letting the water pass. A heavy curtain absorbs sound waves. Theaters often use heavy curtains so the audience cannot hear backstage noise. Dark glass absorbs light waves, which is how some kinds of sunglasses work.

Absorption transfers energy The energy of a wave is transferred to the material into which the wave is absorbed. This is dramatically illustrated by the destructive power of hurricanes. Wind blowing over water waves exerts force on the wave. The force of the wind transfers energy from the wind to the wave, causing the amplitude of the waves to grow large. When a large wave hits the shore, all of its energy is released quickly as the wave is absorbed by everything in its path.

A sharp boundary creates strong reflections.

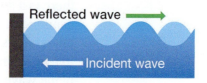

A soft boundary absorbs wave energy and may not produce much reflection.

Figure 14.18: *A hard wall reflects a water wave. A sloped sponge absorbs the wave.*

Diffraction through a small opening turns plane waves into circular waves.

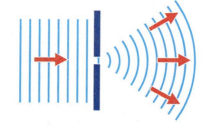

Diffraction Waves can bend around obstacles and go through openings. The process of bending around corners or passing through openings is called **diffraction**. We say a wave is *diffracted* when it is changed by passing through a hole or around an edge. Diffraction usually changes the direction and shape of the wave. For example, diffraction turns a plane wave into a circular wave when the wave passes through a narrow opening (Figure 14.19). Diffraction explains why you can hear someone in another room even if the door is open only a crack. Diffraction causes a sound wave to spread out from the crack.

Figure 14.19: *Diffraction allows waves to bend around corners and spread out after passing through small openings.*

Interference and the superposition principle

The superposition principle

Sometimes, there are many different waves in a system at the same time. For example, if you watch the ocean, you can see small waves on the surface of large waves. When more than one wave is present, the total oscillation of any point is the sum of the oscillations from each individual wave. This is called the **superposition principle**. According to the superposition principle, if there are two waves present, A and B, the total oscillation at any point in time, C, is the sum of the oscillations from Waves A and B. In reality, single waves are quite rare. The sound waves and light waves you experience are the superposition of thousands of waves with different frequencies and amplitudes. Your eyes, ears, and brain separate the waves in order to recognize individual sounds and colors.

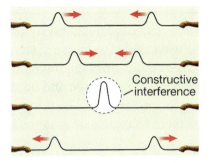

Figure 14.20: *Two wave pulses on the same side add up to make a single, bigger pulse when they meet. This is an example of constructive interference.*

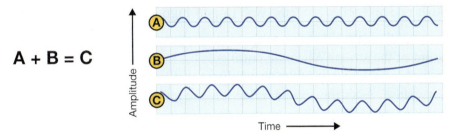

Constructive interference

If two waves add up to create a wave with a larger amplitude, **constructive interference** occurs. Figure 14.20 shows how the constructive interference of two wave pulses makes a single larger pulse at the moment they pass each other. Sometimes on the ocean, two big waves add up to make a gigantic wave that may only last a short time but is taller than ships, which can be very destructive.

Destructive interference

If two wave pulses are started on opposite sides of an elastic string, something different happens. When the pulses meet in the middle, they cancel each other out. One wave wants to pull the string up and the other wave wants to pull it down. The result is that the string is flat at the point the waves meet and both pulses vanish for a moment. This is called **destructive interference**. In destructive interference, waves add up to cancel out or to make a wave with a smaller amplitude (Figure 14.21).

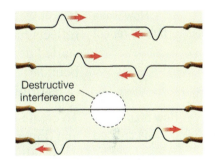

Figure 14.21: *Two equal wave pulses on opposite sides subtract when they meet. The upward movement of one pulse exactly cancels the downward movement of the other. For a moment there is no pulse at all. This is an example of destructive interference.*

14.2 MOTION AND INTERACTION OF WAVES

CHAPTER 14 WAVES

14.3 Natural Frequency and Resonance

Waves can have a natural frequency and create resonance. The natural frequency of a wave depends on the wave and also on the system that contains the wave. In this section, you will learn that resonance in waves comes from the interaction of a wave with reflections from the boundaries of its system. The concepts of resonance and natural frequency apply to a huge range of natural and man-made systems that include waves. The tides of the oceans, the way our ears separate sound, and even a microwave oven are examples of systems that can be explained by waves and resonance.

Boundary conditions and reflections

Resonance and reflections Resonance in waves is caused by reflections from the boundaries of a system. To understand resonance in a wave, consider a stretched elastic string that is fixed to a wall at one end. If you start a wave pulse on the free end of the string, the wave travels to the string's other end and reflects off the wall (top of Figure 14.22).

How wave pulses reflect If you watch carefully, you will observe that a wave pulse launched on the top reflects back on the bottom of the string. When the pulse gets back to where it started, it reflects again, and is back on top of the string. Every reflection from a boundary inverts the wave pulse. After the second reflection, the pulse is traveling in its original form again (Figure 14.22).

How resonance is created To create resonance, you apply a periodic force by shaking the end of the string up and down at regular intervals. Each up and down shake makes a new wave pulse. The timing of the shaking must be just right. To build up a large wave, you wait until a reflected pulse has returned to your hand before launching a new pulse. The new pulse adds to the reflected pulse to make a bigger pulse through constructive interference. The bigger pulse moves away and reflects again. You wait until the reflection gets back to your hand and then shake the string to add a third pulse. The total wave pulse is now three times as large as at the start (bottom of Figure 14.22). Resonance is created when you keep adding pulses so that each new pulse is launched at the exact time a reflected pulse arrives from the far end of the string. After a dozen shakes, the string develops a single large wave motion, and you have created resonance (Figure 14.23). The resonance is created by the addition of new wave pulses with reflections from the boundary at the fixed end of the string.

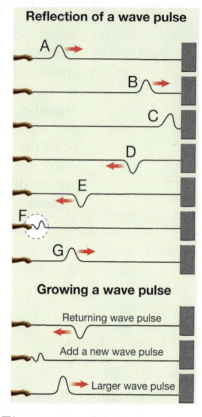

Figure 14.22: *Reflections of a wave pulse on an elastic string.*

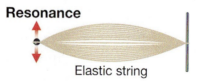

Figure 14.23: *A vibrating string in resonance has a single large wave pattern such as this.*

Standing waves and natural frequency

Standing waves A wave that is confined between boundaries and which appears not to move is called a **standing wave**. In a standing wave, the boundaries reflect the wave back on itself and the wave is trapped or confined in one place. Standing waves can become large if resonance occurs. For example, a jump rope can make a large standing wave if you shake one end up and down at the natural frequency of the rope (Figure 14.24).

The vibrating string A vibrating string is good for investigating standing waves and resonance. A vibrating string oscillates at its natural frequency when it is plucked in the middle and let go. The oscillation looks like the standing wave in Figure 14.24. A jump rope is a large example of a vibrating string. For a lab experiment, an elastic string about a meter long is more appropriate. The top end of the string is fixed by a clamp (Figure 14.25). A periodic force is applied to the bottom end of the string with a small-amplitude shaking motion. Each back-and-forth cycle of the applied force sends a wave pulse up the string.

Matching the natural frequency The experiment consists of watching what happens to the string as the frequency of the periodic force is changed. For most frequencies, the string wiggles around a small amount but nothing much happens. But when the frequency of the applied force gets close to the natural frequency of the string, a large-amplitude standing wave develops. The standing wave forms when the string is in resonance.

Why resonance happens Think of a standing wave in terms of wave pulses created by the oscillation of the applied force. Resonance happens when the period between oscillations of the force is the same as the up-and-back travel time for a wave pulse on the string. Because the wave pulses move quickly, you do not see them. The string develops a standing wave in a second or less. The standing wave has a wavelength of twice the length of the string (Figure 14.25).

Boundaries and natural frequency By itself, a wave does not have a natural frequency. If you had an elastic string that was infinitely long, there would be no natural frequency because a wave you launched from one end would never come back! With all waves, resonance and natural frequency are dependent on reflections from boundaries of the system containing the wave.

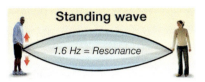

Figure 14.24: *You can make a standing wave by shaking one end of a jump rope up and down with the right frequency. In this example, the right frequency is 1.6 Hz.*

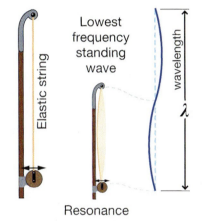

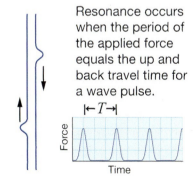

Figure 14.25: *The string is half a wavelength of the standing wave.*

Harmonics of standing waves

Harmonics

As the frequency of the applied force changes, different standing wave patterns appear on a vibrating string (Figure 14.26). The standing wave with the longest wavelength is called the **fundamental**. The fundamental has the lowest frequency of a series of standing waves called **harmonics**. The second harmonic occurs at twice the frequency of the fundamental. The third harmonic occurs at three times the frequency of the fundamental. For example, suppose a vibrating string has a fundamental frequency of 10 hertz. The second harmonic will be at a frequency of 20 hertz and the third harmonic will be at 30 hertz. In the laboratory, you may be able to observe 10 or more harmonics of the same string.

The cause of harmonics

The second harmonic occurs when pulses are launched so each new wave pulse adds up to every *second* reflected pulse. There are two pulses on the string at the time to make the second harmonic (Figure 14.27). The frequency of the second harmonic is twice the fundamental frequency because there are twice as many reflected wave pulses on the string. For the third harmonic, each new pulse adds up to every *third* reflected pulse and there are three pulses on the string at a time. The shapes of the standing wave patterns correspond to the number of wave pulses on the string at one time.

Increasing the resonant frequency

If the elastic string is made tighter, the restoring forces get stronger and the natural frequency of the fundamental increases. The frequency of each harmonic also increases since harmonics occur at multiples of the fundamental. For example, if the frequency of the fundamental is raised to 20 hertz, the second harmonic will occur at a frequency of 40 hertz and the third at 60 hertz.

Almost all systems have harmonics

Almost all systems which have a natural frequency also have harmonics. Each harmonic behaves like another natural frequency. That means the same system can show resonance at the fundamental frequency or any harmonic. In the next chapter, we will see that harmonics are very important to the quality of sound from musical instruments. For another example, car engines have pistons that move rapidly up and down, creating vibrating forces. To avoid resonant vibrations, the whole car must be designed so none of its parts have natural frequencies that match harmonics of the frequencies generated by the moving parts in the engine.

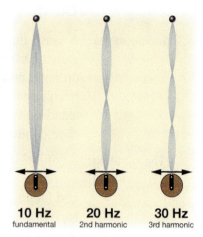

Figure 14.26: *The first three standing wave patterns of a vibrating string. The patterns occur at multiples of the fundamental frequency.*

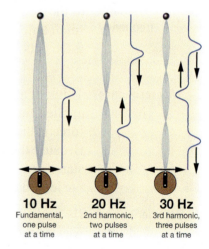

Figure 14.27: *The harmonic number is also the number of wave pulses on the string at one time.*

Energy and waves

Energy of a wave The energy in a wave alternates between two forms, like the energy in an oscillator. For example, with a vibrating string, the energy is potential energy at the maximum amplitude of each cycle. The potential energy comes from stretching the string. The wave has the most potential energy when the string is stretched the most, at maximum amplitude. The kinetic energy comes from the motion of the string. The wave has its maximum kinetic energy as the string swings through its equilibrium position (Figure 14.28). The energy transformations in a water wave are also between potential and kinetic energy as the water's mass rises and falls. All waves propagate by exchanging energy between two forms. For water and elastic strings, the exchange is between potential and kinetic energy. For sound waves, the energy oscillates between pressure and kinetic energy. In light waves, energy oscillates between electric and magnetic fields.

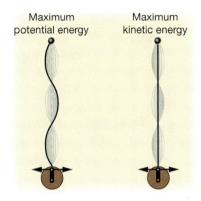

Figure 14.28: *The wave has the maximum potential energy at its most stretched position. The maximum kinetic energy occurs when the string swings through equilibrium.*

Frequency and energy The energy of a standing wave is proportional to the frequency of the wave. Figure 14.29 shows three standing waves with the same amplitude and different frequencies. The wave with the highest frequency has the greatest energy. For a standing wave on a string, the energy increases because the string must move faster to complete more cycles per second at higher frequencies. This result is true for almost all waves.

Amplitude and energy The energy of a wave is also proportional to the amplitude of the wave. Given two standing waves of the same frequency, the wave with the larger amplitude has more energy. With a vibrating string, the potential energy of the wave comes from the stretching of the string. A larger amplitude means the string has to stretch more, and therefore stores more energy.

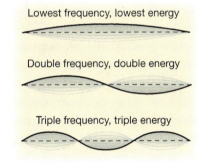

Figure 14.29: *The energy of a wave is proportional to its frequency.*

Why are standing waves useful? Standing waves can be used to store energy at specific frequencies. With the wave on the string, you observed how a small input of energy at the natural frequency accumulated over time to build a wave with much more energy. Musical instruments use standing waves to create sound energy of exactly the right frequency to match musical notes. Radio transmitters and cell phones also use standing waves to create power at specific frequencies. The standing waves are electromagnetic in these applications.

14.3 NATURAL FREQUENCY AND RESONANCE

Describing waves

Wavelength A vibrating string can move so fast that your eye averages out the motion and you see a wave-shaped blur (Figure 14.30). At any one moment the string is really in only one place within the blur. The wavelength is the length of one complete cycle on the string.

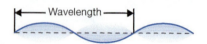

Figure 14.30: *The wavelength of a standing wave is the length of one complete cycle.*

Nodes and antinodes Standing waves have nodes and antinodes. A **node** is a point where the string stays at its equilibrium position (Figure 14.31). An **antinode** is a point where the wave is as far as it gets from equilibrium. A fixed boundary forces the string to always have a node at the boundary. It is also possible to make standing waves with an open boundary that have an antinode at the end of the string.

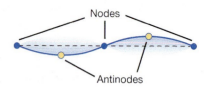

Figure 14.31: *A standing wave may have several nodes and antinodes.*

Frequency and wavelength relationship High frequency waves have a short wavelength compared with low frequency waves. This is generally true of all waves, and comes from the relationship between the speed of a wave, its frequency, and its wavelength. The speed of a wave is not usually affected much by changes in frequency or wavelength. Since speed is frequency times wavelength, for speed to stay the same, the wavelength must go down if the frequency goes up. In fact, if the frequency doubles, the wavelength must be reduced by exactly one-half to compensate. If the frequency is increased by a factor of 10, the wavelength decreases to one-tenth what it was.

Modes A vibrating string has two distinct *modes*. A **mode** is a particular category of wave behavior. One mode of the vibrating string is a rotating wave and the other mode is a transverse wave (Figure 14.32). In the rotating mode, the string spins around in a circular motion. Because the string moves in circles, the wave looks the same from the front and the side. In the transverse mode, the string moves back and forth in two dimensions. The wave looks different from the front and side.

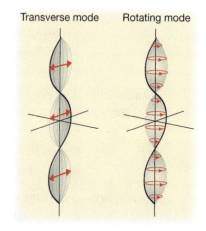

Figure 14.32: *A vibrating string has two possible modes.*

Modes and natural frequency Most systems that can support waves have different modes. In many systems, the different modes have different natural frequencies. The two modes of the vibrating string both stretch the string the same amount and have the same natural frequency. The transverse mode usually occurs when the oscillating force is directly under the fixed end of the string. If the oscillating force is exerted along the string, a slight twist develops and the rotating mode results.

Standing waves in two and three dimensions

Many modes of vibration

Most vibrating objects have more complex shapes than a string. Complex shapes create more ways an object can vibrate. Two- and three-dimensional objects tend to have two or three *families* of modes. Often, each family of modes has its own natural frequency including a fundamental and harmonics. The vibrations of real objects can be complex because more than one mode can be active at one time. For example, when a musician strikes a cymbal, the brass plate vibrates. A careful analysis would show more than 10 different modes of vibration in the motion of the cymbal. Each mode contributes to the sound of the cymbal and the richness of the sound comes from the complexity of the vibrations.

Vibrations of a circular disc

Figure 14.33 shows two families of modes for a vibrating circular disc, like the head of a drum. Two of the different modes in each family are shown. A skillful drummer knows how and where to hit the drum to make mixtures of the different modes and get particular sounds. The radial modes have nodes and antinodes that are circles. The angular modes have nodes and antinodes that are radial lines from the center of the circle. A circular disc has two dimensions because you can identify any point on the surface with two coordinates, angle and radius. Two-dimensional systems usually have two distinct families of vibrating modes. One family of modes has nodes and antinodes along the radius coordinate. The second family of modes has nodes and antinodes along the angle coordinate.

Waves in a water glass

You can also see the two modes of a circular surface with a glass full of water. If you run a moistened finger around the rim of the glass, you can see circular modes. The nodes and antinodes are circles. If you rock the glass side to side, you get a sloshing wave that is the longest wavelength of the angular modes. It is possible to get resonance at a different frequency for each mode.

Vibration of a guitar top

The top of a guitar has a very complex structure of vibrating modes (Figure 14.34). The bands of light and dark are contours of motion away from equilibrium. The lowest frequency mode, at 268 Hz, has the fewest nodes and the simplest pattern. As the frequency increases, the number of nodes increases and the patterns become more complex. Some patterns are harmonics within the same family of modes and others belong to different families of modes.

Radial modes

Angular modes

Figure 14.33: *The two different families of modes are shown for standing waves on a circular disk, like a drum skin.*

268 Hz   980 Hz

873 Hz   1,010 Hz

Figure 14.34: *Vibrating modes of an acoustic guitar top vary with the frequency.*

Chapter 14 Connection

Freak Waves

It was New Year's Day, 1995. The North Sea was being battered by a furious storm. The Draupner oil-drilling platform off the coast of Norway, equipped with a laser designed to measure wave height, measured a monster wall of water 26 meters from trough to crest. Other waves recorded during the same storm measured less than half that height. The wave slammed into the oil rig at 72 kilometers per hour. These measurements mark the first time that a freak wave, more commonly called a *rogue wave*, was recorded by anything other than anecdotal evidence.

In the North Pacific Ocean in 1933, the U.S. Navy oil ship *Ramapo* experienced a monster wave. With the same triangulation method used to determine the height of the world's tallest mountains, the crew calculated the wave's height to be 34 meters.

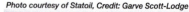

Figure 14.35: *The first laser-recorded rogue wave was measured from this oil rig.*

What they are Rogue waves were long thought to be the stuff of myths and exaggerated eyewitness accounts. But they have earned respect in the scientific community since lasers and other instruments have been confirming their existence.

Rogue waves are defined as waves at least twice the height of other waves in the area at the same time. Such waves often have a steeper slope than other waves—they are almost vertical—and a smaller crest. When they run into a ship, they can break and pummel the ship with about 100 tons of pressure per square meter. The average ship can withstand only 15–30 tons of pressure per square meter.

It's not only the tall crest that presents a huge problem for ships. The deep trough that comes just before the wave, called a "hole in the sea," can cause a ship to plummet precariously for many meters and hit the sea surface with great force that threatens its existence.

How they form Some rogue waves are believed to form by constructive interference. A wave moving in one direction encounters a wave moving in another direction. By coincidence, their crests end up in the same place at the same time, causing a wave with an amplitude as tall as the combined height of the two intersecting waves. But scientists have calculated the occurrence of this random superposition to be only once in every 10,000 years. Clearly there is something else going on in the oceans as well.

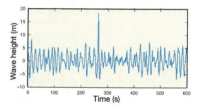

Figure 14.36: *The height of the wave measured on the rig above is graphed here.*

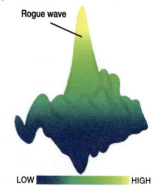

Figure 14.37: *A 3-D graph shows the relatively steep sides and narrow crest of a rogue wave.*

Chapter 14 Connection

Other causes are not so well understood. But areas where an ocean current moves in one direction and prevailing winds blow in the opposite direction are quite prone to experiencing these monsters waves. The Agulhas Current, off the coast of South Africa, is one such place. The Agulhas is also where warm waters from the Indian Ocean meet colder waters from the Atlantic Ocean. A lot of energy is present in this geographic area.

The Kuroshio Current, off the coast of Japan, is prone to rogue waves. So is the Gulf Stream, off the eastern coast of the United States. As it happens, the Gulf Stream runs through the Bermuda Triangle. Perhaps rogue waves are responsible for the disappearance of so many ships in that area.

Some data show that rogue waves often form away from ocean currents, along weather fronts and at the edge of low-pressure areas. Maybe sustained winds during storms build up the water in front of them to enormous proportions.

Wave mathematician Al Osborne has another theory, straight from quantum physics. He says that under certain unstable conditions, some waves can "steal" energy from other waves around them. During a storm at sea, these "robber waves" grow bigger and bigger, while the waves around them shrink. The New Year's Day 1995 data support his work.

Unraveling the mystery
Scientists want to be able to predict these monstrosities, or at least the conditions under which they are likely to form. Ships could then be warned to stay away from danger zones. Ship manufacturers also want to know whether they should redesign ships to withstand more than the customary 15–30 tons of pressure per square meter. There are also questions as to whether they should reinforce weaker parts of ships, such as portholes and hatch covers.

The European Space Agency (ESA) has been contributing a tremendous amount of data on rogues. Unlike optical instruments, their radar satellites can record data even through clouds and in dark skies. The radar data, combined with laser and radar data from oil rigs and depth-sensor data from buoys, are showing scientists that these walls of water at sea are far more common than was believed just a few years ago. The Goma oil field in the North Sea recorded 466 rogue events in 12 years. Wolfgang Rosenthal, a scientist who has helped the ESA study rogues, estimates that 10 of these monsters are in the world's oceans at any given time.

Photo courtesy of the NOAA Photo Library

Figure 14.38: *Huge waves are common near the 100-fathom line in the Bay of Biscay.*

Photo courtesy of Michael Van Woert, NOAA NESDIS, ORA

Figure 14.39: *A sailor captured a rogue in the southern ocean through a porthole in 1998.*

Chapter 14 Assessment

Vocabulary
Select the correct term to complete the sentences.

wave	propagation	amplitude
frequency	wavelength	hertz (Hz)
pulse	transverse	longitudinal
oscillation	crest	trough
wave front	circular wave	plane
continuous	fixed boundary	open boundary
reflection	refraction	absorption
boundary conditions	incident	superposition principle
refracted wave	standing	mode
natural frequency	resonance	fundamental
node	boundary	interference
harmonic	diffraction	antinodes

1. A unit used to measure the frequency of a wave is the _____.
2. A single oscillation can be called a wave _____.
3. The oscillations of a(n) _____ wave are perpendicular to its direction of travel.
4. A material or object that is connected throughout its entirety is said to be _____.
5. The number of oscillations per second made by a point on a wave determines the _____ of the wave.
6. The maximum amount a wave moves from its equilibrium position is its _____.
7. The oscillations move in the same direction as the wave on a(n) _____ wave.
8. The distance from the beginning to end of one wave cycle is a(n) _____.
9. A(n) _____ is a traveling oscillation that moves energy from one place to another.
10. Back-and-forth movement of an object from its equilibrium position is called _____.
11. The crests of _____ waves are represented as straight lines.
12. The high point of a wave is the _____.
13. The leading edge of a wave is called the _____.
14. The movement of a wave away from the point of origin may be referred to as _____.
15. Disturbing water at a single point will create a(n) _____ wave.
16. A(n) _____ is the low point of a wave.
17. The phenomenon of a wave bouncing off a boundary is called _____.
18. When a wave reaches a point at which the surface conditions change, the wave has reached a(n) _____.
19. When the amplitude of a wave is reduced as it passes into a material, _____ of wave energy has taken place.
20. The bending of a wave as it crosses a boundary indicates _____ has taken place.
21. If a boundary does not move in response to a wave, the boundary is a _____ with respect to that wave.
22. The _____ states that if two or more waves are present at the same point in a medium, the total oscillation at that point will be the sum of the oscillations of the waves.
23. A boundary that moves in response to a wave is said to be a(n) _____ for that wave.
24. The process describing the bending of waves around a corner or through an opening is known as _____.
25. _____ determine(s) the way in which a wave behaves as it meets a boundary.
26. A wave approaching a boundary is called a(n) _____.
27. A wave that bends as it crosses a boundary is identified as a(n) _____.
28. Waves generated at the natural frequency and reflected from a boundary will result in _____.

29. The longest standing wave that can occur in an object is called the _____.

30. A point on a standing wave that remains at the equilibrium is called a(n) _____.

31. A category of types of wave behavior is known as a(n) _____.

32. A wave that is confined between boundaries is called a(n) _____ wave.

33. _____ occur(s) when two waves travel through the same space simultaneously.

34. The points on a wave found at maximum distance from equilibrium are known as _____.

Concept review

1. What is the relationship among a wave's frequency, wavelength, and speed?

2. Explain the difference between a pulse and a wave.

3. The wave characteristics, amplitude and wavelength, are both measured in meters. How are they different?

4. The frequency of a wave is the number of vibrations or cycles per second. How is this related to the period of a wave?

5. Write a formula relating the speed of a wave to its period and wavelength.

6. Draw a transverse wave with an amplitude of 2 cm and a wavelength of 4 cm. Label a crest and a trough on the wave.

7. As a wave front strikes a boundary, four interactions may occur. Name and briefly describe each interaction.

8. Read the descriptions below and indicate which type of wave interaction has most likely occurred.
   a. Your friend yells to you from across the park, but you are not able to hear what is said.
   b. As you drive on a hot summer day, you look off in the distance and see what appears to be water on the road.
   c. At sunset on a clear day, the Sun appears to be oval instead of round.
   d. People are talking in the next room and, with the door only slightly opened, you are able to hear their conversation.

9. How are standing waves created? What determines their amplitude?

10. A wave is propagated on a string by the exchange of potential and kinetic energy. Describe the points of maximum potential and kinetic energy for the vibrating string.

Problems

1. In glass, the speed of light is reduced. Calculate the speed of red light with a frequency of 4.33×10^{14} Hz if its wavelength is 4.17×10^{-7} m.

2. A sound wave travels at 340 m/s and has a frequency of 256 Hz. What is its wavelength?

3. A sound wave with a frequency of 512 Hz and a wavelength of 2.99 m is directed toward the bottom of a lake to measure its depth. If the echo of the sound from the bottom is heard 0.25 seconds later, how deep is the lake?

4. A honeybee moves its wings at a frequency of 225 Hz. How much time does it take the honeybee's wings to make one complete cycle?

5. Use the diagram of a spring below to answer the questions which follow.

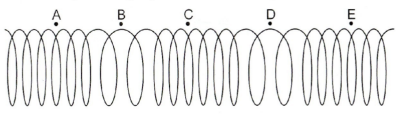

   a. Which type of wave is represented on the spring?
   b. Use the letters to represent an interval of one wavelength.
   c. Compressions are made horizontally. In what direction does the energy travel?

6. Two waves are superimposed in each diagram.
   a. Which pair of waves will completely cancel each other?
   b. Which pair of waves will result in a wave with the largest amplitude?

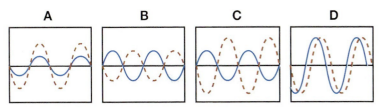

7. Using the diagram of four standing waves in the same medium, answer the questions below.

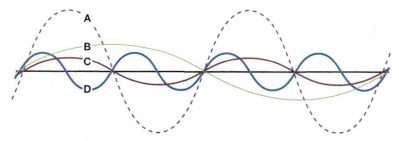

  a. Which two waves have the same wavelength?
  b. Which wave has the lowest frequency?
  c. Which wave has the smallest amplitude?

8. Two waves, A and B, travel in the same direction in the same medium at the same time. Graphs representing their motion are shown below. Answer the questions that follow based on the graphs.

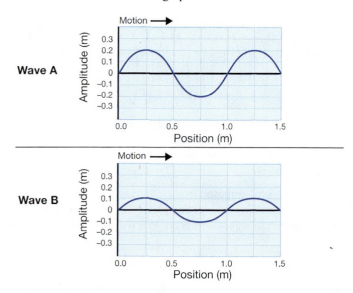

  a. Use a sheet of graph paper to draw the wave that would be produced by the superposition of waves A and B.
  b. What is the amplitude of the resultant wave?
  c. What is the wavelength of the resultant wave?

9. The fundamental harmonic of a standing wave on a vibrating string has a frequency of 40 Hz.
  a. What is the natural frequency of the third harmonic?
  b. How many nodes are in the third harmonic?
  c. How many antinodes are in the third harmonic?
  d. What would happen to the frequency of the third harmonic if the string was loosened? Why?

10. The diagram represents plane wave fronts being produced in a shallow water tank, referred to as a *ripple tank*. The frequency of the generator is 20 Hz. Answer the questions which follow based on this information and the diagram.

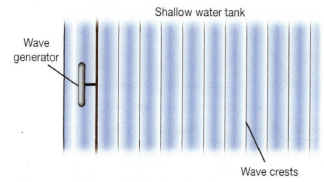

  a. What is the period of the wave?
  b. Use a ruler to determine the wavelength of the waves.
  c. Determine the speed of the wave.

Applying your knowledge

1. The operation of many familiar objects depends upon reflection and/or refraction. Name objects that use one or both of these wave interactions in their operation.

2. Research and write a brief description of noise canceling headsets and their operation.

3. What is a *moiré pattern* and how is it created? Under what circumstances can the formation of this pattern become a problem?

UNIT 5 WAVES AND SOUND

CHAPTER 15

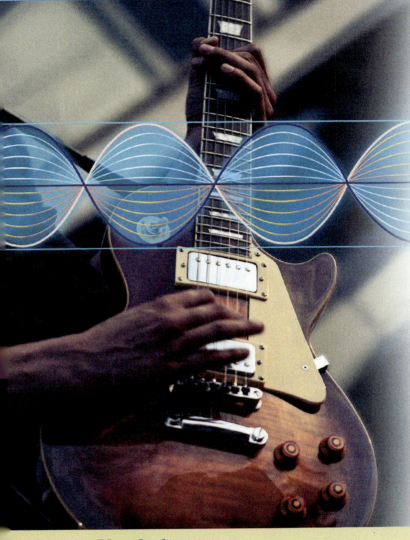

Sound

Objectives:

By the end of this chapter you should be able to:

- ✓ Explain how the pitch, loudness, and speed of sound are related to properties of waves.
- ✓ Describe how sound is created and recorded.
- ✓ Give examples of refraction, diffraction, absorption, and reflection of sound waves.
- ✓ Explain the Doppler effect.
- ✓ Give a practical example of resonance with sound waves.
- ✓ Explain the relationship between the superposition principle and Fourier's theorem.
- ✓ Describe how the meaning of sound is related to frequency and time.
- ✓ Describe the musical scale, consonance, dissonance, and beats in terms of sound waves.

Key Questions:

- What is sound, and how do we hear it?
- Does sound behave like other waves?
- What do frequency and wavelength have to do with pitch and the musical scale?

Vocabulary

acoustics	dissonance	microphone	pressure	speaker
beats	Doppler effect	musical scale	reverberation	stereo
cochlea	Fourier's theorem	note	rhythm	subsonic
consonance	frequency spectrum	octave	shock wave	supersonic
decibel		pitch	sonogram	

307

CHAPTER 15 SOUND

15.1 Properties of Sound

Like other waves, sound has the fundamental properties of frequency, wavelength, amplitude, and speed. Because sound is a significant part of human experience, you probably already know its properties, but you may know them by different names. For example, you almost never hear someone complain about the amplitude of sound. What you hear instead is that the sound is too *loud* or too *soft*.

What is sound?

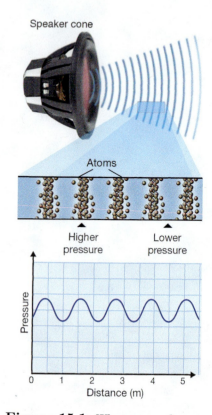

Figure 15.1: *What a sound wave might look like if you could see the atoms. The effect is greatly exaggerated to show the variation. In an actual sound wave, the difference in pressure between the highest- and lowest-pressure regions is much smaller, less than one part in a million. From the graph you can see the wavelength of this sound is about one meter.*

Sound is a wave that carries vibrations Touch the moving cone of a speaker and you can feel it vibrating. Because the speaker is in air, its vibration spreads out through the air as sound. Sound is a vibration that travels as a wave through solids, liquids, or gases. Figure 15.1 shows an illustration of a speaker, a sound wave, and the oscillation of the air. The forward and backward movement of the speaker cone creates the wave that carries the sound through the air from the speaker to your ear.

Sound comes from vibrations too fast to see Anything that vibrates can produce sound. You cannot usually see the actual vibrations. One reason is that the frequency is high. The slowest vibration we can hear is 20 Hz, or 20 vibrations per second. Anything moving back and forth 20 times per second is just a blur to your eyes. The second reason is that even loud sound waves are produced by vibrations with a very small amplitude. For example, the top of an acoustic guitar creates most of the sound of the guitar but the amplitude of the actual vibrations is less than a fraction of a millimeter.

Loudness and pitch Sound has properties of *loudness* and **pitch**. The loudness of a sound depends on the amplitude of vibration. A speaker making a loud sound moves back and forth more than a speaker making a soft sound. The pitch of a sound depends on the frequency of vibration. A speaker making a high-pitched sound like a siren vibrates with a higher frequency than the same speaker making a low-pitched sound like thunder.

Sound moves through matter Sound waves travel faster in liquids and solids than they do in air. You can hear voices through a solid wall because the sound wave in the air pushes on the wall and makes a sound wave in the wall. That sound wave travels through the solid wall and generates a new sound wave in the air on the other side. Sound *cannot* travel through the vacuum of space. Sound waves can only move through matter.

The frequency of sound

Frequency and pressure change

The frequency of sound tells you how fast the **pressure** oscillates. The low humming noise from an electrical transformer has a frequency of 60 hertz (Hz). This means the oscillating pressure of the air goes back and forth 60 times per second. The scream of a fire truck siren may have a frequency of 3,000 Hz. This corresponds to 3,000 oscillations per second in the pressure of the air.

Frequency and pitch

We hear the different frequencies of sound as having different pitches. A low-frequency sound has a low pitch, like the rumble of a big truck. A high-frequency sound has a high pitch, like a whistle or siren. The range of frequencies that humans can hear varies from about 20 Hz to 20,000 Hz. Animals can hear different ranges of sound. For example, bats can hear sound up to frequencies of 200,000 Hz or more.

Most sound has more than one frequency

Most sound that we hear contains many frequencies, not just one. For example, three frequencies can be added to create a complex sound. Remember, we discussed the superposition principle in the last chapter. Complex sound is created by the superposition of many frequencies. In fact, the sound of the human voice contains thousands of different frequencies, all at one time (Figure 15.2).

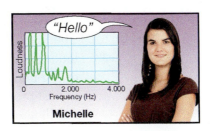

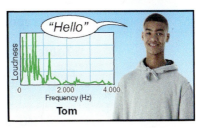

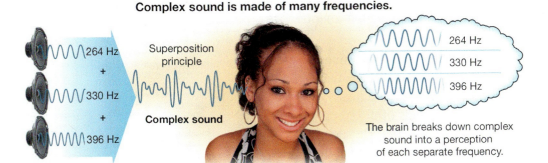

Figure 15.2: *The frequencies in three people's voices as they say the word hello. The highest amplitudes are between 100 and 1,000 Hz. The peaks come from harmonics of each person's fundamental frequency. Women have higher fundamental frequencies than men. The peaks for the two women's voices are farther apart than for the male voice.*

How we hear complex sound

When we hear complex sounds, the nerves in the ear respond separately to each different frequency. The brain interprets the signals from the ear and creates a "sonic image" from the frequencies. The meaning in different sounds is derived from patterns of different frequencies that get louder and softer over time.

The loudness of sound

The decibel scale

The loudness of sound is measured in **decibels** (dB). As you might expect, loudness is determined mostly by the amplitude of a sound wave. The amplitude of a sound wave is one-half of the difference between the highest pressure and the lowest pressure in the wave. Because the pressure change in a sound wave is very small, almost no one uses pressure to measure loudness. Instead we use the decibel scale. Most sounds fall between 0 and 100 on the decibel scale, making it a very convenient number to understand and use.

Table 15.1: Some common sounds and their loudness in decibels

10–15 dB	A quiet whisper 3 feet away
30 dB	Background sound level at a house in the country
40 dB	Background sound level at a house in the city
45–55 dB	The noise level in an average restaurant
65 dB	Ordinary conversation 3 feet away
70 dB	City traffic
90 dB	A jackhammer cutting up the street 10 feet away
110 dB	A hammer striking a steel plate 2 feet away
120 dB	The threshold of physical pain from loudness

The sensitivity of the ear

How the human ear and brain perceive the loudness of sound is affected by the frequency of the sound as well as by the amplitude. The equal loudness curve on the right shows how sounds of different frequencies compare. Sounds near 2,000 Hz seem louder than sounds of other frequencies, even at the same decibel level. For example, the curve shows that a 40 dB sound at 2,000 Hz sounds just as loud as an 80 dB sound at 50 Hz. The human ear is most sensitive to sounds between 300 and 3,000 Hz. The ear is less sensitive to sounds outside this range. Most of the frequencies that make up speech are between 300 and 3,000 Hz.

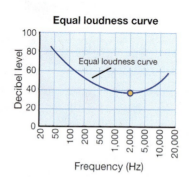

Equal loudness curve

> **The decibel scale**
>
> The decibel scale is a *logarithmic* measure of the amplitude of sound waves. This is different from linear measures you are familiar with. In a logarithmic scale, equal intervals correspond to multiplying by 10 instead of adding equal amounts. For example, every increase of 20 decibels (dB) means the pressure wave has 10 times greater amplitude.
>
> | Logarithmic Scale | Linear Scale |
> |---|---|
> | Decibels (dB) | Amplitude |
> | 0 | 1 |
> | 20 | 10 |
> | 40 | 100 |
> | 60 | 1,000 |
> | 80 | 10,000 |
> | 100 | 100,000 |
> | 120 | 1,000,000 |
>
> The decibel scale is designed to compare a wide range of amplitudes, making it easier to visualize huge changes in numbers. It is a good approximation because the human ear responds to sound in a roughly logarithmic scale. Every 20 dB increase sounds twice as loud.

How sound is created

Vibrations create sound Anything that vibrates with a frequency between 20 and 20,000 Hz will make a sound that humans can hear. The air around you is probably full of overlapping sound waves because so many objects in nature and technology vibrate. When a motor spins, it vibrates and the vibration creates a sound wave that you hear as a low hum. When your heart beats it also makes a vibration that doctors can hear with a stethoscope. Since sound travels through any material—solid, liquid, or gas—sound waves created in one place spread easily.

Voices The human voice is a complex sound that starts in the *larynx* and *voice box*, a small structure at the top of your windpipe. The term *vocal cords* is somewhat misleading because the sound-producing structures are not really cords but are folds of expandable tissue that extend across a hollow chamber known as the voice box. The sound that starts in the larynx is changed as it passes through openings in the throat and mouth (Figure 15.3). Different sounds are made by changing both the vibrations in the larynx and the shape of the openings.

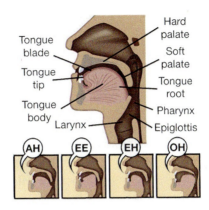

Figure 15.3: *The human voice is created by a combination of vibrating folds of skin in the larynx and the resonant shapes of the throat and mouth.*

Speakers A **speaker** is a device that is specially designed to reproduce sounds accurately. The working parts of a typical speaker include a magnet, a coil of wire, and a cone. When electricity is flowing through it, a coil of wire acts like a magnet. When the electricity flows one way, the magnetism created by the coil is attracted to the central magnet and the speaker cone moves outward (Figure 15.4). When the electricity in the coil is reversed, the coil is repelled by the central magnet and the speaker cone moves inward. The coil moves back and forth with the same frequency, and sound of that frequency is created by the movement of the cone attached to the coil. The oscillating electricity creates sound.

Acoustics Reducing or enhancing the loudness of sound is important in many applications. For example, a library building's interior should be constructed to absorb sound to maintain quiet. A recording studio might want to block sound from the outside from mixing with sound from the inside. **Acoustics** is the science and technology of sound. Knowledge of acoustics is important to many careers, from the engineers who design speakers to the architects who designed your school.

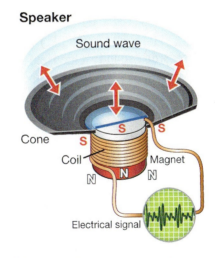

Figure 15.4: *How a speaker produces sound.*

CHAPTER 15 SOUND

Recording sound

Music was rare 100 years ago
We often take for granted that we can listen to our favorite music anytime we wish. This was not true 100 years ago, when the only way to hear music was to be close to the musicians while they were playing. Very few people were able to listen to a variety of musicians so as to even *have* a favorite. The recording of sound was a technological breakthrough that transformed human experience.

The microphone
To record a sound you must store the pattern of vibrations in a way that can be replayed and be true to the original sound. A common way to record sound starts with a **microphone**. A microphone transforms a sound wave into an electrical signal with the same pattern of oscillation (top of Figure 15.5).

Analog to digital conversion
In modern digital recording of sound, a sensitive circuit called an *analog-to-digital converter* measures or *samples* the electrical signal 44,100 times per second. Each measurement consists of a number between 0 and 65,536, corresponding to the amplitude of the signal. One second of digitally recorded sound is a list of 44,100 numbers. The numbers are recorded as data and can be stored on computers, phones, or any digital storage device.

Playback of recorded sound
To play the sound back, the string of numbers is read by a laser and converted into electrical signals again by a second circuit. The second circuit is a *digital-to-analog converter*, and it reverses the process of the previous circuit. The playback circuit converts the string of numbers back into an electrical signal. The electrical signal is amplified until it is powerful enough to move the coil in a speaker and reproduce the sound (bottom of Figure 15.5).

Stereo sound
Most of the music you hear has been recorded in **stereo**. A stereo recording is actually two recordings, one to be played from the right speaker and the other from the left speaker. Stereo sound feels "live" because it creates slight differences in phase of the sound waves reaching your left and right ears. When you listen to a live concert you can hear that a singer is on the left and a guitar player is on the right. The sound from the singer reaching your left ear is slightly out of phase with the sound reaching your right ear. Your brain interprets the difference in phase to provide a sense of depth. A stereo recording can do the same thing. If you close your eyes and listen to a good stereo recording, you can hear different instruments coming from different directions, to the left or right.

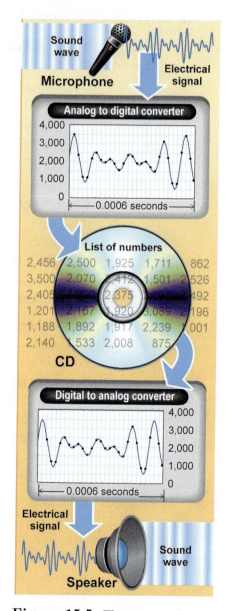

Figure 15.5: *The process of digital-sound reproduction.*

15.2 Sound Waves

Scientifically, sound is one of the simplest and most common kinds of waves. But what a huge influence it has on our everyday experience of life! We know sound is a wave because

1. sound has both frequency that we hear directly and wavelength that can be demonstrated by simple experiments;
2. the speed of sound is frequency times wavelength;
3. resonance happens with sound; and
4. sound can be reflected, refracted, and absorbed and also shows evidence of interference and diffraction.

A close look at a sound wave

Close-up of a sound wave A sound wave is a travelling oscillation of pressure in air or other matter. Anything that vibrates in air creates a sound wave. The wave travels away from the source and eventually reaches our ear, where it vibrates the eardrum and we hear the sound.

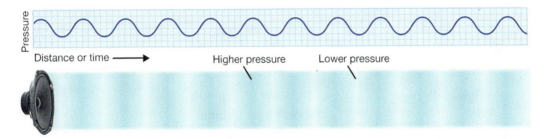

The pressure waves are small The actual oscillations in pressure from a sound wave are very small (Figure 15.6). Table 15.2 gives some examples of the amplitude for different decibel levels. The human ear is remarkably sensitive. For instance, if you were looking at a pile of a million coins, you could not notice one missing. But the human ear can easily hear a pressure wave that is only two parts different out of 100 million. This exquisite sensitivity is why hearing can be damaged by listening to loud sounds for a long time.

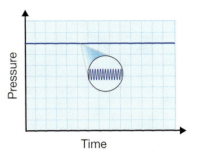

Figure 15.6: *The amplitude of a sound wave is very small. Even a loud 80 dB noise creates a pressure variation of only a few millionths of an atmosphere.*

Table 15.2: Loudness and amplitude of sound waves in the air

Loudness in decibels	Amplitude of pressure wave (fraction of 1 atmosphere)
20 dB	$\dfrac{2}{1,000,000,000}$
40 dB	$\dfrac{2}{100,000,000}$
80 dB	$\dfrac{2}{1,000,000}$
120 dB	$\dfrac{2}{10,000}$

CHAPTER 15 SOUND

The wavelength of sound

Sound is a longitudinal wave
Sound waves are longitudinal waves because the air is compressed in the direction of its motion. You can think of a sound wave like the compression wave on a coiled spring toy. Anything that vibrates creates sound waves as long as there is air or some other matter. Sound does *not* travel in the vacuum of space. Science fiction movies often add sound to scenes of space travel to make them more exciting. In reality, an observer in space would hear total silence because there is no air in space to carry a sound waves.

Range of wavelengths of sound
The wavelength of sound in air is comparable to the size of everyday objects. The chart below gives some typical frequencies and wavelengths for sound at one atmosphere and room temperature. As with other waves, the wavelength of a sound is inversely related to its frequency (Figure 15.7). A low-frequency, 20-hertz sound has a wavelength the size of a large classroom. At the upper range of hearing, a 20,000-hertz sound has a wavelength about the width of your finger.

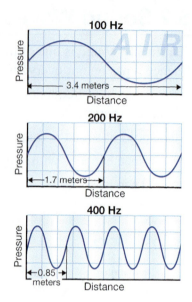

Figure 15.7: *The frequency and wavelength of sound are inversely related. When the frequency increases, the wavelength decreases proportionally.*

Frequency (Hz)	Wavelength	Typical source
20	17 meters	rumble of thunder
100	3.4 meters	bass guitar
500	70 cm (27 in)	average male voice
1,000	34 cm (13 in)	female soprano singer
2,000	17 cm (6.7 in)	fire truck siren
5,000	7 cm (2.7 in)	highest note on a piano
10,000	3.4 cm (1.3 in)	whine of a jet turbine
20,000	1.7 cm (2/3 in)	highest pitched sound you can hear

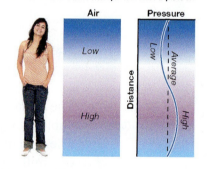

Figure 15.8: *A 200-Hz sound has a wavelength about equal to the average height of a person.*

Why the wavelength of sound is important
Although we usually think about different sounds in terms of frequency, the wavelength can also be important. If you want to make sound of a certain wavelength, you often need to have a vibrating object that is similar in size to the wavelength (Figure 15.8). This is the reason pipes for organs are made in all different sizes. Each pipe is designed for a specific wavelength of sound. The short pipes make short-wavelength sounds that have high frequencies. The long pipes make long-wavelength sounds that have low frequencies.

The Doppler effect

Definition of the Doppler effect

If an object making sound is at rest, stationary observers on all sides will hear the same frequency. Observers will *not* hear the same frequency on all sides of a moving source of sound, like a siren on a racing fire engine. Only people moving with the object or to the side of it hear the sound as if the object were stationary. People in front of the object hear sound of *higher* frequency. People behind the object hear sound of *lower* frequency. The shift in frequency caused by motion is called the **Doppler effect** and it occurs when either the source of the sound or the listener is moving at speeds less than the speed of sound.

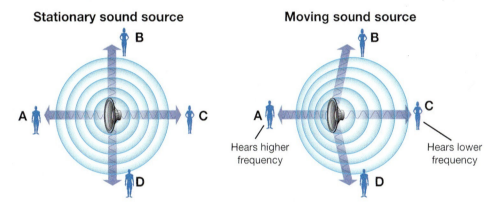

The cause of the Doppler effect

The Doppler effect occurs because an observer hears the frequency at which wave fronts arrive at the ears. Wave fronts are closer together in front of a moving object because the object moves forward between wave crests. This is why Observer A in front hears a higher frequency. The opposite is true for an observer behind. The motion of the object makes more space between successive waves. According to Observer C, the wave fronts get farther apart, and the frequency goes down. The greater the relative speed, the larger the shift in frequency.

Demonstrating the Doppler effect

You can observe the Doppler effect with a sound generator, like a cell phone. When the phone speaker is standing still, it makes a steady sound of a certain frequency. Have a friend whirl the phone overhead in a bag. The sound will no longer have a steady frequency. The frequency shifts up and down with each rotation according to whether the sound source is moving toward or away from your ears.

Doppler radar

The Doppler effect also happens with reflected waves, including light waves. With Doppler radar, an electromagnetic wave is sent out from a transmitter. The wave reflects from moving objects, such as a car. The frequency of the reflected wave increases if the car is moving toward the source and decreases if the car is moving away.

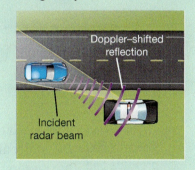

The amount of the Doppler shift is proportional to the speed of the car. The speed of the car can be accurately measured by comparing the original frequency with the frequency of the reflected wave. Doppler radar is used to enforce speed limits, to measure the speed of wind in storms, and in many other applications where speed needs to be measured from a distance.

The speed of sound

Sound is fast, about 340 meters per second

The speed of sound in air is 343 meters per second (660 miles per hour) at 1 atmosphere of pressure and 21°C. The speed increases with temperature and also with pressure. Passenger jets fly slower than sound, usually around 400 to 500 miles per hour. An object is **subsonic** when it is moving slower than sound.

Sonic booms

We use the term **supersonic** to describe motion at speeds faster than the speed of sound. Many military jets are capable of supersonic flight. If you were on the ground watching a supersonic jet fly toward you, there would be silence. The sound would be *behind* the plane, racing to catch up. A **shock wave** forms where the wave fronts pile up (diagram below). The pressure change across the shock wave is what causes a very loud sound known as a *sonic boom* (Figure 15.9).

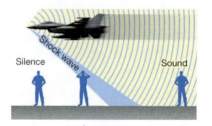

Figure 15.9: *The boundary between sound and silence is called a shock wave. The person in the middle hears a sonic boom as the shock wave passes.*

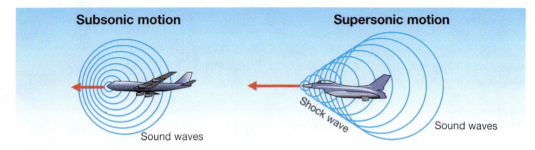

The speed depends on pressure and temperature

The speed of a sound wave in air depends on how fast air molecules are moving. If the molecules are moving slowly, sound does not travel as quickly as when they are moving fast. The kind of molecules also affects the speed of sound. Air is made up of mostly oxygen (O_2) and nitrogen (N_2) molecules. Lighter molecules, like hydrogen (H_2), move faster for a given temperature. Because of the mass difference, sound travels faster in hydrogen than in air.

Material	Sound speed (m/s)
Air	330
Helium	965
Water	1,530
Wood (average)	2,000
Gold	3,240
Steel	5,940

Figure 15.10: *The speed of sound in various materials. Helium and air are at 0°C and 1 atmosphere.*

Sound in liquids and solids

The speed of sound in materials is often faster than in air (Figure 15.10). The restoring forces in solid steel, for example, are much stronger than in a gas. Stronger restoring forces tend to raise the speed of sound. People used to listen for an approaching train by putting an ear to the rails. The sound of an approaching train travels much faster through steel rails than through air.

Standing waves and resonance

Resonance of sound Spaces enclosed by boundaries can create resonance with sound waves. Almost all musical instruments use resonance to make musical sounds. A panpipe is a good example of resonance in an instrument. A panpipe is made of many tubes of different lengths (Figure 15.11); one end of each tube is closed and the other end is open. Blowing across the open end of a tube creates a standing wave inside the tube. The frequency of the standing wave is the frequency of sound given off by the pipe. Longer pipes create longer-wavelength standing waves, and make lower frequencies of sound. Shorter pipes create shorter-wavelength standing waves, and therefore make higher frequencies of sound.

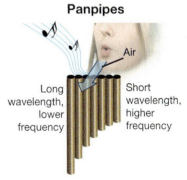

Standing wave patterns The closed end of a pipe is a closed boundary. Remember from the previous chapter that a closed boundary makes a node in the standing wave. The open end of a pipe is an open boundary to a standing wave in the pipe. An open boundary makes an antinode in the standing wave. Figure 15.11 shows different standing waves that have a node at the closed end and an antinode at the open end. Notice that the wavelength of the fundamental mode is four times the length of the pipe. It follows that a pipe will be resonant to a certain sound when its length is one-fourth the wavelength of the sound. Lower frequency sounds have longer wavelengths and are created with longer pipes. Higher frequency sounds have shorter wavelengths and are created with shorter pipes.

Designing a musical instrument Suppose you wish to make a pipe that makes a sound with a frequency of 660 Hz, which is the musical note E. The speed of sound in air is constant and equal to the product of frequency (f) and wavelength (λ). This relationship tells you that 660 Hz times the wavelength must be equal to 343 m/s. The required wavelength is therefore 343 m/s divided by 660 Hz, or 0.52 meters. The length of pipe needs to be one quarter of a wavelength to make a resonance in the fundamental mode. One quarter of 52 centimeters is 13 centimeters. If you make a pipe that is 13 centimeters long with one closed end, it will have a natural frequency of 660 Hz. This is the principle on which virtually all musical instruments are designed. Sounds of different frequencies are made by standing waves. A particular sound is selected by choosing the length of a vibrating object, such as a string or air column, to be resonant at the desired frequency.

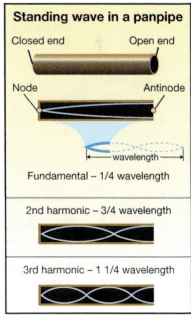

Figure 15.11: *A panpipe is made from tubes of different length. The diagrams show the fundamental and harmonics for standing waves of sound in a panpipe.*

CHAPTER 15 SOUND

Interaction between sound waves and boundaries

Interactions of sound and materials
Like other waves, sound waves can be reflected by surfaces and refracted as they pass from one material to another. Diffraction causes sound waves to spread out through small openings. Carpet and soft materials can absorb sound waves. Figure 15.12 shows examples of sound and materials.

Reverberation
Sound waves reflect from hard surfaces. In a good concert hall, the reflected sound adds to the sound source. You hear a multiple echo called **reverberation**. The right amount of reverberation makes the sound seem livelier and richer. Too much reverberation and the sound is changed from too many reflections. Concert hall designers choose the shape and surface of walls and ceilings to provide a balance between reverberation and the source sound. Some concert halls even have movable panels that can be raised or lowered from the ceiling to help shape the sound.

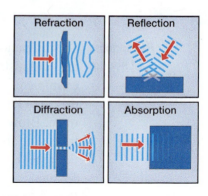

Figure 15.12: *Sound displays all the properties of waves in its interactions with materials and boundaries.*

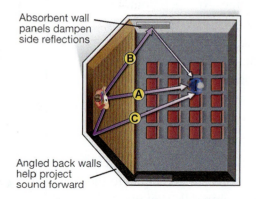

Making a good concert hall

*Direct sound (**A**) reaches the listener along with reflected sound (**B, C**) from the walls. The shape of the room and the surfaces of the walls must be designed so that there is some reflected sound, but not too much.*

Ultrasound

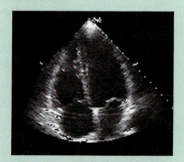

Ultrasound is sound that has very high frequencies, often 100,000 Hz or more. We cannot hear ultrasound, but it can pass through the human body easily. Medical ultrasound instruments use the refraction and reflection of sound waves inside the body to create images. Doctors often take ultrasound pictures of the human body. The ultrasound image pictured above is a heart.

Interference can also affect sound quality
Reverberation causes interference of sound waves. When two waves interfere, the total can be louder or softer than either wave alone. The diagram above shows a musician and an audience of one person. Direct sound reaches the listener along path A. The sound reflected from the walls travels a longer path before reaching the listener. If the distances are just right, one reflected wave might be out of phase with the other. The result is that the sound is quieter at that spot. An acoustic engineer would call it a *dead spot* in the hall. Dead spots are areas where destructive interference causes some of the sound to cancel with its own reflections. It is also possible to make very loud spots where sound interferes constructively. The best concert halls are designed to minimize such interference.

The frequency spectrum and Fourier's theorem

Adding waves Imagine holding a microphone in a noisy room with music playing and people talking. The microphone records a single "wave form" that describes the variation of pressure with time. The recorded wave form is usually complex (Figure 15.13). Yet you could easily distinguish the music from individual voices if you were in the room. The complex wave form recorded by the microphone is the same thing your ears hear. Somehow, this single complex wave form must contain all the sound from the music and voices.

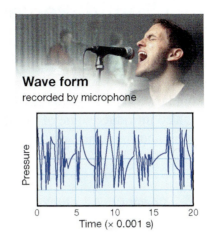

Figure 15.13: *The wave form shows 0.02 s of recorded music.*

Fourier's theorem **Fourier's theorem** says that any wave form can be represented as a sum of single frequency waves. Remember that the superposition principle states that many single waves add up to one complex wave. Fourier's theorem says the opposite is also true: Any complex wave can be made from a sum of single frequency waves. In fact, complex waves are best thought of in terms of *component frequencies*. A complex wave is really a sum of component frequencies, each with its own amplitude and phase. Figure 15.14 shows how a square wave can be built up from component frequencies.

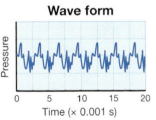 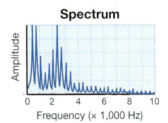

The spectrum shows the frequencies that make up a complex wave form.

Frequency spectrum A **frequency spectrum** is a graph that shows the amplitude of each component frequency in a complex wave. For example, the wave form in the diagram above is from an acoustic guitar playing the note E. The frequency spectrum shows that the complex sound of the guitar is made from many frequencies; in fact, from the evenly spaced peaks, you can identify many harmonics in the sound.

Wave form and spectrum change with time Both the wave form and the spectrum change as the sound changes. The wave form and spectrum in the diagram represent a sample of only 0.02 seconds from the sound. The meaning in sound comes from the changing pattern of frequencies.

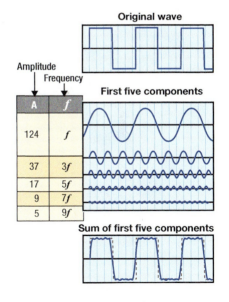

Figure 15.14: *A square wave is a sum of the components of different frequencies and amplitudes.*

15.3 Sound, Perception, and Music

Sound is everywhere in our daily environment. Hearing is one of the most important of our senses, and the ear and brain are constantly perceiving and processing sound. We actively use sound to communicate and we listen to sound for information about what is going on around us. In this section, you will learn about how we hear a sound wave and how the ear and brain construct meaning from sound. This section will also introduce some of the science behind musical sound. Musical sound is a rich language of rhythm and frequency, developed over thousands of years of human culture.

Patterns of frequency

Constructing meaning from sound Think about reading one single word from a story. You recognize the word, but it does not tell you much about the story. When you read the whole story, you put all the words together to get the meaning. The brain does a similar thing with different frequencies of sound. A single frequency by itself does not have much meaning. The meaning comes from patterns in many frequencies together.

Sonograms A **sonogram** is a special kind of graph that shows how loud sound is at different frequencies (Figure 15.15). The sonogram below shows a male voice saying "hello." The word lasts from 0.1 seconds to about 0.6 seconds. You can see lots of sound below 1,500 hertz and two bands of sound near 2,350 and 3,300 hertz. Every person's sonogram is different, even when saying the same word.

A simple version of a sonogram

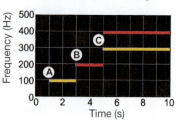

Figure 15.15: *A sonogram shows how the loudness of different frequencies of sound changes with time.*

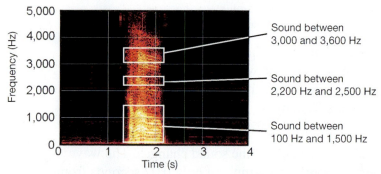

The brighter the sonogram, the louder the sound at that frequency.

> **Reading a sonogram**
>
> A sonogram includes information about frequency, loudness, and time. The vertical axis represents frequency. The example in Figure 15.15 shows frequencies from 0 to 500 Hz. The horizontal axis represents time. The yellow bars represent louder sounds.
>
> The sonogram shows four frequencies of sound over a period of 10 seconds.
>
> 1. A loud sound at 100 Hz that lasts from 1 to 3 seconds.
> 2. A softer sound at 200 Hz that lasts from 3 to 5 seconds.
> 3. A soft sound at 400 Hz and a louder sound at 300 Hz, both starting at 5 seconds.

How we hear sound

Hearing sound — We get our sense of hearing from the **cochlea**, a tiny, fluid-filled structure in the inner ear (Figure 15.16). The inner ear actually has two important functions: providing our sense of hearing and our sense of balance. The three semicircular canals near the cochlea are also filled with fluid. Fluid moving in each of the three canals tells the brain whether the body is moving left-right, up-down, or forward-backward.

How the cochlea works — The perception of sound starts with the eardrum. The eardrum vibrates in response to sound waves in the ear canal. The three delicate bones of the inner ear transmit the vibration of the eardrum to the side of the cochlea. The fluid in the spiral of the cochlea vibrates and creates waves that travel up the spiral. The spiral channel of the cochlea starts out large and gets narrower near the end. The nerves near the beginning are in a relatively large channel and respond to longer-wavelength, lower-frequency sound. The nerves at the small end of the channel respond to shorter-wavelength, higher-frequency sound.

The range of human hearing — The average range of human hearing is between 20 Hz and 20,000 Hz, or 20 kilohertz, abbreviated kHz. The combination of the eardrum, bones, and the cochlea all contribute to the limited range of hearing. You could not hear a sound at 50,000 Hz (50 kHz), even at a very loud volume of 100 decibels. Animals such as cats and dogs can hear much higher frequencies because of more sensitive structures in their inner ears.

Hearing ability changes with time — Hearing varies greatly with people and changes with age. Some people can hear very high frequency sounds and other people cannot. People gradually lose high frequency hearing with age. Most adults cannot hear frequencies above 15,000 Hz, while children can often hear to 20,000 Hz.

Hearing can be damaged by loud noise — Hearing is affected by exposure to loud or high-frequency noise. The nerve signals that carry the sensation of sound to the brain are created by tiny hairs that shake when the fluid in the cochlea vibrates. Listening to loud sounds for a long time can cause the hairs to weaken or break off. Before there were safety rules about noise, people who worked in mines or other noisy places often became partly deaf by the time they retired. It is smart to protect your ears by keeping the volume of music and TVs reasonable and wearing ear protection if you are in a loud place. Many musicians now wear earplugs to protect their hearing when performing.

The Ear

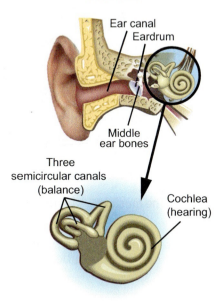

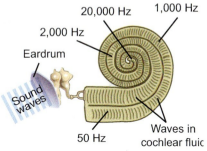

Figure 15.16: *The structure of the inner ear is shown.*

Music

Pitch The pitch of a sound is how high or low we hear its frequency. Pitch and frequency usually mean the same thing. However, because pitch depends on the human ear and brain, sometimes pitch and frequency can be different. The way we hear a pitch can be affected by the sounds we heard before and after. A good musician can create the illusion of a certain note by playing other notes around it.

Rhythm **Rhythm** is a regular time pattern in a sound. Rhythm can be loud and soft: tap-tap-TAP-tap-tap-TAP-tap-tap-TAP. Rhythm can be made with sound and silence, or with different pitches. People respond naturally to rhythm. Cultures are distinguished by their music and the special rhythms used in the music.

The musical scale Music is a combination of sound and rhythm that we find pleasant. Styles of music can be very different, but all music is created from carefully-chosen frequencies of sound. Most of the music you listen to is created from a pattern of frequencies called a **musical scale**. Each frequency in the scale is called a **note**. The range between any frequency and twice that frequency is called an **octave**. Notes that are an octave apart in frequency share the same name. Within the octave there are eight primary notes in the Western musical scale. Each of the eight notes is related to the first note in the scale by a ratio of frequencies. The scale that starts on the note C (264 Hz) is shown in this diagram.

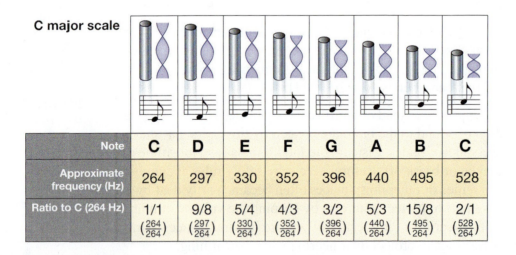

Choosing the notes

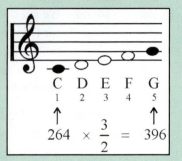

The notes on a musical scale are related to the first note by ratios of frequency. For example, the fifth note has a frequency 3/2 times the frequency of the first note. If the first note is C (264 Hz), then the fifth note has a frequency of 1.5 times 264, or G (396 Hz).

Octaves

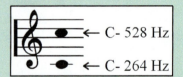

Two notes are an octave apart when the frequency of one note is double the frequency of the other. Notes that are an octave apart are given the same name because they sound similar to the ear. For example, the note C has a frequency of 264 Hz. Frequencies of 132 Hz and 528 Hz are also named C because they are an octave apart from C (264 Hz).

Consonance, dissonance, and beats

Harmony Music can have a profound effect on people's moods. The tense, dramatic soundtrack of a movie is a vital part of the audience's experience. Harmony is the study of how sounds work together to create effects desired by the composer. Harmony is based on the frequency relationships of the musical scale.

Beats An interesting thing happens when two frequencies of sound are close, but not exactly the same. The phase of the two waves changes in a way that makes the loudness of the sound seem to oscillate or **beat**. Sometimes the two waves are in phase, and the total is louder than either wave separately. Other times the waves are out of phase and they cancel each other out, making the sound quieter. The rapid alternation in amplitude is what we hear as beats. Most people find beats very unpleasant to listen to. Out-of-tune instruments playing together make beats. The frequencies in the musical scale are specifically chosen to reduce the occurrence of beats.

Beats come from adding two waves that are slightly different in frequency

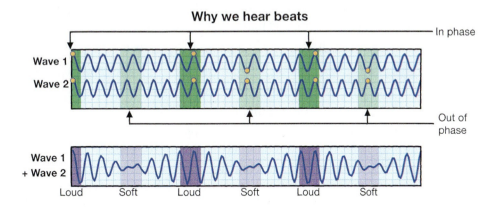

Echolocation and beats

Bats navigate at night using sound waves instead of light waves. A bat's voice is like a sonic flashlight, shining a beam of sound. A bat emits bursts of sound, called chirps, that rise in frequency. When the sound reflects off of an insect, the bat's ears receive the echo. Since the frequency of the chirp is always changing, the echo comes back with a slightly different frequency. The difference between the echo and the chirp makes *beats* that the bat can hear. The beat frequency is proportional to how far the insect is from the bat. A bat can triangulate the insect's position by comparing the echo from the left ear with that from the right ear.

Consonance and dissonance When we hear more than one frequency of sound and the combination sounds good, we call it **consonance**. When the combination sounds bad or unsettling, we call it **dissonance**. Consonance and dissonance are related to beats. When frequencies are far enough apart that there are no beats, we get consonance. When frequencies are too close together, we hear beats that are the cause of dissonance. Dissonance is often used to create tension or drama. Consonance can be used to create feelings of balance and comfort.

Harmonics and the sound of instruments

The same note can sound different
The same note sounds different when played on different instruments. As an example, suppose you listen to the note C (264 Hz) played on a guitar and the same C (264 Hz) played on a piano. A musician would recognize both notes as being C because they have the same frequency and pitch. But the guitar sounds like a guitar and the piano sounds like a piano. If the frequency of the note is the same, what gives each instrument its characteristic sound?

Instruments make mixtures of frequencies
The answer is that the sound from an instrument is not a single pure frequency. The most important frequency is the fundamental note, C (264 Hz) for example. The variation comes from the harmonics. Remember, harmonics are frequencies that are multiples of the fundamental note. We have already learned that a string can vibrate at many harmonics. The same is true for all instruments. A single C from a grand piano might include 20 or more different harmonics.

Recipes for sound
A good analogy is that each instrument has its own "recipe" for the frequency content of its sound. The guitar sound shown in Figure 15.17 has a mix of many harmonics. For this guitar, the fundamental and the 2nd harmonic are about the same size. The 4th, and 6th harmonics are much smaller. The piano recipe has a different mix.

Rise and fall times
The rate at which loudness builds and falls off also influences how we hear a sound. The *rise time* is the time it takes to reach maximum loudness. The *fall time* is the time over which the sound dies away (Figure 15.18). Rise time and fall time are related to resonance and damping in an instrument, and are different for each instrument. Rise and fall times are also different for each harmonic, even from the same instrument. Higher harmonics have faster rise times and shorter fall times.

Synthesized instruments
Today, electronic keyboards and computer programs are available that *synthesize* many different instrument sounds. For example, a keyboard might have buttons that allow you to choose drums, flute, piano, trumpet, or other sounds. The word *synthesize* means "to put together," and that is exactly how electronic instruments work. The sound of each instrument is synthesized by programming a recipe of harmonics and specifying the rise and fall time for each frequency. A good synthesizer might use 64 different frequencies *for each note* to simulate an instrument sound, each with separate rise and fall times.

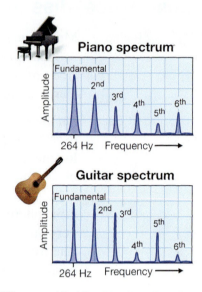

Figure 15.17: *The sound of the note C played on a piano and on a guitar. Notice that the fundamental frequencies are the same but the harmonics have different amplitudes and widths.*

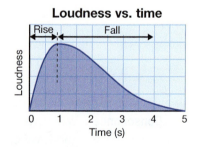

Figure 15.18: *The rise-and-fall times determine the rate at which loudness builds up and falls off for a given frequency of sound.*

Chapter 15 Connection

Sound from a Guitar

The guitar has become a central instrument in popular music. Guitars come in many types, but share the common feature of making sound from vibrating strings. Before 1900, guitars came in two basic varieties. Classical guitars use soft strings, made of nylon today. Folk guitars use steel strings, which are harder on the fingers but much louder. The invention of the electric guitar around 1930 and its cousin, the electric bass, made the voice of the guitar loud enough to be a melody or lead instrument.

Design of the guitar A standard guitar has six strings that are stretched along the neck. The strings have different weights and therefore different natural frequencies. The heaviest string has a natural frequency of 82 Hz and the lightest a frequency of 330 Hz. Each string is stretched by a tension force of about 125 newtons (28 pounds). The combined force from six strings on a folk guitar is more than the weight of a person (750 N or 170 lbs). The guitar is tuned by changing the tension in each string. Tightening a string raises its natural frequency and loosening lowers it.

Each string can make many notes A typical guitar string is 63 centimeters long. To make different notes, the vibrating length of each string can be shortened by holding it down between one of many metal bars across the neck called *frets* (Figure 15.19). The frequency goes up as the vibrating length of the string gets shorter. A guitar with 20 frets and 6 strings can play 126 different notes, some of which are duplicates.

Amplification A vibrating string by itself does not make a loud sound. To make a practical instrument, the vibration from the string needs to amplified. An acoustic guitar amplifies the vibration by coupling the string to the top of the guitar. The guitar top is a large surface that can push much more air around than a thin string. A similar amplification effect can be heard by holding a tuning fork against a hard surface that can vibrate, such as a window pane (Figure 15.20).

Resonance The sound of an acoustic guitar is shaped by sound waves bouncing around inside the guitar, as well as the vibration of the top. Because the shape of the guitar is irregular, there are many resonances. In general, large-bodied guitars have stronger long-wavelength, low-frequency sounds, and are louder. Small-bodied acoustic guitars often lack low frequencies in their range of sounds. As with all instruments, guitar sounds have many harmonics. The highest-quality guitars are prized for both the richness of their sound with its many harmonics and their even balance across high and low frequencies.

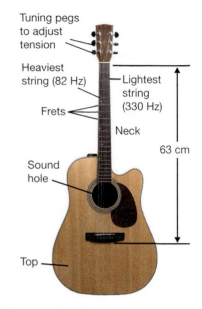

Figure 15.19: *A six-string acoustic guitar.*

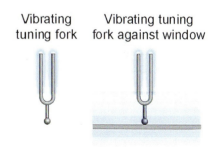

Figure 15.20: *The sound of a tuning fork becomes much louder when the vibrations are amplified by a surface, such as a glass window.*

Chapter 15 Connection

Electric guitars and basses

How electric guitars work
The electric guitar uses electronics to amplify sound from the vibrating strings. Electric guitar *pickups* are made of a coil of wire wound around a set of strong magnets (Figure 15.21). The steel strings are slightly magnetized by being near the magnets in the pickups. As the magnetized strings vibrate up and down, an oscillating electric current is created in the coil by induction (Chapter 23). The electric current is amplified and sent to a speaker to make sound.

Modifying sound electronically
Sound from an electric guitar is not an exact reproduction of the vibrations of the strings. The vibrating signals from the pickups change as they pass through the circuits of the amplifier. Electronic effects chosen by the musician add harmonics, echoes, or emphasize higher or lower frequencies before the electrical signal is turned into sound by the speakers. A common effect is called distortion, which adds the "growl" or "fuzz" to a guitar sound. Figure 15.22 shows wave forms from a clean sound and a distorted sound.

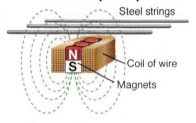

Figure 15.21: *A schematic of a guitar pickup, showing the magnets, coil, and strings, is seen here.*

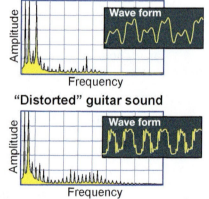

Figure 15.22: *Wave forms from clean and distorted guitar sounds are illustrated here. Notice that both sounds have the same fundamental frequency but the distorted sound has more high-frequency harmonic content.*

The bass guitar
The thump-thump rhythm you hear in the background of many songs comes from an electric bass, which is a guitar designed to play very low frequencies. The strings on a bass guitar are much heavier and longer than on a standard guitar. The extra mass and length both contribute to lower frequencies. The four strings on a traditional bass guitar have natural frequencies from 41 to 98 hertz. Like the guitar, bass players can make different notes on each string by holding the strings between frets to change the vibrating length.

Chapter 15 Assessment

Vocabulary

Select the correct term to complete the sentences.

pressure	frequency	pitch
superposition principle	decibels (dB)	speaker
acoustics	microphone	fundamental
wavelength	stereo	Doppler effect
supersonic	frequency spectrum	shock wave
resonance	node	antinode
dissonance	harmonics	reverberation
notes	sonogram	Fourier theorem
rhythm	musical scale	cochlea
consonance	longitudinal	beats
octave		

1. The loudness of sound is measured in units called _____.
2. An example of a device designed specially to reproduce sounds accurately is a(n) _____.
3. Complex sounds created by the combination of several frequencies can be explained by the _____.
4. _____ is the science and technology of sound.
5. A device commonly used to transform a sound wave into an electrical signal with the same oscillation pattern is a(n) _____.
6. The frequency of vibration of a sound wave determines the _____ of the sound.
7. Sound is transmitted through air as oscillations in the _____ of the air.
8. A combination of two separate recordings—one from the right, one from the left—is called a(n) _____ recording.
9. The _____ of sound is the rate at which the pressure oscillates.
10. The distance between consecutive high pressure areas on a sound wave represents a(n) _____.
11. An object moving faster than sound is referred to as _____.
12. Musical instruments use the principle of _____ to reinforce selected frequencies by enclosing those sound waves between boundaries.
13. Because the compressions that create sound travel in the direction of the wave, a sound wave is considered to be a(n) _____ wave.
14. A shift in the frequency of sound created by motion is called the _____.
15. Wave fronts that pile up in front of a supersonic airplane create a(n) _____.
16. A graph that shows the amplitude of each component frequency in a complex wave is known as a(n) _____.
17. Multiple echoes of a sound is referred to as a(n) _____.
18. The point on a standing wave of highest pressure represents a(n) _____.
19. A statement of _____ says any wave form can be represented as a sum of single frequency waves.
20. The point on a standing wave of lowest pressure represents a(n) _____.
21. Frequencies that are multiples of a note are called _____.
22. A combination of sounds that sounds good is called _____.
23. A combination of sounds that is irritating or unsettling is referred to as _____.
24. A(n) _____ is a range between a given frequency and twice the frequency on a musical scale.
25. Rapid alterations in the phase relationship between two waves of nearly identical frequencies create _____.
26. A set of frequencies mathematically related to one another by specific ratios are considered to be _____.
27. A regular time pattern in a sound is called _____.
28. The _____ is a tiny fluid-filled organ in the inner ear that helps to provide balance.
29. A graph that shows how loud sound is at different frequencies is known as a(n) _____.

30. Frequencies of sound on a musical scale named using the letters A through G are called _____.

31. The lowest frequency at which a body will vibrate is called its _____.

Concept review

1. Sound travels faster in a liquid or a solid than in a gas. Why?

2. If a small piece of space debris was to strike a space station, workers on the inside might hear the sound made by the collision, but workers outside the station would not. Explain.

3. How are pitch and frequency of a sound related?

4. Your parents complain that the music you are listening to is not music. Based on the information in this chapter, defend your choice of music as music.

5. As you tune your clarinet, you hear an oscillating sound your instructor calls beats. What causes them and how can you use them to tune your instrument?

6. Why does an A played on a piano not sound exactly like an A produced by a guitar?

7. A steel string does not produce a loud sound by itself. Explain how an acoustic guitar produces a loud sound when a string is plucked.

8. How does an electric guitar pickup produce an amplified sound?

9. The decibel scale is used to measure the "loudness" of a sound. How is the loudness of a sound related to its amplitude?

10. Most people know that sound is a wave. List at least three statements of evidence to support the idea that sound is a wave.

11. How is the wavelength of sound produced by a musical wind instrument related to the size of the instrument?

12. A patron at a concert claims that she cannot hear clearly certain notes being played unless she moves her head slightly to one side or the other. Explain how this could happen.

13. Parents are concerned about the hearing of their children who wear stereo headsets adjusted to high volume settings. Using Table 15.1 in the chapter, explain why their concerns are justified.

14. When an astronomer observes the Sun, she notices that the light from one edge is slightly shifted toward the red end of the visible spectrum while the opposite edge is slightly shifted toward violet. What causes this shift?

15. While scuba diving, Roy and his partner become separated. Roy taps on his scuba tank to help his partner locate him. Why is this technique not as useful underwater as it might be at the surface?

16. As the temperature increases, the fundamental frequency produced by a flute changes slightly. Does it increase or decrease? Give an explanation for your answer.

17. What causes consonance and dissonance?

Problems

1. Students in a physics class prepare "musical instruments" to play in lab. They cut tubes of half-inch PVC pipe to various lengths. When students blow across the open end and cover the bottom with a thumb, fundamentals of the scale are produced. Approximately what length of pipe will produce the fundamental frequency of 264 hertz, or "middle C"? (*Note*: Assume the speed of sound is 340 m/s.)

2. On a day when the speed of sound is 344 m/s, Tom hollers and hears an echo 2.4 seconds later. How far away is the object that caused the echo?

3. A pipe closed on one end is 2.46 m long. On a day when the speed of sound is 345m/s, what is the fundamental frequency of this pipe?

4. Human hearing depends upon both the frequency and the intensity of the sound. Answer the questions below based upon the graph.

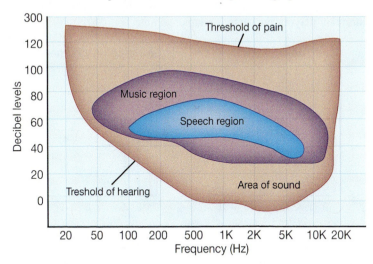

   a. At which frequency is the threshold of pain lowest?
   b. What is the approximate range of frequencies for speech?
   c. What is the decibel range for most speech?
   d. At what frequency are our ears most receptive?

5. The diagram represents wave fronts produced by a source moving at constant velocity through the air. Answer the following questions based upon the diagram and your knowledge of the Doppler effect.

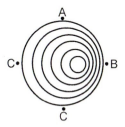

   a. In which direction is the source moving?
   b. If this was a sound source, at which point would the pitch be highest?
   c. If the source was to accelerate, what would happen to the wavelength immediately behind the source?

6. A sonar signal requires 1.31 s to travel to the bottom of the ocean and back to the ship's depth finder in water 1,000 m deep. What is the speed of the sonar signal?

7. The speed of sound in air at 20°C is about 343 m/s. A 1.2-m tube closed on one end has a fundamental frequency of 80 Hz. Is the temperature higher or lower than 20°C? Justify your answer with an explanation and/or calculation.

8. Examine this diagram which illustrates two waves of different frequencies, 50 Hz and 55 Hz, and the sum of these two waves.

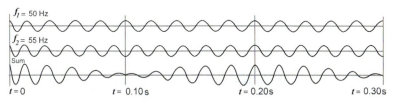

   a. What is represented at $t = 0$ s, $t = 0.1$ s, $t = 0.2$ s, and $t = 0.3$ s on the diagram representing the sum of the two waves?
   b. What is the period between beats? (*Hint*: The period here is the time between two louder or softer sounds.)
   c. What is the beat frequency?

9. Which pair of component waves would most likely produce the complex wave pattern shown below? Assume the times axis is horizontal and the same for all graphs.

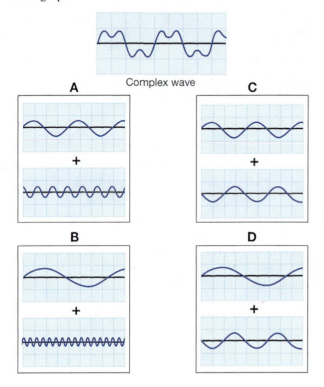

11. The graph on the left shows the wave amplitude of the fundamental harmonic produced by a tuning fork with a frequency of 264 Hz. The graph on the right shows the spectrum of wave amplitudes produced by a musical instrument playing the same fundamental note.

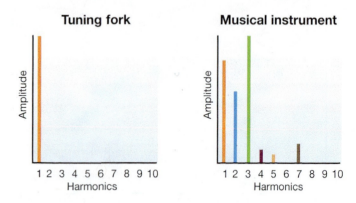

a. What frequencies does the instrument produce?
b. List the frequencies in order of amplitude from greatest to least.

Applying your knowledge

1. Wrap the ends of a 24- to 30-inch piece of thread around your index fingers. Hang a metal hanger from the thread. Have a partner tap the hanger. Put your fingers in your ears and again have a partner tap the hanger with a pencil. Describe what you hear and explain any difference.

2. Some older people are less tolerant of loud music than some younger people. Is there any reason that this should be so? Research online to answer this question and write a short report on your findings.

3. Why does the pitch of a person's voice change when they speak after inhaling helium? Assuming you could survive this, would this happen if the air in the room consisted entirely of helium?

10. Examine this diagrams of four objects whose velocities differ with respect to the sound they emit.

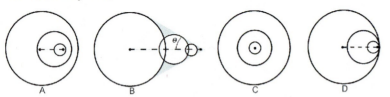

a. Indicate the speed of each object relative to the speed of sound.
b. Indicate for which objects the Doppler effect occurs and for which a shock wave occurs.

UNIT 6 **LIGHT AND OPTICS**

CHAPTER 16

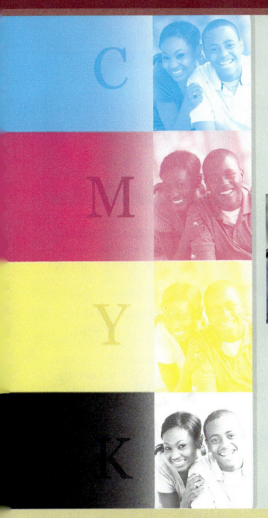

Light and Color

Objectives:

By the end of this chapter you should be able to:

- ✔ Describe at least five properties of light.
- ✔ Describe the meaning of the term *intensity*.
- ✔ Use the speed of light to calculate the time or distance traveled by light.
- ✔ Explain how we perceive color in terms of the three primary colors.
- ✔ Explain the difference between the additive and subtractive color processes.
- ✔ Arrange the colors of light in order of increasing energy, starting with red.
- ✔ Describe light in terms of photons, energy, and color.

Key Questions:

- What are some useful properties of light?
- How do we perceive color?
- How does light fit into the atomic theory of matter?

Vocabulary

additive process	cyan	inverse square law	pigment	spherical pattern
black	fluorescence	light ray	pixel	subtractive process
blue	green	magenta	red	ultraviolet
CMYK color process	incandescence	photoluminescence	RGB color model	white light
color	infrared	photon	rod cell	yellow
cone cell	intensity	photoreceptor	speed of light	

331

CHAPTER 16 — LIGHT AND COLOR

16.1 Properties and Sources of Light

Every time you look at something, light is involved. Whether you are looking at a light bulb, a car, or this book, it is light that brings the information to your eyes. People have wondered for thousands of years how the eye sees. In the past, philosophers suggested light was a fluid that flowed from a candle flame or a "searchlight beam" that comes from the eye. We now know that we see objects by their reflected light. This chapter discusses the properties of light, how we see, and how light is created.

Lighting choices

The "old fashioned" incandescent light bulb patented by Thomas Edison in 1879 was the primary lighting source in the US for well over a century. Despite this, there is a good chance that you've never even seen one! These bulbs have been gradually phased out in our country since Congress passed the "light bulb law" in 2007, which set energy efficiency standards for light bulbs. There are two new types of energy-efficient bulbs: compact fluorescent lamps (CFLs) and light emitting diodes (LEDs). Both types of lighting are more energy efficient than Edison's, and both fit the sockets in all current lighting fixtures. But how do these two types of lights compare, and which should you choose?

Research lighting choices and prepare a chart that compares CFLs to LEDs in terms of power output, price, cost to operate, environmental impact, versatility, and safety.

What is light?

Light is a form of energy Today we know that light—like sound and heat—is a form of energy. Human technologies such as light bulbs, televisions, and cameras use light to do all sorts of useful things. Among the many properties of light are the following.

- Light travels extremely fast and over very short and very long distances.
- Light carries energy and information.
- Light travels in straight lines.
- Light bounces and bends when it comes near or in contact with objects.
- Light has color.
- Light has different intensities, and can be bright or dim.

How do we see? Consider what is physically occurring as you read this page. Light from an electric light bulb, or sunlight, reflects off the page and into your eyes. The reflected light carries information about the page that allows your brain to construct an image of the page. The words and pictures are patterns of light that travel from the page to your eyes. You see because light in the room *reflects* from the page into your eyes. If you turn out the lights in the room, you cannot see the page because the page does not give off its own light. We see most of the world by reflected light.

Electric light

The electric light To see, there must be a source of light that can reflect from objects. For most of human history, people relied on sunlight, moonlight, and fire to provide all the light they needed. Thomas Edison's electric light bulb (1879) is one of the most important inventions in the progress of human development. Chances are that the light you are using to read these words is coming from an electric light.

Incandescent light bulbs The process of making light with heat is called **incandescence**. Incandescent bulbs generate light when electricity passes through a thin piece of metal wire called a *filament*. The filament heats up and gives off light (Figure 16.1). The atoms inside the filament convert electrical energy to heat and then to light. Unfortunately, incandescent bulbs are not very efficient. Only a small fraction of the energy of electricity is converted into light. Most of the energy becomes heat. In fact, the primary function of some incandescent bulbs is to make heat. For example, incandescent heat lamps are used to help chickens' eggs hatch or keep a restaurant's French fries warm.

Fluorescent light bulbs The other common kind of electric light is the *fluorescent* bulb (Figure 16.2). We see many fluorescent bulbs in schools, businesses, and even in homes because they are much more efficient than incandescent bulbs. Compared with a standard incandescent bulb, you get four times as much light from a fluorescent bulb for the same amount of electrical energy. This is possible because fluorescent bulbs convert electrical energy directly to light without generating a lot of heat.

How fluorescent bulbs make light To make light, fluorescent bulbs use high-voltage electricity to ionize electrons from atoms of gas that fill the bulb. Electrical energy is released as light in a process known as **fluorescence**. We cannot directly see the light given off by the atoms in the fluorescent bulbs because it is high-energy ultraviolet, the same kind of light that gives you a sunburn. Another step is needed to get useful light. The ultraviolet light is absorbed by other atoms in a white coating on the inside surface of the bulb. This coating re-emits the energy as white light that we see. Even with the two-step process, fluorescent bulbs are still several times more efficient than incandescent bulbs.

Figure 16.1: *An incandescent light bulb generates light by heating a piece of metal.*

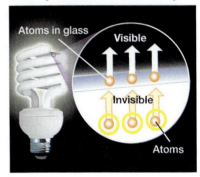

Figure 16.2: *Fluorescent lights generate light by exciting atoms with electricity.*

Light carries energy and power

The intensity of light Light is a form of energy that travels. This is obvious if you stand outside on a hot day. The energy from the Sun is what makes you warm. This energy travels to Earth as light. The **intensity** of light is the amount of energy per second falling on a unit area of surface. For example, on a bright sunny day, 500 joules of light energy may fall on a single square meter of the ground (Figure 16.3). The intensity of this light is 500 watts per square meter (500 W/m²). This intensity equals 500 joules of energy deposited per second on each square meter.

Spherical pattern Like the Sun, most light sources distribute their light equally in all directions, making a **spherical pattern**. If you stuck a bunch of toothpicks into an orange, the toothpicks would point outward in all directions just like light radiates outwards in all directions from a lightbulb or the Sun (Figure 16.4). You can see a bare light bulb from anywhere in a room because the bulb emits light in all directions.

Light intensity follows an inverse square law Because light spreads out in a sphere, the intensity decreases the farther you get from the source. For example, suppose a light bulb gives off 10 watts of light. The light waves form a sphere. The area of a sphere is $4\pi r^2$, where r is the radius. If the radius is 1 meter, the area is about 12.6 m². The light intensity is 10 watts divided by 12.6 m², or 0.8 W/m². If the radius increases to 2 meters, twice as far, the area increases to 50.3 m². The light intensity drops to one fourth because the light that passed through 1 m² at 1 meter radius is spread out over 4 m². This reduction of intensity is known as an **inverse square law** because the intensity is inversely proportional to the square of the distance.

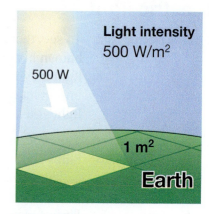

Figure 16.3: *Intensity is the power of light per unit area. In the summer, the intensity of sunlight reaches 500 watts per square meter.*

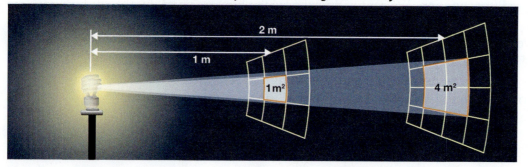

The inverse square law for light intensity

Figure 16.4: *Light emitted from the Sun or from a light bulb travels in a spherical distribution of straight lines.*

Light carries information

Information in images We rely on light to bring us *information* about the world around us. To see the connection between light and information, consider that a video camera captures a picture as a series of colored dots. A standard definition video picture is 720 dots wide by 480 dots high (Figure 16.5). The amount of information in a video picture is roughly the same as the number of dots—346,000 for a video image, which is 720 times 480. The human eye is a very sophisticated light detector, much better than a video camera. The human eye can see more than 100 million dots. Every time you look around, your brain is receiving more than a million dots of information each second and processing the information to give you a picture of what is around you.

Figure 16.5: *A standard definition TV picture is 346,000 dots: 720 dots wide by 480 dots high. A high definition screen of the same size has 922,000 dots: 1,280 dots wide by 720 dots high.*

Information in sound Today, most voice information is also carried by light. In Chapter 15, you read about how music is converted to a string of numbers that can be recorded on a computer. The same string of numbers can be sent on a laser beam by turning the beam on and off very fast. More than 1 billion numbers per second can be sent on a single laser beam. The fiber-optic networks you read about are pipelines for information carried by light (Figure 16.6).

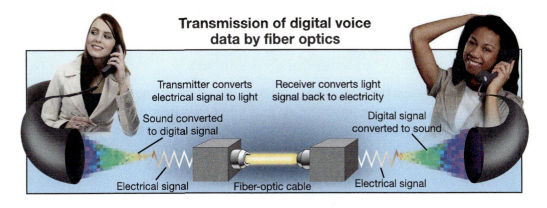

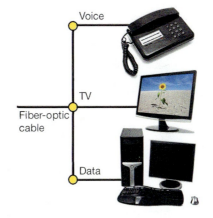

Information in computer data Most of the data transmitted across the Internet is also carried by light. A network of fiber-optic cables crisscrosses the country carrying data from one computer to another. In some cities, a fiber-optic cable comes directly into homes and apartments carrying telephone, television, and Internet signals.

Figure 16.6: *A single fiber-optic cable can carry more than enough information to support television, telephone, and computer data.*

The speed of light

Comparing the speeds of sound and light
Consider what happens when you shine a flashlight on a distant object. You do not see the light leave your flashlight, travel to the object, bounce off, and come back to your eyes. But that is exactly what happens. You do not see the movement of the light because it happens far too fast. If the mirror was 170 m away, almost two football fields, light travels to the mirror and back in about one-millionth of a second (0.000001 s). By comparison, sound travels much slower than light. If you shout, you will hear an echo about one second later from the sound bouncing off a wall 170 m away and back to your ears. Light travels through air almost a million times faster than sound.

The ultimate speed limit
Not only is the speed of light fast, it is as fast as anything gets.

Based on Einstein's theory of special relativity, we believe nothing in the universe can travel faster than the speed of light.

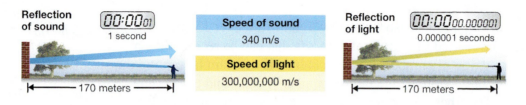

The speed of light, $c = 3 \times 10^8$ m/s
The speed at which light travels through air is approximately 300 million meters per second. This is such a high speed that it is difficult to comprehend. Light is so fast it can travel around the entire Earth 7-1/2 times in 1 second. The **speed of light** is so important in physics that it is given its own symbol, a lowercase c. When you see this symbol in a formula, remember that $c = 3 \times 10^8$ m/s.

The sound of thunder lags the flash of lightning
The speed of light is so fast that when lightning strikes a few miles away, we hear the thunder several seconds after we see the lightning. At the point of the lightning strike, the thunder and lightning are simultaneous. But just a mile away from the lightning strike, the sound of the thunder is already about 5 seconds behind the flash of the lightning.

Accurate measurement of c
Using very fast electronics we are able to measure the speed of light accurately in lab experiments. One technique used to measure the speed of light is to record the time a pulse of light leaves a laser and the time the pulse returns to its starting position after making a round trip. The best accepted experimental measurement for the speed of light in air is 299,792,500 m/s. For most purposes, you do not need to be this accurate and can use a value for c of 3×10^8 m/s.

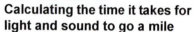

Calculating the time it takes for light and sound to go a mile
Calculate the time it takes light and sound to travel the distance of 1 mile, which is 1,609 meters.
1. You are asked for time.
2. You are given distance and you want to find the speed of sound and light.
3. $t = d \div v$
4. For sound:
   $t = (1{,}609 \text{ m}) \div (340 \text{ m/s})$
   $= 4.73$ s
   For light:
   $t = (1{,}609 \text{ m}) \div (3 \times 10^8 \text{ m/s})$
   $= 0.0000054$ s
   $(5.4 \times 10^{-6}$ s$)$

Reflection and refraction

Light rays, reflection, and refraction
When light moves through a material it travels in straight lines. But when light rays travel from one material to another, the rays may reflect (Figure 16.7) or refract (Figure 16.8). When describing reflection or refraction, it is useful to represent light by one or more imaginary lines traveling in the direction of the light. These imaginary lines are called **light rays**. The light that appears to bounce off the surface of an object is shown by a *reflected ray*. The light that bends as it crosses a surface into a material is shown as a *refracted ray*, as shown. Reflection and refraction cause many interesting changes in the images we see.

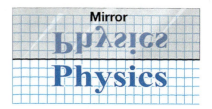

Figure 16.7: *Light rays are reflected in a mirror, resulting in an inverted image.*

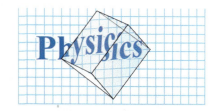

Figure 16.8: *Light rays are refracted (or bent) by a prism, causing the image to be distorted.*

Mirrors
When you look in a mirror, objects that are in front of the mirror appear as if they are behind the mirror. This is because light rays are reflected by the mirror. Your brain perceives the light as if it always traveled in a straight line. You see a reflected image "behind" a mirror because the reflected *light rays* reaching your eye are the same as if the object really was behind the mirror.

A glass of water
When light rays travel from air to water, they refract. This is why the images you see looking through a glass full of water are different from the images you see when the glass is moved away from your eyes. Try looking at some objects through a glass of water; move the glass closer and farther away from the objects. Does it remind you of anything you have ever used, such as a magnifying glass?

Twinkling of stars
Another example of light refraction is the twinkling of a star in the night sky (Figure 16.9). As starlight travels from space into Earth's atmosphere, the rays are refracted. Since the atmosphere is constantly changing, the amount of refraction also changes. The image of a star appears to "twinkle" because the light coming to your eye is constantly being deflected by small amounts. Each small deflection causes the star to appear in a slightly different place.

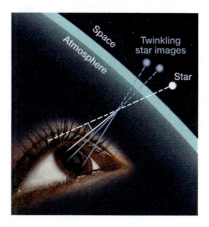

Figure 16.9: *The twinkle of a star is due to the changes in the bending of light as the light moves through the atmosphere.*

16.2 Color and Vision

The rainbow of colors visible to a human eye ranges from deep red, through the yellows, greens, and blues, to deep purples like violet. The order of colors is always the same when white light is separated with a prism: red, orange, yellow, green, blue, and violet. This order of the colors can be remembered by their initials ROY-G-BV, which is pronounced "roy-gee-biv." In this section, we will discuss how light makes different colors and how we perceive color. The discussion may surprise you.

White light, color, and energy

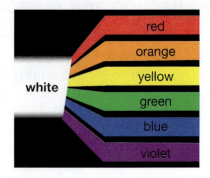

Figure 16.10: *White light is the combination of all the colors.*

White light When all the colors of the rainbow are combined equally, you do not see any particular color. Instead, you see light without any color. An equal combination of all the colors of light makes **white light** (Figure 16.10). White light is a good description of the ordinary light that is all around us most of the time. The light from the Sun and the light from most electric lights are white light.

Where does color come from? What physical property of light accounts for color? In the early 1900s, Albert Einstein proposed a new way of thinking about light. He theorized that **color** was fundamentally a sensation of the *energy content* of light. All of the colors in the rainbow are light of different energies. Red light has the lowest energy we can see, and violet light the highest. As we move through the rainbow from red, to yellow, to blue, to violet, the energy of the light increases.

Energy and light from a flame What do we mean when we talk about the energy of light? Think about the flames coming from a gas stove, a blow torch, or a gas grill. These are very hot flames and they are *blue*. The atoms of gas in the flame have high energy so they give off blue light. The flame from a match or from a burning log in the fireplace is reddish orange. These flames are not nearly as hot as those from gas, so the atoms have a lower energy. The lower-energy light from a match flame appears red or yellow.

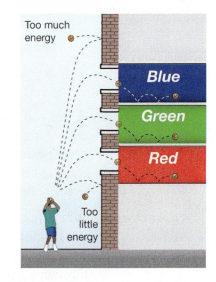

Figure 16.11: *We can think of different colors of light like balls with different kinetic energies. Blue light has a higher energy than green light, like the balls that make it into the top two windows. Red light has the lowest energy, like the ball that can only make it to the lowest window.*

A way to think about light energy You can think of the energy of light being like the kinetic energy of a ball thrown upward: too little energy and we cannot see the light at all. The lowest energy of light we can see appears red (Figure 16.11). Light of intermediate energy appears green, and higher-energy light appears blue. The energy of ultraviolet light is even higher than blue, but a human eye cannot see ultraviolet or higher-energy light. The eyes of other living creatures, and many instruments, can detect light of both higher and lower energy than the human eye can see.

How the human eye sees color

How we see color Scientists have discovered cells in the retina at the back of the eye that contain special cells called **photoreceptors**. Light from an image passes through the lens of the eye and falls on the photoreceptor cells (Figure 16.12). When light hits a photoreceptor cell, the cell releases a chemical that creates an electrical signal that travels down the optic nerve to the brain. In the brain, the signal is translated into a perception of color. The act of *seeing* occurs in the *brain*, not in the eye.

Rods and cones Human eyes have two types of photoreceptors, called *rod cells* and *cone cells*. Cones, or **cone cells**, respond to color and there are three types. One type gives off its strongest signal for red light. Another responds mostly to green light and the third type responds mostly to blue light. Each type of cone responds most strongly to a certain range of light energy. Because there are only three kinds of cones, it is accurate to say we see only three colors of light. We see white light when all three types of cone cells are equally stimulated (Figure 16.13).

Rod cells see black and white The **rod cells**, or rods, respond only to differences in intensity, and not to color. Rods essentially "see" in black, white, and shades of gray. However, rod cells are much more *sensitive* than cone cells and work at very low light levels. At night, colors seem washed out because there is not enough light for your cones to work. When the overall light level is very dim, you are actually seeing black-and-white images from your rod cells.

Black and white vision is sharper than color vision You can think of each photoreceptor cell as being one dot in a total image. The brain assembles all the dots to create the image. An average human eye contains about 130 million rod cells and only 7 million cone cells. Because there are more rod cells, finer details appear sharpest when there is high contrast between light and dark areas. In bright light, each cone cell applies color to many rod cells. The cone cells are concentrated near the center of the retina, making color vision best at the center of the eye's field of view.

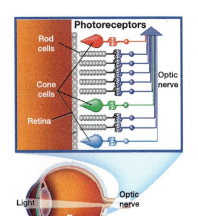

Figure 16.12: *The photoreceptors that send color signals to the brain are in the back of the eye.*

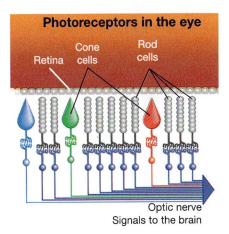

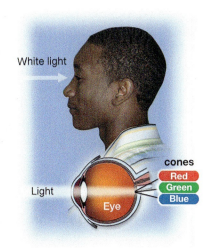

Figure 16.13: *Because white light consists of all the colors, all three types of cones—red, blue, and green—are stimulated.*

How we see colors other than red, green, and blue

How we perceive color
The three color receptors in the eye allow us to see millions of different colors. When the brain receives a signal only from the red cone cells, it thinks *red*. If there is a signal from the green cone cells but neither the blue or red, the brain thinks *green* (Figure 16.14). This accounts for the perception of pure colors.

The additive color process
Now consider what happens if the brain gets a strong signal from both the red and the green cone cells at the same time. The sensation created is different from either red or green. It is what we have learned to call *yellow*. Whether the light is actually yellow, or a combination of red and green, the cones respond the same way and we perceive yellow. If all three cones are sending a signal to the brain at once, we think *white*. The brain makes color by an **additive process** because new colors are formed by the addition of more than one signal from the cone cells.

The additive primary colors
The additive primary colors are **red**, **green**, and **blue**. In reality, our brains are receiving all three color signals just about all of the time. We don't see everything white because the strength may be different for each of the three separate color signals. It is too simple to say that red and green make yellow. If there is a lot of red and only a little green the color is *orange* instead of yellow (Figure 16.15). All the different shades of color we can see are made by changing the relative proportions of red, green, and blue.

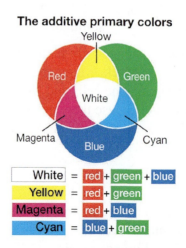

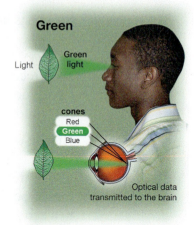

Figure 16.14: *If the brain gets a signal from only the green cone, we see green.*

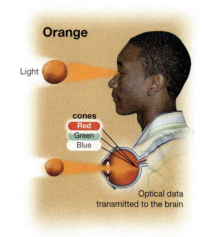

Figure 16.15: *If there is a strong red signal and a weak green signal, we see orange.*

Not all animals see the same colors
To the best of our knowledge, only humans and other closely-related primates, such as chimpanzees and gorillas, have all three types of cone cells—red, green, and blue. Many animals have only rod cells that sense black, white, and shades of gray. Other animals have color sensors that are neither red, blue, or green but respond to different colors. Some birds have four different types of photoreceptors that allows them to sense four primary colors. Marine mammals such as dolphins and whales have no color vision and see only shades of gray. Some insects can sense infrared and ultraviolet light.

How we see the colors of things

We see mostly reflected light
When you see an object, the light that reaches your eyes can come from two different processes.

1. The light can be emitted directly from the object, like a light bulb, glow stick, computer display, or television.

2. The light can come from somewhere else, like the Sun, and be reflected from the object you are seeing.

Most of what you see is from reflected light because most objects do not produce their own visible light. When you look around, you are seeing light originally from another source, usually sunlight or electric light, that is reflected into your eyes.

Figure 16.16: *You see a blue cloth because pigments in the fabric absorb all colors except blue. Blue light gets reflected to your eyes.*

What gives objects their color?
When you look at a blue piece of cloth, you may think the quality of *blue* is in the cloth, *but that is not true*. The reason the cloth looks blue is because the chemicals in the cloth have absorbed all the colors of light *other than blue*. Since blue light is reflected, it is the color that reaches our eyes (Figure 16.16). The blue was never in the cloth. The blue was hidden or mixed in with the other colors in the white light that illuminates the cloth. The cloth unmasked the blue by taking away all the other colors and sending only the blue to your eyes.

The subtractive color process
Colored fabrics and paints create the sensation of color through a **subtractive process**. Chemicals, known as **pigments**, in dyes and paints absorb some colors and allow other colors to be reflected. Pigments work by taking away, or subtracting, colors from white light.

The subtractive primary colors
There are three primary *subtractive* pigments required to make the range of colors visible to the human eye. One pigment absorbs only blue, and reflects red and green. This pigment is called **yellow**. The second pigment absorbs only green, and reflects red and blue. This is a pink-purple called **magenta**. The third pigment is **cyan**, which absorbs only red and reflects green and blue. Cyan is a slightly greenish shade of light blue. Magenta, yellow, and cyan are the three subtractive primary colors (Figure 16.17). By using different proportions of the three pigments, a paint can appear almost any color by varying the amount of reflected red, green, and blue light. All three pigments together absorb all colors and theoretically make **black**, which is the absence of light.

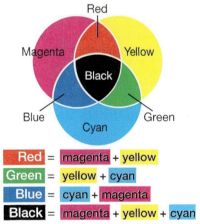

Figure 16.17: *The three subtractive primary colors.*

Why are plants green?

Light is necessary for photosynthesis

In a very unique way, plants absorb physical energy in the form of light and convert it to chemical energy in the form of sugar. The process is called *photosynthesis*. The graph in Figure 16.18 shows the colors of visible light that plants absorb. The *x*-axis on the graph represents color and the *y*-axis represents the amount of light of each color absorbed by plants.

Chlorophyll

Plants are green because of *how* they absorb visible light. The green pigment, chlorophyll a, is the most important light-absorbing pigment. You can see on the graph that chlorophyll a absorbs light at each end of the spectrum. In other words, chlorophyll a *reflects* most of the green light and absorbs blue and red light. Plants are green because they reflect green light and absorb red and blue light. In fact, plants will not grow well if they are placed under pure green light.

Why leaves change color

Notice that chlorophyll b and the orange pigments called carotenoids absorb light where chlorophyll a does not. These extra pigments help plants catch more light. Leaves change color when chlorophyll a breaks down and these pigments become visible. They are the cause of the beautiful bright reds and oranges that you see when the leaves of some plant species change color in the fall.

Plants reflect some light to keep cool

Why do plant pigments not absorb all colors of light? The reason is the same reason you wear light-colored clothes when it is hot outside. Like you, plants must reflect some light to avoid absorbing too much energy and overheating.

Visible light has just the right energy for life

The colors in visible light represent only part of the energy range in sunlight. For example, **ultraviolet** light has energy higher than blue or violet light. **Infrared** light has energy lower than red light. Natural sunlight contains both infrared and ultraviolet light. Human eyes cannot see either of these "colors." Almost all living things on Earth see visible light because this range of colors has just the right amount of energy for living things to use. Ultraviolet light has too much energy. The energy in some ultraviolet light is sufficient to break apart important biological molecules. This is the cause of sunburn. Much of the infrared in sunlight is absorbed by water vapor and carbon dioxide in the atmosphere. Infrared light also has low energy compared to visible light. Unlike ultraviolet light, the energy of infrared light is too low to be significant in the chemical changes that are the basis of life.

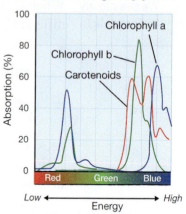

Figure 16.18: *The lines in the graph show which colors of light are absorbed by plant pigments. Chlorophyll a and b absorb blue light and red light, leaving green light to be reflected. This is why plants look green most of the time.*

Color in televisions, video cameras and computer monitors

TV makes its own light
Televisions give off light. They do not rely on reflected light to make color. You can prove this by watching a TV in a dark room. You can see light from the TV even if there are no other sources of light in the room. Computer monitors and movie projectors are similar. All these devices make their own light.

The RGB color process
Televisions and computer monitors make red, green, and blue (RGB) light directly (Figure 16.19). These devices do not use the subtractive method. Use a magnifying glass to look closely at a white area on a TV screen or computer monitor and you will see that what appears white is actually tiny red, green, and blue dots. The dots are called **pixels** and each pixel gives off its own light. The pixels are separated by very thin black lines. The black lines help give intensity to the colors and help make the darker colors darker. From far away, you cannot see the individual pixels. Instead, you see a nice, smooth color picture.

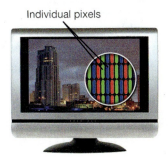

Figure 16.19: *A television makes colors using tiny glowing dots of red, green, and blue.*

Mixing primary colors
By turning on the different color pixels at different intensities, TV sets can mix the three primary colors to get millions of different colors. For example, a light brown tone can be 88 percent red, 85 percent green, and 70 percent blue. A television makes a light brown by lighting the red, green, and blue pixels to these percentages.

Video cameras
A video camera does the opposite of a television. A video camera has red, green, and blue sensors, similar to the cones in your eye (Figure 16.20). The camera records an image by measuring the percentages of red, green, and blue in the light coming through the camera lens. The device inside that actually captures the light is called a CCD (charge-coupled device) sensor. The CCDs in most video cameras are quite small, typically 1 centimeter square or less.

Two complementary color processes
All devices that make their own light use the red, green, and blue, or **RGB color model**. They create different colors by varying the strengths of each of the three primary colors. Anything that relies on *reflected* light to make color uses the **CMYK color process**. The letters *C*, *M*, *Y*, and *K* stand for cyan, magenta, yellow, and black. The CMYK process is used for printing inks, fabric dyes, and paints. In both cases, the colors that you *see* are the colors of light that are *not absorbed* by pigments. The RGB process generates the light you see and the CMYK process subtracts all light *except* the light you see.

Figure 16.20: *Digital cameras have an array of tiny light sensors that are sensitive to red, green, and blue light.*

16.3 Photons and Atoms

You could name many sources of light; sunlight, lightning, firelight, and fluorescent light are a few examples. Common to all of these different sources of light is that they are made of atoms. Virtually all of the light you see is created by atoms. The dyes that absorb light in pigments are also made of atoms in molecules. Both atoms and molecules can create and absorb light. This section is about how atoms create light and how light is described on the atomic scale.

The photon theory of light

Photons   Just as matter is made of tiny particles called atoms, light energy comes in tiny bundles called **photons**. In some ways, photons act like jellybeans of different colors. Each photon has its own color, no matter how you mix them up.

Color and photons   The lowest-energy photons we can see are the ones that appear a dull red in color (Figure 16.21). The highest-energy photons we can see are a blue color tending to deep violet. Low-energy atoms make low-energy photons and high-energy atoms make high-energy photons. As atoms gain energy, the color of the light they produce changes from red, to yellow, to blue, and violet.

White light   White light is a mixture of photons with a wide range of colors or energies. This is a clue that white light is created by atoms that also have a wide range of energies. You can see how this works by watching the filament of a light bulb on a dimmer switch. When the switch is set very low, the filament is relatively cool and the light is very red. As you turn up the dimmer switch, the filament gets hotter and hotter. The hotter it gets, the more green and blue light is created. At full power, the bulb appears a brilliant white, which means it is creating a uniform spread of photons of all colors.

Temperature and energy   For a given temperature, the atoms in a material have a range of energy that go from zero up to a maximum that depends on the temperature. Figure 16.22 shows how the energy is distributed among the atoms in a material for several temperatures. At a temperature of 600°C, you can see that some atoms have just enough energy to make red light (electric stove). At 2,600°C, there are atoms that can make all colors of light, which is why the hot filament of a light bulb appears white. At room temperature (20°C), a rock gives off no visible light at all.

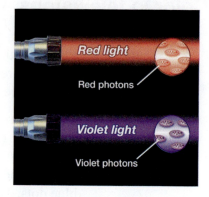

Figure 16.21: *Blue photons have a higher energy than red photons.*

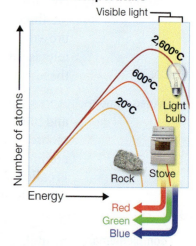

Figure 16.22: *The number of atoms with a given amount of energy depends on the temperature.*

Photons and the intensity of light

Energy, color, and intensity

Intensity measures power per unit area. There are two ways to make light of high intensity. One way is to have high-energy photons. A second way is to have a lot of photons, even if they are low energy (Figure 16.23). In practical terms, to make a red light with an intensity of 100 W/m^2 takes a lot more photons than it does to make the same intensity with blue light. While the color of light depends on the energy of one photon, the intensity of light is a combination of both the number of photons and the energy per photon.

Glow-in-the-dark plastic

Consider an amazing but very common material: glow-in-the-dark plastic. If this material is exposed to light, it stores some energy and later is able to release the energy by giving off its own light. The plastic can only make light if it is "charged up" by absorbing energy from other sources of light. You can test this by holding your hand on some "uncharged" plastic and then exposing it to bright light. If you then bring the plastic into a dark area and remove your hand, you can see that the areas that were covered by your hand are dark while the rest of the plastic glows (Figure 16.24).

Photo-luminescence

The glow-in-the-dark effect comes from atoms of a chemical compound called a *phosphor* in the plastic. When photons of light collide with atoms of the phosphor, the energy from the photons is stored in the atoms. Over a period ranging from minutes to hours, the phosphors release the stored energy as pale green light. The process of releasing stored energy as light is called **photoluminescence**.

An experiment with photon energy

You can use glow-in-the-dark plastic to show that a *single atom interacts with a single photon at a time*. A simple experiment shows that red light does not activate glow-in-the-dark plastic. For an atom of the phosphor to give off a green photon, it must absorb a photon of equal or greater energy. If one red photon is absorbed, the atom does not get enough energy to emit a green photon. If a single atom could absorb two or more red photons, it could get enough energy to emit a green photon. However, the material stays dark and does not glow, no matter how intense a red light is used. *Even many red photons do not cause emission of a green photon.* But even a dim blue light will cause the plastic to glow because one blue photon has *more* energy than a green photon. Each atom in the phosphor absorbs or emits only one photon at a time.

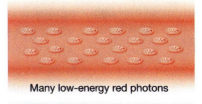
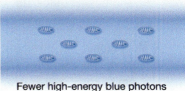

Figure 16.23: *The number and energy of photons determine the intensity of the light.*

Figure 16.24: *The light from the flashlight cannot energize the atoms of the phosphor that your hand blocks. These atoms will not glow because they did not receive any energy from photons from the flashlight.*

Light and atoms

The process of how light is reflected

The light through which you see the image of this page is reflected from another light source. How exactly does this process of reflection occur? The atoms on the surface of the paper in the white areas of the page absorb the light from the room and immediately emit almost all of the light back in all directions. You see a white page because the atoms on the surface of the paper absorb and re-emit light of all colors equally (Figure 16.25). The black letters are visible because light falling on black ink is almost completely absorbed and no light is re-emitted. Where there is no light coming from the page, you see black.

Most atoms absorb and emit light

It turns out that almost all atoms absorb and emit light. What is different about the atoms of the phosphor in glow-in-the-dark plastic is the *time* between the absorption and the emission of the light. For most atoms, the absorption and emission of light happens in less than one-millionth of a second. This is so fast that only the most sensitive instruments can detect the time delay. Phosphorus atoms have a special ability to delay the emission of a photon for a relatively long time.

When thinking about photons and atoms is necessary

Think about photons and atoms the next time you turn on the light in a dark room. The light that you see coming off the wall started at the light bulb. It was absorbed by the atoms in the paint, and its energy was re-emitted as photons of the color that you see. This all happens thousands of times faster than a blink of your eye. It all happens so fast that we can accurately describe the light as reflecting off the surfaces, as we will in the next chapter. When we are interested in the overall path that light travels, we need not worry about the microscopic details. Only when studying the behavior of very small systems, such as individual atoms, do we need to explicitly consider the detailed interaction of atoms and photons.

Light from chemical reactions

Another source of energy that allows atoms to emit light is from chemical changes in materials. Many chemical changes release energy. Some of the energy is absorbed within atoms and subsequently emitted as light. For example, the warm flickering glow from a candle comes from trillions of atoms in the wick giving up photons as they combine with oxygen atoms in the air (Figure 16.26). The light that comes from a glow stick is also made through chemical changes.

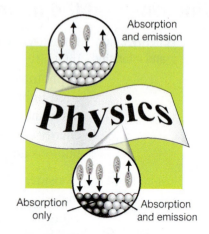

Figure 16.25: *White paper absorbs and immediately re-emits photons of all colors in all directions. Black ink absorbs photons of all colors and emits none, which is why it looks black.*

Figure 16.26: *The light from a candle flame comes from the energy released by chemical changes.*

Chapter 16 Connection

Color Printing

Printing is an ancient art, almost as old as written history. Fragments of whole printed pages have been found in China that are more than 1,400 years old. Early printers carved blocks of wood and pressed them in ink to print a page. To make a printing block, the printer carefully carved out the wood around any areas that were to be inked, such as letters. A new block had to be created for each different page. A new block also had to be created for each different color.

Separating a full color image into the four process colors

High-quality color printing — Today, printing is so common we hardly think about it. Pick up almost any catalog and you will see beautifully-reproduced color. Modern printing presses use the four-color, CMYK process to produce rich, vivid colors from only four inks.

Color separations — To print a color photograph, the image is first converted into four separate images in cyan, magenta, yellow, and black. These are called *color separations*. Each color separation represents what will be printed with its matching C, M, Y, or K color. For example, the cyan separation is printed with cyan ink, the magenta separation with magenta ink, and so on. Figure 16.27 shows a full-color image with the four color separations.

Cyan

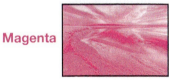

Magenta

Halftone screens make a picture into dots — Each color separation is turned into a *halftone screen*. A halftone turns the image into a series of small dots. The size of the dots is proportional to the darkness of the color. The example on the right has 50 dots per inch. A magazine printer might use 150–180 dots per inch. The dots are necessary to allow space for all four colors. The final image will contain cyan, magenta, yellow, and black dots placed very close together. The dots are so close that they appear to be a continuous color.

Halftone screen

Yellow

Black

The printing press — On the printing press, each of the four separations is printed on the same paper by a separate printing station. Careful alignment of the printing stations ensures that the dots line up and do not overlap each other. If the dots do not line up, the printed image is blurry. You may have seen a color newspaper photograph that is blurry because the dots were not lined up. Newspapers print with coarse halftone screens; you can see the dots easily with a magnifying glass.

Figure 16.27: *To be printed by a full-color press, an image is separated into separate cyan, magenta, yellow, and black images.*

16.3 PHOTONS AND ATOMS   347

Chapter 16 Connection

The four-color printing process

Making all colors from just four

Suppose you want to print the color green using the CMYK process. To see green, you need to remove the red and blue from white light. If you look at Table 16.1, to absorb red and blue, you need a mixture of cyan and yellow ink. If you were mixing paint, this is exactly what you would do. You get green paint by mixing equal quantities of cyan paint and yellow paint.

Table 16.1: The three subtractive primary colors

The color	absorbs	and reflects
Cyan	Red	Blue and green
Magenta	Green	Red and blue
Yellow	Blue	Red and green

Figure 16.28: *Printed color images actually consist of tiny dots spaced very closely together.*

Printing presses do not mix ink

On a printing press, mixing inks to make each color is impossible because there are so many colors to print. Instead, dots of each of the four colors are printed very close together. If you take a powerful magnifying glass you can see the individual cyan, magenta, yellow, and black dots in a printed picture (Figure 16.28).

Black

You see black when no light is reflected. In theory, if you add magenta, cyan, and yellow, you have a mixture that absorbs all light so it looks black. Some electronic printers actually make black by printing cyan, magenta, and yellow together. Because the dyes are not perfect, you rarely get pure black this way. Better printers use a black ink to make black separately.

Color laser and ink-jet printers

Color computer printers also work by putting tiny dots on paper. The dots use the same four colors—cyan, magenta, yellow, and black. With an ink-jet printer, tiny drops of ink are squirted by a print head that moves back and forth across the paper. The image is built up from thousands of tiny dots printed in rows across the paper (Figure 16.29). With a laser printer, the dots are formed by tiny particles of colored plastic which are melted onto the paper. High-resolution color printers print as high as 1,800 dots per inch.

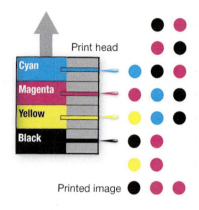

Figure 16.29: *To print a colored image, an ink-jet printer scans its print head back and forth, laying down lines of colored dots by squirting tiny droplets of each of the four colors of ink.*

UNIT 6 LIGHT AND OPTICS

Chapter 16 Assessment

Vocabulary

Select the correct term to complete the sentences.

reflection	refraction	black
fluorescence	intensity	colors
blue	light rays	CMYK color
infrared	photons	RGB color
photoluminescence	additive color	white
red	green	spherical pattern
cyan	magenta	yellow
pigments	speed of light (*c*)	incandescence
pixels	rod cells	cone cells
subtractive color	photoreceptors	ultraviolet
inverse square law		

1. Light is produced by using high-voltage electricity to energize the atoms of a gas in a process called _____.
2. The amount of light falling on a surface per second is a measure of the light _____.
3. The number 3×10^8 m/s, represented by the letter *c*, is the _____.
4. The twinkling of stars is caused by the _____ of the light coming to our eyes through Earth's atmosphere.
5. The process of making light with heat is called _____.
6. A light that appears to bounce off of a surface may have undergone _____.
7. Light from a light bulb radiates in all directions in a(n) _____.
8. The direction in which light travels can be represented by imaginary lines called _____.
9. The intensity of light at a certain distance from a small light source is described by the _____.
10. Cells in the back of the eye that release a chemical in response to incident light are called _____.
11. The additive primary colors of light are _____.
12. The molecules in dyes and paints that absorb some colors and reflect others are known as _____.
13. Photoreceptors that respond to color are called _____.
14. We see the combination of all colors of light as _____.
15. Light of various energies creates all of the visible _____ of light that we see.
16. Cells in the back of the eye that respond only to light-intensity differences are the _____.
17. The subtractive primary pigment that absorbs blue and reflects red and green is _____.
18. The brain combines two or more signals from the cone cells to form color by means of the _____.
19. When no color is reflected from a surface, the surface appears _____.
20. The pigment that reflects blue and red while absorbing only green is _____.
21. The name for the wavelengths of light with slightly less energy than red is _____.
22. Anything that relies on reflected light to make color uses the _____ process.
23. The tiny dots from which images are formed on a TV screen are called _____.
24. Light that has just slightly more energy than the highest-energy visible light is _____ light.
25. The subtractive primary colors are yellow, magenta, and _____.
26. The process by which pigments create color is known as the _____ process.
27. The color picture on a TV is created by combining colors using the _____.
28. The process of releasing stored energy as light is called _____.
29. Tiny bundles of light energy are called _____.

CHAPTER 16 LIGHT AND COLOR

Concept review

1. How is an incandescent bulb different from a fluorescent bulb?
2. In what units is light intensity measured?
3. You wish to read a book in your bedroom. There is enough light for you to read if you use either a 100 W ceiling light or a 15 W desk lamp. Both use the same type of bulb. Explain why you can read with the less powerful desk lamp.
4. List three ways light can be used for communication.
5. Why do we see lightning before we hear thunder?
6. Compare reflection and refraction.
7. What is white light?
8. How could a blacksmith tell the temperature of a fire long before thermometers were invented?
9. Why is it difficult to distinguish among different colors in a dimly-lit room?
10. How can we see many different colors if our eyes contain only three types of cones?
11. Give the color of light that results from combining the colors listed.
    a. red + blue =
    b. blue + green =
    c. red + green =
    d. red + blue + green =
12. Why is mixing pigments called color subtraction?
13. What color results when cyan, magenta, and yellow pigments are mixed?
14. Answer true or false for each.
    a. A green object reflects green light.
    b. A blue object absorbs red light.
    c. A yellow object reflects green light.
    d. A white object absorbs red light.
15. Why is it a good idea to wear a white shirt rather than a black shirt on a hot, sunny day?
16. Why will a plant grow more quickly if it is grown in white light rather than green light?
17. Explain how a television makes pictures of many colors using only three types of pixels.
18. Use the photon theory of light to describe two ways to create high intensity with both high energy and low energy photons.
19. Explain how glow-in-the-dark materials work.

Applying your knowledge

1. The speed of light has been very precisely determined but it was not always known whether light traveled instantaneously from one point to another or had a finite speed. Research and summarize the contributions of some of the scientists who investigated the speed of light.
2. Why is the sky blue? Why are sunsets red? Research these questions and write a brief explanation of each.
3. Stare at the picture of the flag for about one minute. Look away at a blank piece of white paper. Describe what you see. Compare the colors in the printed picture to what you see. How are they related? Find the name of this phenomenon and an explanation for its occurrence by doing an on-line investigation.

4. Thomas Edison's electric light bulb changed our way of life but his light bulb was not the first. Investigate and write a short report on the development of the incandescent light bulb.

UNIT 6 LIGHT AND OPTICS

CHAPTER 17

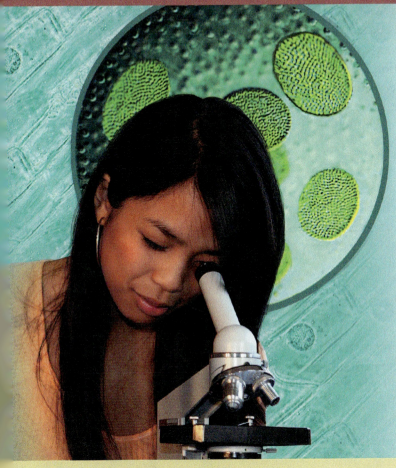

Optics

Objectives:

By the end of this chapter you should be able to:

- ✓ Describe the functions of convex and concave lenses, a prism, and a flat mirror.
- ✓ Describe how light rays form an image.
- ✓ Calculate the angles of reflection and refraction for a single light ray.
- ✓ Draw the ray diagram for a lens and a mirror showing the object and image.
- ✓ Explain how a fiber-optic circuit acts like a pipe for light.
- ✓ Describe the difference between a real image and a virtual image and give an example of each.

Key Questions:

- How do we study and describe the reflection and refraction of light?
- How do lenses and mirrors form images?
- What will a ray diagram tell you about an image?

Vocabulary

angle of refraction	diverging lens	image	magnifying glass	prism	spherical aberration
chromatic aberration	eyepiece	image relay	mirror	ray diagram	telescope
	fiber optics	incident ray	normal line	real image	thin lens formula
converging lens	focal length	index of refraction	object	reflected ray	total internal reflection
critical angle	focal plane	law of reflection	objective	refracting telescope	
diffuse reflection	focal point	lens	optical axis	Snell's law	virtual image
dispersion	geometric optics	magnification	optics	specular reflection	

351

CHAPTER 17 OPTICS

17.1 Reflection and Refraction

Look at your thumb through a magnifying glass. It looks huge. Of course, your hand is the same size it always was, even though what you see is a giant thumb (Figure 17.1). The explanation for why and how magnification occurs is part of the science of *optics*. Optical devices you may have used include mirrors, telescopes, eyeglasses, contact lenses, and magnifying glasses. This section introduces the fundamental ideas behind optics and the formation of images. It is truly amazing how much can be accomplished by clever application of the basic laws for reflection and refraction.

What is optics?

Figure 17.1: *When you look through a magnifying glass, your thumb appears huge.*

Definition of optics The overall study of how light behaves is called **optics**. This chapter will focus on **geometric optics**, which uses the geometry of lines and curves to represent how light passes through lenses or prisms or reflects from surfaces. Geometric optics explains how images are formed by devices like lenses, mirrors, cameras, telescopes, and microscopes. This understanding extends to the study of the eye itself because the human eye forms an image with a lens in exactly the same manner as a magnifying glass.

Light rays As described in Chapter 16, light is always in motion. Unless there is a change in materials, such as air to glass, light moves in straight lines. The path that light follows is often represented in diagrams by *light rays*. Think of a light ray as an imaginary arrow following a thin beam of light. Objects usually reflect or emit light rays in all directions simultaneously. If three people in a room can see the same vase, it is because light rays travel straight from the vase to their eyes. The diagram in Figure 17.2 shows some representative light rays.

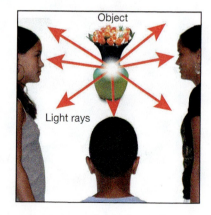

Figure 17.2: *Light rays travel in straight lines to our eyes from the objects we see.*

Basic geometry Light rays might bend when they cross a boundary between materials. For example, a magnifying glass bends light rays when the rays cross from air into glass and back from glass into air again. This is useful because you "see" something where the light rays *appear* to come from. When light rays bend, images may appear in a different place, or appear smaller or larger than the actual objects. A magnifying glass works because the curved surface of a glass lens bends light. Geometric optics applies the laws of reflection and refraction to trace how light rays travel through materials such as air and glass. The results of geometric optics are used to design cameras, telescopes, and other optical instruments.

352 UNIT 6 LIGHT AND OPTICS

Basic optical devices

Optical devices are common Almost everyone has experience with optical devices and their effects. For example, seeing through eye glasses, checking your appearance in a mirror, or admiring the sparkle from a diamond ring all involve optics. Through experiences like these, most of us have seen optical effects created by three basic optical devices: the *lens*, the *mirror*, and the *prism*.

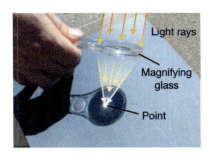

Figure 17.3: *Light rays passing through a magnifying glass converge at a point.*

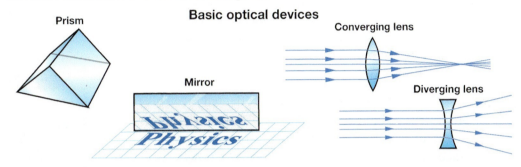

Lenses A **lens** is an optical device with curved surfaces shaped to bend light in a specific way. A **converging lens** bends light so that the light rays come together at a point. This explains how a magnifying glass makes a spot of concentrated light (Figure 17.3). Your eye contains a single converging lens. A **diverging lens** bends light so the light spreads apart instead of coming together at a single point. An object viewed through a diverging lens appears smaller than when it is viewed without the lens.

Mirrors A **mirror** is a familiar optical device—you probably used one this morning. Mirrors reflect light and allow us to see ourselves (Figure 17.4). Flat mirrors show a true-size image. Curved mirrors distort images by causing the light rays to come together or spread apart. At the circus, curved mirrors are used to make you look thin, wide, or upside down. The curved side-view mirrors on a car make any cars behind you look farther away than they really are.

Prisms A **prism** is another optical device that can cause light to change directions. A prism is a solid piece of glass with flat polished surfaces. A common triangular prism is shown in the diagram. Prisms are used to bend and/or reflect light. Many optical devices such as telescopes, cameras, and supermarket laser scanners use prisms of different shapes to bend and reflect light in precise ways.

Figure 17.4: *The image you see in a flat mirror is life-size and oriented as if you were standing in front of yourself.*

Reflection

The image in a mirror
When you look at yourself in a mirror, you see your own image as if your exact twin was standing in front of you. The image appears as far "into" the mirror as you are in front of the mirror (Figure 17.5). If you step back, so does your image.

Incident and reflected rays
Images appear in mirrors because of how light is reflected. In the last chapter, we learned that light is reflected from all surfaces, not just mirrors. But not all surfaces form images. The reason is that there are two types of reflection. Consider a ray of light coming from a light bulb and falling on the mirror. The **incident ray** follows the light from the bulb onto the mirror. The **reflected ray** follows the light after being reflected from the mirror.

Specular reflection
A ray of light that strikes a shiny surface generates a single reflected ray. This type of reflection is called **specular reflection**. In specular reflection, each incident ray results in a single reflected ray (Figure 17.6). Polished surfaces create specular reflection, such as the surface of a mirror. *Images are produced by specular reflection.* If you look closely at a mirror illuminated by a light bulb, somewhere the reflected light forms an image of the light bulb itself. In fact, a surface which has perfect specular reflection is invisible. If you look at that surface, you see reflections of other things, *but you do not see the surface itself.*

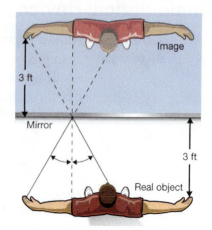

Figure 17.5: *The image you see in a flat mirror is the same distance "behind" the mirror as you are in front of it.*

Diffuse reflection
A surface that is rough, or not shiny, creates **diffuse reflection**. In diffuse reflection, a single ray of light scatters into many reflected rays in many directions (Figure 17.6). Diffuse reflection is caused by the microscopic roughness of a surface. Even a surface that feels smooth to the touch may be rough on the microscopic level. For example, the surface of a wooden board creates a diffuse reflection. In a lighted room, you see the board by reflected light, but you cannot see an image of a light bulb in the board. When you look at a diffuse reflecting surface *you see the actual surface itself.*

Diffuse and specular reflection together
Many surfaces are in between rough and smooth. These kinds of surfaces create both kinds of reflection. For example, a polished wood tabletop can reflect some light in specular reflection, and the rest of the light in diffuse reflection. The specular reflection creates a faint reflected image on the table surface. You also see the table surface itself by diffuse reflection.

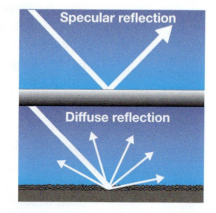

Figure 17.6: *Specular and diffuse reflections.*

The law of reflection

The law of reflection The **law of reflection** relates the direction of a reflected ray to the direction of the incident ray in specular reflection. The law of reflection is easy to remember: Light rays reflect from a mirror at the same angle at which they arrive at the mirror.

The normal line The tricky part of using the law of reflection is defining the angles of the incident and reflected rays. To be consistent, angles are always measured relative to the **normal line**. The normal line is an imaginary line *perpendicular* to the surface of a mirror drawn from the point where the incident ray touches the mirror. You draw the normal line by starting where the incident ray strikes the mirror and drawing a line perpendicular to the mirror's surface (Figure 17.7).

Drawing a ray diagram A **ray diagram** is a type of diagram that shows how light rays interact with mirrors, lenses, and other optical devices. A ray diagram is an accurately-drawn sketch showing the path of one or more light rays. Incident and reflected rays are drawn as arrows on a ray diagram. A mirror is drawn as a solid line. The normal line is drawn as a dotted line perpendicular to a surface.

Measuring the angle of incidence and reflection The angles of incidence and reflection are measured relative to the normal line, as shown in the ray diagram in Figure 17.7. The law of reflection applies to light reflected by curved mirrors as well as flat mirrors (Figure 17.8). With curved mirrors, however, the direction of the normal line changes depending on where the light ray contacts the mirror surface.

Reflection

The angle of incidence is always equal to the angle of reflection.

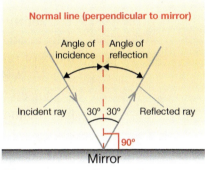

Figure 17.7: *The angle of incidence, θ_i, is equal to the angle of reflection, θ_r.*

Concave mirror

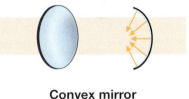

Convex mirror

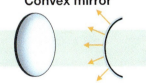

Figure 17.8: *Two common curved mirrors are convex and concave.*

Law of reflection

A light ray strikes a plane mirror at an angle of incidence of 30°. Sketch the incident and reflected rays and determine the angle of reflection.
1. You are asked for a ray diagram and the angle of reflection.
2. You are given the angle of incidence.
3. The law of reflection states the angle of reflection equals the angle of incidence.
4. The angle of reflection is 30°.

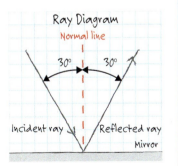

17.1 REFLECTION AND REFRACTION   355

Refraction

Refraction is the bending of light rays

Light rays might bend as they cross a boundary from one material to another, like from air to water. The bending of light rays is known as refraction. You can see a good example of refraction by looking at a straw in a glass of water (Figure 17.9). The straw appears to break where it crosses the surface of the water. Of course, the straw has not actually broken. That visual effect is caused by refracted light rays. The light rays from the straw are refracted or bent when they cross from water back into air before reaching your eyes. Refraction causes the *image* of the straw in the water to appear out of place, not the actual straw itself.

The index of refraction

The ability of a material to bend rays of light is described by its **index of refraction**. The index of refraction is represented by a lowercase letter n. The index of refraction for air is approximately 1.00. Water has an index of refraction of 1.33. A diamond has an index of refraction of 2.42. The high index of refraction is what creates the sparkling quality that makes diamonds so attractive. Table 17.1 lists the index of refraction for some common materials.

The difference in index of refraction

When a ray of light crosses from one material to another, the amount it bends depends on the difference in index of refraction between the two materials. The larger the difference, the greater the "bend." For example, a light ray crossing from air into a diamond at and angle of incidence of 45° is bent by 28°. The difference in index of refraction between air and diamond is 1.42. The same light ray going from air into water bends by just 13 degrees. The light ray bends less because the difference in index of refraction between air and water is only 0.33.

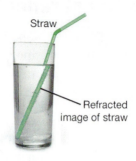

Figure 17.9: *A straw seems to separate at the point it enters the water in the glass. This illusion is created because light is refracted as it travels from water to air.*

Table 17.1: The index of refraction for some common materials

Material	Index of refraction
Vacuum	1.0
Air	1.0001
Water	1.33
Ice	1.31
Glass	1.5
Diamond	2.42

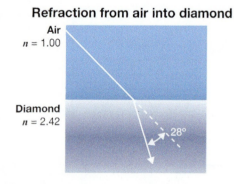
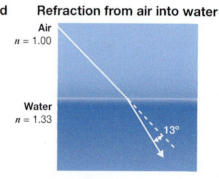

Snell's law of refraction

Snell's law
Snell's law is the relationship that relates the angles of incidence and refraction to the index of refraction of the material on the incident and refracted side of a boundary. The angle of incidence (θ_i) is the angle between the incident ray and the normal, the same as for a mirror. The **angle of refraction** (θ_r) is the angle between the refracted ray and the normal.

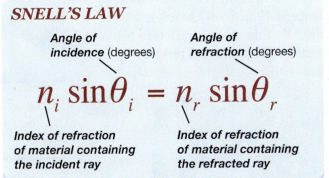
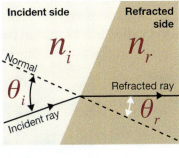

SNELL'S LAW

$$n_i \sin\theta_i = n_r \sin\theta_r$$

- Angle of incidence (degrees)
- Angle of refraction (degrees)
- Index of refraction of material containing the incident ray
- Index of refraction of material containing the refracted ray

The index of refraction of both materials
Two different materials are necessary to create refraction. The *incident* material is the material containing the incident light ray. The index of refraction in the incident material is given the symbol "n-sub-i" (n_i). The *refractive material* contains the refracted ray and its refractive index is "n-sub-r" (n_r). For light entering a prism, the incident material would be air, for which $n_i = 1.0$. The refractive material would be glass, for which $n_r = 1.5$

The direction a light ray bends
The direction in which a light ray bends depends on whether it is moving from a high index of refraction to a lower index or vice versa. A light ray going from a low index of refraction into a higher index bends *toward the normal*. A light ray going from a high index of refraction to a low index bends *away from the normal*.

Snell's law for multiple surfaces
When light passes completely through a refractive material, Snell's law must be used twice to calculate the path of a ray: the first time going into the material and the second time coming back out again. For light *leaving* a glass prism, the incident material would now be glass, for which $n_i = 1.5$. The refractive material would be air, $n_r = 1.0$. Which material is incident or refractive depends on the direction of the light ray and changes from one surface to the next.

Calculating the angle of refraction

A ray of light traveling through air is incident on a smooth surface of water at an angle of 30° to the normal. Calculate the angle of refraction for the ray as it enters the water.

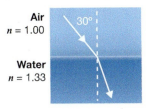

1. You are asked for the angle of refraction.
2. You are told the ray goes from air into water at an angle of incidence of 30°.
3. Snell's law:
   $n_i \sin(\theta_i) = n_r \sin(\theta_r)$
   $n_i = 1.00$ (air), $n_r = 1.33$ (water)
4. Apply Snell's law to find θ_r.
   $1.00 \sin(30°) = 1.33 \sin\theta_r$
   $\sin(\theta_r) = 0.5 \div 1.33 = 0.376$

Use the inverse sine function to find the angle that has a sine of 0.376.
$\theta_r = \sin^{-1}(0.376)$
$= 22°$

Reflection and the critical angle

Total internal reflection

When light goes from a high-index material, such as glass, back into air, it bends away from the normal. Mathematically, it means the angle of refraction is always greater than the angle of incidence. For example, in water ($n = 1.33$) when the angle of incidence is 48 degrees, the angle of refraction is 81 degrees, as shown in the diagram below. When the angle of refraction reaches 90 degrees, the refracted ray is traveling straight along the surface, and *does not actually leave the water*. With water, this occurs when the angle of incidence is 49 degrees. At angles greater than 49 degrees, *there is no refracted ray*. All of the light is reflected back into the water. This is called **total internal reflection**. Total internal reflection occurs when the angle of refraction becomes greater than 90 degrees.

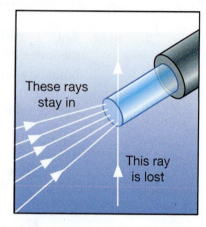

Figure 17.10: *Light entering a glass rod at greater than the critical angle is trapped inside the glass.*

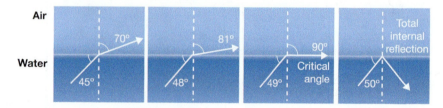

The critical angle

The angle of incidence at which light begins reflecting back *into* a refractive material is called the **critical angle**. The critical angle depends on the index of refraction. The critical angle is 42 degrees for glass and 49 degrees for water.

Fiber optics are pipes for light

Suppose you have a rod of solid glass and you send light into the end at an angle of incidence greater than the critical angle (Figure 17.10). The light reflects off the wall of the rod and bounces back into the glass. It then reflects off the opposite wall as well. In fact, the light always approaches the wall at greater than the critical angle so it always bounces back into the glass. You have constructed a *light pipe*, so to speak: Light goes in one end and comes out the other. If the glass rod is made very thin, it becomes flexible, but still traps light by total internal reflection. **Fiber optics** are made of thin glass fibers and use total internal reflection to carry light, even around bends and corners (Figure 17.11). A bundle of fiber optics can make an *image pipe*, in which each fiber transmits one dot of an image from one end of the bundle to the other end. Image pipes are used for inspections in surgery, for example. Fiber optics are also used for communications and data signals.

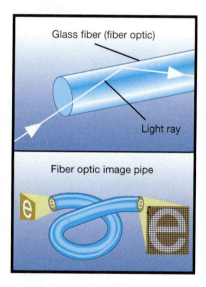

Figure 17.11: *Fiber optics are thin glass fibers that trap light by total internal reflection.*

OPTICS   CHAPTER 17

Dispersion and prisms

Observing the spectrum with a prism
The index of refraction for most materials varies by a small amount depending on the color of the incident light. For example, glass has an index of refraction slightly greater for blue light than for red light. When white light passes through a glass prism, blue is refracted more strongly than red (Figure 17.12). Colors between blue and red are refracted according to their position in the spectrum. Remember, ROYGBV is the order of colors in the spectrum of visible light—red, orange, yellow, green, blue, violet. Yellow is in the middle of the spectrum, so yellow rays are bent about halfway between red rays and blue rays. If white light from a brightly illuminated slit passes through a prism and falls on a screen, the spectrum of colors in white light forms a rainbow on the screen.

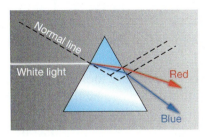

Figure 17.12: *The blue component of white light passing through a prism is refracted more than the red component.*

Dispersion
The variation in refractive index with color is called **dispersion**. The indices of refraction listed in Table 17.1 were measured by observing the bending of a specific color of yellow light. This color was chosen because it is the center of the visible spectrum.

Rainbow
A rainbow is an example of dispersion in nature. Tiny rain droplets act as prisms by separating the colors in the white light rays from the Sun. The different colors of light that reach your eye come from rain droplets at different levels in the sky (Figure 17.13). Next time you see a rainbow, notice that the colors follow ROYGBV from top to bottom.

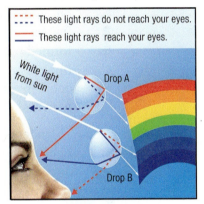

Figure 17.13: *Dispersion in drops of water separates the colors of sunlight in a rainbow.*

Chromatic aberration

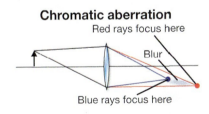

A camera lens forms an image by refracting light. Since camera lenses are made of glass, they suffer from dispersion. When one color is in focus, the other colors are slightly out of focus. If green light in the center of the spectrum is in focus, blue and red light will be slightly blurry. This error is known as **chromatic aberration**. Expensive camera lenses minimize chromatic aberration by using multiple single lenses made from different kinds of glass (Figure 17.14). If at least one of the lenses is a diverging lens, it is possible to make red and blue light focus at the same point. Of course, the colors between red and blue are still slightly blurred but the amount of blur is greatly reduced.

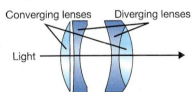

Figure 17.14: *This camera-lens design uses four single lenses and corrects for chromatic aberration.*

17.1 REFLECTION AND REFRACTION

17.2 Mirrors, Lenses, and Images

Think of an *image* as a picture that organizes light in the same way that light is organized somewhere else. For example, the image of a tree duplicates the pattern of light reaching your eye that you would see if you were looking at the actual tree. In fact, the world we see is a world of images created on the retina of the eye by the lens in the front of the eye (Figure 17.15). This section is about images and how images are created by mirrors and lenses. You will learn how to use lenses and mirrors to make images larger or smaller, upright or upside down, or near or far away.

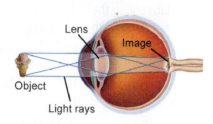

Figure 17.15: *We see because the eye forms images on the retina at the back of the eyeball.*

Images

Objects and images It is helpful to think about optics in terms of *objects* and *images*. **Objects** are real physical things that give off or reflect light rays. **Images** are "pictures" of objects that are formed in space where light rays meet or appear to meet. Images are formed by mirrors, lenses, prisms, and other optical devices (Figure 17.16). Images are organized patterns of light and have no physical substance like matter does. In fact, clever optics can create images that would be impossible to make with real matter.

Rays come together in an image To understand what an image is, consider that each point on a real object gives off light rays in all directions. That is why you can see an object from different directions. Suppose you could collect many rays from the *same* point on an object and bring them back together again. You would have created an *image* of that point. An image is a place where many rays that originated from the same point on an object meet together again. A camera works by collecting the rays from an object so they form an image on the film.

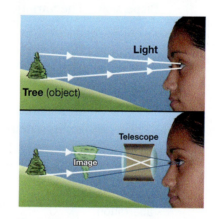

Figure 17.16: *You see the tree because light from the tree reaches your eye. The image of the tree in a telescope is not the real tree, but is a different way of organizing light from the tree. A telescope organizes the light so that the tree appears bigger.*

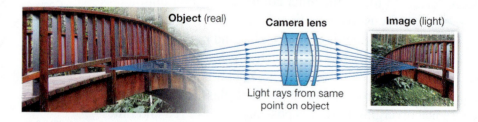

The image in a mirror

Your reflection in a mirror If you stand in front of a flat mirror, your image appears the same distance behind the mirror as you are in front of the mirror. If you move back, the image seems to move back too. If you raise your hand, the image also raises a hand. How does this happen?

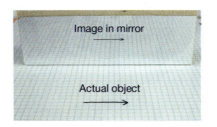

Figure 17.17: *The image in a flat mirror is not left-right reversed.*

The image of an arrow in a mirror The photograph in Figure 17.17 shows a mirror in front of a piece of graph paper that has an arrow drawn on it. The arrow on the graph paper is an *object* because it is a physical source of reflected light. The *image* of the arrow appears in the mirror. Look carefully and you see that the image of the arrow appears the same number of squares *behind* the mirror as the paper arrow is *in front of* the mirror.

A ray diagram of an object in a mirror Figure 17.18 shows a ray diagram of the arrow and mirror. The head of the arrow is a source of light rays. The ray diagram traces three light rays that leave the tip of the arrow and reflect from the mirror. These rays obey the law of reflection. *The reflected rays appear to come from a point behind the mirror.* To see where the rays appear to come from, you extend the actual rays using dashed lines. The image of the tip of the arrow is formed at the point where the dashed lines meet. Remember, an image forms when many rays from an object come together again. The diagram shows that the image appears exactly the same distance behind the mirror as the arrow is in front of the mirror.

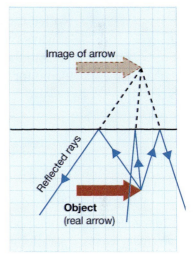

Figure 17.18: *A ray diagram of the arrow in the mirror, showing the location of the virtual image.*

Virtual images The image in a mirror is called a **virtual image** because the light rays do not *actually* come together. They only *appear* to come together. The virtual image in a flat mirror is created by your eyes and brain. Your brain "sees" the arrow where it should be if the light rays reaching the brain had come in a single straight line.

Virtual and real images Because the light rays do not actually meet, a virtual image cannot be projected onto a screen or on film. Virtual images are illusions created by your eye and brain. To show a picture on a screen or record an image on film, you need a *real image*. Real images form where light rays actually come together again. The images formed by a camera lens or a projector lens are real images.

Lenses

A lens and its optical axis

A lens is made of transparent material with an index of refraction different from air. The surfaces of a lens are polished and curved to refract light in a specific way. The exact shape of a lens's surface depends on how strongly and in what way the lens is designed to bend light passing through it. However, nearly all lenses are designed so a light ray passing through the exact center goes straight through and is not deflected. The path along which a light ray is not deflected is called the **optical axis**. Light traveling along the optical axis is not bent by the lens.

Focal point and focal length

A converging lens has convex surfaces that bulge outward (Figure 17.19). Consider a group of light rays that pass through a converging lens parallel to its optical axis. The lens bends these rays so they meet at a point on the other side called the **focal point**. Light can go through a lens in either direction so there are always two focal points, one on either side. The distance from the center of the lens to the focal point is called the **focal length**. The focal length is usually the same for both focal points of a lens.

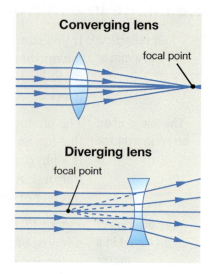

Figure 17.19: *Converging and diverging lenses.*

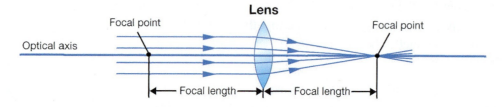

Converging and diverging lenses

The opposite of a converging lens is a *diverging* lens. A diverging lens bends a parallel group of light rays so they spread outward, away from its focal point (Figure 17.19). One or both surfaces of a diverging lens are *concave*.

Lenses follow Snell's law of refraction

Most lenses have surfaces that are parts of a sphere. The normal to a spherical surface is always a radius from the center. When light rays fall on a spherical surface from air, they bend *toward* the normal (Figure 17.20). For a convex lens, the first surface (air to glass) bends light rays toward the focal point. At the second surface (glass to air), the rays bend *away* from the normal. Because the second surface "tilts" the other way, it also bends light rays toward the focal point. Although it is not shown, both surfaces of a diverging lens bend light away from the optical axis. For either type of lens, refraction follows Snell's law.

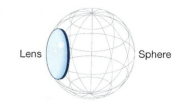

Figure 17.20: *Most lenses have spherical shaped surfaces.*

The image formed by a lens

The focus and focal plane of the lens
The image formed by a lens is created in the same way the image is formed by a mirror. Many rays from the same point on an object are collected and brought to a **focus**. The focus is where the rays from a point on the object come back together to form the equivalent point on the *image*. The surface where the image forms is called the **focal plane** of the lens.

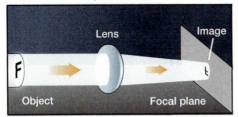
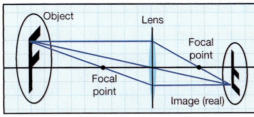

Image in a magnifying glass

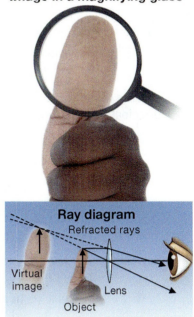

Virtual images in a lens
A lens can form a virtual image just as a mirror does. For example, a convex lens used as a **magnifying glass** creates a virtual image. Rays from the same point on an object are bent by the lens so that they appear to come from a much larger object (Figure 17.21). Virtual images are also created by diverging lenses.

Figure 17.21: *A magnifying glass is a lens that forms a virtual image that is larger-than-life and appears behind the lens.*

Real images in a lens
A converging lens can also form a **real image**. In a real image, light rays from the object *actually* come back together on the focal plane. If you put a piece of paper on the focal plane, you see the image formed where the light rays come together. When the object is very far away from a convex lens, the focal plane is at the focal length. If the object is closer, the image forms farther from the lens.

Magnification
The images from a lens may be smaller than life size, or equal to or larger than life size. The **magnification** of a lens is the ratio of the size of the image divided by the size of the object. For example, a lens with a magnification of 4.5 creates an image that appears 4-1/2 times larger than the real-life object (Figure 17.22).

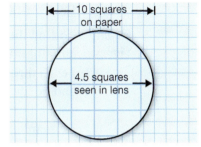

The orientation of images
Images from a lens may be right side up or inverted. The image of a distant object made with a single convex lens is always inverted and "backward" from the shape of the object. The lenses in movie projectors also invert images. The film is put in upside down so that it appears right side up on the screen.

Figure 17.22: *A lens showing an image magnified 4.5 times.*

17.2 MIRRORS, LENSES, AND IMAGES

Drawing ray diagrams of lenses

What a ray diagram tells you
A ray diagram is a good way to determine what type of image is formed by a lens, and whether the image is magnified or inverted. To draw a ray diagram for a lens, you need to know the focal length and the distance of the object from the lens. A vertical arrow is used to represent the object.

The three principal rays
To find the location of the image, there are three rays that are the easiest to draw (Figure 17.23). These three rays follow the rules for how light rays are bent by a lens. The three rules are listed below.

1. A light ray passing through the center of the lens is not deflected at all (A).
2. A light ray parallel to the axis passes through the far focal point (B).
3. A light ray passing through the near focal point emerges parallel to the axis (C).

Ray diagrams
The first step in making an accurate ray diagram is to set up a sheet of graph paper by drawing a straight horizontal line to be the optical axis. The lens itself is drawn as a vertical line crossing the axis. The next step is to draw the two focal points, f, on the axis. The focal points should be the same distance on either side of the lens. It is important that the ray diagram be *drawn to scale*. For example, a scale often used with graph paper is one box equals one centimeter.

Drawing the object
Draw an arrow on the left side of the lens to represent the object. The distance from the arrow to the lens must be drawn to scale. For example, an object 20 centimeters away from a lens would be drawn 20 boxes to the left of the lens in the ray diagram if the scale is one box equals one centimeter.

Finding the image
To find the image, draw three rays from the tip of the arrow. The first ray goes straight through the center of the lens. The second ray goes parallel to the axis and bends through the far focal point. The third ray goes through the near focal point and bends to become parallel to the axis on the far side of the lens. The point where the three rays intersect is where the image of the tip of the arrow will be. Remember, an image forms where many rays from the same point on an object come together again. This is a *real image* because the rays *intersect*. The image is also smaller than life size, backward, and inverted.

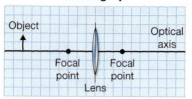
Setting up

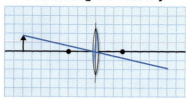
A. Drawing the first ray

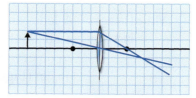
B. Drawing the second ray

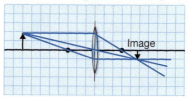
C. Drawing the third ray

Figure 17.23: *The process of drawing a ray diagram.*

Characteristics of images formed by a lens

Ray diagram for a converging lens If an object is placed to the left of a converging lens at a distance greater than the focal length, an inverted image is formed on the right-hand side of the lens. This is a *real* image, because it is formed by the actual intersection of light rays from the object. Real images can be projected on a surface. For example, the image from an overhead projector can be shown on a screen.

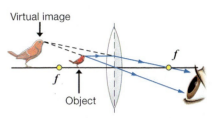

Figure 17.24: *A converging lens becomes a magnifying glass when an object is located inside the focal length of the lens.*

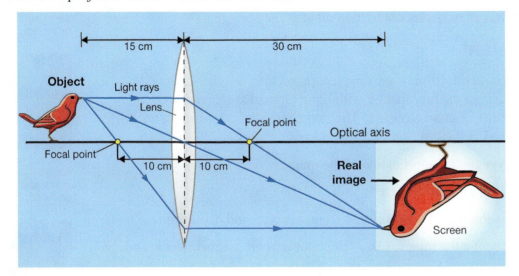

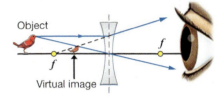

Figure 17.25: *A diverging lens always has the same ray diagram, which shows a smaller image.*

Magnifying glass A converging lens acts as a magnifying glass when the object is closer to the lens than one focal length. The image formed by a magnifying glass is virtual and appears behind the lens (Figure 17.24).

Diverging lens A diverging lens is thicker around the edges and thinner in the center. The image formed by a diverging lens is virtual and right side up. The image is always smaller and appears on the same side of the lens as the object (Figure 17.25). A ray diagram for a diverging lens follows similar rules, except that a parallel incident ray emerges from the lens at an angle away from the axis. A light ray traveling through the center of the lens continues undeflected.

Image summary table The different types of images formed by a single lens are listed in Table 17.2. The magnification varies, depending on the distance of the object from the lens. To find the magnification, you would have to draw a ray diagram.

Table 17.2: Images formed by a lens

Lens	Object	Image
Converging	Beyond focal length	Real
Converging	Inside focal length	Virtual
Diverging	Any	Virtual

17.3 Optical Systems

Most of the optical technology we use is not as simple as a single lens or mirror. For example, a camera may contain several lenses, a prism, and a mirror. The telephoto function in a camera is created by a pair of lenses. When you zoom in and out, the camera changes the separation between the lenses. As the separation changes, the magnification also changes. This section is about optical systems. Optical systems are built from lenses, mirrors, and prisms. An optical system collects light and may use refraction and reflection to form an image, or may process light in other ways.

Describing an optical system

The functions of an optical system

An optical system is a collection of mirrors, lenses, prisms, or other optical elements that performs a useful function with light. Optical systems can be described by a number of characteristics. The first characteristic, forming an image, was discussed in the last section. Other characteristics are

1. the location, type, and magnification of the image;
2. the amount of light that is collected;
3. the accuracy of the image in terms of sharpness, color, and distortion;
4. the ability to change the image, like a telephoto lens on a camera does; and
5. the ability to record images on film or electronically.

The image from a pinhole camera

The more light an optical system collects, the brighter the image it can form. A simple optical system can be made with a pinhole in a box (Figure 17.26). No image forms on the front of the box because rays from many points of the image can reach the same point on the box. An image *does* form inside the box. That image forms because light rays that reach a spot on the back surface are restricted by the pinhole to come from only one spot on the object.

A lens makes a brighter image than a pinhole

The image formed by a pinhole is very dim because the pinhole is small and does not allow much light to come through. The image formed by a lens is brighter because a lens is larger and collects more light. Each point on the image is formed by a cone of light collected by the lens. With a pinhole, the cone is much smaller and therefore the image has a much lower light intensity.

No image forms on the face of the box because light from many points on the object fall on the same point on the box.

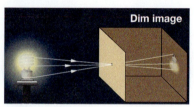

A pinhole forms a dim image by restricting light from each point on the object to a spot the size of the pinhole on the back surface of the box.

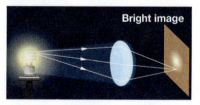

A lens forms a bright image by focusing more light from each point on the object to the equivalent point on the image.

Figure 17.26: *The images formed by a pinhole camera and a lens are different in brightness from a pinhole alone because different amounts of light are collected to form the image.*

The sharpness of an image

Aberrations are imperfect focusing of light
In a perfect image, light from a single point on an object is focused to a perfect point on an image. Real optical systems are never perfect. Light from a point on an object focuses to a small *area* on the image, but not a sharp point. Defects in the image are called *aberrations*, and can come from several sources.

Chromatic aberration
Chromatic aberration is caused by dispersion. Remember, dispersion causes light of different colors to bend by different amounts. With a single glass lens, for example, an image in red light focuses slightly farther away than an image in blue light (Figure 17.27). A multicolored image is slightly blurry because each color focuses at a different distance from the lens. Chromatic aberration can be reduced by using multiple lenses with different refractive indices. A combination of a converging and a diverging lens can bring both red and blue light into focus at the same point. Of course, other colors are still slightly out of focus, but not by much.

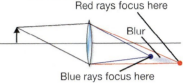

Figure 17.27: *Chromatic aberration occurs when different colors focus at different distances from the lens.*

Spherical aberration
Spherical aberration causes a blurry image because light rays farther from the axis focus at a different point than rays near the axis (Figure 17.28). This error is minimized by blocking all but the center of a lens. For example, a camera lens is restricted by an adjustable hole called the *f-stop*. A larger f-stop value means a smaller opening, which reduces spherical aberration but also reduces the amount of light reaching the film. A better way to reduce spherical aberration is to make the lens a different shape. A lens with a parabolic surface does not have spherical aberration. The best camera and telescope lenses use parabolic surfaces.

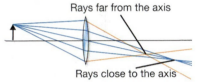

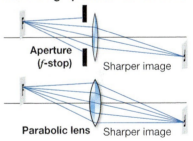

Figure 17.28: *Spherical aberration causes a point to focus to a blur. This error can be reduced by using only the center of the lens, or using a lens with a parabolic surface.*

The diffraction spot size
A third image defect comes from diffraction. Diffraction causes a point on an object to focus as a series of concentric rings around a bright spot. It occurs because a lens collects only a limited amount of light from a point on an object. A larger lens collects more light and the spot size formed by diffraction gets smaller. For this reason, larger lenses produce sharper images.

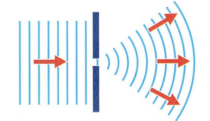

Diffraction through a small opening turns plane waves into circular waves.

The thin lens formula

Calculating image and object distances
Drawing ray diagrams is useful for learning how lenses work, but the process is inaccurate and takes time. The **thin lens formula** is a mathematical way to do ray diagrams with algebra instead of drawing lines on graph paper. Like ray diagrams, the thin lens formula can be used to predict the location, size, and orientation of an image produced by a lens. For optical systems made of several lenses, it is easier to predict the resulting image with the thin lens formula instead of using ray diagrams.

How to use the thin lens formula
The thin lens formula relates the focal length (f) to the distance between the object and the lens (d_o) and the distance from the lens to where an image forms (d_i).

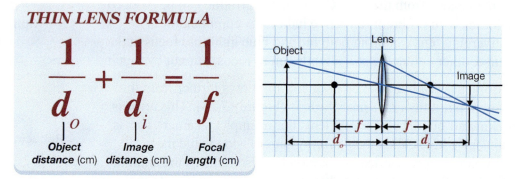

THIN LENS FORMULA
$$\frac{1}{d_o} + \frac{1}{d_i} = \frac{1}{f}$$

Object distance (cm)   Image distance (cm)   Focal length (cm)

Positive and negative signs
To keep track of whether objects or images are to the right or left of lenses, the thin lens formula uses a *sign convention*. The sign convention assumes light goes from left to right. When the object and image appear like the diagram above, all quantities are positive.

1. Object distances are *positive* to the left of the lens and *negative* to the right of the lens.
2. Image distances are *positive* to the right of the lens and *negative* to the left of the lens.
3. Negative image distances or object distances mean *virtual* images or objects. Positive image distances indicate *real* images.
4. The focal length is *positive* for a converging lens and it is *negative* for a diverging lens.

Locating an image with the thin lens formula

Calculate the location of the image if the object is 6 cm in front of a converging lens with a focal length of 4 cm.

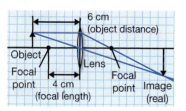

1. You are asked for the image distance.
2. You are given the focal length and object distance.
3. The thin lens formula applies:
   $1/d_i = 1/f - 1/d_o$
4. Solve for d_i:
   $1/d_i = 1/4 - 1/6$
   $1/d_i = 3/12 - 2/12 = 1/12$
   $d_i = 12$ cm

The image forms 12 cm to the right of the lens.

Approximations

The thin lens formula assumes the lenses have no thickness. This is a good assumption when objects and images are far away compared with the thickness of a lens.

Changing the size of an image

Image relay A technique known as **image relay** is used to analyze an optical system made of two or more lenses. The main idea behind image relay is that the image produced by the first lens (1) becomes the object for the second lens (2), and so on. The magnified image you see when you look through a telescope or microscope is produced this way. The ray diagram below shows the image relay analysis for a telescope made with two convex lenses. The first lens (1) forms an image (Image 1) between the lenses. This intermediate image is the object for the second lens (2) which forms the final image (Image 2). Notice that the intermediate image is inverted while the final image is right-side-up.

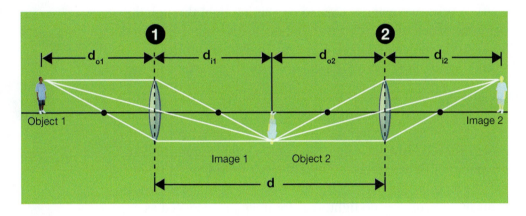

Why multiple lenses are useful Multiple lenses are useful because they allow an optical system to change the size of an image. Remember, the size of the image from a single lens depends on the distance between the object and the lens. If you are looking at a bird through a camera lens, you cannot easily change the distance between you and the bird. It is much easier to change the optical system. A *telephoto lens* is made from several lenses separated by a certain distance. The magnification depends on the distance between the lenses. By adjusting that distance, the image can be made to appear larger or smaller. The same principle applies to the zoom lens on a video camera.

The compound microscope

The microscope that you used in biology class is probably a compound microscope. The optical system of a compound microscope uses two converging lenses. The lens closest to the object, the objective, makes a real, larger, inverted image of the object.

This real image is inside the focal length of the eyepiece lens you look through. Since the image is inside its focal length, the eyepiece lens produces a magnified virtual image of the image from the objective lens. The overall magnification is the magnification of the objective multiplied by the magnification of the eyepiece. For example, a 5× eyepiece lens with a 100× objective lens produces an overall magnification of 500 (5 × 100).

Recording images

The technology of recording images

Before the invention of the camera, the only way to store an image was to draw it or paint it. Since both drawing and painting take skill and time, few images survive from early history. Today, images are so easy to save that you have seen many thousands of them in photographs, on TV, in magazines, and elsewhere.

Recording images on film

There are two basic techniques for recording images. One uses film. Film records an image by using special inks that respond to light. For a black and white photograph, the ink darkens in response to the intensity of light. Where light on the image is intense, the ink becomes dark. Where the image is dark, the ink remains light. Because dark and light are inverted, this image is known as a *negative* (Figure 17.29). A positive image is created by shining light through the negative onto photographic paper, also coated with light-sensitive ink. Light areas on the negative allow light through and darken corresponding areas on the photograph. A color photograph uses three different colors of light-sensitive ink.

Recording images electronically

The second technique for recording an image is electronic. A digital camera uses a tiny sensor called a CCD (Chapter 16), which is located at the focal plane of the camera lens. On the surface of the CCD are thousands of tiny light sensors. There are separate light sensors for red light, blue light, and green light (Figure 17.30). For each sensor, the amount of light is recorded as a number from 0 to 255. For example, if the red sensor records 255, it is seeing the most red light it can handle. A recording of 0 means the sensor sees no light. A color image is recorded as a table of numbers. Each point on the image has three numbers corresponding to the amount of red light, blue light, and green light. The resolution of a digital camera is the number of points, called *pixels*, that can be recorded by the CCD. A *10-megapixel* camera stores 10 million pixels per image. Since each pixel is three numbers, a 10-megapixel image requires 30 million stored numbers.

Recording moving images

A video camera records a sequence of images, one after another, 30 times per second. Because each image requires so many numbers, a single video image typically includes less than 300,000 pixels. The lower resolution is why single images from video cameras look grainy or blurry when they are printed.

Figure 17.29: *Recording an image on film is a two-step process.*

Figure 17.30: *A digital camera records an image as intensities of red, green, and blue light.*

Chapter 17 Connection

The Telescope

The **telescope** is an example of an optical system that has played an important part in human history. When people think of a telescope, most of them think of a refracting telescope. A refracting telescope is usually built like a tube with lenses at each end. Galileo was the first person to use a refracting telescope to learn about the Moon and planets.

The refracting telescope An astronomical **refracting telescope** is constructed of two converging lenses with different focal lengths. The lens with the longest focal length is called the **objective** and the shorter-focal-length lens is the **eyepiece**.

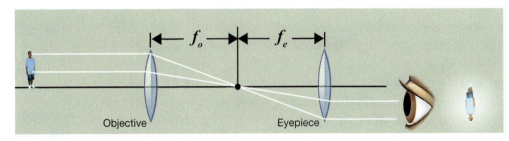

As you can see from the diagram, the ray from the top of the object ends up below the axis as it enters the eye of the observer. This shows that the image from this refracting telescope is inverted. Inverted images are usually fine for looking at objects in space. It does not matter if the image of a star is upside down. But, looking at distant birds or animals upside down does matter.

A telescope with right-side-up images To rectify this problem, the converging eyepiece is replaced by a diverging lens in a terrestrial refracting telescope. This arrangement of lenses produces an image that is right-side-up. The design of the terrestrial telescope also sets the lenses a distance apart equal to the sum of their focal lengths.

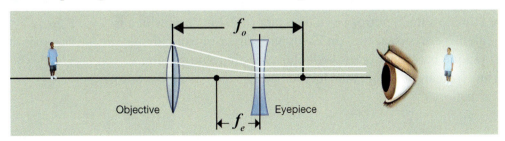

The Sidereal Messenger

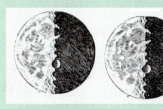

In the early 1600s, Galileo Galilei became the first person to use a telescope to study the sky. Galileo learned that a lensmaker in Holland had put together two lenses to magnify the images of distant objects. Galileo built his own telescope and began studying objects in the sky.

He looked at the Moon, the stars, and the planets. Before Galileo's investigations, people thought the Moon was a smooth ball in the sky. Galileo correctly interpreted the pattern of light and dark as shadows being cast from tall mountains.

When he looked at Jupiter, he discovered four of the moons that orbit the giant planet. This discovery, along with detailed drawings of the moons, was reported in his book which he titled *Sidereal Nuncius*.

Chapter 17 Connection

Newtonian reflecting telescope

Larger telescopes can see more distant objects
The larger the diameter of the objective of a telescope, the more light it can gather to form an image. Consequently, a large-diameter telescope can see faint images in the night sky. Objects appear faint for two reasons. One is that they may actually give off very little light, like a planet that gives off only reflected light from the Sun. The other reason is that the object is far away. A huge galaxy that is far away appears dimmer than a single star that is close due to the inverse square law. A refracting telescope is limited by the size of the lens that can be made. Large glass lenses are heavy and difficult to make.

Newtonian reflecting telescope
All large modern telescopes use a concave mirror instead of an objective lens. The most successful reflecting-telescope design is called a Newtonian telescope, after Sir Isaac Newton, who designed and built the first one. Although we think of the three laws of motion when we think of Newton, he also made important contributions to optics. Besides designing the reflecting telescope, Newton was the first to prove white light is made of many colors.

Mirror arrangement
In the Newtonian telescope, light falls on a curved mirror that focuses the light similar to the way the objective lens focuses light in a refracting telescope. A small secondary mirror directs the light rays that are reflected off the mirror so that they move toward an eyepiece lens (Figure 17.31). The eyepiece is located so that the distance between it and the mirror is equal to the sum of their focal lengths. This combination of mirrors and lenses collects more light than your eye could alone. Newtonian telescopes have been built with mirrors as large as eight meters in diameter, and they can see objects that are so faint they are near the very edge of the observable universe.

The Hubble Space Telescope is a reflector
The Hubble Space Telescope orbiting Earth uses a reflecting telescope (Figure 17.32). The Hubble's main mirror has a diameter of 2.4 meters, which is about 8 feet. Although there are telescopes on Earth with larger mirrors, the Hubble telescope is unique because it orbits well above Earth's atmosphere. Distant objects are much clearer through the Hubble because there is no distortion from atmospheric refraction.

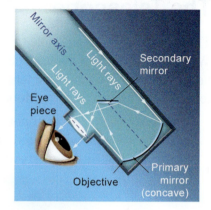

Figure 17.31: *A Newtonian reflecting telescope uses a curved mirror to collect light.*

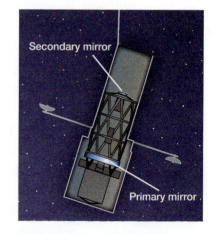

Figure 17.32: *The Hubble Space Telescope uses a reflecting design with a 2.4-meter diameter primary mirror. (NASA)*

Chapter 17 Assessment

Vocabulary

Select the correct term to complete the sentences.

lens	mirror	prism
optics	geometric optics	specular
diffuse	converging	diverging
law of reflection	normal line	ray diagram
magnification	objects	index of refraction
focal point	focal length	optical axis
light ray	thin lens formula	critical angle
Snell's law	real	virtual image
chromatic	refraction	fiber optics
dispersion	magnifying glass	spherical
reflection	diffraction	telescope
pixels	total internal reflection	resolution
	image	focal plane

1. A familiar optical device used to reflect light is a(n) _____.
2. A(n) _____ is an optical device traditionally used to separate light into its colors.
3. The branch of optics that focuses on the creation of images is called _____.
4. A lens that bends light so the light rays come together at a point is a(n) _____ lens.
5. Images produced as light bounces off highly-polished surfaces are the result of _____ reflection.
6. A rule relating the direction of a reflected ray to the direction of the incident ray is known as the _____.
7. The study of how light behaves is called _____.
8. A straight line drawn to indicate the direction in which light travels is known as a(n) _____.
9. A(n) _____ is an accurately-drawn diagram indicating the path of one or more light rays.
10. A surface that is not shiny creates a(n) _____ reflection.
11. An example of an optical device used to bend light in a specific direction is a(n) _____.
12. In a ray diagram, an imaginary line drawn perpendicular to a surface at the point where a light ray strikes is called a(n) _____.
13. The bending of light that takes place as a light ray crosses the boundary from one material to another is called _____.
14. The _____ is a number with no units describing the ability of a material to bend light rays.
15. The angle at which light begins to be reflected back into a material is called the _____.
16. _____ is the separation of light into its component frequencies due to different refractive indices for each color.
17. Light rays that are turned back at the boundary between materials have experienced _____.
18. If the angle of refraction of a light ray exceeds 90°, _____ occurs.
19. Thin glass rods using total internal reflection to carry light signals are called _____.
20. A mathematical relationship used to calculate the path of a light ray as it passes from one material into another is _____.
21. The "picture" of an object formed in space where light rays meet is called a(n) _____.
22. An image that cannot be projected on a screen is known as a(n) _____.
23. A light ray traveling along the _____ of a lens will not be bent by the lens.
24. The ratio of the size of the image produced by a lens to the size of the object is known as the _____ of the lens.
25. Physical things that give off or reflect light are referred to as _____.
26. A convex lens used to create virtual images larger than the object is commonly called a(n) _____.

27. Parallel light rays entering a converging lens meet at a point called the _____.

28. Images that can be brought to focus on a screen are considered to be _____ images.

29. A lens that is thicker around the edges and thinner in the middle is a(n) _____ lens.

30. The surface on which a lens forms an image may be called the _____.

31. The distance from the center of a lens to the focal point is the _____.

32. A mathematical relationship used to predict the size, location, and orientation of an image produced by a lens is known as the _____.

33. Points of light recorded by a CCD are called _____.

34. _____ aberration is the image defect caused by dispersion of light when light is focused by a lens.

35. A common optical instrument used to produce an enlarged image of distant objects is called a(n) _____.

36. The number of points of light recorded by the CCD of a digital camera determines the camera's _____.

37. Light rays striking a lens at different distances from the center of the lens form blurry images due to a lens defect known as _____ aberration.

38. Using a larger lens to focus an image will reduce an image defect caused by _____.

Concept review

1. State the law of reflection. Does this law hold true for specular reflection? Does it hold true for diffuse reflection?

2. Refer to the diagram below. State the correct term for each item listed.

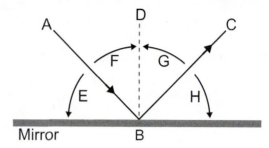

   a. line A–B
   b. line D–B
   c. line B–C
   d. angle F
   e. angle G

3. Use the diagram to answer the following questions.

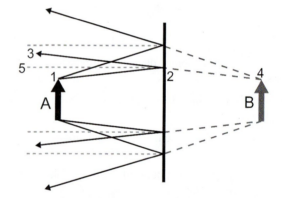

   a. If A is a real object in front of a mirror, is B a real image or a virtual image?
   b. Line 1–2 is the incident ray from the object to the mirror. Identify the corresponding reflected ray.
   c. Line 1–2 is the incident ray from the object to the mirror. Identify the ray that our eye "creates" to allow us to see the virtual image.

4. Refer to Table 17.1 in the text. Which material has a greater ability to bend light—ice or glass? How do you know?

5. Consider the diagram shown. Ray AB represents the incident ray. Identify the refracted ray in each case.

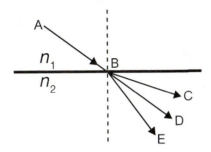

   a. n_1 is greater than n_2
   b. n_1 is equal to n_2
   c. n_1 is less than n_2

6. Explain how fiber optic cables utilize the properties of reflection and refraction.

7. When white light passes through a prism, which color refracts more—yellow or green?

8. According to the sign conventions used with the thin lens formula, object distances to the left of the lens are _____ but image distances to the left of the lens are _____.

9. According to the sign conventions used with the thin-lens formula, negative image distances imply _____ images while positive image distances imply _____ images.

10. The diagram below shows an object to the left of a converging lens and the real image produced to the right of the lens. Use the diagram to identify the following rays.

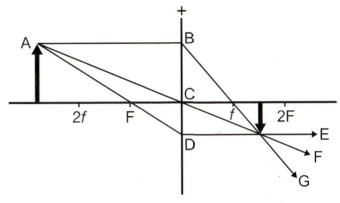

   a. the ray of light that travels from the object to the lens along a line parallel to the optical axis
   b. the ray of light that travels from the object to the center of the lens
   c. the ray of light that travels from the object through the focal point to the lens

11. Large modern reflecting telescopes, which are made using a combination of mirrors, are superior to refracting telescopes, that are made with lenses. Write two or more reasons this is so.

12. What determines the amount of refraction that takes place when a light ray passes from one substance to another?

Problems

1. The diagram shows an object in front of a plane mirror and three rays of light traveling from the object to the mirror. For each ray, draw the corresponding reflected ray, and the virtual ray that we would see coming from the image.

2. The angle between the incident ray striking a mirror and the reflected ray leaving the mirror is 60 degrees. What is the angle of incidence?

3. The phrase "MY MOM" is held in front of a mirror. When you read it in the mirror, what do you read? Try it and find out. Make sure you use capital letters.

Use the diagram below to answer problems 4–6.

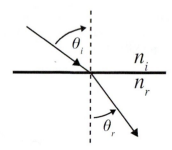

4. An incident ray of light strikes a piece of flint glass ($n = 1.66$) at an angle of incidence of 40 degrees. What is the angle of refraction for this ray inside the glass?

5. Light travels from air into ethyl alcohol. The sine of the incident angle is 0.6427 and the sine of the refracted angle is 0.4726. What is the index of refraction for ethyl alcohol?

6. What angle of incidence would result in an angle of refraction of 30 degrees for light going from air into fused quartz ($n = 1.46$)?

7. An object is placed 10 cm in front of a thin converging lens that has a focal length of 6 cm. Where will the image be located?

8. An object 4 cm in height is placed 12 cm to the left of a thin convex lens with a 6-cm focal length.
   a. Use the thin-lens formula to calculate the location and type of image formed.
   b. Draw a ray diagram showing the position, orientation, type, and height of the image formed.

Applying your knowledge

1. A prism can be used as a reflector as well as a device for dispersing light into its component colors. Research the use of prisms as reflectors. What types of devices use prisms as reflectors? Why are prisms used instead of plane silvered-glass mirrors?

2. It is a bad idea to sprinkle water on tree leaves during the day because the water drops leave brown spots on the leaves. What causes the spots?

3. Why are you able to see better with goggles when swimming underwater?

4. Research the origin of the kaleidoscope. What optical devices are used to construct a kaleidoscope? Describe its basic structure.

UNIT 6 LIGHT AND OPTICS

CHAPTER 18

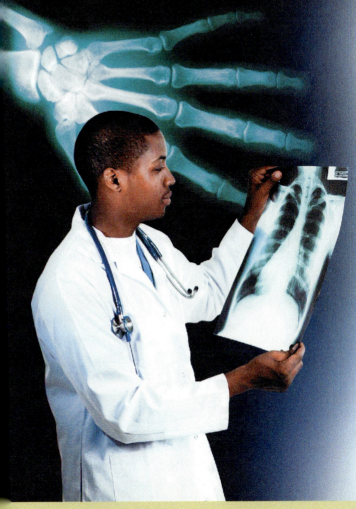

Wave Properties of Light

Objectives:

By the end of this chapter you should be able to:

- ✔ Calculate the frequency or wavelength of light when given one of the two values.
- ✔ Describe the relationship between frequency, energy, color, and wavelength.
- ✔ Identify at least three different types of waves of the electromagnetic spectrum and an application of each.
- ✔ Interpret the interference pattern from a diffraction grating.
- ✔ Use the concept of polarization to explain what happens as light passes through two polarizers.
- ✔ Describe at least two implications of special relativity with regard to energy, time, mass, or distance.

Key Questions:

- What is the electromagnetic spectrum?
- How do diffraction gratings and polarizers work?
- What are some implications of special relativity?

Vocabulary

diffraction grating	gamma ray	polarizer	spectrometer	visible light
electromagnetic spectrum	inference pattern	radio wave	spectrum	X-ray
electromagnetic wave	microwave	rest energy	time dilation	
	polarization	special relativity	transmission axis	

CHAPTER 18 | WAVE PROPERTIES OF LIGHT

18.1 The Electromagnetic Spectrum

Today, we believe light has the properties of a wave, like sound, even though photons also behave partly like particles. Like other waves, light can be described by frequency, wavelength, amplitude, and speed. Although we cannot directly see the wave motion of light, its existence is confirmed by the results of many experiments. The light in the visible **spectrum** is a small part of a whole series of waves called the *electromagnetic spectrum*. This series of waves includes radio waves, microwaves, and X-rays. This chapter explores the broader picture of light as part of the electromagnetic spectrum.

Traveling oscillations of electricity and magnetism

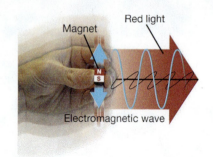

Figure 18.1: *If you could shake a magnet up and down 450 trillion times per second, you could make an electromagnetic wave that you would see as red light.*

Energy fields — Light is a wave. However, this statement prompts a question: a wave of *what*? This is a difficult question to answer. Consider pushing the north poles of two magnets together. They repel each other while still some distance apart. The magnets feel force without actually touching each other because magnets create an *energy field* in the space around them. The energy field is what exerts forces on other magnets. Electricity also creates an energy field. You may have sensed the effects of the electric field in the air during a thunderstorm, or from static electricity on a dry day.

Electromagnetic waves are oscillations of an energy field — The energy field created by electricity and magnetism can *oscillate* and it supports waves that move, just as water supports water waves. These waves are called **electromagnetic waves**, and light is one of them. Anything that creates an oscillation of electricity or magnetism also creates electromagnetic waves. When you move a magnet up and down, you are creating an electromagnetic wave (Figure 18.1). If you could shake the magnet up and down 450 trillion times per second, you would make waves of red light.

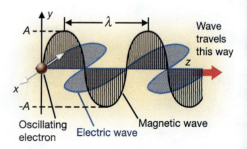

Figure 18.2: *A 3-D view of an electromagnetic wave showing the electric and magnetic portions of the wave. The wavelength and amplitude of the waves are labeled λ and A, respectively.*

Electricity and electromagnetic waves — You can also make electromagnetic waves with electricity. For example, if you could switch the electricity on and off repeatedly in a wire, the oscillating electricity would make an electromagnetic wave. In fact, this is exactly how radio towers make radio waves. Electric currents oscillate up and down the metal towers and create electromagnetic waves of the right frequency to carry radio signals. Electromagnetic waves have both an electric part and a magnetic part (Figure 18.2), and the two parts exchange energy back and forth like a pendulum exchanges potential and kinetic energy back and forth.

Frequency and wavelength of light

The frequency of light The frequency of light waves is incredibly high: 10^{14} is a one with 14 zeros after it. Red light has a frequency of 460 trillion, or 460,000,000,000,000 cycles per second. Light wave frequencies are so high, we use units of terahertz (THz) to measure them. One THz is a trillion hertz (10^{12} Hz), or a million megahertz.

The wavelength of light Because the frequencies are so high, the wavelengths are tiny. Waves of red light have a length of only 0.00000065 meter (6.5×10^{-7} m). Figure 18.3 shows the size of a light wave relative to other small things. Because of the high frequency and small wavelength, we do not normally see the true wavelike nature of light. Instead, we see reflection, refraction, and color.

Frequency 4.6×10^{14} to 7.5×10^{14} Hz

Wavelength 4×10^{-7} to 6.5×10^{-7} meters wavelength

Frequency and color With other waves, we found that energy was proportional to frequency. Higher-frequency waves have more energy than lower-frequency waves. The same is true of light. The higher the frequency of the light, the higher the energy of the wave. Since color is related to energy, there is also a direct relation between color, frequency, and wavelength. Table 18.1 shows the color, frequency, and wavelength of visible light. One nanometer is 10^{-9} meters.

Table 18.1: Frequencies and wavelengths of light

Energy (relative)	Color	1×10^{-6} m	Wavelength (nm)	Frequency (THz)
Low	Red		650	462
	Yellow		580	517
	Green		530	566
	Blue		470	638
High	Violet		400	750

Size

Bee 1×10^{-2} m

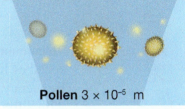
Pollen 3×10^{-5} m

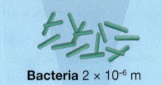
Bacteria 2×10^{-6} m

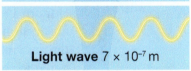
Light wave 7×10^{-7} m

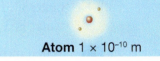
Atom 1×10^{-10} m

Figure 18.3: *A bee is about a centimeter long; a grain of pollen is about 0.03 mm wide; a bacterium is about 2.0×10^{-6} m long; the wavelength of the longest wave we can see is 7×10^{-7} m or 700 nm; and an atom is about 0.1 nm in diameter.*

CHAPTER 18 WAVE PROPERTIES OF LIGHT

The speed of light waves

The speed of light is frequency times wavelength

In the previous chapter, we talked about the speed of light as being incredibly fast (3×10^8 m/s) and being represented by its own symbol, c. Not only light, but all electromagnetic waves travel at that speed in a vacuum. Like other waves, the speed of light is equal to the product of frequency and wavelength.

THE SPEED OF LIGHT (Relationship between frequency and wavelength)

Speed of light (3×10^8 m/s) — $c = f\lambda$ — Wavelength (m)
Frequency (Hz)

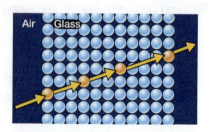

Figure 18.4: *The passage of light through a material takes more time because the light must be absorbed and re-emitted to pass through neighboring atoms.*

To calculate the wavelength of light or any electromagnetic waves, we use the speed of light formula. For example, to calculate the wavelength from the frequency, the formula is rearranged to solve for wavelength, $\lambda = c \div f$. Red light with a frequency of 462×10^{12} Hz, has a wavelength of 649×10^{-9} m:
649×10^{-9} m = (3×10^8 m/s) ÷ (462×10^{12} Hz).

Light travels slower through materials where $n > 1$

The index of refraction (n), from the last chapter, is actually the ratio of the speed of light in a material to the speed of light in a vacuum. When passing through a material like glass, light is absorbed and re-emitted by neighboring atoms (Figure 18.4). Between atoms, light moves at its normal speed. But the process of absorption and emission delays the light. As a result, light travels more slowly through materials than it does in a vacuum. The speed of light in a material is equal to the speed of light in a vacuum divided by the refractive index of the material. For example, water has a refractive index of 1.33. The speed of light in water is $3 \times 10^8 \div 1.33$ or 2.2×10^8 m/s.

Wavelengths are shorter in refractive materials

When moving through a material, the frequency of light stays the same. If 462×10^{12} waves go in per second, then 462×10^{12} waves must come out per second. Otherwise waves would pile up at the boundary between materials. Because the frequency stays the same, the wavelength of light is reduced by the index of refraction. For example, the wavelength of light in water is equal to its wavelength in air divided by 1.33, the index of refraction for water.

Calculating the wavelength of light

Calculate the wavelength in air of blue-green light that has a frequency of 600×10^{12} Hz.

1. You are asked for the wavelength.
2. You are given the frequency.
3. The speed of light is $c = f\lambda$.
4. $\lambda = \dfrac{c}{f} = \dfrac{3 \times 10^8 \text{ m/s}}{600 \times 10^{12} \text{ Hz}}$
   $= 5 \times 10^{-7}$ m

Waves of the electromagnetic spectrum

The electromagnetic spectrum

To a scientist, the word *light* means all the electromagnetic waves that travel at speed, c, not just the ones we can see with our eyes. Visible light is a small part of a whole range of electromagnetic waves called the **electromagnetic spectrum**. On the low-energy end of the electromagnetic spectrum are radio waves with wavelengths billions of times longer than those of visible light. On the high-energy end are gamma rays with wavelengths millions of times smaller than those of visible light.

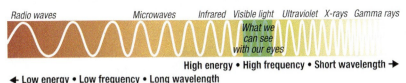

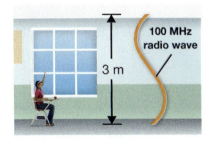

Figure 18.5: *A 100-megahertz radio wave has a wavelength of 3 meters, about the height of a classroom.*

Radio waves

Radio waves are on the low-frequency end of the electromagnetic spectrum. They have wavelengths that range from hundreds of meters down to less than a centimeter (Figure 18.5). Radio broadcast towers are so tall because they have to be at least one-quarter of a wavelength long. Why?

Microwaves

Microwaves have a wavelength between one millimeter and 300 millimeters, or 30 cm. Cell phones and microwave ovens use microwaves (Figure 18.6). The waves in a microwave oven are tuned to the natural frequency of liquid water molecules. The high intensity of microwaves inside a microwave oven rapidly transfers energy to water molecules in food. The heating that comes from absorption of microwave energy occurs through some depth since microwaves can penetrate many centimeters. The statement "microwaves cook from the inside out" is not exactly true. However, microwaves do heat food at more than just the surface, unlike conventional ovens which heat food only at the surface.

Figure 18.6: *Cell phone transmissions are made with microwaves.*

Infrared waves

The infrared or IR region of the electromagnetic spectrum lies between microwaves and visible light. Infrared wavelengths range from 1 millimeter to about 700 nanometers. Infrared waves are often referred to as radiant heat. Although we cannot see infrared waves with our eyes, we can feel them with our skin. Heat from the Sun comes from IR waves in sunlight. Objects that are warmer than their surroundings also radiate energy as IR waves. Figure 18.7 is an infrared view of the Helix nebula, a faraway space object.

Figure 18.7: *An infrared view of the Helix nebula.*

CHAPTER 18 | WAVE PROPERTIES OF LIGHT

Medium- to high-energy electromagnetic waves

Visible light The rainbow of colors of **visible light** is in the medium-energy range of the electromagnetic spectrum with wavelengths between 700 and 400 nanometers. When most people talk about light, they are usually referring to this part of the spectrum. However, when scientists talk about light, they may be referring to any part of the electromagnetic spectrum, from radio waves to gamma rays.

Ultraviolet waves Ultraviolet radiation has a range of wavelengths from 10 to 400 nanometers. Sunlight contains ultraviolet waves. A small amount of ultraviolet radiation is beneficial to humans, but larger amounts cause sunburn, skin cancer, and cataracts. Most ultraviolet light is blocked by the ozone in Earth's upper atmosphere (Figure 18.8). A hole in Earth's ozone layer, such as those observed over the Antarctic, is of concern because it allows more ultraviolet light to reach the surface of the planet, creating problems for humans, plants, and animals.

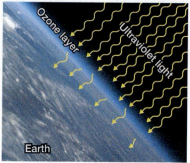

Earth image courtesy NASA/JPL/UCSD/JSC

Figure 18.8: *Most of the ultraviolet light from the Sun is absorbed by Earth's ozone layer.*

X-rays **X-rays** are high-frequency waves that have great penetrating power and are used extensively in medical and manufacturing applications (Figure 18.9). Their wavelength range is from about 10 nanometers to about 0.001 nm, or 10 trillionths of a meter. X-rays are strongly absorbed by calcium and other heavy elements. When you get a medical X-ray, the film darkens where bones are located because the intensity of the X-rays has been absorbed before reaching the film. X-rays allow doctors to quickly determine the extent of an injury such as a broken bone. The X-ray seen here clearly shows a break in the little finger.

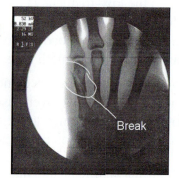

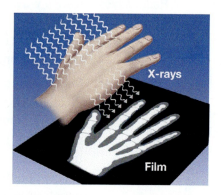

Figure 18.9: *X-rays have a high enough energy to go through your soft tissue, but not bone.*

Gamma rays **Gamma rays** have wavelengths of less than about 10 trillionths of a meter. Gamma rays are generated in nuclear reactions, and are used in many medical applications. Gamma rays can strip atoms of electrons and break chemical bonds, including the chemical bonds holding the molecules in your body together. People who work with gamma radiation have to be protected by lead shielding.

18.2 Interference, Diffraction, and Polarization

In science, we recognize the correct explanation by comparing what we believe with what nature actually does. Saying light is a wave does not mean much unless light actually does the things that waves do. The experimental proof of the wave nature of light is the topic of this section. We know light is a wave because it shows interference, diffraction, resonance, frequency, wavelength, and because light has the same relationships among speed, frequency, and wavelength as other waves.

The interference of light waves

Young's double-slit experiment In 1807, Thomas Young (1773–1829) did the most convincing experiment demonstrating that light is a wave. In the experiment, a beam of light falls on a pair of parallel, very thin slits in a piece of metal. After passing through the slits, the light falls on a screen. You might expect to see two bright stripes on the screen where the light came through the slits, with the rest of the screen in shadow. However, when Young looked at the screen, he saw a pattern of alternating bright and dark bands, called an **interference pattern**.

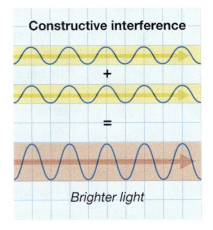

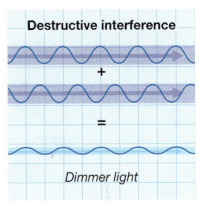

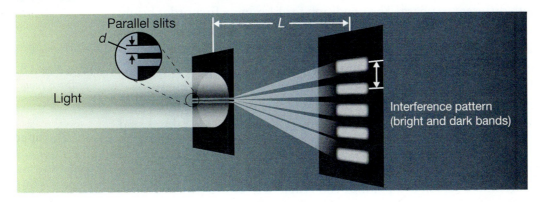

Figure 18.10: *Constructive interference creates brighter light (top). Destructive interference creates dimmer light (bottom).*

Interference of light waves An interference pattern is created by the addition of two waves. The bright bands are at locations where the light waves from both slits meet at the screen in phase. When two waves meet in phase, they add up to make a bigger, brighter wave. In Chapter 14, we called this type of addition constructive interference (top, Figure 18.10). The dark bands appear when the light waves from both slits meet out of phase and subtract from each other. The light intensity is lower because of destructive interference (bottom, Figure 18.10).

CHAPTER 18 WAVE PROPERTIES OF LIGHT

Diffraction gratings and spectrometers

Diffraction grating
A **diffraction grating** creates an interference pattern of light similar to the pattern for the double slit. A grating acts like *many* parallel slits, side by side. When light falls on, or goes through, a diffraction grating, each groove acts like a separate source of light. Because there are many slits, the interference effect is much more dramatic. Compared to the pattern from a double slit, light of a single wavelength passing through a diffraction grating creates narrow, widely separated bright spots with nearly-dark areas in between.

The central spot
Figure 18.11 shows pure red light of a single wavelength passing through a diffraction grating and onto a screen. This kind of light might come from a red laser, which typically makes light with a wavelength of 650 nanometers. A bright spot called the *central spot* appears directly in front of the grating where the light goes straight through.

The first-order bright spot
Another bright spot appears on either side of the central spot. These additional bright spots are called *first order* because they come from waves that are one whole wavelength different in phase. The condition for forming a first order bright spot is that the path length differences for light rays from two adjacent grooves is equal to one wavelength. This condition is satisfied when $\lambda = d \sin \theta$, where d is the spacing between the grooves on the grating (Figure 18.11). Bright spots will also be seen when the path difference for rays from adjacent grooves equals two wavelengths, three wavelengths, or any integer number of wavelengths. These extra spots appear farther from the central spot than the first order, are fainter, and are labeled second order, third order, and so on.

The spectrometer
A diffraction grating spreads out colored light into a spectrum. This happens because each point on the screen is a point of constructive interference for a *different wavelength of light*. A **spectrometer** is a device that measures the wavelength of light. A diffraction grating can be used to make a spectrometer because the wavelength of the light at the first-order bright spot is $\lambda = d \sin \theta$. The groove spacing, d, is known and the angle θ can be measured relatively easily. The wavelength of light is determined from λ and d. Many spectrometers have a printed scale printed that allows you to read the wavelength directly from the pattern of light made by the grating (Figure 18.12).

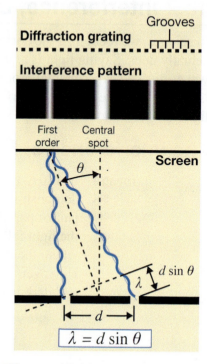

Figure 18.11: *The interference pattern from a diffraction grating.*

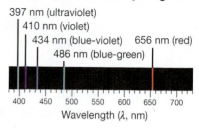

Figure 18.12: *The spectrum of hydrogen seen in a spectrometer.*

Polarization

Polarization of a wave on a spring
Think about shaking a spring or a jump rope to make a wave. If the spring is shaken up and down it makes vertical oscillations. If the spring is shaken sideways it makes horizontal oscillations. The direction of a wave's oscillations is called **polarization**. The wave on the up-and-down spring has a *vertical polarization*. The wave on the sideways-shaken spring has a *horizontal polarization*. Polarization occurs for all transverse waves, which have oscillations perpendicular to the direction of travel.

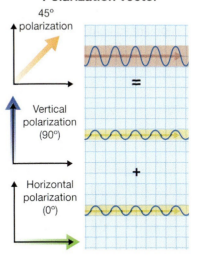

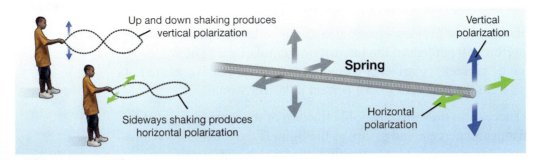

The direction of polarization is a vector
Polarization can be intermediate between horizontal and vertical. For example, a spring that is shaken at a 45-degree angle to the vertical will create a polarization that is also at 45 degrees. In Chapter 7, we learned that a vector can be resolved into components in two directions. The direction of polarization is a vector and can be resolved into components in two directions. You can think of a wave that has 45-degree polarization as the addition of two smaller-amplitude component waves with horizontal and vertical polarizations (Figure 18.13).

Figure 18.13: *Polarization is a vector. A wave with a polarization of 45 degrees can be represented as the sum of two waves. Each of the component waves has a smaller amplitude.*

Polarization of light waves
Polarization is a property of light and demonstrates that light is a *transverse wave*. The oscillation of energy in a light wave is perpendicular to the direction the wave moves. For light, we measure polarization by the orientation of the electrical part of the wave (Figure 18.14). Like a spring, the polarization of a light wave may be resolved into two perpendicular directions we usually call *horizontal* and *vertical*.

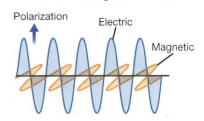

Figure 18.14: *The polarization of a light wave is the direction of the electric part of the wave.*

Unpolarized light
Most of the light that you see is *unpolarized*. That does not mean the light has no polarization. All single photons have a polarization. Instead, unpolarized light is a mixture of all polarizations. We say ordinary light is unpolarized because no single polarization is favored in the whole mixture.

Polarizers

Making polarized light

A **polarizer** is a material that selectively absorbs light depending on polarization. Microscopically, polarizers divide incident light into two polarizations parallel and perpendicular to the **transmission axis** of the polarizer. Light with perpendicular polarization is absorbed by the polarizer. Light with parallel polarization is re-emitted and appears to "pass through" the polarizer. *Light that comes through a polarizer has only one polarization—in the direction of the transmission axis of the polarizer.* After passing through a polarizer, light is polarized because it contains only a single polarization.

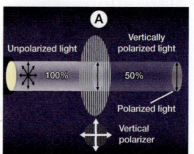

Polarizing unpolarized light

On average, unpolarized light divides equally between parallel and perpendicular polarizations. Therefore, 50 percent of the intensity of unpolarized light will be transmitted through a "perfect" polarizer (Figure 18.15 A). Real polarizers are not perfect, and always transmit less light than a perfect polarizer would.

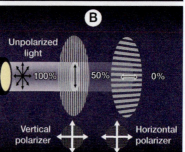

Transmission of light through two polarizers

Two polarizers in succession transmit a variable fraction of the incident light. The first polarizer transmits 50 percent and polarizes the light. If the transmission axis of the second polarizer is perpendicular to the first, the light is completely absorbed and no light is emitted (Figure 18.15 B). If the axis of the second polarizer is at 45 degrees to the axis of the first one, then half the light reaching the second polarizer is transmitted (Figure 18.15 C), which is 25 percent of the incident light. This occurs because the light from the first polarizer is polarized at 45 degrees relative to the transmission axis of the second polarizer.

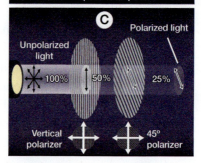

Transmission varies with the angle of polarization

The amount of light passing through a polarizer depends on the angle between the polarization of the light and the transmission axis of the polarizer. For example, suppose the incident light is polarized at 30 degrees to the horizontal. A vertical polarizer transmits 25 percent of this incident light. A horizontal polarizer transmits 75 percent of the light.

Figure 18.15: *The percentage of light transmitted through two polarizers depends on the angle between the transmission axes of the two polarizers.*

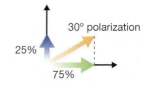
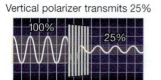
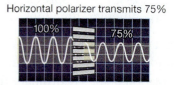

Applications of polarization

Polarized sunglasses Polarizing sunglasses are used to reduce the glare of reflected light (Figure 18.16). This works because light that reflects at low angles from horizontal surfaces is polarized mostly horizontally. This is precisely the light that causes glare. Polarized sunglasses are made to block light waves with horizontal polarization. Because glare is horizontally polarized, it gets blocked by polarized sunglasses much more than other light that is unpolarized. You can still see the light reflected from other objects, but the glare off of a surface, such as water, is blocked.

LCD computer screens The LCD (liquid-crystal diode) screen on a laptop computer uses polarized light to make pictures. The light you see starts with a lamp that makes unpolarized light. A polarizer then polarizes all the light. The polarized light passes through thousands of tiny pixels of liquid crystal that act like windows. Each liquid-crystal window can be electronically controlled to act like a polarizer, or not. When a pixel is *not* a polarizer, the light comes through, like an open window, and you see a bright dot.

Dark dots are made by crossing polarizers The transmission axis of the liquid-crystal is at right angles to the polarization direction of the light leaving the first polarizer. When a pixel becomes a polarizer, the light is blocked and you see a dark dot. The picture is made of light and dark dots. To make a color picture there are separate polarizing windows for each red, blue, and green pixel.

Limits to LCD technology Because the first polarizer blocks half the light, LCD displays are not very efficient, and are the biggest drain on a computer's batteries. LCD displays also suffer from low contrast between light areas and dark areas.

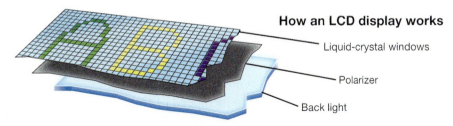

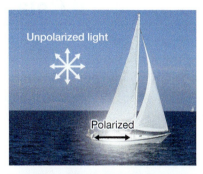

Figure 18.16: *Unpolarized light from the Sun is polarized horizontally when it reflects off the water. Polarized sunglasses block out this light; regular sunglasses do not.*

Try looking at an LCD screen through polarizing sunglasses. Explain what you see when you rotate the sunglasses to change the angle of the transmission axis of the polarizing lenses relative to the LCD screen.

18.3 Special Relativity

Science fiction writers are fond of creating interesting effects such as time travel. It may surprise you, but time travel into the future is actually possible. It just takes a lot of energy; much more energy than we know how to get or control. Albert Einstein's theory of special relativity makes a connection between time and space that depends on how fast you are moving. This section is a brief exploration of the theory of special relativity, which is a fascinating area of physics still being developed.

What special relativity is about

Time runs slower for moving objects.

The relationship between matter, energy, time, and space
The theory of **special relativity** describes what happens to matter, energy, time, and space at speeds close to the speed of light. The fact that light *always* travels at the same speed forces other things about the universe to change in surprising ways. Special relativity does not affect ordinary experience because objects need to be moving faster than 100 million m/s before the effects of special relativity become obvious. However, these effects are observed every day in physics labs.

Mass increases as an object gets close to the speed of light.

Time may move slower
Time moves more slowly for an object in motion than it does for objects that are not in motion. In practical terms, clocks run slower on moving spaceships compared with clocks on the ground. By moving very fast, it is possible for one year to pass on a spaceship while 100 years have passed on the ground. This effect is known as **time dilation**.

Concept of simultaneous changes

Mass may increase
As objects move faster, their *mass increases*. The closer the speed of an object gets to the speed of light, the more of its kinetic energy becomes mass instead of motion. Matter can never exceed the speed of light because adding energy creates more mass instead of increasing an object's speed.

The meaning of simultaneous
The definition of the word *simultaneous* changes. Two events that are simultaneous to one observer may not be simultaneous to another who is moving.

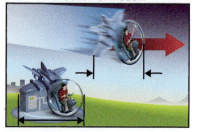
Space contracts in the direction of motion.

Lengths may contract
The length of an object measured by one person at rest will not be the same as the length measured by another person who is moving close to the speed of light. The object does not get smaller or larger, *space itself* gets smaller for an observer moving near the speed of light.

The speed of light paradox

Frame of reference
The theory of special relativity comes from thinking about light. Einstein thought about what light would look like if you could see it when it wasn't moving. Instead of making the light stop, Einstein thought about traveling beside it—going the same speed as light itself. Imagine you could move as fast as light and were traveling right next to the beam from a flashlight. If you looked over, you would see the light standing still, *relative to you*. A similar situation occurs when two people are driving on a road side-by-side at the same speed. The two people look at each other and appear to each other not to be moving relative to each other, because both are traveling at the same speed. The motion you see depends on your frame of reference.

The ways speeds normally add
Consider a person on a railroad train moving at a constant speed of 10 m/s toward you. If you are standing on the track, that person gets 10 meters closer to you every second. The person on the train throws a ball toward you at a speed of 10 m/s relative to their location. In one second, the ball moves forward on the train 10 meters. The train also moves toward you by 10 meters. Therefore, the ball moves toward you 20 meters in one second. The ball approaches *you* with a speed of *20 m/s* as far as you are concerned (Figure 18.17).

How you expect light to behave
Einstein considered the same experiment using light instead of a ball. Suppose the person on the train were to shine a flashlight toward you. You expect the light to approach you faster than it would if the flashlight were stationary. Prior to Einstein, everyone expected the light would come toward you at 3×10^8 m/s plus the speed of the train, like the ball.

The speed of light does not behave as expected
That is not what happens (Figure 18.18). The light comes toward you at a speed of 3×10^8 m/s *no matter how fast the train approaches you!* This experiment was done in 1887 by Albert A. Michelson and Edward W. Morley. They used Earth itself as the "train." Earth moves with an orbital speed of 29,800 m/s. Michelson and Morley measured the speed of light parallel and perpendicular to the Moon. They found the speed to be exactly the same! This result is not what they expected, and was confusing to everyone. Like all unexpected results, it forced people to rethink their assumptions. Einstein's theory of special relativity was the result, and it totally changed the way we understand space and time.

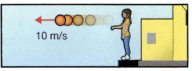

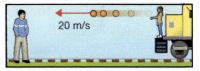

Figure 18.17: *A ball thrown from a moving train approaches you at the speed of the ball relative to the train plus the speed of the train relative to you.*

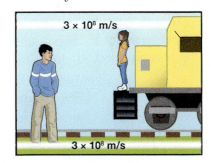

Figure 18.18: *The speed of light appears the same to all observers independent of their relative motion.*

Speed, time, and clocks

Einstein's thinking
With this new idea that the speed of light is constant to all observers, Einstein thought about what this meant for everything else in physics. One of the strangest results that came out of special relativity is that time is dilated, or stretched out, by the motion of an observer. His conclusion about the flow of time is as revolutionary as it is inescapable.

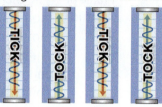

Figure 18.19: *A light clock measures time by measuring how long it takes a pulse of light to move between two parallel mirrors.*

A light clock on a spaceship
Einstein considered a clock that measures time by counting the trips made by a beam of light going back and forth between two mirrors (Figure 18.19). The clock is on a moving spaceship. A person standing next to the clock sees the light go back and forth, straight up and down direction. The time it takes to make one trip is the distance between the mirrors divided by the speed of light.

How the light clock appears on the ground
To someone watching from a frame of reference on the ground while standing still, the path of the light is *not* straight up and down. The light appears to zigzag because the mirrors move with the spaceship (Figure 18.20). The observer on the ground sees the light travel a longer path. This would not be a problem, *except that the speed of light must be the same to all observers, regardless of their motion.*

A stationary observer sees the light zigzag following the motion of the mirrors.

The paradox
Suppose it takes light one second to go between the mirrors. The speed of light must be the same for both people, yet the person on the ground sees the light move a longer distance! How can this be?

Time itself must be different for a moving object
The only way out of this dilemma is that *one second on the ground is not the same as one second on the spaceship*. The speed of light is the distance traveled divided by the time taken. If one second of "ship time" was longer than one second of "ground time," then the problem is resolved. Both people measure a speed of 3×10^8 m/s for the light in the clock. The difference is that one second of "ship time" is *longer* than one second of "ground time."

Figure 18.20: *A clock that counts ticks of light going back and forth in a spaceship. The pulse of a light clock that is moving relative to an observer traces out a triangular path.*

Time slows down close to the speed of light
The consequence of the speed of light being constant is that *time slows down for objects in motion, including people*. If you move fast enough, the change in the flow of time is enormous. For a spaceship traveling at 99.9 percent of the speed of light, 22 years pass on Earth for every year that passes on the ship. The closer the spaceship's speed is to the speed of light, the slower time flows onboard the ship compared to the elapsed time for unmoving observers outside the ship.

Consequences of time dilation

Proof of time dilation
The idea that moving clocks run slower is difficult to believe. Before Einstein's theory of special relativity, time was always considered a universal constant. One second was one second, no matter where you were or what you were doing. After Einstein, we realized this was not true. The rate of time passing for two people depends on their relative motion.

Atomic clocks
One of the most direct measurements of this effect was done in the early 1970s by synchronizing two precise atomic clocks. One was put on a plane and flown around the world, the other was left on the ground. When the flying clock returned home, the clocks were compared. The clock on the plane measured less time than the clock on the ground. The difference agreed precisely with special relativity.

The frequency of light depends on relative motion
Because light is a wave, it has a frequency. Anything that has a frequency acts like a clock because you can count cycles. With this in mind, let's go back to the flashlight on the railroad track. The person moving on the train has a red flashlight making 462×10^{12} waves per second. You see the same 462×10^{12} waves of light in a shorter amount of time because your "second" is shorter than the second of the flashlight. If the train is moving at 70 percent of the speed of light, the waves arrive at your eye at a rate of 630×10^{12} waves per second. This is not red light, it is *blue light*. The effect is more dramatic than the Doppler shift for sound waves. The frequency of light becomes more blue if an object is approaching you and more red if the object is moving away from you.

Discovery of the expanding universe
The technological advances of the last 75 years have allowed astronomers to build larger and larger telescopes. Larger telescopes can see fainter objects because they can collect more light. This has allowed astronomers to see objects much farther away than anything seen before in the thousands of years humans have been observing the night sky. Astronomers discovered that the farther away they looked, the more red the light became. The only possible explanation is that distant galaxies are all moving away from each other. If galaxies are moving away from each other, they must have been closer together in the past. In fact, we now believe the entire universe was once quite small, possibly smaller than a single atom. About 13 billion years ago, the universe came into existence and has been expanding and cooling ever since.

The twin paradox

A well-known thought experiment regarding time dilation is known as the twin paradox. The story goes like this: Two twins are born on Earth. They grow up to young adults and one of the twins becomes an astronaut. The astronaut twin goes on a mission into space. The space ship in which the twin travels moves at a velocity near the speed of light. Because of traveling at this high speed, the clocks on the ship, including the twin's biological clock, run much slower than the clocks on Earth.

Upon returning from a trip that was only a few years by the ship clocks, the traveling twin learns that the twin that stayed behind is much older.

CHAPTER 18 WAVE PROPERTIES OF LIGHT

The equivalence of energy and mass

Relationship between mass and energy

The equation $E = mc^2$ is probably the most recognized symbol of physics. This equation tells us that matter and energy are really two forms of the same thing. If you put enough energy in one place, you can create matter. If you make matter disappear, you get energy. The law of conservation of energy becomes a law of mass-energy conservation. The amount of energy it takes to create a kilogram of matter is calculated using Einstein's formula.

EINSTEIN'S MASS-ENERGY FORMULA

Energy (J) — $E = mc^2$ — Speed of light (m/s)

Mass (kg)

Why matter cannot travel as fast as light

If a particle of matter is as rest, it has a total amount of energy equal to its **rest energy**. The rest energy is the rest mass (m_0) of the particle multiplied by the speed of light squared ($E_{rest} = m_0 c^2$). If work is done to a particle by applying force, the energy of the particle increases. At speeds that are far from the speed of light, all the work done increases the kinetic energy of the particle. As the speed approaches the speed of light, however, the work does two things. Some of the work goes to increasing the speed of the particle. *Some of the work goes to increasing the particle's mass.* The closer the speed gets to the speed of light, the larger the proportion of work that goes to increasing mass. It would take an infinite amount of work to accelerate a particle to the speed of light, because at the speed of light the mass of a particle also becomes infinite. In physics, *infinity* has a well-defined meaning. It means *larger than anything possible*.

Einstein's thinking

Although Einstein's argument is beyond the scope of this book, he was able to deduce the equivalent of mass and energy by thinking about the momentum of two particles moving near the speed of light. Remember from Chapter 12, momentum involves mass and speed. Kinetic energy also involves mass and speed. The speed of light must be the same for all observers regardless of their relative motion. Energy and momentum must be conserved. The only way to resolve these two constraints is for mass to increase as the speed of an object gets near the speed of light. The increase in mass comes from energy.

Calculating the equivalents of mass and energy

A nuclear reactor converts 0.7 percent of the mass of uranium to energy. If the reactor uses 100 kg of uranium in a year, how much energy is released?

One gallon of gasoline releases 1.3×10^8 joules. How many gallons of gasoline does it take to release the same energy as the uranium?

1. You are asked for energy in joules and gallons of gas.
2. You are given the mass, a percent converted to energy, and joules per gallon for gasoline.
3. Einstein's formula: $E = mc^2$
4. The amount of mass converted to energy is
   $m = 0.007 \times 100$ kg = 0.7 kg
   Energy released is
   $E = (0.7 \text{ kg})(3 \times 10^8 \text{ m/s})^2$
   $= 6.3 \times 10^{16}$ J

To calculate the equivalent in gasoline divide by the energy per gallon:

$$N = \frac{6.3 \times 10^{16} \text{ J}}{1.3 \times 10^8 \text{ J/gal}}$$

$= 4.8 \times 10^8$ gal

A single kilogram of uranium releases as much energy as 480 million gallons of gasoline.

Simultaneity

The meaning of "simultaneous"
When we say that two events are simultaneous, we mean they happen at the same time. Since time is not constant for all observers, whether two events are simultaneous depends on the relative motion of the observers. In special relativity, simultaneity is defined by the time it takes light to get from one place to another.

Two events that appear simultaneous
The example that Einstein gave to help explain this concept starts with two lightning strikes hitting the front and back of a moving train. Imagine you are watching the train from a distance. You are the same distance from the front and back of the train when the lightning strikes (Figure 18.21). You see the two bolts of lightning hit the train at the same time. To you, the two events, which are the lightning strikes, are *simultaneous* because it takes the same amount of time for light from either event to reach you.

The events are *not* simultaneous to a moving observer
For a person sitting on the train, however, it is a different situation. Suppose the person is in the center of the train. If the train were at rest, the person would also observe two simultaneous lightning strikes. But, the train *is* moving and as a result, the two lightning strikes are *not* simultaneous to the person on the train. The person on the train sees the light hit the front of the train first, and then the back of the train afterward. Because the velocity of the train has no effect on the speed of light, the light from both lightning strikes travels at the same speed. But the person is moving *toward* the point where lightning struck the front of the train. Light from the front of the train has a shorter distance to travel before reaching the person. The person is moving *away* from the point where lightning struck the rear of the train. Light from the rear has a longer distance to go to reach the person.

The definition of *simultaneous* in special relativity
No information can travel faster than the speed of light. This means the *definition of simultaneous is when light from two events reaches the observer at the same time*. Einstein's example shows that two events that are simultaneous to one observer may not be simultaneous to another observer moving relative to the first. This effect is real and not just a trick of language. In collisions between subatomic particles, whether or not two events are simultaneous often affects the outcome of experiments.

Simultaneity depends on the relative motion of your frame of reference.

Figure 18.21: *The two lightning strikes are simultaneous to an observer at rest, but an observer on the train sees the lightning strike the front of the train first.*

Chapter 18 Connection

Holography

The stunning 3-D images that are produced by a hologram are captivating. A normal picture looks the same from any angle. A hologram looks different from different angles, just like a real scene does. A well-made hologram appears to have depth and perspective as if the actual three-dimensional scene was embedded in the picture. In many ways, the three-dimensional scene *is* embedded in the hologram, except it is done with the *phase* of light and not with solid objects. As amazing as holograms are, they are not quite up to what science fiction special effects make them! There are limits to what a hologram can do.

Figure 18.22: *The brightness, level of detail, front-to-back order, and size give you cues that your brain interprets, placing the rock formation in front of the mountain range.*

How do we see 3-D?

Shading and vanishing point An artist can create the sensation of depth in a picture. Things farther away can be drawn smaller and light can be shaded differently on different sides of objects. By making things change size as they get "farther away" in the picture and by using color and shading, a good artist can make a flat picture look like a three-dimensional scene as viewed from one place. A photograph creates the illusion of depth in a similar way. Differences in size, shape, and shading capture the sensation of depth (Figure 18.22).

Figure 18.23: *An early 3-D stereo viewer shows different images from a celluloid photograph to each eye, simulating the effect of stereo vision.*

Two views A true 3-D scene looks different when seen from different angles. There have been several inventions that recreate aspects of 3-D images. Most are based on the fact that we have two eyes and we see things simultaneously from two different angles. Your left eye has a view of the world, and your right eye has the same view, but from a slightly different angle. Your brain combines the two slightly different images from each eye to see 3-D. Your earliest experience with 3-D images may have come from a *stereo viewer* where you put flat round disks into the top and looked through the two eyepieces (Figure 18.23). As you clicked through the images on the disk, you saw pictures with amazing depth. The stereo viewer is derived from a technique of taking two pictures of the same scene with two cameras spaced the same distance apart as our eyes. The illusion of 3-D from a stereo viewer is limited, however, because it only shows a scene from a single perspective. You cannot "move around" the image to see objects from different sides, like you can in a real 3-D scene such as a model.

394 UNIT 6 LIGHT AND OPTICS

Chapter 18 Connection

A real object is different than a picture

Wave fronts A flat object gives off a flat wave front and you see the same image from every angle. A real object gives off light in a wave front that is shaped like the actual object. For example, the wave front of light reflecting from a vase is similar in shape to the vase itself (Figure 18.24). Depending on the angle, you see different parts of this wave front. The light from the left side of the vase travels out towards the left. People on the left side of the vase can see its left side. Those on the right cannot see the left side because they do not receive that part of the wave front.

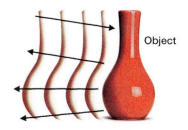

Figure 18.24: *The shape of the wave from a 3-D object is similar to the shape of the object.*

How a hologram works A hologram duplicates the three-dimensional shape of the wave front that is coming from the real object. A hologram is made by using interference of waves to create constructive and destructive interference that makes the wave look different from different angles, similar to a diffraction grating. Even though a hologram itself is flat, when light strikes a hologram it bounces off with the same wave front as would be coming from a real object (Figure 18.25). When you see the wave front, your brain thinks that it is looking at a real object. You can change your angle and see the image that you would see if you were looking at the real object. You can move over to the left side of the wave front and see what people on the left would have seen.

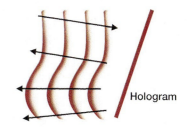

Figure 18.25: *A hologram looks three dimensional because it recreates the original shape of the wave front from the original object.*

Making a hologram To make a hologram you need to use the interference from a beam of light that has been divided into two beams. Half of the beam bounces off the object that will be in the hologram. This is the *image beam*. The other half of the beam goes right to the film. This is the *reference beam*. The interference of the image beam and the reference beam creates an interference pattern on the film (Figure 18.26). When the film is developed it looks nothing like the original object. That is because the film recorded the interference of the two beams, not the image itself. To recreate the image, you must shine light on the hologram from the same direction as the original reference beam. The light that bounces off the hologram recreates the original image beam, including the depth and three-dimensional sensation. The hologram looks different from different angles because you see a different interference pattern at different angles, just as you do with a diffraction grating.

Figure 18.26: *The making of a hologram.*

Chapter 18 Assessment

Vocabulary

Select the correct term to complete the sentences.

X-rays	spectrum	microwaves
index of refraction	electromagnetic wave	spectrometer
gamma rays	radio wave	transmission axis
diffraction grating	special relativity	polarization
polarizer	destructive interference	infrared
ultraviolet	rest energy	visible light
speed of light	time dilation	
wavelength	constructive	

1. Energy waves with a wavelength less than 10 trillionths of a meter belong to the part of the electromagnetic spectrum identified as _____.

2. Wavelengths of light between 700 and 400 nanometers comprise the _____ portion of the electromagnetic spectrum.

3. The entire range of light to which a physicist refers is called the electromagnetic _____.

4. The ratio of the speed of light to the frequency equals the _____ of light.

5. An electromagnetic wave with the lowest possible energy would be classified as a(n) _____.

6. Waves that are strongly absorbed by the calcium of our bones are _____.

7. Waves used to transmit cell phone messages as well as heat food are the _____.

8. A wave created by the periodic oscillation of electric or magnetic fields is called a(n) _____.

9. Ozone protects Earth's inhabitants by absorbing most _____ waves that might otherwise cause sunburn, skin cancer, and cataracts.

10. The energy waves we feel as heat are the part of the electromagnetic spectrum identified as _____ waves.

11. The letter c represents a special quantity in physics, the _____.

12. The ratio of the speed of light in a material to the speed of light in a vacuum is equal to the _____ for the transmitting material.

13. When two light waves meet in a medium in phase, the resulting light is brighter due to _____ interference.

14. A(n) _____ is a device used to measure the wavelengths of light.

15. A series of thin, parallel grooves on glass used to produce an interference pattern of light is called a(n) _____

16. A light wave on which the vibration of the electric field is in a single plane has experienced _____.

17. The plane in which a polarizer allows the electric part of a passing electromagnetic to vibrate is called the _____.

18. _____ occur(s) at the point where two light waves meet out of phase.

19. A partially-transparent material with the ability to restrict the vibrational orientation of an electromagnetic wave may be called a(n) _____.

20. The concept that time passes more slowly for an object in motion than an object at rest is known as _____.

21. The total amount of energy contained in a motionless particle can be referred to as its _____.

22. The theory that describes the relationship between matter, energy, time, and space at speeds close to the speed of light is Einstein's theory of _____.

Concept review

1. Give an example of an energy field.
2. Describe an electromagnetic wave.
3. List the different types of electromagnetic energy, from highest energy to lowest energy.
4. Which color of visible light has the highest frequency? Which has the lowest frequency?
5. True or false: All electromagnetic waves travel at the same speed in a vacuum.
6. Use the formula $c = f\lambda$ and write equations to solve for f and for λ.
7. Explain how microwaves cook food.
8. We often say that light slows down when it goes through glass but speeds up again when it leaves the glass. Explain how the light speeds back up.
9. Why must the frequency of light be the same regardless of the material through which it is traveling?
10. Explain how Thomas Young demonstrated the wave nature of light.
11. Where is the first-order bright spot seen when light goes through a diffraction grating?
12. Why is a rainbow seen when white light goes through a diffraction grating?
13. What does it mean to say a wave is polarized?
14. Explain what a polarizer does to unpolarized light.
15. Why does an LCD display use so much energy?
16. How was time dilation proven?
17. Describe the "twin paradox."
18. Arnold the astronaut is being paid for the time he spends in space. After a long voyage traveling at speeds approaching the speed of light, he returns to Earth. What is his reaction when he opens his pay envelope? Give an explanation for your answer.

Problems

1. A red keychain laser has a wavelength of 650 nm (6.50×10^{-7} m). Calculate the frequency of the laser.
2. Calculate the frequency of a radio wave with a wavelength of 1.5 m.
3. An alien is able to tune his visor so that he can see infrared waves ($\lambda = 1 \times 10^{-5}$ m) and ultraviolet waves ($\lambda = 1 \times 10^{-8}$ m). What are the frequency settings for this device?
4. Blue light has a frequency of 6.40×10^{14} Hz. When it is traveling through Lucite, its velocity is 2×10^{8} m/s. What is its wavelength?
5. The index of refraction (n) is found by dividing the speed of light in a vacuum (c) by the speed of light in a material (v). If light travels at a velocity of 1.239×10^{8} m/s in a diamond, calculate its index of refraction.
6. Zircon has an index of refraction of 1.92. What is the speed of light when it is traveling through zircon?
7. A fiber-optic cable is made of flint glass with an index of refraction of 1.66. If this cable is 10,000 m long, how much time would be required for light to pass through its length?
8. When light passes through two slits in a diffraction grating that are separated by a measured distance (d), a bright spot will form on a screen located a measured distance (L) from the grating. The first-order bright spot will form a distance (x) from the central bright spot. This distance is a function of the wavelength (λ) of the light. The ratio is $\frac{\lambda}{d} = \frac{x}{L}$. Algebra changes this to $\lambda = \frac{dx}{L}$. Light shines through a diffraction grating with $d = 2.5 \times 10^{-4}$ m. Dots appear on a screen 2.0 m away and the distance to the first-order bright spot (x) is 0.25 m. What is the wavelength of the light?
9. Calculate the maximum number of joules of energy contained in a stationary 4.0-kilogram mass.

Applying your knowledge

1. When white light is observed through a diffraction grating, the light source is visible in the center of two visible spectra to the right and left of the white light. Which color of the visible spectrum will appear closer to the white central spot: red or violet? Why?

2. Two flashlights do not produce an interference pattern when they simultaneously illuminate a screen. Research the difference between coherent and incoherent light sources to explain why this is so.

3. Research colors that occur in nature due to interference rather than absorption and reflection.

4. In some movie theaters, polarized light is used to produce 3-D movies. After doing research on the production of these movies, explain how the 3-D illusion is created.

UNIT 7 ELECTRICITY AND MAGNETISM

CHAPTER 19

Electricity

Objectives:

By the end of this chapter you should be able to:

✔ Describe the difference between current and voltage.
✔ Describe the connection between voltage, current, energy, and power.
✔ Describe the function of a battery in a circuit.
✔ Calculate the current in a circuit using Ohm's law.
✔ Draw and interpret a circuit diagram with wires, a battery, a bulb, and a switch.
✔ Measure current, voltage, and resistance with a multimeter.
✔ Give examples and applications of conductors, insulators, and semiconductors.

Key Questions:

- What is electricity, and how does an electric circuit work?
- What are current and voltage and how are they measured?
- How are voltage, current, and resistance related?

Vocabulary

ammeter	electric circuit	electrical symbols	open circuit	short circuit
amperes (amps, A)	electric current	electricity	potentiometer	switch
battery	electrical conductivity	multimeter	resistance	volt (V)
circuit diagram	electrical conductor	ohm (Ω)	resistor	voltage
closed circuit	electrical insulator	Ohm's law	semiconductor	wire

CHAPTER 19 ELECTRICITY

19.1 Electric Circuits

Imagine your life *without* electricity. There would be no TVs, computers, refrigerators, or light bulbs. All of the products that are made by machines—from clothes to newspapers—would have to be made using muscle power, water power, or wind. The use of electricity has become so routine that many of us never stop to think about what happens when we switch on a light or turn on a motor. This section is about electricity and electrical circuits. Electric circuits, usually made of wires, are how we connect and control electricity and devices that use electricity.

Electricity

What is electricity? In common use, the word **electricity** refers to the presence of **electric current** in wires, motors, light bulbs, and other devices. Electricity is usually invisible and is a form of energy that comes from the motion of tiny particles inside and between atoms. This chapter and the next will introduce the practical use of electricity. The electrons, which are the moving particles responsible for most electric current, will be discussed in Chapter 21.

Electric current Electric current is similar to a current of water, but electric current flows in solid metal wires so it is not visible. A good way to think about electric current is in terms of its ability to carry energy and power as it flows. An electric current can do work just as a current of water can. For example, a waterwheel turns when a current of water exerts a force on it (Figure 19.1). A waterwheel can be connected to a machine such as a loom for making cloth, or to a millstone for grinding wheat into flour. Before electricity was available, waterwheels were used for many machines. Today, the same tasks are done using energy from electric current. Look around right now and you can probably see wires carrying electric current into a house or a building or a device that does work for you using electricity.

Electricity can be powerful and dangerous Electric current can carry a lot of power. A five-horsepower electric motor the size of a basketball can do as much work as five big horses or 15 strong people. A 1.5-horsepower electric saw can cut wood much faster than a person with a hand saw (Figure 19.2). Because it carries so much power, electric current can also be dangerous. Touching a live *high-voltage* electric wire can do significant damage to your body, or even be fatal. The more you know about electricity, the easier it is to use it safely.

Figure 19.1: *A waterwheel uses a current of water to turn a wheel and do useful work.*

Hand saw cuts a board in 60 seconds

Electric saw cuts a board in 10 seconds

Figure 19.2: *Electricity uses an electric current to power electric motors.*

Electric circuits

Electricity travels in circuits
An **electric circuit** is a conducting path that connects electrical devices and through which electric current flows. A good example of an electric circuit is the one in an electric toaster. Bread is toasted by a heating element that converts electrical energy to heat. The circuit has a switch that turns on when the lever on the side of the toaster is pulled down. When the switch is on, electric current flows in one prong of the plug from the socket in the wall, through the toaster, and back out the other prong of the plug.

A circuit of pipes distributes water through a house.

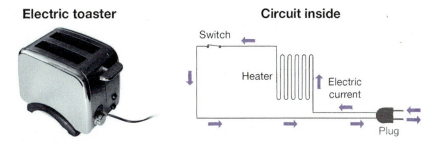

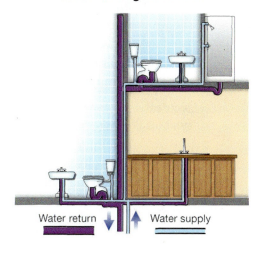

Wires are like pipes for electricity
Wires in electric circuits are similar in some ways to pipes and hoses that carry water (Figure 19.3). Wires act like pipes for electric current. Current flows into the house on the supply wire and out on the return wire. The big difference between wires and water pipes is that you cannot get electricity to leave a wire the way water leaves a pipe. If you cut a water pipe, the water comes out. If you cut a wire, the electricity immediately stops flowing.

A circuit of wires distributes electric current through a house.

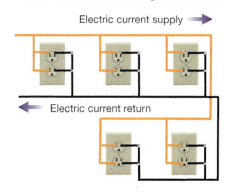

Examples of circuits in nature
Circuits are not confined to appliances, wires, and devices built by people. People's first experience with electricity was in the natural world. Some examples of natural circuits are as follows.

- The nerves in your body are an electrical circuit connecting muscles to messages from the brain.
- The tail of an electric eel makes an electric circuit when it stuns a fish with a jolt of electricity.
- Earth makes a giant circuit when lightning carries electric current between the clouds and the ground.

Figure 19.3: *In a house or other building, we use pipes to carry water and wires to carry the flow of electric current.*

Circuit diagrams and electrical symbols

Circuit diagrams Circuits are made up of wires and electrical components such as *batteries*, *light bulbs*, *motors*, and *switches*. When designing a circuit, it is often necessary to draw the arrangement of the electrical components. This is most easily done with a **circuit diagram**. When drawing a circuit diagram, symbols are used to represent each part of the circuit. These **electrical symbols** are quicker and easier to draw than realistic pictures of the components.

Electrical symbols A circuit diagram is a shorthand method of describing a real circuit. The electric symbols used in circuit diagrams are standard, so anyone familiar with electricity can interpret them. Figure 19.4 shows some common electric components and their electrical symbols. The picture below shows an actual circuit. See if you can match the symbols with each part of the actual circuit.

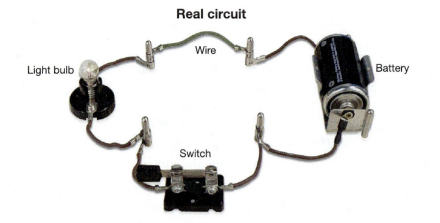

Real circuit

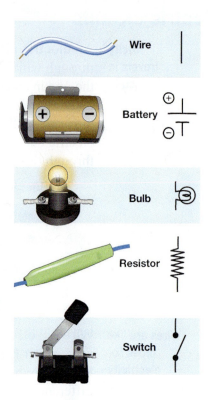

Figure 19.4: *These electrical symbols are used when drawing circuit diagrams.*

Resistors A **resistor** is an electrical component that uses the energy carried by electric current. In many circuit diagrams, any electrical device that uses energy is shown with a resistor symbol. A light bulb, heating element, speaker, or motor can be represented with a resistor symbol. When you analyze a circuit, many electrical devices may be treated as resistors for the purpose of determining the electrical current flowing in the circuit.

Open and closed circuits

Batteries The positive end of a battery is a source of electric current. The negative end of the battery is a return of electric current. When a battery is connected to an electric circuit, current flows out of the positive end, through the circuit, and back to the negative end.

Open and closed circuits Unlike water, electric current only flows through a **closed circuit**. A closed circuit is a complete and unbroken conducting path between the source of the current and the return of the current. A circuit with a switch turned to the "off" position, or a circuit with any break in it, is called an **open circuit**. Electricity cannot travel through an open circuit.

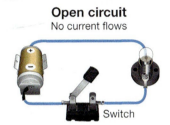
Open circuit
No current flows
Switch

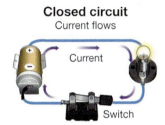
Closed circuit
Current flows
Current
Switch

Switches **Switches** are used to turn electric current on and off. Turning the switch off creates an open circuit by making a break in the wire. The break stops the flow of current because electricity cannot normally travel through air.

Breaks in circuits A common problem found in circuits is that an unintentional break occurs. If there are any breaks, the circuit is open and electric current cannot travel through the circuit. If you look inside an incandescent light bulb that has burned out, you will see that the thin wire that glows inside the bulb, called the *filament*, is broken. This creates an open circuit and the bulb will not light.

Short circuits A **short circuit** is not the same as either an open or closed circuit. A short circuit is usually an accidental extra path through which a current flows. A short circuit provides an easy, but sometimes dangerous, shortcut through which current travels to avoid other electrical components in the circuit. Short circuits will be covered in more detail in a later section when we talk about *parallel* and *series* circuits.

Why electricity does not leak from open circuits

Electric current is usually the flow of electrons around atoms in a metal, like copper. In copper and other conductors, some of the atoms split up into free electrons and ions. The electrons are tiny and can move to create an electric current. Atoms and ions are mostly empty space and there is plenty of room for the electrons to flow around them.

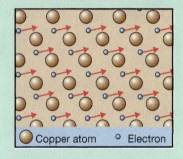
Copper atom   Electron

Electric current does not leak from a wire because very strong forces maintain an exact balance between the density of electrons and the density of copper ions. One electron can move away from its ion only if another electron comes in to keep the balance. If any did leak out, there would be more ions than electrons. The forces between ions and electrons are so great that this does not happen except in extreme circumstances.

CHAPTER 19 ELECTRICITY

19.2 Current and Voltage

Current and voltage are the two *most important* concepts to understanding electricity. Current is what actually flows through wires, carries energy, and does work. Like water, electric current only flows when there is a difference in energy. Water current flows downhill from high to low energy. The difference in energy that makes water flow is measured in height. Electric current also flows because of a difference in energy. *Voltage* measures the difference in energy between two places in a circuit. Current flows in response to differences in voltage just like water flows in response to differences in height. Current is what flows and does work. Voltage differences are what make current flow.

Current

Measuring electric current
As discussed in the previous section, electricity is the flow of electric current. The flow of water in a hose might be measured in gallons per minute. Electric current is measured in units called **amperes**, or **amps (A)** for short. The unit is named in honor of Andre-Marie Ampere (1775–1836), a French physicist who studied electricity and magnetism.

Current flows from positive to negative
The direction of electric current is from positive to negative. If you examine a battery, you will find one end marked with a plus, typically the end with the raised metal dimple. The other end of a battery is marked negative.

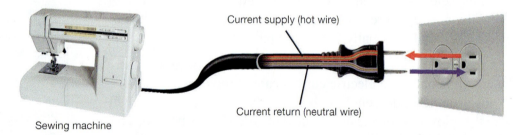

Current in equals current out
The amount of electric current entering any electrical device always equals the amount leaving the device. When you plug an appliance into the wall, there are two wires that carry current. The current source is on one wire, the hot wire, and the current return is on the other wire, the neutral wire. You can think about this rule conceptually like marbles flowing through a tube. When you push one in, one comes out. The rate at which the marbles flow in equals the rate at which marbles flow out (Figure 19.5).

Conventional Current

The direction of current is from the positive to the negative end of a battery. This definition of current is called *conventional current* and was proposed by Ben Franklin in the 1700s. Scientists later discovered that the particles that carry electricity in a wire actually travel from negative to positive. However, we still use Franklin's definition today. You will learn more about the particles that make current later in this unit.

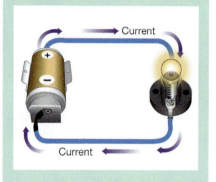

Figure 19.5: *Marbles can flow through a tube even though the number of marbles in the tube stays the same. When one marble goes in, another comes out.*

UNIT 7 ELECTRICITY AND MAGNETISM

Voltage

Energy and voltage

Voltage is a measure of electric *potential energy*, just like height is a measure of gravitational potential energy. When one point in a circuit has a higher voltage, that point is at a higher potential energy than a point at lower voltage. **Voltage** is measured in **volts (V)**. Like other forms of potential energy, voltage *differences* represent energy that can be used to do work. For electricity, the way to extract the energy is to let the voltage cause current to flow through a circuit.

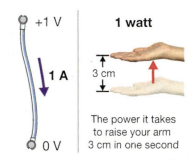

Figure 19.6: *The power equivalent of one amp at one volt.*

What voltage means

A voltage difference of 1 volt means 1 amp of current does 1 joule of work in 1 second. Since 1 joule per second is a watt, you can interpret voltage as measuring the available electrical power per amp of current (Figure 19.6). The average voltage in your home electrical system is 120 volts, which means each amp of current carries 120 watts of power. The higher the voltage, the more power is carried by each amp of electric current.

Using a meter to measure voltage

Humans cannot see or feel small voltages, so an electric meter is necessary to find the voltage in a circuit. The most common type of meter is a **multimeter**, which can measure voltage or current. To measure voltage, the probes attached to the meter are touched to two places in a circuit. The meter reads positive voltage when the red (positive) probe is at a higher voltage than the black (negative) probe.

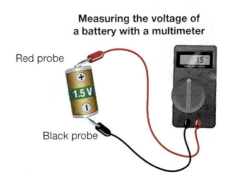

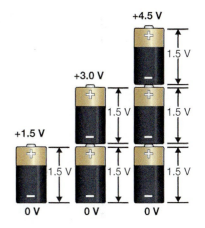

Figure 19.7: *The positive end of a 1.5-volt battery is 1.5 volts higher than the negative end. If you connect batteries positive-to-negative, each battery adds 1.5 volts to the total. Three batteries make 4.5 volts. Each unit of current coming out of the positive end of the three-battery stack has 4.5 joules of energy.*

Meters measure voltage difference

A multimeter measures the *difference* in voltage between two places in a circuit. If the meter is connected across a battery, it will read the voltage difference between the positive and negative terminals of the battery. *All measurements of voltage are actually measurements of voltage difference.* For example, a reading of +1.5 volts means the red probe is touching a point in the circuit that is 1.5 volts higher in potential energy than the point the black probe is touching. In fact, you can make larger voltage differences by stacking batteries (Figure 19.7). You can also make a multimeter voltage negative by touching the red lead to the *lower* voltage.

Batteries

Batteries A **battery** uses chemical energy to create a voltage difference between its two terminals. When the current leaves a battery, it carries power. If the battery has a voltage of 1.5 volts, then 1 amp of current carries 1.5 joules per second, or 1.5 watts of power. The current can give up its power when it passes through an electrical device such as a light bulb. Every second that a bulb is lit, some energy is taken from the current and is transformed into light and heat energy. The current returns to the battery where it gets more energy.

A pump is like a battery because it brings water back to a higher energy.

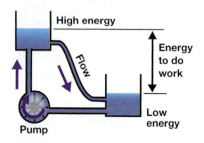

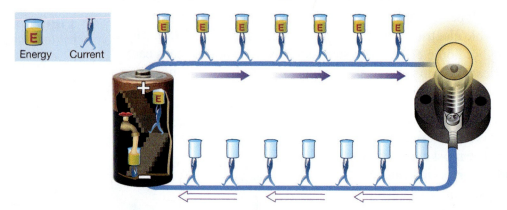

Figure 19.8: *A battery raises electric current back up to higher energy, similar to the way a pump pushes water back up to high energy so it can do work again.*

Batteries are like pumps Two water tanks connected with a pump make a good analogy for a battery in a circuit (Figure 19.8). The pump gives potential energy to the water as it lifts it up to the higher tank. As the water flows down, its potential energy is converted into kinetic energy. In a battery, chemical reactions provide the energy to pump the current from low voltage to high voltage. The current can then flow back to low voltage through a circuit and use its energy to turn motors and light bulbs.

Battery voltage A fully-charged battery adds energy proportional to its voltage. The positive end of a 1.5-volt battery is 1.5 joules higher in energy than the negative end. That means every amp of current that leaves the positive end has 1.5 joules more energy than it has after traveling through the circuit. The voltage of a battery depends on how the battery is constructed and what chemicals it uses. Nickel-cadmium (NiCd) batteries are 1.2 volts each. Lead-acid batteries, like the one in a car, are usually 12 volts. Different voltages can also be made by combining multiple batteries.

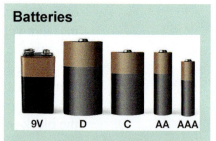

Batteries

Batteries come in different sizes as well as different voltages. One D-size battery stores more energy than a AAA-size battery, even though the voltage is the same. That means a D battery can keep 1 amp of current flowing for much longer than a AAA battery can.

Measuring current in a circuit

Measuring current with a meter
Electric current can be measured with a multimeter. However, to measure current, the circuit must be set up differently than for measuring voltage. If you want to measure current in a circuit, you must make it flow *through* the meter. That usually means you must break the circuit somewhere and rearrange wires so the current passes through the meter. For example, the diagram below shows a circuit with a battery and bulb. The meter has been inserted into the circuit to measure current. If you trace the wires, the current comes out of the positive end of the battery, through the light bulb, *through the meter*, and back to the battery. The meter in the diagram measures 0.37 A of current. Some electric meters, called **ammeters**, are designed specifically to measure only current.

Circuit breakers

Wires can get dangerously hot if they carry more current than they are designed to carry. Electric circuits in your house contain *circuit breakers* that prevent more than the allowable 15 or 20 amps of current from flowing through the wires. A circuit breaker uses temperature-sensitive metal that expands with heat. When the current gets too high, the metal gets hot, bends, melts, and breaks the circuit.

Most overloads are caused by using too many electric appliances on the same circuit at one time. If the total current drawn from all of the appliances exceeds the rated current on the breaker, the circuit breaker trips and breaks the circuit before the wires get hot enough to cause a fire. If this happens, you must unplug some appliances before you reset the circuit breaker.

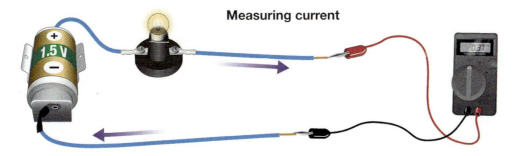

Measuring current

Setting up the meter
If you use a multimeter, you must set its dial to measure current. Multimeters can measure two different types of electric current, called alternating current (AC) and direct current (DC). You will learn about the difference between alternating and direct current in a later section. For circuits with light bulbs and batteries, you must set your meter to the direct current setting.

Be careful measuring current
The last important thing about measuring current is that the meter itself can be damaged if too much current passes through it. Your meter may contain a *circuit breaker* or *fuse*. Circuit breakers and fuses are two kinds of devices that protect circuits from too much current by creating a break in the circuit and stopping the current. If your meter does not work, a likely explanation is that the fuse may have blown from too much current and created an open circuit. A circuit breaker can be reset, but a blown fuse must be replaced.

19.3 Electrical Resistance and Ohm's Law

You can apply the same voltage to different circuits and different amounts of current will flow. For example, when you plug in a desk lamp, about one amp of current flows. If a hair dryer is plugged into the same outlet, with the same voltage, *10 amps* of current will flow. The amount of current that flows in a circuit is determined by the *resistance* of the circuit. Resistance is the subject of this section.

Electrical resistance

Current and resistance **Resistance** is a measure of how easily electric current moves through an object or electrical device. A device with low resistance, such as a copper wire, allows current to easily move through it. An object with high resistance, such as a rubber band, does not allow current to pass through it easily.

A water analogy The relationship between electric current and resistance can be compared to water flowing out of a bottle through an opening (Figure 19.9). If the opening is large, the resistance is low and lots of water flows out quickly. If the opening of the bottle is small, there is a lot of resistance and the water flow is slow.

Circuits The total amount of electrical resistance in a circuit determines the amount of current flowing in the circuit for a given voltage. The more resistance the circuit has, the less current that flows. Every device that uses electrical energy adds resistance to a circuit. For example, if you connect several light bulbs in a series as shown, the resistance in the circuit increases and the current decreases, making each bulb dimmer than a single bulb would be.

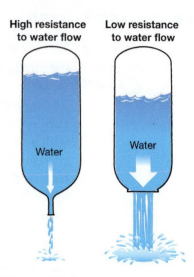

Figure 19.9: *Just as with water, only a small amount of electric current passes through the "opening" of a circuit element if its resistance is high. A large amount of current passes through the element if its resistance is low.*

One bulb
Single resistance
Full current

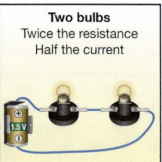

Two bulbs
Twice the resistance
Half the current

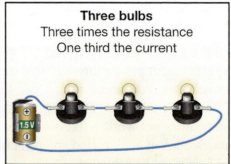

Three bulbs
Three times the resistance
One third the current

Measuring resistance

The ohm — Electrical resistance is measured in units called **ohms (Ω)**. This unit is abbreviated with the Greek letter *omega* (Ω). When you see Ω in a sentence in physics, think or read "ohms." For a given voltage, the greater the resistance, the lower the current. If a circuit has a resistance of one ohm, then a current of one ampere flows when a voltage of one volt is applied. As an example, a 100-watt light bulb has a resistance of 145 Ω. When the light bulb is connected to the 120-volt circuit in your house, a current of 0.83 amps flows.

Figure 19.10: *You can measure the resistance of a light bulb with a multimeter.*

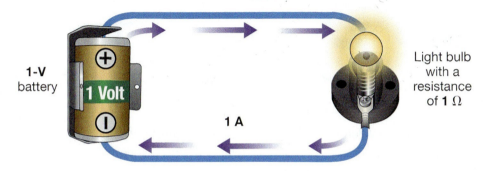

Resistance of wires — The wires used to connect circuits are made of metals such as copper or aluminum and have very low resistance. The resistance of wires is usually so low, less than 0.001 ohms, that you can ignore resistance from the wires compared with other devices in a circuit. Practically, this means the voltage in a wire is the same everywhere along the wire. The exception is when there are large currents. If the current is large, the resistance of wires may be important.

Current, voltage, and resistance — You can use a multimeter to measure the resistance of wires, light bulbs, and other devices (Figure 19.10). Set the dial on the meter to the resistance setting and touch the meter leads to each end of the device. The meter will display the resistance in ohms (Ω), kilohms (1,000 Ω), or megohms (1,000,000 Ω). Light bulbs and other electrical devices typically have resistances between a few ohms and a few hundred ohms. Electronic components and insulators may have resistances of kilohms (kΩ) or megohms (MΩ).

> **How resistance is measured**
>
> A multimeter measures resistance by passing a precise amount of current through an electrical device. The meter then measures the voltage it takes to cause the current to flow. The resistance is calculated from the voltage and current. The currents used to measure resistance are typically very small, 0.001 amps or less.

19.3 ELECTRICAL RESISTANCE AND OHM'S LAW

CHAPTER 19 ELECTRICITY

Ohm's law

Ohm's law The mathematical relationship that relates current, voltage, and resistance in a circuit is called **Ohm's law**. Ohm's law states that electric current is directly proportional to voltage and is inversely proportional to resistance. If you know the voltage and resistance in a circuit, you can calculate the current by dividing the voltage by the resistance.

OHM'S LAW

Current (amps, A) $I = \dfrac{V}{R}$ Voltage (volts, V)
Resistance (ohms, Ω)

Calculating the current flowing in a circuit

A light bulb with a resistance of 3 ohms is connected in a circuit that has a single 1.5-volt battery. Calculate the current that flows in the circuit. Assume the wires have negligible resistance.

Circuit

If you know.	Equation	Gives you.
voltage and resistance	$I = V \div R$	current (I)
current and resistance	$V = I \times R$	voltage (V)
voltage and current	$R = V \div I$	resistance (R)

The connection between current and voltage Ohm's law makes the connection between the voltage applied to a circuit and the amount of current that flows. Devices such as motors are designed with a specific resistance that allows them to draw the proper current when connected to the proper voltage. For example, a particular 1.5-V electric motor requires 0.75 amps. The motor is designed with a resistance of 2 Ω. When connected in a circuit with a 1.5-volt battery, the motor draws the right current (0.75 A = 1.5 V ÷ 2 Ω).

Diagram

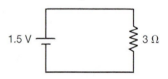

1. You are asked for the current.
2. You are given the voltage and resistance.
3. Ohm's law relates current, voltage, and resistance
4. Solve:
   $I = (1.5 \text{ V}) \div (3 \text{ Ω})$
   $= 1.5 \text{ A}$

The light bulb draws 1.5 amps of electric current.

Current through motor

$I = \dfrac{1.5 \text{ V}}{1 \text{ Ω}} = 1.5 \text{ A}$

Current through multimeter

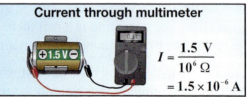

$I = \dfrac{1.5 \text{ V}}{10^6 \text{ Ω}}$
$= 1.5 \times 10^{-6} \text{ A}$

High resistance means low current When measuring voltage in a circuit, you do *not* want current to pass through the multimeter itself. Multimeters are designed to have high resistance for voltage measurement, typically 1 million ohms (1 MΩ) or more. According to Ohm's law, the current passing through the multimeter when measuring 1.5 V is only 1.5×10^{-6} A. This small extra current is not likely to affect the circuit.

The resistance of electrical devices

Resistance of common devices
The resistance of electrical devices ranges from very small (0.001 Ω) to very large (10^7 Ω). Each device is designed with a resistance that allows the right amount of current to flow when connected to the voltage and source for which the device was designed. For example, a 100-watt light bulb has a resistance of 145 ohms. When connected to 120 volts from a wall socket, 0.83 amps flows and the bulb lights (Figure 19.11).

Resistances match operating voltage
If you connect a 1.5 V battery to an ordinary 100-watt bulb, it will not light; there is not enough current. According to Ohm's law, 1.5 volts can push only 0.01 amps through 145 ohms of resistance. This amount of current is much too small to make the bulb light. A 100-watt bulb is designed with the appropriate resistance to draw the right amount of current at 120 volts. Most electrical devices are designed to operate correctly at a certain voltage.

The resistance of skin
Your skin has a fairly large resistance. A typical resistance of dry skin is 100,000 ohms or more. You can safely handle a 9-volt battery because the resistance of your skin is high. According to Ohm's law, 9 V ÷ 100,000 Ω is only 0.00009 amps. This is not enough current to be harmful. On average, nerves in the skin can feel a current of around 0.0005 amps. You can get a dangerous shock from 120 volts from a wall socket because 120 V divided by 100,000 Ω is 0.0012 amps, which is more than twice the amount you can feel.

Water lowers skin resistance
Wet skin has much lower resistance than dry skin. Because of the lower resistance, the same voltage will cause more current to pass through your body when your skin is wet. The combination of water and 120-volt electricity is especially dangerous because the high voltage and lower resistance make it possible for large, or possibly fatal, currents to flow.

Changing resistance
The resistance of many electrical devices varies with temperature and current. For example, a light bulb's resistance increases when there is more current through the bulb. This change occurs because the bulb gets hotter when more current passes through it. The resistance of many materials, including those in light bulbs, increases as temperature increases. A graph of current versus voltage for a light bulb shows a curve (Figure 19.12). A device with constant resistance would show a straight line on this graph.

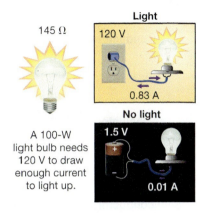

Figure 19.11: *The resistance of a light bulb is chosen so the bulb draws the correct current when connected to 120 volts from a wall socket. The bulb will not light if connected to a battery because there is not enough current.*

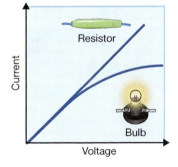

Figure 19.12: *The current versus voltage graph for a resistor is a straight line with a constant slope. The graph for a light bulb is curved with a decreasing slope because resistance increases with current.*

Conductors and insulators

Conductors Current passes very easily through metals, such as copper. A material such as copper is called an **electrical conductor** because it can *conduct*, or carry, electric current. The electrical resistance of wires made from conductors is very low. Most metals are good conductors.

Insulators Other materials, such as glass and plastic, do not allow current to flow easily through them. These materials are classified as **electrical insulators** because they insulate against or block the flow of current. Things made from insulators usually have very high resistance.

Semiconductors Some materials are in between conductors and insulators. These materials are named **semiconductors** because their ability to carry current falls between that of conductors and insulators. Computer chips, LEDs, and some types of lasers are made with semiconductors.

Electrical conductivity No material is a perfect conductor or insulator. Some amount of current will flow through any material if a voltage is applied. **Electrical conductivity** describes a material's ability to pass electric current. Materials with high conductivity, such as metals, allow current to flow easily and are good conductors. Materials with low conductivity block current from flowing and are insulators.

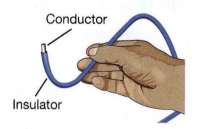

Figure 19.13: *A wire uses both conductors and insulators. The conductor carries the current. The insulator keeps the current from reaching you when you touch the wire.*

Electrical properties	Category	Example materials	
High conductivity / **Low resistance** ↑	Conductors	• silver • gold • copper	• aluminum • tungsten • iron
↕	Semiconductors	• carbon • germanium	• silicon
↓ **Low conductivity** / **High resistance**	Insulators	• air • paper • ice	• glass • rubber • plastic (most)

Applications of conductors and insulators Both conductors and insulators are necessary materials in human technology. For example, a wire has one or more conductors on the inside and an insulator on the outside (Figure 19.13). An electrical cable may have many conductors, each separated from the others by a thin layer of insulator.

Breakdown voltage

Even the best insulators will become conductors when the voltage gets high enough. For example, air is normally a good insulator. One centimeter of air will remain an insulator up to about 10,000 volts. When the voltage exceeds 10,000 V/cm (the breakdown voltage), atoms in the air split apart and release electrons that can carry current. Air becomes a conductor, as it does in a lightning strike.

Resistors

Resistors are used to control current
Resistors are electrical components that are designed to have a well-defined resistance that remains the same over a wide range of current. Resistors are used to control the amount of current in circuits. They are found in many common electronic devices such as computers, televisions, telephones, and stereos.

Fixed resistors
Fixed resistors have a resistance that cannot be changed. If you have ever looked at a circuit board inside a computer or other electrical device, you have seen fixed resistors. Resistors are small cylindrical or rectangular components. They can be colored or black, and have resistance values printed on them.

Potentiometers
Potentiometers are fixed resistors that can be adjusted to have a range of resistance. If you have ever turned a dimmer switch or volume control, you have used a potentiometer. When the resistance of a dimmer switch is increased, the current decreases, and the bulb is dimmed. Inside a potentiometer is a circular resistor and a sliding contact called a wiper, as shown below. The wiper moves when you turn the knob and is connected to a wire (B). The resistance between the wires at A and C always stays the same. As you turn the knob, the resistance between A and B changes. The resistance between B and C also changes, but in the opposite direction. A potentiometer, like this one, can have a resistance between 0 and 10 ohms, depending on the position of the wiper and which terminals are connected to the circuit.

Variable resistors
A variable resistor is a special configuration of a potentiometer where the wiper is directly connected to one end of the potentiometer. By turning the dial, the resistance changes between the two ends of the device because of the external connection. Variable resistors are used to vary the amount of current in a circuit.

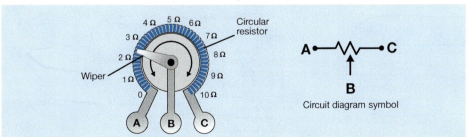

The inside of a potentiometer

Chapter 19 Connection

Hybrid Gas/Electric Cars

According to most specialists on the subject, world oil supplies are decreasing and the environmental impact of automobile pollution is increasing. The most promising near-term technology for reducing both problems is the hybrid gas/electric car. Hybrid cars and vehicles use a diesel- or gasoline-powered engine and an electric motor (Figures 19.14 and 19.15). Several models now on the road get gas mileage close to 60 miles per gallon, twice the fuel economy of ordinary cars.

Automotive technology

Electric cars — Electric motors can generate large forces very fast, are quiet, and produce no pollution. Despite these advantages, electric cars have not been widely accepted. The biggest problem is that one kilogram of gasoline yields more than 30 times the energy stored in one kilogram of batteries. To store enough energy to be practical, electric cars are heavy and have a range of only 50 to 100 miles before the batteries need to be recharged. Recharging is slow, and takes many hours.

Conventional gas-powered cars — The gasoline engine in a car is ideal for transportation in many ways. Gasoline is a compact fuel in that it contains a lot of energy in a relatively small and light form that is easy to store and transport. Some disadvantages are the pollution from refining oil into gasoline and a decreasing supply of oil.

Hybrid cars — Hybrid cars combine the advantages of gasoline as a fuel, with many of the advantages of electric power. In the most promising hybrids on the market today, a small, efficient gasoline engine is combined with a battery-powered electric motor. These two power sources work together to achieve good performance and excellent fuel efficiency.

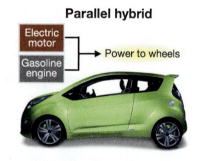

Figure 19.14: *A series hybrid like this bus has a diesel engine that powers a generator directly, which produces electricity to power an electric motor that moves the bus.*

Figure 19.15: *A parallel hybrid can be powered by either the gasoline engine or the electric motor, or by both. Today's hybrid cars are parallel hybrids and use advanced aerodynamics to increase efficiency.*

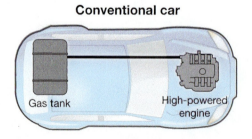
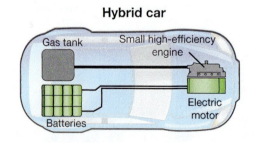

Chapter 19 Connection

Hybrid car technology

Conventional car engines — Ordinary gasoline combustion car engines are built with extra power capacity to provide rapid acceleration when passing or starting. Cruising at constant speed requires less than 15 percent of a typical engine's maximum power. Fuel efficiency suffers because large engines are inefficient at the low power levels used for 95 percent of driving. Conventional engine designs are a compromise between the high power needed for acceleration and the low power needed for normal driving.

Gas-engine use in a hybrid car — In a hybrid car, the small, high-performance gasoline engine is used for normal driving. The electric motor supplies extra power for fast starts and for rapid acceleration. The gas engine can be relatively powerful while producing ultra-low levels of emissions. One hybrid model has an engine that weighs a mere 124 pounds yet produces 67 horsepower.

Hybrid cars do not need to be "plugged in" overnight to be recharged like pure electric cars. The gasoline engine includes an electric generator that continuously recharges the batteries while you are driving.

Electric-motor use in a hybrid car — The high-efficiency electric motors in hybrids can be up to 90 percent efficient. At low speed and when idle, the gas engine shuts off and the electric motor takes over. Drivers have been surprised—and pleased—by the hybrid car's quietness when operating in full-electric mode.

Regenerative braking — One of the most innovative developments in hybrid technology is *regenerative braking*. Conventional braking systems dissipate a car's kinetic energy through friction, resulting in heat. In a hybrid car, however, the electric motor also works as a generator. Regenerative braking uses some of the car's kinetic energy to turn the motor into a generator that charges the batteries during slowing or stopping. This process recycles kinetic energy that would otherwise be lost.

Battery technology — The first hybrid cars used heavy lead-acid batteries, similar to the ones used in conventional cars, to start the engine. New lithium-ion batteries are lighter and smaller than lead-acid batteries. Hybrid cars will continue to improve and replace conventional cars as battery technology gets better.

Current and voltage in hybrid cars

Conventional car

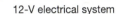

12-V electrical system

Hybrid car

96-V electrical system

The electrical system in conventional cars uses 12 volts. The battery is only used to start the engine, and is designed to supply high current of 50 amps or more for a short time.

Some hybrid cars use 96-volt electrical systems. At 96 V, each amp of current carries 96 watts of power instead of 12 watts like a conventional car. Hybrids typically use multiple banks of specialized batteries to provide the higher voltage. Unlike ordinary car batteries, hybrid car batteries are designed to supply steady current for long periods of time.

19.3 ELECTRICAL RESISTANCE AND OHM'S LAW

CHAPTER 19 ELECTRICITY

Chapter 19 Assessment

Vocabulary

Select the correct term to complete the sentences.

electricity	electric current	voltage
resistance	Ohm's law	battery
open	closed	short
switch	circuit diagram	electrical conductivity
potentiometer	wire	volt
electrical symbols	amperes (amps)	multimeter
ohms	resistor	ammeters
electrical conductors	semiconductors	conductor
electric circuit		

1. Anything which provides a complete path over which electricity can travel can be called a(n) _____.

2. A complete and unbroken path between a source of current and the return of current is referred to as a(n) _____ circuit.

3. A device that can be used to open and close a circuit is a(n) _____.

4. A path, generally created unintentionally, over which electric current can flow is called a(n) _____ circuit.

5. Symbols are used to represent various components of an electrical circuit when drawing a(n) _____.

6. The presence of electric current in a wire is referred to as _____.

7. A common electrical component that uses a specific amount of energy carried by electric current is known as a(n) _____.

8. If a break accidentally occurs in a complete conducting path, a(n) _____ circuit is created.

9. Simple drawings used to represent the components of a circuit are called _____.

10. Electrical energy flows through wires as _____.

11. The circuit component generally responsible for providing the current path is a(n) _____.

12. _____ is a measure of the electrical potential energy carried by the current in a circuit.

13. A device that is used to measure voltage, electrical current, and resistance is known as a(n) _____.

14. Units that are used to represent quantities of electric current are _____.

15. Meters designed specifically to measure only electrical current are called _____.

16. An example of a device that uses chemical energy to establish a voltage between two points is a(n) _____.

17. Voltage or potential difference in a circuit is measured in units called _____.

18. A common name for a variable resistor is a(n) _____.

19. Electrical resistance is measured in units of _____.

20. The mathematical relationship between current, resistance, and voltage in a circuit is given by _____.

21. Materials, such as copper and silver, through which electric current passes easily are referred to as _____.

22. Materials that are neither good conductors nor good insulators are called _____.

23. The opposition to the flow of electrical current through an object or electrical device is _____.

24. A material's ability to pass electric current through it easily is called _____.

Concept review

1. Explain the similarities and differences between electric and water current.
2. Why are electrical symbols used when making circuit diagrams?
3. Draw the electrical symbols for each of the following.
    a. resistor
    b. bulb
    c. battery
    d. switch
    e. wire
4. A circuit contains a battery, switch, and resistor. Draw its circuit diagram.
5. Explain the difference between an open circuit and a closed circuit.
6. Explain the function of a circuit breaker or fuse.
7. What provides the "push" to current in a circuit?
8. How is a battery's voltage related to the amount of energy it supplies to the current in a circuit?
9. You install two batteries in a flashlight such that their positive ends are facing each other. Will the flashlight work? Explain your answer.
10. What is the energy source for a battery?
11. Give an example of a material with high electrical resistance and one with low resistance.
12. Which three quantities in a circuit can be measured with a multimeter?
13. You are doing a classroom investigation with circuits. Your lab partner gathers the supplies for the investigation and you notice the multimeter has only one probe. Can you use this meter to measure current? Can you use it to measure voltage? Explain your answers.
14. A battery is connected to a light bulb, creating a simple circuit. What happens to the current in the circuit if
    a. the bulb is replaced with a bulb having a higher resistance?
    b. the bulb is replaced with a bulb having a lower resistance?
    c. the battery is replaced with a battery having a higher voltage?
15. List the unit in which each of the following is measured.
    a. current
    b. voltage
    c. resistance
16. According to Ohm's law, how is current related to resistance in a circuit? How is current related to voltage?
17. Why is it important to make sure your hands are dry before you plug wires into electrical outlets?
18. Explain why the current versus voltage graph is a straight line for a resistor but is curved for a light bulb.
19. What is the difference between a conductor and an insulator?
20. Identify each material as a conductor, a semiconductor, or an insulator.
    a. carbon
    b. gold
    c. glass
    d. copper
    e. air
21. If a material has a low resistance, does it have a high or low electrical conductivity?
22. Explain why electrical wires in homes are covered with a layer of plastic.
23. How is a potentiometer different from a fixed resistor?
24. When you measure the voltage and current in a simple circuit of a battery, wires, and a resistor, you discover that the current is lower than you wanted. In order to raise the current, you exchange the resistor for a new one. Should the new resistor have a higher value or a lower value than the original resistor?
25. You are given a 20 Ω resistor and a 100 Ω resistor. Which resistor allows current to flow more easily in a circuit?

Problems

1. A hair dryer draws a current of 10 A when it is plugged into a 120-V circuit. What is the resistance of the hair dryer?
2. A light bulb with a resistance of 2 Ω is connected to a 1.5-V battery. What is the current in the circuit?
3. A battery supplies 0.25 A of current to a motor with a resistance of 48 Ω. What is the voltage of the battery?
4. A portable CD player has a resistance of 15 Ω and requires 0.3 A of current. How many 1.5-V batteries must be used in the CD player?

5. A flashlight bulb has a resistance of approximately 6 Ω. It works in a flashlight with two AA alkaline batteries. How much current does the bulb draw?

6. Household circuits in the United States commonly run on 120 V of electricity. Circuit breakers are frequently installed to open a circuit if it draws more than 15 A of current. What is the minimum amount of resistance that must be present in the circuit to prevent the circuit breaker from activating?

7. Julie's hair dryer has a resistance of 9 Ω when first turned on.
   a. How much current does the hair dryer draw from the 120-V outlet in her bathroom?
   b. What happens to the resistance as the hair dryer continues to run for a long time?

8. Which device has the higher resistance?
   a. A 100-watt light bulb connected to 120 volts from a wall socket drawing 0.83 amps of current
   b. A toaster oven connected to 120 volts from a wall socket drawing 10 amps of current

9. Using electrical symbols, draw a circuit diagram of a closed circuit that includes the following: a switch, wires, a battery, a bulb, and a resistor.

Applying your knowledge

1. Investigate the following questions about superconductivity: What is superconductivity? Under what circumstances does it occur? What materials are most commonly used as super conductors? Of what value is superconductivity?

2. You have been given a fixed resistor, but you do not know its resistance. How could you build a circuit to figure out the resistor's value using the resistor, a 5-volt battery, an ammeter, and some wire? Explain what measurements you would take, and how you would calculate the resistance.

UNIT 7 ELECTRICITY AND MAGNETISM

CHAPTER 20

Electric Circuits and Power

Objectives:

By the end of this chapter you should be able to:

✔ Recognize and sketch examples of series and parallel circuits.
✔ Describe a short circuit and why a short circuit may be a hazard.
✔ Calculate the current in a series or parallel circuit containing up to three resistances.
✔ Calculate the total resistance of a circuit by combining series or parallel resistances.
✔ Describe the differences between AC and DC electricity.
✔ Calculate the power used in an AC or DC circuit from the current and voltage.

Key Questions:

- How do series and parallel circuits work?
- What are network circuits, and how do you analyze them?
- How are electrical energy and power measured?

Vocabulary

alternating current (AC)	kilowatt (kW)	network circuit	short circuit
circuit analysis	kilowatt-hour (kWh)	parallel circuit	voltage drop
circuit breaker	Kirchhoff's current law	power factor	
direct current (DC)	Kirchhoff's voltage law	series circuit	

20.1 Series and Parallel Circuits

A simple electric circuit contains one electrical device, a battery, and a switch. Flashlights use this type of circuit. However, most electrical systems, such as a stereo, contain many electrical devices connected together in multiple circuits. This section talks about the two fundamental ways to connect multiple devices in a circuit. Series circuits have only one path for current to flow. Parallel circuits have at least two branches that create multiple paths for current to flow. A complex system like a stereo has networks of both series and parallel circuits.

Series circuits

Series circuits contain one path   A **series circuit** contains only one single path for current to flow. Since there is only one path, the current is the same at all points in a series circuit. For example, the circuit below has three identical bulbs connected in series. Because there is only one path for the current, the same current goes through each bulb making each bulb equally bright.

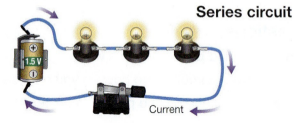
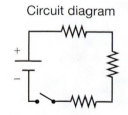

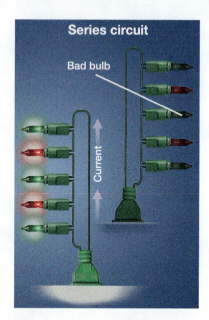

Figure 20.1: *Some types of holiday lights are wired in series. When one bulb burns out, the circuit is broken and all the lights go out.*

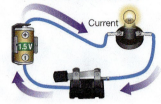

Figure 20.2: *Switches are usually placed "in series" with other elements in a circuit so they can stop the flow of current in the whole circuit.*

Stopping current   If there is a break at any point in a series circuit, the current will stop everywhere in the circuit. Some types of holiday lights are wired with the bulbs in series. When one bulb burns out, the current stops and none of the bulbs will light (Figure 20.1). Connecting bulbs in series uses the least amount of wire, so strings of lights connected in series can be less expensive to manufacture.

Using series circuits   There are times when devices are connected in series for specific purposes. On-off switches are placed in series with the other components in most electrical devices (Figure 20.2). When a switch is turned to the "off" position, it breaks the circuit and stops current from reaching all of the components in series with the switch. Dimmer switches placed in series with light bulbs adjust the brightness by changing the *amount* of current in the circuit.

ELECTRIC CIRCUITS AND POWER   CHAPTER 20

Current and resistance in a series circuit

Adding resistance in series Each resistance in a series circuit adds to the total resistance of the circuit. Think about adding pinches to a hose (Figure 20.3). Each pinch adds more resistance and reduces the current of water flowing in the hose. The total resistance of a hose is also the sum of the resistances from each pinch.

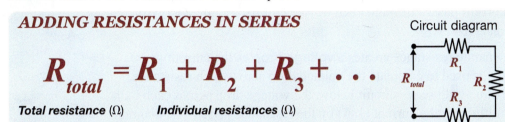

ADDING RESISTANCES IN SERIES

$$R_{total} = R_1 + R_2 + R_3 + \ldots$$

Total resistance (Ω)   Individual resistances (Ω)

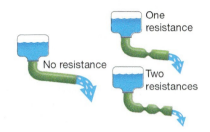

Figure 20.3: *Resistances in series add up like pinches in a hose. Each one reduces the current in the circuit.*

Calculating current If you know the voltage and the total resistance in a circuit, you can calculate the current using Ohm's law. To calculate the total resistance for a series circuit you *add* all the individual resistances because all the current must pass through all of the resistances. Since the current is the same everywhere in a series circuit, use Ohm's law to calculate the current by dividing the voltage applied to the circuit by the total resistance of the circuit.

Resistance in wires and batteries Every part in a circuit has some resistance, even the wires and batteries. However, light bulbs, resistors, motors, and heaters usually have much greater resistance than wires and batteries. Therefore, when adding resistances, we can almost always leave out the resistance of wires and batteries.

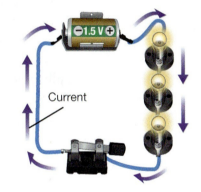

Calculating the current in a series circuit

How much current flows in a circuit with a 1.5-volt battery and three 1-ohm bulbs in series (Figure 20.4)?

1. You are asked to calculate current.
2. You are given the voltage and resistances.
3. Use Ohm's law, $I = V \div R$, and add the resistances in series.
4. Solve:
   Resistance = $R_1 + R_2 + R_3 = 1\,\Omega + 1\,\Omega + 1\,\Omega = 3\,\Omega$
   Current, $I = (1.5\text{ V}) \div (3\,\Omega) = 0.5$ A

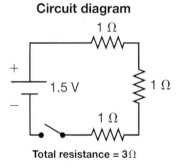

Figure 20.4: *Adding resistances in a series circuit of three 1-Ω bulbs. The total resistance is 3Ω.*

20.1 SERIES AND PARALLEL CIRCUITS

CHAPTER 20 ELECTRIC CIRCUITS AND POWER

Voltage in a series circuit

Energy You have learned that energy is not created or destroyed, but can be transformed from one form to another. As current flows along a circuit, each resistance—a bulb, motor, or any other electrical device—transforms some of the electrical energy into other forms of energy. Voltage represents the available energy per amp of current. As the current passes through each device that uses some energy or power, the *voltage gets lower*.

The voltage drop We often say each separate resistance creates a **voltage drop** as the current passes through it. Ohm's law is used to calculate the voltage drop across each resistance. For example, in the three-bulb series circuit below, the voltage drop across each bulb (1 V) is found by multiplying the current (1 A) by the resistance (1 Ω). The diagram below shows how the power carried by the current is used. The current leaves the battery at a potential of 3 V, so each amp carries 3 watts of electrical power. The first bulb uses one watt and the voltage drops by 1 volt ($V = IR = 1\text{ A} \times 1\text{ }\Omega$). The next bulb uses the same power and the voltage drops by 1 volt again.

Batteries and cells

Although the terms *battery* and *cell* are often used interchangeably, cells are the building blocks of batteries. A, AA, AAA, B, C, and D batteries each contain a single 1.5-V cell. A chemical reaction inside the cell supplies electric current.

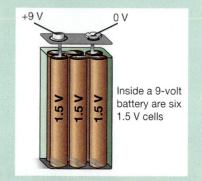

Inside a 9-volt battery are six 1.5 V cells

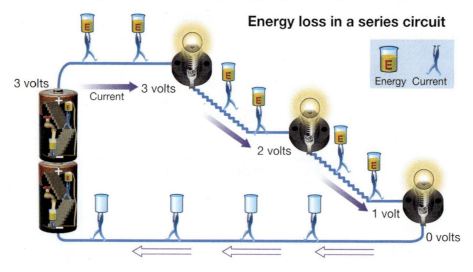

Energy loss in a series circuit

If a battery contains multiple cells, they may be connected in series or parallel. If a higher voltage is needed, the cells are connected in series. Six 1.5-volt cells in series make up the 9-volt battery, as shown.

If a battery must supply a large current, a larger cell may be used or multiple cells may be connected in parallel. Parallel cells make a battery with the voltage of one cell, but with a greater current capacity.

Kirchhoff's voltage law Over the entire circuit, the power used by all of the resistances must equal the power supplied by the battery. This means the total of all the voltage drops must add up to the total voltage supplied by the battery. This is known as **Kirchhoff's voltage law**, after German physicist Gustav Robert Kirchhoff (1824–1887).

Parallel circuits

Multiple paths for current Unlike series circuits, **parallel circuits** contain more than one path for current to flow. Every parallel circuit contains at least one point where the circuit divides, providing multiple paths for the current. Sometimes these paths are called *branches*. The current through a branch is also called the *branch current*.

Kirchhoff's current law Because there are multiple branches, the current may not be the same at all points in a parallel circuit. When analyzing a parallel circuit, keep in mind that the current always has to go somewhere. At every branch point, the current flowing out must equal the current flowing in. This rule is known as **Kirchhoff's current law**. The total current in the circuit is the sum of the currents in all the branches.

Three bulbs in parallel For example, suppose you have three 3-Ω light bulbs connected in parallel. Because each bulb has a direct wire to the battery with no voltage drops, each bulb has 3 volts of potential difference. That means that each 3-Ω bulb draws a current of 1 amp. If there are three bulbs drawing 1 amp each, then the battery must supply a total of 3 amps. At the first branch point, 1 amp splits off to the first bulb and 2 amps continue to the next two bulbs. After passing through the bulbs, the current combines again and 3 amps flows back to the battery. *Note*: This example uses 3-Ω bulbs instead of 1-Ω bulbs used in the previous example.

Why are birds not electrocuted?

16,000 V

If high-voltage wires are so dangerous, how do birds sit on them without being instantly electrocuted? First, the bird's body has a higher resistance than the electrical wire. The current tends to stay in the wire because the wire is a lower-resistance path.

The most important reason, however, is that the bird has both feet on the wire. That means the voltage is the same on both feet and no current flows through the bird.

If a bird had one foot on the wire and the other touching the electric pole, then there would be a voltage difference across the bird's feet. A lot of electricity would pass through the bird. You don't see birds standing with one foot on a wire and the other foot elsewhere because any bird who makes that mistake doesn't survive to be seen!

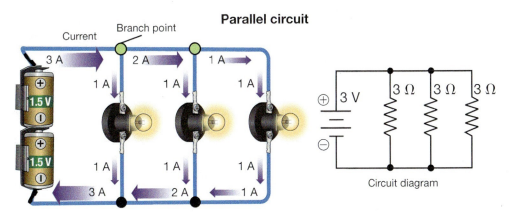

Kirchoff's current law
All the current flowing into a branch point in a circuit has to flow out.

CHAPTER 20   ELECTRIC CIRCUITS AND POWER

Voltage and current in a parallel circuit

Voltage In a parallel circuit, the *voltage* is the same across each branch because each branch has a low resistance path back to the battery. The amount of current in each branch in a parallel circuit is *not* necessarily the same. The resistance in each branch determines the current in that branch. Branches with less resistance have larger amounts of current than branches with more resistance.

Advantages of parallel circuits Parallel circuits have two big advantages over series circuits.

1. Each device in the circuit is at the full battery voltage.
2. Each device in the circuit may be turned off independently without stopping the current flowing to other devices in the circuit.

Remember, in a series circuit, each additional resistance reduces the current in the whole circuit. And a break anywhere in a series circuit stops the current to all devices on the circuit. Parallel circuits use more wires but are used for most of the wiring in homes and businesses because of the two advantages above.

Short circuits A **short circuit** is a parallel path in a circuit with zero or very low resistance. Short circuits can be made accidentally by connecting a wire between two other wires at different voltages. Short circuits can be dangerous because they can draw large amounts of current. For example, suppose you connect a length of wire across a circuit, creating a short circuit as shown below.

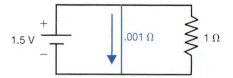

$$I = \frac{1.5\ \text{V}}{0.001\ \Omega}$$
$$= 1{,}500\ \text{amps}$$

Why a short circuit is dangerous The white wire in the circuit makes a short circuit because it has a very low resistance of 0.001 Ω. According to Ohm's law, the current through the offending wire should be 1.5 V divided by 0.001 Ω, or 1,500 amps! This much current would melt the wire in an instant and probably burn you as well if the battery shown could deliver that much current. Short circuits are always a concern when working around electricity. Fuses or circuit breakers are used for protection to prevent the high current of a short circuit.

Calculating the current in a parallel circuit

Two bulbs with different resistances are connected in parallel to batteries with a total voltage of 3 volts. Calculate the total current from the batteries.

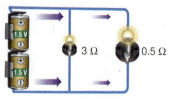

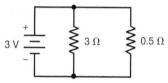

1. You are asked for the current.
2. You are given the voltage and resistance.
3. Use Ohm's law: $I = V \div R$.
4. For the 3-Ω bulb:
   $I = (3\ \text{V}) \div (3\ \Omega) = 1\ \text{A}$
   For the 0.5-Ω bulb:
   $I = (3\ \text{V}) \div (0.5\ \Omega) = 6\ \text{A}$

The battery must supply the current for both bulbs, which adds up to 7 amps.

Resistance in parallel circuits

More branches means more current In series circuits, adding extra resistance increases the total resistance of the whole circuit. The opposite is true in parallel circuits. Adding resistance in parallel provides another path for current, and *more* current flows. When more current flows at the same voltage, the total resistance of the circuit must *decrease*.

Similarity to checkout lines A similar result occurs in the check-out area of a grocery store. If only one register is open, every person must pass through the same lane, and the rate of people leaving the store is slow. If a second register opens, half of the people can pass through each lane. The average amount of time each person has to spend waiting in line can be cut in half, and the total flow of people from the store is higher.

Example of a parallel circuit Figure 20.5 shows the parallel circuit of three 3-Ω bulbs used in a previous example. The total current in this circuit is 3 amps. The voltage is 3 volts. According to Ohm's law, the total resistance of the circuit, R, equals voltage, V, divided by the total current, I. Substituting values gives 3 V divided by 3 A, or 1 Ω. Three 3-Ω resistors add up to a total resistance of 1 Ω. This happens because every new path in a parallel circuit allows more current to flow for the same voltage. The resistance of the whole circuit decreases.

The formula for adding parallel resistances If you work backward from current and voltage, you can derive a formula for the total resistance of a parallel circuit. The formula allows you to calculate the total current that flows in the circuit using Ohm's law and the total resistance. At the end of the chapter, you will see that all the outlets in your home are connected in parallel. Every device you plug in adds a parallel resistance and draws more current.

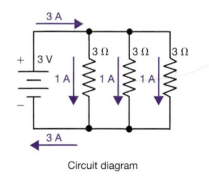

Figure 20.5: *A parallel circuit with three equal resistances divides the current three ways.*

Calculating the resistance of a parallel circuit

A circuit contains a 2-ohm resistor and a 4-ohm resistor in parallel. Calculate the total resistance of the circuit.

1. You are asked for the resistance.
2. You are given the circuit diagram and resistances.
3. Use the rule for parallel resistances.
4. Solve:

$$\frac{1}{R_{total}} = \frac{1}{2\ \Omega} + \frac{1}{4\ \Omega} = \frac{3}{4\ \Omega}$$

$$\frac{1}{R_{total}} = \frac{4}{3}\ \Omega \approx 1.33\ \Omega$$

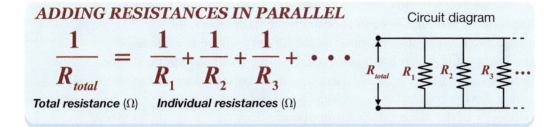

ADDING RESISTANCES IN PARALLEL

$$\frac{1}{R_{total}} = \frac{1}{R_1} + \frac{1}{R_2} + \frac{1}{R_3} + \cdots$$

Total resistance (Ω)   Individual resistances (Ω)

CHAPTER 20 — ELECTRIC CIRCUITS AND POWER

20.2 Analysis of Circuits

Electric circuits perform so many useful tasks that it is impossible to give examples of every type of circuit. There are circuits in computers, refrigerators, cars, televisions, cell phones, and every device that uses electricity. Some electrical engineers design circuits. Designing a good circuit is like a puzzle: You might need a specific current and voltage output, and there may be a specific current and voltage input. This section is a brief introduction to designing and analyzing circuits that contain only resistors. In later chapters you will be introduced to some other kinds of electrical devices that appear in circuits, such as capacitors, transistors, and diodes. Even though they are presented for resistor circuits, the three rules of this section apply to all circuits, even the most complex ones.

The three circuit laws

Why circuit analysis is useful — Before you can design a circuit, you need to know how to figure out what a circuit does. All circuits work by manipulating currents and voltages. The process of **circuit analysis** involves calculating the currents and voltages in a circuit and how they affect each other. There are three basic laws of circuit analysis. Used with the formulas for combining resistors, these three laws can be used to solve any circuit problem.

Ohm's law — Ohm's law, $I = \dfrac{V}{R}$, relates current, voltage, and resistance in a circuit. It can be used to find one quantity when the other two are known (Figure 20.6, top).

Kirchhoff's current law — Kirchhoff's current law states that the total current flowing into any junction in a circuit equals the total current flowing out of the junction (Figure 20.6, middle).

Kirchhoff's voltage law — Kirchhoff's voltage law states that the total of all voltage drops and voltage gains around any loop of a circuit must be zero (Figure 20.6, bottom). The voltage law is the trickiest to use because you have to choose a direction you believe the current will flow around the loop. You do not have to choose the right direction; the current comes out negative if you guess wrong. You do, however, have to choose and stick with your choice until the problem is solved. A voltage *increase* in the direction of the assumed current flow is counted as *positive*. A voltage *decrease* in the direction of current flow is counted as *negative*. The signs reverse if the voltage change is opposite to the assumed direction of current.

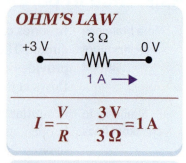

OHM'S LAW

$$I = \dfrac{V}{R} \quad \dfrac{3\text{ V}}{3\,\Omega} = 1\text{ A}$$

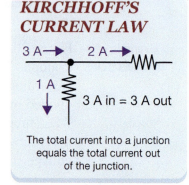

KIRCHHOFF'S CURRENT LAW

The total current into a junction equals the total current out of the junction.

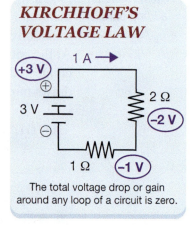

KIRCHHOFF'S VOLTAGE LAW

The total voltage drop or gain around any loop of a circuit is zero.

Figure 20.6: *The three circuit laws.*

426 UNIT 7 ELECTRICITY AND MAGNETISM

A voltage divider circuit

Changing voltage in a circuit Nearly all electronic devices use a voltage divider somewhere in the circuit. This is because electrical devices work within a certain range of voltage. For example, the input circuit for a microphone is designed for a voltage up to 0.1 volts. The circuit for a speaker operates at 48 volts (Figure 20.7). If you want to record a speaker output directly into a microphone input, you need to divide the voltage by a factor of 48 divided by 0.1, or 480.

A simple voltage divider Consider the series circuit with two resistors shown in Figure 20.8. The total resistance of this circuit is 10 ohms ($9\,\Omega + 1\,\Omega$). The total current is 1 amp from Ohm's law ($10\,V \div 10\,\Omega$). What is the voltage at point A?

Analyzing the circuit We can use Kirchhoff's voltage law to find the answer. Assume the current flows according to the colored loop. The battery starts at $+10$ V. Resistor R_1 drops the voltage by 9 V. This voltage drop is calculated from Ohm's law and the resistance, $V = 1\,A \times 9\,\Omega$. Resistor R_2 drops the voltage by 1 V ($1\,A \times 1\,\Omega$). Around the whole loop, the sum of the voltage gains and drops is zero ($+10 - 9 - 1 = 0$).

The circuit divides by 10 The voltage at point A is 1 volt because it is the battery voltage, which is 10 V, minus the voltage drop across R_1, which is 9 V. The circuit is a *voltage divider*. The voltage at point A is the battery voltage divided by 10. If the battery had been 20 V, the voltage at point A would be 2 volts. No matter what the battery voltage, the voltage at point A will always be one-tenth as great.

Input and output voltage Think of the battery voltage as the *input voltage*, and the voltage at point A as the *output voltage* (Figure 20.8, bottom). The output voltage (V_o) is always lower than the input voltage (V_i) by the ratio $R_2 \div (R_1 + R_2)$. For the example circuit, the ratio was 1 to 10, but it could be any number depending on which resistor values you choose. For example, if R_2 was $10\,\Omega$ and R_1 was $4{,}790\,\Omega$, the output voltage would be the input voltage divided by 480.

VOLTAGE-DIVIDER CIRCUIT

$$\text{Output voltage (V)}\ V_o = \left(\frac{R_2}{R_1 + R_2}\right) V_i\ \text{Input voltage (V)}$$

where $\frac{R_2}{R_1+R_2}$ is the Resistor ratio.

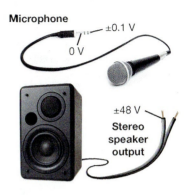

Figure 20.7: *Voltages from typical music electronics.*

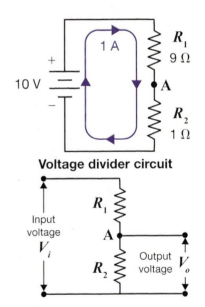

Figure 20.8: *The top diagram shows a circuit that divides the battery voltage by 10. The same circuit divides any voltage by a ratio of $R_2 \div (R_1 + R_2)$.*

CHAPTER 20 — ELECTRIC CIRCUITS AND POWER

Solving circuit problems

Using a systematic approach

It is best to take a systematic approach when analyzing circuits. The following steps are a general guide that will help you solve circuit problems.

1. Identify what the problem is asking you to find. Assign variables to the unknown quantities.
2. Make a large, clear diagram of the circuit. Label all of the known resistances, currents, and voltages. Use the variables you defined to label the unknowns.
3. You may need to combine resistances to find the total circuit resistance. Use multiple steps to combine series and parallel resistors. Table 20.1 is a good guide.
4. If you know the total resistance and current, use Ohm's law in the form $V = IR$ to calculate voltages or voltage drops. If you know the resistance and voltage, use Ohm's law in the form $I = V \div R$ to calculate the current.
5. An unknown resistance can be found using Ohm's law as $R = V \div I$ if you know the current through the resistor and the voltage drop across it.
6. Use Kirchhoff's current and voltage laws as necessary.

There is often more than one way to solve circuit problems. Finding solutions to complex circuits requires creativity, logical thinking, and practice.

Table 20.1: Comparing series and parallel circuits

	Series	Parallel
	R_1, R_2, R_3 in series	R_1, R_2, R_3 in parallel
	One path for current	Multiple paths for current
	Same current at every point	Current may be different in each branch
	Voltage drops across each resistance	Same voltage across each resistance
	Break anywhere stops current in entire circuit	Break might only stop current in one branch of circuit
Total resistance	$R_{total} = R_1 + R_2 + ...$	$\dfrac{1}{R_{total}} = \dfrac{1}{R_1} + \dfrac{1}{R_2} + ...$
	Adding series resistances decreases the total current.	Adding parallel resistances increases the total current.

Calculating the resistor needed to limit current in a series circuit

A bulb with a resistance of 1 ohm is to be used in a circuit with a 6-volt battery. The bulb requires 1 amp of current. If the bulb is connected directly to the battery, it will draw 6 amps and burn out instantly. To limit the current, a resistor is added in series with the bulb. What size resistor is needed to make the current 1 amp?

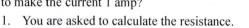

1. You are asked to calculate the resistance.
2. You are told it is a series circuit and given the voltage, total current, and one resistance.
3. Use Ohm's law, $R = V \div I$, and add the resistance in series.
4. Solve:
   Total resistance = 6 V ÷ 1 A = 6 Ω

Since the bulb is 1 Ω, the additional resistor must be 5 Ω to give a total 6 Ω of resistance.

Network circuits

Resistors in both series and parallel
In many circuits, resistors are connected both in series and in parallel. Such a circuit is called a **network circuit**. A network circuit contains more than one current path and can have resistors in series or parallel on one or more of those paths (Figure 20.9).

Solving a network circuit problem
There is no single formula for adding resistors in a network circuit. Often, you must simplify the circuit step by step, using either the series or parallel resistor formula to combine two or more resistors. Keep combining resistors until you are left with one resistance that is equivalent to the original combination.

Calculating currents and voltages
From the total resistance and voltage, you can use Ohm's law to calculate the total current. You can then work back through the circuit, calculating voltage drops at intermediate points, as necessary. For very complex circuits, electrical engineers use computer programs that can rapidly solve a large number of simultaneous equations for the circuit using Kirchhoff's laws.

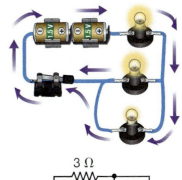

Circuit diagram

Figure 20.9: *An example of a network circuit with three bulbs.*

Calculating the currents and voltages in a three-resistor network circuit

Three bulbs, each with a resistance of 3 Ω, are combined in the circuit shown in Figure 20.9. Three volts are applied to the circuit. Calculate the currents in each of the bulbs. From your calculations, do you think all three bulbs will be equally bright?

1. You are asked to calculate the currents.
2. You are given the circuit diagram, voltages, and resistances.
3. Use Ohm's law, $R = V \div I$, and the series and parallel resistance formulas.
4. First, reduce the circuit by combining the two parallel resistances.

$$\frac{1}{R_{total}} = \frac{1}{3\,\Omega} + \frac{1}{3\,\Omega} = \frac{2}{3\,\Omega} \rightarrow R_{total} = \frac{3}{2}\,\Omega = 1.5\,\Omega$$

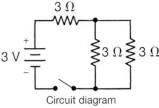

Calculate the total resistance of 4.5 Ω by adding up the remaining series resistances.

Calculate the total current using Ohm's law: $I = 3\text{ V} \div 4.5\,\Omega = 0.67$ A.

The two bulbs in parallel have the same resistance, so they divide the current equally; each one gets 0.33 amps.

The single bulb in series gets the full current of 0.67 amps, but the other two bulbs get only 0.33 amps each. That means the bulbs in parallel will be much dimmer since they only get half the current.

20.3 Electric Power, AC, and DC Electricity

In the last chapter, the volt was defined by the amount of energy carried by each amp of current in 1 second. A current of 1 amp flowing through a voltage difference of 1 volt does 1 joule of work in 1 second. Since 1 joule per second is 1 watt, this definition really links voltage, current, and *power*. If you look carefully at a stereo, hair dryer, or other household appliance, you find that most devices list a "power rating" that tells how many watts the appliance uses (Figure 20.10). In this section you will learn what these power ratings mean, and how to calculate the electricity cost of using various appliances of different power ratings.

Figure 20.10: *The back of an electrical device often tells you how many watts it uses.*

Electric power

The three electrical quantities You have learned about three important electrical quantities:

Current	Current is the flow of electricity. Current is measured in amperes.
Voltage	Voltage measures the potential energy difference between two places in a circuit. Voltage differences produce current. Voltage is measured in volts.
Resistance	Resistance measures the ability to resist current. Resistance is measured in ohms.

Paying for electricity Electric bills sent out by utility companies do not charge by the volt, the amp, or the ohm. And you may have noticed that electric appliances in your home usually include another unit—the *watt*. Almost every appliance has a label that lists the number of watts it uses. You may have purchased 60-watt light bulbs, a 1,000-watt hair dryer, or a 1,500-watt toaster oven.

Measuring power The watt (W) is a unit of power. Remember from Chapter 11 that power is the rate at which energy moves or is used. Since energy is measured in joules, power is measured in joules per second. One joule per second is equal to 1 watt (Figure 20.11). A 100-watt light bulb uses 100 joules of energy *every second*. This means the bulb transforms 100 joules of electrical energy into heat and light every second. The longer the bulb is turned on, the more electrical energy it uses, and the higher your monthly electricity bill.

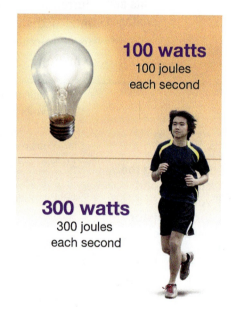

Figure 20.11: *One watt is an energy flow of 1 joule per second. A 100-watt light bulb uses 100 joules every second. A person running uses about 300 watts, or 300 joules every second.*

Power in electric circuits

Power in a circuit A voltage of 1 volt means 1 amp of current can do 1 joule of work each second. Since 1 joule per second is 1 watt, this definition of a volt is really a formula for calculating power from current and voltage. If the voltage and current in a circuit are multiplied together, the result is the power used by the circuit.

ELECTRICAL POWER

Power (W) $P = VI$ Voltage (V), Current (A)

Voltage	×	Current	=	Power
joules / (amperes × second)	×	amperes	=	joules / second

Watts and kilowatts One watt is a pretty small amount of power. In everyday use, larger units are more convenient to use. For example, a 1,500-watt toaster oven may be labeled "1.5 kW" instead of "1,500 W." A **kilowatt (kW)** is equal to 1,000 watts.

Horsepower The other common unit of power often seen on electric motors is the horsepower. One horsepower is 746 watts. Electric motors you find around the house range in size from a small electric fan at 1/25th of a horsepower, which is about 30 watts, to an electric saw that develops two horsepower, or 1,492 watts.

Voltage and current Power depends on both current and voltage. You can get 1,000 watts of power from 1,000 amps at 1 volt, or from 10 amps at 100 volts. It would take a very large wire to carry 1,000 amps. That is why most electrical systems that move a lot of power also operate at high voltages, typically over 100 volts.

Calculating the power used by a small light bulb

A light bulb with a resistance of 1.5 Ω is connected to a 1.5-volt battery in the circuit shown here. Calculate the power used by the light bulb.

1. You are asked to find the power used by the light bulb.
2. You are given the voltage of the battery and the bulb's resistance.
3. Use Ohm's law, $I = \dfrac{V}{R}$, to calculate the current; then use the power equation, $P = VI$, to calculate the power.
4. Solve: $I = 1.5\text{ V} \div 1.5\text{ Ω} = 1\text{ A}$
   $P = 1.5\text{ V} \times 1\text{ A} = 1.5\text{ W}$; the bulb uses 1.5 watts of electric power.

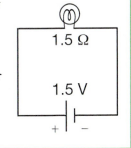

Electric cars

Many people believe that eventually all cars will be electric because they give off little or no pollution. Electric cars are challenging to build because of the power required by a car. An average automobile gasoline engine delivers about 100 horsepower, or about 75,000 watts.

Suppose in your electric car you wanted to use 12-volt batteries, like the ones used to *start* cars today. To make 75 kilowatts of power at 12 volts, you need a current of 6,250 amps. By comparison, most people's entire homes use less than 100 amps.

The solution is to use more efficient motors and higher voltages. The higher the voltage, the more power can be carried by each amp of current. For example, some electric buses use 96-volt systems.

CHAPTER 20 | ELECTRIC CIRCUITS AND POWER

Paying for electricity

Your electric bill   What do you buy from the electric utility company? If you look at your electric bill, you will see that the utility company charges you for the number of kilowatt-hours of electricity you use. One **kilowatt-hour (kWh)** means that a kilowatt of power has been used for 1 hour. Since power multiplied by time is energy, a kilowatt-hour is a unit of *energy*. One kilowatt-hour is 3.6×10^6 joules.

Kilowatt-hours   Electric companies charge for the number of kilowatt-hours used during a set period of time, often a month (Figure 20.12). Your home is connected to a meter that counts kilowatt-hours used. The meter is read in person or remotely so the electric company knows how much to charge you.

Estimating the cost   If you know the cost per kilowatt-hour your utility company charges, you can estimate the cost of running an appliance for a period of time.

Figure 20.12: Most people pay an electric bill monthly that charges for the kilowatt-hours of energy used.

Calculating the currents and voltages in a three-resistor network circuit

Your electric company charges 14 cents per kilowatt-hour. A coffee maker has a power rating of 1,050 watts. How much does it cost to use the coffee maker 1 hour per day for a month?

Coffeemaker
AC 120 V   8.75 A
1050 W Heating Element

1. You are asked to find the cost of using the coffee maker.
2. You are given the power in watts and the time.
3. Use the power formula $P = VI$ and the fact that 1 kWh = 1 kW × 1 hr.
4. Solve: Find the number of kilowatts of power that the coffee maker uses.
   1,050 W × 1 kW/1,000 W = 1.05 kW

Find the kilowatt-hours used by the coffee maker each month. Assume 30 days per month.
1.05 kW × 1 hr/day × 30 days/month = 31.5 kWh per month

Find the cost of using the coffee maker.
31.5 kWh/month × $0.14/kWh = $4.41 per month

Table 20.2: Typical power ratings

Appliance	Power (watts)
Electric stove	5,000
Electric heater	1,500
Hair dryer	1,000
Iron	800
Washer	750
Light	100
Small fan	50
Clock radio	10

Alternating (AC) and direct (DC) current

DC current The current from a battery is always in the same direction. One end of the battery is positive and the other end is negative. The direction of current flows from positive to negative. This is called **direct current (DC)**. Most of the experiments you do in the lab use DC current.

AC current Imagine putting a battery on a rotating wheel. You could touch wires against the wheel so the voltage applied to the circuit alternated back and forth between positive and negative (Figure 20.13). If the voltage alternates, so does the current. When the voltage is positive, the current in the circuit is clockwise, as shown in the diagram above the graph. When the voltage is negative, the current goes in the opposite direction. This type of current is called **alternating current (AC)**.

AC is used for most electric power AC current is used for almost all high-power applications because it is easier to generate and to transmit over long distances. For example, all the power lines you see carry AC current. The outlets in the walls of your homes or classroom also carry AC.

The frequency of AC electricity With DC electricity, the voltage and current tell the whole story. AC electricity is more complicated. The frequency at which the voltage alternates is an important characteristic of AC electricity. In the United States, most AC power alternates at a frequency of 60 Hz. This means the voltage on the same wire switches back and forth between positive and negative 60 times each second.

The waveform of AC electricity The second important characteristic of AC electricity is the shape of the voltage versus time graph. Since it looks like a wave, this shape is normally called the *waveform*. Figure 20.14 shows the waveform of common household AC electricity. This waveform is called a sine wave because it is related to the sine function from trigonometry.

Peak and average voltages The 120 volt AC (VAC) electricity used in homes and businesses alternates between peak values of +170 V and −170 V at a frequency of 60 Hz. This kind of electricity is called 120 VAC because +120 V is an *average* positive voltage and −120 V is the average negative voltage. AC electricity is usually identified by this average voltage, not the peak voltage.

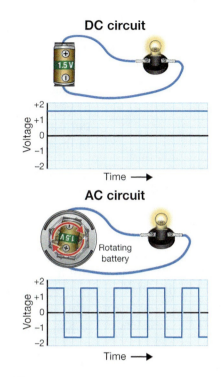

Figure 20.13: *You could make an AC circuit by rotating a battery and connecting the circuit to a disc with sliding electrical contacts.*

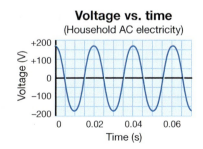

Figure 20.14: *The voltage versus time graph for common household AC electricity.*

Voltage, current, and power for AC circuits

Power in resistive circuits The power in an AC circuit with only resistances is calculated in the same way as the power in a DC circuit. The difference is that *average* values are used for voltage and current. For example, a 100-watt light bulb uses a power of 100 watts. The average AC voltage from the wall socket is 120 V. That means the average current through the bulb is 100 watts divided by 120 volts, or 0.82 amps. Light bulbs and heaters are pure resistances, but motors and transformers are not.

Peak current and average current The peak current through a bulb or any other AC device is much higher than the average current. From Ohm's law, you can calculate the resistance of a 100-watt bulb by dividing 120 V by 0.83 A, with the result $R = 145\ \Omega$. The peak current is the peak voltage, or 170 V divided by 145 Ω, or 1.17 A. The peak current is 41 percent higher than the average current.

Motors and devices that are not pure resistances For a circuit containing a motor, the power calculation is different than it is for a pure resistance, like a light bulb. Because motors store energy and act like generators (Chapter 23), the current and voltage are not in phase with each other. When the voltage reverses to a motor, the current cannot respond immediately. This means the current is always a little behind the voltage (Figure 20.15). Because the current is not in sync with the voltage, the average power in a motor is *less* than the average current times the average voltage.

The power factor Electrical engineers use a **power factor** (*pf*) to calculate power for AC circuits. For a circuit with only resistors, the power factor is 100 percent and the power is equal to the average current times the average voltage. For a motor or transformer, the power factor is typically 80 percent or less, depending on the design of the device. The power in an AC circuit is given by the formula below, which is similar to the power formula $P = VI$ except that it includes the power factor.

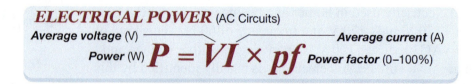

ELECTRICAL POWER (AC Circuits)

Power (W) $P = VI \times pf$
Average voltage (V) — Average current (A) — Power factor (0–100%)

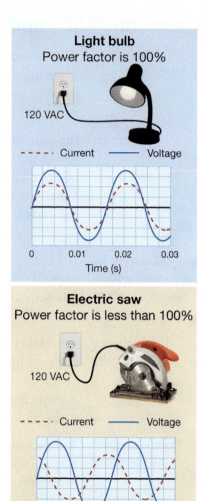

Figure 20.15: *Power in an AC circuit may be less than current times voltage because the current and voltage may be out of phase, as with an electric saw.*

Chapter 20 Connection

Wiring in Homes and Buildings

You use electric current in your house every day. When you plug in an electric appliance, you connect it to a circuit of wires in the walls. The wires eventually connect to power lines outside your house that bring the current from an electricity-generating source.

Wire sizes and circuit breakers

The 120 VAC electricity comes into a typical home or building through a circuit breaker panel. The **circuit breakers** protect against wires overheating and causing fires. The wires in a house are of different sizes to carry different amounts of current safely (Figure 20.16). For example, a circuit made with 12-gauge wire can carry 20 amps. This circuit is protected with a circuit breaker rated for 20 amps that opens the circuit automatically if more than 20 amps of current flows through it.

Wire gauge	Current (amps)
12	20
14	15
16	10
18	7

Figure 20.16: *Different gauges of wire can carry different amounts of current safely.*

Hot, neutral, and ground wires

Each wall socket has three wires feeding it. The hot wire carries 120 VAC. The neutral wire stays at zero volts. When you plug something in, current flows in and out of the hot wire, through your appliance, and back through the neutral wire. The ground wire is for safety and is at 0 V like the neutral wire. If there is a short circuit in your appliance, the current flows through the ground wire rather than through you or whatever caused the short circuit.

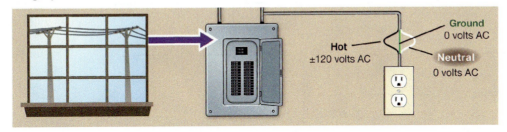

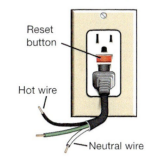

Figure 20.17: *A ground fault interrupt outlet might be found in bathrooms and kitchens where water may be near electricity.*

Ground fault interrupt (GFI) outlets

Electrical outlets in bathrooms, kitchens, or outdoors are now required to have ground fault interrupt (GFI) outlets installed (Figure 20.17). A GFI outlet contains a circuit that compares the current flowing out on the hot wire and back on the neutral wire. If everything is working properly, the two currents should be exactly the same. If they are different, some current must be flowing to ground through another path, such as through your hand. The ground fault interrupter detects any difference in current and immediately breaks the circuit. GFI outlets are excellent protection against electric shocks, especially in wet locations.

20.3 ELECTRIC POWER, AC, AND DC ELECTRICITY

Chapter 20 Connection

Home wiring circuits

Why parallel circuits are used The electric circuits in homes and buildings are parallel circuits. Two properties of parallel circuits make them a better choice than series circuits.

- Each outlet has its own current path. This means one outlet can have something connected and turned on, with current flowing, while another outlet has nothing connected or something turned off.

- Every outlet has the same voltage because the "hot" side of every outlet is connected to the same wire that goes to the main circuit breaker panel.

Why series circuits are not If outlets and lights were wired in series, turning off anything electrical in the circuit would break the whole circuit. This is not practical; you would have to keep everything on all the time just to keep the refrigerator running. Also, in a series circuit, everything you plugged in would use some energy and would lower the voltage available to the next outlet.

What happens if you plug too many things into a socket?

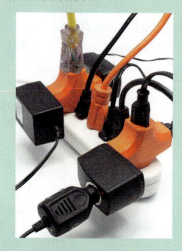

In a parallel circuit, each connection uses as much current as it needs. If you plug in a coffee maker that uses 10 amps and a toaster oven that uses 10 amps, a total of 20 amps needs to come through the wire.

If you plug too many appliances into the same circuit or outlet, you will eventually use more current than the wires can carry without overheating. Your circuit breaker will click open and stop the current. You should unplug things to reduce the current in the circuit before resetting the circuit breaker.

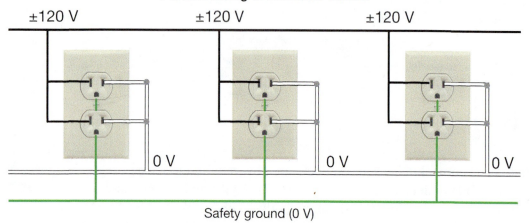

Parallel wiring of electrical outlets

Multiple parallel circuits Each room in a house typically has its own parallel circuit, protected by the appropriate-sized circuit breaker. For example, all the outlets in one bedroom might be on one circuit and the outlets in the living room might be on a separate circuit. By dividing the circuits up, many electrical devices can be connected without drawing too much current through any one wire.

Chapter 20 Assessment

Vocabulary

Select the correct term to complete the sentences.

series circuit	parallel circuit	short circuit
network circuits	circuit analysis	power
Kirchhoff's voltage law	voltage drop	direct current (DC)
alternating current (AC)	kilowatt	Kirchhoff's current law
horsepower	power factor	circuit breaker
watt	kilowatt-hour	

1. The energy per electrical charge that is lost as current flows through each resistance is called a(n) _____.
2. The current flowing into a circuit branch must equal the current flowing out of a current branch is a statement of _____.
3. A parallel path in a circuit with zero or very low resistance is called a(n) _____.
4. A circuit containing two or more resistances but only one path for current to follow is referred to as a(n) _____.
5. The scientific law which states that "energy used by all of the resistors in a circuit must equal the energy supplied by the battery" is _____.
6. A circuit which contains more than one path over which current flows may be called a(n) _____.
7. Circuits which contain resistors connected in both series and parallel are named _____.
8. The process of determining the value and the relationships between currents and voltages in a circuit is known as _____.
9. The unit of power equivalent to 746 watts is the _____.
10. Current which flows in only one direction is called _____.
11. A unit of energy equivalent to 1,000 watts of power consumed for one hour is a(n) _____.
12. An example of a device that protects wires against overheating is a(n) _____.
13. Current which changes direction in a circuit several times each second is referred to as _____.
14. The rate at which energy is consumed is known as _____.
15. When 1 joule of energy is used every second, the rate of energy consumption is measured as 1 _____.
16. One thousand watts of power is equivalent to 1 _____.
17. The numerical expression used by electrical engineers to calculate the power for alternating current (AC) circuits is known as the _____.

Concept review

1. Explain what a series circuit is and when one might be used.
2. A circuit contains a battery and two bulbs connected in series. What happens if one of the bulb burns out?
3. What happens to the total circuit resistance as more bulbs are added to a series circuit? Why?
4. A circuit contains a battery and two bulbs connected in parallel. What happens if one of the bulbs burns out?
5. The current in a _____ circuit is the same at every point. The current in a _____ circuit can be different at different points.
6. What happens to the total circuit resistance as more branches are added in a parallel circuit? Why?
7. Sketch a network circuit containing resistors. Identify the parallel branches and the series connections.
8. What is a short circuit? How is a short circuit created?
9. Identify and explain the three circuit laws.
10. In a voltage divider circuit, the output voltage is always _____ than the input voltage.
11. A 100-watt light bulb converts 100 _____ of electrical energy into heat and light energy each _____.
12. List three units in which power can be measured.
13. Explain what happens to the electrical energy used by a fan.

14. The unit kilowatt-hour is used to measure _____, not power.
15. Explain how the electric company determines how much to bill its customers each month.
16. What is the difference between AC and DC current?
17. The _____ is always between 0 and 100 percent in an AC circuit since the power in a motor is always _____ than the average current times the average voltage.

Problems

1. Calculate the total resistance of each combination of resistors.
   a. three 4-Ω resistors in series
   b. a 5-Ω resistor and a 2-Ω resistor in series
   c. two 4-Ω resistors in parallel
   d. a 6-Ω resistor and a 9-Ω resistor in parallel

2. Calculate the total resistance and the current for the circuit shown.

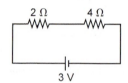

3. Calculate the unknown resistance in the circuit shown.

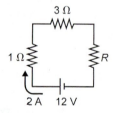

4. Calculate the total resistance, total current, and current in each branch for each of the circuits shown.

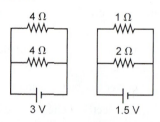

5. Calculate the output voltage for a voltage divider circuit that contains a 12-Ω resistor and a 6-Ω resistor if the input voltage is 24 V.

6. Calculate the total resistance and total current in the network circuit shown.

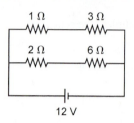

7. A hair dryer has a power rating of 1,000 W when connected to a 120-V outlet. How much current does it draw?

8. A television draws 0.8 A of current when connected to a 120-V outlet. Calculate its power.

9. A digital alarm clock has a power rating of 10 W.
   a. Calculate the number of kilowatt-hours of energy used by the alarm clock in one day.
   b. Calculate the cost of using the alarm clock for one year if 1 kilowatt-hour of energy costs $0.15.

Applying your knowledge

1. At first, circuits in a home were protected by devices called fuses. Now circuit breakers are used. After researching the characteristics of each device, explain how each device works and why circuit breakers are now commonly used in place of fuses.

2. In a normal 120-volt household circuit which is wired in parallel, a 100-watt bulb burns brighter than a 60-watt bulb. If the two bulbs are placed in a series circuit at the same voltage, is the result the same? You may need to do some research to help answer this question. Use the information you find to explain your answer.

3. Given three identical light bulbs and a battery, sketch as many different electrical circuits as you can using all of the components in each circuit design.

UNIT 7 ELECTRICITY AND MAGNETISM

CHAPTER 21

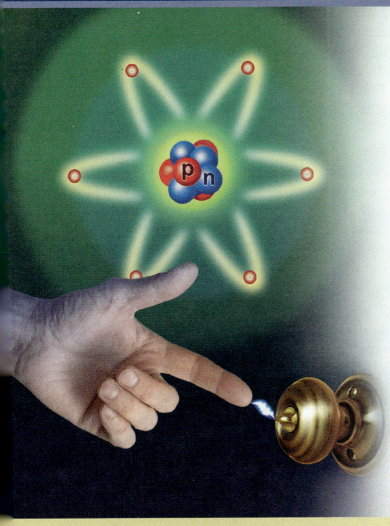

Electric Charges and Forces

Objectives:

By the end of this chapter you should be able to:

- ✓ Describe and calculate the forces between like and unlike electric charges.
- ✓ Identify the parts of the atom that carry electric charge.
- ✓ Apply the concept of an electric field to describe how charges exert forces on other charges.
- ✓ Sketch the electric field around a positive or negative point charge.
- ✓ Describe how a conductor shields electric fields from its interior.
- ✓ Describe the voltage and current in a circuit with a battery, a switch, a resistor, and a capacitor.
- ✓ Calculate the charge stored in a capacitor.

Key Questions:

- How do electric charges interact?
- What is Coulomb's law?
- What are capacitors, and how do they work?

Vocabulary

capacitance	Coulomb's law	electrons	force field	neutrons	shielding
capacitor	discharged	electroscope	gravitational field	parallel-plate capacitor	static electricity
charge by friction	electric field	farad (F)	induction		test charge
charge polarization	electric forces	field	microfarad (μF)	positive charge	
coulomb (C)	electrically neutral	field lines	negative charge	protons	

CHAPTER 21 | ELECTRIC CHARGES AND FORCES

21.1 Electric Charge

The last two chapters have developed the practical properties of electricity without digging into what electricity is on a fundamental level. This chapter will fill in the missing foundation by exploring the properties of electric charge. Electric charge is a fundamental property of all matter. The forces between positive and negative electric charge are what hold atoms together and also are what makes current flow in a circuit. This section will also discuss **electric forces**, fields, and static electricity.

Positive and negative charge

Two types of charge If you have ever felt a shock when touching a doorknob or removing clothes from the dryer, you have experienced the effect of electric charge (Figure 21.1). Electric charge, like mass, is a fundamental property of matter. An important difference between mass and charge is that there are two types of charge, called **positive charge** and **negative charge**.

Neutral objects All ordinary matter contains both positive and negative charges. You do not usually notice the charge because most matter contains the same number of positive and negative charges. When the number of positive charges equals the number of negative charges, the total charge is *zero*. An object with zero total charge is **electrically neutral**. Electrically neutral objects still contain charge! However, positive and negative are exactly balanced so the *total* charge is zero. Your pencil, textbook, and even your body are all electrically neutral, as is almost all matter.

Net charge Objects can lose or gain electric charges. If an object has more negative charges than positive charges, the net, or total, charge is negative (Figure 21.2). If it has more positive charges than negative charges, it has a net positive charge. The net charge is also sometimes called *excess* charge because a charged object has an excess of either positive or negative charges.

Transferring charge A tiny charge imbalance resulting in net positive or negative charge on an object is the cause of **static electricity**. If two neutral objects are rubbed together, charges may be rubbed off one object and onto the other. One object acquires a slight excess of negative charge and the other an equal excess of positive charge. This is what happens to clothes in the dryer and to your socks when you walk on a carpet. The shock you feel is the excess charge moving between you and the charged object to restore neutral charge balance.

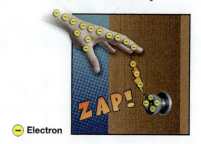

Figure 21.1: *Touching a metal object on a dry day can give you a shock from static electricity.*

Figure 21.2: *If an object has an unequal number of positive and negative charges, it has a net charge.*

The source of electric charge

Charge is a property of the particles that make up the atom

Ordinary matter is made of atoms, and electric charge is a property of the particles inside each atom. The nucleus at the center of the atom contains all of the positive charge in particles called **protons**. The nucleus also contains neutral particles called **neutrons**. Outside the nucleus are negative charges called **electrons**. The charge on the proton and electron are exactly equal and opposite. Since a complete atom has equal numbers of protons and electrons, atoms have zero net electric charge. Chapter 28 has a detailed discussion of the atom. For the purpose of understanding electricity, you need to know that electric charge is a property of electrons and protons within all atoms. Electrons are smaller than protons and have a negative charge. Protons have positive charge, are much more massive, and typically move slower than electrons.

Lightning and electric charge

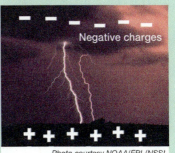

Photo courtesy NOAA/ERL/NSSL

Lightning is caused by a huge buildup of charge. As particles in a cloud collide, charges are transferred between them. Positive charges build up on smaller particles and negative charges on bigger ones. Gravity and wind push positively charged particles toward the top of the cloud and negatively charged particles to the bottom.

Negative charge on the bottom of a cloud causes the ground to become positive. When enough charges have been separated by the storm, the cloud, air, and ground act like a giant circuit. All the accumulated negative charges flow from the cloud to the ground, heating the air along the path to as hot as 20,000°C, so that it glows like a bright streak of light.

Structure of an atom

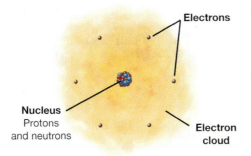

Nucleus
Protons and neutrons

Electrons

Electron cloud

	Mass (kg)	Charge (coulombs)
Electron	9.109×10^{-31}	-1.602×10^{-19}
Proton	1.673×10^{-27}	$+1.602 \times 10^{-19}$
Neutron	1.675×10^{-27}	0

The coulomb is the unit of charge

The unit of electric charge is the **coulomb (C)**. The name is chosen in honor of Charles-Augustin de Coulomb (1736–1806), the French physicist who succeeded in making the first accurate measurements of the force between charges. One coulomb is a *very* large amount of charge. One coulomb of charge is equal to the charge of about 6×10^{18} protons or electrons. The static electricity described on the previous two pages typically results from an excess charge of less than one-millionth of a coulomb.

Label charge with a plus or minus

A quantity of charge should always be identified with a positive or a negative sign. The units of coulombs apply to both positive and negative charge. For example, -2×10^{-6} C describes a negative charge of two-millionths of a coulomb.

CHAPTER 21   ELECTRIC CHARGES AND FORCES

Electric forces

Attraction and repulsion

Electrical forces exist between electric charges. Because there are two kinds of charges, positive and negative, the electrical force between charges can be attractive or repulsive. This makes electrical forces different from gravity. There is only one kind of mass, so gravity creates only *attractive* forces.

Like charges repel and unlike charges attract

Whether two charges attract or repel each other depends on whether they are the same sign or different. A positive and a negative charge will attract each other. Two positive charges will repel each other. Two negative charges will also repel each other. The force between charges is illustrated in the diagram.

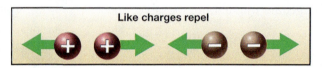
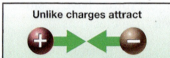

Parts of an electroscope

The forces between the two kinds of charge can be observed with an **electroscope**. An electroscope contains two very thin "*leaves*" of metal that can swing from a central rod connected to a metal ball (Figure 21.3). Charges can flow freely between the ball and the leaves. An insulator holds the rod in place and keeps charges from getting to the outside of the electroscope.

Charging an electroscope

Suppose a positively-charged rod touches the metal ball of an electroscope. Some negative electrons are attracted to the rod. The metal ball and leaves of the electroscope are left with a net positive charge. Since both leaves have the same positive charge, the leaves repel each other and spread apart.

Testing an unknown charge with an electroscope

Once an electroscope is charged, it can be used to test other charged objects. The leaves spread farther apart if another positively charged rod is brought near the metal ball. This happens because the positive rod attracts some negative electrons from the leaves toward the ball, increasing the positive charge on the leaves. If a negatively charged rod is brought near the ball, the opposite effect occurs. A negatively charged rod repels negative electrons from the ball into the leaves where they neutralize some of the positive charge. The positive charge on the leaves is reduced and the leaves get closer together (Figure 21.3, bottom).

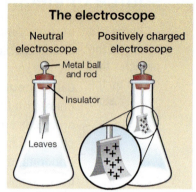

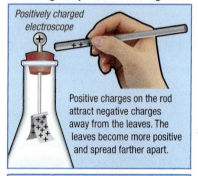

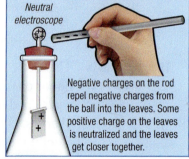

Figure 21.3: *An electroscope can be used to observe the forces between positive and negative charges.*

Current and charge

Current is the flow of charge

We can now be more specific and say current is the movement of electric charge through a circuit. You can think of electric current much as you would think of a current of water. If a faucet is on, you can measure the rate of water flow by finding out how much water comes out in one minute. You might find that the current, or flow, is 10 gallons per minute. In a circuit, you can measure the current by counting the amount of electric charge that flows through a wire in one second. *One ampere is a flow of 1 coulomb per second.* Higher current means more charge flows per second. For example, a current of 10 amperes means that 10 coulombs of charge flow through the wire every second.

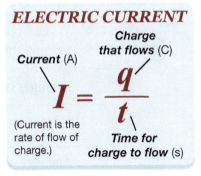

ELECTRIC CURRENT

$$I = \frac{q}{t}$$

(Current (A) is the rate of flow of charge.) Charge that flows (C). Time for charge to flow (s).

Current and the motion of charges

The direction of current was historically defined as the direction that positive charges move. However, both positive and negative charges can carry current. In conductive liquids, such as salt water, both positive and negative charges carry current. In solid metal conductors, only the electrons can move, so current is carried by the flow of negative electrons. However, no matter whether positive charges or negative charges are actually moving, positive current is defined to flow from higher voltage to lower voltage, or positive to negative (Figure 21.4).

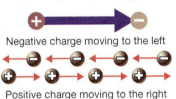

Two ways to make current flow from positive to negative

Negative charge moving to the left

Positive charge moving to the right

Figure 21.4: *Electron flow and the direction of current.*

Electron flow and the drift velocity

Not all electrons carry current, and even the electrons that do don't follow a straight path. Electrons in a wire bounce around at high speed colliding with fixed atoms. When a voltage is applied, electrons are attracted to the positive voltage. However, the random bouncing speed is so high, that even very high voltages deflect the electrons only slightly. Applied voltages cause a slow *drift velocity* to be added to the random bouncing and the drift velocity is what creates current.

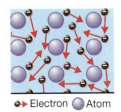

 Electron Atom

Calculating the current from the flow of charge

Two coulombs of charge pass through a wire in 5 seconds. Calculate the current in the wire.

1. You are asked to find the current.
2. You are given charge and time.
3. Use the equation $I = q \div t$.
4. Solve: $I = (2\ \text{C}) \div (5\ \text{s})$
   $= 0.4\ \text{C/s}$ or $0.4\ \text{A}$

The source of current carrying electrons

Batteries do not provide most of the electrons that flow in a circuit. Current flows because the voltage from a battery makes electrons that are *already in the wire* move. This is why a light bulb goes on as soon as you connect a circuit. A copper wire contains many electrons bouncing around randomly between atoms. Without an applied voltage, as many electrons bounce one way as the other. There is no net flow of electrons and no electrical current. The electrons only acquire an average drift velocity and create electric current when a voltage is applied.

21.1 ELECTRIC CHARGE

Conductors and insulators

Electrons are too small to be seen
When you look at a wire in an active circuit, you do not see the electric current passing through it. If you *could* see atoms, however, you would find that a metal like copper is a three-dimensional matrix of atoms that are vibrating, but on average, staying fixed in place. The fixed atoms are surrounded by a sea of moving electrons. The electrons in this "sea" are the current carriers. A good conductor like copper contributes one free electron to the "sea" per copper atom. Each atom of copper has 29 electrons and 28 of these stay attached to their atoms. The 29th electron is the one that is free to carry current.

Electrons in insulators are trapped
All materials contain electrons. The electrons in insulators, however, are not free to move; they are tightly bound inside atoms. Since the atoms are fixed in place, the electrons in insulators are also fixed in place. Current cannot flow through an insulator because there are no movable charges with which to make the current.

Electrons in semiconductors
A *semiconductor* has a *few* free electrons, but, on average, many fewer than one per atom. This is not nearly as many as a conductor has. The diagram below illustrates the internal structures of conductors, insulators, and semiconductors.

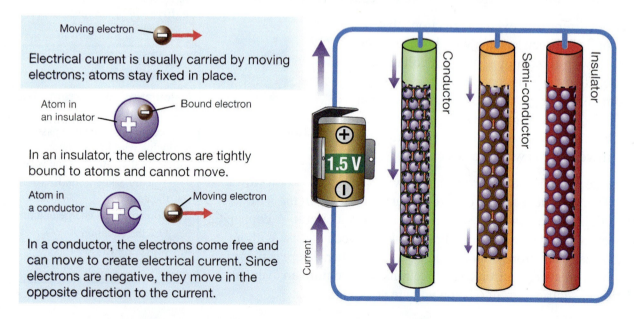

Electrical current is usually carried by moving electrons; atoms stay fixed in place.

In an insulator, the electrons are tightly bound to atoms and cannot move.

In a conductor, the electrons come free and can move to create electrical current. Since electrons are negative, they move in the opposite direction to the current.

Superconductivity

Certain materials become superconductors when they are cooled to very low temperatures. For example, the metal alloy niobium-zirconium becomes a superconductor at –262°C. A superconductor carries electrical current with no resistance. A current flowing in a loop of superconducting wire will keep flowing forever, even without a battery or any source of voltage!

Electrical resistance is caused by the transfer of momentum in collisions from moving electrons to stationary atoms. As an electron moves, it leaves a "wake" of slightly greater positive charge behind it. This positive charge can attract another electron. The small attraction causes electrons to travel together in pairs. When electrons are paired, any momentum lost by one is gained by the other. The pair together lose no momentum as they bounce off fixed atoms. That is why the resistance drops to zero. In a superconductor, the current is carried by pairs of electrons moving together. The pairing only occurs at very low temperature because the pair-bond between electrons is very weak. Warmer atoms have more energy and the electron pair-bond cannot form.

Static electricity, charge polarization, and induction

Charging by friction When two neutral objects are rubbed together, charge can be transferred from one to the other and the objects become oppositely charged. This is called **charging by friction**. Objects charged by this method will attract each other. For example, a balloon will become negatively charged if rubbed on hair or fur. After losing a few electrons, the hair will have a net positive charge and will then be attracted to the balloon. This tends to work best with fine, straight hair on a dry day.

Polarization A charged balloon will stick to a neutrally-charged wall or other insulating surface. When a negatively-charged balloon is near a wall, electrons inside atoms in the wall are repelled. Since the wall is made of insulating material, the repelled electrons are not free to travel between atoms. The electrons *can* move within each atom, so they spend more time on the side of the atom that is farthest from the balloon. The atoms undergo **charge polarization**; one end is positive and the other is negative (Figure 21.5). The balloon is both attracted to the positive side of each atom and repelled by the negative side. The force of attraction is stronger because the positive side of each atom is closer to the balloon than the negative side.

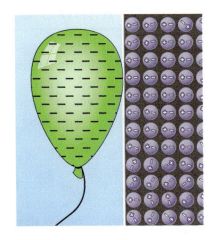

Figure 21.5: *When a charged balloon is brought near a wall, the atoms in the wall become polarized. The balloon is then attracted to the wall and will stick to it.*

Conductors Atoms in a material only become polarized if the material is an insulator. In a conductor, electrons are free to move from atom to atom so the entire object becomes polarized. A negatively-charged balloon brought near a conducting object polarizes the object, and an attractive force results. However, as soon as the balloon touches the conducting object, some of the balloon's excess electrons move onto the object. Both become negatively charged and repel each other. This is why a balloon sticks to a wood door but not a metal doorknob.

Charging by induction Because charges can flow, a charged object like a rubbed balloon can be used to charge an electroscope by **induction**. To charge by induction, the electroscope is first connected by a wire to a large neutral object. When the balloon comes near, the charge on the balloon induces an opposite charge to flow through the wire onto the electroscope. The wire is then removed quickly so the charge on the electroscope cannot flow back where it came from. The electroscope stays charged after the balloon is removed (Figure 21.6). The leaves spread apart because the added charges repel *each other* and spread out onto the electroscope leaves.

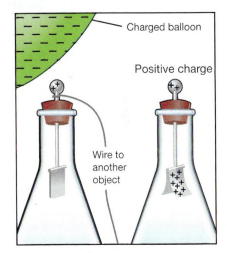

Figure 21.6: *Charging an electroscope by induction.*

21.2 Coulomb's Law

Coulomb's law describes the strength of the force between two charges. Coulomb's law is one of the fundamental relationships in the universe because atoms are held together by the electrical attraction between positive protons and negative electrons. The strength of the Coulomb force determines how close the electrons and protons get, and therefore determines the size of atoms. Since we are made of atoms, Coulomb's law indirectly determines our size as well.

Coulomb's law relates charge, distance, and force

The strength of electric forces The force between two charges gets stronger if the charges are closer together. The force also gets stronger if the amount of either charge is larger. **Coulomb's law** relates the force between two single charges (q_1, q_2) separated by a distance (r).

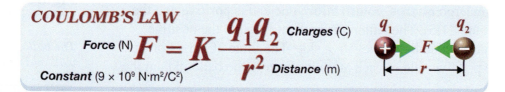

COULOMB'S LAW
Force (N) $F = K \dfrac{q_1 q_2}{r^2}$ Charges (C)
Constant (9×10^9 N·m²/C²) Distance (m)

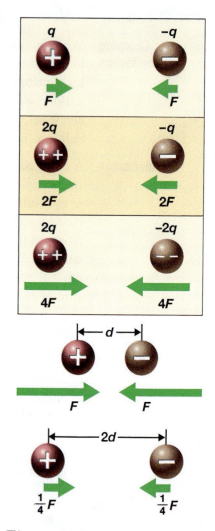

Force and charge The force between charges is directly proportional to the magnitude, or amount, of each charge (Figure 21.7). Doubling one charge doubles the force. Doubling both charges quadruples the force.

The strength of the coulomb force The coulomb constant, K, describes the strength of the coulomb force. The value of K is 9×10^9 N·m²/C². This means two 1-coulomb charges that are separated by one meter exert a force of 9,000,000,000 N, or 9 *billion* newtons on each other. Compared to other forces in nature, the coulomb force is very strong.

Force and distance The force between charges is inversely proportional to the square of the distance between them. Doubling the distance reduces the force by a factor of 2^2, decreasing the force to one-fourth of its original value. Tripling the distance reduces the force to one-ninth of its original value. This relationship is called an *inverse square law* because force and distance follow an inverse-square relationship. The force between charges decreases very quickly as charges are moved apart.

Figure 21.7: *The force between charges depends on their magnitudes and the distance between them.*

The force between charges

Point charges — Coulomb's law can be used to directly calculate the force between two point charges. Charges are considered "point-like" if they are physically much smaller than the distance between them. For example, two marbles a meter apart may be treated as points because the diameter of a marble is much smaller than 1 meter. Two baseballs 10 centimeters apart would *not* be considered as points, because the size of a baseball is not that different from the 10-centimeter separation.

The direction of the force — The direction of the force between two charges lies along a line between them. Figure 21.8 illustrates this concept for both attractive and repulsive forces. Like all forces, coulomb forces always occur in pairs according to Newton's third law, If the charges are both positive or both negative, the direction of the forces will be away from each other. If one is positive and the other is negative, the direction of the forces will be toward each other.

The magnitude of force between charges — Forces between charges are so strong it is hard to imagine them. For example, a cubic millimeter of carbon the size of your pencil point contains about 77 coulombs of positive charge in its protons and exactly then same amount of negative charge in its electrons. If you could separate the positive and negative charge by 1 meter, the attractive force between the charges would be 50 thousand, *billion* newtons. This is about the same force as the weight of a billion trucks (Figure 21.9), all from 1 cubic *millimeter* of pure charge!

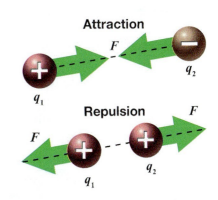

Figure 21.8: *The force between two charges is directed along a line connecting their centers.*

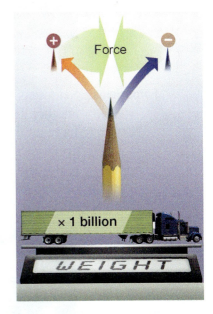

Figure 21.9: *If you could separate the positive and negative charge in a pencil point, the force pulling the charge back together would exceed the weight of a billion trucks.*

Calculating the force between two static charges

Two spheres are given equal but opposite charges of one ten-thousandth (0.0001) of a coulomb each. Calculate the force between the charges when they are separated by one-tenth (0.1) of a meter. Compare the force with the weight of an average 70-kg person.

1. You are asked to calculate the force and compare it to a person's weight.
2. You are given the charges and separation, and the mass of the person.
3. Use Coulomb's law, $F = \dfrac{-Kq_1 q_2}{r^2}$, for the electric force and $F = mg$ for the weight.
4. Solve:
   $F = (9 \times 10^9 \text{ N·m}^2/\text{C}^2)(0.0001\text{C})(0.0001\text{C}) \div (0.1 \text{ m})^2$
   $= 9{,}000 \text{ N}$

The weight of a 70-kg person: $F = mg = (70 \text{ kg})(9.8 \text{ N/kg}) = 686 \text{ N}$

The force between the charges is $9{,}000 \div 686$, or 13.1 times the weight of an average person.

Fields and forces

Fields Newton's law of universal gravitation describes the strength and direction of the force of gravity, but not how the force gets from one body to the next. How does the force of gravity act between two objects that are not touching? The answer is through a **field**. A field is a useful model to describe the way forces are transmitted through space between objects. All mass in the universe creates a **gravitational field** that exerts forces on other masses. The gravitational force that attracts the Moon to Earth results from the interaction of each of their gravitational fields.

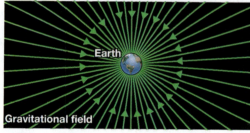

The mass of Earth creates a gravitational field.

The gravitational field exerts a force on the Moon.

Force fields Scientists believe forces between two objects do not act directly from one object to another. Instead, one object creates a **force field**. The force field then creates the force on the second object. *This applies to all forces.* You can sometimes, like with gravity, calculate the force between objects directly, skipping the force field. In reality, however, forces between objects are always exchanged through fields. Charge creates an **electric field** that creates forces on other charges. Think of a *field* as the *way* forces are transmitted through space between objects.

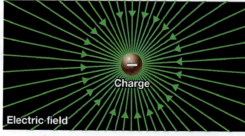
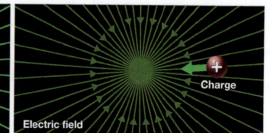

Charge creates an electric field.

The electric field exerts force on other charges.

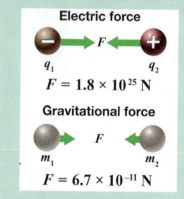

Gravity is far weaker than electric forces

Electric force

$F = 1.8 \times 10^{25}$ N

Gravitational force

$F = 6.7 \times 10^{-11}$ N

Consider two 1-kilogram iron spheres. When they are separated by 1 meter, the gravitational attraction between them is 6.7×10^{-11} N, about the weight of a speck of dust!

If you separated the positive and negative charges in a kilogram of iron, the force between them would be 1.8×10^{25} N, close to the weight of Earth.

We feel gravity because Earth has tremendous mass. We do not often feel the full strength of the electric force because it is so strong that positive and negative charges in matter are almost never separated.

Drawing the electric field

An electric field exists around a charge
The force between charges is always transmitted through the electric field. First, a charge called the *source* charge creates an electric field in the space around it. Second, the electric field exerts forces on other charges.

Strength of an electric field
The strength and direction of the electric field depends on both charge and distance. A larger source charge makes a stronger electric field. The electric field is strongest close to the source charge and gets weaker with increasing distance from the source charge.

Direction of an electric field
The direction of the electric field depends on the sign of the source charge. To determine the direction, imagine placing an imaginary positive **test charge** in the region of the field. Because it is imaginary, the test charge itself does not change the electric field. The electric field points in the direction of the force felt by the positive test charge. Electric field lines therefore point *toward* negative charges and *away* from positive charges (Figure 21.10). Because of this convention, a positive charge placed in an electric field will feel a force in the direction of the field, and a negative charge will feel a force opposite the direction of the field.

Field lines
It is sometimes convenient to diagram the field in a particular region. Vectors called **field lines** are used to indicate the direction of a field. The direction of a field line indicates the direction of the force exerted on a positive charge placed in the field. The strength of the field is indicated by the spacing between the field lines. The field is strong in the region where the field lines are close together and is weak where the lines are far apart.

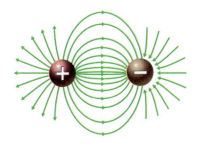

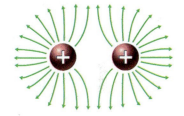

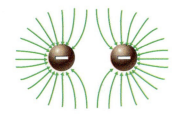

Figure 21.10: *The electric field lines flow from a positive to a negative charge. For like charges, they flow away from each other.*

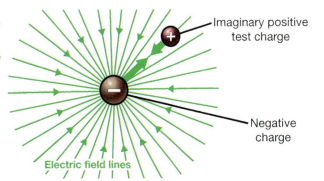

The field lines show the force on a positive test charge.

Field lines always point away from positive charge and toward negative charge.

The spacing of the lines indicates the strength of the electric field.

CHAPTER 21 ELECTRIC CHARGES AND FORCES

Electric fields and electric force

Measuring electric fields How is an electric field measured? What are the units of electric field? How do you calculate Coulomb forces from the electric field? The comparison between gravitational and electric fields helps answer these questions.

Gravitational field strength The strength of the gravitational field determines the strength and direction of gravitational force on an object. We call this force the object's *weight*. On Earth's surface, the gravitational field creates 9.8 N of force on each kilogram of mass. The equation relating force (F) and mass (m) is written $F = mg$, where g is the strength of the gravitational field in units of force per mass, or 9.8 N/kg.

Electric field strength The strength of the electric field determines the amount of force a charged object feels near another charged object. The object that creates the field is the *source* charge. The charge you place to test the force is the *test* charge. The force (F) on the test charge is equal to the amount of charge (q) multiplied by the electric field (E), or $F = qE$.

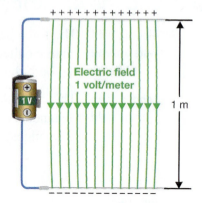

Figure 21.11: *A potential difference of 1 volt across a distance of 1 meter makes an electric field of 1 volt per meter (V/m).*

Field	Units	Equation	Interpretation
Gravity	$\vec{g} = \frac{\text{newtons}}{\text{kilograms}} \left(\frac{N}{kg} \right)$	$\vec{F} = m\vec{g}$	The force ($\vec{F}$) on an object is the mass (m) times field strength (g)
Electricity	$\vec{E} = \frac{\text{newtons}}{\text{coulomb}} \left(\frac{N}{C} \right)$	$\vec{F} = q\vec{E}$	The force ($\vec{F}$) on an object is the charge (q) times field strength (E)

Units of electric field With gravity, the strength of the field is in newtons per kilogram (N/kg) because the field describes the amount of force per kilogram of mass. With the electric field, the strength is in *newtons per coulomb* (N/C) for a similar reason. The electric field describes the amount of force *per coulomb of charge*. For example, a 10-C test charge feels 10 times as much force as a 1-C test charge in the same field.

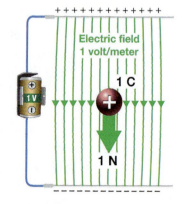

Figure 21.12: *An electric field of 1 V/m exerts a 1-N force on a 1-C test charge.*

Alternate units for the electric field The electric field can be expressed in more practical units. Remember, 1 volt is 1 joule per coulomb. A joule is equal to a newton-meter. By combining the relationships between units you can prove that 1 newton per coulomb is the same as 1 *volt per meter*. This is a prescription for how to make an electric field in the lab. A voltage difference of 1 volt over a space of 1 meter makes an electric field of 1 V/m (Figure 21.11). That same field exerts a force of 1 newton on a 1-coulomb test charge (Figure 21.12).

450 UNIT 7 ELECTRICITY AND MAGNETISM

Accelerators and electric shielding

How to make an electron beam accelerator
An electric field can be produced by maintaining a voltage difference across any insulating space, such as plastic, air, or a vacuum. Many electrical devices use electric fields created in this way. For example, suppose a flat plate is given a negative voltage and a metal screen is given a positive voltage (Figure 21.13). The voltage difference creates an electric field between the plates. Any electrons between the plates feel a force from the electric field. This device is an *accelerator* for electrons. Electrons are repelled from the plate and attracted to the screen. Because the screen has holes, many of the electrons pass right through it. It is easily possible to make a beam of electrons move at a speed of 1 million m/s with such a device. Electron beams generated this way are used in X-ray machines, televisions, computer displays, and many other technologies.

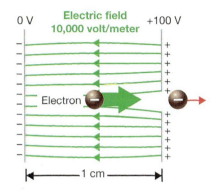

Figure 21.13: *Electric fields can create beams of high-speed electrons.*

Conductors can block electric fields
In a conductor, charges are free to move under the influence of any electric field. When a hollow conductor is placed in an electric field, an interesting thing happens. If the field is positive, negative charges move toward it until the field is neutralized. If the field is negative, electrons move away, leaving enough positive charge behind to neutralize the field. *On the inside of the conductor, the field is zero!*

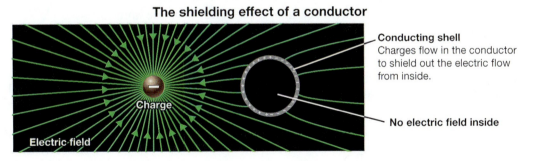

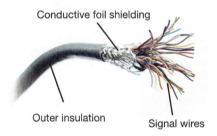

Figure 21.14: *A shielded computer cable blocks electrical interference with conductive foil.*

Shielding out electrical interference
Electric fields are created all around us by electric appliances, lightning, and even static electricity. These stray electric fields can interfere with the operation of computers and other sensitive electronics. Many electrical devices and wires that connect them are enclosed in conducting metal shells to take advantage of the **shielding** effect. For example, if you unwrap a computer network wire, you will find a bundle of smaller wires wrapped by aluminum foil. The aluminum foil is a conductor and shields the wires inside from electrical interference (Figure 21.14).

21.3 Capacitors

The circuits introduced in the last two chapters contained only resistances. In resistor circuits, the current stops flowing immediately when the voltage is removed. The subject of this section is a device called a *capacitor* which holds charge, like a bucket holds water. If the voltage is removed from a circuit containing a capacitor, the current keeps going for a while, until the capacitor is empty of charge. Almost all electric appliances, including televisions, cameras, and computers, use capacitors in their circuits. Capacitors are also a useful tool for investigating the relationship between electric charge, voltage, and current.

A capacitor is a storage device for electric charge

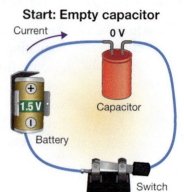

Figure 21.15: *When the switch is closed, current flows and charges the capacitor. As the capacitor charges, its voltage builds to oppose the current. When the capacitor is fully charged, it has the same voltage as the battery and no more current flows.*

A capacitor stores charge A **capacitor** is a device that stores charge. Think of a capacitor like a bottle with a hose attached. Current flowing into a capacitor fills the bottle with electric charge until it is full. Once a capacitor is full of charge, current stops flowing until some charge is emptied out again. Current flowing from the capacitor empties it of charge. The current will keep flowing out until the capacitor is empty of charge.

The comparison to a bottle of water is not totally accurate because current cannot flow *through* a capacitor. Also, capacitors fill equally with positive and negative charges so their net charge is zero. We will soon see how a capacitor is made and this distinction will become clearer.

Charging a capacitor A capacitor can be charged by connecting it to a battery or any other source of voltage that can also provide some electric current. As the capacitor fills up, the voltage across its terminals increases. When the capacitor is full, its voltage is equal to the battery voltage. Current stops flowing into a full capacitor because there is no longer any voltage difference between the capacitor and the battery (Figure 21.15).

Discharging a capacitor A capacitor can be **discharged** by connecting it to any closed circuit that allows current to flow. A low-resistance circuit will discharge the capacitor quickly because low resistance means high current and high current carries away charge faster. A high-resistance circuit will discharge a capacitor more slowly because high resistance limits the current, and therefore limits the rate at which charge flows out of the capacitor.

Current and voltage in capacitor circuits

Figure 21.16: *Capacitors and the circuit symbol for a capacitor.*

Capacitors in circuits The symbol for a capacitor in a circuit is shown in Figure 21.16, along with diagrams of actual capacitors. Capacitors can be connected in series or parallel in circuits, just like resistors. Unlike resistors, the current and voltage for a capacitor have a more complex relationship.

Current The current flowing into or out of a particular capacitor depends on four things.

- the amount of charge already in the capacitor
- the voltage applied to the capacitor by the circuit
- any resistance that limits the current flowing in or out of the capacitor
- the *capacitance* of the capacitor

A simple capacitor circuit Consider the circuit with an empty capacitor, switch, resistor, and battery shown in Figure 21.17. When the switch is closed, the current in the circuit is greatest because the capacitor is empty. In fact, the current is exactly what it would have been if the capacitor were not there. If the voltage is 1.5 V and the resistance is 15 Ω, the current initially in the circuit is 0.1 amps.

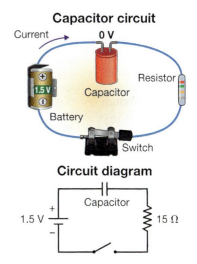

Figure 21.17: *A simple capacitor circuit with a resistor, capacitor, switch, and battery.*

Current and voltage while charging a capacitor The current in the circuit starts high and gets lower over time. The current gets lower because the battery voltage attracts negative charge to one side of the capacitor. The negative charge attracts an equal amount of positive charge to the other side. As charge fills up the capacitor, it creates a voltage difference between the two terminals of the capacitor. Charge continues to fill up the capacitor until the capacitor voltage is equal and opposite to the battery voltage.

Current and voltage change with time As the capacitor charges, the current in the circuit decreases. This is because the current flow is proportional to the voltage difference between the battery and the capacitor. As the voltage on the capacitor increases, the voltage difference gets lower and the current flow gets lower in response. The graphs in Figure 21.18 show how the current and voltage change in the circuit.

Voltage of a charged capacitor A fully-charged capacitor contains equal amounts of positive and negative charge. Because both kinds of charge are present but separated, a capacitor develops a voltage across its two terminals, like a battery. The voltage across a fully-charged capacitor is equal and opposite to the voltage applied to the circuit.

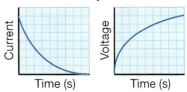

Figure 21.18: *Current and voltage graphs for a charging a capacitor circuit.*

How a capacitor works inside

Parallel plate capacitors The simplest type of capacitor is called a **parallel plate capacitor** (Figure 21.19). It is made of two conductive metal plates that are close together. When a parallel plate capacitor is charged, one plate is negative and the other is positive. To keep the positive and negative charges from coming together, a sheet of plastic or other insulating material is placed between the charged plates.

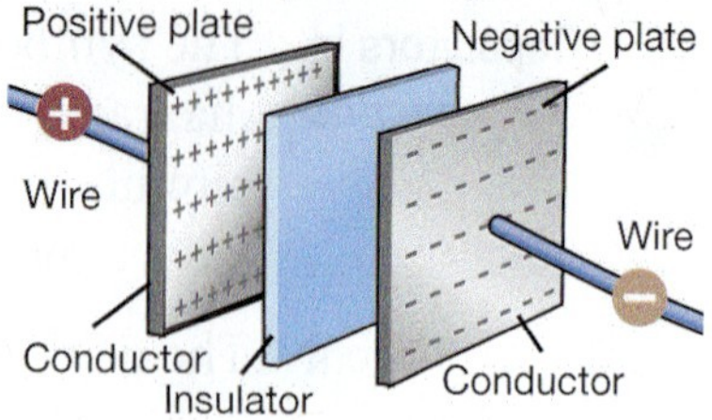

Making a simple capacitor A cardboard milk container and aluminum foil can be used to make a capacitor. The cardboard is an effective insulator. The aluminum can be shaped so that it covers the inside and outside surfaces (Figure 21.20). Wires can be connected to the aluminum to conduct charge into and out of the capacitor.

The amount of charge in a capacitor The amount of charge a capacitor can store depends on several factors.

- The amount of charge is proportional to the voltage applied.
- The electrical properties of the material between the positive and negative plates affect how much charge is stored.
- The area of the two plates affects charge. Larger areas can hold more charge.
- The separation distance between the plates affects charge storage.

The limit to a capacitor Unlike a bottle full of water which has a fixed volume, the amount of charge stored in a capacitor increases when more voltage is applied. For example, the same capacitor charged to 10 volts has twice as much charge as when it is charged to 5 volts. The ultimate limit to how much charge can be stored in a capacitor is the strength of the insulating material between the conducting plates. As the voltage increases, eventually the electric field in the insulator gets so large that the insulator breaks down and becomes a conductor itself. A miniature version of lightning happens inside the capacitor and the positive and negative charges flow through and neutralize each other.

Cylindrical capacitors A practical parallel plate capacitor is made of a flexible insulating material with metal foil on either side for the "plates." If the insulator and foil are thin, a relatively large area can be rolled up into a small cylinder. This allows each plate to have a large area while the whole capacitor fits into a small space. Some other types of capacitors, like batteries, use chemical reactions to store charge.

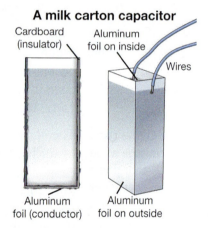

Figure 21.19: *A diagram of a parallel plate capacitor*

Figure 21.20: *Using a milk carton and aluminum foil to make a capacitor. Caution: Charging it against a carpet on a dry day can give a shock.*

Capacitance

The ability to store charge The ability of a capacitor to store charge is called **capacitance** (C). Capacitance is measured in **farads (F)**. A 1-farad capacitor can store 1 coulomb of charge when the voltage across its plates is 1 volt. One farad is a large amount of capacitance, so the **microfarad (μF)** is frequently used in place of the farad. There are 10^6 microfarads in 1 farad. Most capacitors in electronic devices are measured in microfarads or even smaller units. The equation for capacitance tells how much charge a capacitor stores per volt.

CAPACITANCE

Charge (C) $\quad q = CV \quad$ Voltage (V)

Capacitance (C/V)

Capacitor

More capacitance means more charge A capacitor with a large capacitance can hold more charge at the same voltage compared with a capacitor with a small capacitance. For example, if a 1-farad capacitor is connected to a 1.5-volt battery, it will store a charge of 1.5 coulombs when fully charged. If a 2-farad capacitor is connected to a 1.5-volt battery, it will hold 3 coulombs of charge.

Factors affecting capacitance There are three factors that affect a capacitor's ability to store charge. The size of the capacitor is the most obvious factor. The greater the area of a capacitor's plates, the more space for storing charge, and the greater the capacitance. The distance between the plates also affects the capacitance. The closer together the plates are, the greater the capacitance. The third factor is the type of insulator placed between the plates. The capacitance is the least when there is nothing between the plates, and can be many times greater when the right insulator is used.

Using a capacitor Batteries are used to supply energy for use over a long period of time, while capacitors are used for quick bursts of energy. Cameras use capacitors to supply energy to the flash attachment. When turned on, it takes a few seconds for the camera batteries to charge the capacitor. Pressing the shutter button completes the circuit. A large current flows through the flash lamp for a short time producing a very bright flash, until the capacitor's charge is gone.

Calculating capacitance and voltage for a capacitor

A capacitor holds 0.02 coulombs of charge when fully charged by a 12-volt battery. Calculate its capacitance and the voltage that would be required for it to hold 1 coulomb of charge.

1. You are asked to find the capacitance and the voltage needed to hold 1 C of charge.
2. You are given the voltage and corresponding charge.
3. Use $C = q \div V$ to calculate the capacitance.
4. Solve:
   $C = (0.02\ C) \div (12\ V)$
   $= 0.001667\ F$ or $1,667\ \mu F$

Rearrange $C = \dfrac{q}{V}$ to get $V = \dfrac{q}{C}$ and calculate the voltage required to store a charge of 1 C on the capacitor.
$V = (1\ C) \div (0.001667\ F)$
$= 600\ V$

The capacitor would hold 1 coulomb of charge at a voltage of 600 volts. Most capacitors would be destroyed by a voltage this high.

Chapter 21 Connection

Rival Projector Technologies

Today's data projectors use computer-controlled technologies to create and project both still and moving images. Two major designs make up the majority of all projectors used, from home entertainment systems to projectors used for business meetings—LCD and DLP® technology. Both projector designs have benefits and drawbacks, and use quite different methods for projecting images.

Liquid Crystal Display LCD stands for *liquid-crystal display*, and projectors using this design use three separate LCD panels to control the light that goes into the final projected image. Many LCD projectors use a metal-halide bulb, which produces a huge amount of light without a tremendous amount of electricity. The bright light created by the bulb travels through a prism, or a series of prisms, that separates out the red, green, and blue wavelengths of light. Each color of light then passes through an LCD panel that controls the actual picture itself. The image is broken down into a grid of picture elements, called pixels. Each LCD screen is comprised of a grid of tiny dots or squares, each representing one pixel. If the pixel is turned on, that dot or square becomes black, blocking light so it cannot get through. If it is off, the pixel remains transparent and light is allowed to pass through.

Recombining the colors of light Each LCD panel has the exact same number and arrangement of pixels. The amount of red, blue, and green light in each pixel is controlled using the on-off method. The amount of color required for each pixel is controlled by the rate at which each pixel is turned on and off, and these pixels can be switched back and forth very quickly. This allows for great control over each pixel's color and brightness. Once light has passed through all three LCD panels, the red, green, and blue images are recombined and then directed out the focusing lens and onto the projection surface all together. When all three separate color images are perfectly overlaid on the screen, a complete full-color image is produced. With all the prisms and separate LCD panels involved, this design tends to be a bulky and heavy. However, this method can produce rich colors and may be easier on the eyes than its direct competitor, DLP technology.

DLP and the DLP® logo are trademarks of Texas Instruments.

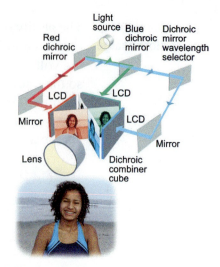

Figure 21.21: *LCD projector technology splits light up into separate colors, then recombines them to create high-resolution images.*

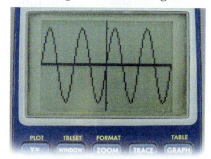

Figure 21.22: *Transparent LCD screens, similar to this graphing calculator screen, but with many more pixels, are used in LCD projectors. Darkened pixels block light from passing through the LCD screen, and control the amount of red, green, and blue light in the projected image.*

Chapter 21 Connection

Digital Light Processing

DLP technology processes image signals digitally. This technology uses tiny mirrors mounted on a microchip, and a series of color filters to control the light it projects. Light from a very bright bulb shines through a color wheel which is a small disk that has three or more color filters—red, green, and blue. As the color wheels spins, different colors of light are produced, depending on the filter that the light passes through. The DLP chip is a digitally-controlled micro-mirror device (DMD) that reflects the colored light from the filter through the focusing lens and onto the screen. As with an LCD system, the image is broken down into pixels, but the way the pixels are controlled is much different. In a DLP system, each pixel has its own tiny micro mirror located directly on the DLP chip. The DLP system reads the digital data of an image and times the reflecting angle of each micro mirror with the color of light passing through the filter. The amount of red, green, and blue light needed to create the desired color of each pixel at each moment is controlled by moving all the mirrors at once, one color at a time.

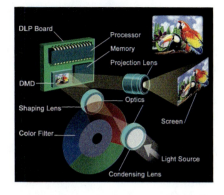

Figure 21.23: *DLP systems use a very bright bulb, a spinning color wheel, up to 2 million micro mirrors, focusing lenses, advanced electronics, and careful computer-controlled timing to produce high-quality images.*

For example, if a pixel requires the full amount of red light available to make a particular color, then it reflects the red light falling on it for the entire time period that the color wheel lets red light through for each rotation. If it needs less than the full amount of red, it will change the mirror's angle back and forth. This causes some of the red light for that pixel to be shown on the screen, and some to be absorbed inside the projector. This happens for every pixel in the image. Every single mirror reflects the correct amount of colored light coming through the wheel as it spins, one color at a time. When the color changes, all the mirrors reflect the correct amount of that color, and the process constantly repeats. All of this is controlled by the DLP system, and it happens so fast we see a crisp, projected image instead of a jerky, flickering series of red, blue, and green images. The more mirrors there are on the DMD chip, the more pixels in the image, and the higher its resolution.

Power savings and reduced size

Recent advances in *light-emitting diode* (LED) technology have introduced some of the first DLP projectors to use LEDs as a light source, which have a similar brightness and efficiency to metal halide, at a much-reduced size. Instead of being the size of a thick student textbook, several compact DLP projectors are on the market today that are the size of a deck of cards.

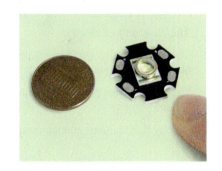

Figure 21.24: *The small yellow and silver piece on the black disc shown next to a penny is an LED element. The black disc helps dissipate excess heat generated by the LED when it operates.*

21.3 CAPACITORS   **457**

Chapter 21 Assessment

Vocabulary

Select the correct term to complete the sentences.

charge	electrically neutral	static electricity
electric force	field lines	point charges
charging by friction	electroscope	protons
neutrons	electrons	gravitational
charged	induction	Coulomb's law
capacitor	parallel plate capacitor	microfarad (µF)
coulomb (C)	electric field	capacitance
charge polarization	shielding	test charge
farad (F)	discharged	inverse square
force field	parallel plate	negative

1. The attraction between positively- and negatively-charged bodies is an example of _____.

2. A tiny imbalance in either positive or negative charge on an object is the cause of _____.

3. Positively charged particles in the nucleus of an atom are named _____.

4. The basic unit of electric charge is the _____.

5. The extremely light and negatively-charged particles found outside the nucleus of an atom are _____.

6. The fundamental property of matter identified as positive or negative is called _____.

7. A body that contains equal amounts of negative and positive charge is said to be _____.

8. Atoms in which electrons have moved toward one side have experienced _____.

9. The slightly more massive neutral particles found in the nucleus of an atom are the _____.

10. A separation of charge by rubbing neutrally-charged objects together is referred to as _____.

11. Electrons have a(n) _____ charge.

12. When an electroscope is charged by _____ it can acquire a charge opposite to the object causing the charging.

13. Any object with an excess of negative or positive charges is referred to as _____.

14. A device that is used to observe the force between charges is a(n) _____.

15. The small positive charge used to determine the direction of an electric field is called a(n) _____.

16. The force fields holding Earth and the Moon in their orbits are _____ fields.

17. The law relating two single charges separated by a distance is named _____.

18. Vectors used to indicate the direction of a field are called _____.

19. One charge exerts force on a second charge by creating a(n) _____.

20. Force and the distance between charges are related by a(n) _____ law.

21. Charged bodies that are small compared to the distance between them are considered to be _____.

22. A metal casing surrounding wires prevents electrical interference from stray fields by _____.

23. The basic unit for measuring capacitance is the _____.

24. When the current flowing from a capacitor in a closed circuit is reduced to zero the capacitor is considered to be _____.

25. The simplest type of capacitor is called a(n) _____.

26. A type of device used to store charge is called a(n) _____.

27. The ability of a capacitor to store charge is called _____.

28. An amount of capacitance equivalent to 1×10^{-6} farads is one _____.

Concept review

1. Why don't we usually notice the electric charge in everyday objects?
2. What does it mean to say an object is electrically neutral?
3. If a neutral object loses negative charges, what will its net charge be?
4. Two negative charges will _____ each other. Two positive charges will _____ each other. A positive charge and a negative charge will _____ each other.
5. Draw a model of an atom, showing all three types of its particles.
6. A negatively-charged balloon is held near the top of a neutral electroscope but doesn't touch it. What is the charge on the electroscopes leaves?
7. A circuit has a current of 2 A. How many coulombs of charge will pass by a point in the circuit in
   a. 1 second?
   b. 5 seconds?
   c. 1 minute?
8. Which type of charges are free to move in a metal?
9. Electric current is defined as the direction in which _____ charges move, or opposite the direction in which _____ charges move.
10. Explain how electrons move through a wire.
11. Explain what happens when a positively-charged balloon is used to charge a neutral electroscope by induction. What is the electroscope's charge?
12. Coulomb's law is an inverse-square law because the _____ and the _____ follow an inverse-square relationship.
13. Two protons are 1 m apart. What happens to the force between them if they are moved 2 m apart?
14. What happens to the strength of an electric field as you get farther away from a charge that creates the field?
15. An electric field shows the direction of the force felt by an imaginary _____ test charge placed in the field.
16. List two units that can be used to measure electric fields.
17. How is a gravitational field similar to an electric field? How are the two types of fields different?
18. Which forces are stronger—electric or gravitational?
19. Explain why the electric field inside a conductor is zero.
20. What is a parallel plate capacitor?
21. A battery is attached to a parallel plate capacitor as shown below.

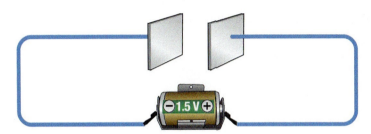

   a. Which plate will become positive and which will become negative?
   b. Will the positive plate gain electrons, gain protons, lose electrons, or lose protons?
   c. Will the negative plate gain electrons, gain protons, lose electrons, or lose protons?
22. What happens to the current in a circuit as a capacitor is discharging? Why?
23. How much charge can a 3-F capacitor store if connected to a 1-V battery?
24. What three factors affect a capacitor's ability to store charge?

Problems

1. Twenty coulombs of charge pass by a point in a circuit in 30 seconds. What is the current?

2. Calculate the force between a pair of 1-C charges that are 0.01 m apart.

3. Calculate the force between a 2-C positive charge and a 3-C negative charge separated by a distance of 0.5 m.

4. The repulsive force between two identical charges is measured to be 0.1 N when the charges are 0.2 m apart. Calculate the magnitude of each charge.

5. A capacitor in a camera flash is charged by four 1.5-V batteries. It holds a maximum of 0.015 C of charge. What is its capacitance?

6. A capacitor holds 0.06 C of charge when fully charged by a 9-V battery.
   a. Calculate its capacitance.
   b. How much voltage would be required to store 2 C of charge?

Applying your knowledge

1. The hazards of lightning are obvious. Other sources of static electricity can create hazardous situations.
   a. What are some sources of static electricity?
   b. What is the main hazard associated with static charge?
   c. How can static electricity be controlled?

2. How was Coulomb able to measure the force between charges? Research and write a short report detailing his technique.

3. What is a Leyden jar and what is its relationship to capacitors?

UNIT 7 ELECTRICITY AND MAGNETISM

CHAPTER 22

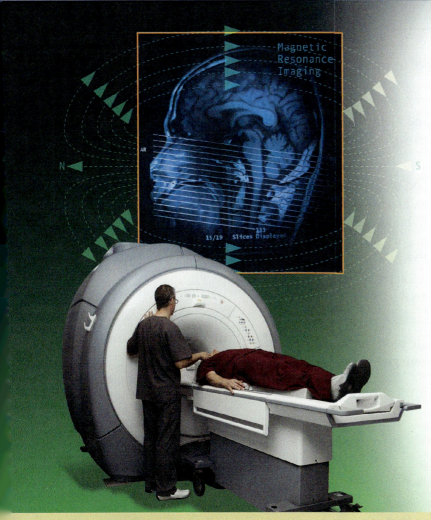

Magnetism

Objectives:

By the end of this chapter you should be able to:

✔ Describe the force between two permanent magnets.
✔ Sketch the magnetic field of a single permanent magnet.
✔ Predict the direction of the force on a magnet placed in a given magnetic field.
✔ Explain why ferromagnetic materials always attract magnets of either pole.
✔ Describe the theory behind why a compass works.
✔ Use a compass to find the direction of true north.

Key Questions:

- What are the characteristics common to all magnets?
- How do magnets interact with different materials?
- How does a compass work, and what is magnetic declination?

Vocabulary

compass	gauss	magnetic domain	magnetic north pole	paramagnetic material
demagnetization	hard magnet	magnetic field	magnetic south pole	permanent magnet
diamagnetic material	magnet	magnetic field lines	magnetism	soft magnet
ferromagnetic	magnetic declination	magnetic forces	magnetization	

461

CHAPTER 22 — MAGNETISM

22.1 Properties of Magnets

Magnetism has fascinated people since the earliest times. Until the period of the Renaissance, many people thought magnetism was a form of life because it could make some types of rocks move. Today we use magnets to stick notes on the refrigerator or pick up paper clips. Magnets are also part of electric motors, computer disk drives, burglar alarm systems, and many other common devices. This chapter develops some concepts that help explain some of the properties of magnets and magnetic materials.

What is a magnet?

Magnets and magnetic materials If a material is *magnetic*, it has the ability to exert forces on magnets or other magnetic materials. Some materials are actively magnetic, and we call them **magnets**. Other materials are attracted to nearby magnets but do not show magnetism otherwise. Iron and steel are in the second category because they are attracted by magnets but are not magnetic unless *magnetized*. A magnet on a refrigerator is attracted to the magnetic material, steel, that is part of the refrigerator's door. However, a steel nail is not attracted to the refrigerator door because steel is not magnetic unless magnetized.

Permanent magnets A **permanent magnet** is a material that keeps its magnetic properties, even when it is not close to other magnets. Bar magnets, refrigerator magnets, and horseshoe magnets are good examples of permanent magnets.

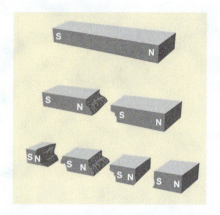

Figure 22.1: *If a magnet is cut in half, each half will have both a north pole and a south pole.*

Bar magnets

Horseshoe magnet

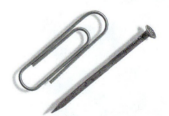
Materials that can be magnetized

Poles All magnets have two opposite poles, called the **magnetic north pole** and **magnetic south pole**. If a magnet is cut in half, each half will have its own north and south pole (Figure 22.1). It is impossible to have just a north or a south pole by itself. The north and south poles are comparable to the two sides of a coin. Every coin has two sides, just as every magnet has a north and a south pole.

The difference between magnetic poles and electric charge

Electric charges can be separated into a single type. For example, you can have a single negative charge or a single positive charge. Magnetic poles cannot be separated. It is not possible to have a magnetic north pole without a magnetic south pole. This is a fundamental difference between magnetism and electricity. There are some scientists looking for single magnetic poles, called *monopoles*, but none have ever been found.

The magnetic force

Attraction and repulsion

When near each other, magnets exert **magnetic forces** on each other. Magnetic forces can both attract and repel each other. Whether the magnetic force between two magnets is attraction or repulsion depends on the alignment of the poles (Figure 22.2). If two opposite poles are facing each other, the magnets will attract each other. If two like poles face each other, the magnets repel each other.

Most materials are not affected by magnetic forces

Magnetic forces can pass through most materials with no apparent decrease in strength. For example, the force of one magnet can drag another magnet from which it is separated by a piece of wood or your hand! Plastics, wood, and most insulating materials are not affected by magnetic forces. Conducting metals, like aluminum, also allow magnetic forces to pass through, but may change the forces. Iron and a few metals near iron on the periodic table have strong magnetic properties. Iron and iron-like metals that block magnetic forces are discussed later in this chapter.

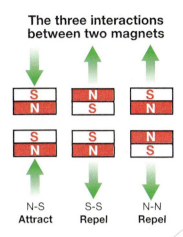

Figure 22.2: *The force between two magnets depends on how the poles are aligned.*

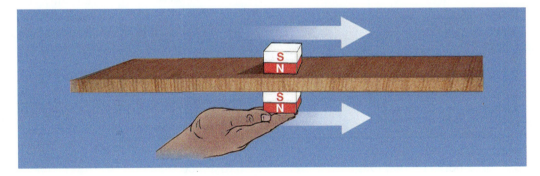

Comparing magnetic and electric forces

Magnetic forces are used in many applications because they are relatively easy to create and can be very strong. There are many ways to make large magnets with forces that are strong enough to lift a car (Figure 22.3). Small magnets are everywhere. For example, some doors are sealed with magnetic weatherstripping that blocks out drafts. There are several patents for magnetic zippers, and many handbags, briefcases, and cabinet doors close with magnetic latches. Although electric forces are stronger than magnetic forces, they are harder to create and control. It is much more difficult to keep electric charges separated than it is to make a magnet.

Figure 22.3: *Powerful magnets are used to lift cars in a junkyard.*

22.1 PROPERTIES OF MAGNETS   **463**

CHAPTER 22 MAGNETISM

The force between two magnets

Distance and the magnetic force

The strength of the force between magnets depends on the distance between them. When magnets are close together, the force between them is strong. As magnets are moved farther apart, the attractive or repulsive force gets weaker. Because every magnet has two poles, the force between two magnets decreases with distance much faster than the force between two electric charges.

Why magnetic force decreases rapidly with distance

Consider two magnets attracting each other. The total force between the magnets is the sum of the forces between *all four* magnetic poles. When two magnets are separated by a distance that is large compared with their size, both pairs of north and south poles are about the same distance away, as shown below. As a result, the attractive and repulsive forces nearly cancel each other out.

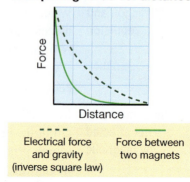

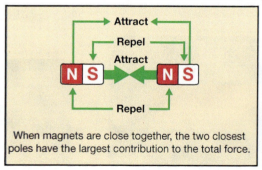

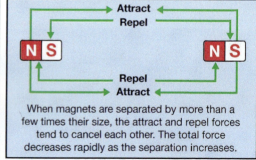

Figure 22.4: *The magnetic force decreases with distance much faster than does either gravity or the electric force. Gravity and the electric force between point charges both obey an inverse square law. The force between two magnets does not.*

Magnetic force does not obey an inverse square law

Mathematically, the force between two magnets decreases with distance much more rapidly than an inverse square law (Figure 22.4). If you try to squeeze two repelling magnets together, you can easily feel how quickly the force gets strong as the magnets get very close together. Separate a pair of small magnets by a few centimeters, and you can hardly feel a force at all.

Torque between two magnets

Two magnets near each other often experience a twisting force, or *torque*. This is also a result of having two poles. One pole is attracted and the other is repelled. The combination of attractive and repulsive forces on the same magnet creates a torque. Figure 22.5 shows how the torque on a test magnet changes, depending on its position relative to a fixed-source magnet.

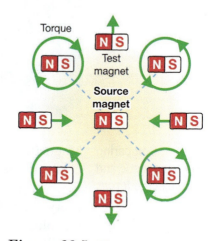

Figure 22.5: *The torques on a test magnet are shown at different positions around a fixed magnet.*

464 UNIT 7 ELECTRICITY AND MAGNETISM

The magnetic field

Describing magnetic force The force between two magnets is transmitted through a **magnetic field**, in a similar way as the Coulomb force is transmitted through an electric field or gravitational force through a gravitational field. The word *field* in physics means that there is a quantity, such as force, associated with every point in space. There are many other kinds of fields. For example, the "odor field" near a sewer would be strongest nearest the sewer, and get weaker farther away.

Using a test magnet to trace magnetic field lines All magnets create a magnetic field in the space around them, and the magnetic field creates forces on other magnets. Imagine you have an infinitesimal *test magnet* (Figure 22.6) that you are moving around a *source magnet*. The north pole of your test magnet feels a force everywhere in the space around the source magnet. For the purpose of tracing the magnetic field, we ignore the south pole of the test magnet. To keep track of the force, imagine drawing an arrow in the direction in which the north pole of your test magnet is pulled or pushed as you move it around. The arrows that you draw show the magnetic field. If you connect all the arrows, you get lines called **magnetic field lines**. Notice from the drawing that magnetic field lines often form loops that close on each other through the source magnet.

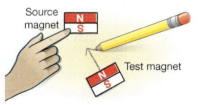
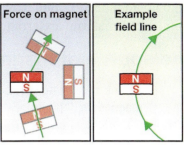

Figure 22.6: *Drawing a field line with a test magnet is done by moving the test magnet to different locations around the source magnet.*

Understanding magnetic field lines How do you interpret a drawing of a magnetic field? The number of field lines in a certain area indicates the relative strength of the magnetic field in that area. The closer the lines are together, the stronger the field. The arrows on the field lines indicate the direction of the force on the *north pole* of a test magnet (Figure 22.7). Magnetic field lines always point away from a magnet's own north pole and toward its south pole.

The interaction of fields Magnetic fields are typically more complex than electric or gravitational fields because there are no "point" magnets. All magnetic fields come from objects that have a three-dimensional distribution of magnetism. Calculations with magnetic fields are complicated, and there is no simple formula like Coulomb's law. However, experienced engineers and physicists are able to calculate magnetic fields and forces so precisely that it is possible to predict magnetic effects inside a single atom. For example, the medical technique of magnetic resonance imaging (MRI) is based on magnetic stimulation of atoms in the human body.

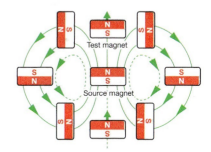

Figure 22.7: *The magnetic field is defined in terms of the force exerted on the north pole of another magnet.*

22.2 Magnetic Properties of Materials

It seems unusual that magnets can attract and repel other magnets but can only *attract* objects such as steel paper clips and nails. The reason this is true and the explanation for magnetism itself lie inside the atoms that make up matter. This section takes a microscopic look inside materials to explain their magnetic properties. **Magnetism**, like charge and mass, is a fundamental property of the particles that make up all atoms of matter. Whether or not we observe magnetism in a material depends on how the atoms in the material are arranged.

The source of magnetism

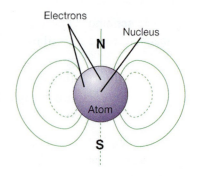

Figure 22.8: *Some atoms have magnetic fields that derive mainly from the motion of electrons surrounding the nucleus.*

Electrons and magnetism — The sources of nearly all magnetic effects in matter are the electrons in atoms. There are two ways in which electrons create magnetism. First, the electrons move around the nucleus and their motion makes the entire atom a small magnet (Figure 22.8). Second, electrons themselves act as though they are magnets.

Why magnetic properties vary in materials — All atoms have electrons, so you might think that all materials should be magnetic. In fact, there is great variability in the magnetic properties of materials. The variability comes from the arrangement of electrons within the atoms of different elements. The electrons in some atoms align to cancel out one another's magnetic influence. In other atoms, the electrons are aligned in a way that strengthens the overall magnetic field. While all materials show some kind of magnetic effect, the magnetism in most materials is too weak to detect without highly-sensitive instruments.

Diamagnetic materials — If an external magnetic north pole is brought near a **diamagnetic material**, the material develops its own slight north pole in response. Because the poles are similar, a diamagnetic material will slightly *repel* any external magnet. Diamond, copper, and silver are diamagnetic.

Paramagnetic materials — Individual atoms in **paramagnetic materials** *are* magnetic because the magnetism of individual electrons does not completely cancel out. However, the atoms themselves are randomly arranged so the overall magnetism of a sample with many atoms is zero (Figure 22.9). However, if an external north pole is brought near a paramagnetic material, the material develops a *south* pole in response. A paramagnetic material will always weakly *attract* an external magnet, regardless of its polarity. Aluminum is a paramagnetic metal.

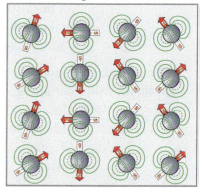

Figure 22.9: *Atoms in a paramagnetic material act like magnets. However, the atoms are randomly arranged so that the net magnetic effect is zero. If placed in a magnetic field, the atoms align so that the material is weakly magnetic.*

Ferromagnetism

Ferromagnetic materials
A small group of paramagnetic metals, including iron, nickel, and cobalt, have very strong magnetic properties. These metals are the best known examples of **ferromagnetic** materials. The electrons in each atom of a ferromagnetic material, such as iron, align so their magnetic fields do not cancel each other. Individual atoms of ferromagnetic materials do not act independently like they do in paramagnetic materials. Instead, atoms with similar magnetic orientations line up with neighboring atoms in groups called **magnetic domains**. Because many atoms align with each other, the magnetic strength of a domain is greatly multiplied in comparison to that of a single atom.

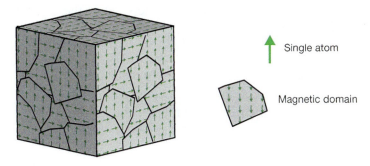

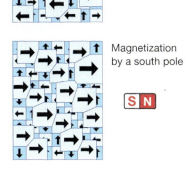

The alignment of domains to a north magnetic pole
Magnetic domains can grow or shrink in response to magnetic fields. For example, if you bring the north pole of a magnet near iron, the south pole of each atom in the iron is attracted to the magnet's north pole (Figure 22.10). All south-pointing magnetic domains quickly grow larger as atoms are attracted by the external magnet. Other magnetic domains shrink if they have magnetic poles pointed the wrong way. The iron becomes magnetized because it develops a south pole in response to the external north pole. For example, a paper clip sticks to a magnet because the magnetic domains in the steel that are attracted to the magnet have grown and the domains that are repelled by the magnet have shrunk.

Domains realign to attract any pole
If you bring the magnet's south pole near the paper clip, the opposite happens. Domains grow that have north poles facing the external magnet's south pole. This is why a magnet will always attract ferromagnetic materials, regardless of whether the magnet's north or south pole approaches the object. Domains in a ferromagnetic material can easily realign themselves to be attracted to either pole.

Figure 22.10: *Magnetic domains in a ferromagnetic material will always orient themselves to attract a permanent magnet. If a north pole approaches, domains grow that have south poles facing out. If a south pole approaches, domains grow that have north poles facing out.*

CHAPTER 22 MAGNETISM

Properties of ferromagnetic materials

Creating permanent magnets
If a magnet near a ferromagnetic material is removed, the domains in the material tend to go back to their random orientation. However, if an object made of a magnetic material is repeatedly magnetized or is placed in a very strong magnetic field, the domains become so well aligned that they stay aligned even after the external magnet is removed. This is how permanent magnets are made.

Hard magnets
Ferromagnetic materials differ in their ability to become and remain magnetized. Materials that make good permanent magnets are called **hard magnets**. The domains in these materials tend to remain aligned and are useful in devices that require permanent magnets. Because it is difficult to change the orientation of magnetic domains in hard magnets, creating permanent magnets out of these materials requires the application of a strong external magnetic field.

Materials for permanent magnets
Steel, which contains iron and carbon, is a common and inexpensive material used to create hard magnets. Typical bar and horseshoe magnets are made of steel. Most magnets however, are no longer made of steel. Stronger magnets are made from ceramics containing nickel and cobalt, or the rare-earth metal neodymium. Using these materials, it is possible to manufacture magnets that are very small, very strong, and harder to demagnetize than steel magnets.

Soft magnets
Materials that lose their magnetism quickly are called **soft magnets**. Soft magnets are easy to magnetize with even a weak bar magnet. You can see both the **magnetization** and **demagnetization** of small iron nails or paper clips using a magnet (Figure 22.11). If you use the north end of a bar magnet to pick up a nail, the nail becomes magnetized with its south pole toward the magnet. Because the nail itself becomes a magnet, it can be used to pick up other nails. If you separate that first nail from the bar magnet, the entire chain demagnetizes and falls apart.

How a magnet may be demagnetized
Even hard magnets can be demagnetized. If a magnet is vibrated or repeatedly struck, the domains can become unaligned and the magnetism will weaken. Forcing two like poles together is enough to demagnetize some steel magnets. High temperatures can also demagnetize magnets, because the energy of high-temperature atoms can exceed the energy that keeps the atoms aligned in magnetic domains.

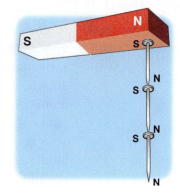

Figure 22.11: *Iron nails become temporarily magnetized when placed near a magnet.*

There are no liquid magnets

The atoms in a liquid are free to move around. That means they cannot be fixed into place with their magnetic poles aligned. For this reason, there are no liquid permanent magnets. Permanent magnetism is a property of solids alone. In fact, if you heat a magnetic solid up, it loses its magnetism. The temperature at which this happens is called the *Curie temperature*. This temperature varies for different materials and is approximately 360°C for nickel, 770°C for iron, and 1,100°C for cobalt.

22.3 Earth's Magnetic Field

The biggest magnet on Earth is the planet itself. Earth has a magnetic field that has been useful to travelers for thousands of years. Compasses, which contain small magnets, interact with our planet's magnetic field to indicate direction. Certain animals, including migratory birds, can feel Earth's magnetic field and use their magnetic sense to locate north or south. This section is about the magnetic field of Earth.

Discovering and using magnetism

Lodestone — As early as 500 BCE, people discovered that some naturally occurring materials—such as *lodestone* and *magnetite*—have magnetic properties. The Greeks observed that one end of a suspended piece of lodestone pointed north and the other end pointed south, helping sailors and travelers to find their way. This discovery lead to the first important application of magnetism—the **compass** (Figure 22.12).

The Chinese "south pointer" — The invention of the compass is also recorded in China, in 220 BCE. Writings from the Zheng dynasty tell stories of how people would use a "south pointer" when they went out to search for jade, so that they wouldn't lose their way home. The pointer was made of lodestone. It looked like a large spoon with a short, skinny handle. When balanced on a plate, the "handle" was aligned with magnetic south.

The first iron needle compass — By 1088 CE, iron refining had developed to the point where the Chinese were making a small needle-like compass. Shen Kua recorded that a needle-shaped magnet was placed on a reed floating in a bowl of water. Chinese inventors also suspended a long, thin magnet in the air, realizing in both cases that the magnet ends were aligned with geographic north and south. Explorers from the Sung dynasty sailed their trading ships all the way to Saudi Arabia using compasses among their navigational tools. About 100 years later, a similar design appeared in Europe and soon spread through the region.

Compasses and exploration — By 1200 CE, explorers from Italy were using a compass to guide ocean voyages beyond the sight of land. The Chinese also continued exploring with compasses, and by the 1400s, they were traveling to the east coast of Africa. The compass, and the voyages that it made possible, led to many interactions among cultures.

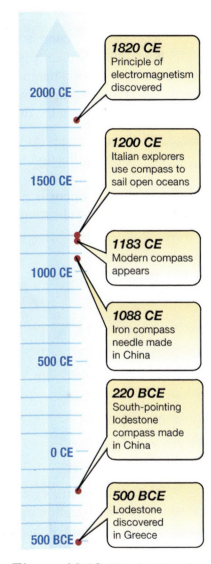

Figure 22.12: *The timeline shows the developments that lead to the modern compass.*

How does a compass work?

A compass is a magnet
A compass needle is a magnet that is free to spin. The needle spins until it lines up with any magnetic field that may be present (Figure 22.13). Because the north pole of a magnet is attracted to the south pole of another magnet, the north pole of a compass needle always points toward the south pole of a permanent magnet. This is in the direction of the field lines.

North and south poles
The origin of the terms *north pole* and *south pole* of a magnet come from the direction that a magnetized compass needle points. The end of the magnet that pointed towards geographic north was called the north pole of the magnet and opposite pole was called south. The names were decided long before people truly understood *why* a compass needle worked.

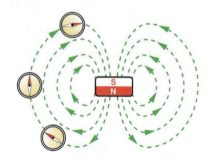

Figure 22.13: *A compass needle aligns itself with the magnetic field lines. The north pole of the compass needle points toward the south pole of the bar magnet.*

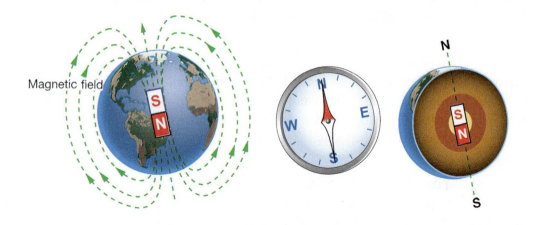

Earth is a magnet
Earth's *geographic* north and south poles are the points at which Earth's axis of rotation intersects the planet surface. When you use a compass, the north-pointing end of the needle points toward a spot near, but not exactly at, Earth's geographic north pole. Earth's *magnetic* poles are defined by the planet's magnetic field. That means the *south magnetic pole* of the planet is near the north geographic pole. Earth has a planetary magnetic field that acts as if the core of the planet contained a giant magnet oriented as shown in the diagram.

> **Some animals have biological compasses**
>
> Many animals, including species of birds, frogs, fish, turtles, and bacteria, can sense Earth's magnetic field. Migratory birds are the best known examples. Scientists have found magnetite, a magnetic mineral, in bacteria and in the brains of birds. Tiny crystals of magnetite may act like compasses and allow the animals to sense Earth's magnetic field.

The source of Earth's magnetism

The strength of Earth's magnetic field

Earth's magnetic field is very weak compared with the strength of the magnetic field on the surface of the ceramic magnets you might have in your classroom. For this reason, you cannot trust a compass to point north if any magnets are close by. The **gauss** is a unit used to measure the strength of a magnetic field. A small ceramic permanent magnet has a field of a few hundred up to 1,000 gauss at its surface. By contrast, the magnetic field averages about 0.5 gauss at Earth's surface. Of course, the field is much stronger nearer the core of the planet.

Earth's magnetic core

While Earth's core is magnetic, we know it is not a permanent magnet. Studies of earthquake waves reveal that Earth's core is made of hot, dense molten iron, nickel, and possibly other metals that slowly circulate around a solid inner core (Figure 22.14). The motion of the molten core creates electric currents, which produce a magnetic field like that made by an *electromagnet* (Chapter 23).

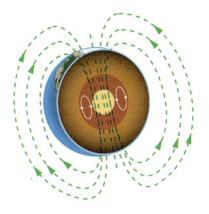

Figure 22.14: *Scientists believe that the motion of molten metals in Earth's outer core create its magnetic field.*

Reversing poles

Historical data shows that the strength of Earth's magnetic field can vary and the location of the north and south magnetic poles can switch places. Studies of magnetized rocks in Earth's oceanic crust provide evidence that the poles have reversed many times over the last tens of millions of years (Figure 22.15). The reversal has happened every 500,000 years on average. The last field reversal occurred roughly 750,000 years ago, so Earth is overdue for another change of magnetic polarity.

The next reversal

Today, Earth's magnetic field is losing approximately seven percent of its strength every 100 years. We do not know whether this trend will continue, but if it does, the magnetic poles could reverse in the next 2,000 years. During a reversal, Earth's magnetic field would not completely disappear. However, the main magnetic field that we use for navigation would be replaced by several smaller fields with poles in different locations.

Movements of the magnetic poles

The location of Earth's magnetic pole is always changing slowly, even between full reversals. Today, Earth's magnetic south pole is about 1,000 kilometers, or 600 miles, from the geographic north pole. This is the pole to which the north pole of a compass needle points. When navigating with a compass, it is necessary to take this difference, called *declination*, into account.

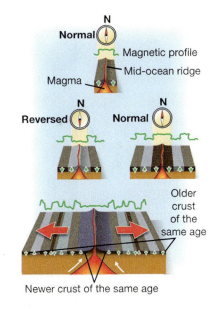

Figure 22.15: *When oceanic crust is made on the seafloor at mid-ocean ridges, the crustal rock records Earth's polarity.*

Magnetic declination

Compensating for pole differences
Earth's geographic north pole, or *true north*, is where the planet's axis of rotation intersects its surface. Because Earth's magnetic south pole is not at the same place, a compass does *not* point *directly* to the geographic north pole. Depending on where you are, a compass will point slightly east or west of true north. The difference between the direction a compass points and the direction of true north is called **magnetic declination**. Magnetic declination is measured in degrees and is indicated on topographical maps.

Finding true north with a compass
Most good compasses contain an adjustable ring with a degree scale and an arrow that can be turned to point toward the destination on a map (Figure 22.16). The ring is turned the appropriate number of degrees to compensate for the declination. If you were using the map shown below, you would have to adjust your compass by 16 degrees to find true north.

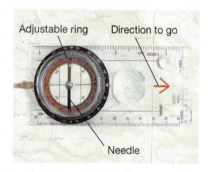

Figure 22.16: *This compass has an adjustable ring that is rotated to set the direction you want to go. After correcting for the declination, you rotate the whole compass until the north-pointing end of the needle lines up with zero degrees on the ring. The large arrow points in the direction you want to go.*

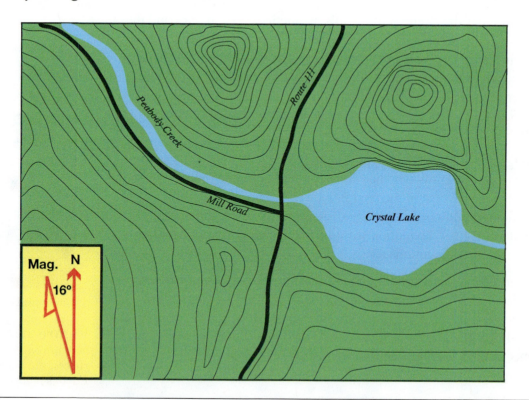

Orienteering

What was once a necessary survival skill is today a challenging competition. Competitors in orienteering are given a compass, a destination, and a map. They must find their way to the destination using wilderness navigation skills to get around many obstacles, such as rivers or mountains.

Chapter 22 Connection

Magnetic Resonance Imaging

Imagine how medical diagnosis worked before X-rays, ultrasound, or *magnetic resonance imaging* (MRI). A doctor of 100 years ago might have had to perform surgery just to find out what was wrong. Today, we enjoy the benefits of advanced medical technologies that allow doctors to see inside the body without having to perform surgery that might be more dangerous than the illness.

What an MRI scanner does — MRI is a powerful diagnostic technology. An MRI scanner makes a three-dimensional map of the *inside* of the body. As the name implies, MRI technology uses magnets and resonance to create images (Figure 22.17).

Magnetic field of the nucleus — The nucleus of every atom has a magnetic field (Figure 22.18). The nuclear magnetic field has a different strength for different elements. For example, a hydrogen nucleus has a different magnetic field than a carbon nucleus. Even though the nuclear magnetic field is very small, it is important in MRI.

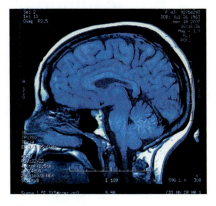

Figure 22.17: *An MRI image.*

Energy and magnetic orientation — Consider what happens to a magnet in a strong magnetic field. The magnet aligns so that its own north pole points in the direction of the magnetic field. If you want to turn the magnet opposite to the field, you have to apply a force, and thus do work. Applying force and doing work means exchanging energy. As a result, a magnet that *opposes* the field has higher energy than a magnet that is aligned *with* the field. Left alone, magnets tend to settle into the lowest energy position and align themselves with the magnetic field around them.

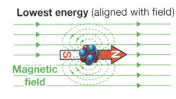 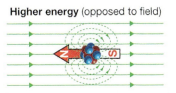 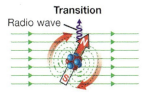

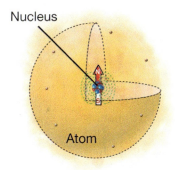

Figure 22.18: *The nucleus of an atom has a magnetic field of its own.*

Energy exchange by a nucleus in a magnetic field — As you learned in the last unit, atoms exchange energy through electromagnetic radiation, including light. Because the nucleus of an atom acts like a small magnet, there is an energy difference depending on whether the nucleus is aligned with, or opposed to, an applied magnetic field. If a nucleus absorbs the right amount of energy, it will flip its magnetic orientation from aligned to opposed. If the same nucleus is left in the opposed, high energy position, it quickly flips back to being aligned with the field and gives off the change in energy as radio waves.

22.3 EARTH'S MAGNETIC FIELD

Chapter 22 Connection

The MRI scanning machine

What an MRI scanner does

An MRI scanner uses the magnetism of the nucleus to map the locations of different parts of the body. The large cylinder you slide into is a powerful magnet (Figure 22.19). The magnet creates a strong and very uniform magnetic field. It also creates a preferred orientation for the nuclei of atoms in the body. When a body is inside an MRI scanner, the nuclei of every atom tends to line up with the magnetic field.

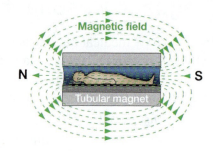

Figure 22.19: *For an MRI scan, the patient lies in the center of a tubular magnet. The magnetic field inside the magnet is very uniform. MRI scanners use electromagnets instead of permanent magnets.*

Each nucleus is a small resonant oscillator

A nucleus is in equilibrium with the magnetic field when its own magnetic poles are aligned with the field. The interaction between the magnetism of the nucleus and the external field creates a restoring force. Remember, from Chapter 13, restoring forces tend to push systems back toward equilibrium. Because the nucleus has mass, it also has inertia. The combination of a restoring force and inertia makes each nucleus into a tiny oscillator that can flip back and forth in the magnetic field (Figure 22.20). The resonant frequency of the nuclear oscillator depends on the ratio of the mass of the nucleus to its magnetic strength. Since each element has a different mass and field strength, each element also has a different resonant frequency.

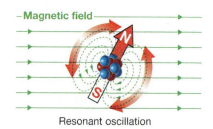

Figure 22.20: *The magnetic field's restoring force turns an atom's nucleus into a resonant oscillator.*

MRI uses radio waves at the natural frequency of the nucleus

The MRI scanner measures the absorption and emission of radio waves that are tuned to the natural frequency of different nuclei. For example, one set of frequencies might be used to map the density of hydrogen atoms in a body. Since hydrogen is in water, the body has a lot of hydrogen. A different frequency is used to map calcium atoms, which are found in bones. A third frequency is used for carbon atoms, and so on. Because the radio waves can pass right through the body, the MRI scan is able to make three-dimensional images of the body's interior. These images are so detailed that doctors can see individual veins and arteries clearly. The procedure is painless and virtually risk-free to the patient. One of the only dangers is that the strong magnetic fields might exert forces on metal parts inside a person's body. Some people who have had broken bones repaired with metal pins, have had knee or hip joint replacements, or have pacemakers for heart arrhythmia cannot be examined safely with an MRI scanner.

Chapter 22 Assessment

Vocabulary
Select the correct term to complete the sentences.

magnets	north pole	south
magnetization	demagnetization	magnetic fields
compass	magnetic field lines	diamagnetic
paramagnetic	ferromagnetic	soft magnets
magnetic declination	magnetic domains	hard magnets
permanent magnet	gauss	

1. The forces between two magnets are created by their _____.
2. Magnetic field lines of a bar magnet always point away from a magnet's _____.
3. Objects which attract magnetic materials or may repel magnets are called _____.
4. Lines drawn to represent the direction of the magnetic field surrounding a magnet are called _____.
5. A material that keeps its magnetic properties even when separated from other magnets is considered a(n) _____.
6. The end of a suspended bar magnet that points toward the south geographic pole should be labeled _____ pole.
7. Materials in which neighboring atoms readily form groups called magnetic domains are considered _____.
8. The process by which iron metal forms south poles in response to an external north pole is known as _____.
9. Materials that make good permanent magnets are called _____.
10. Areas of ferromagnetic material in which atoms of similar magnetic orientation are aligned are referred to as _____.
11. Materials in which the electrons are oriented so their individual magnetic fields cancel each other out are _____.
12. Materials that lose their magnetism quickly are called _____.
13. Magnets that are heated, vibrated, or dropped may undergo _____.
14. Materials in which the magnetism of individual electrons does not completely cancel out that of other electrons is called _____.
15. A magnet free to spin until it lines up with the magnetic field at its location is called a(n) _____.
16. The difference between true north and the direction a compass points is known as _____.
17. A unit used to measure the strength of a magnetic field is the _____.

Concept review

1. List three objects made of magnetic materials.
2. A north pole of a magnet will _____ the north pole of a second magnet. A south pole will _____ a south pole. A north pole will _____ a south pole.
3. Explain how the force between two magnetic poles depends on the distance between them.
4. What does the spacing of magnetic field lines tell you?
5. Magnetic fields have direction. How is the direction determined?
6. How are magnetic poles and electric charges similar? How are they different?
7. What is the source of magnetism in materials?
8. Compare diamagnetic, paramagnetic, and ferromagnetic materials.
9. If all materials contain moving electrons, why is the magnetism in most materials too weak to measure?
10. What are magnetic domains?
11. Compare hard magnets and soft magnets.
12. How can a permanent magnet be created?
13. Name two ways a magnet can become demagnetized.
14. Where is the magnetic north pole located with respect to Earth's geographic poles?
15. How is a compass used to determine direction?
16. What do scientists believe to be the source of Earth's magnetic field?
17. What evidence do scientists have to support the theory that the location of Earth's magnetic poles is constantly changing?
18. What is magnetic declination?

Problems

1. The magnet shown in the picture on the left is dropped and breaks into three pieces. Copy the diagram of the pieces and label the north and south poles on each piece.

2. A compass is located at point X near a bar magnet as shown. Which diagram shows the proper direction of the compass needle?

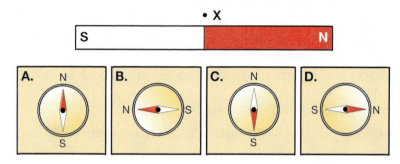

3. Which diagram below best represents the magnetic field near a bar magnet?

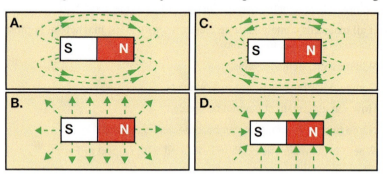

4. Which magnet(s) pictured below experience(s) a clockwise torque?

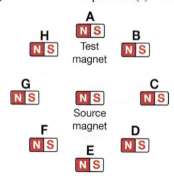

5. Which pair of objects will *not* experience an attractive magnetic force?

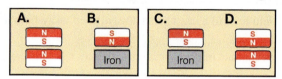

6. Two cross-country skiers refer to a topographic map and note that the magnetic declination is 16° east in the area where they are skiing. If they were to ski 10 km north according to their compasses, how far would they be from a point that is 10 km true north of their position?

7. The diagram shows magnetic field lines that result when a piece of soft iron is placed between unlike magnetic poles. At which point—A, B, C, or D—is the strength of the magnetic field greatest?

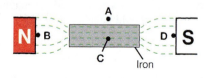

Applying your knowledge

1. If you hang a dollar bill by one end and bring a strong magnet toward it, the bill will move toward one pole of the magnet. Why? Of what commercial value is this?

2. The bubble in a carpenter's level will move when a strong magnet is brought near the fluid of the level. Why does this happen? In what direction does the bubble move?

UNIT 7 ELECTRICITY AND MAGNETISM

CHAPTER 23

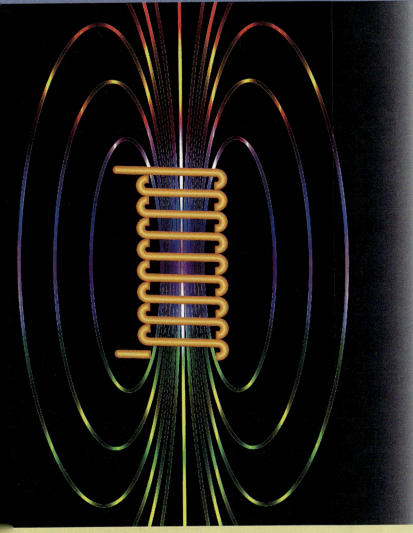

Electricity and Magnetism

Objectives:

By the end of this chapter you should be able to:

- ✔ Predict the direction of the force on a moving charge or current-carrying wire in a magnetic field by using the right-hand rule.
- ✔ Explain the relationship between electric current and magnetism.
- ✔ Describe and construct a simple electromagnet.
- ✔ Explain the concept of commutation as it relates to an electric motor.
- ✔ Explain how the concept of magnetic flux applies to generating electric current using Faraday's law of induction.
- ✔ Describe three ways to increase the current from an electric generator.

Key Questions:

- What is the relationship between electricity and magnetism?
- What is an electromagnet, and why is it important to motors?
- How does Faraday's law of induction explain how generators work?

Vocabulary

coil	Faraday's law of induction	induced current	right-hand rule
commutator		magnetic flux	solenoid
electromagnet	generator	polarity	tesla

477

CHAPTER 23 ELECTRICITY AND MAGNETISM

23.1 Electric Current and Magnetism

For a long time, it was thought that electricity and magnetism were unrelated. As scientists began to understand electricity better, they searched for relationships between electricity and magnetism. The breakthrough discovery was made in front of a class of students. In 1819, Hans Christian Oersted, a Danish scientist and professor, tried an experiment in front of his class. In the experiment, Oersted placed a compass needle near a wire through which he could make electric current flow by closing a switch. When the switch was closed, the compass needle moved just as if the wire was a magnet. We now know that magnetism is created by the motion of electric charge and that electricity and magnetism are really two forms of the same basic force.

The magnetic field of a wire carrying current

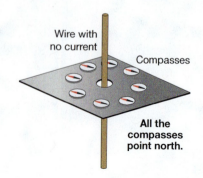

Figure 23.1: *An apparatus for investigating the magnetic field around a straight wire.*

An experiment with a wire and compasses Wires carrying electric current create magnetic fields. The proof is in the following experiment. A long straight wire is connected to a battery with a switch. The wire passes through a board with a hole in it. Around the hole are many compasses which can detect any magnetic field (Figure 23.1).

The response of the compasses to electric current When the switch is off, the compasses all point north. As soon as the switch is closed, the compasses point in a circle (Figure 23.2). The compasses point in a circle as long as electric current is flowing in the wire. If the current stops, the compasses point north again. If the current is reversed in the wire, the compasses again point in a circle, but in the opposite direction.

The magnetic field of a wire The experiment with the compasses shows that a wire carrying electric current makes a magnetic field around it. The magnetic field lines are concentric circles with the wire at the center. As you may have guessed, the direction of the field depends on the direction of the current in the wire. The **right-hand rule** can be used to show which way the magnetic field lines point. When your thumb is in the direction of the current, the fingers of your right hand wraparound the wire in the direction of the magnetic field.

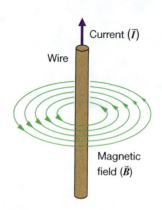

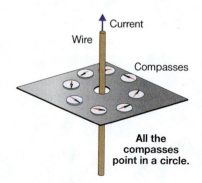

Figure 23.2: *The compass needles all form a circle when the current is switched on in the wire.*

478   UNIT 7 ELECTRICITY AND MAGNETISM

Magnetic forces and electric currents

The force between two wires
Two wires carrying electric current exert forces on each other, just like two magnets. The forces can be attractive or repulsive depending on the direction of current in both wires.

Observing the force between wires
As an example, consider two parallel wires. When the current flows in the same direction in both wires, the wires attract each other. If the currents go in opposite directions, the wires repel each other. For the amount of current in most electric circuits, the forces are small but can be seen with careful experiments. For example, if the wires are 1 meter long and each carries 100 amps of current, the force between them is 0.1 newton when the wires are 1 centimeter apart.

Force on a current in a magnetic field

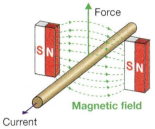

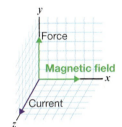

The force is perpendicular to both the current and the magnetic field.

Figure 23.3: *A current-carrying wire in a magnetic field experiences forces.*

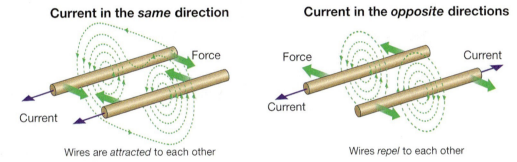

Force on a current in a magnetic field
The force between wires comes from the interaction of the magnetic field with moving current moving in the wire. A similar effect can be seen with a wire in any magnetic field. For example, Figure 23.3 shows a wire between two magnets. The wire feels a force when current is flowing. *The force is perpendicular to both the current and the magnetic field.* This is a new kind of behavior for a force. Magnetic forces between currents require you to work in all three dimensions because force, current, and the magnetic field are at right angles to each other.

The right-hand rule
The direction of the force can be determined by the right-hand rule. If you bend the fingers of your right hand as shown in Figure 23.4, your thumb, index, and middle finger indicate the directions of the magnetic field, force, and current, respectively. It does not matter which finger or thumb is assigned to which of the three quantities. The rule works correctly as long as you use your *right* hand.

The right hand rule
(for force, current and field)

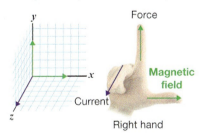

Figure 23.4: *The right-hand rule is used to find the direction of the force between a current and a magnetic field.*

The magnetic field of loops and coils

Making a strong magnetic field from current

The magnetic field around a single wire is too small to be of much use. However, there are two techniques to make strong magnetic fields from current in wires.

- Many wires are bundled together, allowing the same current to create many times the magnetic field of a single wire.
- Bundled wires are made into **coils** which concentrate the magnetic field in their center.

Bundling wires

When wires are bundled, the total magnetic field is the sum of the fields created by the current in each individual wire. By wrapping the same wire into a coil, current can be "reused" as many times as there are turns in the coil (Figure 23.5). This works because the current that creates the magnetic field is the *total current* crossing a surface perpendicular to the wire. It does not matter if the current is in one wire or 50 wires. For example, suppose a wire carries 1 amp of electric current. A coil with 50 turns of this wire creates the same magnetic field as a single wire with 50 amps of current. A coil with 50 turns is preferable since it is easier and safer to work with 1 amp than it is to work with 50 amps.

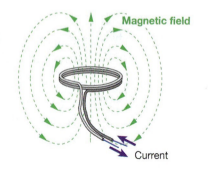

Figure 23.5: *A coil of wire creates a magnetic field as shown.*

Coils and solenoids

A coil concentrates the magnetic field at its center. When a wire is bent into a circular loop, field lines on the inside of the loop squeeze together. Field lines that are closer together indicate a higher magnetic field. Field lines on the outside of the coil spread apart, making the average field lower outside the coil than inside. The **solenoid** is the most common form of a coil with multiple turns of wire (Figure 23.6).

Where coils are used

A coil takes advantage of both techniques for maximizing the magnetic field created by current, bundling wires, and making bundled wires into coils. Coils are used in electromagnets, speakers, electric motors, electric guitars, and almost every kind of electric appliance that has moving parts. As you will read in the next section, magnetic forces are the simplest way to make electric currents do mechanical work. Coils are the most efficient way to make a strong magnetic field with the least amount of current. This explains why coils are found in so many electric machines and appliances.

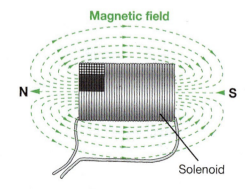

Figure 23.6: *A solenoid is a tubular coil of wire with many turns. The upper left corner of the solenoid in the diagram has been cut away to show the arrangement of wires.*

The true nature of magnetism

Coils and permanent magnets
The magnetic field of a coil and the magnetic field of a cylindrical magnet are identical (Figure 23.7). Both have a north and south pole and both have the same shape of field lines. In fact, if all you could measure was the magnetic field itself, you could not tell whether it was made by current flowing in a coil or by a permanent magnet.

Permanent magnets and atomic currents
Permanent magnets and electromagnets create the same magnetic field because *magnetism is fundamentally an effect created by electric current*. Remember, from the last chapter, that the magnetism of individual atoms creates the magnetic field of a permanent magnet. Now we can see why atoms themselves are magnetic. The electrons moving around the nucleus carry electric charge. Moving charge makes electric current so the electrons around the nucleus create currents within an atom. These currents are what create the magnetic fields that determine the magnetic properties of individual atoms.

How atomic currents add up
When the current from many atoms is all in the same direction (Figure 23.8), the atoms act like a single large loop of current. The magnetic fields from each atom add up to make a single, larger magnetic field, as if the atoms were replaced by a coil carrying the loop current. The magnetic field of a permanent magnet is the sum of the individual fields of trillions and trillions of individual atoms.

Not all atoms are magnetic
All matter is not normally magnetic for two reasons. First, atoms heavier than hydrogen contain many electrons. The current from each electron in the same atom is *not* usually in the same direction. In fact, in most atoms the currents created by individual electrons almost completely cancel each other out. For every electron making a clockwise current, there is one making a counterclockwise current. As a result, the majority of atoms have zero net current, and therefore do not have a magnetic field.

Why only some materials are magnetic
The second reason most matter is not magnetic is that the currents in neighboring atoms do not line up. One single atom might be magnetic, but it may be neutralized by the magnetic atom next to it which is pointing in the opposite direction. Only the ferromagnetic elements, such as iron, nickel, and cobalt, have both magnetic atoms and a tendency for neighboring atoms to align their magnetic poles.

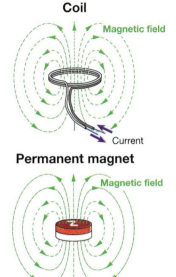

Figure 23.7: *The magnetic field of a coil is identical to the field of a disk-shaped permanent magnet.*

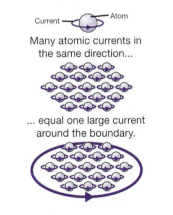

Figure 23.8: *Atoms with their currents oriented the same way act like one large coil of wire carrying current in the same direction.*

CHAPTER 23 ELECTRICITY AND MAGNETISM

The magnetic force on a moving charge

Making charges orbit

In Chapter 8, you learned that forces perpendicular to velocity create circular motion. This is also true for charges in magnetic fields. A charge moving perpendicular to a magnetic field moves in a circular orbit.

Magnetic force on a moving charge The magnetic force on a wire is really due to force acting on moving charges in the wire. A charge moving in a magnetic field feels a force perpendicular to both the magnetic field and to the direction of motion of the charge.

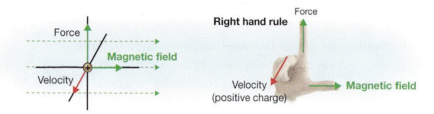

The tesla, a unit of magnetic field strength The strength of the force depends on the strength of the field, the charge, and the speed at which the charge is moving. A magnetic field that has a strength of 1 **tesla** (1 T) creates a force of 1 newton (1 N) on a charge of 1 coulomb (1 C) moving at 1 meter per second (1 m/s). This relationship is how the unit of magnetic field strength is defined. Since one coulomb per second is the same as one amp, the relationship is usually expressed in terms of amps and meters instead of coulombs and meters per second. In this form, a 1-T magnetic field exerts 1 N of force on a current of 1 A flowing for 1 m perpendicular to the magnetic field.

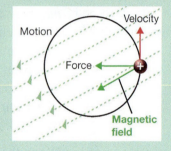

A charge moving at an angle to a magnetic field moves in a spiral.

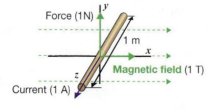

Relationship of current, force, and magnetic field strength units

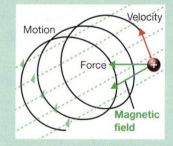

Magnetic forces are used to deflect the electron beam in an old-fashioned cathode-ray television picture tube and in other applications that use moving electric charge.

Tesla, gauss, and typical field strengths One tesla is a *very* strong magnetic field. The gauss, another common unit for magnetic field strength, is smaller; 10,000 gauss equals 1 tesla. Earth's magnetic field is about one-half gauss, or 5×10^{-5} T. The field strength near a strong permanent magnet can be as high as 0.3 T, or 3,000 gauss. The strongest electromagnetic fields ever produced have reached 100 teslas for a fraction of a second.

Calculating magnetic fields and forces

Magnetic fields and forces are complex calculations Magnetic fields are three-dimensional and are created by magnetic domains or currents distributed in three-dimensional objects. Except for a few simple cases, the calculations are beyond the scope of this book. The calculation of magnetic forces is even harder because forces come from interactions between the field and magnetic domains or currents throughout the volume of a object. These types of calculations are more easily done with calculus or with computer programs.

The field of a straight wire Two cases which do use simple formulas are a long straight wire and the center of a circular loop. The field of a straight wire is proportional to the current in the wire and inversely proportional to the radius from the wire. The field lines make concentric circles with a direction given by the right-hand rule.

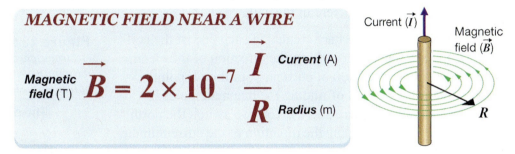

MAGNETIC FIELD NEAR A WIRE

Magnetic field (T) $\vec{B} = 2 \times 10^{-7} \dfrac{I}{R}$ Current (A) / Radius (m)

The field at the center of a coil The magnetic field at the center of a coil comes from the whole circumference of the coil. As a result, the field is a factor of π times larger ($\pi \approx 3.14$) than the field of a straight wire. If the coil has N turns, the magnetic field at the center is multiplied by N. For example, a coil of 25 turns will have a magnetic field 25 times stronger than a coil of one turn.

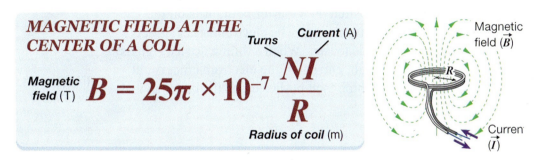

MAGNETIC FIELD AT THE CENTER OF A COIL

Magnetic field (T) $\vec{B} = 25\pi \times 10^{-7} \dfrac{NI}{R}$ Turns / Current (A) / Radius of coil (m)

Calculating the magnetic field at the center of a coil

A current of 2 amps flows in a coil made from 400 turns of very thin wire. The radius of the coil is 1 centimeter. Calculate the magnetic field strength in teslas in the center of the coil.

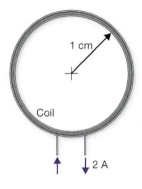

1. You are asked for the magnetic field strength in teslas.
2. You are given the current, radius, and number of turns.
3. Use the formula for the field of a coil: $B = (25\pi \times 10^{-7})(NI \div R)$
4. Solve:

$$B = \dfrac{(25\pi \times 10^{-7})(400)(2\ \text{A})}{(0.01\ \text{m})}$$

$$= 0.628\ \text{T}$$

CHAPTER 23 ELECTRICITY AND MAGNETISM

23.2 Electromagnets and the Electric Motor

Magnets made with electric current are called electromagnets, and they are the subject of this section. Electromagnets are very useful in many technologies because they can be made much stronger than permanent magnets. Electromagnets can also be switched on and off. The magnetic poles of an electromagnet can also be reversed easily. This section will also show you how electromagnets work in electric motors, generators, and many familiar devices.

Electromagnets

A coil of wire   **Electromagnets** are devices that create magnetism through electric current flowing in wires. The simplest electromagnet uses a coil of wire, often wrapped around an iron core (Figure 23.9). Because iron is magnetic, it concentrates and amplifies the magnetic field created by the current in the coil.

The poles of an electromagnet   The north and south poles of an electromagnet are located at each end of the coil (Figure 23.9). Which end is the north pole depends on the direction of the electric current. When your fingers curl in the direction of current, your thumb points toward the magnet's north pole. This method of finding the magnetic poles is another example of the *right-hand rule* (Figure 23.10). You can switch the north and south poles of an electromagnet by reversing the direction of the current in the coil. This is a great advantage over permanent magnets.

Current and the strength of an electromagnet   Electromagnets have several advantages over permanent magnets and are more useful in many applications. By changing the amount of current in an electromagnet, you can easily change its strength or turn its magnetism on and off. Electromagnets can also be designed to be much stronger than permanent magnets by using a large amount of electric current.

The electromagnet in a toaster   Toasters use electromagnets. The sliding switch on a toaster both turns on the heating circuit and sends current to an electromagnet. The electromagnet attracts a spring-loaded metal tray to the bottom of the toaster. When a timing device signals that the bread has been toasting long enough, current to the electromagnet is cut off. This releases the spring-loaded tray, which then pushes up on the bread so that it pops out of the toaster.

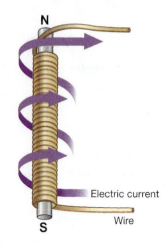

Figure 23.9: *A simple electromagnet can be made by wrapping a coil of wire around a rod of iron or steel.*

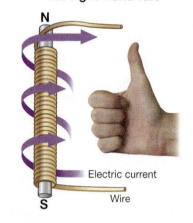

Figure 23.10: *Another right-hand rule: When your fingers curl in the direction of the current, your thumb points toward the magnet's north pole.*

Building an electromagnet

Wire and a nail You can easily build an electromagnet from wire and a piece of iron, such as a nail. Wrap the wire snugly around the nail many times and connect a battery as shown in Figure 23.11. When current flows in the wire, the nail becomes a magnet. Use the right-hand rule to identify which end of the nail is the north pole and which is the south pole. To reverse north and south, reverse the connection to the battery, making the current flow the opposite way.

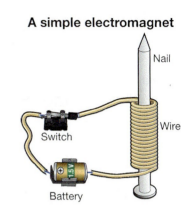

Figure 23.11: *An electromagnet can be made from a nail, wire, and a battery.*

Increase the electromagnet's strength You might expect that more current would make an electromagnet stronger. This is correct, and it can be achieved in two ways:

1. You can apply more voltage by adding a second battery.
2. You can add more turns of wire around the nail, while using only one battery.

Why adding turns works The second method works because the magnetism in your electromagnet comes from the *total* amount of current flowing *around* the nail (Figure 23.12). If there is 1 ampere of current in the wire, each loop of wire adds 1 ampere to the total amount that flows around the nail. Ten loops of 1 ampere each creates a magnetic field that is the equivalent of a magnetic field generated by one turn of wire with 10 amperes flowing through it. By adding more turns, you get a stronger magnetic field.

Resistance Of course, nothing comes for free. By adding more turns you also increase the resistance of the coil. Increasing the resistance makes the current a little lower. Lower current decreases the magnetic field instead of increasing it. More resistance also generates more heat. A good electromagnet strikes a balance between having enough turns to make the magnet strong and not having so many turns that the resistance is too high.

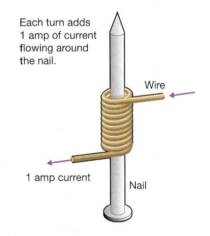

Figure 23.12: *Adding turns of wire increases the magnetic field strength from the current flowing in the wire. The same current flows in each turn, but each turn increases the magnetic field strength.*

Factors affecting the field The magnetic field created by a simple electromagnet depends on three factors:

- the amount of electric current in the wire;
- the amount and type of material in the electromagnet's core; and
- the number of turns in the coil.

In more sophisticated electromagnets, the shape, size, material in the core, and winding pattern of the coil each can be changed to control the strength of the magnetic field produced.

The principle of the electric motor

Imagine a spinning disk with magnets

An electric motor uses electromagnets to convert electrical energy into mechanical energy. One set of magnets rotates around the axis of the motor and turns the shaft. The other set of magnets is stationary and does not rotate. The stationary set of magnets pushes and pulls on the rotating magnets to turn the motor.

To see how this works, imagine a disk that is free to spin. Around the edge of the disk are permanent magnets arranged so their north and south poles alternate facing outward. Figure 23.13 is a diagram of this rotating disk.

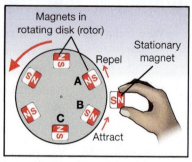

How to make the disk spin

The south pole of another magnet is now brought near the disk. The new magnet attracts the north pole of Magnet B and repels the south pole of Magnet A. These forces make the disk spin a small distance counterclockwise.

Reversing the magnet is the key

To keep the disk spinning, the external magnet must be reversed as soon as Magnet B passes by. Once the magnet has been reversed, Magnet B will now be repelled and Magnet C will be attracted. As a result of the push-pull, the disk continues to rotate counterclockwise.

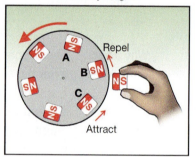

The principle of an electric motor

The disk will keep spinning as long as the external magnet is reversed every time the next magnet in the disk passes by. This is the operating principle of the electric motor. Stationary magnets reverse their *polarity* so they alternately push or pull on a second, rotating assembly of magnets. In the example, permanent magnets are used for the rotating part and a single permanent magnet is used for the stationary part. In an actual electric motor, one or both sets of magnets, rotating and stationary, could be electromagnets.

Figure 23.13: *Using a single magnet to spin a disk of magnets. Reversing the magnet in your fingers attracts and repels the magnets in the rotor, making it spin.*

Knowing when to reverse the magnet

The disk is called the *rotor* because it can rotate. The key to making the rotor spin smoothly is to reverse the *polarity* of the stationary magnet when the disk is in the right position. The term **polarity** means the orientation of the north and south poles. The polarity reversal should occur just as a rotor magnet passes by a stationary magnet. If the reversal comes too early, the rotor magnet is repelled before it reaches the stationary magnet. If the reversal is too late, the stationary magnet attracts the rotor magnet backwards after it has passed. For the highest efficiency, the switching of the stationary magnet must be synchronized with the passage of each magnet in the rotor.

Commutation

How electromagnets are used in electric motors

In electric motors, an electromagnet provides a clever way to reverse the polarity of the stationary magnet without moving anything. The switch from north to south is done by reversing the electric current in the coil. The sketch below shows how the electromagnet keeps the rotor spinning.

First, the electromagnet repels magnet A and attracts magnet B.

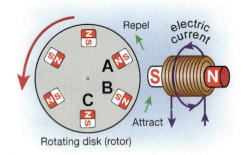

Then, the electromagnet switches so it repels magnet B and attracts magnet C.

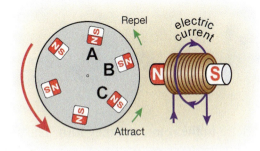

Blender

Laptop computer

The commutator is a kind of switch

The electromagnet must switch from north to south as each magnet in the rotor passes by. The process of reversing the current in the electromagnet to synchronize with the rotor is called **commutation**. The switch that makes it happen is called a **commutator**. As the rotor spins, the commutator switches the direction of current back and forth in the electromagnet. This makes the side of the electromagnet facing the rotor change polarity from north to south and back again. The electromagnet alternately attracts and repels the magnets in the rotor, and the motor turns.

Drill

The three things you need to make a motor

Electric motors are very common (Figure 23.14). All types of electric motors have three key components. The components are

- a rotating element, the rotor with magnets that may be permanent or electromagnets;
- at least one stationary permanent magnet or electromagnet; and
- if there are electromagnets, a commutator that switches the electromagnets from north to south at the right place to keep the rotor spinning.

Figure 23.14: *There are electric motors all around you, even where you do not see them. Can you identify where an electric motor is used in each of the devices shown above?*

23.2 ELECTROMAGNETS AND THE ELECTRIC MOTOR

How a battery-powered electric motor works

Inside a small electric motor

If you take apart an electric motor that runs on batteries, it does not look like the motor you built in the lab. The same three mechanisms are there; the difference is in the arrangement of the electromagnets and permanent magnets. The picture below shows a small battery-powered electric motor and what it looks like inside with one end of the motor case removed. The permanent magnets are on the outside, and they stay fixed in place.

AC motors

Almost all of the electric motors you find around your house use AC electricity. Remember, AC means alternating current, so the current switches back and forth as it comes out of the wall socket. This makes it easier to build motors.

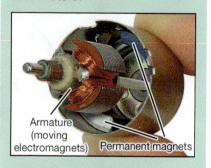

Electromagnets and the armature

The electromagnets are in the rotor, and they turn. The rotating part of the motor, including the electromagnets, is called the *armature*. The armature in the picture above has three electromagnets, corresponding to the three coils, A, B, and C, in the sketch below.

Most AC motors use electromagnets for both the rotating magnets on the armature, and also for the stationary magnets around the outside. The attract-repel switching happens in both sets of electromagnets.

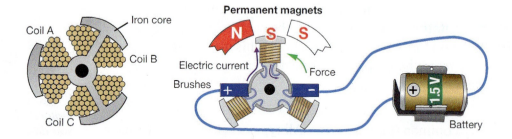

How the switching happens

The wires from each of the three coils are attached to three metal plates—the commutator—at the end of the armature. As the rotor spins, the three plates come into contact with positive and negative *brushes*. Electric current flows through the brushes into the coils. As the motor turns, the plates rotate past the brushes, switching the electromagnets from north to south by reversing the positive and negative connections to the coils. The turning electromagnets are attracted and repelled by the permanent magnets and the motor turns.

23.3 Induction and the Electric Generator

Motors transform electrical energy into mechanical energy. Electric generators do the opposite. They transform mechanical energy into electrical energy. In this section, you will learn how generators produce electricity. You will also see an important example of the principle of *symmetry* in physics.

Electromagnetic induction

Magnetism and electricity A current flowing through a wire creates a magnetic field. The reverse is also true. If you move a magnet near a coil of wire, an electric current will be produced. This process is called **electromagnetic induction** because a moving magnet *induces* electric current to flow.

Symmetry in physics Many laws in physics display *symmetry*. When a physical law is symmetric, a process described by the law works in two directions. Earlier in this chapter, you learned that moving electric charge creates magnetism. The symmetry is that changing magnetic fields also cause electric charges to move. Nearly all physical laws display symmetry of one form or another.

Current flows as the magnet moves into the coil Figure 23.15 shows an example of an experiment demonstrating electromagnetic induction. In the experiment, a magnet can move in and out of a coil of wire. The coil is attached to an ammeter that measures the electric current produced. When the magnet moves into the coil of wire, *as it is moving,* electric current is induced in the coil and the ammeter swings to the left. The current stops if the magnet stops moving.

Current reverses as the magnet moves out When the magnet is pulled back out again, *as it is moving,* current is induced in the opposite direction. The ammeter swings to the right as the magnet is moving out. Again, if the magnet stops moving, the current also stops.

Current is induced only by changing magnetic fields Current is produced only if the magnet is moving because a *changing* magnetic field is what creates current. Moving magnets induce current because they create changing magnetic fields. If the magnetic field does not change, such as when the magnet is stationary, the current is zero. If the magnetic field is *increasing,* the **induced current** is in one direction. If the field is *decreasing,* the induced current is in the opposite direction.

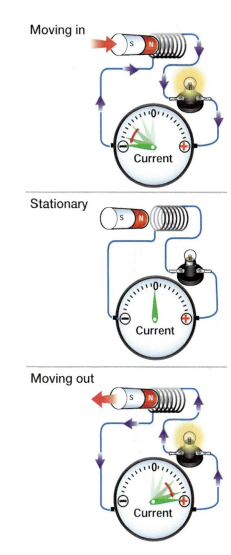

Figure 23.15: *A moving magnet produces a current in a coil of wire.*

Magnetic flux

Not all moving magnets induce current Suppose an inventor overhears the story about moving magnets making electricity. The inventor gets a magnet and sets up a coil of wire and an ammeter. The inventor shakes the magnet up and down and moves it every which way (Figure 23.16). But, the experiment does not work. No current is measured by the ammeter.

Magnetic field lines A moving magnet induces current in a coil *only if* the magnetic field of the magnet passes through the coil. If a magnet is far away, its field does not pass through the coil and no amount of motion will cause current to flow. The closer the magnet is, the more of its magnetic field is "linked" with the coil, and the stronger the induced current.

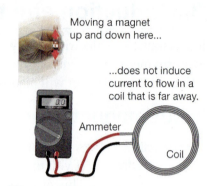

Figure 23.16: *Moving a magnet far from a coil does not induce a current in the coil.*

Magnetic flux Recall that magnetic field lines point outward from the north pole of a bar magnet and loop around to the south pole. If a coil of wire is placed near the north pole of a bar magnet, many of the magnet's field lines pass through the coil. The **magnetic flux** *through* the coil is a measure of the amount of the magnetic field lines encircled by the coil (Figure 23.17). The magnetic flux depends on the size of the coil, its orientation, and the magnetic field strength. A coil can enclose more flux if it is larger, has more turns, or if the magnetic field is stronger. Note, the flux has to go *through* the coil, not past it. A coil that is parallel to the field does *not* capture any flux because the field lines are not enclosed by the coil.

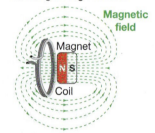

Changing flux causes current to flow It is the *change in magnetic flux* through a coil that induces current to flow. You can think of magnetic flux as the number of magnetic fields *linking* the magnet and the coil. If the magnetic flux increases, current flows in one direction. If the flux decreases, current flows in the opposite direction.

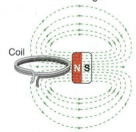

Faraday's law of induction The amount of current depends on the *rate* at which the magnetic flux changes. The greater the rate of change of flux, the greater the current. This relationship is known as **Faraday's law of induction**. When you move a magnet into or out of a loop of wire during the process of electromagnetic induction, you create a current by changing the magnetic flux through the loop. The faster you move the magnet, the greater the rate of change of the flux, and the greater the current.

Figure 23.17: *The magnetic flux through a loop of wire depends on the size of the loop and the strength of the magnetic field.*

Faraday's law of induction

Faraday's law

Faraday's law says the current in a coil is proportional to the rate at which the magnetic field passing through the coil, the flux, changes. To make sense of this law, consider a coil of wire rotating between two magnets (Figure 23.18).

The induced current in a coil rotating through a magnetic field

Look at the illustration showing positions A–F. When the coil is in position A, the magnetic flux points from left to right. As the coil rotates to position B, the number of field lines that go through the coil decreases. As a result, the flux starts to decrease and current flows in a negative direction.

At position C, the largest negative current flows because the *rate of change* in flux is greatest. The graph of flux versus time has its greatest slope at position C, and that is why the current is largest. Remember, current is proportional to the *rate of change* of flux, not the amount of flux. In fact, at position C, no magnetic field lines are passing through the coil at all and therefore the flux through it is *zero*.

As the coil rotates to position D, flux is still decreasing and becoming more negative. Current flows in the same direction, but decreases proportionally to the decreasing rate of change; the slope of flux versus time levels out. At position E, the flux through the coil reaches its most negative value. The slope of the flux versus time graph is zero and the current is zero. As the coil rotates to position F, the flux starts increasing and current flows in the opposite direction.

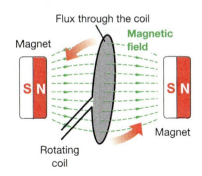

Figure 23.18: *Farady's law can be seen when a coil rotates in a magnetic field. The flux through the coil is a measure of the magnetic field passing through the area enclosed by the coil.*

Energy conservation and Faraday's law

The electric current produced by induction does not create electrical energy from nothing. The induced current makes its own magnetic field that opposes the rotation of the coil. Because of the coil's own field, it takes force to make the coil rotate. The electrical energy created by a rotating coil can never exceed the work done to make the coil turn.

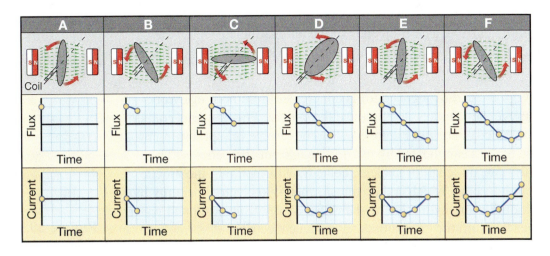

23.3 INDUCTION AND THE ELECTRIC GENERATOR

CHAPTER 23 ELECTRICITY AND MAGNETISM

Generating electricity by induction

How a generator makes electricity

A **generator** is a device that uses induction to convert mechanical energy into electrical energy. An effective laboratory generator can be made from a spinning disk with magnets on it. As the disk rotates, first a north pole and then a south pole pass the coil. When a north pole is approaching, the current flows one way. When the north pole passes and a south pole approaches, the current flows the other way. As long as the disk is spinning, there is a changing magnetic flux through the coil and electric current is induced to flow.

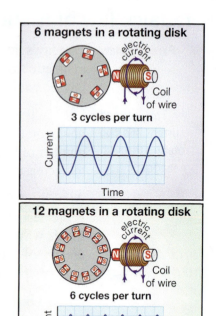

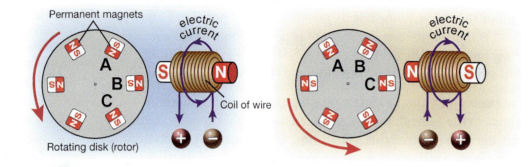

Alternating current

Because the magnet near the coil alternates from north to south as the disk spins, the direction of the current reverses every time a magnet passes the coil. *This creates an alternating current.* The frequency of the alternating current is the frequency at which magnets pass the coil (Figure 23.19). Generators are the source of alternating current that is supplied to your home.

Figure 23.19: *The frequency of the AC current depends on how many times per second the field reverses.*

Energy for generators

The electrical energy created by a generator is not created from nothing. Work must be done to move the magnets that produce the current. Power plants contain a rotating machine called a *turbine* (Figure 23.20). The turbine is kept turning by a flow of expanding air or steam heated by gas, oil, coal, or nuclear energy. The energy stored in the gas, oil, coal, or nuclear fuel is transformed into the kinetic energy of the turning turbine, which is then transformed into electrical energy by the generator. Windmills and hydroelectric dams use energy from wind and water to turn the turbines that turn the generators and produce electricity.

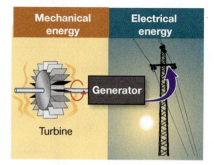

Figure 23.20: *A power plant generator.*

Transformers

Electricity is transmitted at high voltage

From the perspective of physics, it makes sense to distribute electricity from a generator to homes using very high voltage. For example, high-voltage power lines carry current at 13,800 volts. Since power is current times voltage, that means each amp of current carries 13,800 watts of power. The problem is, you would not want 13,800 volts coming to your wall outlet. With a voltage this high, you probably would not *survive* plugging in an appliance!

Electric power transformers

The voltage from the outlet in your wall is 120 VAC. The device that steps down the voltage from 13,800 volts down to 120 volts is called a **transformer**. Transformers are extremely useful because they efficiently change voltage and current, while providing the same total power. For example, a transformer can take 1 amp at 13,800 volts and convert it to 115 amps at 120 volts (Figure 23.21).

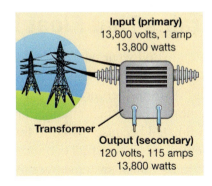

Figure 23.21: *A high-power transformer can reduce the voltage and keep the power constant.*

Transformers operate by electromagnetic induction

The transformer uses electromagnetic induction, similar to a generator. Figure 23.22 shows the basic design of a transformer. You may have seen one inside a doorbell or an AC adapter. The two coils are called the *primary* and *secondary* coils. The input to the transformer is connected to the primary coil. The output of the transformer is connected to the secondary coil. The two coils are wound around a ferromagnetic core. The core concentrates and amplifies the magnetic flux, which "couples" the two coils.

How the two coils work

Consider what happens when current is increasing in the primary coil. This creates an increasing magnetic flux through the secondary coil. The increasing flux in the secondary coil induces current to flow through any circuit connected to the output of the transformer. The useful characteristic of transformers is that *the two coils can have a different number of turns*. For example, suppose the primary coil has 100 turns and the secondary coil has 10 turns. Because the secondary has fewer turns, each amp of current that flows in the primary induces 10 times the current to flow in the secondary! The ratio of the voltages is inverse to the ratio of currents. If 100 V is applied to the primary, only 10 V appears on the secondary.

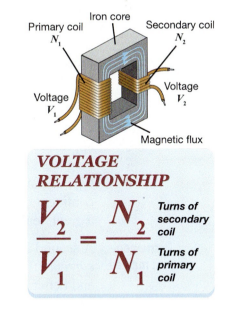

VOLTAGE RELATIONSHIP

$$\frac{V_2}{V_1} = \frac{N_2}{N_1}$$

Turns of secondary coil
Turns of primary coil

Figure 23.22: *The voltages and turns of wire in a transformer are related.*

Transformers only work with AC

Transformers only work with AC current. Remember, the rate of change of magnetic flux is what induces current to flow. The current must go up and down in the primary coil in order to keep the magnetic flux changing in the secondary coil.

Chapter 23 Connection

Trains That Float by Magnetic Levitation

Friction and wear take an expensive toll on vehicles. The cost is both in fuel efficiency and in maintenance. Mass transit vehicles, such as trains, operate seven days a week. If you drove an ordinary car 500 miles per day, every day, you would reach the 100,000 mile life of a typical car in seven months! Magnetic levitation, or *maglev*, trains float several centimeters above the track. Wear is almost eliminated because there are few moving parts that carry heavy loads. Friction is greatly reduced, thereby increasing fuel economy. Many engineers believe maglev technology will become the standard for mass transit systems over the next 100 years.

Magnetically levitated trains Maglev train technology uses electromagnetic force to lift the train 10–20 centimeters above the track (Figure 23.23). Depending on the design, magnets are used in the train, track, or both train and track. The track and train repel each other through powerful magnetic fields. The train "floats" on a nearly frictionless cushion of magnetic force. Although air friction is still present, friction between the wheel and rail is eliminated. Because of reduced friction, maglev trains reach high speeds using less power than a normal train. In 1999, in Japan, a prototype five-car maglev carrying 15 passengers reached a record speed of 552 kilometers (343 miles) per hour. Maglev trains are now being developed and tested in Germany and the United States as well.

Figure 23.23: *A maglev train track has electromagnets in it that both lift the train and pull it forward.*

Electromagnetic maglev Two different approaches are being used to develop maglev technology. One approach uses electromagnets on either the train or the track or both. Powered magnets are necessary to get the high lift needed for the fastest speeds. This type of maglev is being tested for long-distance runs between cities or across continents. Unfortunately, powered tracks are very expensive to construct and use a great deal of electricity.

Figure 23.24: *The Magplane has a track but rises above it when moving.*

Permanent-magnet maglev A second approach uses permanent magnets in the train and relies on the *eddy current* effect in the track to create lift. This approach is simpler, more reliable, and better suited for urban areas where extreme high speed is not necessary. The best developed example of the permanent-magnet approach is the Magplane (Figure 23.24). Developed by Dr. D. Bruce Montgomery at Massachusetts Institute of Technology, the Magplane borrows from aircraft technology as it banks and rolls into turns using spoilers and wing surfaces for control.

Chapter 23 Connection

How the Magplane levitates

Eddy currents

To understand the eddy current effect, consider the following experiment. A magnet is dropped down a copper tube and a cardboard tube of the same size. The magnet in the copper tube falls slower than the magnet in the cardboard tube. As the magnet falls, an increasing magnetic field in induced in the copper tube as the magnet falls past it. In accordance with Faraday's law, an induced current flows around the copper, making a temporary electromagnet. The magnetic field of the induced current pushes on the falling magnet, slowing it down (Figure 23.25). The currents induced in the copper tube are called *eddy currents*. An eddy current is a circular current flowing in a solid conductor in a changing magnetic field.

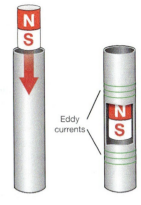

Figure 23.25: *The falling magnet is slowed by the magnetic field it creates by inducing eddy currents in the tube.*

How eddy currents lift the Magplane

The underside of the Magplane has powerful permanent magnets. The surface of the Magplane track is made from curved copper sheets (Figure 23.26). As the Magplane moves forward, eddy currents are induced in the copper sheets on the track. The eddy currents create a magnetic field that pushes against the field of the permanent magnets and lifts the train off the track. The faster the train moves, the more powerful the induced eddy currents become, and the higher the train floats off the track. The Magplane is a very elegant technology because it requires no moving parts and no power supplies for either the train or the track.

The drive system of Magplane

Of course, the train has to be kept moving or the eddy currents will not flow. In the center of the track are a set of electromagnet coils arranged as a linear motor. The coils push and pull on a set of permanent magnets along a line running down the center of the Magplane's underside. As with an ordinary electric motor, the polarity of the electromagnets in the track can be changed to drive the Magplane forward and can accelerate or decelerate the vehicle.

Stopping and starting

When picking up or discharging passengers, the Magplane rides on wheels, like a bus. The wheels are also used at very slow speeds during starting and stopping. Once the Magplane gets going above a speed of 10 miles per hour, the eddy current forces from the track become strong enough to gently lift the train off its wheels. During most of its operation, the Magplane is truly flying over the track with a much smoother and quieter ride than any aircraft.

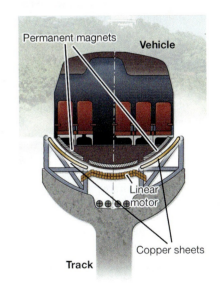

Figure 23.26: *The Magplane uses permanent magnets to levitate the vehicle.*

CHAPTER 23 ELECTRICITY AND MAGNETISM

Chapter 23 Assessment

Vocabulary

Select the correct term to complete the sentences.

gauss	right-hand rule	tesla
solenoid	magnetic field	induced current
Faraday's law of induction	commutator	generator
magnetic flux	polarity	transformer
electromagnet	coil	induction

1. One tesla of magnetic field strength is equivalent to 1×10^4 _____.

2. Turns of wire bundled to concentrate a magnetic field in their center are referred to as a(n) _____.

3. An electromagnetic device in the form of a tube-shaped coil of wire is called a(n) _____.

4. The rule used to indicate the relative direction of force, current, and magnetic field is known as the _____.

5. Two units of magnetic field strength are the gauss and _____.

6. The area surrounding a magnet in which a magnet's effect may be measured is known as the _____.

7. Switching the north and south pole orientation of a motor's stationary magnet changes its _____.

8. Changing the current direction in a motor's electromagnet is accomplished by the _____.

9. A coil of wire wrapped around a rod of iron or steel becomes a(n) _____ as current passes through the coil.

10. Producing current in a coil of wire by moving a magnet near the coil is known as electromagnetic _____.

11. An example of a device that converts mechanical energy to electrical energy is called a(n) _____.

12. Current produced by changing the magnetic field through a coil is known as _____.

13. A measure of the amount of magnetic field linking a magnet and a coil in its field is referred to as _____.

14. _____ says that the current in a coil is proportional to the rate of change of the amount of the magnetic flux passing through the coil.

Concept review

1. What is created by a moving electric charge?

2. When the right-hand rule is used to find the direction of the magnetic field near a current carrying wire, your thumb points in the direction of the _____ and your fingers wrap in the direction of the _____.

3. Two parallel wires carry current. When the currents run in the same direction, the wires _____ each other; when the currents run in opposite directions, the wires _____ each other.

4. Where is a magnetic field concentrated in a coil?

5. When a charge moves through a magnetic field, it experiences a force. On what three things does the size of the force depend?

6. When a charge moves through a magnetic field, how can the direction of the force be determined?

7. A charge that enters perpendicular to a magnetic field will move in a(n) _____; a charge that enters at an angle to a magnetic field will move in a(n) _____.

8. Name two units used for measuring the strength of a magnetic field. Which unit represents a stronger field?

9. What is an electromagnet?

10. Why are electromagnets frequently used instead of permanent magnets?

11. List three devices that contain electromagnets.

12. How can you determine where an electromagnet's north pole is located?

13. Make a diagram of an electromagnet. Indicate the direction of the current and the location of the north and south poles.

14. How can you make an electromagnet stronger?
15. An electric motor uses electromagnets to convert _____ energy into _____ energy.
16. List the three main components of an electric motor.
17. What happens to the poles of the electromagnet in a motor as the rotor spins? Why must this happen?
18. What is electromagnetic induction?
19. What happens if the strength of a magnetic field around a wire changes?
20. If a magnet is held still near a coil of wire, will it cause current to flow? Why or why not?
21. A permanent magnet is held near a loop of wire, creating a magnetic flux through the loop. List two ways the flux through the wire can be changed.
22. A generator converts _____ energy into _____ energy.
23. Explain the function of a transformer.
24. Why does a transformer work only with alternating current?

Problems

1. A wire is oriented vertically in a region of space. Current flows from the top of the wire toward the bottom. Use the right hand rule to determine whether the magnetic field is directed clockwise or counterclockwise around the wire.
2. Imagine that a magnetic field exists on the page of your textbook, directed from right to left. A positive charge moves from the top of the page toward the bottom of the page. Use the right hand rule to determine the direction of the force exerted on the moving charge.
3. Calculate the size of the magnetic field that exists at a distance of 5 cm from a straight wire carrying 1 A of current.
4. A coil of wire with a radius of 2 cm is made up of 25 turns of wire. If the coil carries 3 A of current, what is the strength of the magnetic field at its center?
5. The electromagnet in a motor has its north pole facing the rotor at the instant shown in the diagram. In which direction is the rotor spinning?

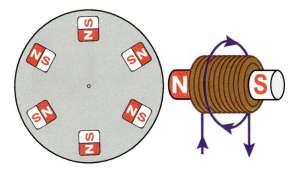

6. Some electric toothbrushes contain rechargeable batteries that are charged by placing the toothbrush on a plastic charging base. Because both the bottom of the toothbrush and the base are made of an insulating material, current does not flow from the base to the toothbrush. How do you think the toothbrush battery gets recharged?

7. A transformer has 9,000 turns in its primary coil and 30 turns in its secondary coil. If the voltage in the secondary coil is 120 V, what is the voltage in the primary coil?

8. The diagram shows a wire that is being moved vertically through a magnetic field. In what direction will the induced electric current flow?

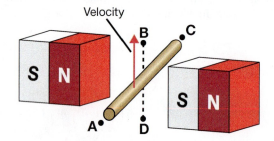

9. The diagram represents a coil of wire connected to a battery. What is the direction of the magnetic field at point P in the coil?

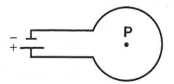

Applying your knowledge

1. The diagram represents a coil of wire surrounding an iron core. A metal ring slips over the coil. If the coil is connected to a source of AC, the ring may be lowered over the coil and will be supported in place. If the current is switched on very rapidly while the ring is in place, the ring may jump vertically off of the coil. Explain why this happens.

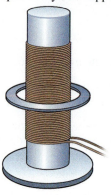

2. Electromagnetic induction can be used to produce large currents. Small but significant currents can be produced due to the piezoelectric effect. Review this concept in Chapter 13. List some ways in which the piezoelectric effect is used.

UNIT 7 ELECTRICITY AND MAGNETISM

CHAPTER 24

Electronics

Objectives:

By the end of this chapter you should be able to:

✔ Describe how a diode and a transistor work in terms of current and voltage.
✔ Explain the difference between a *p*-type and an *n*-type semiconductor.
✔ Construct a half-wave rectifier circuit with a diode.
✔ Construct a transistor switch.
✔ Describe the relationship between inputs and outputs of the four basic logic gates.
✔ Construct an adding circuit with logic gates.

Key Questions:

- What is a diode, and why is it a basic element of electronics?
- How is a transistor used in a circuit?
- What are the basic types of logic circuits and why are they useful?

Vocabulary

amplifier	bit	diode	logic gate	*p-n* junction
amplify	central processing unit (CPU)	emitter	memory	*p*-type
analog signal		forward bias	*n*-type	program
AND	collector	gain	NAND	rectifier
base	conductivity	hole	NOR	reverse bias
bias voltage	depletion region	integrated circuit	OR	transistor
binary	digital signal	logic circuits		

499

24.1 Semiconductors

It is almost impossible to do anything today without being affected by electronics. Electronic devices are in telephones, computers, video games, cars, watches—a virtually endless list. Contemporary explorers on land and sea carry *electronic global positioning system* (GPS) receivers to keep from getting lost. Electronic devices use semiconductors to precisely control current and voltage in complex circuits. This chapter introduces some of the more important semiconductor components used in all electronic devices, including diodes, transistors, and integrated circuits.

Diodes

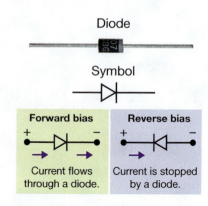

Figure 24.1: *A diode and its circuit symbol. The stripes indicate the negative side when the diode is forward biased.*

What a diode does A **diode** is a one-way "valve" for electric current. Current can only flow one way through a diode and not the other way. Diodes are a basic building block of all electronics and are used to control the direction of current flowing in circuits. A common diode looks like a small cylinder with stripes (Figure 24.1).

Forward and reverse bias When a diode is connected in a circuit so current flows through it, we say the diode is **forward biased**. When the diode is reversed so it blocks the flow of current, the diode is **reverse biased**. If you plot the current through the diode versus the voltage across the diode, the graph will look like Figure 24.2.

The bias voltage In a forward-biased diode, the current stays at zero until the voltage reaches the **bias voltage** (V_b), which is 0.6 V for common silicon diodes. You can think of the bias voltage as the amount of energy difference it takes to switch on the diode. Once the voltage gets higher than the bias voltage, diodes have low resistance.

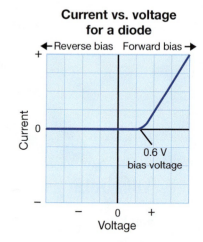

Figure 24.2: *The current versus voltage graph for a common silicon diode.*

Diodes and AC adapters An important application for diodes is in AC-to-DC adapters. Most electronic devices need DC electricity, including stereos, laptop computers, and battery chargers. The adapter uses a circuit of diodes to turn AC into DC. Diodes are ideal for this type of circuit because they allow current to flow in only one direction.

Transistors

What is a transistor?
Transistors are solid, semiconductor devices. Using a small current or voltage, a transistor has the ability to control a large amount of current flow. Transistors can be the size of the image in Figure 24.3, or they can be extremely tiny. A basic laptop computer has hundreds of *billions* of transistors built into its integrated circuits.

A transistor is a flow control valve for current
Transistors are devices that allow you to *control* the current, not just block it in one direction. A transistor is like a variable flow valve for electric current. A good analogy for a transistor is a pipe with an adjustable gate. When the gate is closed, the pipe has very high resistance and not much water flows. When the gate is open, the pipe has low resistance and water flows easily. The gate controls the flow of current by controlling the resistance of the pipe.

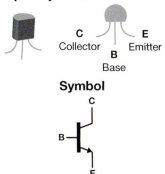

Figure 24.3: *An early transistor's design is reflected in its symbol diagram.*

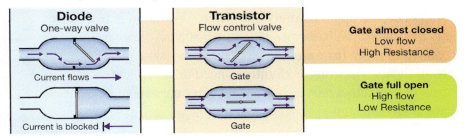

Types of transistors
Two of the main categories of transistors are the bipolar junction transistor (BJT) and the field effect transistor (FET). All transistors have three terminals. A signal applied between one pair of terminals controls the signal between another pair of terminals.

Bipolar junction transistor
The BJT regulates its collector current proportionally to the base current. The three terminals of the BJT are called the **collector, emitter,** and **base.** The main path for current is between the collector and emitter. The base controls how much current flows, just like the gate controls the flow of water in the pipe. Figure 24.3 shows the electrical symbol for a BJT transistor.

Field effect transistor
The FET is controlled by the input voltage. FETs also have three terminals, called the source, drain, and gate, which are analogous to the BJT's emitter, collector, and base. The voltage at the gate is what controls the current from the drain to the source. The higher the gate voltage, the more drain current flows.

CHAPTER 24 ELECTRONICS

Conductivity and semiconductors

Conductivity The relative ease with which electric current flows through a material is known as **conductivity**. Conductors, like copper, have a very high conductivity. Insulators, like rubber, have very low conductivity.

Semiconductors can change their conductivity Semiconductors are materials that are neither conductors nor insulators but are somewhere between the two in conductivity. The conductivity of a semiconductor depends on its conditions. For example, at low temperatures and low voltages a semiconductor acts like an insulator. When the temperature and/or the voltage is increased, the conductivity increases and the material acts more like a conductor. The ability to pass current in one direction and block current in the opposite direction comes from the ability of semiconductors to change their electrical conductivity in response to a change in voltage.

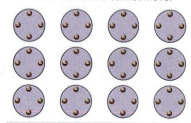

Why metals are conductors Metals are good conductors because approximately one electron per atom is free to separate and move independently. With only a small voltage "push" from a battery, electrons in a conductor move from atom to atom throughout the material. In an insulator, virtually all the the electrons are tightly bonded to atoms and cannot move. Since the electrons cannot move, they cannot carry current, and that explains why an insulator *is* an insulator.

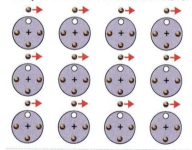

Electrons in a semiconductor The electrons in a semiconductor are also bound to atoms, but the bonds are relatively weak. The energy from a battery or heat is enough to free a few electrons, which can then move and carry current. The density of free electrons is what determines the conductivity of a semiconductor. If there are many free electrons to carry current, the semiconductor acts more like a conductor. If there are few free electrons, the semiconductor acts like an insulator.

Silicon is the most common semiconductor Silicon is the most commonly used semiconductor. Atoms of silicon have 16 electrons. Twelve of the electrons are bound tightly inside the atom. Four electrons are near the outside of the atom and only loosely bound. In pure silicon, the atoms are arranged so that each of the four outer electrons is paired with another electron from each of four neighboring atoms (Figure 24.4). At room temperature, a fraction of the electrons in silicon has enough energy to break free from their electron pairs and carry current. This small population of free electrons is what makes silicon a semiconductor.

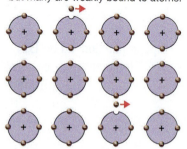

Figure 24.4: *Electrons' motion varies in insulators, conductors, and semiconductors.*

Changing the conductivity of semiconductors

Adding impurities to a semiconductor
Anything that changes the number of free electrons has a huge effect on conductivity in a semiconductor. For example, adding an impurity of one boron atom per 10 million silicon atoms increases the conductivity by 20,000 times. Useful semiconductors are created by adding impurities to adjust conductivity.

n-type semiconductors
Phosphorus atoms have five outer electrons compared with silicon's four. When a phosphorus atom tries to bond with four silicon atoms, four of its five outer electrons pair up with the neighboring silicon atoms. The extra electron does not pair up and is free to carry current. Adding a phosphorus impurity to silicon *increases* the number of electrons that can carry current. Silicon with a phosphorus impurity makes an **n-type** semiconductor. Current in an *n*-type semiconductor is carried by electrons with *negative* charge (Figure 24.5).

p-type semiconductors
When a small amount of *boron* is mixed into silicon the opposite effect happens. A boron atom has three outer electrons, one less than silicon. When a boron atom tries to bond with silicon, it needs another electron so it can pair up with its four neighbors. The boron atom captures an electron from a neighboring silicon atom.

When an electron is taken by a boron atom, the silicon atom is left with a positive charge. The silicon atom with the missing electron is called a **hole** because it needs to be filled with another electron (Figure 24.6). The positive silicon atom attracts an electron from one of *its* neighbors, and the *hole moves*. The new hole takes an electron from *its* neighbor and *the hole moves again*. In fact, as electrons jump from atom to atom, the positive hole moves in the opposite direction and can carry current. Silicon with a boron impurity is a **p-type** semiconductor. The current in a *p*-type semiconductor is carried by holes with *positive* charge.

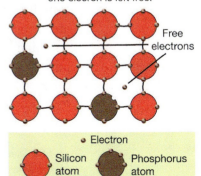

Figure 24.5: *An* n-*type semiconductor.*

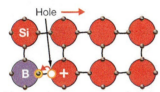

A boron electron creates a hole by taking an electron from a silicon atom.

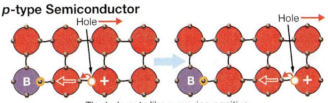

The hole acts like a moving positive charge as electrons jump from atom to atom.

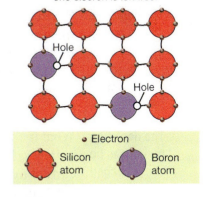

Figure 24.6: *In a* p-*type semiconductor, each boron atom creates a "hole" where an electron is needed to match pairs with silicon.*

The p-n junction

The p-n junction A ***p-n junction*** forms where *p*-type and *n*-type semiconductor materials meet. Initially, the *n* side has free electrons and the *p* side has holes. The holes attract electrons so some negative electrons from the *n* side flow over to the *p* side and combine with the positive holes. As the electrons move, the *n* side becomes positively charged and the *p* side becomes negatively charged. The charge difference grows until it is large enough to keep any more electrons from crossing over. For silicon, this equilibrium is reached when the *n* side is 0.6 volts more positive than the *p* side.

Semiconductors and crystals

A crystal is a solid in which all the atoms are perfectly organized in neat rows and columns in three dimensions. Semiconductors are made from almost perfectly-pure crystals of silicon, or at least as perfect as human technology can achieve.

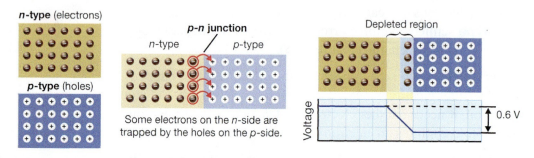

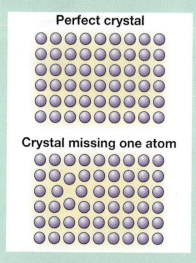

The depletion region When an electron fills a hole, a neutral silicon atom is left. As a result, the material right near the *p-n* junction has neither electrons or holes. This area is called the **depletion region**. The depletion region has *no movable charges* that can carry current because the electrons and holes have combined to make neutral silicon atoms. As a result, *the depletion region becomes an insulating barrier to the flow of current*.

The depletion region can be changed What makes a transistor work is that the depletion region is affected by external voltages. It can grow and become a stronger insulator, or disappear and allow current to flow. The depletion region can switch from insulating to conducting very quickly because it is thin, typically 0.5 millionths of a meter or less, and because electrons are small and fast. Computer chips are made from semiconductors with *p-n* junctions. The fundamental limit to how fast computers work is the speed at which the depletion region can change from an insulator to a conductor. In today's fastest computer technology, the switch of a *p-n* junction can occur in a trillionth of a second or less.

When an atom of silicon is out of place, it acts like an impurity, because the electrons in neighboring atoms cannot pair up one-to-one. If too many atoms are out of place, the precise effects of the added impurities are changed and the *p-n* junction may not work as it should.

The physics of diodes

Reverse biased p-n junction The depletion region of a *p-n* junction is what gives diodes, transistors, and all other semiconductors their useful properties. Suppose an external voltage is applied in a direction that attracts electrons on the *n* side. The same voltage also attracts holes on the *p* side. Both electrons and holes are drawn away from the junction and the depletion region gets larger. Even as the voltage increases, no current can flow because it is blocked by the larger, insulating depletion region.

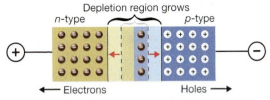

A reverse-biased *p-n* junction

Forward biased p-n junction Now suppose the opposite voltage is applied. Both electrons and holes are repelled *toward* the depletion region. As a result, the depletion region gets smaller. The larger the opposing voltage gets, the smaller the depletion region becomes. When the applied voltage becomes greater than 0.6 V, the depletion region *goes away completely*. Once the depletion region is gone, electrons are free to carry current across the junction and the semiconductor becomes a conductor.

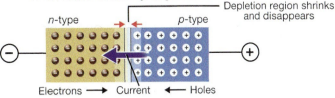

A forward-biased *p-n* junction

A p-n junction is a diode The two paragraphs above can be summarized as follows:

- the *p-n* junction *blocks* the flow of current from the *n* side to the *p* side;
- the *p-n* junction *allows* current to flow from the *p* side to the *n* side if the voltage difference is more than 0.6 volts.

In short, a *p-n* junction is a *diode*. Current is only allowed to flow in one direction across the junction. Transistors have two such *p-n* junctions.

Transistors

A transistor is made from two *p-n* junctions back to back. The three terminals are connected to the three regions as shown in the diagram below. The layer of *p*-type semiconductor in an actual transistor is much thinner than the *n*-type layers on either side.

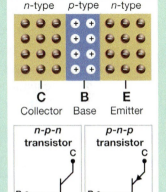

With a transistor, a small current flowing into the base can control a large current from the collector to emitter by changing the conductivity of the depletion region. An *n-p-n* transistor (shown) has a *p*-type layer sandwiched between two *n*-type layers. A *p-n-p* transistor is the reverse—an *n*-type semiconductor between two layers of *p*-type.

24.2 Circuits with Diodes and Transistors

Electric circuits made with diodes and transistors can do much more than circuits with only resistors and capacitors. For example, your CD player uses a transistor circuit to *amplify* the electrical signal from a CD until it carries enough current to drive earphones or speakers. This section presents a few examples of important and useful types of circuits made with transistors and diodes.

A rectifier circuit turns AC electricity into DC

A single diode AC-DC converter A diode can convert alternating-current electricity to direct current. Consider what happens when an AC voltage is applied to a diode. The diode conducts current in one direction only. When the AC cycle is positive, the voltage passes through the diode because the diode is conducting and has low resistance. When the AC cycle is negative, the voltage is blocked by the diode because the diode has a very high resistance to current in the reverse direction. A single diode is called a *half-wave rectifier* since it converts half of the AC cycle to DC.

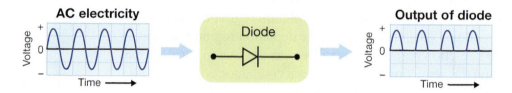

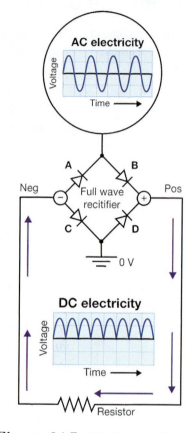

Figure 24.7: *A bridge-rectifier circuit uses the entire AC cycle by inverting the negative portions.*

A four-diode AC-DC converter There is a way to convert the whole AC cycle to DC with four diodes arranged in the circuit shown in Figure 24.7. This circuit is called a *full-wave* or *bridge rectifier*. A version of the full-wave rectifier circuit is in most AC-to-DC adapters you have used.

How the four-diode circuit works When the AC cycle is positive, current flows through Diode B, through the resistor, and back through Diode C. The current flows out of the positive terminal of the bridge rectifier and back to the negative terminal. When the AC cycle is negative, current flows through Diode D, through the resistor, and back through Diode A. On either part of the AC cycle, the resistor receives current from the positive terminal that flows back to the negative terminal. The result is a "bumpy" DC current. The bumps can be smoothed out with a capacitor.

A transistor switch

Transistors as electronic switches In many electronic circuits, a small voltage or current is used to switch a much larger voltage or current. Transistors work very well for this application because they behave like switches that can be turned on and off *electronically* instead of using manual or mechanical action.

Turning a transistor on and off Consider an n-p-n transistor. Because there are two p-n junctions, the transistor normally blocks current in both directions. The current in the p layer is carried by positive holes. When positive current flows into the base, electrons are drawn in from the n-type regions and the whole p layer becomes a conductor. It typically takes only a tiny amount of base current—10 millionths of an amp—to turn a transistor from an insulator into a conductor.

The resistance of a switch You can think of a regular mechanical switch as a device that goes from very high resistance to very low resistance. When the switch is open, the resistance is greater than a million ohms. When the switch is closed, the resistance drops to 0.001 ohms or less. Transistor switches work because very small currents to the base change the resistance of the transistor by almost as great an amount as a mechanical switch. For example, when the current into the base is zero, a transistor has a resistance of 100,000 ohms or more. When a tiny current flows into the base, the resistance drops to 10 ohms or less. The resistance difference between "on" and "off" for a transistor switch is not as great as for a mechanical switch, but it is good enough for many useful circuits.

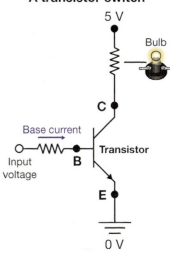

Figure 24.8: *A transistor can be used as a switch to turn on a light bulb.*

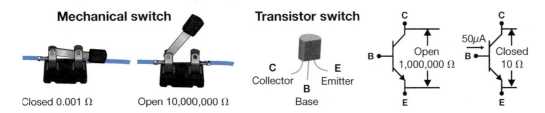

A transistor switch circuit An example of a transistor switch circuit is shown in Figure 24.8. Closing the switch causes a small current to flow into the base of the transistor. This turns the transistor "on" and current flows through to light the bulb.

A transistor amplifier

What does an amplifier do? One of the most important uses of a transistor is as an **amplifier**. Amplifier circuits are used to change the amplitude of an input signal without changing the shape of the signal. A common use of amplifier circuits is in volume control. When you increase or decrease the volume on a television, an amplifier circuit is used to make the sound louder or softer.

A two-transistor amplifier circuit

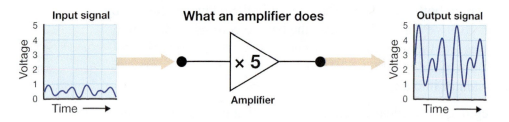

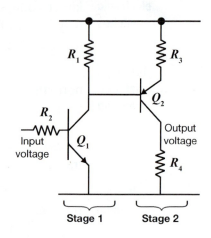

Gain and amplifiers The amount of increase or decrease from the input to the output signal is called the *gain*. The gain is the ratio of the output voltage to the input voltage. For example, an amplifier with a gain of five would change a 1-volt input signal into a 5-volt output signal, leaving the shape of the voltage versus time graph the same. The output voltage of an amplifier is equal to the input voltage multiplied by the gain.

How is a signal changed? Amplifier circuits can be designed to *increase* the voltage or current, which is called *amplifying* the signal, or they can be designed to *decrease* the voltage or current, which is called *attenuating* the signal. A signal can also be *inverted* by the amplifier circuit, meaning that when the input voltage goes up, the output voltage goes down, and vice versa.

Figure 24.9: *Stage 1 of the amplifier circuit has a gain of -5; the signal is amplified by five and inverted. Stage 2 inverts the signal again, so the gain of the entire circuit is +5.*

A two-transistor amplifier circuit Figure 24.9 shows a two-transistor amplifier circuit. The current in the collector for transistor Q1 can be calculated by multiplying the input current by Q1's current gain. The output voltage from stage 1 is inverted from the input voltage, and is connected as the input voltage to stage 2. In stage 2, the current through R4 is the same magnitude as the current in R3. The current in R3 is set by the input voltage to stage 2. The ratio of the resistors R1R4/R2R3 equals the total gain of this circuit. By adjusting resistor ratios, you can change the gain of the amplifier.

Electronic logic

Circuits that make decisions Many electronic circuits are designed to perform certain functions only if a number of input conditions are met. For example, the circuit that starts a car might only work when (a) the car is in park, (b) the brake is on, and (c) the key is turned. The circuit that applies power to start the engine must evaluate three conditions before it turns itself on. How do electronic circuits make decisions like this?

Electronic logic Logic circuits are designed to compare inputs and produce a specific output when all the input conditions are met. Logic circuits assign voltages to the two logical conditions of TRUE (3 V) and FALSE (0 V). For example, for 3-volt circuits, 3 V is considered TRUE and 0 V is considered FALSE. Using this "electronic logic," the problem of the car-starter circuit is summarized by the table below. There are three inputs corresponding to the three conditions that must be satisfied. There is one output which starts the car if TRUE and does not start the car if FALSE.

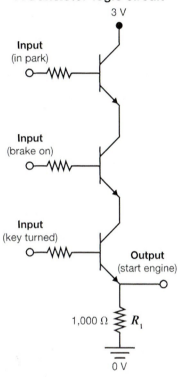

A transistor-logic circuit

Figure 24.10: *A circuit with three transistors can solve a three-input AND logical decision.*

INPUT Car in park	INPUT Brake on	INPUT Key turned	OUTPUT Start engine
0 V	0 V	0 V	0 V
3 V	0 V	0 V	0 V
3 V	3 V	0 V	0 V
3 V	0 V	3 V	0 V
0 V	3 V	3 V	0 V
3 V	3 V	3 V	3 V

A circuit that solves the three-input logic problem The circuit with three transistor switches shown in Figure 24.10 does exactly what the table prescribes. This circuit operates like a big voltage divider. The three transistors in series act like one resistor and R_1 is the other. If *any* of the three transistors has a high resistance compared with R_1, then the output is close to zero volts. If all three transistors are "on," their total resistance is much less than R_1 and the output voltage is close to 3 volts. The only way for the output to be 3 V is when all three transistors are on, which only happens if all three inputs are TRUE. The circuit of Figure 24.10 is an example of an AND logic circuit. An AND circuit compares its inputs and makes the output TRUE only if the first input is TRUE *and* the second is TRUE *and* the third is TRUE.

> **Voltages between 0 and 3V**
>
> Logic circuits treat a range of voltages as TRUE or FALSE. For example, in 3-V logic, any voltage less than 1 V is FALSE and any voltage greater than 1.5 V is TRUE.

24.2 CIRCUITS WITH DIODES AND TRANSISTORS

24.3 Digital Electronics

We seem to live in a so-called "digital age" where everything is better if it is "digital." Supposedly, CDs have better sound because they are digital. Commercials claim digital TV has better picture quality. As with most advertising, the truth is not quite so clear cut. You do not directly hear or see digital signals. You hear and see analog signals. This section will help you understand the important difference between the digital world of computer-based electronics and the analog world of our senses.

Analog and digital signals

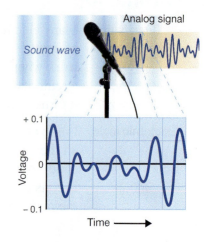

Figure 24.11: *A microphone creates an analog signal, shown by the voltage versus time graph.*

Signals and information A signal is anything that carries information. Today the word *signal* usually means a voltage, current, or light wave that carries information. In electronics, signals are usually voltages and the information is contained in the way the voltages vary with time. For example, a voice is a sound wave. A microphone converts the variations in air pressure from the sound wave into variations in voltage in an electrical signal (Figure 24.11).

Analog signals The voltage versus time graph from a microphone is an example of an **analog signal**. The voltage in an analog signal can have continuous values. For example, a particular microphone might produce a voltage from –0.1 V to +0.1 V. The signal from the microphone is a continuous voltage between –0.1 V and +0.1 V (Figure 24.11). The information in an analog signal is contained in both the value of the signal and the way the signal changes with time.

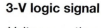
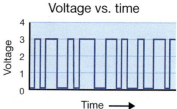

Figure 24.12: *A digital signal is a sequence of transitions between* high *and* low *voltages. The voltage versus time graph shown is a 3-V logic signal.*

Digital signals A **digital signal** can only be on or off. For the digital signals in many computers, "on" is 3 volts, "off" is 0 volts. A 3-volt digital signal has only two values: 0 V or 3 V. A digital signal is very different from an analog signal. The information in a digital signal is coded in the sequence of changes between 0 V and 3 V. Figure 24.12 shows an example of a digital signal, like the one from a CD player.

Comparing analog and digital signals At first glance, the claim that digital signals are better seems impossible. How can a signal that can only be on or off tell you as much as a signal that can have all the values in between? The answer is that digital signals can send billions of ones and zeros per second, and are extremely stable. If a digital signal degrades, it is easily fixed. Analog signals cannot be recovered once they degrade. Digital signals are also easier to store, process, and reproduce than analog signals.

Digital information

Information in analog signals
All circuits use real voltages and currents that are analog variables. The difference is in how the *information* in a signal is used by a circuit. In an analog circuit, the voltage or current *is* the information. For example, an electronic thermometer makes a voltage proportional to temperature. The higher the temperature, the higher the voltage.

Information in digital signals
In a digital circuit, the information is not in the voltages or currents directly, but instead is coded in the patterns of change between high and low voltages. For example, with a digital thermometer, the temperature could be represented by a number between 0 and 99. The digital temperature signal assigns a code to each digit as shown in Table 24.1. (*Note*: For more on **binary** numbers, see the connection on page 514.)

Table 24.1: Binary Coded Decimal (BCD) for the digits 0–9

#	Code	#	Code	#	Code	#	Code	#	Code
0	0000	2	0010	4	0100	6	0110	8	1000
1	0001	3	0011	5	0101	7	0111	9	1001

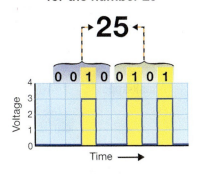

Figure 24.13: *Base-10 numbers (25) can be written in binary code.*

Digital representation of a number
To represent a temperature of 25°C as a digital signal requires the code for "2" followed by the code for "5." If the electronics uses 3-V logic, then 3 V is a "1" and zero volts is a "0." The number 25 is represented by a sequence of eight voltages in the order shown in Figure 24.13.

Bits and representation of letters
In a digital signal, a **bit** is a place in the signal that can be either a "0" or a "1." All 10 numbers can be represented by four bits in decimal notation. The letters in the alphabet also have codes. Since there are more letters, it takes more bits to create a unique code for each letter, including the lowercase letters. A few of the codes for the alphabet are shown in Table 24.2. A digital signal for the word *face* is shown in Figure 24.14.

Table 24.2: The American Standard Code for Information Interchange (ASCII)

#	Code	#	Code	#	Code	#	Code
A	0100 0001	D	0100 0100	a	0110 0001	d	0110 0100
B	0100 0010	E	0100 0101	b	0110 0010	e	0110 0101
C	0100 0011	F	0100 0110	c	0110 0011	f	0110 0110

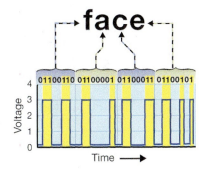

Figure 24.14: *The binary code for the word* face *is shown.*

Computers

Circuits as information processors
You can think about a circuit as a processor of analog information. Currents and voltages applied to the input of the circuit result in other currents and voltages at the output. The amplifier or the AC-DC rectifier are examples of circuits that process an input to create a specific kind of output.

Circuits are task-specific
Analog circuits are designed to do a particular task, such as amplifying a signal. To make a circuit do something different, you have to build a different circuit. This is often inconvenient. For example, you can make an analog circuit start the car if three conditions are met. It would take a completely different circuit if only two out of the three conditions were met.

Why computers are useful
A computer is an electronic device for processing digital information. Computers use **programs** that direct the processing of information. To make a computer do something different, you need a different program. *The electronic circuits stay the same.* This is a tremendous advantage and is the reason digital electronics is a foundation of human technology.

The memory system
All computers have three basic parts (Figure 24.15). The **memory** stores digital information. The information might be words, numbers, pictures, or programs that tell the computer what to do with the letters, numbers, or pictures. The average classroom computer can store about 800 billion bits of information. For comparison, this chapter has about 6,000 words averaging seven letters each plus a space. It takes 380,000 bits to store the words in the chapter as ASCII codes. The average computer memory can hold the words in 2 million chapters of this length.

The CPU
Information from memory is processed by the **central processing unit (CPU)**. The CPU of a computer is a huge circuit using more than a billion transistors. The program tells the CPU what to do with each string of digital data that comes in from memory.

The input/output system
The third component of a computer is the input-output system, or I/O. The I/O system includes the keyboard, display, sound interface, network interface, and other sub-systems that allow the CPU and memory to interact with you or other electronic devices, such as other computers.

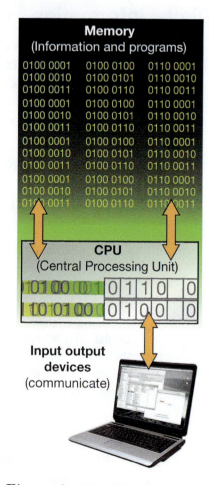

Figure 24.15: *The basic components of a computer are its CPU, memory, and input-output devices.*

Logic circuits

Computers use logic circuits Inside a computer are complex electronic circuits that take a signal of ones and zeros and create a different signal of ones and zeros. A modern CPU chip has a circuit with 780 *million* transistors in it. Since the ones and zeros are represented by voltages, the electronic circuits inside also work with voltages. You have already read about one type of digital circuit: the AND circuit used to solve the car-starter problem.

The four basic logic gates Circuits called **logic gates** are the basic building blocks of computers and almost all digital systems. The fundamental logic gates are called **AND**, **OR**, **NAND**, and **NOR**. As their names imply, these gates compare two input voltages and produce an output voltage based on the inputs. The tables show the output of each of the four logic gates for every combination of inputs.

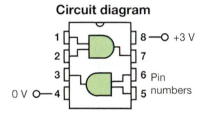

Figure 24.16: *Some integrated circuits use AND gates as shown.*

Inputs		Output
A	B	
0	0	0
0	1	0
1	0	0
1	1	1

AND

Inputs		Output
A	B	
0	0	0
0	1	1
1	0	1
1	1	1

OR

Inputs		Output
A	B	
0	0	1
0	1	1
1	0	1
1	1	0

NAND

Inputs		Output
A	B	
0	0	1
0	1	0
1	0	0
1	1	0

NOR

Chips and integrated circuits Logic gates are built with many transistors in **integrated circuits**, commonly known as "chips." Figure 24.16 shows a picture of a chip that has two AND circuits. This chip operates with 3-V signals and must be supplied with 3 volts to pin 8 and 0 volts to pin 4. The inputs and outputs are pins 1–3 and 5–7.

An example of a logic circuit As an example of a logic circuit, suppose a computer wants to "recognize" a four-bit number. The computer memory stores the number 3, which has a code of 0011. The number 3 is entered from the keyboard. An AND gate returns a one only if both signals are one. A NOR gate returns a one only if both signals are zero. A second OR compares the output of the first two and returns a one if either of its inputs is one. The output of this circuit will be four ones, with 3 V on each, only if the number entered from the keyboard exactly matches the number in the computer's memory. The CPU of a computer contains hundreds of millions of such circuits.

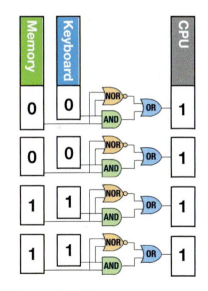

Figure 24.17: *A logic circuit compares two four-bit electronic numbers.*

Chapter 24 Connection

Electronic Addition of Two Numbers

Computers in science are useful for many reasons. One reason is to do complicated calculations such as calculating the magnetic field from a coil at places other than its center. To do calculations, computers represent all numbers in zeros and ones using the **binary** number system. Once numbers are in binary, electronic logic circuits can be used to add, subtract, multiply and divide them.

The decimal numbering system In a base-10 whole number, each digit represents a power of 10. The digit farthest to the right equals the number of tens to the zero power ($10^0 = 1$). The digit to the immediate left is the number of tens to the first power ($10^1 = 10$). The next digit is the number of tens to the second power and so on. The quantity represented by the number is calculated by adding the digits multiplied by the appropriate power of 10. For example, the number 115 equals $1 \times 10^2 + 1 \times 10^1 + 5 \times 10^0$.

The binary number system It would take 10 different voltages to represent 10 digits. Making circuits work on 10 voltages is difficult, so all numbers in computers are represented in binary. Binary numbers work the same as decimal numbers except each digit represents a power of 2 instead of a power of 10. The diagram below shows how the number 115 is represented in binary.

Adding binary numbers To add two numbers, each digit is lined up just as you would with decimal numbers (Figure 24.18). A zero plus one equals one. One plus one equals 10, with the one being carried to the next place to the left. The usual rules of arithmetic hold for binary numbers. For example, with decimal numbers, one plus one equals two. The same is true in binary numbers, except the calculation is written $1 + 1 = 10$. In binary, the number 10 is really *two* in the decimal system.

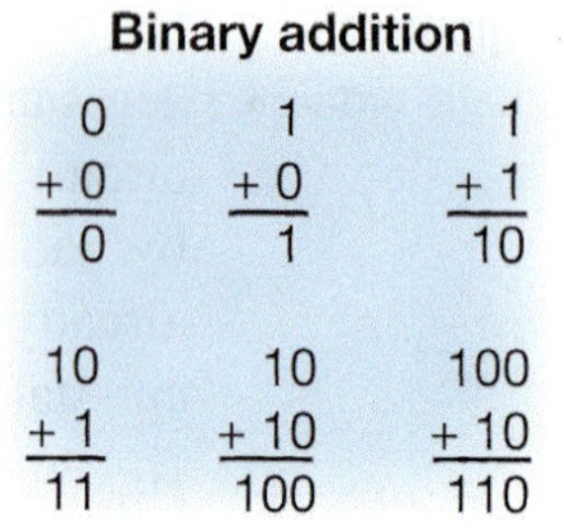

Binary addition

```
  0      1      1
+ 0    + 0    + 1
———    ———    ———
  0      1     10

 10     10    100
+ 1    +10    +10
———    ———    ———
 11    100    110
```

Figure 24.18: *Adding two numbers in binary.*

Floating point numbers

Integers have no decimal point. In computer technology, decimals are stored in a form called *floating point* notation. A floating-point number has two parts: The first is the multiplier and the second is the power of two.

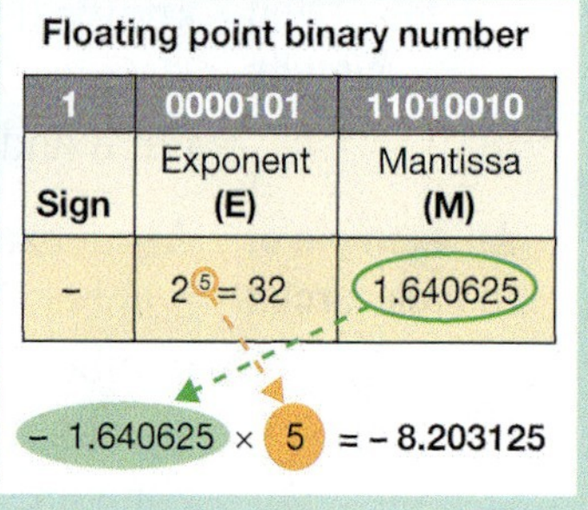

Chapter 24 Connection

An electronic adding circuit

Components of an adding circuit

Two one-bit binary numbers can be added with four logic gates: two AND gates, a NOR gate, and a NAND gate, as shown in Figure 24.19. To see how this circuit works, consider the process of addition. Two single-digit binary numbers can have four different sums, as shown below. The sum has at most two digits.

A circuit to add two one-bit binary numbers

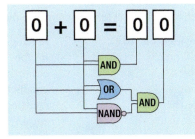

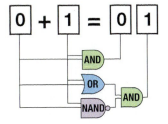

The logic for the two's place

To see how the circuit is designed, consider the logical relationship between the two input numbers and each of two digits of the result. The two's place should be a one only if the first number is one and the second number is one. This is done electronically with an AND gate.

The logic for the one's place

The one's place of the result is a little harder to figure out. The one's place should be a zero if both inputs are zero, *or* if both inputs are one. The one's place should be a one if either input number is one, but not both. This cannot be done with a single AND, NAND, OR, or NOR gate. It can be done with the combination of three gates shown in Figure 24.19. The diagram below shows how the logic in the circuit works. This type of diagram is called a *truth table*. If you experiment with digital electronics, you will find other ways to make an adding circuit, with other gates, such as an XOR gate.

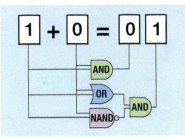

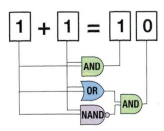

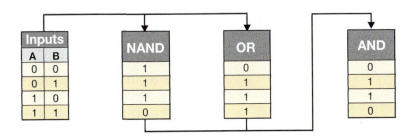

Figure 24.19: *A one-bit binary adder circuit can be made with four logic gates.*

24.3 DIGITAL ELECTRONICS

Chapter 24 Assessment

Vocabulary

Select the correct term to complete the sentences.

forward biased	reverse bias	bias voltage
p-type	n-type	depletion region
hole	collector	emitter
base	conductivity	p-n junction
logic circuits	rectifier	diode
transistor	amplifier	gain
analog	digital	AND
OR	NAND	binary
CPU	programs	memory
bit	integrated circuits	semiconductors
NOR		

1. The terminal that controls the amount of current to flow through a transistor is the _____.

2. A diode placed in a circuit so current will flow through it is described as _____.

3. The relative ease with which current flows through a material is known as _____.

4. A semiconductor through which current is carried by electrons with a negative charge would be classified as a(n) _____ semiconductor.

5. A common device that acts as a one-way "valve" for electric current flow is a(n) _____.

6. To control the amount of current that flows in a circuit an electronic "variable flow valve" called a(n) _____ can be used.

7. The amount of energy difference required to cause energy to flow through a diode is referred to as _____.

8. The current in a(n) _____ semiconductor is carried by holes with a positive charge.

9. Diodes which block current flow are said to have a(n) _____.

10. In a transistor, the area near the p-n junction that has neither electrons nor holes is called the _____.

11. The interface between p-type and n-type semiconductor material in a transistor is the _____ of a transistor.

12. The main path for current through a transistor is between the _____ and _____.

13. The location on a p-type semiconductor vacated by an electron is identified as a(n) _____.

14. An electronic device capable of converting any portion of alternating current to direct current may be called a(n) _____.

15. The ratio of the input signal to the output signal for an amplifier is referred to as _____.

16. A transistor used to change the size of an input signal to produce a larger output signal is being used as a(n) _____.

17. Circuits considered the basic building blocks of computers are called _____.

18. The logic gate that only returns a zero when both inputs are zero is a(n) _____. The rest of the time, its output is one.

19. A signal which can have a continuous range of values is called a(n) _____ signal.

20. A(n) _____ logic gate can only have an output value of zero when both inputs are one. The rest of the time, its output is one.

21. In a computer, digital information is stored in the _____ of the computer system.

22. A system used to represent information coded in combinations of zeroes and ones is known as a(n) _____ system.

23. The system in a computer responsible for processing information from memory is the _____.

24. In a digital signal, a place that can be either a "0" or a "1" is called a(n) _____.

25. A(n) _____ logic gate returns a one only when both inputs are zero.

26. Signals that code information as either "on" or "off" are referred to as _____ signals.

27. Logic gates are built from a combination of transistors called chips or _____.

28. Computers are internally directed to process information by _____.
29. The logic gate that returns a "1" only when both inputs are "1" is a(n) _____.

Concept review

1. Compare the two ways in which a diode can be connected in a circuit.
2. Why is it important to know a diode's bias voltage?
3. How does the graph of current versus voltage for a diode compare to the current versus voltage graph for a resistor?
4. Explain the function of the base connection on a transistor.
5. Under what conditions does a semiconductor act like an insulator?
6. How is the number of free electrons in a semiconductor related to its conductivity?
7. What effect do impurities have on semiconductors? Why?
8. Compare n-type and p-type semiconductors.
9. Why does the n side of a p-n junction become positively charged?
10. Why does the depletion region of a semiconductor act like an insulator?
11. How can the size of the depletion region be changed?
12. Explain how a half-wave rectifier and a full-wave rectifier differ.
13. Compare and contrast a transistor switch with a mechanical switch.
14. Give an example of a device that uses a logic circuit.
15. Which type of signal can only be on or off?
16. Explain how different numbers and letters can be represented using only zeroes and ones.
17. Describe the main components of a computer.
18. List the four main types of logic gates.

Problems

1. The diagram shows the input voltage for an AC circuit.

   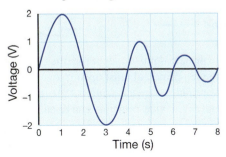

   a. A half-wave rectifier is added to the circuit. Sketch the output voltage versus time graph for the rectifier.
   b. A full-wave rectifier is added to the circuit. Sketch the output voltage versus time graph.

2. An amplifier has a gain of three. The input signal is shown. Sketch a graph of the output signal.

   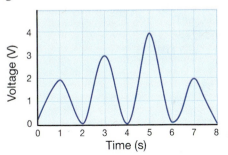

3. Give the decimal number that corresponds to each binary number.
   a. 0001
   b. 0010
   c. 0111

4. This logic circuit represents who can open your school locker. Which choice gives the conditions that will allow the locker to be opened?

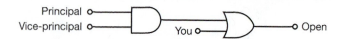

   a. You and the principal must open it together, but not alone.
   b. You and the vice-principal must open it together, but not alone.
   c. You and the principal or vice-principal together can open it together.
   d. The principal or the vice-principal can open it, but not together.

5. Which choice gives the input conditions for the output to be true?

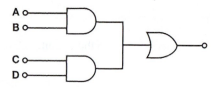

   a. A and B, but not C and D
   b. A or B, and C or D
   c. A and B and C and D
   d. A and B or C and D

Applying your knowledge

1. The silicon point-contact diode was invented in 1906. In 1947, the first transistor was produced; prior to this, vacuum tubes were used to control current. Research why it took scientists so long to make the transition to transistors, which are essentially three-wire diodes.

2. Research the history of the computer. Create a timeline of significant milestones in computer history.

UNIT 8 **MATTER AND ENERGY**

CHAPTER 25

Energy, Matter, and Atoms

Objectives:

By the end of this chapter you should be able to:

- ✓ Describe the relationship between atoms and matter.
- ✓ Find an element in the periodic table.
- ✓ Identify how elements, compounds, and mixtures differ.
- ✓ Convert temperatures between Fahrenheit, Celsius, and Kelvin scales.
- ✓ Understand the concept of absolute zero temperature.
- ✓ Describe the phases of matter and explain solid, liquid, and gas in terms of energy and atoms.
- ✓ Describe the concepts of heat and thermal energy and apply them to real-life systems.
- ✓ Perform basic calculations with specific heat.

Key Questions:

- What makes up all matter?
- How is temperature related to phases of matter?
- How are temperature and thermal energy related?

Vocabulary

absolute zero	Celsius scale	heat of fusion	molecule	specific heat
boiling point	condensation	heat of vaporization	periodic table of elements	temperature
British thermal unit (BTU)	evaporation	ionized		thermistor
	Fahrenheit scale	Kelvin scale	random motion	thermocouple
calorie	heat	melting point	relative humidity	thermometer

519

25.1 Matter and Atoms

We experience a tremendous variety of matter. You can easily list several hundred different examples of matter in your home. Your list might include wood, many kinds of plastic, concrete, plaster, glass, copper wires, steel pots, paint, water, leaves, salt, tile, and even the air you breathe. Humans have wondered about the diversity of matter since early in recorded history. A growing tree creates wood from water, air, and trace elements in soil. The observation that wood is created from other forms of matter suggests that wood is not a fundamental substance but is made from simpler things. Is it possible that all the different forms of matter are made from a few ingredients? If so, what are the basic ingredients of matter?

Figure 25.1: *Steam, water, and ice are three forms of the same substance that can be converted into each other.*

Three big questions

Question 1 Think about cutting matter into pieces. When you cut a piece of wood, you have two smaller pieces of wood. It seems logical that there should be a "smallest" piece of wood that is still "wood." Since large pieces of wood are made of smaller pieces of wood, the same should be true of all matter, but there must be a limit.

What is the smallest piece of matter?

Question 2 Matter does not always stay the same. If you freeze water it becomes ice. Water is liquid and flows. Ice is hard and solid. Water and ice act like very different materials. But when you heat ice, it becomes water again. It is natural to assume that ice and water are two forms of the same substance (Figure 25.1).

Why can the same kind of matter assume different forms, like solid or liquid?

Question 3 Think about what happens when you heat wood in a candle flame. The wood turns into black, powdery ashes. If you take the ashes out of the flame, they do not turn back into wood as they cool down. Ashes appear to be a completely different substance from wood (Figure 25.2).

How can one kind of matter (like wood) turn into another kind of matter with very different properties (like ashes)?

Figure 25.2: *Wood and ashes are different substances. The conversion from wood into ashes does not reverse when the ashes cool down.*

Physics and chemistry The search for answers to the first question is part of physics. The science of chemistry is the search for answers to the second and third questions. We don't have complete answers to any of the three questions for all types of matter.

Matter is made of tiny particles

Brownian motion
A speck of dust dropped into a glass of water swirls around as it floats on the water surface. If the dust speck is *very* small and you look at it with a powerful microscope, its motion is *not* what you might imagine. The dust speck moves in a jerky irregular way. It appears to be bounced like a bumper car where the dust speck is being bounced around by impacts from smaller, invisible but fast moving objects. The jerky movement of a dust speck in water is an example of Brownian motion and provides a clue to identifying the smallest particle of matter.

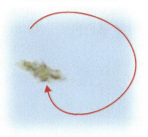

Figure 25.3: *A large floating dust speck moves smoothly because it is much larger than the smallest particles of water.*

The idea of atoms
In 430 BCE, the Greek philosopher Democritus and his teacher Leucippus proposed that matter must be made of small particles they called *atoms*. They had no proof that such particles existed, but it made sense that there should be a smallest particle of matter. Few believed them, and for the next 2,300 years atoms were just an idea. In 1803, the English scientist John Dalton revived the idea of atoms, but he also lacked proof. In 1905, Albert Einstein finally proved that matter was made of tiny particles by explaining Brownian motion.

Why Brownian motion occurs
A large speck of dust moves smoothly as it floats around. This is because a large speck of dust is so much larger than a single particle of water is hit by thousands of such particles, that a single collision is not noticeable (Figure 25.3). However, if the dust speck is *very* small, collisions with single particles of water are visible because the speck is hit by only a few particles at a time and its mass is not so much larger than that of a single particle of water. Moving water particles bump into the dust speck and cause Brownian motion (Figure 25.4). Why does the dust speck move around? Why doesn't it stay still? The answer is in this chapter and involves temperature.

Figure 25.4: *A tiny dust speck shows Brownian motion because of collisions with particles of water which have a comparable mass.*

A human-sized example
To get an idea of the scale of things, imagine throwing marbles at an inflatable swimming pool toy floating in the water. If you keep throwing marbles, the inflatable toy will slowly move. The motion of the inner tube will be small because each marble weighs a lot less than the tube. Next, think about throwing marbles at a paper cup floating on the water. The cup visibly moves under the impact of each marble. This is because the weight of the cup is not that much greater than the weight of a single marble. Brownian motion proves that matter exists in microscopic particles that are smaller than a tiny dust speck. We call these particles atoms and molecules.

Atoms and molecules

The smallest piece of matter Suppose you want to make the smallest possible piece of gold. You can keep cutting a piece of gold into smaller and smaller pieces until you cannot cut it any more. That smallest piece is one atom. A single atom is the smallest amount of gold you can have.

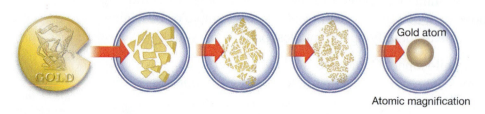

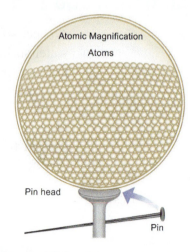

Figure 25.5: *The head of a pin contains more than 10^{20} atoms.*

Atoms We now know that all the matter you are familiar with is made of atoms. Atoms make up everything that we see, hear, feel, smell, and touch. We don't experience atoms directly because they are so small. The head of a pin contains 10^{20} atoms (Figure 25.5). Aluminum foil is thin but is still more than 200,000 atoms thick (Figure 25.6). A single atom is about 10^{-10} meters in diameter. That means you can lay 10,000,000,000 (10^{10}) atoms side by side in a one meter length.

Molecules A **molecule** is a group of two or more atoms that are joined together. If you could look at water with a powerful microscope you would find each particle of water is made from one oxygen atom and two hydrogen atoms.

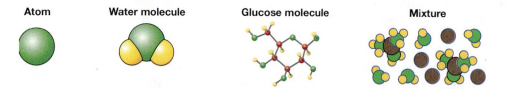

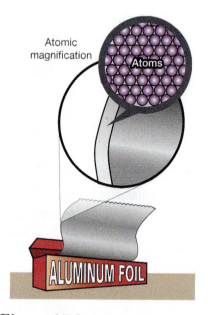

Figure 25.6: *A sheet of thin aluminum foil is 200,000 atoms thick.*

Matter is mostly molecules and mixtures Most matter you encounter is made of molecules, or mixtures of molecules. For example, glucose is a common sugar found in food. Each molecule of glucose has 6 carbon atoms, 6 oxygen atoms, and 12 hydrogen atoms. Grape soda is a mixture that contains water molecules, sugar molecules, molecules that make the purple color, and other molecules that create grape flavor and make the soda fizz.

Elements

Explaining the diversity of matter

You can make millions of colors by mixing different amounts of red, green, and blue. Is it possible that millions of different kinds of matter are really mixtures of a few simpler things? The ancient Greeks thought so. Their theory proposed that all matter was made of four fundamental elements: *air*, *fire*, *water*, and *earth*. According to the Greek theory of matter, everything could be made by combining different amounts of the four elements. For example, wood contained certain proportions of water, earth, air, and fire. When wood was burned, the smoke was the fire and air. The ash left over was earth. Gold was a different mixture with more earth and less water. The theory was based on simple observations.

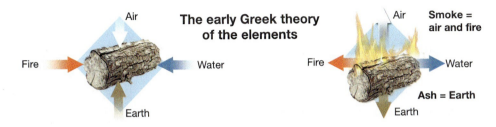

Elements

The Greeks had the right idea, but not the right elements. Today we know that nearly all the matter in the world is made from 92 different elements. Water is made from the elements hydrogen and oxygen. Air contains mostly nitrogen and oxygen. Steel is mostly iron and carbon, with a few exotic elements mixed in, like vanadium and chromium. Some rocks are mostly silicon and oxygen. We get different kinds of matter from combinations of these naturally-occurring elements.

Atoms and elements

Each of the 92 naturally-occurring elements has a unique type of atom. All atoms of a given element are similar to each other. If you could examine a million atoms of carbon, you would find them all to be similar. But carbon atoms are different from iron atoms or oxygen atoms. The atoms of an element are similar to atoms of the same element but different from atoms of other elements. You will find more details about atoms in Chapter 28.

One atom

One single atom is the smallest particle of an element that retains the identity of the element. Since all atoms of the same element are similar, every atom of carbon is identifiable as carbon. Similarly, every atom of gold is identifiable as gold.

The search for the elements

The search for the true elements has been a human goal for thousands of years. Fortunes and medicine were among many historical reasons people were interested. If the difference between gold and lead was just a change in recipe, then it might be possible to make lead into gold! The person who found the right recipe would be rich. The search for the secret of turning lead into gold developed into the "false science" of alchemy during the Middle Ages. Other early experimenters believed sickness was an imbalance in the elements of the body. Various treatments were devised to adjust the amount of "ill humours" present in a sick body. The experiments and observations of the alchemists and healers led directly to our modern understanding of chemistry. *Chemistry* is the science of how substances interact with each other, like wood burning in air. The foundation of chemistry is the study and description of the properties and interactions between the elements.

CHAPTER 25 — ENERGY, MATTER, AND ATOMS

The periodic table of elements

The periodic table
The **periodic table of elements** shows the elements in order from atomic number 1 (hydrogen) to 92 (uranium). Elements with atomic numbers greater than 92 can be made in a laboratory but are not normally found in nature. The periodic table also groups the elements by their chemical properties. All the elements in the same column have similar chemical properties. For, example all the elements in column 18 are noble gases that do not form molecules with other atoms. The elements in the middle columns are all metals.

Mass and atomic number
The mass of an atom is proportional to its atomic number. The lightest atom is hydrogen at the top left. The heaviest atom shown is uranium, shown in the bottom row of four elements.

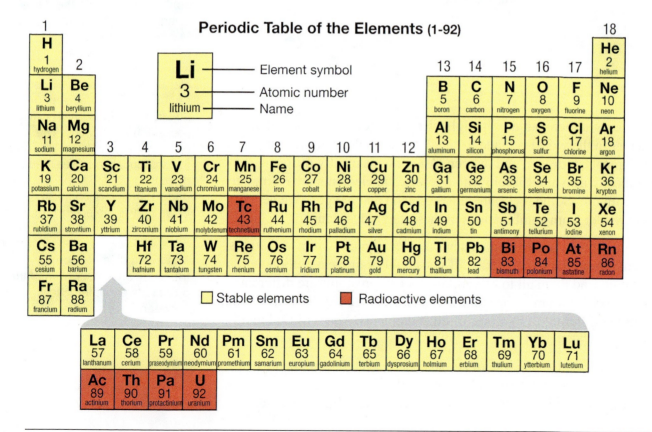

Elements past 92

Np 93	Pu 94	Am 95	Cm 96	Bk 97	Cf 98
Es 99	Fm 100	Md 101	No 102	Lr 103	Rf 104
Db 105	Sg 106	Bh 107	Hs 108	Mt 109	Ds 110

There are more elements than the 92 that make up the matter we find on Earth. For example, element 99 is called einsteinium. It does not occur naturally on Earth but has been made in laboratories. At the time of this writing, scientists had created elements 93 to 118 in research laboratories.

We don't find much of elements 93 to 118 in nature because all of these elements are *radioactive* and break down into other elements. Eventually all the elements heavier than lead (number 82) break down into lighter elements. Chapter 29 introduces radioactivity and nuclear reactions.

524   UNIT 8 MATTER AND ENERGY

The diversity of matter

Compounds Salt is a solid crystal that dissolves in water and is commonly used to flavor foods. Pure sodium is a soft, silvery, and highly-reactive metal. Pure chlorine is a toxic yellow-green gas. Salt is actually a chemical combination of the elements sodium and chlorine, even though the properties of salt are very different from the properties of sodium or chlorine alone (Figure 25.7). The incredible diversity of matter we experience is created by combinations of the 92 basic elements into compounds like salt. A compound is made of more than one element. Water is also a compound since it is made of molecules containing hydrogen and oxygen. Pure elements are rare. Most matter exists in the form of compounds.

Element
One single kind of atom

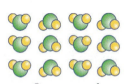
Compound
Molecules containing more than one kind of atom

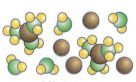
Mixture
Combination of different compounds and/or elements

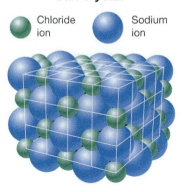

Figure 25.7: *Salt is a combination of sodium atoms and chlorine atoms.*

Molecules and properties of materials The properties of matter depend on how the atoms are arranged into molecules and mixtures. For example, one carbon atom with four hydrogen atoms makes a methane molecule. Methane is a flammable gas that is used to produce heat. Ten carbon atoms and 22 hydrogen atoms make a decane molecule. Decane is an oily liquid which you may have used as charcoal lighter fluid. Other combinations of carbon and hydrogen make wax and plastic. You would be surprised how many different substances can be made from just carbon and hydrogen!

How one material changes into another You can rearrange the same atoms into different molecules and get completely different materials (Figure 25.8). Wood and wood ash are made from the same elements, just arranged in different molecules. The heat that turns wood into wood ash is actually providing energy to break bonds between atoms, combine some with oxygen from the air, and rearrange the atoms into different molecules. That is how one substance can turn into something else.

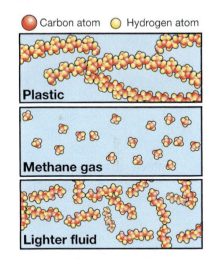

Figure 25.8: *Material properties depend more on molecules than on atoms.*

25.1 MATTER AND ATOMS **525**

25.2 Temperature and the Phases of Matter

The concepts of warm and cold are familiar to everyone. What causes ice to feel cold and fresh coffee to feel hot? The simple answer is *temperature*. Ice feels cold because its temperature is less than the temperature of your skin. Coffee feels hot because its temperature is greater than your skin temperature. Temperature is related to energy and determines whether matter takes the form of solid, liquid, or gas.

Temperature scales

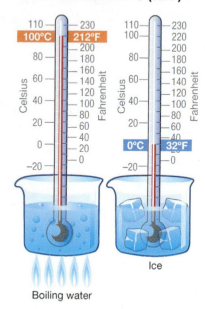

Water boils at 100°C (212°F) and freezes at 0°C (32°F)

Figure 25.9: *The thermometers compare the Celsius and Fahrenheit temperature scales.*

Fahrenheit There are two commonly used temperature scales. On the **Fahrenheit (F) scale**, water freezes at 32 degrees and boils at 212 degrees (Figure 25.9). There are 180 Fahrenheit degrees between the freezing point and the boiling point of water. Temperature in the United States is commonly measured in Fahrenheit. For example, 72°F is a common room temperature.

Celsius The **Celsius (C) scale** divides the difference between the freezing and boiling points of water into 100 degrees. Water freezes at 0°C and boils at 100°C. Most science and engineering temperature measurement is in Celsius because 0 and 100 are easier to use in calculations than 32 and 212. Most other countries use the Celsius scale for all descriptions of temperature, including daily weather reports.

CONVERTING BETWEEN FAHRENHEIT AND CELSIUS

$$T_{Fahrenheit} = \frac{9}{5} T_{Celsius} + 32 \quad \bigg| \quad T_{Celsius} = \frac{5}{9}(T_{Fahrenheit} - 32)$$

Fahrenheit to Celsius To convert from Fahrenheit to Celsius, subtract 32 then multiply by five-ninths. Subtracting 32 is necessary because water freezes at 32°F and 0°C. The factor of $\frac{5}{9}$ is applied because the Celsius degree is larger than the Fahrenheit degree.

Celsius to Fahrenheit To convert from Celsius to Fahrenheit, multiply by $\frac{9}{5}$ then add 32. If you travel to other countries, you will want to learn the difference between the two temperature scales. A weather report that says 21°C in London, England, predicts a warm day. A weather report predicting 21°F in Minneapolis, Minnesota, means clothing to protect against the cold weather. The United States is one of the few countries still using the Fahrenheit scale.

Measuring temperature

Human temperature sense
Our sense of temperature is not very accurate. Humans can sense when something is warm or cold, but cannot sense exact temperature. If you walk into a 65°F room from being outside on a winter day, the room feels warm. The same room will feel cool if you come in from outside on a hot summer day.

Thermometers
A **thermometer** is an instrument that measures temperature. The common alcohol thermometer (Figure 25.10) uses the expansion of liquid alcohol. As the temperature increases, the alcohol expands and rises up a long, thin tube. The temperature is measured by the height to which the alcohol rises. The thermometer can read small changes in temperature because the bulb at the bottom has a much larger volume than the tube.

How thermometers work
There are many ways to make a thermometer. All thermometers are based on some physical property, such as color or volume, that changes with temperature. A **thermistor** is a device that changes its electrical resistance as the temperature changes. Some electronic thermometers sense temperature by measuring the resistance of a thermistor. A **thermocouple** is another electrical sensor that measures temperature. A thermocouple is made by joining two metals of different elements. A small electrical voltage is created where the metals touch. The voltage depends on the temperature. An electronic thermometer that uses a thermocouple measures the voltage and converts it to temperature. Some kinds of chemicals change color at different temperatures.

> **Writing temperatures**
> Temperatures in Fahrenheit and Celsius are measured in degrees indicated with a little circle and the capital letters *F* or *C*. The temperature 72°F reads "72 degrees Fahrenheit." Likewise, 21°C reads "21 degrees Celsius."

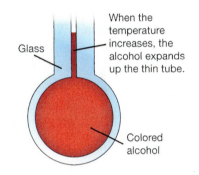

Figure 25.10: *How an alcohol thermometer works. The alcohol is often colored red to make it easier to see. Older thermometers used mercury instead of colored alcohol. Mercury is no longer used because it is toxic.*

Converting from Celsius to Fahrenheit

A friend in Paris sends you a recipe for a cake. The French recipe says to bake the cake at a temperature of 200°C for 45 minutes. At what temperature should you set your oven, which reads temperature in Fahrenheit?

1. You are asked for the temperature in Fahrenheit.
2. You are given the temperature in Celsius.
3. Use the conversion formula.

$$T_F = \frac{9}{5} T_C + 32$$

4. $T_F = \left(\frac{9}{5}\right)(200) + 32 = 392$ degrees

CHAPTER 25 ENERGY, MATTER, AND ATOMS

What is temperature?

Temperature and energy Think about Brownian motion again. Why is a tiny dust speck constantly moving around when floating in still water? Why doesn't it just sit still? What forces are acting to make the dust speck move if the water is standing still? The answer is that atoms in matter are never at rest but are always in constant, vibrating motion, even in a solid. Temperature is fundamentally a measure of the average kinetic energy of individual atoms. Imagine you have a microscope powerful enough to see individual atoms in a solid at room temperature. You would see that the atoms are in constant motion. The atoms in a solid material act like they are connected by springs (Figure 25.11) where each atom is free to move a small amount.

Average motion We already know the relationship between motion and kinetic energy for a single object, or atom. For a *collection* of atoms, the situation is different. The kinetic energy of a collection has two distinct parts. The kinetic energy you already know comes from the motion of the whole collection. If all the atoms were moving identically, the velocity of each individual atom would be the same as the velocity of the whole collection. However, this is not what really occurs.

Random motion Each atom in a collection can also have **random motion**. Random motion is motion that is scattered equally in all directions. In pure random motion, the average change in position for the whole collection is zero because as many atoms are moving one way as are moving the opposite way.

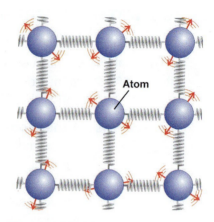

Figure 25.11: *Atoms in a solid are connected by bonds that act like springs. The atoms vibrate and the temperature measures their average energy of vibration.*

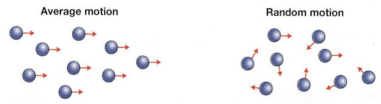

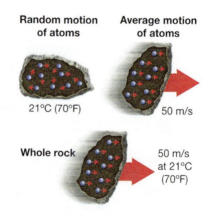

Figure 25.12: *A collection of atoms can have both average motion and random motion. That is why a thrown rock has both a velocity and a temperature.*

Temperature and random motion **Temperature** measures the kinetic energy *per atom* due to random motion. Temperature is not affected by any kinetic energy associated with average motion. That is why throwing a rock does not make it hotter (Figure 25.12). When you throw a rock, you give each atom in the rock the same average motion because all the atoms move together. *Temperature affects only the random motion of atoms.* When you heat a rock with a torch, each atom moves around more, but the whole rock stays in the same place because temperature does not affect average motion.

Absolute zero and the limits of temperature

Absolute zero There is a limit to how cold matter can get. As the temperature is reduced atoms move more and more slowly. When the temperature gets down to **absolute zero**, the atoms have the lowest energy they can have and the temperature cannot get any lower. You can think of absolute zero as the temperature where atoms are completely frozen, like ice, with no motion. Technically, atoms can never become absolutely motionless, but the distinction does not matter for most situations. Absolute zero occurs at –273°C (–459°F). It is not possible to have a temperature lower than absolute zero.

Quantum effects Technically, according to the quantum theory, atoms can never stop moving completely. Even at absolute zero some tiny amount of energy is left. Here, the "zero point" energy might as well be exactly zero because the rules of physics prevent the energy from ever going any lower. Explaining what happens when atoms are cooled to absolute zero is an area of active research.

The Kelvin scale The **Kelvin (K) scale** is useful for many scientific calculations because it starts at absolute zero. For example, the pressure in a gas depends on how fast the atoms are moving. The Kelvin scale is used because it measures the actual energy of atoms. A temperature in Celsius measures only the energy *relative* to 0°C. Kelvin measurements are written without the degree symbol (250 K).

Converting to Kelvin The Kelvin unit of temperature is related to the Celsius unit. Add 273 to the temperature in Celsius to get the temperature in Kelvins. For example, a temperature of 21°C is equal to 294 K (21 + 273).

CONVERTING CELSIUS TO KELVIN

$$T_{Kelvin} = T_{Celsius} + 273$$

High temperatures Temperature can be raised almost indefinitely. As the temperature increases, exotic forms of matter appear. For example, at 10,000 K, atoms start to come apart and become a plasma. In a plasma, the atoms themselves are broken into separate positive ions and negative electrons. Plasma conducts electricity and is formed in lightning and inside stars. Figure 25.13 compares the temperatures at which several natural events occur.

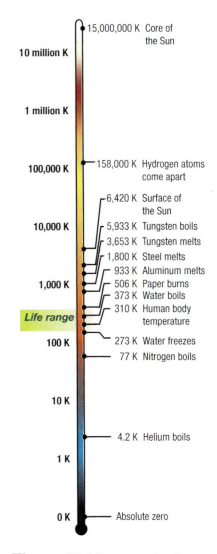

Figure 25.13: *A sample of temperatures in the universe. Most of our lives occur in a narrow 100°C range around the freezing point of water (223 K–323 K). The diagram is not shown to scale.*

The phases of matter

Solid, liquid, and gas   The three most common phases of matter are called solid, liquid, and gas. Matter in the solid phase holds its shape and does not flow. Ice is a good example of a solid. Matter in the liquid phase has constant volume but can flow and change its shape. Water is a good example of a liquid. Matter in the gas phase flows like liquid, but also can expand or contract to fill its container. Air is a good example of a gas.

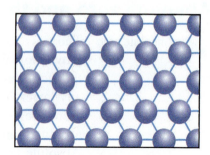

Figure 25.14: *Atoms or molecules in a solid stay bonded together.*

Solid   Matter at low temperatures is often solid. Atoms or molecules in a solid stay together because their thermal energy is too low to break the bonds between them (Figure 25.14). Imagine a marching band marching in place *with every one holding hands*. People move, but each person stays in the same place relative to others. Everyone in a marching band moves together, like the atoms in a solid.

Liquid   The liquid phase occurs at a higher temperature than the solid phase. Liquids flow because the atoms have enough energy to move around by temporarily breaking and reforming bonds with neighboring atoms (Figure 25.15). Imagine a room full of people dancing. The crowd generally stays together, *but people can switch partners if they want to do so*, like the atoms in a liquid.

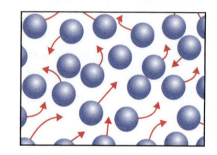

Figure 25.15: *Atoms in a liquid stay close together but can move around and bond with other atoms.*

Gas   The gas phase occurs at a higher temperature than the liquid phase. Atoms in a gas have enough energy that bonds between neighbors are completely broken (Figure 25.16). Gas expands because atoms can move independently. Imagine many people running fast in different directions. Every person is moving independently with a lot of space between people, like the atoms in a gas.

Plasma   At temperatures greater than 10,000 K, the atoms in a gas start to break apart. In the plasma state, matter becomes **ionized** as electrons break loose from atoms. Because the electrons are free to move independently, plasma can conduct electricity. Lightning is a good example of plasma. The Sun is another example.

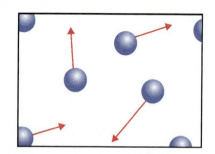

Figure 25.16: *Atoms in a gas move fast and are separated by relatively large spaces.*

Changing from solid to liquid

Melting The **melting point** is the temperature at which a material changes phase from solid to liquid. Melting occurs when the kinetic energy of individual atoms equals the attractive force between atoms. Different materials have different melting points because the bonds between atoms have different strengths. Water melts at 0°C (32°F). Iron melts at a much higher temperature, about 1,500°C (2,800°F). The difference in melting points tells us the attractive force between iron atoms is much greater than the attractive force between water molecules.

Heat of fusion The **heat of fusion** is the amount of energy it takes to change 1 kilogram of material from solid to liquid or vice versa at 1 atmosphere or pressure and at 0°C. The energy goes to make or break bonds between atoms. Table 25.1 gives some representative values of the heat of fusion (h_f) for common materials. Note how large the values are. It takes 335,000 joules of energy to turn 1 kilogram of ice into liquid water!

Phase changes take energy When thermal energy is added or subtracted from a material, either the temperature changes, or the phase changes, but usually not both at the same time. Think about heating a block of ice that has a temperature of –20°C. As you add heat energy, the temperature increases. Once it reaches 0°C, *the temperature stops increasing* as ice starts to melt and form liquid water. As you add more heat, more ice becomes water but the temperature stays the same. The heat goes to changing the phase from solid to liquid. Once all the ice has become liquid water, the temperature starts to rise again as more heat is added. This can be observed easily with an ordinary thermometer in an experiment.

Table 25.1: Heat of fusion for common materials

Material	Heat of fusion (J/kg)
Water	335,000
Wax	175,000
Aluminum	321,000
Iron	267,000
Silver	88,000

CALCULATING THE ENERGY FOR CHANGING FROM SOLID TO LIQUID

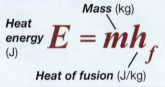

Heat energy (J) $E = mh_f$ Mass (kg), Heat of fusion (J/kg)

Calculating the energy to melt ice

How many joules does it take to melt a 30-gram ice cube at 0°C?

1. You are asked for heat energy (E) in joules.
2. You are given mass (m) and that the material is ice.
3. Use the phase change equation: $E = mh_f$
4. $E = (0.03 \text{ kg})(335,000 \text{ J/kg})$
   $= 10,050$ J

Start with ice at –20°C.

Add heat energy at a constant rate.

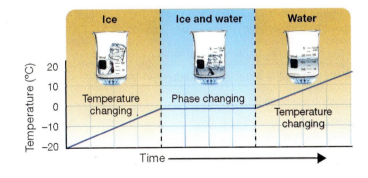

CHAPTER 25: ENERGY, MATTER, AND ATOMS

Changing from liquid to gas

Boiling The **boiling point** is the temperature at which the phase changes from liquid to gas. Water boils at 100°C (212°F) at 1 atmosphere of pressure. The steam rising from a pot of boiling water is water molecules in the gas phase. Even metals such as iron can become a gas, although it takes a much higher temperature. Iron boils above 2,900°C (5,200°F).

Heat of vaporization Just as with melting, it takes energy for an atom to go from liquid to gas. The **heat of vaporization** is the amount of energy it takes to convert 1 kilogram of liquid to 1 kilogram of gas. In a gas, all the bonds between one atom and its neighbors are completely broken. Some representative values for the heat of vaporization are given in Table 25.2. It takes 2,256,000 joules to turn a kilogram of water into a kilogram of steam! That is why stoves and steam irons require so much electricity. The heat of vaporization is much greater than the heat of fusion because breaking bonds between atoms or molecules takes much more energy than exchanging bonds. In a liquid, molecules move around by *exchanging* bonds with neighboring molecules. The energy needed to break one bond is recovered when the molecule forms a new bond with its neighbor.

Energy and phase change If you add heat to a pot of boiling water at 100°C, the temperature of the water stays right at the boiling point. The heat you add goes to changing water atoms from the liquid phase to the gas phase. The temperature of the boiling water will stay at 100°C until all the liquid has been converted to steam (Figure 25.17). Heat can change a material's temperature or its phase, but not both at the same time.

Table 25.2: Heat of vaporization for common materials

Material	Heat of vaporization (J/kg)
Water	2,256,000
Alcohol	854,000
Liquid nitrogen	201,000
Lead	871,000
Silver	2,336,000

CALCULATING THE ENERGY FOR CHANGING LIQUID TO GAS

$$E = mh_v$$

Heat energy (J) = Mass (kg) × Heat of vaporization (J/kg)

Figure 25.17: *It takes 2,256,000 joules to turn 1 kilogram of liquid water at 1 atmosphere and 100°C into steam.*

Calculating the heat needed to boil water into steam

A steam iron is used to remove the wrinkles from clothes. The iron boils water in a small chamber and vents steam out the bottom. How much energy does it require to change one-half gram (0.0005 kg, or about half a teaspoon) of water into steam?

1. You are asked for the heat energy required (E) in joules.
2. You are given that the material is water, and the mass (m) is 0.0005 kg.
3. The heat of vaporization equation applies: $E = mh_v$.
4. $E = (0.0005 \text{ kg})(2,256,000 \text{ J/kg}) = 1,128$ joules

A typical steam iron takes about 5 seconds to boil this amount of water.

Evaporation and condensation

Evaporation

Evaporation occurs when molecules go from liquid to gas at temperatures *below* the boiling point. Evaporation happens because temperature measures the *average* random kinetic energy of molecules. Some have energy above the average and some below the average. If an energetic energy molecule is right on the surface, it may get "bumped out" and become a gas. High-energy molecules at a liquid surface are the source of evaporation.

Evaporation cools liquids

Evaporation takes energy away from a liquid because the molecules that escape are the ones with the most energy. The average energy of the molecules left behind is lowered. Evaporation cools the surface of a liquid because the fastest molecules escape and carry energy away. That is why we sweat on a hot day. The evaporation of sweat from your skin cools your body (Figure 25.18).

Condensation

Condensation occurs when molecules go from gas to liquid at temperatures below the boiling point (Figure 25.19). Water vapor in the air often condenses on cool surfaces. Water molecules with less-than-average energy may stick to a cool surface, forming drops of liquid water. Condensation raises the temperature of a gas because atoms in a gas have more energy than atoms in a liquid. Low energy gas atoms condense into liquid leaving higher-energy atoms in the gas.

Air contains water vapor

Ordinary air contains some water vapor. Evaporation adds water vapor to the air. Condensation removes water vapor. The percentage of water vapor in the air is a balance between evaporation and condensation. When air is *saturated*, it means the processes of evaporation and condensation are exactly balanced. If you try to add more water vapor to saturated air, it condenses into liquid water again.

Relative humidity

The **relative humidity** tells how close the air is to saturation. When the relative humidity is 100 percent, the air is completely saturated. That means any water vapor that evaporates from your skin is condensed right back again, which is why you feel hot and sticky when the humidity is high. The body's natural cooling mechanism cannot work effectively because the air is already saturated with water vapor. The opposite is true in the dry air of desert climates. Hot desert air has a very low relative humidity, allowing water to evaporate rapidly. This is why dry heat feels more bearable than humid heat.

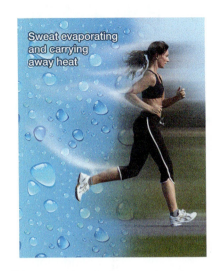

Figure 25.18: *Evaporation is how we sweat. When water evaporates from the skin it carries away energy and cools the body.*

Figure 25.19: *Dew is formed by condensation. Water vapor in the air condenses to form liquid droplets. Condensation warms the air since the heat is given up when the water goes from gas to liquid.*

25.3 Heat and Thermal Energy

To change the temperature of matter, you need to add or subtract energy in the form of heat. When you want to get your house warm in the winter, you add heat. If you want to cool your house in summer, you remove heat. This section makes the connection between temperature, heat, and energy. By the end of the section, you should be able to determine how much heat to add or subtract when you want to change the temperature by a certain number of degrees.

The relationship between heat, energy, and temperature

What causes hot and cold? You can warm your cold hands by holding a hot cup of coffee (Figure 25.20). What flows from the hot coffee into your hands to make them warm? For a long time, people thought a fluid called *caloric* was responsible. These people thought of caloric as "liquid heat." According to their theory, hot coffee was hot because it had more caloric than cold coffee. Your hands warm up because caloric flows from the hot cup to your colder hands. It seemed like a good explanation, *except it was wrong*. If you carefully measure the mass of an object when it is hot and when it is cold, you find the mass is exactly the same. If the caloric theory was correct, the hot object, with more caloric, should have more mass than the cold object.

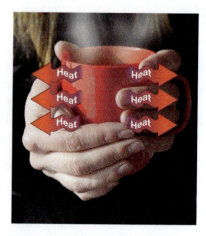

Figure 25.20: *You can warm your by holding a cup of hot coffee because heat flows from hot to cold, carrying energy to your hands.*

Thermal energy Today we know that *energy* is what flows from hot to cold and warms your hands. Energy has no intrinsic mass of its own, so hot and cold objects have the same mass. Thermal energy is energy stored in materials because of differences in temperature. Remember, temperature measures the random kinetic energy of *each atom*. The thermal energy of an object is the total amount of random kinetic energy for *all* the atoms in the object.

Temperature and thermal energy Temperature is not the same as thermal energy. Imagine heating a cup of coffee to a temperature of 100°C. Next, think about heating up 1,000 cups of coffee to 100°C (Figure 25.21). The final temperature is the same in both cases but the amount of energy needed is very different. It takes more energy to heat up 1,000 cups than to heat up a single cup. Temperature measures the average energy per kilogram of matter. Thermal energy is the total energy in a sample of matter and depends on both the temperature and the amount of matter you have.

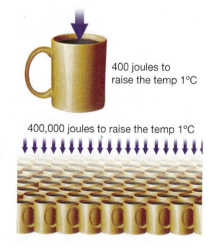

Figure 25.21: *It takes 400 joules to raise the temperature of one cup of coffee 1 degree. It takes 400,000 joules to raise the temperature of 1,000 cups of coffee the same 1 degree.*

Heat and thermal energy

Heat — **Heat** is what we call thermal energy that is moving. Heat flows naturally from hot to cold and moves thermal energy from higher temperatures to lower ones. When you leave a cup of hot coffee on the table, it cools down. Heat flows from the hot coffee to the cooler air in the room. The thermal energy of the hot coffee is decreased by the heat that flows. The thermal energy of the air is increased by the heat that flows. Everything balances; the increase in thermal energy of the air is exactly the same as the decrease in thermal energy of the coffee (Figure 25.22).

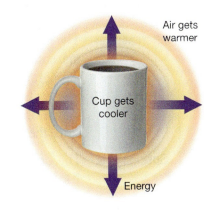

Figure 25.22: *Heat carries thermal energy from hot to cold. The cup gets cooler and the air in the room gets warmer.*

Joules — The joule (J) is the unit of heat, or thermal energy, used for physics and engineering (Chapter 10). Heat is a form of energy and the joule is the same unit used for all other forms of energy. There are also other commonly-used units to measure heat. Each unit was developed to measure heat in a different application. For physics calculations, you may have to convert from the other units to joules.

Calories — The **calorie** is a unit of heat often used in chemistry. One calorie is the amount of heat required to raise the temperature of 1 gram of water by 1 degree Celsius. The calorie is a larger unit of energy than the joule. There are 4.184 joules in 1 calorie. The calorie is different from the *food* calorie (Chapter 11).

British thermal units (BTU) — Air conditioners and furnaces are rated in **British thermal units (BTU)**. One BTU is the amount of heat required to raise the temperature of 1 pound of water by 1 degree Fahrenheit. A typical single-family home-heating furnace can produce 10,000 to 100,000 BTU per hour. One BTU equals 1,055 joules.

Units of heat and thermal energy	Practical application		
1 joule = Energy to push with a force of 1 newton for 1 m	1 kg of water at 10°C (50°F)	Heat added 41,840 J 10,000 cal 39.7 BTU	1 kg of water at 20°C (68°F)
1 BTU = 1,055 joules			
1 calorie = 4.184 joules			

25.3 HEAT AND THERMAL ENERGY

Specific heat

Differences in materials The same amount of heat causes a different change in temperature in different materials. For example, if you add 4 joules of heat to 1 gram of water, the temperature goes up by about 1 degree Celsius. If you add the same 4 joules of heat to 1 gram of gold, the temperature goes up by 31°C! The different temperature increases occur because different materials have different abilities to store thermal energy (Figure 25.23).

Specific heat The **specific heat** is the quantity of heat it takes to raise the temperature of 1 kilogram of material by 1 degree Celsius. Water is an important example. The specific heat of water is 4,184 J/kg·°C. It takes 4,184 joules to raise the temperature of 1 kilogram of water by 1 degree Celsius. The specific heat of gold is 129 J/kg·°C. Since the specific heat of gold is lower than that of water, a given change in energy produces a greater temperature change in gold than in water. Table 25.3 gives the specific heat of some common materials.

The heat equation The heat equation is used to calculate how much heat (E) it takes to make a temperature change ($T_2 - T_1$) in a mass (m) of material with specific heat (c_p).

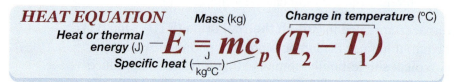

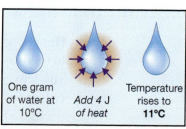

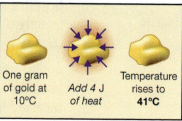

Figure 25.23: *The same amount of heat causes the temperature of gold to rise much more than the temperature of water.*

Table 25.3: Specific heat of common materials

Material	Specific heat (J/kg·°C)
Air	1,006
Water	4,184
Aluminum	900
Steel	470
Silver	235
Oil	1,900
Concrete	880
Glass	800
Gold	129
Wood	2,500

Note: Specific heat often changes with temperature and pressure.

Calculating the heat required to reach a temperature

One kilogram of water is heated in a microwave oven that delivers 500 watts of heat to the water. One watt is a flow of energy of 1 joule per second. If the water starts at 10°C, how much time does it take to heat up to 100°C?

1. You are asked for the time (t) to reach a given temperature (T_2).
2. You are given the mass (m) of water, power (P), and initial temperature (T_1). The specific heat of water is 4,184 J/kg·°C.
3. The heat equation, $E = mc_p(T_2 - T_1)$, gives the heat required. Power is energy over time: $P = E/t$.
4. First, calculate the heat required: $E = (1 \text{ kg})(4{,}184 \text{ J/kg·°C})(100°C - 10°C) = 376{,}560$ joules. Next, recall that 500 watts is 500 joules per second. At 500 J/s, it takes $376{,}560 \div 500 = 753$ seconds, or about 12.6 minutes.

Why the specific heat is different for different materials

Why specific heat varies
One reason for the variation in specific heat is the mass of each atom. Elements with heavy atoms have a lower specific heat compared to elements that have lighter atoms. This is because temperature measures the energy *per atom*. Heavy atoms mean fewer atoms per kilogram. Energy that is divided among fewer atoms means more energy per atom, and therefore more temperature change.

An example: silver and aluminum
Suppose you add 4 joules of heat to a gram of silver and 4 joules to a gram of aluminum. Silver's specific heat is 235 J/kg·°C and 4 joules is enough to raise the temperature of the silver by 19°C. Aluminum's specific heat is 900 J/kg·°C, so 4 joules only raises the temperature of the aluminum by 5°C. The silver has fewer atoms than the aluminum because silver atoms are heavier than aluminum atoms. When heat is added, each atom of silver gets more energy than each atom of aluminum because there are fewer silver atoms in a gram. Because the energy per atom is greater, the temperature increase in the silver is also greater.

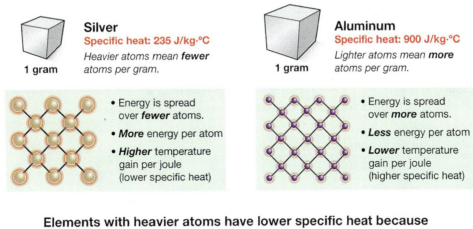

Elements with heavier atoms have lower specific heat because temperature depends on the energy *per atom*.

Thermodynamics

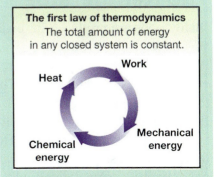

Thermodynamics is a branch of physics that deals with the relationship between thermal energy and other forms of energy. The first law of thermodynamics is a restatement of the law of conservation of energy, including thermal energy.

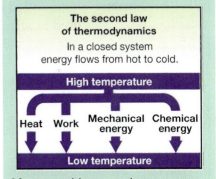

Many machines, such as a car engine, get useful work as energy flows from higher-temperature to lower-temperature regions. The second law says there must be a temperature difference for a machine to extract useful work from thermal energy.

Chapter 25 Connection

The Refrigerator

Many foods are produced far away from where you buy them. The fact that you can buy fruits, vegetables, dairy products, eggs, and meat almost everywhere all year round is due mainly to the invention of the refrigerator. A refrigerator moves thermal energy out of the inside of the appliance, making things inside it colder. The thermal energy is pumped to the outside of the refrigerator, making the room warmer (Figure 25.24). If you think about it, a refrigerator should not work! Heat usually flows from hot to cold, not the other way around. However, if you turn off the power to the refrigerator, the situation reverts to normal. Heat flows from the warmer room to the inside of the refrigerator and things will not stay cold. A refrigerator takes power, usually from electricity, to make heat flow opposite to its normal direction.

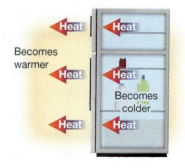

Figure 25.24: *A refrigerator moves heat from the colder inside to the warmer outside.*

Gas cools when it expands When you open a can of soda, you may notice that the escaping gas feels cold. When a gas at high pressure expands to low pressure, the gas cools. The cooling effect happens because energy of vibration is transformed into energy of motion as the molecules move rapidly away from each other. The cooling effect of an expanding gas is what makes a refrigerator work.

The principle of a refrigerator In the working parts of a refrigerator, a high-pressure liquid flows through a tiny hole, called an expansion valve, and becomes a low-pressure liquid. The liquid used in refrigerators typically drops from 40°C to –20°C between the high-pressure and low-pressure sides of the valve. A refrigerator moves heat by circulating a fluid that changes from a warm, high-pressure liquid to a cold, low-pressure gas-and-liquid mixture.

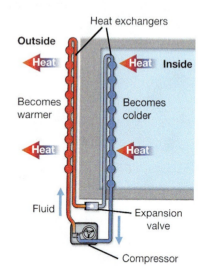

Figure 25.25: *The working parts of a refrigerator. A fluid flows through the tubes and changes from liquid, to gas, and back again at different places.*

The compressor and heat exchangers A refrigerator has two heat exchangers called the *evaporator coil* and the *condenser coil* and a special pump called a *compressor* (Figure 25.25). The evaporator coil is colder than the inside of the refrigerator and absorbs heat from inside. The condenser coil is hotter than room-temperature air and ejects the heat into the room. The compressor compresses the cool gas into a hot gas and moves it along through the condenser. The best way to learn how a refrigerator works is to follow some fluid through the cycle from the compressor through the heat exchangers' coils and back.

Chapter 25 Connection

The refrigeration cycle

Expansion and cooling The cooling effect begins with expansion of a liquid at high pressure through the expansion valve. The liquid cools as it passes through the hole and enters the low-pressure part of the cycle. The cold, low-pressure liquid passes through the evaporator coils. Heat flows into the cold liquid inside the evaporator coils and starts evaporating because the mixture is colder than the air in the refrigerator. After passing through the evaporator coils, all the liquid has evaporated into gas.

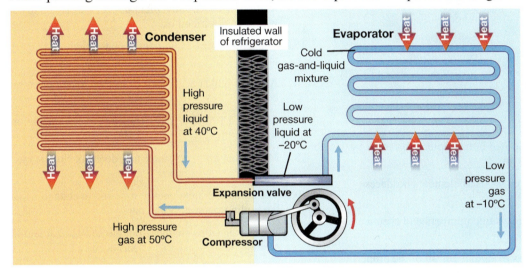

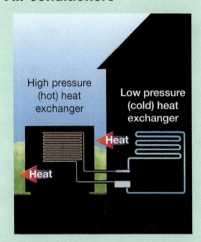

Air conditioners

An air conditioner is just like a refrigerator, except that the whole house becomes the inside of the refrigerator. The cold, low-pressure heat exchanger is inside the house. The hot, high-pressure heat exchanger is outside the house. An air conditioner moves thermal energy from inside to outside, making the air in the house cooler and the air outside warmer.

Ejecting heat into the room The cold, low-pressure gas is raised back to high pressure by the compressor. Raising the pressure makes the temperature go up so that the high-pressure gas is warmer than room-temperature air. The high-pressure gas loses heat to the room and condenses into a liquid as it flows through the condenser coils. The high-pressure liquid flows back through the expansion valve, where it cools. The fluid in the refrigerator continuously cycles from cold and low pressure to hot and high pressure as it carries thermal energy.

Conservation of energy As clever as it is, a refrigerator still obeys the law of conservation of energy. The heat energy absorbed by the room equals the heat energy taken out of the refrigerator plus the electrical energy you have to supply to make the compressor work. If you leave a refrigerator door open, the room actually gets warmer overall.

25.3 HEAT AND THERMAL ENERGY   **539**

Chapter 25 Assessment

Vocabulary

Select the correct term to complete the sentences.

atom	compound	elements
mixture	molecule	periodic table
Kelvin	Celsius	thermometer
Fahrenheit	specific heat	temperature
calorie	absolute zero	random
melting point	heat of vaporization	boiling point
evaporation	ionized	condensation
thermal energy	relative humidity	British thermal unit (BTU)
heat	solid	plasma
liquid	heat of fusion	gas

1. A pure substance that cannot be broken down into simpler substances by chemical or physical means is known as an _____.

2. The chemical combination of two or more types of elements produces a(n) _____.

3. A table of elements grouped according to atomic numbers and chemical properties is called the _____.

4. The smallest piece of an element, such as iron, that has all of the properties of the element is called a(n) _____.

5. The chemical combination of one atom of oxygen with two atoms of hydrogen produces one _____ of water.

6. Salt and pepper stirred together could be referred to as a(n) _____.

7. When a change from liquid to gas phase at a temperature below the boiling point occurs it is known as _____.

8. Water freezes at 32° and boils at 212° on the _____ scale.

9. A device commonly used to measure temperature is called a(n) _____.

10. The temperature at which atoms have their lowest energy is called _____.

11. Matter becomes ionized at high temperatures in the _____ phase of matter.

12. The amount of heat energy required to change 1 kilogram of a solid to liquid is known as the _____.

13. The scale used for most scientific calculations that start at absolute zero is the _____ scale.

14. Motion in which movement of particles is equally scattered in all directions would be described as _____.

15. The temperature at which the phase changes from liquid to gas is called the _____.

16. An atom which acquires a charge by losing or gaining an electron has become _____.

17. There is a 100-degree difference between the melting and boiling point of water on the _____ scale.

18. Matter that holds its shape and does not flow is identified as a(n) _____.

19. The amount of heat energy required to change 1 kilogram of liquid to gas at the boiling point is called the _____.

20. When molecules change from gas to liquid at temperatures above the melting point, _____ has occurred.

21. The temperature at which a substance changes from a solid to a liquid phase is identified as the _____ of the substance.

22. Matter in the _____ phase can flow but retains its volume.

23. _____ is the measure of the kinetic energy per atom due to random motion.

24. The amount of moisture in the air compared to the amount it can hold at a certain temperature is _____.

25. Matter that can flow and fill any container that encloses it is called a(n) _____.

26. The amount of heat energy required to raise the temperature of 1 pound of water 1 degree Fahrenheit is 1 _____.

27. The quantity of heat energy required to raise the temperature of 1 kilogram of a substance 1 degree Celsius is known as the _____ of the substance.

28. Thermal energy that moves is referred to as _____.

29. The unit of heat energy required to raise the temperature of 1 gram of water 1 degree Celsius is the _____.

30. Energy stored in bodies due to temperature differences is known as _____.

Concept review

1. As you observe the movement of a smoke particle under a microscope, you see irregular, "jerky" motion called Brownian movement. What is the cause of this motion?

2. Compare and contrast the terms *compound* and *molecule*.

3. Compare and contrast the terms *atom* and *element*.

4. A student from Germany flies to visit a family in Oswego, New York, in the United States. He hears the captain report the weather on arrival as "clear and sunny with temperatures in the low 20s." He changes on the plane into shorts and a T-shirt. Explain his behavior.

5. The operation of all thermometers is based on some physical property that changes with temperature. Describe the operation of two different thermometers and the properties on which they are based.

6. Fahrenheit and Celsius degrees are most commonly used to indicate temperature on household thermometers. Which unit is greater—Celsius or Fahrenheit? How can this be determined?

7. Temperature is a measure of the kinetic energy or the motion of the molecules of an object. What type of molecular motion is responsible for temperature differences?

8. Compare and contrast the Celsius and Kelvin scales of temperature.

9. What causes compounds such as water to change phase from solid to liquid or gas to plasma?

10. Heat is added to ice at its melting point and to the same amount of water at its boiling point. Describe the difference in the amount of heat needed to change the phase of water in each case.

11. Why does evaporation cause the temperature of a liquid to decrease?

12. The graphs represent temperature versus time graphs for 1-kilogram masses of various materials under various heating or cooling circumstances. Match each graph with a statement that could correctly represent the circumstances.

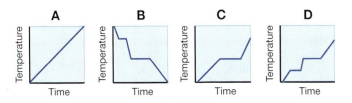

a. a material changing from solid to liquid to gas
b. a material with a heat of fusion greater than its heat of vaporization
c. a material with a specific heat that is greatest in the liquid phase
d. a material which is present in only one phase

13. The diagrams represent ionic compounds in their phases. Label each diagram as solid, liquid, or gas. Summarize the characteristics of each phase.

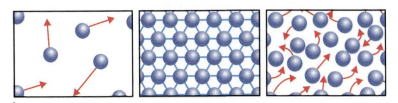

14. Which would cause a more severe burn to your skin—exposure to 1 kilogram of steam at 100°C or 1 kilogram of liquid water at 100°C? Explain your answer.

15. Why is it more comfortable to exercise on a day with low relative humidity?

16. Which contains more heat, a 100-milliliter beaker of water at 98°C or a bathtub full of water at 0°C? Explain your answer.

17. The specific heat of gold is less than the specific heat of aluminum. Explain why this is so.

18. Explain why a manual bicycle pump is warm after inflating a tire.
19. Ed thinks that he can cool his kitchen by leaving the refrigerator door open. What is wrong with this thinking?

Problems

1. How much heat is required to change the temperature of 0.5 kilograms of aluminum from 25°C to 37°C?
2. A material with a mass of 0.25 kilograms absorbs 322.5 joules of energy as its temperature rises from 20°C to 30°C. What is the material?
3. A 0.15-kg piece of glass is dropped into 100 milliliters of water which has a temperature is 20°C. The temperature of the water rises 4°C. What was the original temperature of the glass? Assume no loss of heat to the air or the container.
4. Answer questions a–h based upon the diagram which represents a cooling curve for 10 kilograms of a substance as it cools from a vapor at 160°C to a solid at 20°C. Energy is removed from the material at a rate of 200 kilojoules per minute.

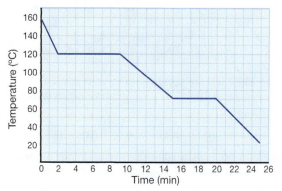

a. Calculate the heat of fusion of the substance.
b. Calculate the specific heat of the substance as a liquid.
c. What is the melting point of the substance in degrees Kelvin?
d. What is the boiling point of the substance in degrees Celsius?
e. What is the freezing point of the substance in degrees Fahrenheit?
f. Calculate the number of BTUs of heat energy removed as the substance changes from the gas phase to the liquid phase.
g. What is the difference in temperature in degrees Celsius between the boiling point and the freezing point of this substance?
h. What is the difference in temperature in degrees Kelvin between the boiling and freezing points of the substance?

Applying your knowledge

1. Explain the relevance of water's unusually high specific heat to life on Earth.
2. Research the specific heat of hydrogen gas. Compare the specific heat of hydrogen gas to the specific heat of liquid water. Offer an explanation for the difference between the two.
3. Plasma makes up about 99 percent of the matter in the universe but only occurs naturally in lightning and parts of flames. Research how man-made plasmas are used on Earth.
4. Which graph best represents the relationship between the Celsius temperature of an ideal gas and the average kinetic energy of its molecules? What is the approximate temperature represented by the point where the line crosses the x-axis?

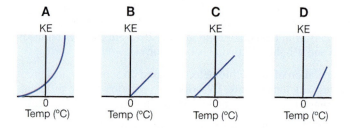

UNIT 8 MATTER AND ENERGY

CHAPTER 26

Heat Transfer

Objectives:

By the end of this chapter you should be able to:

✔ Explain the relationship between temperature and thermal equilibrium.
✔ Explain how heat flows in physical systems in terms of conduction, convection, and radiation.
✔ Apply the concepts of thermal insulators and conductors to practical systems.
✔ Describe free and forced convection and recognize these processes in real-life applications.
✔ Identify the relationship between wavelength, color, infrared light, and thermal radiation.
✔ Calculate the heat transfer in watts for conduction, convection, and radiation in simple systems.
✔ Explain how the three heat-transfer processes are applied to evaluating the energy efficiency of a house or building.

Key Questions:

- How is thermal energy transferred by conduction?
- How is thermal energy transferred by convection?
- How is heat transfer by radiation different from conduction and convection?

Vocabulary

blackbody	forced convection	R-value	thermal insulator
blackbody spectrum	free convection	thermal conductivity	thermal radiation
conduction	heat transfer	thermal conductor	windchill factor
convection	heat transfer coefficient	thermal equilibrium	

26.1 Heat Conduction

Almost everyone has accidentally picked up a hot pot from the stove and *quickly* realized it was too hot to hold. A pot feels hot because heat flows from the pot into your skin, causing your skin temperature to rise uncomfortably fast. The science of how heat flows is called **heat transfer**. There are three ways heat transfer works: conduction, convection, and radiation. This section is about conduction, which is the transfer of heat by direct contact of particles of matter.

Heat flow and thermal equilibrium

The cause of heat transfer — The rate of heat transfer is proportional to the difference in temperature. Consider standing outside when the air temperature is 68°F (20°C). You probably will not feel extremely cold because the temperature difference between your skin (75°F) and the air is only seven degrees. Your body loses heat relatively slowly. If the air temperature were to drop to 40°F (4°C), you would quickly feel cold. You get cold because five times more heat flows out of your body each second when there is a 35-degree difference compared with a 7-degree difference.

Heat transfer in living things — Heat flow is necessary for life. This is because all biological processes release energy. Your body regulates its temperature through the constant flow of heat. The inside of your body averages 98.6°F. Humans are most comfortable when the air is around 75°F because the rate of heat flow out of the body matches the rate at which the body generates heat internally. If the air is 50°F, you get cold because heat flows too rapidly from your skin to the air. If the air is 100°F, you feel hot partly because heat flows from the air to your body and partly because your body cannot get rid of its internal heat fast enough (Figure 26.1).

Thermal equilibrium — Two bodies are in **thermal equilibrium** with each other when they have the same temperature. In thermal equilibrium, no heat flows because the temperatures are the same (Figure 26.2). In nature, heat *always* flows from hot to cold until thermal equilibrium is reached. For example, if you leave a cup of hot coffee on the table, heat flows from the cup to the room until everything is the same temperature. Putting the hot coffee in an insulated cup only slows the process down. Heat still flows, only slower. Hot coffee in an insulated cup takes *longer* to reach thermal equilibrium with the room, but the end result is the same.

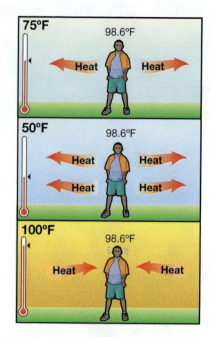

Figure 26.1: *Heat flow depends on the temperature difference between you and your surroundings.*

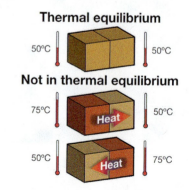

Figure 26.2: *No heat flows when objects are at the same temperature.*

Heat conduction

Conduction

Conduction is the transfer of heat *through* materials by the direct contact of matter. Imagine two blocks of metal at different temperatures. What happens if you bring the blocks together so that they touch? The warmer block will cool down and the colder block will warm up. Heat flows from the warmer block to the colder block by conduction (Figure 26.3). The heat keeps flowing until both blocks are at the same temperature.

Heat flow by conduction

Figure 26.3: *Conduction is the flow of heat through direct contact.*

Thermal conductors

Dense metals like copper and aluminum are very good **thermal conductors**. Think about holding one end of a copper pipe while the other end is in a hot flame. Copper is a good thermal conductor so heat flows rapidly from one end of the pipe to the other. You cannot hold on for long because the pipe gets hot so quickly.

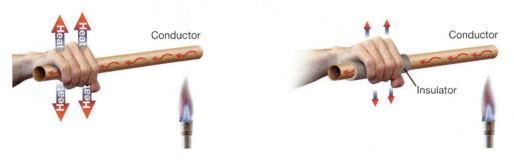

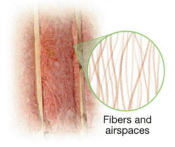

Thermal insulators

A **thermal insulator** is a material that conducts heat poorly. Foam is a good example. You can comfortably hold a hot copper pipe surrounded by a centimeter of foam. Heat flows very slowly through the plastic so that the temperature of your hand does not rise very much. Foam gets its insulating ability by trapping spaces of air in bubbles (Figure 26.4). We use thermal insulators to maintain temperature differences without allowing much heat to flow from a hotter to a colder object.

What affects a material's ability to conduct heat?

All materials conduct heat at some rate. Solids usually are better heat conductors than liquids, and liquids are better conductors than gases. The ability to conduct heat often depends more on the structure of a material than on the material itself. For example, solid glass is a thermal conductor when it is made into windows. When glass is spun into fine fibers and made into fiberglass insulation, the combination of glass fibers and trapped air makes a thermal insulator.

Figure 26.4: *Fiberglass insulation and foam derive their insulating ability from trapping air between fibers or in bubbles.*

CHAPTER 26 HEAT TRANSFER

Thermal conductivity

Thermal conductivity

The **thermal conductivity** of a material describes how well the material conducts heat. Materials with high thermal conductivity are good thermal conductors, such as copper, aluminum, and other metals. Materials with low thermal conductivity are thermal insulators, such as fiberglass and foam. Table 26.1 gives the thermal conductivity of some common materials.

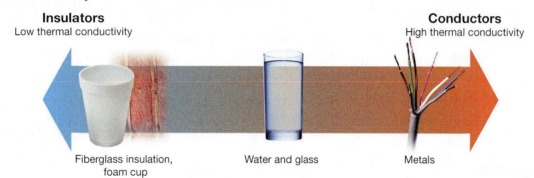

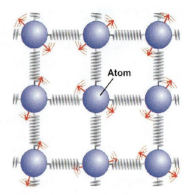

Figure 26.5: *The bonds between atoms act like springs that allow atoms to vibrate and transmit the vibration to neighboring atoms.*

Conduction in solids and liquids

Heat conduction in solids and liquids works by transferring energy through bonds between atoms or molecules. The bonds between neighboring atoms or molecules act like springs (Figure 26.5). If you shake one end of a spring, eventually the motion is transferred to the other end. Thermal motion is transferred along bonds in a similar way. The result is that energy flows from hotter atoms to cooler atoms. Solid materials are the best conductors because solids have a high density of atoms connected by strong bonds.

Conduction in a gas

Heat conduction in a gas works through collisions between atoms. Hotter atoms in a gas move faster. A fast atom slows down a little when it collides with a slower atom. A slow atom speeds up a bit when it collides with a faster atom. Each collision transfers a little thermal energy. Some of the kinetic energy of hotter gas atoms is transferred, one collision at a time, to cooler atoms. Gases are thermal insulators because atoms are much farther apart and collisions are less effective at transferring energy than bonds between atoms.

Conduction in liquids

Liquids conduct heat better than gases but not as well as solids. The atoms in a liquid are close together, like in a solid. However, the bonds are more like those in a gas than a solid.

Table 26.1: Thermal conductivity of common materials

Material	Thermal cond. (W/m·°C)
Diamond	2,650.
Copper	401.
Aluminum	226.
Steel	43.
Rock	3.
Glass	2.2
Ice	2.2
Liquid water	0.58
Wood	0.11
Wool fabric	0.038
Fiberglass insulation	0.038
Foam	0.025
Air	0.026

546 UNIT 8 MATTER AND ENERGY

The heat conduction equation

Predicting heat flow In many cases, we want to know how much heat per second flows. For example, to keep your house at the same temperature, any heat that flows out the windows must be replaced by energy you pay for. In physics, power is the flow of energy per second. Power is measured in *watts*; a watt is a joule per second (Chapter 11). We will use the symbol P_H for the power transferred by heat. The power transferred by heat conduction (P_H) depends on four factors (Figure 26.6):

- the area through which the heat flows (A);
- the length the heat has to travel (L);
- the temperature difference ($T_1 - T_2$); and
- the thermal conductivity of the material (κ).

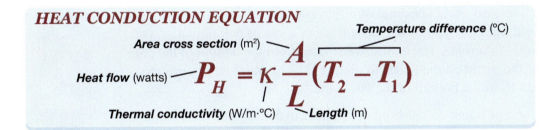

A metal bar connects two beakers of water at different temperatures. Heat flows through the bar.

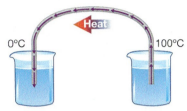

Four variables affect the amount of heat that flows:

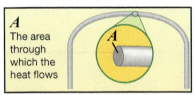

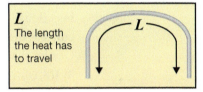

Calculating heat transfer through a metal bar

A copper bar connects two beakers of water at different temperatures (Figure 26.6). One beaker is at 100°C and the other is at 0°C. The bar has a cross section area of 0.0004 m² and is one-half meter (0.5 m) long. How many watts of heat are conducted through the bar from the hot beaker to the cold beaker? The thermal conductivity of copper is 401 W/m·°C.

1. You are asked for the power transferred as heat.
2. You are given the area, length, temperature difference, and thermal conductivity.
3. The heat conduction equation is $P_H = kA(T_2 - T_1)/L$.
4. Solve the problem:
   $P_H = (401 \text{ W/m·°C})(0.0004 \text{ m}^2)(100°C)/(0.5 \text{ m}) = 32$ watts

Figure 26.6: *The variables in the heat conduction equation can be understood in an experiment in which a metal bar connects beakers of water at unequal temperatures.*

26.2 Convection

Convection is the transfer of heat by the motion of liquids and gases. A candle flame provides a good example of convection. If you hold your hand a half meter directly above a candle flame, you will quickly feel heat (Figure 26.7). (*Note*: Do *not* hold your hand above the candle flame for more than a few seconds!) Hold your hand the same distance to the side and you feel no heat at all. You feel heat directly above the flame because hot air rises after being heated by the flame. The heat carried by the motion of the air is an example of convection.

The causes of convection

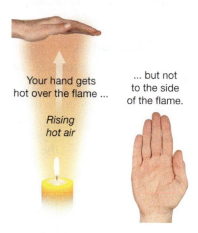

Figure 26.7: *The rising column of hot air over a candle flame is an example of convection.*

Why convection occurs Convection in a gas occurs because gas expands when heated. When a gas expands, the mass is spread out over a larger volume so the density decreases. Hot gas with lower density is lighter than surrounding cooler gas and floats upward. Convection occurs because currents flow when hot gas rises and cool gas sinks.

Convection in liquids Convection in liquids also occurs because of differences in density. Hot liquid is less dense than cold liquid. If you watch the surface of a pot of boiling water, you can see the convection currents. The hottest water rises from the bottom of the pot. Cooler water near the surface sinks. The circulating flow of water is very effective at transferring heat from the bottom of the pot to the surface.

Free convection When the flow of gas or liquid comes from differences in density and temperature, it is called **free convection**. The heat rising above a candle flame is an example of free convection. The water circulating in a boiling pot of water is also an example of free convection.

Forced convection When the flow of gas or liquid is circulated by pumps or fans it is called **forced convection**. Many homes and buildings are heated by forced convection. A furnace heats water in a boiler. Pumps circulate water through the boiler to get it hot, then circulate the hot water to rooms where the heat is needed.

Convection systems Many heat transfer systems use a combination of free and forced convection. For example, a common home-heating system uses copper tubes and pumps to circulate hot water. Forced convection carries the heat from the boiler to finned copper tubes in each room. Free convection transfers the heat from the finned tubes to the air in the room (Figure 26.8).

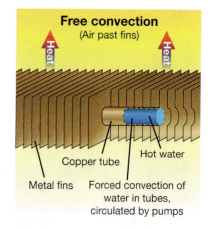

Figure 26.8: *Homes heated with hot-water boilers use both free convection and forced convection.*

Convection depends on speed and surface area

Faster flow increases heat transfer
Think about standing outside on a cold day. You get cold much faster when the wind is blowing. The faster the wind blows, the more effectively heat is carried away from your body. Motion increases heat transfer by convection in all fluids. Figure 26.9 shows the heat transferred by oil flowing in a tube in the engine of a car. The amount of heat transferred to the oil depends on the flow speed.

The windchill factor
The Antarctic explorers who invented the **windchill factor** did a very simple experiment in convection. They measured how long it took for a gallon of water to freeze in air at different temperatures and wind speeds. They found a temperature of 0°F with a 30-mile-per-hour wind froze the water in the same time as a lower temperature (−26°F) with no wind. A windchill of −26°F means your skin loses heat at the same rate as if the air temperature were actually −26°F with no wind. The chart below shows the windchill for different temperatures and wind speeds.

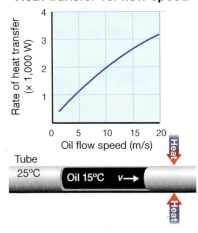

Figure 26.9: *The rate of heat transfer changes dramatically with the speed of flow.*

Wind chill equivalent temperatures (°F)

Wind speed (mph)	Air temperature (°F)						
	−20	−10	0	10	20	30	40
10	−41	−28	−16	−4	9	21	34
20	−48	−35	−22	−9	4	17	30
30	−53	−39	−26	−12	1	15	28
40	−57	−43	−29	−15	−1	13	27

Fins increase heat transfer
Almost all devices made for convection have fins (Figure 26.10). Convection transfers heat between a surface and a moving fluid. If the surface contacting the fluid is increased, the rate of heat transfer also increases. Fins provide a tremendous increase in surface area and greatly increase the rate of heat transfer. Motorcycle engines, car radiators, and home-heating elements are examples of devices that use fins to enhance heat transfer by convection.

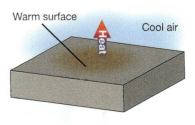

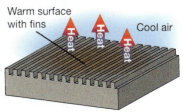

Figure 26.10: *Fins dramatically increase the surface area, improving heat transfer by convection.*

26.2 CONVECTION 549

Convection in the environment

Convection and weather Currents caused by convection are responsible for much of our weather. Warm air rises off of Earth's surface. As the warm air rises higher, it cools. The cooler air sinks back down, creating a circulation pattern.

Sea breezes The sea breezes that form near coastlines are a good example of circulation patterns created by convection. During the day, the land is warmed by the Sun more than water because rocks and earth have a lower specific heat than water. Warm air over land rises because of convection and is replaced by cooler air from the ocean. A daytime sea breeze blows from the ocean inward (Figure 26.11). In the evening the circulation pattern reverses. At night, the ground cools rapidly but the ocean remains warm because water has a high specific heat. Warm air rises over the water and is replaced with cooler air from over the land. The nighttime breeze blows from the land out to the sea (Figure 26.12).

Convection in the ocean Much of Earth's climate is regulated by giant convection currents in the oceans. Dense, cold water from melting ice near the poles sinks to the ocean floor and flows toward the equator. Warmer water from the equator circulates back toward the poles near the ocean surface. The weather pattern known as El Niño causes heavy storms in some years. El Niño is caused by an oscillation in the flow of convection currents in the Pacific Ocean.

Figure 26.11: *During the day, a sea breeze is created when warm air over the land rises because of convection and is replaced by cooler air from the ocean.*

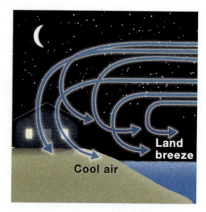

Figure 26.12: *At night, temperatures reverse and a land breeze occurs—land cools more rapidly than the ocean.*

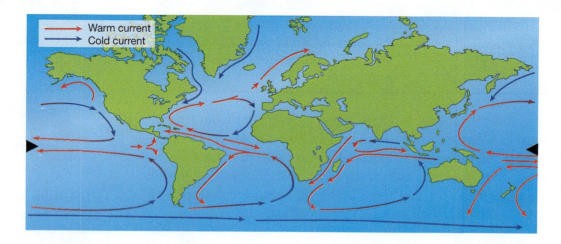

A model for convection

Temperature difference
The rate of heat transfer is proportional to the temperature difference. For example, a 0°C (32°F) wind blowing on bare skin will carry away much more heat than a 20°C (68°F) wind because the colder wind creates a larger temperature difference between the air and skin.

The heat transfer coefficient, h
A simple model for convection includes the contact area (A) and the temperature difference ($T_2 - T_1$). The complicated effects of flow speed and surface conditions are grouped together in the **heat transfer coefficient**, h. A value of $h = 1$ W/m²·°C means that 1 watt of heat is transferred from each square meter of area when the temperature difference is 1°C. Most practical heat transfer systems have a value of h between 10 and 10,000 W/m²·°C (Table 26.2). The convection equation shows how to calculate the power of heat transferred (P_H).

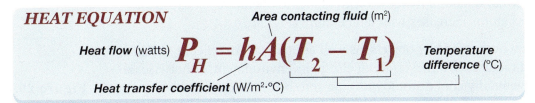

HEAT EQUATION
Heat flow (watts) — $P_H = hA(T_2 - T_1)$ — Temperature difference (°C)
Area contacting fluid (m²)
Heat transfer coefficient (W/m²·°C)

Table 26.2: Heat-transfer coefficients for common flow conditions

Condition	Range of heat transfer coefficients (W/m²·°C)
Free convection	
Gases	5–25
Oil	10–60
Water	100–1,000
Forced convection	
Gases	10–300
Oil	50–2,000
Water	100–20,000

Many things influence the heat transfer coefficient
The heat transfer coefficient, h, depends on *many* factors. Surface roughness is one factor. Whether the flow is smooth or turbulent is another very important factor. Whether the fluid is a liquid or gas is another important factor. In almost all real cases, h is measured in experiments, and not accepted as a constant.

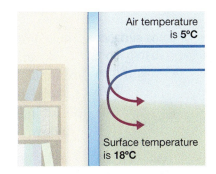

Figure 26.13: *Cold air absorbs heat by convection when it blows on the outside surface of a window.*

Calculating the heat lost through a glass window

The temperature of the surface of a window is 18°C (64°F). A 5°C (41°F) wind is blowing on the window fast enough to make the heat transfer coefficient 100 W/m²·°C. How much heat is transferred between the window and the air if the area of the window is 0.5m² (Figure 26.13)?

1. You are asked for the rate of heat transfer.
2. You are given the temperature difference, area, and heat transfer coefficient.
3. Use the convection equation $P_H = hA(T_2 - T_1)$.
4. Solve:
   $P_H = (100$ W/m²·°C$)(0.5$ m²$)(13$°C$) = 650$ watts

This is a lot of heat to lose out your window! Storm windows and double-pane glass help reduce convection by preventing the outside air from contacting the inner window.

CHAPTER 26 HEAT TRANSFER

26.3 Radiant Heat

If you stand in sunlight, you can feel warm even on a cold day. The heat you feel from the Sun comes from the energy of light waves soaking into your skin. *Radiation* is heat transfer by electromagnetic waves. Virtually all of the energy that makes Earth warm comes from the Sun as radiation. Radiation plays an important role in heat transfer here on Earth, too, as you will discover in this section.

Properties of thermal radiation

Definition of thermal radiation **Thermal radiation** is electromagnetic waves, including light, produced by objects because of their temperature. All objects with a temperature above absolute zero give off thermal radiation. Your own body gives off thermal radiation and that is why you and all other living things "glow" when viewed with an infrared night-vision camera. Thermal radiation comes from the thermal energy of atoms. The power in thermal radiation increases with higher temperatures because the thermal energy of atoms increases with temperature (Figure 26.14).

Objects emit and absorb radiation Thermal radiation is absorbed by objects, as well as emitted. An object constantly receives thermal radiation from everything else in its environment. Otherwise, all objects would eventually cool down to absolute zero by radiating their energy away. The temperature of an object rises if it absorbs more radiation. The temperature falls if more radiation is given off. The temperature adjusts until there is a balance between radiation absorbed and radiation emitted.

Some surfaces absorb more energy than others The amount of thermal radiation absorbed depends on the surface of a material. Black surfaces absorb almost all the thermal radiation that falls on them. For example, black asphalt pavement gets very hot in the Sun because it effectively absorbs thermal radiation. A silver mirror surface reflects most thermal radiation, absorbing very little. A mirrored screen reflects the Sun's heat back out your car window, helping your car stay cooler on a hot day (Figure 26.15).

Radiation can travel through space Thermal radiation can travel through the vacuum of space. Conduction and convection cannot carry heat through space because both processes require matter to transfer heat. Radiation is different. Because the energy is carried by electromagnetic waves, radiation does not require matter to provide a path for the heat to flow. Radiant energy also travels very fast—at the speed of light!

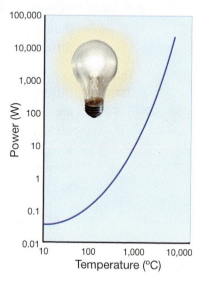

Figure 26.14: *The higher the temperature of an object, the more thermal radiation it gives off.*

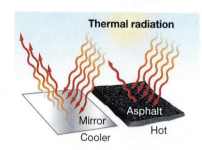

Figure 26.15: *Black or dark colored surfaces absorb most of the thermal radiation they receive. Silver or mirrored surfaces reflect most of the thermal radiation they receive.*

Thermal radiation and infrared light

Thermal radiation is mostly invisible infrared light

At room temperature, we do not see objects by their thermal radiation. We see objects by the light they reflect from other sources. We do not see the thermal radiation because it occurs at infrared wavelengths invisible to the human eye. The power versus wavelength graph shows how light from thermal radiation is spread over a range of wavelengths. A rock at room temperature does not "glow" because the curve for 20°C does not extend into visible wavelengths. All of the thermal radiation at 20°C is invisible infrared light. Up to a few thousand degrees Celsius, most of the energy in thermal radiation is in infrared light.

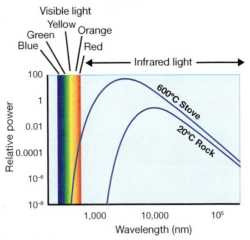

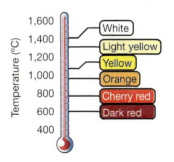

Figure 26.16: *Objects glow different colors at different temperatures. The glow comes from the visible part of the spectrum.*

Why hot objects glow

As objects heat up, they start to give off visible light, or glow. At 600°C, objects glow dull red, like the burner on an electric stove. The curve for 600°C on the power versus wavelength graph shows some radiation emitted in red and orange visible light, though most of the energy is still radiated as invisible infrared. More than 99.9 percent of the radiant heat from a hot stove element is in infrared light.

Temperature and color

As the temperature rises, thermal radiation produces shorter-wavelength, higher-energy light. At 1,000°C, the color is yellow-orange, turning to white at 1,500°C (Figure 26.16). If you carefully watch a bulb on a dimmer switch, you see its color change as the filament gets hotter. The bright white light from a bulb is thermal radiation from an extremely hot filament, near 2,600°C (Figure 26.17).

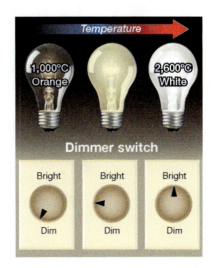

Figure 26.17: *A dimmer switch and an incandescent bulb illustrate the effect of temperature on the color of thermal radiation.*

Infrared thermometers

Between room temperature and 600°C, the power of thermal radiation increases more than 100 times. You may have had a doctor place a thermometer in your ear that measures your temperature. Ear thermometers work by measuring infrared thermal radiation.

CHAPTER 26 — HEAT TRANSFER

The blackbody spectrum

Perfect absorption of light
To a physicist, an object is perfectly *black* when it absorbs all radiation that falls on its surface. If all radiation is absorbed, then any radiation coming off of the surface can only be thermal radiation. A perfect **blackbody** is a surface that reflects nothing and emits pure thermal radiation. Most objects reflect some radiation and therefore emit less thermal radiation than a perfect blackbody. To a physicist, the white-hot filament of a bulb is a near-perfect blackbody. All light from the filament is thermal radiation; almost none of it is reflected from other sources.

The blackbody spectrum
The graph of power versus wavelength for a perfect blackbody is called the **blackbody spectrum**. The curve for 2,600°C shows that radiation is emitted over the whole range of visible light. White light is a mixture of colors, and incandescent light bulbs need the high temperature to make white light. The Sun has a surface temperature of 5,500°C. From the blackbody spectrum for 5,500°C, you can see the Sun produces visible light, infrared light, and quite a lot of harmful, high-energy, ultraviolet light as well.

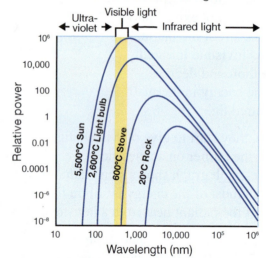

Figure 26.18: *Sirius is a hot blue-white star in the constellation Canis Major. Sirius is one of the brightest stars in the night sky.*

The temperature and color of stars
Stars have different surface temperatures. Stars bigger than our Sun burn hotter and radiate more blue-white light. Sirius, in the constellation of Canis Major, is a hot, young star about two times the size of the Sun (Figure 26.18). Sirius is about 22 times brighter than our Sun. Because its temperature is hotter, Sirius also appears bluer than the Sun. Some older stars, called red giants, are also larger but cooler than the Sun. The star Betelgeuse is a red giant. Light from these stars is redder than sunlight because the surface temperature of red giant stars is lower.

Stars and the blackbody spectrum

A star is a near-perfect blackbody. According to the blackbody spectrum, the distribution of energy across different wavelengths (colors) depends strongly on the temperature. Therefore, a star's color tells us the temperature of the star. The temperature and brightness tell us the size of the star. From the temperature and size, astronomers can calculate the mass of the star, how old it is, and how long it is likely to keep shining. For example, we know the Sun is around 4 billion years old and we can expect sunshine for another 4 billion years.

UNIT 8 MATTER AND ENERGY

A model for radiation

The Stefan-Boltzmann formula

The total power emitted as thermal radiation by a blackbody depends on temperature (*T*) and surface area (*A*). The Stefan-Boltzmann formula allows us to calculate the power. Virtually all real surfaces emit less than the full blackbody power, with the actual emission varying from 10 to 90 percent of full blackbody power. The power emitted as thermal radiation power increases *very quickly* with temperature, as the fourth power of the temperature (T^4; Figure 26.19). The power of four means that when you double the temperature, the power emitted in thermal radiation increases by 2^4, or *16 times*. At temperatures over 500°C, radiation almost always transfers more heat than convection or conduction (Figure 26.20).

Use Kelvins for radiation calculations

The Kelvin temperature scale is used in the Stefan-Boltzmann formula because thermal radiation depends on the temperature *above absolute zero*. When you are given a problem, remember to add 273 if the temperature is in degrees Celsius.

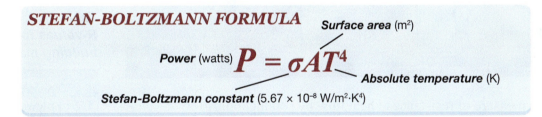

STEFAN-BOLTZMANN FORMULA

Power (watts) $P = \sigma A T^4$ Surface area (m²), Absolute temperature (K)

Stefan-Boltzmann constant (5.67×10^{-8} W/m²·K⁴)

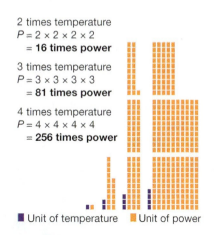

Figure 26.19: *Thermal radiation increases as the fourth power of the temperature. This means every time the temperature doubles, the power of radiation increases by 2^4, or 16 times.*

Calculating the radiation power from a small light bulb filament

The filament in an incandescent light bulb has a diameter of 0.5 millimeters and a length of 50 millimeters. The surface area of the filament is 4×10^{-8} m². If the temperature is 3,000 K, how much power does the filament radiate?

1. You are asked for the power radiated.
2. You are given the size, surface area, and temperature in Kelvins.
3. Use the Stefan-Boltzmann formula: $P = \sigma A T^4$.
4. Solve:
   $P = (5.7 \times 10^{-8} \text{ W/m}^2\cdot\text{K}^4)(4 \times 10^{-8} \text{ m}^2)(3{,}000 \text{ K})^4 = 0.2$ watts

This is not much light! Incandescent bulbs are not very efficient at converting electrical energy to light.

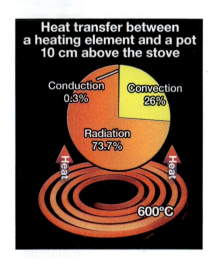

Figure 26.20: *The circle graph compares heat transfer by convection and radiation at different temperatures.*

Chapter 26 Connection

Energy-Efficient Buildings

Heat transfer is very important to the design of homes and buildings. In winter, the temperature inside is greater than the temperature outside. Heat flows out, and any that is lost must be replaced with heat we pay for, produced by electricity, oil, wood, or gas. In summer, the situation is reversed. We want to keep the heat outside. Heat that leaks in must be pumped back out with air conditioning, using expensive electricity. An efficient house does not lose too much heat in winter or gain too much heat in summer.

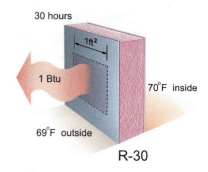

Figure 26.21: *The R-value is based on temperature change and time elapsed.*

R-value — The **R-value** is a measure of resistance to heat flow by conduction. A low R-value means heat flows quickly. A high R-value means heat flows slowly. A wall rated at R-1 means it takes one hour for one BTU (1,054 J) of heat to pass through one square foot when the temperature difference is 1°F. One BTU is not much heat; it takes about 25 BTU to heat a cup of coffee. A typical cold-climate, home-heating system produces more than 50,000 BTU per hour. Such houses require R-30 in the roof, which means it takes 30 hours for one BTU of heat to pass through each square foot when the temperature difference is 1°F (Figure 26.21).

The R-value of a wall or roof — The walls and roofs of houses built today are insulated with materials such as fiberglass or foam. Table 26.3 lists the R-value for some common building materials. The total R-value of a wall is equal to the sum of the R-values for each layer. For example, a roof with one-half inch of plywood, nine inches of fiberglass insulation, and one-half inch of plaster board would have a total R-value greater than R-31.

Table 26.3: R-values for common building materials

Material	R-values
1/2-in plywood	0.62
1/2-in plaster board	0.45
8-in concrete	1
4-in wood (pine)	4
1-in foam board	7
3.5-in fiberglass ins.	13
6-in fiberglass ins.	19
9-in fiberglass ins.	30

Chapter 26 Connection

Reducing heat loss

Heat loss through windows

Convection is the chief cause of heat loss through thin materials, such as glass windows. To reduce convection, modern windows have double or triple panes of glass (Figure 26.22). The outer pane of glass stops the wind from reaching the inner pane of glass. By separating the inner and outer panes of glass, heat must travel through the insulating air space between them. This greatly reduces heat transfer. Older homes without double-pane windows may use *storm windows*. A storm window also provides a second pane of glass closer to the outside.

Other sources of convection

Cracks around doors and siding are also a large source of heat loss by convection. Weatherstripping is an effective way to seal openings around doors and windows and reduce air exchange between inside and outside. To slow down air flowing through siding, many homes are now built with a wind barrier, or house wrap.

Radiation

In Southern climates, preventing heat transfer by radiation is an important design goal. Radiation passes through glass windows easily and the energy is absorbed by interior surfaces. One strategy for reducing radiation heat gain is to use low emissivity (low-E) windows. A low-E window has two or three panes of glass. The inside surface is treated with a special coating that reflects infrared light, but allows most visible light to pass through (Figure 26.23). You can still see through the windows, but much of the heat is blocked.

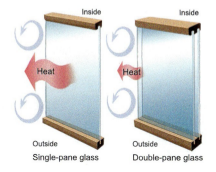

Figure 26.22: *Double-pane glass windows reduce heat loss by convection because the outside air does not contact the inner window pane.*

Energy use for heating and cooling

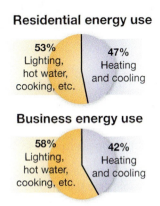

Heating and air conditioning use a lot of energy. These two uses are responsible for 47 percent of all the energy used by homes and 42 percent of that used by businesses. By many estimates, as much as half of this energy could be saved by designing buildings to be better at controlling heat transfer. Every state has a *building code* which includes rules for constructing homes and commercial buildings to meet standards of energy efficiency. Insulation, double-pane windows, low-E glass, and weatherstripping are part of many modern building-code requirements.

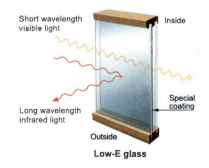

Figure 26.23: *A special coating on the glass of a low-E window transmits most visible light but reflects infrared light.*

26.3 RADIANT HEAT

Chapter 26 Assessment

Vocabulary

Select the correct term to complete the sentences.

infrared	windchill factor	thermal insulator
thermal equilibrium	forced convection	R-value
convection	blackbody spectrum	thermal conductivity
thermal conductors	blackbody	heat transfer
thermal radiation	free convection	heat transfer coefficient
conduction	insulators	conductors

1. The ability of a material to conduct heat is measured as _____.

2. A material that conducts heat poorly is referred to as a(n) _____.

3. The transfer of heat through a material by the direct contact of matter is called _____.

4. No heat flows because their temperatures are the same when two bodies are in _____.

5. The science of how heat flows is called _____.

6. Dense metals like copper, aluminum, gold, and silver make good _____.

7. For calculating heat flow, the effects of flow speed and surface conditions are combined in the _____.

8. A comparison of a body's heat loss at various wind speeds and heat loss in still air was used to establish the _____.

9. The transfer of heat due to the motion of fluids is called _____.

10. When the flow of gas or liquid is caused by pumps or fans, the resulting transfer of heat is known as _____.

11. Heat transfer causing gases to flow resulting from density differences is called _____.

12. All thermal radiation occurring at 20°C is in invisible _____ wavelengths.

13. Electromagnetic waves produced by objects because of their temperature are referred to as _____.

14. A surface that reflects nothing and emits pure thermal radiation is known as a perfect _____.

15. The measure of the resistance to heat flow by conduction is known as _____.

16. The graph of power as a function of wavelength for a perfect blackbody is called the _____.

Concept review

1. Answer the following.
   a. Name a factor affecting the rate of heat transfer.
   b. Describe the direction of heat transfer between two objects.
   c. Explain how thermal equilibrium affects heat transfer.

2. Two objects, A and B, are in contact with one another. Initially the temperature of A is 300 K, and temperature of B is 400 K. Which diagram indicates the correct direction of heat flow?

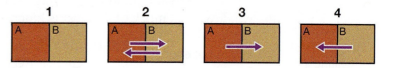

3. When you step out of the shower, the tile floor feels cold but the bath mat feels warm even though they are both at room temperature. Why?

4. Wood is a poorer thermal conductor than glass but wooden buildings are insulated using glass fibers. Why?

5. Compare the ability of solids, liquids, and gases to conduct heat.

6. A copper rod conducts heat from one area of higher temperature to one of lower temperature at a certain rate. If the diameter of the rod is doubled, how would this affect the rate of heat conduction?

7. All objects at a temperature above absolute zero constantly emit thermal radiation. Why do objects not cool to absolute zero by emitting all their thermal energy?

8. Ammonia boils at −33°C. Referring to the chart of wind chill equivalent temperatures, estimate the highest temperature and accompanying wind speed that could be used to prevent ammonia from boiling.

Wind chill equivalent temperatures (°F)

Wind speed (mph)	Air temperature (°F)						
	−20	−10	0	10	20	30	40
10	−41	−28	−16	−4	9	21	34
20	−48	−35	−22	−9	4	17	30
30	−53	−39	−26	−12	1	15	28
40	−57	−43	−29	−15	−1	13	27

9. Explain why it might be easier to sail toward the shore in a sailboat during the day rather than at night.

10. At room temperature, wood feels warmer than metal. At what temperature will both seem to be the same temperature? Explain.

11. You are asked to shovel snow from a driveway on a sunny winter day. You have plans for the afternoon. Your brother suggests that you scatter the black ashes from the wood stove on the driveway. Will this effect the amount of time required to shovel the driveway? Explain.

12. Astronauts in an orbiting space station are trying to celebrate a fellow astronaut's birthday with burning candles on a birthday cake. The candles will not stay lit. Why?

13. There are some people who believe that water can be boiled over an open flame in a paper cup or a foam cup. Is this true? Support your answer with an explanation from this chapter.

14. Many a young person who has been told their tongue would stick to cold metal has tested the warning only to discover that it really happens! Why does this happen? Does this happen with a wooden object? Explain.

15. Examine the diagram. Both sunlight and light from a light bulb contain harmful ultraviolet radiation. Why are you more likely to get sunburn from the Sun than from a light bulb?

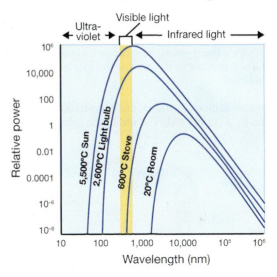

The blackbody spectrum
Power vs. Wavelength

CHAPTER 26 HEAT TRANSFER

Problems

1. As a dimmer switch is turned, the temperature of the filament in an incandescent bulb increases from 100°C to 1,592°C. By how many times is the thermal radiation increased?

2. The filament of a 100-watt bulb is 5.0 centimeters long. What must be the diameter of the filament if the bulb has an operating temperature of 2,800 K?

3. A bar of copper and a bar of aluminum of the same area and thickness are in thermal contact. The temperature at the outer surface of the copper is 80°C while the temperature of the aluminum is maintained at 20°C at its outer surface. When heat conduction through the bimetal object reaches a steady state, what is the temperature at the interface?

4. When the equation for heat flow by conduction, $P_H = \kappa(A/L)(T_2 - T_1)$, is used by engineers, the ratio L/κ is called the R-value. Rearrange the equation to include the R-value. Use the new equation to determine the effect a four-fold increase in R-value of the insulation would have on the heat loss in a home.

5. Two insulated containers, A and B, containing identical amounts of water are connected by a copper rod. The initial temperatures of A and B are 100°C and 20°C, respectively.

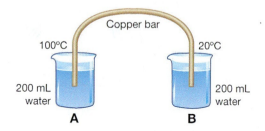

   a. Assuming all heat lost by A is transferred to B through the copper rod as A's temperature decreases to 80°C, what change occurs in the heat flow?
   b. If an aluminum bar 0.25 meters long and 0.0075 meters in diameter is substituted for the copper bar, calculate the initial rate of energy flow. The thermal conductivity for aluminum is 226 W/m²·°C.

6. If you place your hand in a beaker of water at room temperature (20°C), what is the rate of heat flow? Assume your hand has a surface area of 0.08 m² and the heat transfer coefficient of the water is 200 W/m²·°C.

7. The temperature inside a room is maintained at 20°C while the outside temperature is –20°C. What is the rate at which heat is transferred to a window pane by convection? The window dimensions are as follows: length: 0.67 m; width: 0.50 m; thickness: 0.25 cm. The heat transfer coefficient is 5.0 W/m²·°C. Assume the window temperature to be the average of the inside and outside air temperature.

8. In the previous problem, it was assumed that the temperature of the glass was the average of the inside and outside temperatures. Find the actual temperature of the inside and the outside of the glass window from the previous problem. Use κ = 0.82 W/m²·°C for glass.

9. Suppose a person snow skied in a bathing suit when the temperature was 5°C. How much heat energy would be lost due to radiation from the skin of the skier in an hour of skiing? Assume the area of the exposed skin is 1.5 m² and typical body temperature is 37°C.

Applying your knowledge

1. Global climate change may cause a change in certain convection currents in the North Atlantic. Research this topic and briefly describe how global climate change might ultimately affect Europe.

2. Good conductors of heat are generally good conductors of electricity. Briefly explain why good thermal conductors are good electrical conductors. Compare the electrical and thermal conductivities of five metals such as aluminum, copper, iron, gold, and silver. Do they occur in the same relative order? Are these properties temperature-dependent?

UNIT 8 MATTER AND ENERGY

CHAPTER 27

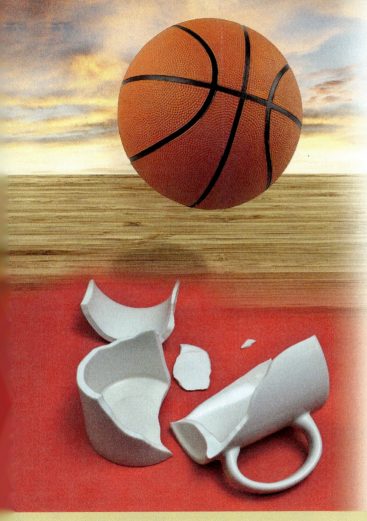

The Physical Properties of Matter

Objectives:

By the end of this chapter you should be able to:

- Perform calculations involving the density of solids, gases, and liquids.
- Apply the concepts of force, stress, strain, and tensile strength to simple structures.
- Describe the cause and some consequences of thermal expansion in solids, liquids, and gases.
- Explain the concept of pressure and calculate pressure caused by the weight of fluids.
- Explain how pressure is created on a molecular level.
- Understand and apply Bernoulli's equation to flow along a streamline.
- Apply the gas laws to simple problems involving pressure, temperature, mass, and volume.

Key Questions:

- What is the relationship between force, strength, and stress?
- What is Bernoulli's equation, and how does it explain fluid behavior?
- How do we describe the behavior of gases?

Vocabulary

airfoil	Charles's law	fluid mechanics	safety factor	tension
alloy	composite material	gas constant (R)	strain	thermal expansion
Bernoulli's equation	compression	ideal gas law	streamline	turbulent flow
Boyle's law	cross-section area	laminar flow	stress	viscosity
brittle	ductile	modulus of elasticity	tensile strength	
buoyancy	fluid	pascal (Pa)		

27.1 Properties of Solids

Solid materials normally hold their shape and do not flow like liquids. Solids have a wide range of properties that make them useful. Some solids are strong and heavy, like steel. Some solids are light and bouncy, like rubber. Other solids are hard and brittle, like glass. This section will explore some of the properties of solids.

Density of solid materials

Materials have a wide range of densities

Solid materials have a wide range of densities (Table 27.1). One of the densest metals is platinum, with a density of 21,450 kg/m³. Platinum is twice as dense as lead and almost three times as dense as steel. A ring made of platinum has three times as much mass as a ring of the exact same size made of steel. Rocks have lower densities than metals, between 2,200 and 2,700 kg/m³. As you might expect, the density of wood is less than rock, ranging from 400 to 600 kg/m³.

The definition of density

The density of a material is the ratio of mass to volume. Density is a physical property of the material and stays the same no matter how much material you have. For example, a 1-meter steel cube on a side has a mass of 7,800 kilograms (Figure 27.1). A steel nail has a volume of 1.6 millionths of a cubic meter, 1.6×10^{-6} m³, and a mass of 12.5 grams (0.0125 kg). The densities are the same.

The density formula

The formula for density is mass divided by volume. Most engineers and scientists use the Greek letter *rho* (ρ) to represent density.

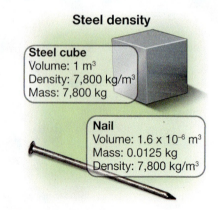

Figure 27.1: *The density of a steel nail is the same as the density of a solid steel cube.*

DENSITY

$$\rho = \frac{m}{V}$$

Density (kg/m³)   Mass (kg) / Volume (m³)

Use	if you know	and want to find
$\rho = m \div V$	mass and volume	density
$m = \rho \times V$	volume and density	mass
$V = m \div \rho$	mass and density	volume

Table 27.1: Densities of common materials

Material	Density (kg/m³)
Platinum	21,450
Lead	11,340
Steel	7,800
Titanium	4,500
Aluminum	2,700
Glass	2,700
Granite	2,640
Concrete	2,300
Sandstone	2,200
Plastic	2,000
Brick	1,600
Rubber	1,200
Liquid water	1,000
Ice	920
Oak (wood)	600
Pine (wood)	440
Cork	120

UNIT 8 MATTER AND ENERGY

The strength of materials

The meaning of "strength"
The concept of physical "strength" means the ability of an object to hold its form even when force is applied. We use solid materials for many applications because of their strength. It would be foolish to build a bridge from gas or liquid because these forms of matter have no strength! However, solid materials vary widely in strength. You would not build a bridge from wax, even though wax is a solid. To evaluate the strength of an object, consider the following questions as illustrated in Figure 27.2.

- How much does the object bend or deform under applied force?
- How much force can the object take before it permanently bends or breaks?

Separating design from material properties
The strength of an object can be further broken down into *design* and *materials*. As an example, think about breaking two sticks of a strong wood, like oak. A small force can break a thin stick. A thicker stick takes more force to break, even though the wood itself is the same (Figure 27.3). To evaluate the properties of oak *as a material*, it is necessary to separate out the effects of design, such as shape and size. The strength of a material is described in terms of *stress* instead of force.

Force and stress
The **stress** in a material is the ratio of the force acting through the material divided by the **cross-section area** through which the force is carried. The cross section area is the area perpendicular to the direction of the force. Dividing force by its cross-section area separates out most of the effects of size and shape from the strength properties of a material itself. The Greek letter sigma (σ) is used for stress. Stress (σ) is force (F) divided by cross section area (A). The units of stress are newtons per square meter (N/m²) in SI or pounds per square inch (PSI) in the English system. These are the same units as *pressure*.

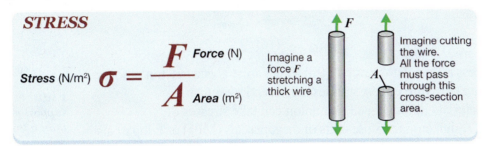

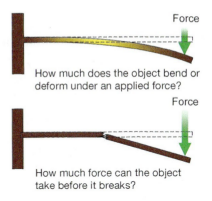

Figure 27.2: *Two questions that we use to define the physical strength of an object relate to forces applied to the object.*

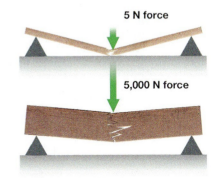

Figure 27.3: *It takes a much larger force to break a beam of oak than to break a thin stick.*

The breaking strength

Stress and breaking Materials break when the stress within them reaches a limit that depends on the material and the stress "history" of the object. Soft solids like wax break at a low value of stress. Materials like steel break at a much higher stress level. The breaking strength of a material is the maximum stress the material can normally withstand. Stress explains why things get stronger as they get larger or thicker. A larger-diameter wire has more cross section area and therefore can handle more force before it reaches the stress level at which it breaks (Figure 27.4).

Tensile strength The **tensile strength** is the stress at which a material may break under a **tension** force. Tension forces are "stretching" forces. Strong materials like steel have high tensile strength. Weak materials like wax and rubber have low tensile strength. Materials like wood and plastic have intermediate values of tensile strength. Table 27.2 lists the tensile strength of some common materials.

Table 27.2: Tensile strength of common materials

Material	Tensile strength (MPa)
Titanium	900
Steel (alloy)	825
Steel	400
Aluminum (alloy)	290
Aluminum (pure)	110
Oak (wood)	95
Pine (wood)	60
Nylon plastic	55
Rubber	14

Strength in bending The tensile strength also describes how materials break in bending. Imagine you have a rubber bar. When you bend the bar, it stretches on one side and squeezes together in **compression** on the other side. The bar breaks when the stress on the stretched side reaches the tensile strength of the material. The same idea applies to bars or beams, made of wood or steel.

Units for stress The metric unit of stress is named the **pascal (Pa)**. One pascal is one newton of force per square meter of area (1 N/m²). Most stresses are much larger than one pascal. Strong materials like steel and aluminum can take stresses of 100 million pascals. Design engineers use units of megapascals (MPa) or kilopascals (kPa). One megapascal is 10^6 Pa and one kilopascal is 1,000 Pa or 10^3 Pa.

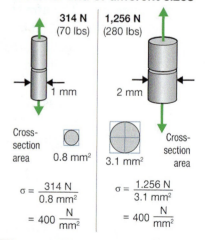

Figure 27.4: *A thicker wire can support more force at the same stress than a thinner wire because the cross section area is greater.*

Designing things to be strong enough

Designing for success
This may seem odd, but the way you design something to work is to design it *not to fail*. You can design something strong enough *not* to break only when you understand the forces and conditions that *will* make it break. Analysis of failure is a *very* important part of the process of design. Engineers use properties of materials, such as tensile strength, to evaluate their designs against failure.

Safety factors
The **safety factor** is the ratio of how strong something *is* compared with how strong it has to be. For example, suppose you need to choose a steel wire to support a weight of 1,000 newtons. A safety factor of 10 means you choose the wire to have a breaking strength of 10,000 newtons, 10 times stronger than it has to be. The safety factor allows for things like rust that might weaken the wire over time or things you did not consider in the design, like accidental heavy loads.

Some kinds of failure give warning
The *way* something fails is also important. Wooden ladders are not as strong as aluminum ladders of the same weight. But wooden ladders break slowly, making splintering and cracking noises that warn you to get down. Aluminum ladders are stronger but they break quickly with no warning. Because they can fail quickly, aluminum ladders are designed with a higher safety factor than wood ladders.

Example problem

The road must be held from above by 3 supports.

A total of 4.5 million N of force is required to hold up the road.

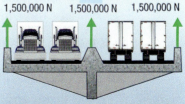

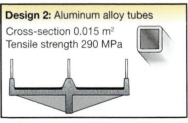

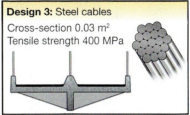

Figure 27.5: *Three designs to support a road from above.*

Evaluating three designs for a bridge

Three designs (Figure 27.5) have been proposed for supporting a section of road. Each design uses three supports spaced at intervals along the road. Evaluate the strength of each design. The factor of safety must be five or higher even when the road is bumper-to-bumper on all four lanes with the heaviest possible trucks.

Design 1:
The stress is 1,500,000 N ÷ 0.015 m² = 100 MPa. The tensile strength of the steel is six times greater (600 MPa) than the stress in the tube, so the safety factor is six.
Design 1 is acceptable for strength.

Design 2:
The stress is 1,500,000 N ÷ 0.015 m² = 100 MPa. The tensile strength of the aluminum alloy is only 2.9 times greater (290 MPa) than the stress in the tube, so the safety factor is only 2.9.
Design 2 is *not* acceptable because the safety factor is too low.

Design 3:
The stress is 1,500,000 N ÷ 0.03 m² = 50 MPa. The tensile strength of the steel in the cables is eight times greater (400 MPa) than the stress in the cable, so the safety factor is eight.
Design 3 is acceptable for strength.

CHAPTER 27 THE PHYSICAL PROPERTIES OF MATTER

Elastic properties of solids

Elasticity of materials
Elasticity measures the ability of a material to stretch. In Chapter 6, you studied Hooke's law which states that the distance a spring stretches is proportional to the applied force. A more general form of Hooke's law applies to three-dimensional solid materials. In the general form of Hooke's law, *stress* and *strain* take the place of force and distance.

Strain is the amount a material deforms
The **strain** is the amount a material has been deformed, divided by its original size. For example, imagine stretching a rubber rod by 1 centimeter (0.01 m). If the rod started with a length of 1 meter, the strain is 0.01, or 1 percent. The Greek letter epsilon (ε) is usually used to represent strain. Because strain is a ratio of two lengths, it is a number without units.

Table 27.3: Modulus of elasticity of common materials

Material	Modulus of elasticity (MPa)
Steel	200,000
Titanium	180,000
Aluminum	70,000
Oak (wood)	14,000
Pine (wood)	9,000
Nylon plastic	2,000
Rubber	100

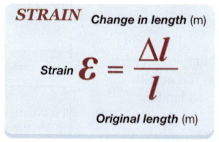
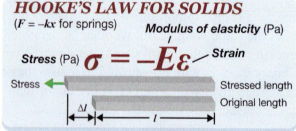

Hooke's law for solids
Hooke's law for solids states that the strain in a material is proportional to the applied stress. The **modulus of elasticity** plays the role of the spring constant for solids. Rubber has a very low modulus of elasticity. A small stress in rubber creates a relatively large strain. Steel has a very high modulus of elasticity. Steel is often used for structures because a lot of stress produces very little strain. Table 27.3 gives the modulus of elasticity for some common materials.

Elastic and brittle materials
A material is elastic when it can take a large amount of strain before breaking. Rubber is a good example of an elastic material. You can easily stretch rubber tubing to three times its original length, a strain of 200 percent. A **brittle** material breaks at a very low value of strain. Glass is a good example of a brittle material (Figure 27.6). You cannot stretch glass even one-tenth of a percent (0.001) before it breaks. Concrete, rock, and some plastics are also brittle.

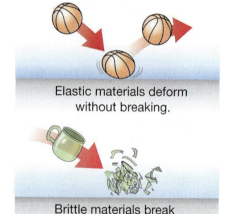

Figure 27.6: *Elastic materials deform when they are stressed. Brittle materials break.*

Thermal expansion

Atoms and thermal expansion
Almost all solid materials expand as the temperature increases. As temperature increases, the vibrational energy of atoms also increases. This makes each atom take up a little more space, causing thermal expansion.

The coefficient of thermal expansion
The coefficient of **thermal expansion** describes how much a material expands for each change in temperature. A thermal-expansion coefficient of 10^{-4} per degree Celsius means each 1°C rise in temperature causes an object to expand by 0.0001 times its original length. Table 27.4 gives the thermal expansion coefficient for several common materials.

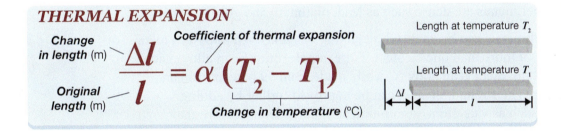

THERMAL EXPANSION

$$\frac{\Delta l}{l} = \alpha (T_2 - T_1)$$

Change in length (m): Δl
Original length (m): l
Coefficient of thermal expansion
Change in temperature (°C)
Length at temperature T_2
Length at temperature T_1

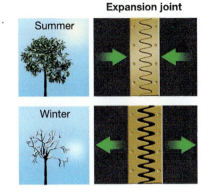

Figure 27.7: *Concrete bridges always have expansion joints to allow the concrete to expand and contract with changes in the temperature.*

Materials contract as well as expand
The thermal expansion coefficient works both ways. If the temperature *decreases*, objects *contract*. The amount of contraction or expansion is equal to the temperature change times the coefficient of thermal expansion. Bridges and many other structures include specific design features that allow for thermal expansion or contraction (Figure 27.7).

Thermal stress
Thermal stress is stress caused by differences in temperature. Thermal expansion is a type of *strain* caused by temperature changes. Thermal *stress* occurs when materials are *not* allowed to expand or contract as much as they normally would in response to changes in temperature. For example, pouring boiling water into a glass often causes the glass to crack. The glass cracks from thermal stress. The inside surface of the glass touching the hot water heats up quickly and tries to expand. The rest of the glass does not heat up as quickly and expands at a lower rate. The difference in expansion causes stress which cracks the glass if the stress becomes greater than the tensile strength of glass. Hooke's law can be used to calculate the stress produced by thermal expansion or contraction.

Table 27.4: Coefficient of thermal expansion for common materials

Material	Coefficient of thermal expansion ($\times 10^{-5}$ per °C)
Steel	1.2
Brass	1.8
Aluminum	2.4
Glass	2.0
Copper	1.7
Concrete	1
Nylon plastic	8
Rock	7
Rubber	16
Wood	3

Types of solid materials

Plastics Plastics are solids formed from long-chain molecules (Figure 27.8). Different plastics can have a wide range in physical properties such as strength, elasticity, thermal expansion, and density. Many common plastics melt and can be shaped easily when they are liquid. Other plastics are liquid at room temperature and harden when heated.

Figure 27.8: *Plastics are made of long-chain molecules.*

Ductile metals Metals are strong and relatively easy to form. Metals that bend and stretch easily without cracking are called **ductile**. Copper and some kinds of steel are considered ductile metals. Other metals are hard and brittle, like cast iron. Some metals are very soft, like lead. Some metals are light, like aluminum or magnesium. Other metals are dense such as lead, platinum, and gold.

Alloys The properties of metals can be changed by mixing elements. An **alloy** is a metal that is a mixture of more than one element. For example, brass is an alloy made from copper and tin. Steel is an alloy typically made from 99 percent iron and 1 percent carbon. The strength of steel can be improved further by adding other elements, such as vanadium, chromium, and molybdenum. More than 100 different alloys of steel are in common use. Virtually all common objects made from metal are made from alloys rather than pure elements.

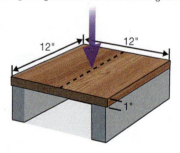

Figure 27.9: *Some properties of wood vary with the direction of the grain.*

Material properties can change with direction Many materials have different properties in different directions. Wood is a good example of this kind of material. Wood has a *grain* that is created by the way trees grow. Wood is very difficult to break against the grain, but easy to break along the grain (Figure 27.9). Plywood is strong in two directions because it is made from layers. The grain alternates direction in each layer.

Composite materials **Composite materials** are made from strong fibers supported by much weaker plastic. Like wood, composite materials tend to be strongest in a preferred direction. Fiberglass and carbon fiber are two examples of useful composite materials (Figure 27.10). Individual fibers can be stronger than steel but are too flexible to be useful. To make useful materials, fibers are bundled or woven and then embedded in plastic. Modern composite materials can be lighter and stronger than steel. However, composite materials are only strong in the direction of the fibers and are weak in the direction across the fibers.

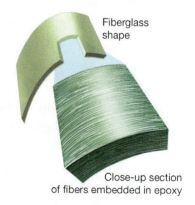

Figure 27.10: *Fiberglass and carbon fiber are composite materials.*

27.2 Properties of Liquids and Fluids

Liquids are a type of **fluid**. Fluids can change shape and flow when forces are applied to them. Gas is also a fluid because gas can change shape and flow. Liquids can be thin like water or thick like syrup. Liquids can be light like alcohol, or dense like mercury. This section introduces some of the properties of fluids. The concepts apply to liquids and gases.

Density and buoyancy

Density The density of a liquid is the ratio of mass to volume, just like the density of a solid. Because the atoms have more energy in a liquid, they tend to be slightly farther apart. Because the atoms are slightly farther apart, the density of a liquid is almost always less than the density of the same material as a solid. Water is an exception to this rule. Table 27.5 gives the density of some common liquids.

Why objects sink Because liquids can flow, objects of higher density sink through a liquid of lower density. Liquids of higher density also sink in liquids of lower density. Water has a higher density than oil. If you pour water into oil, the water sinks to the bottom.

Why objects float Objects of lower density float on liquids of higher density. Wood floats on water because wood has a lower density than water. A steel boat floats because the boat is mostly full of air. The *average* density of the boat is less than the density of water. Fill a steel boat with water and it sinks because the average density rises.

Ice is less dense than water Water is an exception to the rule about the density of solid and liquid phases of a substance. Ice floats because it is *less* dense than water. If ice were denser than water, it would sink. Ice is less dense because electrical forces between water molecules force them into a hexagonal pattern with an unusually large amount of empty space when they freeze into a solid (Figure 27.11). The empty space between molecules is why ice is less dense than liquid water.

Buoyancy An object submerged in a liquid feels an upward force called **buoyancy** (Figure 27.12). The buoyancy force is exactly equal to the *weight of liquid displaced* by the object. Objects float if the buoyancy force is greater than their own weight. Objects sink if the buoyancy force is less than their own weight. If you think carefully about weight and volume, you will recognize that buoyancy derives from differences in the density of things.

Table 27.5: Densities of common liquids

Material	Density (kg/m³)
Mercury	13,560
Glycerin	1,264
Water	1,000
Oil	888
Alcohol	789

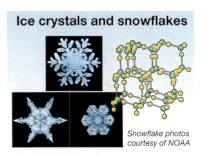

Figure 27.11: *Water molecules form crystals, like ice and snowflakes.*

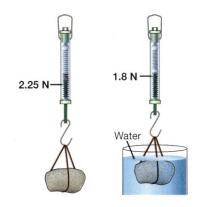

Figure 27.12: *Submerged objects seem lighter due to buoyancy.*

CHAPTER 27   THE PHYSICAL PROPERTIES OF MATTER

Pressure

Force and fluids Think about what happens when you push down on a balloon. The downward force you apply creates forces that act in other directions as well as down. For example, sideways forces push the sides of the balloon out. This is very different from what happens when you push down on a bowling ball. The solid ball transmits the force directly down. Because fluids can easily change shape, forces applied to fluids create more complex effects than forces applied to solids.

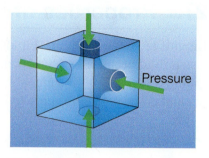

Figure 27.13: *Pressure exerts equal force in all directions in liquids that are not moving. If you put a box with holes underwater, pressure makes water flow in from all sides.*

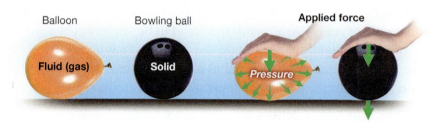

Pressure Forces applied to fluids create pressure instead of stress. Like stress, pressure is a ratio of force per unit area. Unlike stress, however, pressure acts in *all directions*, not just the direction of the applied force (Figure 27.13). Pressure is caused by forces acting on and within fluids. For example, gravity creates pressure in a pool of water. The air flowing around a wing creates pressure that lifts an airplane.

Pressure is an important concept The concept of pressure is central to understanding how fluids behave internally and also how fluids interact with surfaces, such as containers. The motion of fluids depends on pressure and density like the motion of solids depends on force and mass. Pressure exerts forces on all surfaces that come in contact with a fluid because of Newton's third law of action and reaction.

Units of pressure Pressure is force per unit area, like stress. A pressure of 1 N/m² means a force of 1 newton acts on each square meter. Like stress, the metric unit of pressure is the N/m², named the *pascal* (Pa). One pascal is equal to a pressure of 1 newton of force per square meter of area (N/m²). The English unit of pressure is pounds per square inch (PSI). One PSI describes a pressure of 1 pound of force per square inch of area (lb/in²). One pascal is less than one PSI (Figure 27.14).

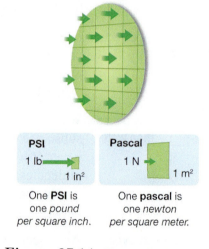

Figure 27.14: *Two units for pressure are Pa (N/m²) and PSI (lb/in²).*

UNIT 8 MATTER AND ENERGY

Pressure caused by gravity

Pressure from the weight of a liquid
Gravity is one cause of pressure because fluids have weight. The pressure increases the deeper you go beneath the surface of a fluid because the weight of fluid above you increases with depth. The rate at which pressure increases depends on the density of the fluid. Heavy fluids, like water, create more pressure than light fluids, like air, at an equal depth.

The atmosphere
Air is a fluid and Earth's atmosphere has a pressure. At sea level, a column of air 1 meter square has a mass exceeding 10,000 kilograms and exerts a pressure of 101,000 N/m². Air has a low density, but the atmosphere is deep, exceeding 80,000 meters (Figure 27.15). You are not crushed because the pressure from air *inside* your lungs pushing *out* is the same as the pressure *outside* pushing *in*.

The ocean
Water is much denser than air, so pressure under the ocean surface increases rapidly with depth (Figure 27.16). At a depth of 1,000 meters, the pressure is nearly 10 million N/m², or almost 100 times the pressure of the atmosphere. Submarines are engineered and built to withstand these deep-ocean pressures.

Pressure is equal at equal depths
The pressure at the same depth is the same everywhere in any liquid that is not moving. It does not matter what the shape of the container is. The formula below gives the pressure in a fluid that is at rest.

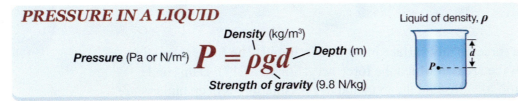

PRESSURE IN A LIQUID

Pressure (Pa or N/m²) $P = \rho g d$ — Depth (m), Density (kg/m³), Strength of gravity (9.8 N/kg), Liquid of density, ρ

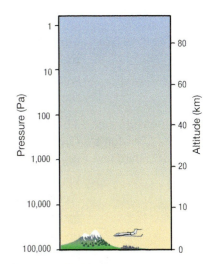

Figure 27.15: *The pressure of the atmosphere decreases with altitude. Atmospheric pressure comes from the weight of air.*

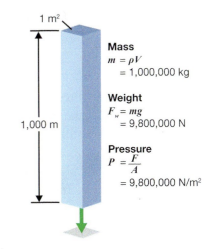

Figure 27.16: *The pressure at any point in a liquid is created by the weight of liquid above that point.*

Calculating pressure in the ocean

Calculate the pressure 1,000 meters below the surface of the ocean. The density of water is 1,000 kg/m³. The pressure of the atmosphere is 101,000 Pa. Compare the pressure 1,000 meters deep with the pressure of the atmosphere.

1. You are asked for the pressure and to compare it to one atmosphere.
2. You are given the density and depth.
3. Use the pressure formula $P = \rho g d$ and add the atmospheric pressure to water pressure.
4. $P = (1,000 \text{ kg/m}^3)(9.8 \text{ N/kg})(1,000 \text{ m}) + 101,000 \text{ Pa}$
   = 9,800,000 Pa + 101,000 Pa = 9,901,000 Pa, or 99 times atmospheric pressure.

27.2 PROPERTIES OF LIQUIDS AND FLUIDS

CHAPTER 27 THE PHYSICAL PROPERTIES OF MATTER

Pressure, molecules, and force

The molecular explanation
On the microscopic level, pressure comes from collisions between atoms or molecules. The molecules in gases and liquids are not bonded tightly to each other as they are in solids. Molecules move around and collide with each other and with the solid walls of a container.

Pressure and Newton's third law
Think about water in a jar. The water exerts pressure against the inside of the jar. On a microscopic level, water molecules are moving around and they bounce off the jar. It takes force to make a molecule reverse its direction and bounce the other way. That force is applied *to* the molecule *by* the inside surface of the jar. According to Newton's third law, an equal and opposite reaction force is exerted *by* the molecule *on* the jar. The reaction force is what creates the pressure acting on the inside surface of the jar. Trillions of molecules per second are constantly bouncing against every square millimeter of the inner surface of the jar. Pressure comes from the collisions of those atoms.

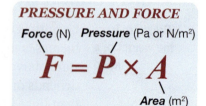

PRESSURE AND FORCE

Force (N)   Pressure (Pa or N/m²)

$$F = P \times A$$

Area (m²)

Figure 27.17: *Pressure creates a force on any surface immersed in a liquid. The force is equal to the pressure times the area of the liquid on which the force acts.*

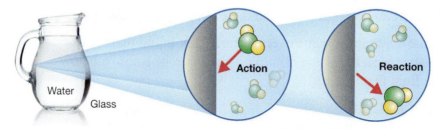

Pressure creates forces on surfaces
Pressure creates a force on any surface immersed in liquid. The force is equal to the pressure times the area on which the force acts (Figure 27.17). Dams and submarines are built to withstand the tremendous pressure forces deep underwater.

Figure 27.18: *The pressure inside the tires holds the car up. When the tire pressure is too low, the shape of the tire changes because more area is needed to exert enough force to hold up the car.*

Calculating pressure in car tires

A car tire is at a pressure of 35 PSI. Four tires support a car that weighs 4,000 pounds. Each tire supports 1,000 pounds. How much of the tire's surface area is holding up the car (Figure 27.18)?

1. You are asked for area.
2. You are given force and pressure.
3. Force is pressure times area, so area is force divided by pressure.
4. $A = F \div P = (1{,}000 \text{ lbs}) \div (35 \text{ PSI}) = 28.5 \text{ in}^2$

This is about equal to a contact area measuring 5 by 5.7 inches for each tire.

Motion of fluids

Applying the laws of motion to fluids

The study of the motion of fluids is called **fluid mechanics**. Fluid mechanics is a complex subject because fluids can change shape. To understand fluid motion, we cannot think of the whole fluid at once because different parts might be moving differently (Figure 27.19). Instead, we focus our attention on a small sample, so small that we can assume all the fluid in the sample is moving together, like a solid block. The principles of fluid motion can be discovered by applying the physical laws we already know to the motion of the small sample of fluid. Fluids move according to the same fundamental laws of motion as solid objects.

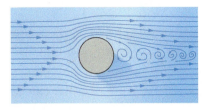

Figure 27.19: *Water flowing around a pole reverses. The water reverses direction immediately behind the pole and forms swirls called* eddies.

The speed of fluids

Moving fluids usually do not have a single speed throughout their volume. The speed is often different at different places. For example, think about thick syrup flowing down a plate held at an angle. The syrup near the plate sticks to the surface and moves very slowly. The syrup farthest from the plate's surface moves faster (Figure 27.20).

Fluids flow because of pressure differences

Fluids flow because of differences in pressure. Think about water flowing in a garden hose. The pressure is high where the hose is connected to the faucet. The pressure decreases along the hose and is lowest at the outlet.

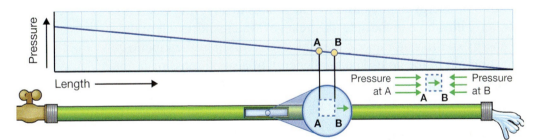

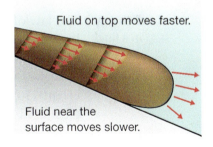

Figure 27.20: *The arrows represent the speed of syrup flowing down a plate. Friction slows the syrup touching the plate. The top of the syrup moves faster because the drag from friction decreases away from the plate's surface.*

Applying Newton's laws to a fluid

Consider a cube-shaped sample of water in the hose. The pressure is higher at point A than at point B. The pressure difference makes a net force pushing the cube of water onward. The fluid moves according to Newton's laws, just like all matter. The water flows because the pressure at one end of the hose is higher than the pressure at the other end. Fluids flow because of unequal pressures just like solids move because of unequal forces.

27.2 PROPERTIES OF LIQUIDS AND FLUIDS

CHAPTER 27 THE PHYSICAL PROPERTIES OF MATTER

Energy in fluids

Pressure and energy

Pressure and energy are related. Remember, our definition of energy was the stored ability to exert force and do work. Fluid in a container has energy because any pressure created by the fluid pushes on the sides of the container with forces that can do work. One joule of work is done when a pressure of 1 pascal pushes a surface of 1 square meter a distance of 1 meter. The volume swept out by the expanding fluid is 1 cubic meter (Figure 27.21). One joule per cubic meter is a *potential energy density* just like 1 kilogram per cubic meter is a *mass density*.

Pressure is potential energy

Differences in pressure create potential energy in fluids just like differences in height create potential energy from gravity. A pressure difference of 1 N/m² is equivalent to a potential energy density of 1 J/m³ (Figure 27.22). This potential energy can be used to do work by allowing the fluid to expand. This is the principle behind the steam engine, gas-powered engines, and turbines.

Pressure and work

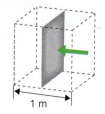

A pressure of 1 N/m² pushing one square meter does one joule of work for each meter.

Figure 27.21: *Pressure does work as fluids expand. A 1-pascal pressure does 1 joule of work pushing 1 square meter a distance equal to 1 meter.*

PRESSURE AND POTENTIAL ENERGY

Potential energy (J) $E = PV$ — Volume (m³)
— Pressure (Pa or N/m²)

Total energy in a fluid

The law of conservation of energy applies to fluids as it does to anything else. Imagine a tank of water with a hole in the side. A stream of water squirts out of the hole. The pressure energy of the water inside the tank is converted to the kinetic energy of water squirting out of the hole. The total energy of a small mass of fluid is equal to its potential energy from gravity plus its potential energy from pressure plus its kinetic energy. The formula below gives the total energy.

Pressure and energy density

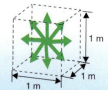

A pressure difference of one newton per square meter ...

... is equivalent to ...

... one joule of potential energy per cubic meter.

Figure 27.22: *Differences in pressure can create potential energy which can do work.*

ENERGY OF A SMALL MASS OF FLUID

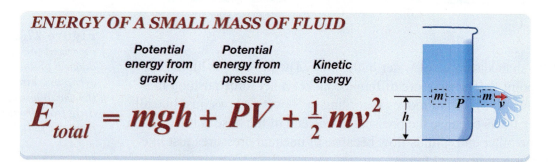

$$E_{total} = mgh + PV + \tfrac{1}{2}mv^2$$

Potential energy from gravity · Potential energy from pressure · Kinetic energy

THE PHYSICAL PROPERTIES OF MATTER   CHAPTER 27

Energy conservation and Bernoulli's equation

Deriving Bernoulli's equation The law of conservation of energy is called **Bernoulli's equation** when applied to a fluid. To get Bernoulli's equation, we set the energy inside and outside the container to be equal as shown in Step 1. In Step 2, we combine mass and volume.

Streamlines moving around a car

Step 1: Set energy equal inside and outside
Energy inside the container = Energy outside the container
$$mgh_1 + P_1 V + \tfrac{1}{2} m v_1^2 = mgh_2 + P_2 V + \tfrac{1}{2} m v_2^2$$

Step 2: Replace mass (m) and volume (V) by density (ρ).
$$\rho g h_1 + P_1 + \tfrac{1}{2}\rho v_1^2 = \rho g h_2 + P_2 + \tfrac{1}{2}\rho v_2^2$$
$$= constant$$

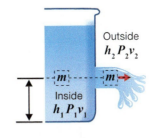

Figure 27.23: *These streamlines show the flow of air around a car. Fluid flows along streamlines. Fluid does not flow across streamlines when represented this way.*

The three variables Bernoulli's equation relates the three variables of height, pressure, and speed by energy conservation. If one variable increases, at least one of the other two must decrease. For example, if speed goes up, pressure goes down.

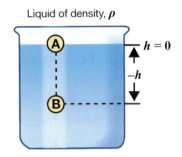

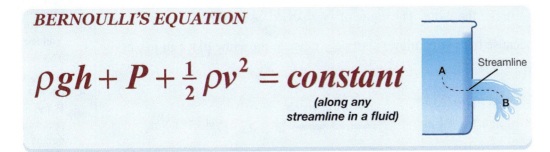

BERNOULLI'S EQUATION

$$\rho g h + P + \tfrac{1}{2}\rho v^2 = constant$$
(along any streamline in a fluid)

At point Ⓐ   At point Ⓑ
$\rho g h + P = 0$ $\rho g(-h) + P = 0$
                    $P = \rho g h$

Figure 27.24: *In an unmoving fluid we choose h = 0 at the surface where the pressure (P) is also zero. The value of $\rho g h + P$ at point A is zero. Bernoulli's equation says that at point B the value of $\rho g h + P$ must also be zero.*

A model for fluid flow **Streamlines** are imaginary lines drawn to show the flow of fluid. We draw streamlines so that they are always parallel to the direction of flow (Figure 27.23). Bernoulli's equation tells us that the quantity $\rho g h + P + \tfrac{1}{2}\rho v^2$ is the same anywhere along a streamline. If the fluid is not moving, v is 0, and Bernoulli's equation gives us the relation between pressure and depth, which is negative height (Figure 27.24).

27.2 PROPERTIES OF LIQUIDS AND FLUIDS

CHAPTER 27 THE PHYSICAL PROPERTIES OF MATTER

Applying Bernoulli's equation

Airfoils The wings of airplanes are made in the shape of an **airfoil**. Air flowing along the top of the airfoil (A) moves faster than air flowing along the bottom of the airfoil (B). The speed is different because the shape of the airfoil forces the air on the top of the wing to take a longer path than the air under the wing.

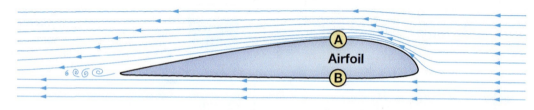

Lift forces According to Bernoulli's equation, if the speed goes up, the pressure goes down. When a plane is moving, the pressure on the top surface of the wings is lower than the pressure beneath the wings. The difference in pressure is what creates the lift force that supports the plane in the air.

Choosing streamlines By picking the right streamlines, we can use Bernoulli's equation to calculate the lift force on the wing. Streamline A goes over the top surface of the wing. Streamline B goes under the wing. The value of $\rho gh + P + \frac{\rho v^2}{2}$, must be the same at points A and B because both streamlines start very close to the same place in front of the wing.

At point A   **At point B**   Solve for the pressure difference
$\rho gh + P_A + \frac{1}{2}\rho v^2 = \rho gh + P_B + \frac{1}{2}\rho v^2$ → $P_B - P_A = 3{,}781 \text{ N/m}^2$

Calculating lift force When air above the wing is moving at 150 m/s and air below the wing is moving at 125 m/s, the difference in pressure is 3,781 N/m². An airplane with 20 m² of wing surface would experience a lift force of 75,600 newtons, almost 17,000 pounds. The faster a plane moves through the air, the greater the difference in speeds above and below the wing, and the greater the lift force. This is why planes must reach a minimum speed before they can take off.

Calculating the speed of water from a faucet

Water towers create pressure to make water flow. At what speed will water come out if the water level in the tower is 50 meters higher than the faucet?

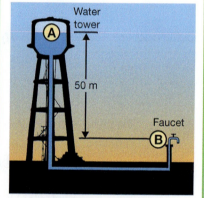

1. You are asked for the speed.
2. You are given height.
3. The Bernoulli equation relates speed and height in a fluid.
4. Choose a streamline from A to B and note that speed and pressure at A are zero, while height and pressure at B are zero.

$(\rho gh)_A = \left(\frac{1}{2}\rho v^2\right)_B$

$v = \sqrt{2gh}$

$v = \sqrt{2(9.8 \text{ m/s}^2)(50 \text{ m})}$

$= 31 \text{ m/s}$

Fluids and friction

Viscosity

Viscosity measures a fluid's resistance to flow. Thick fluids like syrup have a high viscosity. It takes a large pressure difference to make syrup flow fast. Thin liquids like water have a low viscosity. Even a small pressure difference can produce a large flow of water because the viscosity of water is so low.

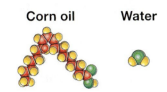

Figure 27.25: *Molecules of different sizes are one of the reasons liquids have different viscosities*

The cause of viscosity

Viscosity is caused by forces that act between atoms and molecules in a liquid. Corn oil has a high viscosity because corn oil is made of large molecules that interfere as they slip over each other. Water has a low viscosity because water molecules are small and move around each other easily (Figure 27.25).

The effect of temperature

The viscosity of liquids decreases when temperature increases. Thick fluids like corn oil flow much easier when they are hot because the viscosity is greatly reduced. The reduction in viscosity with raised temperatures comes from the increase in molecular motion. Warmer molecules in rapid motion have enough energy to jostle around each other more easily. Oil for a car engine is specially made to have the right viscosity at different temperatures.

Laminar flow

Friction in fluids also depends on the type of flow. In **laminar flow**, the streamlines are smooth and parallel. Fluids in laminar flow do not mix across streamlines. Water running from a faucet at a very low rate produces laminar flow (Figure 27.26). The water runs out in a clear, smooth stream. Laminar flow usually creates the lowest amount of friction.

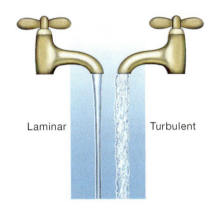

Figure 27.26: *Water running from a faucet can be laminar or turbulent, depending on the flow rate.*

Turbulent flow

When fluid moves fast, the flow often becomes turbulent. In **turbulent flow**, the streamlines are broken up into very disorganized patterns. There is constant churning and mixing of fluid in turbulent flow. A faucet at high volume produces turbulent flow (Figure 27.26). The stream of water appears foamy from the rapid mixing. Turbulent flow creates much higher friction than laminar flow.

CHAPTER 27 — THE PHYSICAL PROPERTIES OF MATTER

27.3 Properties of Gases

Gases are fluids because gases can change shape and flow when forces are applied. Gases are different from liquids because they can expand and contract, greatly changing their density while the density of liquids and solids remains nearly constant. This property of expansion makes gases uniquely suited to work in *engines* which convert heat into motion and mechanical work. This section introduces some of the properties of gases.

Density and buoyancy in air and other gases

Density — Gases have much lower densities than liquids because the atoms in a gas are much farther apart than in a liquid. The density of air is about one kilogram per cubic meter. Air feels "light" because air is 1,000 times less dense than water.

Air is not "nothing" — Air may seem like "nothing," but all the oxygen our bodies need and the carbon needed by plants comes from air. As a tree grows, the soil does not sink down to supply mass for the tree. All of the carbon atoms in wood come from carbon dioxide in the air.

Air is a mixture of gases — Air is the most important gas to living things on Earth. Earth's atmosphere is a mixture of nitrogen, oxygen, water vapor, argon, and a few trace gases (Figure 27.27). Molecules of nitrogen and oxygen account for 97.2 percent of the mass of air. The amount of water vapor depends on the temperature and relative humidity.

Sinking in a gas — Because gas can flow and has a very low density, objects of higher density sink quickly in a gas. For example, if you drop a penny, it falls through the air easily because the density of the penny is 9,000 times greater than the density of air.

Floating in a gas — Objects of lower density can float on a gas of higher density. A helium balloon floats because the average density of the balloon and helium inside is less than the density of air (Table 27.6).

Buoyancy — An object submerged in a gas feels an upward buoyancy force. You do not notice buoyancy forces from air because the density of ordinary objects is so much greater than the density of air. Just as with liquids, the buoyancy force is exactly equal to the weight of gas displaced by an object. Whether an object sinks or floats depends on whether the buoyancy force is greater or less than the object's weight.

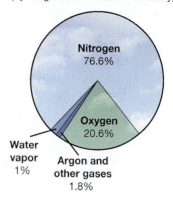

Figure 27.27: *Air is a mixture of gases.*

Table 27.6: Densities of common gases at 1 atm and 27°C.

Gas	Density (kg/m^3)
Carbon dioxide	1.8
Argon	1.6
Oxygen	1.3
Air	1.2
Nitrogen	1.1
Helium	0.16

Pressure and Boyle's Law

Pressure and temperature affect density
The density of a gas depends on pressure and temperature. If the pressure increases, the density may also increase. If the pressure decreases, the density may also decrease. This is very different from liquids or solids. The density of a liquid or solid stays almost the same when the pressure is changed. Depending on the pressure and temperature, the density of a gas can vary from near zero, as in outer space, to densities greater than solids if the temperature is high enough.

Boyle's law
The pressure goes up if you squeeze a specific mass of gas into a smaller volume while keeping temperature constant. This relationship is known as **Boyle's law**. Pressure increases because the same number of molecules are squeezed into a smaller space. The molecules hit the walls more often because there are more molecules per cubic unit of volume. The pressure also increases because there are more collisions. The formula for Boyle's law relates the pressure and volume of gas. If mass and temperature are kept constant, the product of pressure and volume stays the same. The subscripts 1 and 2 in the formula indicate the pressure and volume at two different conditions.

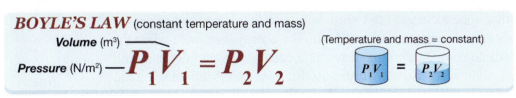

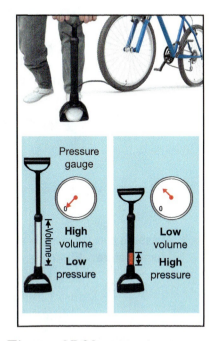

Figure 27.28: *A bicycle pump changes the volume of air to increase the pressure.*

Calculating the pressure increase from a change in volume

A bicycle pump creates high pressure by squeezing air into a smaller volume (Figure 27.28). If air at atmospheric pressure (14.7 PSI) is compressed from an initial volume of 30 cubic inches to a final volume of 3 cubic inches, what is the final pressure?

1. You are asked for pressure.
2. You are given initial and final volume.
3. Apply Boyle's law: $P_1V_1 = P_2V_2$
4. Solve for P_2: $P_2 = (V_1/V_2) \times P_1 = (30 \div 3) \times 14.7 = 147$ PSI

Note: The tire-pressure gauge will read 132.3 PSI because most pressure gauges measure the pressure *difference* (147 − 14.7) between inside the gauge and the atmosphere outside. Boyle's law and the other gas laws use "absolute pressure," which is pressure relative to zero. Gauges read "gauge pressure," which is pressure greater than the pressure of the atmosphere.

CHAPTER 27: THE PHYSICAL PROPERTIES OF MATTER

Temperature and pressure

Charles's Law The volume of a gas is affected by temperature. If mass and pressure are kept constant, then volume increases when temperature increases. The volume goes down when the temperature goes down. The relationship known as **Charles's Law** relates volume and temperature in a gas (Figure 27.29).

PRESSURE-TEMPERATURE RELATIONSHIP

Pressure (N/m²) $\dfrac{P_1}{T_1} = \dfrac{P_2}{T_2}$ (Volume and mass = constant)
Temperature (K)

$\dfrac{P_1}{T_1} = \dfrac{P_2}{T_2}$

CONVERTING CELSIUS TO KELVIN

$$T_{Kelvin} = T_{Celsius} + 273$$

CHARLES'S LAW

$$\dfrac{V_1}{T_1} = \dfrac{V_2}{T_2}$$

Volume (m³)
Temperature (K)

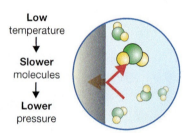

Figure 27.29: *Charles's Law shows the relationship between the temperature and volume of a gas.*

Pressure and temperature Pressure increases with temperature because temperature is a measure of the kinetic energy of moving molecules. High-temperature molecules move faster than low temperature molecules. Because the change in momentum is greater, a faster molecule creates more force when it bounces off of a surface. The increase in force creates a corresponding increase in pressure (Figure 27.30).

Use temperature in Kelvins The temperature that appears in gas laws *must be in Kelvins*. The speed of gas molecules depends on their energy compared with the energy they have when the temperature is absolute zero. At absolute zero, molecules are essentially standing still. Because the Kelvin scale starts at absolute zero, it measures the total kinetic energy of gas molecules relative to zero energy.

Calculating the pressure increase at high temperatures

A can of hair spray has a pressure of 300 PSI at room temperature (21°C or 294 K). The can is accidentally moved too close to a fire, and its temperature increases to 800°C (1,073 K). What is the final pressure in the can?

1. You are asked for pressure.
2. You are given initial and final temperatures.
3. Apply Charles's Law: $P_1 \div T_1 = P_2 \div T_2$
4. Solve for P_2: $P_2 = (T_2 \div T_1) \times P_1 = (1{,}073 \div 294) \times 300 = 1{,}095$ PSI

This is why you should *never* put spray cans near heat. The pressure can increase so much that the can explodes.

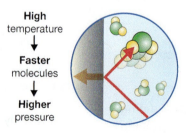

Figure 27.30: *Faster molecules create higher pressure because they exert larger forces.*

The ideal gas law

The ideal gas law The **ideal gas law** combines the pressure, volume, and temperature relations for a gas into one single equation which also includes the mass of the gas. In physics and engineering, mass (m) is used to describe a quantity of gas. In chemistry, the ideal gas law is traditionally written in terms of the number of moles, or atoms, of gas (n) instead of mass.

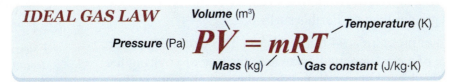

IDEAL GAS LAW
Pressure (Pa) — $PV = mRT$ — Temperature (K)
Volume (m³) / Mass (kg) / Gas constant (J/kg·K)

The gas constant Each different kind of gas has a unique **gas constant (R)**. You need to select the appropriate value for R when using the ideal gas law. Table 27.7 lists the gas constants for several common gases. The gas constants are different because the size and mass of gas molecules are different. The gas constant for air is an average based on the proportions of oxygen and nitrogen in air.

Using the combined gas law The values for the gas constant in Table 27.7 are in metric units. Therefore, to use these values with the ideal gas law, pressure should be in pascals, volume in meters cubed, mass in kilograms, and temperature in Kelvins. Since the law applies to the total amount of gas, pressure needs to be absolute pressure, not gauge pressure. Absolute pressure is gauge pressure plus the pressure of the atmosphere.

Table 27.7: Gas constants for common gases

Gas	Gas constant (R, J/kg·K)
Air	287
Argon	208
Nitrogen (N_2)	297
Oxygen (O_2)	260
Carbon dioxide	189
Helium	2,078
Water vapor	462
Methane	518
Propane	189

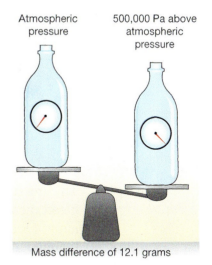

Figure 27.31: *Two bottles of air at different pressures have different masses. The bottle at higher pressure has more mass because there are more air molecules in it.*

Calculating the mass of air from the ideal gas law

Two soda bottles contain the same volume of air at different pressures (Figure 27.31). Each bottle has a volume of 0.002 m³. The temperature is 21°C. One bottle is at a gauge pressure of 500,000 pascals. The other bottle is at a gauge pressure of zero. Calculate the mass difference between the two bottles.

1. You are asked for a mass difference.
2. You are given the volume, temperature, gauge pressure, and the gas is air.
3. Use the ideal gas law, $PV = mRT$, with $R = 287$ J/kg·K
4. Convert gauge pressure to absolute pressure by adding 101,000 Pa.
   First bottle: $m = PV/RT = (601{,}000 \times 0.002) \div (287 \times 294) = 0.0143$ kg
   Second bottle: $m = PV/RT = (101{,}000 \times 0.002) \div (287 \times 294) = 0.0024$ kg
   The difference is 0.0121 kg, or 12.1 grams.

Chapter 27 Connection

The Deep Water Submarine *Alvin*

Most of Earth's surface lies under the oceans. Deep beneath the ocean surface are undersea mountains and volcanoes, strange forms of life, and many clues to the past and present condition of our planet. Exploring the deep ocean requires courage and very sophisticated engineering. The exploration submarine *Alvin* is famous for research done during deep dives (Figure 27.32). Scientists aboard *Alvin* have made many remarkable discoveries, including forms of life that live near deep hot spots where there is no light and pressures are 400 times greater than on Earth's surface.

Photo courtesy OAR/National Undersea Research Program (NURP); Woods Hole Oceanographic Inst.

Figure 27.32: *The* Alvin *deep-water submarine can dive to depths of 4,500 meters below the ocean surface.*

Pressure force At 4,500 meters, the water pressure is 44 million N/m². This extreme pressure is equivalent to the weight of a car supported on an area the size of your big toe! *Alvin* is 7.1 meters long and 3.7 meters tall, but the spherical pressure hull inside where scientists work is only two meters in diameter. The force acting on the two-meter sphere is equal to the pressure times the area. For *Alvin*, the force on one side of the pressure hull is 31 million pounds (138 million N)!

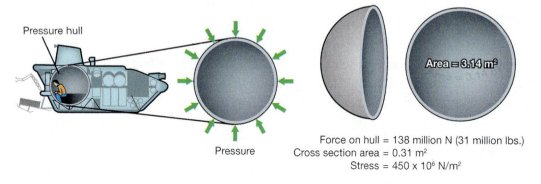

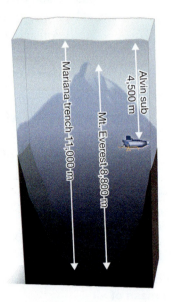

Stress in the hull The pressure hull is spherical because a sphere is the shape that can withstand the greatest compression. To withstand the pressure, the hull is made from titanium alloy, 4.9 centimeters thick (almost two inches). The tensile strength of the titanium alloy is greater than 900×10^6 N/m². The cross-section area of the hull is 0.31 m². The stress in the hull is 450×10^6 N/m². The titanium alloy used in *Alvin*'s hull is one of the strongest materials ever developed. A hull of ordinary steel or aluminum would be crushed by the forces exerted by the ocean's enormous pressure. This is why only robot probes are used to explore deeper parts of the ocean (Figure 27.33).

Figure 27.33: *The deepest part of the ocean floor lies more than 8,000 meters below the surface. Even the* Alvin *cannot reach this depth. The deepest places have only been seen with robot probes.*

Chapter 27 Connection

Air for life support and buoyancy control

Staying level *Alvin* and other submarines control their depth by changing their buoyancy. Aboard the submarine is a chamber that can be filled with air or water. The amount of air and water is adjusted with pumps until the average density for the whole submarine is the same as the density of water. When the average densities are matched, the submarine achieves neutral buoyancy and neither rises or sinks. To rise, some water is pumped out of the tank and replaced with air. The average density decreases and the submarine rises because of the positive buoyancy force. To dive, water is pumped into the tank and air is released. Negative buoyancy occurs when the average density becomes greater than the density of water, and the submarine sinks.

Submarine
Mass: 10,000 kg
Volume: 10.00 m³
Average density: 1,000 kg/m³
(Buoyancy tank 1/2 full)

Life support Air for breathing is kept in tanks at very high pressure. At 20 breaths per minute, an average adult inhales 0.08 m³ of air each minute. The interior volume of *Alvin*'s hull has a volume of just over 2 m³. The normal three-person crew would breathe all of the air in the hull in just eight minutes. A seven-hour mission with a crew of three requires at least 100 m³ of air. This volume can be stored in a tank with a volume of 0.5 m³ by raising the pressure to 200 times atmospheric pressure, or 200 *atmospheres*. Air tanks for diving typically store air at pressures near or exceeding 200 atmospheres (Figure 27.34).

The *Alvin* *Alvin* has made more than 4,700 dives since its commissioning in 1964, and is considered the most productive research submarine in the world. Among its accomplishments over the last 50 years, Alvin has explored undersea volcanoes, discovered strange new life forms at hydrothermal vents, and confirmed sea-floor spreading up close in the Mid-Atlantic Ridge. Even though Alvin is the world's oldest research submersible, its complete overhaul and upgrade in 2013 keeps it on the cutting edge of scientific research.

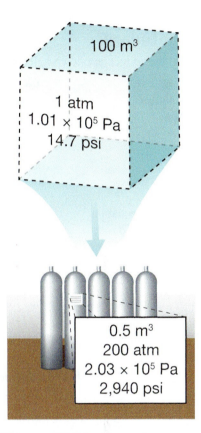

Figure 27.34: *Air can be compressed to high pressures and stored in small volumes.*

27.3 PROPERTIES OF GASES   **583**

Chapter 27 Assessment

Vocabulary

Select the correct term to complete the sentences.

stress	Boyle's law	fluid
tensile strength	strain	safety factor
volume	pressure	airfoil
elastic	elasticity	ideal gas law
ductile	Charles's law	laminar flow
alloy	brittle	pascal (Pa)
fluid mechanics	modulus of elasticity	composite material
streamlines	thermal expansion	buoyancy
Bernoulli's equation	density	compression
gas constant (R)	cross section	turbulent flow
tension		

1. A metal that stretches and bends easily without cracking is _____.
2. The amount a metal expands for each change in temperature is described by its coefficient of _____.
3. The ability of a material to stretch is measured by the physical property of _____.
4. The stress at which a material breaks under tension is called _____.
5. The ratio of the mass of a material to its volume is known as _____.
6. The ratio of the force acting through a material divided by the cross-section area through which is is exerted is referred to as _____.
7. The ratio of the strength of a material to the amount of strength needed in a certain application is known as the _____.
8. Squeezing a material creates _____.
9. The proportionality constant relating stress and strain in solids is the _____.
10. The amount a material has been deformed divided by its original size is called _____.
11. The force applied when stretching a material is _____ force.
12. The area perpendicular to the force applied to an object is the _____ area.
13. A material that can withstand a large amount of strain before breaking would be described as _____.
14. A mixture of two or more metal elements is called a(n) _____.
15. Material made by combining strong fibers with weaker material is known as _____.
16. The space occupied by an object is referred to as its _____.
17. Materials that break at very low values of strain are identified as _____.
18. Forces applied to fluids create _____ on and within the fluid.
19. The basic metric unit of stress or pressure is the _____.
20. Lines drawn in a diagram to represent the flow of fluid are called _____.
21. The upward force experienced by an object submerged in a fluid is known as _____.
22. The study of the motion of fluids is called _____.
23. A material that can change shape and flow when force is applied is referred to as a(n) _____.
24. The law of conservation of energy applied to fluids is known as _____.
25. The motion of fluid in disorganized and broken streamlines is described as _____.
26. A surface that creates lift as it passes through air is known as a(n) _____.
27. Fluid motion represented by smooth and parallel streamlines is called _____.
28. If the mass and temperature of a gas are held constant, the product of pressure and volume are constant according to _____.
29. If the mass and volume of a gas remain the same, the temperature and pressure of the gas are directly proportional, according to _____.

30. In the ideal gas law, the proportionality constant based upon the type of gas is known as the _____.

31. The law which relates temperature, pressure, and volume of a gas is referred to as the _____.

Concept review

1. Which has greater density, a sewing needle made of steel or a heavy bar of steel from which the needle is made?

2. Liquid mercury has a density of approximately 13,600 kg/m^3. Use Table 27.1 to decide which material(s) will float in mercury.

3. When inquiries are made about the strength of a material, what two questions must be answered to best describe a material's physical strength?

4. Both stress and pressure are calculated as the ratio of force to area. How do the definitions for *stress* and *pressure* differ?

5. Explain why road beds and other structures experiencing high stress are often made of concrete containing steel rods or cables.

6. The load capacity of a bridge is often stated on a sign at the bridge. How can vehicles weighing two or three times the stated capacity of the bridge travel across the bridge with no apparent effect at one time and yet the bridge fails at another time?

7. Golf balls are commonly constructed by covering a rubber core with a rubber-cord winding and a durable cover. Why doesn't it make sense to make any part of the golf ball of glass?

8. Rubber and steel are both elastic, yet rubber is not used in making automobile springs or as structural supports in bridges and buildings. Explain why rubber is not used.

9. Glass marbles can be fractured by heating them in a frying pan for several minutes and then dropping them into cold water. Explain what happens.

10. Concrete sidewalks, bridges, and roadways are built in sections. Provide an explanation for this.

11. To which board would a karate expert apply a "chop" to most easily break the board? Explain.

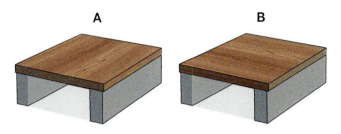

12. Steel is more dense than water but steel ships float in water. Explain why.

13. With regard to motion, force is to the mass of a solid as _____ is to _____ of a fluid.

14. A diver lies on the bottom of a pool. Compared to the average pressure exerted on the diver in this position, would he experience more, less, or the same average pressure if he were standing on the bottom of the pool? Explain.

15. People who drive on the loose sand of the beaches in the Outer Banks of North Carolina are advised to lower the pressure of their tires to 20 lb/in^2 or less. Explain why this is done.

16. In the diagram of the airfoil, is the internal pressure greater at location A or B? Explain your choice.

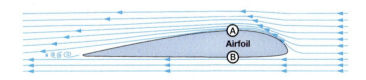

17. Blimps weighing several tons fly over sports games carrying TV cameras and people aloft in a gondola suspended under the inflated part of the blimp. Yet, a single person is not able to float in air. Explain why.

18. Two identical, rigid, air-tight containers are filled with air at 20°C. Container A is filled at a pressure of 2 atmospheres while Container B is filled at 4 atmospheres of pressure. The containers are placed on a balance as shown. On which side is Container A?

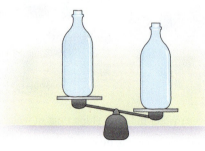

Problems

1. A 355-milliliter can of diet cola has a mass of 0.349 kilograms. Will it float in water? Show calculations to support your answer.

2. The winners of a bridge-building contest find that their model bridge will hold 2.45 kg of mass before breaking. The cross-sectional area of the bridge supports is 2 cm². How much stress was applied to the supports of the bridge?

3. A bridge built with a safety factor of five bears a sign declaring the safe load limit to be 7 tons. How much weight can the bridge hold before collapsing?

4. The steel-alloy string on a certain musical instrument is 0.75 mm in diameter. A musician, tightening the string to tune the instrument, stretches the steel string until it breaks. How much force was applied to the string?

5. The main steel span of the Golden Gate Bridge in San Francisco is 1,280 meters long. If the temperature changes from 10°C to 20°C during the day, how much longer is this span at the end of the day?

6. A solid lead object weighs 45.2 newtons. Calculate the buoyant force on it when it is placed in a container of oil.

7. As you breathe, air is forced into your lungs by a pressure difference between atmospheric pressure often stated as "760 mm of Hg" and an internal lung pressure of "759 mm of Hg." These pressure designations refer to the pressure present at the bottom of columns of mercury 760 or 759 mm in depth. What is the equivalent pressure in pascals that pushes air into your lungs?

8. Water is sprayed from a fire hose into a burning building 20 meters above the ground.
   a. With what speed does the water leave the fire hydrant on the ground?
   b. What is the pressure of the water at the hydrant?

9. As a scuba diver descends below the surface in a lake from a depth of 30 meters, a bubble with a volume of 1 liter is exhaled. Just before the bubble breaks at the surface, what is its volume? Assume the temperature of the water is constant at all depths.

10. On a warm summer day, the temperature is 25°C at 1 atmosphere of pressure. You buy your young cousin a helium-filled balloon. It has a volume of 0.25 m³. Your cousin accidentally releases the balloon. To what volume does the balloon expand if it rises to a height at which the temperature is –40°C and the pressure is 1.01×10^4 Pa?

11. If 15 grams of water are placed in a 1.5-liter pressure cooker and heated to a temperature of 400°C, what is the pressure inside the container?

12. Lines which attach to the foil of the sail on a kiteboard will support a weight of 2,200 newtons. If the lines that support 2,200 newtons are 2 mm in diameter, how much weight would be supported by a line of the same material with a diameter of 1.5 mm?

13. Grain silos are generally reinforced with metal bands wrapped around the structures. The supporting bands are closer together near the bottom than near the top. Why is the silo reinforced in this way?

Applying your knowledge

1. Archimedes is best known for the principle of buoyancy named Archimedes' Principle. He was a well-known scientist of the third century BCE. Research the device known as Archimedes Screw. What was its original function? What are some of its modern adaptations?

2. A high school baseball player wearing baseball shoes with metal cleats walks into a home in which the kitchen floor is covered with vinyl flooring. The home owner is concerned. Explain the basis for the concern.

3. The metric unit for pressure is the pascal, named for French scientist Blaise Pascal. Research his contributions to the field of fluid dynamics. Briefly describe Pascal's law of fluid dynamics and list some applications of the law.

4. Daniel Bernoulli reached his conclusions on fluid dynamics based on his examination of the work of Evangelista Torricelli. Briefly describe Torricelli's law and explain it in terms of Bernoulli's law.

UNIT 9 THE ATOM   CHAPTER 28

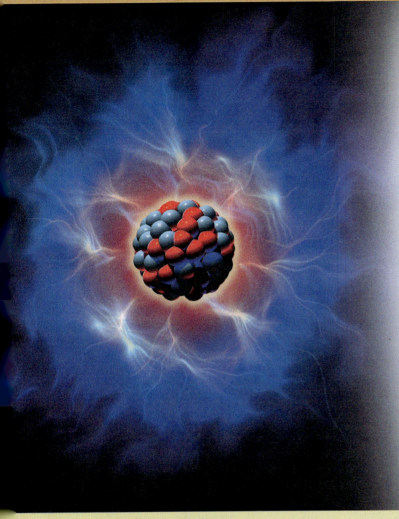

Inside the Atom

Objectives:

By the end of this chapter you should be able to:

✔ Describe the structure of an atom.
✔ Describe the four forces acting inside an atom.
✔ Use the periodic table to obtain information about the atomic number, mass number, atomic mass, and isotopes of different elements.
✔ Predict whether a certain nucleus is stable or unstable and explain why.
✔ Distinguish between and provide examples of chemical reactions and nuclear reactions.
✔ Describe how atomic spectral lines can be explained by energy levels and quantum states.
✔ Explain quantum theory as it relates to light and electrons.
✔ Describe the major developments in quantum theory and identify the scientists associated with each.

Key Questions:

- What is the structure of an atom?
- How are electrons described by quantum states?
- What is quantum theory, and how does it explain electron behavior?

Vocabulary

atomic mass	energy level	Pauli exclusion principle	quantum numbers	strong nuclear force
atomic mass unit	isotope	photoelectric effect	quantum physics	uncertainty principle
atomic number	mass number	Planck's constant	quantum state	wave function
chemical reaction	nuclear reaction	probability	radioactive	weak force
electromagnetic force	nucleus	quantum	spectral line	

28.1 The Nucleus and Structure of the Atom

This section introduces the structure of the atom. The atom was first thought to be the smallest particle of matter that could exist. Today, we understand that atoms themselves are made from even smaller particles. Knowing the structure of the atom makes it possible to explain many properties of matter, just as knowing the structure of DNA makes it possible to explain many processes in biology.

Three particles make up the atom

Charge of the three particles — Atoms are made of three kinds of particles: electrons, protons, and neutrons (Table 28.1). Protons are particles with positive electric charge. Electrons are particles with negative electric charge. Neutrons are neutral and have zero charge. The charge on the electron and proton are equal and opposite. This is important because a proton and an electron together inside an atom have a total charge of exactly zero. Matter is electrically neutral because of the exact and complete charge cancellation between protons and electrons.

Table 28.1: Charge and mass of particles in the atom

	Mass (kg)	Charge (coulomb)
Electron	9.109×10^{-31}	-1.602×10^{-19}
Proton	1.673×10^{-27}	$+1.602 \times 10^{-19}$
Neutron	1.675×10^{-27}	0

Particles in the atom

	Relative charge	Relative mass
Electron	−1	1
Proton	+1	1,835
Neutron	0	1,837

Mass of the three particles — Electrons are tiny and light. While scientists know the mass of the electron accurately, we do not know how small it really is. We do know that protons and neutrons are much larger and more massive. The mass of the proton is 1,835 times the mass of the electron. Neutrons have a bit more mass than protons, but the two masses are so close that we usually assume they are the same. Because the mass of a proton is tiny by normal standards, scientists use **atomic mass units** (amu). One amu is 1.661×10^{-27} kg, or slightly less than the mass of a proton. One electron has a mass of 0.0005 amu.

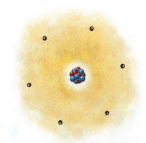

Carbon atom	Charge
6 electrons	−6
6 protons	+6
6 neutrons	0
Total charge	0

Figure 28.1: *The total positive and negative charges of a neutral atom equals zero.*

Atoms are neutral — The positive charge in a neutral atom equals its negative charge. A *neutral* atom has a total charge of zero (Figure 28.1). Since the number of electrons equals the number of protons in an atom, it tends to stay neutral because electric forces are very strong. Any atom with excess protons usually attracts electrons until it becomes neutral again.

Structure of the atom

The nucleus — The neutrons and protons are grouped together in the **nucleus**, which is at the center of the atom (Figure 28.2). There are no electrons in the nucleus, only protons and neutrons. The nucleus is extremely small in comparison to the atom as a whole. If the atom were the size of your classroom, the nucleus would be the size of a single grain of sand in the center of the room.

The electron cloud — The electrons are found outside the nucleus. Because electrons move so fast and have so little mass, we tend to speak of the electron "cloud" rather than talk about the exact location of each electron. Think about a swarm of bees buzzing in a "cloud" around a beehive. It is not easy to precisely locate any one bee, but you can easily see that, on average, the bees are confined to a cloud of a certain size around the hive. On average, electrons are confined to a similar cloud around the nucleus.

Structure of an atom

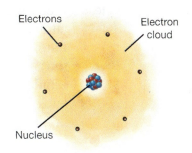

Figure 28.2: *Protons and neutrons are found in the nucleus of an atom. Electrons are outside the nucleus in the "electron cloud." Note: The diagram is not drawn to scale.*

Mass and the nucleus

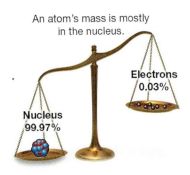

An atom's mass is mostly in the nucleus.

Most of an atom's mass is concentrated in the nucleus. The number of electrons and protons is the same, but electrons are so light they contribute very little mass. For example, a carbon atom has six protons, six electrons, and six neutrons. The mass of the nucleus is 12 amu. The mass of the electrons is only 0.003 amu. So 99.97 percent of the carbon atom's mass is in the nucleus and only 0.03 percent is in the electron cloud.

Electrons and the size of atoms

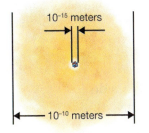

An atom is the size of its electron cloud. The nucleus is tiny by comparison.

The size of an atom depends on how far the electrons spread out. When we talk about the "size" of an atom, what we really mean is how close atoms get to each other. Unless the atoms are chemically bonded together, the electron cloud of one atom does not normally overlap the electron cloud of another. For this reason, the size of an atom is more accurately the size of its electron cloud, which is 10,000 times larger than the nucleus.

CHAPTER 28 | INSIDE THE ATOM

Forces in the atom

Electromagnetic forces
Electrons are bound to the nucleus by **electromagnetic force**. The force is the attraction between positive protons and negative electrons. A good analogy is Earth orbiting the Sun. The gravity of the Sun creates a force that pulls Earth toward it. Earth's momentum causes it to orbit rather than fall into the Sun. Similarly, an electron in the outer part of an atom is attracted to protons in the nucleus. The momentum of the electron causes it to "orbit" the nucleus rather than falling into it (Figure 28.3).

Strong nuclear force
Neutrons are the "glue" that holds the protons in the nucleus together. We already know that the Coulomb force is large. That means the positively-charged protons in the nucleus of an atom repel each other with great force. The nucleus stays together because there is another force even stronger than the electric force. We call it the **strong nuclear force**. The strong nuclear force attracts neutrons and protons to each other, independent of electric charge. If there are enough neutrons, the attraction from the strong force overcomes repulsion from the Coulomb force and the nucleus stays together. In atoms heavier than helium, there is at least one neutron for each proton.

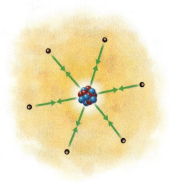

Figure 28.3: *The negative electrons are attracted to the positive protons in the nucleus of the atom.*

Helium nucleus   **Electrical force**   **Strong nuclear force**

Weak force
There is a third force acting in the nucleus called the **weak force.** The weak force is weaker than both electric force and strong nuclear force. If you leave a solitary neutron outside the nucleus, the weak force eventually causes it to break up into a proton and an electron. The weak force doesn't play a big role in a stable atom, but is a factor in special cases when atoms break apart.

Gravity
The fourth force in nature, gravity, is insignificant inside the atom because it is much weaker than even the weak force. It takes a relatively large mass to create enough gravity to make a significant force. We know particles inside an atom do not have enough mass for gravity to be an important force in their interactions. But there are many unanswered questions. Understanding how gravity works inside atoms is an unsolved mystery in physics.

The four forces of nature
Strong nuclear force
Electromagnetic force
Weak force
Gravity

Every process in the universe *that we understand* can be explained in terms of the four fundamental forces—strong, electromagnetic, weak, and gravity. But there are many things that we do not yet understand. A physics book written in the year 2100 CE might well list other forces. Maybe you will discover one of them!

Elements and atoms

Elements — The variety of matter we find in nature here on Earth is made from 92 different types of atoms called *elements*. Water is made from the elements hydrogen and oxygen. Some rocks are mostly silicon and oxygen. The atoms of the 92 elements are created from the same three basic particles: electrons, protons, and neutrons.

Atomic number — All atoms of the same element have the same number of protons in the nucleus. For example, every atom of helium has two protons in its nucleus. Every atom of iron has 26 protons in its nucleus (Figure 28.4). The **atomic number** of each element is the number of protons in its nucleus. The periodic table arranges the elements in increasing atomic number. Atomic number 1 is hydrogen with 1 proton. Atomic number 92 is uranium with 92 protons.

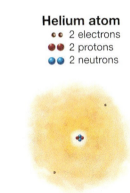

Helium atom
- 2 electrons
- 2 protons
- 2 neutrons

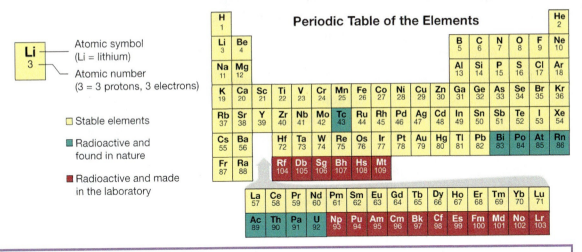

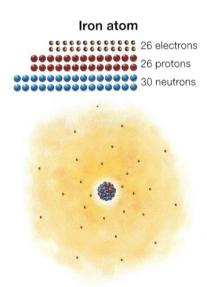

Iron atom
- 26 electrons
- 26 protons
- 30 neutrons

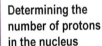

Determining the number of protons in the nucleus

How many protons are in the nucleus of an atom of vanadium (V)?
1. You are asked for the number of protons.
2. You are given that the element is vanadium.
3. The number of protons is the atomic number.
4. The atomic number of Vanadium is 23 so there are 23 protons in the nucleus of a vanadium atom.

Figure 28.4: *One type of helium atom has 2 electrons, 2 protons, and 2 neutrons. One type of iron atom has 26 electrons, 26 protons, and 30 neutrons.*

28.1 THE NUCLEUS AND STRUCTURE OF THE ATOM

Isotopes

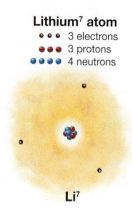

Lithium⁷ atom
- 3 electrons
- 3 protons
- 4 neutrons

Figure 28.5: *The mass number is equal to the total number of protons plus neutrons in the nucleus. A lithium 7 atom has 3 protons and 4 neutrons in its nucleus.*

Isotopes — There are different ways to form the nucleus of a lithium atom. Each form is called an **isotope**. All the isotopes of lithium have three protons but they have different numbers of neutrons. Different isotopes exist for atoms of each element. Some isotopes occur naturally. For example, lithium has two naturally-occurring isotopes. Other isotopes can be created in a laboratory.

Mass number and naming isotopes — The **mass number** is the total number of protons and neutrons in the nucleus. Different isotopes of the same element have different mass numbers. For example, there are two natural isotopes of lithium. Lithium 6 (Li^6) has a mass number of 6 with 3 protons and 3 neutrons in the nucleus. Lithium 7 (Li^7) has a mass number of 7 with 3 protons and 4 neutrons in the nucleus (Figure 28.5). Mass numbers are written above the element symbol in the periodic table. They are also written to the upper right of the symbol, as shown in Figure 28.5. When physicists say "lithium seven" (Li^7), they are talking about atoms of lithium with 7 particles in the nucleus—3 protons and 4 neutrons.

Stable and unstable — Not all isotopes are stable. If an isotope has too many or too few neutrons, the nucleus eventually breaks up and we say the atom is **radioactive**. In a *stable* isotope, the nucleus stays together. In an *unstable* isotope, the nucleus is radioactive and eventually breaks up. Radioactivity is discussed more in Chapter 29.

Stable

Nucleus

Radioactive and unstable

Chemical properties — Different isotopes of the same element have the same chemical properties. The chemical properties are the same because electrons determine all the properties of elements *except* mass. Changing the number of neutrons affects only the nucleus, not the electrons. The chemical identity makes some radioactive isotopes very useful for research and medicine. For example, suppose some of the phosphorus in a DNA molecule is replaced with a radioactive isotope of phosphorus. By looking for the radioactive phosphorus, it is possible to follow exactly how specific parts of DNA function in a living system.

Radioactive isotopes

Most of the matter you see is not radioactive. Almost all of the naturally-produced radioactive isotopes broke apart long ago in Earth's history. Uranium is one of the few rare radioactive isotopes still found in nature. We can still find U^{238} because it takes a very long time to decay into other elements.

Radioactive isotopes can be created in a laboratory for medical use. For example iodine 131 (I^{131}) is used to track how well blood is flowing in the body.

Atomic mass

Atomic mass Elements in nature usually have a mixture of isotopes. For example, the periodic table lists an **atomic mass** of 6.94 for lithium. That does *not* mean there are 3 protons and 3.94 neutrons in a lithium atom. On average, 94 percent of lithium atoms are Li^7 and 6 percent are Li^6 (Figure 28.6). The *average* atomic mass of lithium is 6.94 because of the mixture of isotopes. The table below gives the mass numbers and average atomic masses for the stable isotopes of elements 1–26.

Units of atomic mass The atomic mass of an atom is usually given in atomic mass units (amu). One amu is 1.66×10^{-27} kg, and is defined as $1/12$ ($\frac{1}{12}$) the mass of a carbon-12 atom. To determine the mass of a single atom, you multiply the atomic mass in amu by 1.66×10^{-27} kg/amu. For example, an "average" lithium atom has a mass of $(6.94 \text{ amu}) \times (1.66 \times 10^{-27} \text{ kg/amu}) = 1.15 \times 10^{-26}$ kg.

Atomic mass for stable isotopes of elements 1–26

Atomic number	Element symbol	Element name	Mass numbers of stable isotopes	Average atomic mass
1	H	Hydrogen	1, 2	1.008
2	He	Helium	3, 4	4.003
3	Li	Lithium	6, 7	6.941
4	Be	Beryllium	9	9.012
5	B	Boron	10, 11	10.81
6	C	Carbon	12, 13	12.01
7	N	Nitrogen	14, 15	14.07
8	O	Oxygen	16, 17, 18	16.00
9	F	Fluorine	19	19.00
10	Ne	Neon	20, 21, 22	20.18
11	Na	Sodium	23	22.99
12	Mg	Magnesium	24, 25, 26	24.31
13	Al	Aluminum	27	26.98
14	Si	Silicon	28, 29, 30	28.09
15	P	Phosphorous	31	30.97
16	S	Sulfur	32, 33, 34, 36	32.06
17	Cl	Chlorine	35, 37	35.45
18	Ar	Argon	36, 38, 40	39.95
19	K	Potassium	39, 41	39.10
20	Ca	Calcium	40, 42, 43, 44, 46, 48	40.08
21	Sc	Scandium	45	44.96
22	Ti	Titanium	46, 47, 48, 49, 50	47.88
23	V	Vanadium	51	50.94
24	Cr	Chromium	50, 52, 53, 54	52.00
25	Mn	Manganese	55	54.94
26	Fe	Iron	54, 56, 57, 58	55.85

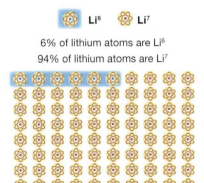

Figure 28.6: *The periodic table lists the average atomic mass for each element. For most elements, the averages include several different isotopes.*

CHAPTER 28 — INSIDE THE ATOM

A graph of protons versus neutrons

Making a stable nucleus
A nucleus is unstable if there are too few or too many neutrons. If there are too few neutrons, the electrical repulsion between protons tears the nucleus apart. If there are too many neutrons, the nucleus tends to eject one or more of them.

The ratio of protons to neutrons
The chart of stable isotopes shows a graph of protons versus neutrons. Each dark blue square represents a stable nucleus. For carbon with six protons, you can see two blue squares representing six and seven neutrons. The chart tells you that carbon has two stable isotopes, C^{12} and C^{13}. For light elements, the number of neutrons and protons is about equal. As the elements get heavier, more neutrons than protons are required to keep the nucleus stable. Only two stable isotopes have fewer neutrons than protons. Can you find them?

Determining the number of neutrons in a nucleus

How many neutrons are in the nucleus of an atom of titanium 49 (Ti^{49})?

1. You are asked for the number of neutrons.
2. You are given that the isotope is titanium 49.
3. The number of neutrons is the mass number minus the atomic number.
4. Titanium is atomic number 22. If 22 of the 49 particles in the Ti^{49} nucleus are protons, then there must be 49 − 22 or 27 neutrons.

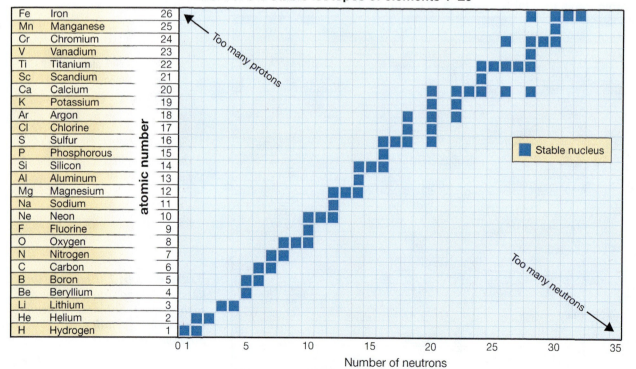

Chart of the stable isotopes of elements 1–26

UNIT 9 THE ATOM

Reactions inside and between atoms

Molecules Most atoms in nature are found combined with other atoms into molecules. A molecule is a group of atoms that are chemically bonded together. For example, water (H_2O) is a molecule of two hydrogen atoms and one oxygen atom.

Chemical reactions A **chemical reaction** rearranges the same atoms into different molecules. For example, the chemical reaction between methane (CH_4), or natural gas, and oxygen (O_2), rearranges one carbon atom, four hydrogen atoms, and four oxygen atoms. The same nine atoms that make up the methane and oxygen molecules are rearranged into carbon dioxide (CO_2) and water molecules. Chemical reactions rearrange atoms into new molecules but do not change atoms into other kinds of atoms.

Nuclear fusion

Fusion is a type of nuclear reaction that combines small atoms to make larger atoms. The energy produced in stars like the Sun comes from fusion reactions.

The interior temperature of the Sun is about 15 million degrees Celsius. At this temperature, nuclei are moving fast enough that they can almost touch despite the electric forces pushing them apart. If two nuclei get close enough, the strong force causes a fusion reaction.

On Earth, we would need to generate about 100 million degrees Celsius to create fusion of hydrogen for producing energy. The higher temperature is necessary because reactors on Earth do not have the Sun's immense gravity to force atoms together. Many countries are working together on fusion research. Someday we may get "clean" energy from power plants using nuclear fusion.

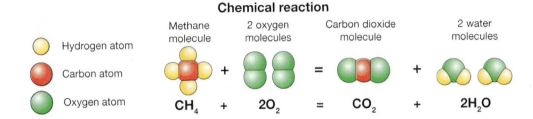

Nuclear reactions A **nuclear reaction** is a process that changes the nucleus of an atom. Because the nucleus is affected, a nuclear reaction can change atoms of one element into atoms of a different element. For example, two helium nuclei can be combined to create one lithium nucleus and an extra proton. There is even a nuclear reaction that turns lead into gold! The ability to change one element into another is one important way nuclear reactions are different from chemical reactions.

Nuclear reaction

$He^4 + He^4 = Li^7 + p^1$

2 helium-4 nuclei — Lithium-7 nucleus — Proton

- Proton
- Neutron
- Helium-4 nucleus
- Lithium-7 nucleus

28.1 THE NUCLEUS AND STRUCTURE OF THE ATOM

28.2 Electrons and Quantum States

Nearly every property of matter we experience is determined by the behavior of electrons in atoms. The color of paints comes from how electrons absorb light. Oxygen is vital to life because of how it makes chemical bonds with other elements. Chemical bonds are formed between electrons of different atoms. The size of atoms is determined by how far electrons range from the nucleus. The exception is mass; mass derives from the nucleus. Just about everything else is determined by electrons.

The birth of quantum physics

The discovery of quantum physics The electrons in an atom obey a very strange set of rules. The Danish physicist Niels Bohr (1885–1962) was the first person to put the clues together correctly, and in 1913, proposed a theory that described the electrons in an atom. A brilliant scientist, Bohr is often called the father of **quantum physics**. Quantum physics is the branch of science that deals with extremely small systems such as an atom.

The spectrum An unusual feature of light was the clue that lead to the discovery of quantum physics. When a substance is made into a gas, and electricity is passed through the gas, light is given off, like in a neon sign. When this light was examined carefully, it was found that the light did not include all colors. Instead they saw a few very specific colors, and the colors were different for different elements (Figure 28.7). This characteristic pattern of colors is called a *spectrum*. The colors of clothes, paint, and everything else around you come from this unusual property of substances to emit *or absorb* light of only certain colors (Figure 28.8). Since the energy of light depends on the color, the lines in a spectrum meant that substances could only emit light of certain energies.

Spectrometers and spectral lines Each individual color is called a **spectral line** because each color appears as a line in a spectrometer. A spectrometer is a device that spreads light into its different wavelengths, or colors. The diagram below shows a spectrometer made with a prism. The spectral lines appear on the screen on the right.

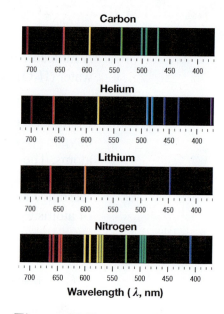

Figure 28.7: *Spectra differ for different elements.*

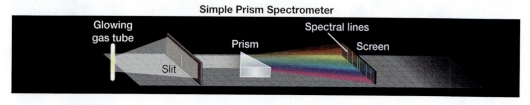

Figure 28.8: *Cloth looks blue because molecules in the dye absorb all colors of light except blue.*

The hydrogen spectrum

Johann Balmer's discovery
The spectrum shown below is from hydrogen. When hydrogen gas is heated, it gives off a unique pattern of colors. The first serious clue to an explanation of the spectrum was discovered in 1885 by Johann Balmer, a Swiss high school teacher. He showed that the wavelengths of the light given off by hydrogen atoms could be predicted by a mathematical formula now known as Balmer's formula.

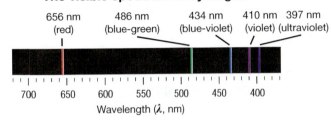

The visible spectrum of hydrogen

656 nm (red), 486 nm (blue-green), 434 nm (blue-violet), 410 nm (violet), 397 nm (ultraviolet)

$$\frac{1}{\lambda} = R\left(\frac{1}{2^2} - \frac{1}{n^2}\right)$$

Where R is the Rydberg constant and is equal to $1.0972 \times 10^{-7} \, \text{m}^{-1}$

$n = 3, 4, 5, \ldots$

Balmer's formula

n	λ (nm)	Color
3	656	Red
4	486	Blue-green
5	434	Blue-violet
6	410	Violet
7	397	Ultraviolet

Using the Balmer formula
In Balmer's formula, n is an integer greater than two. For example, if we choose $n = 4$, then the formula predicts a wavelength of 486 nm. This exactly matches the blue-green line in the hydrogen spectrum. Choosing $n = 3, 4, 5, 6$, and 7 gives the correct wavelengths of other spectral lines of hydrogen. Balmer identified patterns in the spectrum, but could not explain what caused it.

What the formula implied
Something inside an atom corresponds to the integer numbers—$n = 3, 4, 5$, and so on—from Balmer's formula. The mechanism in a hydrogen atom that creates light acts like the numbers are "click stops" on a rotary switch. The switch can be set to any integer, such as 2 or 3, but not to any number in between, such as 2.5. Of course, there is no switch in the atom, but you may find the true explanation even stranger.

The discovery of helium

Atoms can absorb light at the same wavelengths that they emit light. When a bright light containing all wavelengths is passed through a gas, dark spectral lines indicate which wavelengths of light are absorbed. The Sun's spectra shows such dark spectral lines.

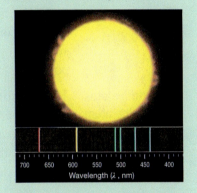

The element helium is a light gas that is very rare on Earth. In fact, helium was not discovered on this planet. It was discovered in the Sun, hence the name. In Greek, *helios* means "sun." Astronomers saw a series of lines in the spectrum of the Sun that did not match any known element on Earth. Researchers were then able to find it on Earth because they knew what to look for.

Quantum states

The quantum meaning of the word "state"

Neils Bohr proposed that electrons in the atom were limited to certain **quantum states**. In quantum physics, the word *state* means the *complete* description of a system. If you know the quantum state of an electron, you know *everything* you can know about that electron—its energy, how it is moving, where it is, and its spin. If you could know the *state* of a used car, in the quantum sense, you would know much more than its every scratch or speck of dirt. You would know the location, motion, and energy of every single atom in the car. This would be enough information to *exactly* duplicate the car, at the same temperature, with every detail the same. This is too much to know for any macroscopic object, like a car. This is why "transporter beams" exist only in science fiction. However, for a single atom, it *is* possible to know its state, and even more strange is that the possible states of an atom are restricted to only specific values.

Quantum states in the atom

The quantum states in an atom have certain allowed values of energy, momentum, position, and spin. A graph showing the energy of an electron within an atom looks like a hilly surface with peaks and valleys. Each quantum state represents a valley on the energy graph big enough to hold a single electron. An electron can be found in one valley or another, but never in between.

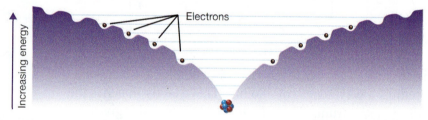

Quantum numbers

The number n in the Balmer formula is one of four **quantum numbers** that describe which quantum state an electron is in. To understand quantum numbers, think of an atom like a parking garage for electrons. Each parking space in the garage is a quantum state that can hold one electron. Quantum numbers are the code describing each space where an electron can be parked. To locate an electron, you need to know the numbers of its code. The code has four numbers—n, l, m, and s—and each number can only have values corresponding to actual parking spaces called *quantum states*. Every quantum state in the atom is identified by a unique combination of the four quantum numbers.

n, l, m, s

The four quantum numbers of an electron

Every electron in an atom can be completely described by the values of its four quantum numbers: n, l, m, and s. The first quantum number (n) can be any integer bigger than zero.

The second quantum number (l) must be a positive integer from zero to $n - 1$. For example, if $n = 1$, the only possibility is $l = 0$. If $n = 2$, then l can be 0 or 1.

The third quantum number (m) is an integer that can go from $-l$ to $+l$. For example, if $l = 3$, m can have any of seven values between -3 and $+3$. Can you list them? The fourth quantum number (s) can only be either $+1/2$ or $-1/2$.

Each possible combination of values for the four quantum numbers represents one quantum state. For example, one of the two quantum states in the first energy level has quantum numbers: $n = 0$, $l = 0$, $m = 0$, and $s = +1/2$. The other state in the first level has $n = 0$, $l = 0$, $m = 0$, and $s = -1/2$.

From the four quantum numbers, it is possible to calculate everything about the electron, including its energy, angular momentum, position, and spin.

Energy levels and spectra

Energy and quantum states
The energy of an electron depends on its quantum state. Quantum states that keep the electron far from the nucleus have more energy than states of electrons that are closer to the nucleus. If an electron moves to a quantum state closer to the nucleus, energy is released, often as light.

Energy levels
The quantum states in an atom are grouped into **energy levels**. All the quantum states in each level have approximately the same energy. A good analogy is a multilevel parking garage. Each floor of the garage has a limited number of parking spaces for cars. Each parking space can hold one car. Each energy level is like one floor of the garage. Each quantum state in an energy level is like a parking space for one electron. The diagram in Figure 28.9 shows how the quantum states are arranged in the first five energy levels. The first level has two states, and can therefore hold two electrons. The second and third levels have eight quantum states. The fourth and fifth levels have 18 states.

Energy levels explain spectral lines
Bohr explained that spectral lines are produced by electrons moving between different energy levels. An electron in a hydrogen atom dropping from the third level to the second level gives off an amount of energy exactly equal to the red line in the hydrogen spectrum. An electron falling from the fourth level to the second level gives off more energy, creating the blue-green line in the spectrum. All of the spectral lines described by the Balmer formula correspond to electrons falling from higher levels to the second energy level. Bohr's model developed into the quantum theory of the atom.

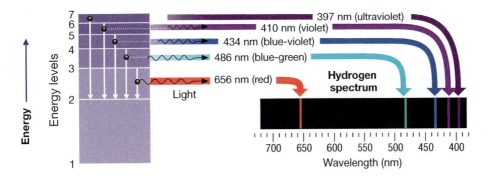

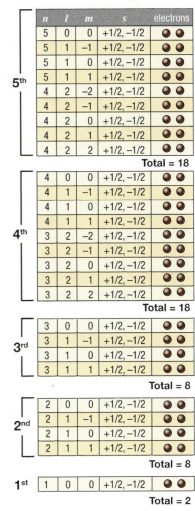

Figure 28.9: *The quantum numbers are shown for electrons in the first five energy levels.*

The Pauli exclusion principle & the periodic table

A quantum state can hold one electron

According to the quantum theory, two electrons in an atom can never be in the same quantum state at the same time. This rule is known as the **Pauli exclusion principle** after Wolfgang Pauli, the physicist who discovered it. The exclusion principle prevents all the electrons in an atom from falling immediately to the lowest energy level. Once all the quantum states in the first level are occupied by electrons, the next electron has to go into a higher energy level.

Patterns of electrons in the periodic table

The rows of the periodic table correspond to the number of quantum states in each energy level. The first energy level has two quantum states. Hydrogen (H) has one electron and helium (He) has two electrons. These two elements are the only ones in the top row of the periodic table because there are only two quantum states in the first energy level. The next element, lithium (Li), has three electrons. Lithium begins the second row because the third electron goes into the second energy level. The second energy level has eight quantum states and there are eight elements in the second row of the periodic table, ending with neon. Neon (Ne) has 10 electrons, which exactly fill all the quantum states in the first and second levels. Sodium (Na) has 11 electrons, and starts the third row because the 11th electron goes into the third energy level.

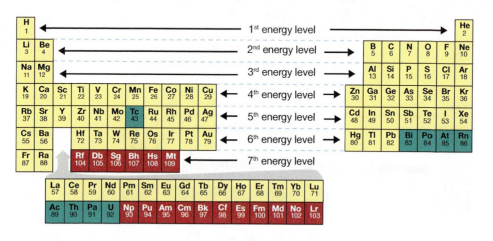

If atoms are mostly empty space, why can't I walk through a wall?

The nucleus takes up only one thousandth of a trillionth (10^{-15}) of the volume of the atom. It is often said that the rest is "empty space" occupied only by a few dozen electrons. However, the so-called empty space is not empty.

To walk through a wall, your atoms and the wall's atoms would have to overlap. That means twice as many electrons would have to be in the same space. According to the exclusion principle, two electrons cannot occupy the same quantum state at the same time.

During the overlap, electrons from one atom must go into higher unoccupied energy states of the other atom. But, putting an electron into a higher energy state takes energy. The need for the extra energy is why we do not walk through walls. You actually *can* walk through a wall. However, it takes so much energy to do it that neither you or the wall would survive!

The "shape" of quantum states

Orbitals In chemistry, the quantum states for electrons in an atom are called *orbitals*. The name comes from an older idea that electrons moved in orbits around the nucleus, like planets around the Sun. Today, we know quantum states are *not* similar to orbits (Figure 28.10), but the name *orbital* is still commonly used.

Molecules have a shape The shape of a molecule is important to what the molecule does. For example, the two hydrogen atoms in a water molecule make an angle of 104 degrees. The 104-degree angle is created by the shape of the quantum states of the oxygen atom in the middle of the molecule. Many substances dissolve in water because the 104-degree angle puts both hydrogen atoms on one side of the oxygen atom. Many medicines work because the shape of a molecule fits precisely with another molecule found in the body, like pieces in a puzzle.

Orbital shapes for n = 1, 2, 3

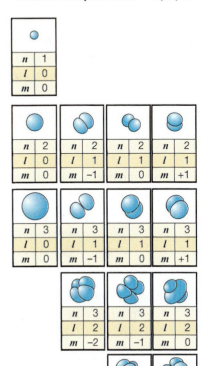

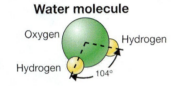
Water molecule
Oxygen — Hydrogen
Hydrogen 104°

Tetrahedron
(four-sided pyramid)

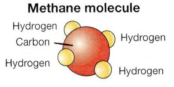
Methane molecule
Hydrogen — Carbon — Hydrogen
Hydrogen Hydrogen

The shape of orbitals Each "orbital" shape shows the most likely locations for a pair of electrons with matching quantum numbers n, l, and m. The $l = 0$ shapes are spherical. The $l = 1$ shapes fall along the x, y, and z axes. The $l = 2$ shapes are more complex (Figure 28.10). The orbital shapes overlap in an atom with many electrons. The shape of the electron cloud of an atom comes from the shapes of all the orbitals that contain electrons.

Outer orbitals form the shape of molecules The structure of molecules comes from the shapes of the orbitals of each atom making up the molecule. When atoms bond, they tend to align along the orbitals that hold the outermost electrons. For example, carbon has six electrons. The first two go in the inner, $n = 1$, quantum states. The last four electrons occupy quantum states with $n = 2$, and are available for bonding with other atoms. Carbon combines with four hydrogen atoms to make *methane*. The four hydrogen atoms in a methane molecule line up with the corners of a tetrahedron. The shape comes from the orbitals occupied by the four electrons in the $n = 2$ states.

Figure 28.10: *The shapes and relative sizes of the orbitals are shown for quantum numbers n = 1, 2, and 3. Each shape is two quantum states corresponding to s = +1/2 and −1/2.*

28.3 The Quantum Theory

In the microscopic world of atoms and particles, familiar rules such as Newton's laws of motion do not tell the whole story. The *quantum theory* describes what happens to matter and energy when things get as small as the size of single atoms. The structure of the atom and the behavior of electrons are described by the quantum theory. This section will describe some of the basic ideas. As you read, keep in mind that quantum theory is far from complete. Quantum physics is a relatively young field of science and many discoveries are yet to be made.

Classical physics	Quantum physics
Matter is particles	Matter is both a particle and a wave
Newton's laws of motion	Quantum mechanics
Coulomb's law and wave theory of light	Photon theory of light

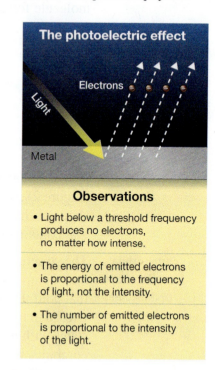

Figure 28.11: *The table shows some of the differences between classical and quantum physics.*

The discovery of a "new" physics

Discovering new knowledge Nature always behaves the "right" way. The clash between an unexpected observation and our imperfect knowledge leads us to say, "That was not supposed to happen!" In science, "pushing the limits" means trying to understand the *unexpected* result. Starting with Newton's laws (1685), "classical physics" was very successful at explaining things (Figure 28.11). The quantum theory started when classical physics disagreed with the results of new experiments.

Two outstanding puzzles The quantum theory began between 1899 and 1905 with Max Planck and Albert Einstein. Planck was trying to understand why light given off by hot materials follows the blackbody spectrum (Chapter 26). Einstein was thinking about the photoelectric effect. Neither phenomenon could be explained by classical physics.

The photoelectric effect When light falls on the surface of some metals, electrons are emitted from the surface. This is called the **photoelectric effect**. If the light is made brighter, the metal absorbs more energy. Classical physics predicts that electrons coming off the metal should have more kinetic energy when the light is made brighter. But that is *not* what happens. Classical physics gives the wrong answer.

Results of experiments Experiments on the photoelectric effect showed that the *frequency* of the light is the most important variable. With low-frequency red light, which has a long wavelength, no electrons are emitted, even if the light is very bright. As the light's frequency increases, electrons start to be emitted. The kinetic energy of the emitted electrons depends on the frequency of the light. Once the frequency threshold is passed, the higher the frequency, the more energy the emitted electrons have (Figure 28.12).

The photoelectric effect

Observations

- Light below a threshold frequency produces no electrons, no matter how intense.
- The energy of emitted electrons is proportional to the frequency of light, not the intensity.
- The number of emitted electrons is proportional to the intensity of the light.

Figure 28.12: *Variables affecting the photoelectric effect.*

The quantum theory of light

The photon — In 1899, Max Planck proposed that light existed in small bundles of energy called *photons*. The smallest amount of light you could have is a single photon. Bright light consists of billions of photons per second, while dim light has very few photons per second. Planck's idea was very different from the wave theory of light. You could make a wave as small as you want by reducing the amplitude. You could not split a photon. You could make light of 1 photon, 10 photons, or 10 trillion photons, but you could never make half a photon.

The energy of a photon — According to Planck, the energy of a single photon is related to its frequency by the formula $E = hf$, where h is **Planck's constant** (Figure 28.13). Higher frequency means higher photon energy. Planck's constant is $h = 6.626 \times 10^{-34}$ J·s. Therefore, the energy of a single photon is also small. A typical flashlight produces 10^{20} photons per second! Like atoms, photons are such small quantities of energy that light appears as a continuous flow of energy under normal circumstances.

Einstein explains the photoelectric effect — Quantum theory was confirmed in 1905 when Albert Einstein published his explanation of the photoelectric effect. Einstein proposed that an atom can absorb only one photon at a time. An electron needs a minimum amount of energy to break free from an atom. If the energy of the photon is too low, there is not enough energy to free an electron and no photoelectric effect is observed. Making brighter light does not help. Brighter light has more photons, but none with enough energy to free an electron. If the frequency of light gets higher, at a particular frequency, one photon has just enough energy to free an electron. Even if the light is made very dim, you get exactly one electron for each photon of light.

Wavelength and kinetic energy — If light of an even higher frequency is used, there is more than enough energy in each photon to free an electron. Part of the photon energy goes to freeing the electron and the rest becomes kinetic energy of the electron. Making the frequency still higher and more blue increases the amount of "leftover" energy available to become kinetic energy. Einstein's explanation matched the data collected in experiments. His explanation of the photoelectric effect was strong evidence that the quantum theory of light was correct.

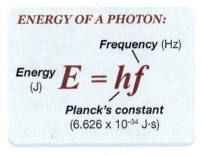

Figure 28.13: *You can think of a photon as a bundle of energy with a frequency. The photon energy is given by Planck's formula.*

Quantum theory

What "quantum" means

Planck's theory became a **quantum** description of light. To a physicist, if something is *quantized,* it can only exist in whole units, not fractions of units. For example, the number of students assigned to a class is quantized. There can be 25 students or 26 or 32, or any other whole number, but there cannot be 25.3 students. Light is quantized and one photon is the smallest unit, or *quantum,* of light. A quantum of something often means the smallest amount that can exist.

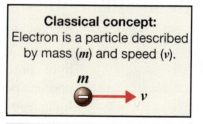

Waves and particles

In the quantum theory, *all* matter and energy are quantized when you get down to the scale of atoms. That means matter and energy have both wavelike and particle-like properties. Light acts like a wave from far away. But up close, light acts "particle-like" because the wave is made of individual photons. An electron acts like a particle when it is both free to move and far from other electrons. However, if an electron is confined in a small space, as in an atom, it behaves like a wave.

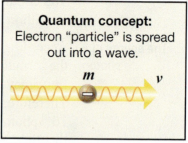

The wavelength of a particle

The wavelength of a particle (λ) depends on its mass (m) and speed (v), according to the DeBroglie formula (Figure 28.14). The wavelengths of particles tend to be extremely small. An electron moving at a million meters per second has a wavelength of only 7×10^{-10} meters. The short wavelength is why an electron looks like a particle most of the time. The wave properties only become apparent when the electron is confined to a space near the size of its wavelength, such as an atom.

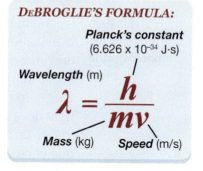

Properties of wave/particles

You might think quantities like position and velocity can be applied to an electron as if it were a tiny baseball. For example, the electron could be at exactly 1 meter moving at exactly 100 m/s. Quantum theory gives a completely different picture. When you try to look at extremely small details, quantum theory tells us that the electron spreads out into a wave. You cannot say where on the wave the electron "is." The electron has no exact value of position.

Figure 28.14: *The DeBroglie formula gives the wavelength of a moving particle. The wavelength of an electron is close to the size of an atom.*

We do not usually see particles act like waves

The wavelike properties of matter are not normally seen unless you look at very fine details. If the smallest important detail of a system you are trying to study is much larger than the quantum wavelength ($\lambda = h/mv$), then you can use Newton's laws or the wave theory of light and you will get the right answer. If important details are less than or equal to a factor of 10, compared with the quantum wavelength, then you must use quantum theory to get an accurate answer.

The uncertainty principle

The uncertainty principle
Quantum theory puts limits on how precisely we can know the value of quantities such as position, momentum, energy, and time. In classical physics, you could say an electron is exactly at a certain place, at a certain time, moving at a certain speed, and with a specific energy. In the quantum world, this is not possible. The **uncertainty principle** places a limit on how precisely these four quantities can be measured. Planck's constant (h) shows up again, as it often does in quantum theory.

Understanding the uncertainty principle
The uncertainty principle arises because the quantum world is so small. When you see a car, your eye collects trillions of photons that bounce off the car. Photons are so small compared with a car that the car is not affected by your looking at it. To "see" an electron you also have to bounce a photon off of it, or interact with the electron in some way. Because the electron is so small, even a single photon moves it and changes its motion. That means the moment you use a photon to locate an electron, you push it so you no longer know precisely how fast it is going, or in what direction. In fact, any process of observing in the quantum world changes the very system you are trying to observe. The uncertainty principle works on pairs of variables because measuring one always effects the other in an unpredictable way (Figure 28.15).

The meaning of the uncertainty principle
The uncertainty principle has some very strange implications. In the quantum world, anything that *can* happen, *does* happen. Put more strongly, unless something is specifically *forbidden* from happening, it *must* happen. For example, suppose you could create a particle out of nothing, then make it disappear again. Suppose you could do this so fast that it was within the energy and time limit of the uncertainty principle. You could break the law of conservation of energy if you did it quickly enough and in a very small space. *We believe this actually happens.* Physicists believe the so-called "vacuum" is not truly empty when we consider details so small that the uncertainty principle prevents us from seeing them. There is considerable experimental evidence that supports the belief that particles of matter and antimatter are continually popping into existence and disappearing again, out of pure nothing. This implies that the vacuum of empty space may never truly be empty of *everything*. It may have intrinsic energy of its own, even when there is no ordinary matter or energy present.

Uncertainty principle

The uncertainty in position (Δx) multiplied by the uncertainty in momentum (Δp) can never be less than $h/2\pi$.

$$\Delta x \times \Delta p \geq \frac{h}{2\pi}$$

The uncertainty in energy (ΔE) multiplied by the uncertainty in time (Δt) can never be less than $h/2\pi$.

$$\Delta E \times \Delta t \geq \frac{h}{2\pi}$$

Figure 28.15: *Two pairs of variables are related by the uncertainty principle.*

A question of size

The uncertainty principle does not normally affect us because we cannot see details as small as Planck's constant. However, the uncertainty principle does put a limit on how small we can make computer circuits, magnetic disk drives, and other devices that rely on extremely small details.

Probability and the quantum theory

Quantum theory and probability
Calculations in quantum physics do not result in knowing what *will* happen, but instead give the **probability** of what is likely to happen. This is a very strange concept. For example, take the motion of a ball tossed in the air. According to Newton's laws, you can calculate exactly where the ball will be at every moment of its motion. If the ball were an electron, this calculation would not be possible. You could calculate that there is a 98 percent *chance* the electron is at a particular place and time. But there is a two percent chance it is somewhere else! The result of any calculation in quantum physics is the *probability* of something occurring.

The meaning of probability
To understand probability, consider tossing a penny. There are two ways for the penny to land: heads or tails (Figure 28.16). The term *probability* describes the chance of getting each possible outcome of a system. There is an equal probability that the penny lands heads or tails. With a single penny, there is a 50 percent probability of getting heads and a 50 percent probability of getting tails. Suppose you flip a penny 100 times and record the number of "heads." The graph of your results looks like Figure 28.16. The graph tells you that there is a 5.5 percent chance that you will get exactly 50 heads out of 100 coin tosses. If you repeated the experiment 1,000 times, you would expect 55 experiments to come up with exactly 50 heads and 50 tails. While you can never accurately predict the outcome of one toss of the penny, what you *can* do is to make accurate predictions about a collection of many tosses, and this is how quantum theory works.

The wave function
Quantum theory uses probability to predict the behavior of large numbers of particles. In quantum physics, each quantum of matter or energy is described by its **wave function**. The wave function mathematically describes how the probability for finding a quantum of matter or energy is spread out in space (Figure 28.17). For example, quantum physics allows you to calculate the probability of one electron being in a certain place. If you observe a trillion identical electrons, you can say with great precision how many will be found at that place. But quantum theory still cannot tell you where any *single* electron is. Because of its basis in probability, quantum theory can only make accurate predictions of the behavior of large systems with many particles.

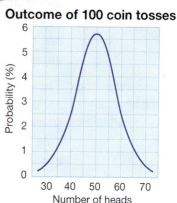

Figure 28.16: *The graph shows the probability curve for the outcome of 100 tosses of a penny.*

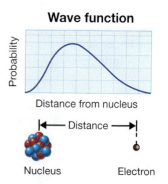

Figure 28.17: *The wave function describes how the probability of a quantum of matter or energy is spread out in space.*

Chapter 28 Connection

The Laser

The word *laser* is an acronym for light amplification by stimulated emission of radiation. A laser is a device that depends on both quantum mechanics and optics. The light from lasers has special properties that make possible such technologies as compact discs, laser surgery, and fiber-optic communications. Since the development of the first laser in 1960, laser-based products and services have grown to a multibillion-dollar industry.

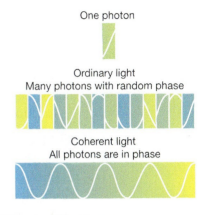

Figure 28.18: *Phase differences distinguish coherent light from light, which is not coherent.*

Coherence of laser light   The special characteristics of laser light are that it is *coherent* and *monochromatic*. The word *coherent* means that all the photons of light are lined up in such a way that they have the same phase. To understand coherence, consider light as a wave that is broken up into many small pieces (Figure 28.18). Each "piece" is a single photon. In ordinary light, the photons are scrambled so that each is independent of the others. Each photon contributes to the wave pattern but the photons do not add up to a single wave. In coherent light, the photons are aligned in phase so that they create a single, continuous wave pattern.

Lasers are monochromatic   White light is a mixture of all colors. By comparison, monochromatic light contains only one single frequency. Because all the photons have the same frequency, they also all have the same wavelength. For example, a common red laser has a wavelength of 650 nanometers.

The three components of a laser   Although there are many different types of lasers, they all have three main parts: a pump, laser material, and a laser cavity. The pump is the source of energy that starts the laser process. The energy of the pump can come from many different sources. The pump in a compact disc player's laser uses a continuous low-voltage electric current. The helium-neon lasers used in grocery store scanners use high-voltage electric current. Some very high-power lasers use a chemical process to pump energy into the laser.

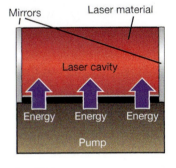

Figure 28.19: *The three main components of a laser are the laser cavity, the laser material, and the pump. The laser cavity is made by sandwiching the laser material between two mirrors.*

Laser materials   Many materials, such as crystals of ruby, nitrogen gas, and even colored gelatin, have the properties needed to be a laser material. The special atomic configuration of a laser material gives it the ability to store energy for release as light. The laser cavity is a space between partially-reflecting mirrors where light can bounce back and forth (Figure 28.19). The release of stored energy in the laser material makes the light stronger with each bounce.

Chapter 28 Connection

How lasers make light

Emission and absorption of photons

As energy is absorbed by an atom in a laser material, one electron moves to a higher energy level. The energy is released as a photon when the electron falls back down to its normal energy level. In normal materials, absorption and release happen almost simultaneously. In laser materials, there is a time delay between the absorption of the energy and the subsequent release of a photon.

Stimulated emission

The atoms and molecules used in a laser have an internal structure that traps an electron in a higher level. The electron remains trapped in the higher level until it is liberated by a photon with just the right energy. When a *stimulating photon* with the right energy does come along, the atom emits a photon with the exact same energy and phase as the stimulating photon (Figure 28.20). It is this property of atoms to emit photons that match other photons that makes the laser possible. The process of light emission triggered by a photon is called *stimulated emission*.

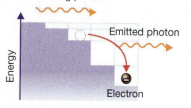

Figure 28.20: *Stimulated emissions cause electrons to emit light as they lose energy.*

A standing wave forms in the laser cavity

A laser material has many photons in many atoms in its system. The *laser cavity* ensures that the emission from each atom is synchronized with a resonant light wave in the laser. The simplest laser cavity is made of two parallel mirrors (Figure 28.21). Light traveling between the mirrors forms a standing wave. Every photon in the standing wave is in phase with every other photon. To get light out of the laser, one of the mirrors is made with a reflectivity less than 100 percent. A fraction of the photons bouncing back and forth pass through the partially-reflecting mirror and create the laser light.

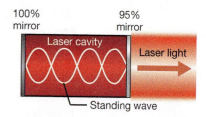

Figure 28.21: *The standing wave in the laser cavity.*

Amplification of photons

To see how *amplification* works, consider one photon starting out across the cavity. This photon hits an energized atom, which releases a second photon matching the first (Figure 28.22). Now, two photons hit two other energized atoms, releasing two more matching photons, for a total of four. As long as there are atoms with electrons in higher energy levels, any photon moving in the cavity triggers a chain reaction releasing many photons, all with identical phase and energy. The process of multiplying photons through stimulated emission is called light amplification by stimulated emission of radiation, or simply, *laser*.

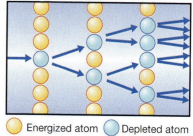

Figure 28.22: *The process of light amplification.*

Chapter 28 Assessment

Vocabulary

Select the correct term to complete the sentences.

nucleus	electron	proton
neutron	atomic mass	atomic mass unit
electromagnetic force	strong nuclear force	weak force
elements	atomic number	mass number
isotopes	radioactive	nuclear reaction
quantum	spectral lines	spectrum
quantum state	energy level	quantum numbers
Pauli exclusion principle	Planck's constant	wave function
probability	orbitals	uncertainty principle
photoelectric effect	photon	chemical reaction
spectrometer	quantum physics	molecule

1. Any process that changes the nucleus of an atom is called a(n) _____.
2. The total number of particles in the nucleus of an atom is the _____.
3. The average mass of all isotopes of an element is known as the _____.
4. The matter on Earth is made from 92 different types of atoms called _____.
5. A nuclear force that causes a neutron to break apart into an electron and proton is referred to as the _____.
6. An element can be identified by its _____, the number of protons in its nucleus.
7. Atoms containing the same number of protons but different number of neutrons in their nuclei are called _____.
8. Atoms with too many neutrons are unstable and are described as _____.
9. Electrons are bound to the nucleus of an atom by _____.
10. Neutrons and protons are held together by the _____.
11. The positive atomic particle found in the nucleus of an atom is the _____.
12. The compact center of an atom in which protons and neutrons are located is the _____.
13. The unit used by scientists to express the mass of atoms and atomic particles is the _____.
14. The light, negatively charged atomic particle found outside the nucleus is the _____.
15. The atomic particle considered the "glue" for holding protons together in the nucleus is a(n) _____.
16. A process that rearranges the same atoms into different molecules is called a(n) _____.
17. Bright lines of color produced by a spectrometer are called _____.
18. The claim that two electrons can never concurrently occupy the same quantum state is an idea expressed by the _____.
19. Quantum states for electrons in an atom are called _____.
20. The values for energy, momentum, position, and spin determine the _____ of an electron in an atom.
21. A scientist might call a rainbow in the sky a complete _____.
22. A device that spreads light into different wavelengths or colors is a(n) _____.
23. Quantum states in an atom are grouped into _____.
24. The set of numbers describing the quantum state of an electron is known as _____.
25. The smallest amount of something that can exist is called a(n) _____.
26. The proportionality constant used to relate the energy of an electron to its frequency is named _____.
27. The phenomenon that occurs when light shines on a metal and causes the emission of electrons is called the _____.
28. The explanation for the behavior of large numbers of particles based on probability is known as _____.

29. A quantum of light is a(n) _____.

30. The mathematical description for the probability of how a quantum of matter or energy is distributed in space is called the _____.

31. The _____ places a limit on the precision with which the four parameters of a quantum state can be determined.

32. The relative chance for getting any possible configuration in a system can be described as _____.

Concept review

1. Summarize the properties of the major atomic particles by completing the chart below.

Particle name	Relative charge	Relative mass	Mass (kg)	Charge (C)
Electron				
Proton				
Neutron				

2. If electrons are in constant motion, explain why atoms tend to remain neutral rather than lose electrons.

3. The electric forces between protons of the nucleus are large, positive, and mutually repulsive. Why does this force not cause the positively-charged protons to fly apart?

4. List the four forces of nature in order from strongest to weakest, name the particles affected by the forces, and describe how they might affect an atom structurally.

5. Using the terms *mass number* and *atomic number*, describe the difference between the various isotopes of carbon.

6. If the mass number of an atom is the sum of protons and neutrons in the atom, why is the atomic mass of an atom of magnesium given as 24.31?

7. Using the chart of stable isotopes in this chapter, state whether or not C^{14} is stable and explain your answer.

8. Hydrogen reacts with oxygen to form water. Helium reacts to form lithium. One is a nuclear reaction, the other a chemical reaction. Identify each and state a major difference between chemical and nuclear reactions.

9. Which particle in the atom is responsible for most characteristics of matter?

10. How is the color of clothing related to the electronic structure of the atom?

11. What does a physicist mean by the term *quantum state*?

12. To locate an electron in an atom, you need to know the four _____ of its address: n, l, m, and s.

13. When an electron in one quantum state moves to a state closer to the nucleus, does it gain or lose energy? Explain.

14. Use the analogy of a multilevel parking garage to explain the terms *quantum state* and *energy level*.

15. When heated, an unknown gas emits light with wavelengths of 589 nm and 704 nm, creating colored lines at these frequencies in a bright-line spectrum as observed in a spectrometer. The spectra of four gases are shown. Which of the four is the unknown gas?

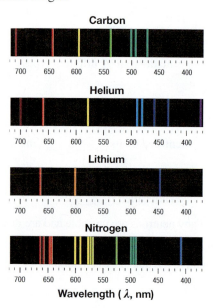

16. What is the most important variable in photoelectric effect experiments?
17. How are the wavelength and energy of a photon related?
18. When do electrons exhibit particle-like properties? When do they exhibit wavelike properties?
19. State the *l* quantum number for each orbital shape below.

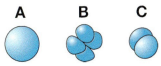

20. Explain how the uncertainty principle relates to the idea that the exact position of an electron can never be determined precisely.
21. What does the wave function of a quantum of matter describe?
22. List the contributions of the following scientists to the quantum theory.
    a. Johann Balmer
    b. Neils Bohr
    c. Wolfgang Pauli
    d. Max Planck
    e. Albert Einstein
    f. Louis DeBroglie

Problems

1. List the element that contains each number of protons.
    a. 22
    b. 36
    c. 79
2. Calculate the mass of a C^{12} atom in kilograms.
3. Calculate the number of neutrons in the isotope of uranium U^{235}.
4. A fictitious element, pennium (Pe), has two isotopes, Pe^{25} and Pe^{31}. In an average sample of pennium, 8 of 10 atoms are Pe^{31}. What is the average atomic mass for pennium?
5. Refer to the chart of stable isotopes to answer the following questions. Use the appropriate symbols.
    a. Which two isotopes have fewer neutrons than protons?
    b. Give the number(s) of neutrons not found in stable isotopes.
    c. Which element on the chart has the most stable isotopes? How many stable isotopes does it have?
    d. Which element's stable isotope contains no neutrons?
    e. Name the stable isotopes of argon.
6. Sulfur's common isotopes are stable. Using this information and the chart of the atomic masses of stable isotopes in the chapter, solve the following questions.
    a. Assuming that sulfur's most common isotope occurs more frequently than others, what is the most common isotope of sulfur?
    b. How many protons and neutrons does this isotope have?
7. Johann Balmer predicted that other spectral lines in the ultraviolet frequency would be found if *n* in his formula represented numbers larger than seven. Use Balmer's formula, $\frac{1}{\lambda} = 1.0972 \times 10^7 \, m^{-1} \left(\frac{1}{2^2} - \frac{1}{n^2} \right)$, to predict the wavelength of light that might be emitted if $n = 8$.
8. Johann Balmer predicted that other wavelengths of spectral lines in the infrared region would be identified if the 2 in the fraction $\frac{1}{2}$ was replaced by the number 3. Use Balmer's formula, substituting $\frac{1}{3}$ for $\frac{1}{2}$, to predict the wavelength of light produced by a hydrogen atom when $n = 4$. Was his prediction correct? Explain.
9. A photon of ultraviolet light with wavelength 5×10^{-7} meters is directed at a zinc metal surface. It causes an emission of electrons from the surface of the zinc.
    a. Calculate the energy of the incident photon.
    b. Assuming it takes 1×10^{-19} joules to remove the electron from the surface of the metal, how much kinetic energy does the emitted electron have?
10. Find the wavelength of an electron traveling at 1.23×10^6 m/s.
11. Find the wavelength of a 2,000-kg car traveling at 100 km/h.

12. A tetrahedron is a pyramid with four triangular faces. It has one black face, one orange face, and two white faces. If it is tossed into the air, what is the probability of
    a. an orange face landing as the base?
    b. an orange face *not* landing as the base?
    c. an orange or black face landing as the base?

Applying your knowledge

1. The discovery of the elementary particles from which the atom is made was not a single event. Developing an understanding of the atom started centuries ago and continues today. Pick one of the particles—the electron, proton, or neutron—and write a brief history of its discovery, including the scientists involved.

2. What do the acronyms PRK, LASEK, and LASIK represent? Compare and contrast the three procedures. What type of laser is used? Why?

3. Radioactive materials are used to provide electrical energy. Research other uses for radiation in the home, science, industry, agriculture, and law enforcement.

UNIT 9 THE ATOM

CHAPTER 29

Nuclear Reactions and Radiation

Objectives:

By the end of this chapter you should be able to:

✓ Describe the causes and types of radioactivity.
✓ Use energy concepts to explain why radioactivity occurs.
✓ Use the concept of half-life to predict the decay of a radioactive isotope.
✓ Write the equation for a simple nuclear reaction.
✓ Describe the processes of fission and fusion.
✓ Describe the difference between ionizing and non-ionizing radiation.
✓ Use the graph of energy versus atomic number to determine whether a nuclear reaction uses or releases energy.

Key Questions:

- Why does radioactivity occur?
- What types of radiation exist in the environment?
- How is a fusion reaction different from a fission reaction?
- How are nuclear and chemical reactions different?

Vocabulary

alpha decay	detector	fusion reaction	ionizing	nuclear waste
antimatter	dose	gamma decay	neutrino	radiation
background radiation	energy barrier	Geiger counter	non-ionizing	radioactive decay
beta decay	fallout	half-life	nuclear chain reaction	rem
CAT scan	fission reaction	ionization		

29.1 Radioactivity

You would be very surprised if you saw a bus spontaneously turn itself into two cars and a van. A radioactive atom does something almost as strange. If left alone, a radioactive atom spontaneously turns into other kinds of atoms. The process releases energy and can produce elements that would not otherwise be found in nature, like radon gas. Although radioactivity is a natural process, it can also be created by human technology. This section talks about radioactivity.

The discovery of radioactivity

The discovery of radioactivity The word *radioactivity* was first used by Marie Curie in 1898. A brilliant and tenacious experimenter, Curie noticed that minerals containing uranium gave off some kind of invisible energy that could expose photographic film. She coined the term *radioactivity* to describe the property of certain substances to give off invisible "radiations" that could be detected by photographic plates.

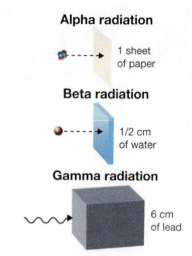

Figure 29.1: *Different thicknesses of different materials are required to stop alpha, beta, and gamma radiation of equal energies.*

Three kinds of radioactivity Many scientists pursued the mystery of radioactivity. They quickly learned that there were three different kinds of **radiation** given off by radioactive materials. The scientists called them "rays" because the radiation carried energy and moved in straight lines, like light rays. Alpha rays came from uranium and could be stopped easily by a thin sheet of material such as paper. Beta rays had more penetrating power. Gamma rays were hardest of all to stop. Many centimeters of a dense material, like lead, were needed to stop gamma rays (Figure 29.1).

Radioactive decay We now know that radioactivity comes from the nucleus of the atom. If the nucleus has too many neutrons, or is unstable for any other reason, the atom undergoes **radioactive decay**. The word *decay* means to "break down," and in radioactive decay, the nucleus breaks down and forms a different nucleus. Almost all elements have some isotopes that are radioactive and other isotopes that are not radioactive. For example, three isotopes of carbon occur naturally. All have six protons in the nucleus but C^{12} has six neutrons, C^{13} has seven neutrons, and C^{14} has eight neutrons. Both C^{12} and C^{13} have stable nuclei and are not radioactive. C^{14} has one too many neutrons and is radioactive. On average, every 5,700 years, half of the unstable C^{14} nuclei decay into stable nitrogen-14 (N^{14}), giving off beta radiation in the process (Figure 29.2). Today, we know that both alpha and beta radiations are particles emitted by the decay of a nucleus.

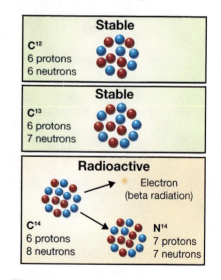

Figure 29.2: *Radioactive carbon-14 (C^{14}) decays into nitrogen-14 (N^{14}) and an electron.*

The three types of radioactivity

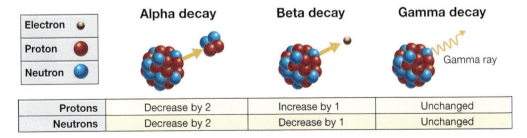

	Alpha decay	Beta decay	Gamma decay
Protons	Decrease by 2	Increase by 1	Unchanged
Neutrons	Decrease by 2	Decrease by 1	Unchanged

Alpha radiation — In **alpha decay**, the nucleus ejects two protons and two neutrons. Check the periodic table and you can see that two protons and two neutrons are the nucleus of a helium-4 (He^4) atom. Alpha radiation is actually fast-moving He^4 nuclei. When alpha decay occurs, the atomic number is reduced by two because two protons are removed. The atomic mass is reduced by four because two neutrons go along with the two protons. For example, uranium-238 undergoes alpha decay to become thorium-234.

Beta radiation — **Beta decay** occurs when a neutron in the nucleus splits into a proton, an electron, and a third particle called a *neutrino*. The proton stays behind in the nucleus, but the high-energy electron is ejected from the nucleus and is the source of beta radiation. This fast electron is still called a beta particle because it is the source of the beta rays observed by Henri Bequerel in 1895. The neutrino is also ejected from the nucleus, but neutrinos are weakly-interacting particles and normally have very little effect on matter. During beta decay, the atomic number increases by one because one new proton is created. The atomic mass stays about the same because neutrons and protons have nearly identical mass.

Gamma radiation — **Gamma decay** is how the nucleus gets rid of excess energy. Gamma decay is not truly a decay reaction in the sense that the nucleus becomes something different. In gamma decay, the nucleus emits a high-energy photon but the number of protons and neutrons stays the same. The nucleus decays from a state of high energy to a state of lower energy. Gamma ray photons are energetic enough to break apart other atoms, making gamma rays dangerous to living things. For this reason, working with gamma rays require heavy shielding. Alpha and beta decay are often accompanied by gamma radiation from the same nucleus.

How a smoke detector works

Smoke detectors contain a tiny amount of americium-241 (Am^{241}), a radioactive isotope that emits alpha radiation. When an alpha particle hits a molecule of air, it knocks off an electron, which ionizes the air. The positive ion and negative electron are collected by positively- and negatively-charged metal plates attached to the battery in the smoke detector. The flow of ions and electrons creates a small electric current that is measured by the electronics of the smoke detector.

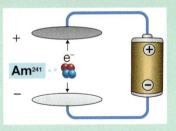

When smoke is in the air, particles of smoke interrupt the flow of ions and electrons. The electric current collected by the metal plates drops. The circuit in the smoke detector senses the drop in current and sounds the alarm.

Radioactive decay releases energy

Energy and radioactivity
Radioactive decay gives off energy. The energy comes from the conversion of mass into energy. If you started with 1 kilogram of C^{14}, it would decay into 0.999988 kg of N^{14}. The difference of 0.012 grams is converted directly into energy via Einstein's formula $E = mc^2$. Because the speed of light (c) is such a large number, a tiny bit of mass generates a huge amount of energy. In fact, the energy released by the decay of one kilogram of C^{14} is equivalent to that released by 27,000 kilograms of gasoline! Fortunately, the decay of C^{14} takes a long time and the energy is released slowly, over thousands of years.

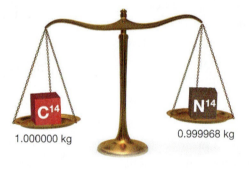

Mass difference
1.000000 kg
− 0.999988 kg
0.000012 kg

Energy released
$E = mc^2$
$= (0.000012 \text{ kg})(3 \times 10^8 \text{ m/s})^2$
$= 1.1 \times 10^{12}$ joules

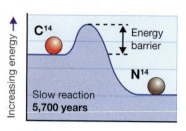

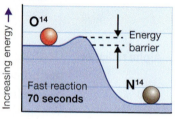

Figure 29.3: *The lower the energy barrier, the more likely the atom is to decay quickly.*

Energy barriers
The radioactive decay of C^{14} does not happen immediately because it takes a small input of energy to start the transformation from C^{14} to N^{14}. The energy needed to start the reaction is called an **energy barrier** (Figure 29.3). The decay of C^{14} is slow because the energy barrier is high compared with the energy of the reaction. The decay of radioactive O^{14} is much faster because the energy barrier is lower compared with the energy of the reaction. Classically, a nucleus should not be able to get over the energy barrier to decay. However, the uncertainty principle of quantum physics allows atoms to violate energy conservation for a short time and get over the barrier to a lower energy. Physicists call this *tunnelling*.

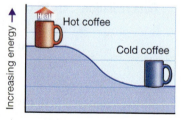

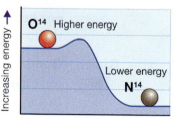

Why radioactivity occurs
Radioactivity occurs because everything in nature tends to move toward lower energy. A ball rolls downhill to the lowest point. A hot cup of coffee cools down. Both are examples of systems that move from higher energy to lower energy over time. The same is true of the nucleus. A radioactive nucleus decays because the neutrons and protons have lower overall energy in the final nucleus than they had in the original one (Figure 29.4).

Figure 29.4: *Systems in nature tend to move toward lower energy. Radioactive decay allows the nucleus to change so as to reach a lower energy.*

The half-life

Chance and radioactivity

Radioactive decay is a process of *chance*. That means it is possible to predict the average behavior of lots of atoms, but impossible to predict what any one particular atom will do. Flipping a coin is a good analogy. You cannot predict whether a specific toss will come up heads or tails. But, you can make a good prediction of the average outcome of 10,000 tosses. Since the chances are 50/50, you could expect 5,000 of the outcomes to be heads and 5,000 to be tails. The larger your sample, the more accurate a prediction you can make about the average. Since even small samples of ordinary materials contain many more then 10,000 atoms, it is possible to accurately predict the average rate of decay.

The half-life

One very useful prediction we can make is the **half-life**. The half-life is the time it takes for one half of the atoms in any sample to decay. For example, the half-life of carbon-14 is about 5,700 years. If you start out with 200 grams of C^{14}, 5,700 years later, only 100 grams will still be C^{14}. The rest will have decayed to nitrogen-14 or N^{14} (Figure 29.5). If you wait another 5,700 years, half of the 100 remaining grams of C^{14} will decay, leaving 50 grams of C^{14} and 150 grams of N^{14}. Wait a third interval of 5,700 years, and you will be down to 25 grams of C^{14}. *One half of the atoms decay during every time interval of one half-life.*

The half-life of different isotopes varies greatly

The half-life of radioactive materials varies greatly. Uranium-238 (U^{238}) has a half-life of 4.5 billion years. It was created in the nuclear reactions of exploding stars, the remains of which condensed to form the solar system. We can still find uranium-238 on Earth because the half-life is so long. The isotope fluorine-18 (F^{18}) has a half-life of 110 minutes. This isotope is used in medicine. Hospitals have to make it when they need it because it decays so quickly. Any natural F^{18} decayed billions of years ago. Carbon-15 has a half-life of 2.3 seconds. Scientists who make C^{15} in the laboratory have to use it immediately.

Radioactive decay series

Most radioactive materials decay in a series of reactions called an *activity series*. For example, radon gas comes from the decay of uranium in the soil (Figure 29.6). Radon itself decays into lead in a chain of three alpha decays and two beta decays. Radon is a source of indoor air pollution in some houses that do not have adequate ventilation. Many people test for radon before buying a house.

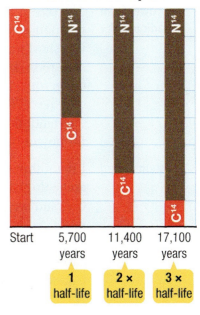

Figure 29.5: *After 5,700 years, half of a 200-gram sample of C^{14} has decayed to N^{14}.*

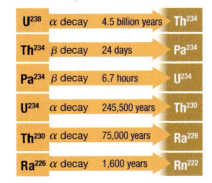

Figure 29.6: *Six decay reactions produce radon-222 (Ra^{222}) from uranium (U^{238}).*

29.1 RADIOACTIVITY

CHAPTER 29 NUCLEAR REACTIONS AND RADIATION

Applications of radioactivity

Power from radioactivity
The power released by radioactive decay depends on the energy released by the reaction and on the half-life. Isotopes with a short half-life give off lots of energy in a short time and can be extremely dangerous. Isotopes with a long half-life such as carbon-14 give off small amounts of power, and usually are much less dangerous. For example, spreading out the decay energy of one kilogram of C^{14} over its 5,700-year half-life yields an average power of about six watts. This is one-tenth the power of a 60-watt light bulb. Many satellites use radioactive decay for power because energy can be produced for a long time without refueling.

Carbon dating
Living things contain a large amount of carbon. The isotope carbon-14 is used by archeologists to determine age. Earth is far older than the 5,700 year half-life of C^{14}, and we only find C^{14} in the environment because it is constantly being produced in the upper atmosphere by cosmic rays. Cosmic rays are high energy particles that come from stars and elsewhere in the universe. The ratio of carbon-14 to carbon-12 in the environment is a constant that is determined by the balance between production and decay of C^{14}. As long as an organism is alive, it constantly exchanges carbon with the environment. The ratio of C^{14} to C^{12} in the organism is the same as in the environment.

Why carbon dating works
When a living organism dies, it stops exchanging carbon with the environment. All the carbon-12 in the organism remains because C^{12} is a stable isotope. Almost no new carbon-14 is created because most cosmic rays do not reach the ground. As the carbon-14 decays, the ratio of C^{14} to C^{12} slowly gets smaller with age. By measuring this ratio, an archeologist can tell how long it has been since the material was alive. Carbon dating works reliably up to about 10 times the half-life, or 57,000 years. After 10 half-lives, there is not enough carbon-14 left to measure accurately. An important limitation is that carbon dating only works on material that has once been living, such as bone or wood.

Determining how much Carbon^{14} is left after five half-lives

A sample of 1,000 grams of the isotope C^{14} is created. The half-life of C^{14} is 5,700 years. How much C^{14} remains after 28,500 years?

1. You are asked for the amount of C^{14} left after 28,500 years.
2. You are given the half-life is 5,700 years.
3. One half the C^{14} decays every half-life.
4. The total, 28,500 years, is 5 times the half-life. The amount of C^{14} is reduced by half every 5,700 years.

Start: 1,000 grams
5,700 years: 500 grams
11,400 years: 250 grams
17,100 years: 125 grams
22,800 years: 62.5 grams
28,500 years: 31.2 grams
Answer: = 31.2 grams

29.2 Radiation

The word *radiation* means the flow of energy through space. There are many forms of radiation. Light, radio waves, microwaves, and X-rays are forms of electromagnetic radiation. The energy in alpha and beta radiation comes from moving particles. Ultrasound is the radiation of very high-frequency sound through different materials, such as the human body. Many people mistakenly think of radiation as only associated with nuclear reactions. This section will explore the topic of radiation and how radiation affects matter.

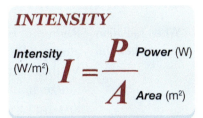

Intensity

Definition of intensity The intensity of radiation measures how much power flows per unit of area. On a clear day, the intensity of sunlight on Earth's surface is about 1,300 watts per square meter (W/m²). A good flashlight produces an intensity of about 100 W/m² in the brightest part of the beam at a distance of one half-meter.

The inverse square law When radiation comes from a single point, the intensity decreases inversely as the square of the distance. This is called the inverse square law, and it applies to all forms of radiation. If you get two times farther away from a radiation source, the intensity is reduced to one fourth of the source value. If you get 10 times farther away, the intensity is reduced to one hundredth of what it was, because one divided by 10^2 is 0.01. Figure 29.7 shows how the inverse square law works.

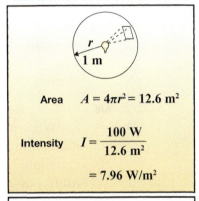

Radiation spreads out as it travels, lowering the intensity.

The inverse square law comes from geometry The inverse square law is a property of geometry. Think of a 100-watt light bulb at the center of a sphere. The intensity at the surface of the sphere is 100 watts divided by the area of the sphere which is $4\pi r^2$. At twice the distance, the area of the sphere is four times greater. Since the same amount of power is spread over a larger area, the intensity goes down by one fourth. Mathematically, the area increases as the square of the radius, so the intensity decreases as the square of the radius. The increase in area is the basis for the inverse square law.

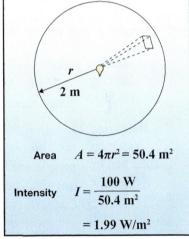

Figure 29.7: *Radiation intensity obeys the inverse square law. The power of a source is spread over a spherical area, as shown.*

When is radiation harmful?

When radiation is harmful

Radiation becomes harmful when it has enough energy to remove electrons from atoms. The process of removing an electron from an atom is called **ionization**. Visible light is not harmful because each photon does not have enough energy to ionize an atom. Visible light is an example of **non-ionizing** radiation, and so is low-frequency ultraviolet light. High-frequency ultraviolet (UV) light is harmful because an ultraviolet photon has enough energy to eject an electron from an atom. UV light is an example of **ionizing** radiation. Figure 29.8 shows the difference between the two categories of radiation. Electromagnetic waves with frequencies higher than UV have enough energy in their photons to ionize electrons in your body and cause damage.

Dangerous effects of ionizing radiation

Ionizing radiation can be harmful because it can break chemical bonds in DNA and other important biological molecules. The human body constantly repairs itself from microscopic damage caused by naturally-occurring, low-intensity radiation. However, too much radiation overwhelms the body's ability to heal itself. For example, too much UV light from the Sun causes sunburn and can damage your eyes. Prolonged exposure to high levels of ionizing radiation has been linked to cancer and other serious diseases.

Limit your exposure to ionizing radiation

Limit your exposure to the Sun's ultraviolet rays to avoid a higher risk of skin cancer. If you ever work with radioactive materials, keep your distance, and use shielding materials, such as lead, to block radiation. Moving three times farther away reduces your exposure by 89 percent because of the inverse square law.

Measuring radiation absorbed by people

Ionizing radiation absorbed by people is measured in a unit called the **rem**. The total amount of radiation received by a person is called a **dose**. Radiation doses in people are measured in rems instead of watts because different kinds of radiation are absorbed differently by body tissues. For example, gamma rays are more dangerous than alpha particles because the skin stops alpha particles. Gamma rays are more dangerous because they can penetrate the skin and reach internal organs. The safe limit for people who work with radiation is five rems per year. One watt of gamma radiation results in a dose of one rem. It takes 20 watts of alpha radiation to produce a dose of one rem. The average person absorbs about 0.3 rems per year from radiation sources in the environment.

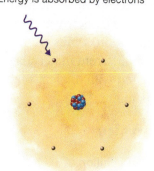

Non-ionizing radiation
Energy is absorbed by electrons

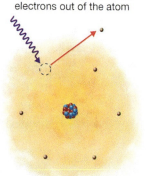

Ionizing radiation
Energy is enough to knock electrons out of the atom

Figure 29.8: *The magnitude of the energy of radiation determines whether it will be ionizing or non-ionizing.*

Sources of radiation

Background radiation Ionizing radiation is a natural part of our environment. There are two chief sources of radiation to which you may be exposed to: **background radiation** and radiation from medical procedures, such as X-rays. Background radiation comes from the environment. Background radiation results in an average dose of 0.3 rems per year for someone living in the United States. Background radiation levels can vary from place to place. They can also vary based on the types of activities you do. Spending time at high altitude tends to increase your exposure to background radiation. Spending time underground where radon gas tends to accumulate from the decay of natural uranium also increases exposure.

Sources of background radiation The chart in Figure 29.9 shows the average breakdown of background radiation in the United States. Uranium and other radioactive elements are found naturally on Earth. The decay of radioactive materials in rocks and soil is a source of natural radiation. Radioactive radon gas is present in the atmosphere in very small quantities. Cosmic rays are another source of radiation. Cosmic rays are high-energy particles that come from outside our solar system. More than 10,000 cosmic ray particles pass through your body every second. Flying in a commercial plane increases your exposure to ionizing cosmic radiation because you are above much of the atmosphere. The atmosphere acts as a shield for this radiation. Human technology also contributes to radiation in the environment. The testing of nuclear weapons since 1945 has distributed a small amount of radioactive material all over Earth. Radioactive material from nuclear weapons is called **fallout**, and contributes about 2 percent of the background radiation.

Radiation from the Sun The Sun is a source of beneficial visible-light radiation and harmful ultraviolet radiation. Many scientists are concerned that the thinning of Earth's ozone layer will allow more harmful UV radiation to reach the planet's surface. Sunbathing also increases your exposure to UV radiation.

Radiation from medical procedures Medical X-rays generate low doses of radiation. A typical chest X-ray produces a dose of 0.02 rem. Comparatively, a single X-ray produces less than 10 percent of the dose you get every year from environmental sources of radiation. That does not mean there is no risk from X-rays. However, the potential health benefit of an X-ray is usually worth whatever small increased risk there might be.

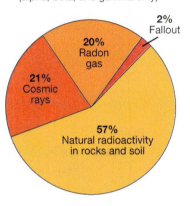

Figure 29.9: *Natural radioactivity is the major source of background radiation in the environment.*

X-ray machines

X-rays and medical applications

X-rays are photons like visible light photons only with higher frequency and much more energy. A typical medical X-ray photon has almost 100,000 times more energy than a photon of red light, and more that 10,000 times more than an ultraviolet photon from the Sun. Unlike UV radiation, which does not have enough energy to get past human skin, X-rays have enough energy to penetrate deeply into the human body. There are two important medical uses for X-rays. *Diagnostic X-rays* make images of the body whereas *therapeutic X-rays* destroy unhealthy cells, such as cancer cells.

Diagnostic X-rays

Diagnostic X-rays are used to produce images of bones and teeth on X-ray film. X-ray film turns black when exposed to X-rays. Images on X-ray film are made by passing X-rays through the body and onto the film. Bones and teeth contain elements of higher atomic numbers, such as calcium. The inner electrons of heavier atoms such as calcium have energy levels comparable to the energy of an X-ray photon. Electrons in calcium atoms are strong absorbers of X-rays because their energy levels match the X-ray photons. By comparison, the soft tissues of the body contain mostly light elements such as carbon, hydrogen, and oxygen. The energy levels of the lighter elements are much lower, so most X-rays pass through them. Bones appear white on X-ray film because the calcium atoms in bones absorb more of the X-ray photons than does the surrounding tissue (Figure 29.10). Modern diagnostic X-ray machines use highly-sensitive film, so only very low doses of X-rays are needed to produce images.

Therapeutic X-rays

Therapeutic X-rays are used to destroy diseased tissue, such as cancer cells (Figure 29.11). Low levels of X-rays do not destroy cells, but high levels do. A typical therapeutic X-ray procedure uses focused multiple beams of X-rays. Each beam by itself is too weak to destroy cells. The beams are made to overlap at the place where the doctor wants to destroy diseased cells. The energy in the overlap region is much higher because of the intensity of the multiple beams. This process can destroy diseased cells without killing adjacent healthy tissue.

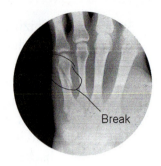

Figure 29.10: *This X-ray shows a broken bone in one finger. Bones are white on X-ray film because they absorb X-rays more than other tissue.*

Figure 29.11: *X-rays are used in therapeutic treatments for cancer.*

CAT scans

CAT scans use X-rays Advances in computer technology have made it possible to use diagnostic X-rays not only to produce pictures on X-ray film, but also to produce three-dimensional images of bones and other structures within the body. The three-dimensional image is called a computerized axial tomography scan, or **CAT scan**.

How a CAT scan is made To produce a CAT scan, a computer controls an X-ray machine as it takes pictures of the body from different angles. Each image is like an ordinary X-ray. The three-dimensional image is constructed by the computer by combining pictures of the same location taken from different angles.

Three-dimensional images from multiple views To see how this process works, imagine you are controlling a robot rover mapping an unknown part of another planet. Your only way to see the planet is with a TV camera mounted on the rover. You stop the rover and take a picture that shows two rocks in the distance. From your camera position, the rocks appear separated by four meters (Figure 29.12). But one view is not enough information to reconstruct a map showing the rocks' location. A second view from a different angle solves the problem. From the second position, the rocks appear 7 meters apart. There is only one way the rocks can be arranged so they appear 4 meters apart in one view and 7 meters apart in a second view, taken from a known position relative to the first. Two views allow you to uniquely determine the positions of the two rocks.

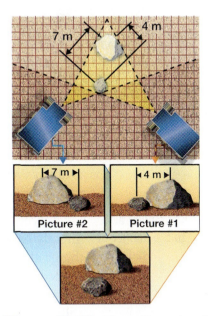

Figure 29.12: *Three-dimensional information can be determined by combining two views of the same objects.*

CAT scans compared with X-rays CAT scans use a similar multi-image principle to the one described for the rover. A single X-ray exposure is like a single TV picture. It shows how far apart the bones and other body parts are, but not how deep in the body they are. By taking multiple X-ray images from multiple angles and combining the image information with a computer application, incredibly detailed, three-dimensional renderings of the body can be produced. X-ray imaging by computer is far more sensitive than X-ray film, so images of virtually all parts of the body can be obtained. For example, CAT scans can show detailed, three-dimensional images of blood vessels in the heart or tumors in the brain, all without surgery (Figure 29.13).

CAT versus MRI Although the machines may look similar from the outside, a CAT scan is very different form an MRI scan. CAT scan technology uses X-rays and is better at imaging hard structures, such as bones. MRI uses magnetic resonance and radio waves, which are non-ionizing radiation and consequently have less health risk.

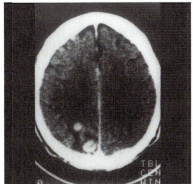

Figure 29.13: *This view of the human brain from a CAT scan is combined with other views to produce a three-dimensional image.*

CHAPTER 29   NUCLEAR REACTIONS AND RADIATION

Measuring radiation

Observing what you cannot see

People who work with radiation use radiation **detectors** to tell when radiation is present and to measure its intensity. You are already familiar with one radiation detector—your own eyes. Unfortunately, ionizing radiation is invisible to the eye. For example, the ultraviolet light that tans your skin on a sunny day is invisible ionizing radiation. You may think the bright light you see is what gives you the suntan or sunburn. In fact, what you see is only the visible portion of sunlight. The ultraviolet portion of sunlight is invisible. Fortunately, UV radiation from the Sun is always accompanied by bright visible light as well. The presence of bright visible light from the Sun tells you the harmful UV radiation is there as well.

Invisibility adds to the danger of radiation

Other types of ionizing radiation cannot be so easily detected by the human body, directly or indirectly. High temperatures are dangerous but people work with hot things all the time because you can feel heat in time to pull away before getting burned. In the experimental development of X-rays, researchers were badly hurt because you cannot feel harmful levels of ionizing radiation. Radiation damage to cells does not become evident for hours or days after the exposure. Thus, the extent of a sunburn is often not known until the day after. One of the great dangers of ionizing radiation is that you cannot tell you are being hurt until it is too late and the damage has been done. People who work around ionizing radiation use many types of detectors to warn them of unsafe conditions.

The Geiger counter

The **Geiger counter** is a type of radiation detector invented to measure X-rays and other ionizing radiation. A Geiger counter is a gas-filled tube with electrodes, which are connected to a power source that creates a voltage of several hundred volts between them (Figure 29.14). Normally, the gas in the tube is neutral and no electric current flows. However, if the tube is exposed to ionizing radiation, positive and negative ions are created in the gas by the radiation. These ions carry electric current, which can be measured. The more radiation that is present, the larger the current becomes. The amount of current in the Geiger counter is a measure of how much radiation is present. Geiger counters are routinely used to measure all types of ionizing radiation.

Figure 29.14: *A Geiger counter detects radiation by electrically collecting ions of gas.*

29.3 Nuclear Reactions and Energy

A nuclear reaction is any process that changes the nucleus of an atom. Radioactive decay is one form of nuclear reaction. Because the nucleus is affected, a nuclear reaction can change one element into another. It is even theoretically possible to create a nuclear reaction that turns lead into gold! The ability to change elements is one important way nuclear reactions are different from chemical reactions. Remember, a chemical reaction can rearrange atoms into different molecules but cannot change atoms of one element into atoms of another element.

Nuclear reactions

Equations for nuclear reactions

Nuclear reactions and chemical reactions are written in a similar way. The mass number of each isotope is written by the element symbol. Individual particles such as electrons, e^-, protons, p^+, and neutrons, n^0, may also appear in a reaction. The example below is a reaction that combines helium and carbon to make oxygen.

A nuclear reaction

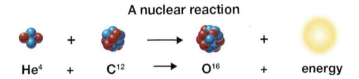

$$He^4 + C^{12} \longrightarrow O^{16} + \text{energy}$$

Mass is not conserved in nuclear reactions

If you could take apart a nucleus and separate all of its protons and neutrons, the separated protons and neutrons would have more mass than the nucleus. This bizarre fact is explained by Einstein's formula $E = mc^2$, which tells us that mass can be converted to energy, and vice versa. The mass of a nucleus is reduced by the energy that is released when the nucleus comes together. Nuclear reactions can convert mass into energy.

Energy is higher in nuclear reactions

Nuclear reactions involve much more energy than chemical reactions. The energy in a nuclear reaction is much greater because nuclear reactions involve the strong nuclear force. Chemical reactions only involve electrical forces between electrons distant from the nucleus. The electrical force acting on an electron far from the nucleus is much smaller than the strong force acting on a proton or neutron in the nucleus itself. The difference in the magnitude of these forces is the reason chemical reactions are much less energetic than nuclear reactions.

Turning lead into gold

For almost 2,000 years, people have sought a way to turn lead and other metals into gold. The many clever schemes people invented were always unsuccessful.

With today's understanding of nuclear physics, it is now possible to make lead into gold. We don't do it because the process is much more expensive than gold itself.

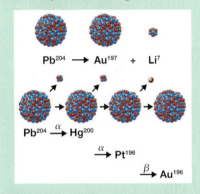

Gold (Au^{197}) has 79 protons and 118 neutrons. The closest stable isotope of lead is lead-204 (Pb^{204}), with 82 protons and 122 neutrons. One possible reaction to turn lead into gold would be to split away a lithium-7 (Li^7) nucleus of three protons and four neutrons. A second possibility would be to make the lead undergo two alpha decays and a beta decay to the isotope Au^{196}. Neither occurs in nature because lead is stable.

The source of energy in nuclear reactions

Energy of the nucleus
When separate protons and neutrons come together in a nucleus, energy is released. Think about balls rolling downhill (Figure 29.15). The balls roll down under the force of gravity and potential energy is released. Protons and neutrons are attracted by the strong nuclear force and also release energy as they come together. The more energy that is released, the lower the energy of the final nucleus. The energy of the nucleus depends on the mass and atomic number. The nucleus with the lowest energy is iron-56 with 26 protons and 30 neutrons. See the graph below. Protons and neutrons assembled into the nuclei of carbon or uranium have higher energy, and are, therefore, higher on the graph.

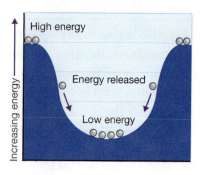

Figure 29.15: *The force of gravity pulls balls down to the lowest point in a valley. Energy is released and reaches its lowest value in the valley.*

Energy of the nucleus vs. atomic number
Relative values with reference to iron-56 (Fe^{56}) with energy set at 0 J/kg

Different nuclei have different energies

Two opposing forces are at work in the nucleus. The electromagnetic force causes protons to repel each other. The strong nuclear force causes protons and neutrons to attract each other. In the iron-56 nucleus, the two effects are balanced. Lighter nuclei are less tightly bound because there are too few neutrons and protons contributing to the strong attractive force. Heavy nuclei are less tightly bound because the repulsion from the protons competes with the attraction from the strong force.

Nuclear energies are very large
The graph compares the energy of the nucleus in 1 kilogram of matter for helium through uranium. Note that the units of energy are hundreds of trillions (10^{12}) of joules per kilogram of material! Nuclear reactions often involve huge amounts of energy as protons and neutrons are rearranged to form different nuclei. A nuclear reaction releases energy when it produces a nucleus with lower energy. A nuclear reaction uses energy when it creates a nucleus that has a higher energy.

Fusion reactions

Fusion reactions
A **fusion reaction** is a nuclear reaction that combines, or fuses, two smaller nuclei into a larger nucleus. Fusion reactions can release energy if the final nucleus is lower on the energy of the nucleus graph (Figure 29.16). For example, a kilogram of carbon-12 contains 104 trillion joules (TJ) of nuclear energy according to the graph. Two carbon atoms have a total of 12 protons and 12 neutrons. These same particles can combine to make one nucleus of magnesium-24 (Mg^{24}). According to the graph, a kilogram of Mg^{24} has 48 TJ of nuclear energy. If the protons and neutrons in a kilogram of carbon are rearranged to form magnesium, about 56 trillion joules of energy is released.

Fusion reactions need very high temperatures
It is difficult to make fusion reactions occur because positively charged nuclei repel each other. The attraction from the strong nuclear force has a very short reach. Two nuclei must get very close for the attractive strong nuclear force to overcome the repulsive electric force. One way to make two nuclei get close is to make the temperature very high. The temperature must be high enough to strip all the electrons from around the nucleus. If temperature is raised even higher, nuclei slam together with enough energy to touch, which allows the strong force to take over and initiate a fusion reaction. The hydrogen fusion reactions in the core of the Sun occur at a temperature of about 15 million degrees Celsius.

Density and fusion power
It takes high density as well as high temperature to make significant energy from fusion reactions. A single fusion reaction makes a lot of energy for a single atom. But a single atom is tiny. To produce enough power to light a single 100-watt bulb requires 10^{14} fusion reactions per second. The density of atoms must be large enough to get a high rate of fusion reactions.

Fusion in the Sun
The Sun and other stars make energy from fusion reactions. The temperature at the core of the Sun is about 15 million degrees Celsius. The density is so high that a tablespoon of material weighs more than a ton. The primary fusion reaction that happens in the Sun combines hydrogen nuclei to make helium, which converts protons and two electrons into two neutrons along the way. All of the energy reaching Earth from the Sun comes ultimately from fusion reactions in the Sun's core.

Energy release by a fusion reaction

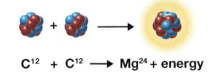

$C^{12} + C^{12} \longrightarrow Mg^{24} + \text{energy}$

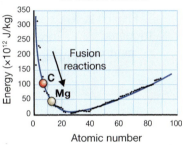

+104 TJ	Energy of carbon (C) nucleus
− 48 TJ	Energy of magnesium (Mg) nucleus
+ 56 TJ	Energy released by fusion of carbon into magnesium

Figure 29.16: *In a fusion reaction, such as the one shown, more energy is released than is used to cause the reaction.*

Fission reactions

Fission reactions
A **fission reaction** splits a large nucleus into smaller pieces. For elements heavier than iron, breaking the nucleus into smaller pieces releases nuclear energy (Figure 29.17). For example, a kilogram of uranium-235 has about 123 trillion joules (TJ) of nuclear energy. A fission reaction splits the uranium nucleus into two pieces. Both pieces have a lower atomic number and are lower on the energy of the nucleus graph. The average energy of the nucleus for a combination of molybdenum-99 (Mo^{99}) and tin-135 (Sn^{135}) is 25 TJ/kg. The fission of a kilogram of U^{235} into Mo^{99} and Sn^{135} releases the difference in energies, or 98 trillion joules. This amount of energy from a golf-ball-sized piece of uranium is enough to drive an average car 19 million miles!

Fission is triggered by neutrons
A fission reaction typically happens when a neutron hits a nucleus with enough energy to make the nucleus unstable. Fission breaks the nucleus into two smaller pieces and often releases one or more extra neutrons. Some of the energy released by the reaction appears as gamma rays and some as kinetic energy of the smaller nuclei and the extra neutrons. For example, a possible fission reaction for uranium-235 (U^{235}) is shown in the diagram in Figure 29.17.

Chain reactions
A **nuclear chain reaction** occurs when the fission of one nucleus triggers fission of many other nuclei. In a chain reaction, the first fission reaction releases two or more neutrons. The two neutrons can hit two other nuclei and cause fission reactions that release four neutrons. The four neutrons hit four new nuclei and cause fission reactions that release eight neutrons. The number of neutrons increases rapidly. The increasing number of neutrons causes more nuclei to have fission reactions and enormous energy is released.

Fission products are radioactive
The small nuclei produced by fission are called *fission products*. Fission products usually have too many neutrons to be stable. For example, Mo^{99} is radioactive and has two radioactive decays before it reaches a stable nucleus of ruthenium-99 (Ru^{99}). First, molybdenum-99 (Mo^{99}) decays to technetium-99 (Tc^{99}), which then decays to ruthenium-99. The decay from Tc^{99} to Ru^{99} has a half-life of 211,000 years. The term **nuclear waste** includes used fuel from nuclear reactors that contains radioactive isotopes such as Mo^{99} that have long half-lives.

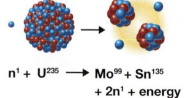

Energy release by a fission reaction

$n^1 + U^{235} \rightarrow Mo^{99} + Sn^{135} + 2n^1 +$ energy

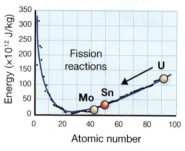

Energy of the nucleus vs. atomic number

+123 TJ	Energy of uranium (U) nucleus
− 25 TJ	Average energy of nuclei of molybdenum (Mo) and Tin (Sn)
+ 98 TJ	Energy released by fission of uranium into Mo and Sn

Figure 29.17: *Fission of uranium releases some energy because elements of lower atomic number are lower on the graph of energy versus atomic number.*

Rules for nuclear reactions

Conservation of energy and momentum
Nuclear reactions obey rules called *conservation laws*. You have already studied two of the conservation laws. The law of conservation of energy applies to nuclear reactions just as it does to any other process in physics. The total amount of energy before the reaction has to equal the total amount of energy after the reaction. The law of conservation of momentum also applies. The total linear and angular momentum before the reaction must be the same after the reaction.

Energy stored as mass
Energy stored as mass must be included in order to apply the law of conservation of energy to a nuclear reaction. The energy of a particle's mass is added to the usual potential and kinetic energy. The energy from mass is given by Einstein's formula $E = mc^2$ where m is the mass of the particle. Even small amounts of mass contain tremendous energy because c^2 is such a large number.

Conservation of charge
Nuclear reactions must conserve electric charge. The total amount of electric charge before the reaction must equal the total electric charge after the reaction.

Conservation of baryons and leptons
There are conservation laws that apply to the type of particles before and after a nuclear reaction. Protons and neutrons belong to a family of particles called *baryons*. Both protons and neutrons have a *baryon number* of one. The total baryon number before and after the reaction must be the same. For the reactions in this book, that means the total number of protons plus neutrons must stay the same before and after the reaction. Electrons come from a family of particles called *leptons*. An electron has a *lepton number* of one. Another conservation law states that the total lepton number must stay the same before and after the reaction.

Conservation laws

Reactants ⟶ Products
$He^4 + C^{12} \longrightarrow O^{16}$ + energy

Mass and energy	The total mass plus energy of reactants equals total mass plus energy of products.
Linear momentum	Total linear momentum of reactants equals total linear momentum of products.
Angular momentum	Total angular momentum of reactants equals total angular momentum of products.
Electric charge	Total electric charge of reactants equals total electric charge of products.
Baryons	Total baryon number of reactants equals total baryon number of products.
Leptons	Total lepton number of reactants equals total lepton number of products.

Using the rules for nuclear reactions

The following nuclear reaction is proposed for combining two atoms of silver to make an atom of gold. This reaction cannot actually happen because it breaks the rules for nuclear reactions. List two rules that are broken by the reaction.

$Ag^{107} + Ag^{107}$ + energy ⟶ Au^{197}

Hint: The atomic number of silver is 47. The atomic number of gold is 79.

1. You are asked for the rules that would be broken if this reaction was to happen.
2. You are given the reaction and the atomic mass of each isotope.
3. Check the rules:

   Two silver nuclei have a total charge of 94 (47 + 47). One gold nucleus has a charge of 79. The reaction violates the rule of conservation of charge.

   Two silver nuclei have a total of 214 protons and neutrons (107 + 107). A gold nucleus has 197 protons and neutrons. The reaction also breaks the rule about the total number of baryons (protons and neutrons).

CHAPTER 29 NUCLEAR REACTIONS AND RADIATION

Antimatter, neutrinos, and other particles

Other particles of matter — The matter you meet in the world ordinarily contains protons, neutrons, and electrons. But, there are other particles of matter besides these three. Cosmic rays contain particles called *muons* and *pions*. Thousands of particles called *neutrinos* from the Sun pass through you unnoticed every second.

Matter and antimatter — Every particle of matter has an **antimatter** "twin." Antimatter is the same as regular matter except its properties like electric charge are reversed. An antiproton is just like a normal proton except it has a negative charge. An antielectron, or *positron*, is like an ordinary electron except that it has positive charge. Some nuclear reactions create antimatter.

Antimatter reactions — When antimatter meets an equal amount of normal matter, both the matter and antimatter are converted to pure energy. For example, if an antiproton meets a normal proton, the two particles are immediately converted to pure energy. Antimatter reactions release thousands of times more energy than ordinary nuclear reactions. If a grain of sand weighing 0.002 kg made of ordinary matter was to collide and react with a grain of sand made from 0.002 kg of antimatter, the resulting event would release 400 trillion J of energy, which is enough to power a large city for almost a week (Figure 29.18).

Neutrinos — In section 1, you read about beta decay, in which a neutron turns into a proton and an electron. When beta decay was first discovered, physicists were surprised to find the total energy of the resulting proton and electron was less than the energy of the disintegrating neutron. Where was the missing energy going? The famous Austrian physicist Wolfgang Pauli proposed that there must be a very light, previously undetected neutral particle that was carrying away the missing energy. We now know the missing particle is a type of **neutrino**. In beta decay, a neutron becomes a proton, an electron, and an *antineutrino* (Figure 29.19). Neutrinos and antineutrinos react only weakly with matter and are therefore extremely difficult to detect. A neutrino can easily pass through Earth without any measurable interaction with any other particle. Despite the difficulty of detection, several carefully-constructed neutrino experiments have detected neutrinos coming from nuclear reactions in the Sun.

Figure 29.18: *A bit of antimatter the size of a grain of sand would release enough energy to power a city if it combined with an equal amount of normal matter.*

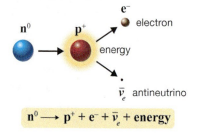

Figure 29.19: *Beta decay of a neutron produces a proton, an electron, and an electrion neutrino, which is one of the three types of neutrinos. The overban on the Greek letter u (ν) means it is an antineutrino.*

Chapter 29 Connection

Nuclear Power

Schematic of a nuclear power plant

In a conventional nuclear power plant, the U^{235} fission reaction is used to generate electricity (Figure 29.20). The process of getting electricity from nuclear reactions takes many steps. First, nuclear reactions in uranium produce heat in the reactor core. The heat is carried by high-pressure hot water into the steam generator. Heat in the steam generator boils water and makes steam. The steam turns a turbine. In the last step, the turbine is connected to an electric generator that makes electricity. The steam is condensed and pumped back to the steam generator.

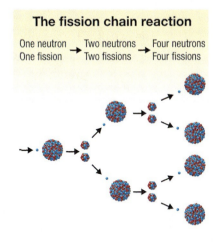

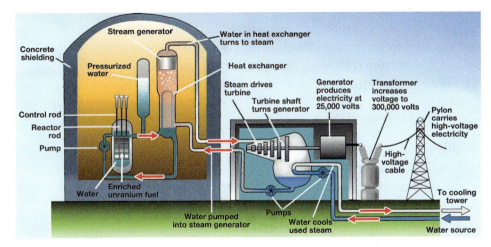

Figure 29.20: *The U^{235} fission chain reaction releases extra neutrons that make more fission reactions.*

Producing uranium fuel

To get enough U^{235} to power a nuclear reactor, naturally-occurring uranium-bearing ore must first be mined. Most natural uranium is the isotope U^{238}, which is less reactive than U^{235}. Uranium must be processed to increase the fraction of U^{235} before it can be used as fuel in a reactor. That process is called *enrichment*.

The reactor core

The reactor core contains long fuel rods of enriched uranium in a heavy pressure vessel filled with water. Between the fuel rods are *control rods*, made from cadmium, an element which absorbs neutrons (Figure 29.21). Neutrons absorbed by the control rods cannot create more fission reactions. The deeper the control rods are placed in the reactor core, the fewer neutrons are available and the slower the chain reaction proceeds. In an emergency, the control rods are released and drop into the reactor, shutting down the chain reaction completely.

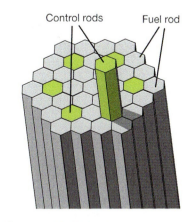

Figure 29.21: *The reactor core contains fuel rods and control rods. The control rods absorb neutrons and slow down the chain reaction.*

29.3 NUCLEAR REACTIONS AND ENERGY

Chapter 29 Connection

Fusion and the future of nuclear energy

The use of nuclear energy Today, about 20 percent of U.S. electric power comes from nuclear fission plants like the one described. Other countries, such as France, use fission power for as much as 70 percent of their electricity. Nuclear fission reactors do not burn fossil fuels such as oil or natural gas. This means no carbon dioxide or other air pollutants are produced. Unfortunately, nuclear fission *does* produce dangerous nuclear waste (see sidebar) and uses technology that can be used to make nuclear weapons. As with many technologies, there is a trade-off between risk and benefit.

Fusion power In the future, nuclear energy may come from *fusion* power instead of fission. Fusion reactions have all the benefits of fission, but generate little or no nuclear waste. The easiest fusion reaction to produce on Earth is the fusion of deuterium and tritium (reaction below). Deuterium (H^2) is an isotope of hydrogen with one neutron. Tritium (H^3) is an isotope of hydrogen with two neutrons.

$$H^2 + H^3 \rightarrow He^4 + n^1 + \text{energy}$$

Fusion fuel comes from seawater The only by-products of this reaction are helium and a neutron. No radioactive nuclei are produced, so there is no dangerous spent nuclear fuel that needs to be isolated from the environment. Fusion fuel is abundant and relatively inexpensive. Deuterium is found naturally in seawater. Tritium does not occur naturally. However, tritium can be produced by combining the neutron from the deuterium-tritium reaction with lithium (reaction below). Lithium is a very common element in Earth's crust. This secondary reaction allows a fusion power plant to *generate* part of its own fuel!

$$Li^6 + n^1 \rightarrow H^3 + H^4 + \text{energy}$$

Fusion power is technically difficult Nuclear fusion is in many ways an ideal energy source since it uses inexpensive, abundant fuel and it generates little harmful waste. However, it has proven very difficult to build a fusion reactor capable of generating electricity. One difficulty is heating the deuterium and tritium fuel to more than 50 million degrees Celsius. Fusion fuel, in the form of a hot plasma, must be contained in a magnetic force field and kept isolated from solid materials. The tokamak reactor (Figure 29.22) is the most successful experimental fusion reactor yet constructed.

Nuclear waste

Generating power from fission creates nuclear waste. When the U^{235} nucleus fissions, many radioactive isotopes are created, some of which remain dangerously radioactive for hundreds or even thousands of years. However, the volume of nuclear waste generated is small. A reactor powering a large city for a year generates spent radioactive fuel with a volume of about 8 m³. However, this waste is dangerous and must be isolated from the environment for a long time. How and where to store this waste is a subject of debate in the scientific community and in society as a whole.

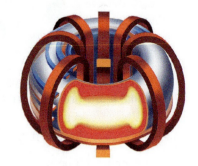

Figure 29.22: *A tokamak fusion reactor confines the hot plasma in a doughnut-shaped chamber. Temperatures in fusion experiments typically reach 80 million °C, five times hotter than the core of the Sun.*

Chapter 29 Assessment

Vocabulary

Select the correct term to complete the sentences.

radioactive	alpha decay	beta decay
gamma decay	radiation	isotopes
radioactive decay	energy barrier	intensity
inverse square law	shielding	fission reaction
CAT scans	ionizing	non-ionizing
ultraviolet	fusion reaction	Geiger counter
rem	nuclear waste	neutrons
antimatter	X-rays	neutrino
background radiation	dose	fallout
detectors	half-life	chain reaction

1. The presence of too many _____ in the nucleus of an atom creates an unstable nucleus.

2. Radioactive decay that occurs when a neutron in the nucleus splits to produce a proton and an electron is called _____.

3. Decay in which a nucleus breaks down to form a different nucleus is known as _____.

4. Atoms of the same element with different numbers of neutrons are called _____.

5. Radioactive decay that emits energy without changing the number of protons or neutrons is called _____.

6. The amount of time required for half of the atoms in a sample to decay is the _____ of the isotope.

7. The energy necessary to start radioactive decay is called a(n) _____.

8. As a result of _____, two protons and two neutrons are ejected from the nucleus.

9. Atoms which spontaneously turn into other kinds of atoms are said to be _____.

10. Energy emitted from the nucleus of spontaneously disintegrating atoms may be called _____.

11. Radiation which is capable of removing an electron from an atom is referred to as _____.

12. Radioactive material produced by the detonation of nuclear weapons is called _____.

13. A unit that is used to measure the ionizing radiation absorbed by the body is the _____.

14. Radiation we are exposed to from environmental sources is known as _____.

15. Sunburn, damage to the eyes, and skin cancer are conditions that may result from exposure to _____ light.

16. The measure of power that flows per unit of area is defined as _____ of radiation.

17. The total amount of radiation received by a person is called a(n) _____.

18. Materials like lead used to block radiation provide _____.

19. The decrease of radiation intensity with the square of the increase in distance from the source is described as a(n) _____.

20. One type of detector developed to measure X-rays and other forms of ionizing radiation is the _____.

21. Photons with enough energy to penetrate deeply into the human body are known as _____.

22. Three-dimensional pictures of the body created using computer-controlled X-ray beams are called _____.

23. Radiation, such as visible light, with insufficient energy to remove an electron from an atom is described as _____.

24. Because ionizing radiation is invisible to the eye, people who work with radiation monitor their exposure with _____.

CHAPTER 29   NUCLEAR REACTIONS AND RADIATION

25. In beta decay, neutrons disintegrate to produce a proton, an electron, and a(n) _____.

26. A nuclear reaction that combines two smaller nuclei into a larger nucleus is a(n) _____.

27. The used fuel from nuclear reactors is generally referred to as _____.

28. Matter which is the same as regular matter, but with some properties like charge reversed is called _____.

29. A nuclear reaction which releases energy as heavy nuclei are split into smaller pieces is a(n) _____.

Concept review

1. Marie Curie referred to radioactivity as an invisible radiation that exposed photographic plates. What causes materials to be radioactive and what part of the atom is affected?

2. Complete the chart for three kinds of radioactivity decay.

Decay	Proton change	Neutron change	Ejected particle	Penetrating ability
Alpha				
Beta				
Gamma				

3. In what ways are energy changes which occur in chemical reactions similar to those which occur in radioactivity? How are they different?

4. Scientists cannot predict when an individual nucleus will decay but can predict when half of the nuclei present will have decayed. Explain how.

5. Carbon-14 can be used to estimate the age of once-living organisms to an age of about 57,000 years. Why is there a limit to the age which can be determined?

6. Light is projected on a screen. As the distance to the screen is increased, what happens to the light's intensity and coverage area?

7. Radiation can take many forms. Why are some forms harmful? Give two examples which are harmful.

8. X-rays are highly-energetic photons. They can be used to destroy unhealthy cancer cells. How is this done without killing healthy cells?

9. A CAT scan can yield three-dimensional pictures of the internal structure of the human body. Explain how this is done.

10. Infrared radiation can be harmful but damage can be avoided by simply moving away from the source. What makes the damage from UV light, X-rays, and other forms of ionizing radiation more difficult to avoid?

11. In a chemical reaction, balanced equations are written using the law of conservation of mass. Can this same law be applied to nuclear reactions? Explain your answer.

12. Why is the energy released from a nuclear reaction so much greater than the energy from a chemical reaction?

13. Fission and fusion are sources of tremendous amounts of energy. Neither is without drawbacks. Compare the "positives" and "negatives" of the two nuclear reactions, fusion and fission, as a means of supplying energy for mankind by naming at least one positive and one negative for each method.

Reaction	Positive	Negative
Fission		
Fusion		

14. Use the graph and the periodic table to number the elements in parts a–f in order of the increasing energy of their nuclei.

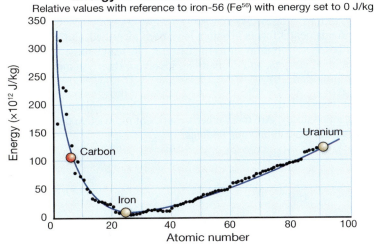

a. carbon
b. iron
c. magnesium
d. lithium
e. lead
f. krypton

15. Describe what occurs if a proton collides with an antiproton.

16. Can S^{34} be created by combining He^3 and P^{31}? If the nuclei named are the only particles involved, explain why this reaction will not take place.

17. Fission and fusion produce energy but differ in the manner in which the energy is released. In what way do they differ?

18. How was the law of conservation of energy responsible for the discovery of the neutrino?

Problems

1. A 1,000-kilogram mass of TNT releases about 4.2×10^9 joules of energy when it explodes. How much of this mass is converted to energy?

2. A proton and an antiproton collide. How much energy does this collision release? (*Note*: The mass of a proton is 1.673×10^{-27} kg.)

3. A substance with a mass of 0.25 kilograms undergoes radioactive decay. What mass of the original isotope will remain after five half-lives?

4. Radon has a half-life of 3.8 days. How long does it take for 16 grams of radon to be reduced to 2 grams of radon?

5. Chris holds a flashlight 1 meter from a wall, illuminating a circle 25 centimeters in diameter. He moves the bulb 3 meters from the wall.
   a. What is the diameter of the new circle of illumination?
   b. Compare the new intensity with the original intensity.

6. A miner is unaware that he was exposed to a 0.005-watt gamma radiation source 0.25 meters from his work station. Upon detection of the radiation exposure, the station is closed down.
   a. What was the intensity of radiation received by this miner?
   b. What was the intensity for a miner 2 meters away?

7. In one year, an individual working at a nuclear power plant receives 35 watts of alpha radiation and 2 watts of gamma radiation.
   a. What is her dose of radiation for the year?
   b. Should she be permitted to return to work?

8. Using the graph of nucleus energy versus atomic number found in the chapter and your knowledge of nuclear reactions, indicate which pairs of nuclei would be most likely to release energy by fission and which would release energy by fusion.
   a. He^4 and C^{12}
   b. U^{235} and Sn^{135}
   c. C^{12} and C^{12}

9. Identify the type of each reaction.
   a. $Ca^{66} \rightarrow Zn^{66} + e^{13}$
   b. $Pb^{204} \rightarrow Rn^{222} + He^{4}$
   c. $n^{1} \rightarrow H^{1} + e^{0}$

10. The graph shows the amount of nitrogen-13 remaining as it undergoes radioactive decay.

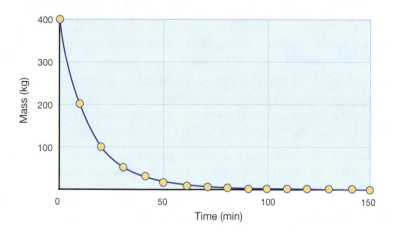

   a. How much nitrogen-13 remains after 40 minutes?
   b. How long is one half-life for nitrogen-13?

Applying your knowledge

1. Are protons, neutrons, and electrons the true "building blocks" of the atom or are there smaller fundamental particles? Write a brief report discussing what you can discover about these particles.

2. Four fundamental forces in nature have been identified. Current theory describes these forces in terms of "carriers." Research these fundamental forces, their carriers, and the particles involved. Briefly summarize the information.

3. Three types of radioactive decay are commonly recognized—alpha, beta, and gamma decay. Are there other varieties of radioactive decay? Write a brief report describing your findings.

UNIT 9 THE ATOM

CHAPTER 30

Frontiers in Physics

The universe is a wonderful and complex place. We are far from understanding how nature works in all of its detail. This chapter presents several interesting ideas in physics that are areas of active research. The discussion is a broad overview only, touching on a few interesting points. Many entire books have been written on each of the ideas discussed on each page of this chapter. Scientists are doing experiments and debating with each other whether these ideas are the right ones to describe nature, or not. Only more research will tell. As you read this chapter, do not worry about understanding all of what you read. Even scientists who work with these ideas do not completely understand all of them. The purpose of the chapter is to give you a sense of how truly interesting and surprising the universe is, and also to encourage you to further explore some of the concepts discussed here.

Vocabulary

baryon	dark matter	general relativity	meson
big bang	elementary particles	gravitational mass	quark
black hole	equivalence principle	inertial mass	red shift
boson	escape velocity	lepton	reference frame
curved space	event horizon		

637

30.1 The Origin of the Universe

The universe is, well, *everything*. All matter and all energy are part of the universe. It seems very strange to think the universe *had* a beginning, partly because it is such a huge stretch of the mind to imagine the entire universe itself. Like the rest of this chapter, a book written 50 years from now may present a different story from the ideas in the next few pages. But today that seems improbable. The evidence for the origin of the universe is very strong.

The expanding universe

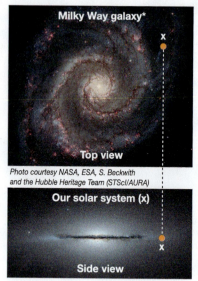

Figure 30.1: *The Milky Way is a typical spiral galaxy, which is one of several known types of galaxies.*

Milky Way galaxy — When we look out into space with powerful telescopes, we see the stars of our own Milky Way galaxy. The Milky Way contains about 200 billion stars in a spiral galaxy (Figure 30.1). Some of the stars are like the Sun. Some are hotter, some are cooler, some older, some younger. Just recently we have discovered that many stars have planets like the planets in our solar system.

The universe contains billions of galaxies — Beyond our galaxy, with its 200 billion stars, are other galaxies. We can see billions of galaxies, of which our own Milky Way is just one example. Some galaxies are much larger, some are smaller. This is the universe on its largest scale.

The laws of physics seem to be the same — In the chapter about atoms, you learned that each element has a characteristic *spectrum*. When we look at the light from distant galaxies, we see the spectra of familiar elements. That tells us that other galaxies also have hydrogen, helium, lithium, oxygen, and the same elements as our own galaxy. Because the lines have the same pattern, it also tells us that the laws of physics are the same.

Red shift — The spectra of each element observed in a distant galaxy has the same pattern as here on Earth, *except* shifted toward the red or longer wavelengths. The appearance of spectral-line patterns at longer wavelengths is called a **red shift**. For example, the spectrum of helium from one distant galaxy is red-shifted 30 nanometers compared to the spectra of helium from the Sun (Figure 30.2). The spectra are red shifted *because galaxies are moving away from each other*. The Doppler effect and special relativity cause light to be red shifted if the source of light and the observer are moving away from each other at speeds that are of the same order of magnitude as the speed of light. Galaxies that are farther away have a larger red shift so they are moving away even faster. *The universe is expanding.*

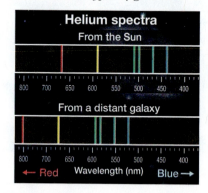

Figure 30.2: *Light from distant galaxies show spectra that are red-shifted. Notice the same pattern of spectral lines from helium is at a longer wavelength in light from a distant galaxy.*

The big bang

An expanding universe implies a beginning
If the universe is expanding, then it must have been smaller in the past. It seems reasonable to ask how small was the early universe, and how long has it been since the universe was small. The best evidence indicates that the age of the universe is 13 billion years, plus or minus a few billion years. This is roughly four times older than the age of the Sun. This range of ages for the universe agrees with other estimates, such as the ages of the stars we can see.

The big bang
It also appears that the universe was once very small, possibly smaller than a single atom. Thirteen billion years ago, a cataclysmic explosion occurred and the universe started growing from a tiny point into the incredible vastness we now observe. In jest, someone called this beginning the "**big bang**" and the name stuck. We have no idea why the universe came into existence or what came before the big bang. It is not clear these questions can even be answered by science.

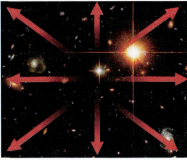

Figure 30.3: *The observed expansion of the universe is strong evidence for the big bang theory.*

Evidence for the big bang
We do see evidence for the big bang itself. The fact that galaxies are expanding away from each other is a strong argument for the big bang. As far as we can look into the universe, we find galaxies are expanding away from each other (Figure 30.3). We do not see galaxies coming toward each other.

The cosmic background radiation
When you light a match, the flame bursts rapidly from the first spark and then cools as it expands. When the big bang exploded, it also created hot radiation. This radiation has been expanding and cooling for 13 billion years. The radiation is now at a temperature only 2.7 K above absolute zero throughout the universe. This *cosmic background radiation* is evidence of the big bang (Figure 30.4).

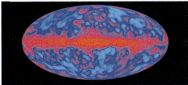

Figure 30.4: *The COBE satellite measured this image of the cosmic background radiation. (NASA)*

Ratios of the elements
We have other evidence that supports the big bang theory. The proportion of hydrogen to helium is consistent with the physics of the big bang (Figure 30.5). Elements heavier than hydrogen and helium are formed in stars. When stars reach the end of their life cycle, they spread heavy elements such as carbon, oxygen, and iron out into the universe. If the universe was significantly older, there would be more heavy elements present than just hydrogen and helium.

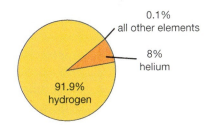

Figure 30.5: *The universe is mostly hydrogen with a small amounts of helium and other elements.*

A one-page history of the universe

The first second — Immediately after the big bang, the universe was a hot, dense fireball of expanding energy (Figure 30.6). At an age of 0.01 seconds, the estimated temperature of the fireball was 100 billion degrees Kelvin (10^{11} K). This temperature was so high that protons could not stay together. The universe consisted of exotic particles and photons.

Protons and neutrons form at 4 minutes — As the universe expanded, it cooled as its energy spread out over a larger volume. About four minutes after the big bang, the universe had expanded and cooled enough that protons and neutrons could stick together to form the nuclei of atoms. Because atoms were still flying around with high energy, heavy nuclei were broken apart immediately. In fact, only 1 helium atom survived for every 12 hydrogen atoms. Almost no elements heavier than helium were created. When we look at the matter in the universe today, we see this ratio of hydrogen to helium left by the big bang, with the exception of elements formed much later in stars.

Matter and light decouple in 700,000 years — For the next 700,000 years, the expanding universe was like the inside of a star—hot ionized hydrogen and helium. At the age of 700,000 years, the universe had expanded enough to become transparent to light. At this point, the light from the fireball was freed from constant interaction with hot matter. The light separated from matter and became the cosmic background radiation we can detect today.

Stars and galaxies form — When the universe was about 1 billion years old, it had expanded and cooled enough for galaxies and stars to form. At this point, the universe probably began to look similar to how it looks today. The Sun and solar system formed about 4 billion years ago, by which time the universe was 12 billion years old.

Unresolved questions — While scientists feel relatively confident about the overall "universal" picture, they are not confident about the details. There are many puzzling observations yet to be fully explained. From the gravitational behavior of galaxies, astronomers believe that as much as 90 percent of the matter in the universe is invisible "**dark matter**." The identity of dark matter is unknown. There are recent observations that suggest the expansion of the universe is accelerating. This is a puzzle because, if anything, the expansion should be decelerating as the combined effect of gravity from the matter in the universe slows down the expansion. Acceleration implies a pressure from an unknown "dark energy."

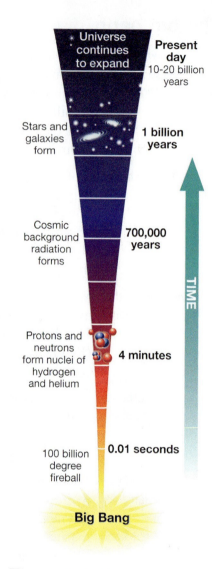

Figure 30.6: *The history of the universe spans over 16 billion years.*

The future of the universe

Two different futures
Our current theories suggest two very different scenarios for the future of the universe (Figure 30.7). We do not know which one is the correct description. It is possible the correct answer is neither of these alternative views. This is an area of active research in the field of *cosmology*.

Continued expansion and cooling
The universe could continue to expand and cool. This option is called the *open universe*. The Sun will continue to shine pretty much as it is for another 4 billion years. Eventually it will become a white dwarf star, which is a small hot ball of electrons, protons, and neutrons. Over time, the white dwarf will cool by radiation and become relatively cold. After 100 billion years, a similar scenario will happen to all of the stars in the universe. The universe becomes a much-less energetic place than it is today. Eventually, only radiation will remain.

Reversal followed by Big Crunch
If there is enough matter in the universe, the combined effect of gravity will eventually slow and reverse the expansion. This option is called the *closed universe*. When the expansion stops, the universe would then start collapsing on itself. Over many billions of years, the universe would slowly heat up again. Eventually, the universe might collapse to a small point again in a "big crunch." Some cosmologists believe that the universe is a repeating cycle of big bang and big crunch.

Density and dark matter
Which of the two options lies in the future depends on how much mass there is in the universe. This is something we are trying hard to find out. The amount of matter in stars and galaxies is what we can see because this matter gives off light. This "light" matter is less than five percent of the matter needed to "close" the universe. Astronomers are confident that dark matter exists because of its gravitational effects on matter we can see. For example, you could calculate the mass of Earth by observing the orbit of the Moon. We know from the motion of galaxies that there is much more dark matter than light matter in stars. The total mass of dark matter appears to exceed that of light matter, but is still not enough to close the universe. New research has discovered a mass for neutrinos, and the possibility of forms of "dark energy." Whether the universe is open or closed is not known.

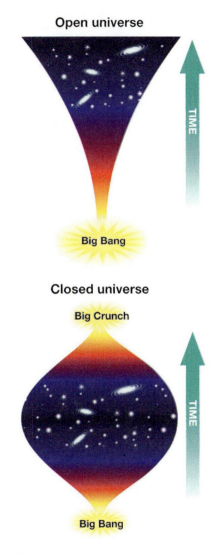

Figure 30.7: *Two different scenarios for the evolution of the universe by physicists.*

30.2 Gravity and General Relativity

Einstein's theory of **general relativity** describes gravity in a fundamentally different way than Newton's law of universal gravitation. According to Einstein, the presence of mass changes the shape of space-time itself. In general relativity, an object in orbit is moving in a straight line through **curved space**. The curvature of space itself causes a planet to move in an orbit. The force we call gravity is an effect created by the curvature of space and time.

Inertial mass and gravitational mass

The two interpretations for g In Chapter 6, we discussed two interpretations for g. One interpretation is that g represents the strength of the gravitational field, or 9.8 newtons per kilogram of mass. The second interpretation is that an object in free fall experiences an acceleration of 9.8 m/s². We argued that both interpretations were identical because the mass, m, that appears in Newton's second law is the same mass, m, that appears in Newton's law of universal gravitation (Figure 30.8).

Inertial mass The mass that appears in the second law is a measure of an object's inertia. Inertia means resistance to acceleration. This mass is often called **inertial mass** because an object with more mass has more inertia and requires more force to accelerate it.

Gravitational mass The mass that appears in the law of gravitation is a measure of an object's ability to act *through* and be *acted on* by gravity. **Gravitational mass** measures an object's susceptibility to the force of gravity. Two masses attract each other with a force that is proportional to the product of their gravitational masses. This law, called the law of universal gravitation, is analogous to the force that two charges exert on each other that is proportional to the product of the two charges.

The paradox and the solution What is the problem, you say? Is it not the same "m" in both formulas? The problem is that, at first glance, *inertia has nothing to do with gravity*. Why should an object's resistance to acceleration be the exact same property that determines an object's interaction with gravity? Why should it be the same "m" in both formulas? The results of every experiment tells us that *the "m" is the same*. But why? The coincidence seemed too perfect for Einstein. His brilliant theory of general relativity showed how inertia and gravity are intimately connected and provided a beautiful explanation for why inertial mass and gravitational mass should be the same.

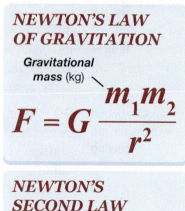

NEWTON'S LAW OF GRAVITATION

$$F = G \frac{m_1 m_2}{r^2}$$

Gravitational mass (kg)

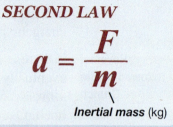

NEWTON'S SECOND LAW

$$a = \frac{F}{m}$$

Inertial mass (kg)

The paradox
Why does the same property of matter determine both the effect of gravity and the resistance to acceleration?

Figure 30.8: *General relativity helped resolve the paradox between gravitational and inertial mass.*

The equivalence principle

Different perspectives on the same motion

Consider the example of a boy and a girl who jump into a bottomless canyon, where there is no air friction. On the way down, they play catch and throw a ball back and forth. If the girl looks at the boy, she sees the ball go straight to him. If the boy looks at the girl, he sees the ball go straight to her. However, an observer at rest watching them fall sees the ball follow a curved zigzag path back and forth (Figure 30.9). Who is correct? What is the real path of the ball?

The boy and girl perceive no gravity

Both are correct. Imagine enclosing the boy and girl in a windowless box falling with them. From inside their box, they see the ball go straight back and forth. To the boy and girl in the box, the ball follows the exact same path it would *if there were no gravity*. Remember, gravity would cause the ball to follow a curved path.

The reverse situation

Next, imagine the boy and girl are in the same box throwing the ball back and forth in deep space *where there is no gravity*. This time, assume that the box is accelerating upwards. When the boy throws the ball to the girl, the ball does not go straight to her, but drops in a parabola toward the floor. This happens because the floor is accelerating upward and pushing the girl with it. The girl moves up while the ball is moving toward her. But, from her perspective, she sees the ball go down. If the girl calculates the path of the ball, she finds it to be a parabola *exactly like what it would follow if there was a force of gravity pulling it downward*.

Reference frames and the equivalence principle

In physics, the box containing the boy and girl is called a **reference frame**. Everything they can do, measure, or see is inside their reference frame. The **equivalence principle** says that *no experiment the boy or girl do can distinguish whether they are feeling the force of gravity or they are in a reference frame that is accelerating*. The only way to tell is to look outside the box.

The meaning of the equivalence principle

An experiment done in a reference frame at rest on Earth's surface finds a gravitational force of 9.8 N/kg. Suppose a person does the same experiment in a spaceship in deep space that is accelerating at 9.8 m/s^2. The experiment *also finds a gravitational force of 9.8 N/kg, even though there is no gravity*. According to the equivalence principle, *any* result from *any* experiment is exactly the same whether the experiment is done in a place with a gravitational force of 9.8 N/kg or in a reference frame that is accelerating at 9.8 m/s^2.

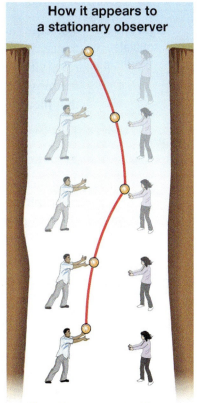

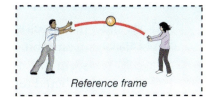

Figure 30.9: *The ball's path appears different to an outside observer than it does in the accelerating reference frame.*

CHAPTER 30 — FRONTIERS IN PHYSICS

Curved space-time

Light and the equivalence principle
The equivalence principle applies to any experiment, even an experiment that measures the path taken by light. In Chapter 18, we discussed the theory of special relativity, which says the speed of light is the same for all observers, whether they are moving or not. In order to make the equivalence principal true for experiments that measure the speed of light, two things must be true.

1. Space itself must be curved.
2. The path of light must be deflected by gravity, even though light has no mass.

Flat space
To understand what we mean by curved space, consider rolling a ball along a sheet of graph paper. If the graph paper is flat, the ball rolls along a straight line. A flat sheet of graph paper is like "flat space." In flat space, parallel lines never meet, all three angles of the triangle add up to 180 degrees, etc. Flat space is what you would consider "normal."

Curved space
The presence of a large mass, such as a star, creates curved space in the region close to the star. Figure 30.10 shows an example of a graph paper made of rubber which has been stretched down in one point. If you roll a ball along this graph paper, it bends as it rolls near the "well" created by the stretch. From directly overhead, the graph paper still looks flat. If you look straight down on the graph paper, the path of the ball appears to be deflected by a force pulling it toward the point at which it has been stretched. You might say the ball experiences the force of gravity which deflected its motion. And, you would be right. The effect of curved space is identical to the force of gravity.

Orbits and curved space
In fact, close to a source of gravity, straight lines become circles. A planet moving in an orbit is actually moving in a straight line through curved space. This may seem like a strange way to think, but all of the experimental evidence we have gathered tells us it is the *right* way to think.

Proving general relativity
The event that made Einstein famous was his prediction that light from distant stars should be bent by the curvature of space near the Sun. People were skeptical because, according to Newton's law of gravitation, light is not affected by gravity. In 1919, an expedition was launched to see if Einstein was right by observing a star near the Sun during a solar eclipse. Einstein was right.

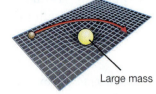

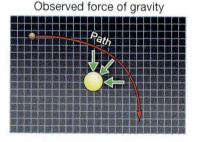

Figure 30.10: *Large amounts of mass cause space to become curved. An object following a straight path in curved space bends the same way as if it were under the influence of gravity.*

Black holes

General relativity predicts black holes

One of the strangest predictions of general relativity is the existence of **black holes**. To understand a black hole, consider a rocket trying to leave Earth. If the rocket does not go fast enough, Earth's gravity eventually pulls it back. The minimum speed a rocket must have to overcome gravity is called its **escape velocity**. The stronger gravity becomes, the higher the escape velocity.

The escape velocity of a black hole

If gravity becomes strong enough, the escape velocity can reach the speed of light. A *black hole* is an object with such strong gravity that its escape velocity equals or would theoretically exceed the speed of light. When the escape velocity equals the speed of light, nothing can get out because nothing can go faster than light. In fact, even light cannot get out, because in general relativity, light is affected by gravity. The name *black hole* comes from the fact that anything that falls in never comes out. Since no light can get out, the object is "black" (Figure 30.12).

Black holes are extremely compact matter

To make a black hole, a very large mass must be squeezed into a very tiny space. For example, to make Earth into a black hole, you would have to squeeze the mass of the entire planet down to the size of a marble as wide as your thumb. For a long time, nobody took black holes seriously because they seemed so extreme that they could never actually be real.

We see black holes by what is around them

But then astronomers started finding them. You might think it would be impossible to see a black hole—and it is. But, you *can* see what happens *around* a black hole. In Chapter 10, you learned that an object loses potential energy as it falls. When an object falls into a black hole, it loses so much energy that a significant fraction of its mass turns into energy. Any matter that falls into a black hole gives off so much energy it creates incredibly "bright" radiation as it spirals into the black hole. Astronomers believe the core of our own Milky Way galaxy contains a black hole with a mass of more than 1 million times the mass of the Sun.

The event horizon

Black holes have very strange properties. The **event horizon** is the sphere around a black hole with a diameter at which the escape velocity just equals the speed of light. Once something crosses the event horizon, it is lost to our universe forever. Special relativity (Chapter 18) made a connection between time and the speed of light. Because of this connection, *time slows to a stop at the event horizon.*

Earth image courtesy NASA

Figure 30.11: *On Earth, light travels in nearly straight lines because light's escape velocity from Earth is much less than the speed of light.*

Figure 30.12: *Light from a black hole cannot escape because its escape velocity is higher than the speed of light.*

30.3 The Standard Model

Around the middle of the 20th century, the three-particle atomic model that includes protons, electrons, and neutrons was very successful. Some people made half-serious jokes about physics having no purpose because "everything has been discovered." However, just as scientists were getting confident, a particle was discovered with a negative charge like the electron, but 200 times the mass of an electron. More particles were discovered, and then more, and soon there were more than 100 elementary particles. A common goal of science is to explain complexity with a few basic principles. The proton, electron, and neutron were able to explain the elements. Now, physicists started looking for a similar way to simplify what they began to call the *particle zoo*.

A walk through the particle zoo

Sorting particles into categories
When you walk through a zoo, many animals are sorted by shared characteristics. Mammals are warm-blooded, have fur, and bear live young; birds have feathers; and reptiles lay eggs and are cold-blooded. Particles also fall into categories, although the properties we use to sort them are not as familiar as fur or feathers. Table 30.1 shows a partial list of **elementary particles** sorted into four basic types. Notice that some come in three varieties with positive, negative, and zero electric charge. The sigma particles (Σ^+, Σ^-, Σ^0) are of this type.

Baryons and mesons
Protons and neutrons are examples of **baryons**. Baryons are relatively heavy particles, with their mass greater or equal to one proton. Each baryon is made of three even smaller objects called *quarks*. The pion particle is a **meson**. Mesons are made of two quarks. Both baryons and mesons respond to the strong nuclear force. Remember, the strong nuclear force holds the nucleus of an atom together by creating an attraction between neutrons and protons.

Leptons
The electron is a **lepton**. Leptons are particles that have no internal structure. That means leptons are not made from anything else and are truly elementary by themselves. The lightest leptons are called *neutrinos*. There are three kinds of neutrinos, one for each of the other types of leptons. Like the electron, the muon and tau particles have negative charge.

Bosons
The last group consists of the **bosons**. Bosons are particles that carry *forces*. The particle of light we met in Chapter 16, the photon, is a boson. The first two bosons have zero mass. The next two have a relatively large mass.

Table 30.1: A partial list of elementary particles

Name	Symbol	Mass (amu)*
Baryons		
Neutron	n	1.009
Proton	p	1.007
Lambda	Λ	1.20
Sigma	$\Sigma^+ \Sigma^- \Sigma^0$	1.28
Omega	Ω	1.80
Mesons		
Pion	$\pi^+ \pi^- \pi^0$	0.15
Kaon	$K^+ K^- K^0$	0.53
D	$D^+ D^- D^0$	2.0
J/Psi	J/ψ	3.3
B	$B^+ B^- B^0$	5.7
Leptons		
Neutrino	$\nu_e \, \nu_\tau \, \nu_\mu$	$<5 \times 10^{-11}$
Electron	e	5.5×10^{-4}
Muon	μ	0.11
Tau	τ	1.9
Bosons		
Photon	γ	0.0
Gluon	g	0.0
W	$W^+ W^-$	86.0
Z	Z	98.0

Note: Masses are average for families of particles.

Matter and antimatter

The quantum equations include antimatter

An ordinary proton has a positive charge, and a mass of 1 atomic mass unit (amu). An ordinary electron has a negative charge equal and opposite to the charge of the proton. In 1928, Paul Dirac (1902–1984), an English physicist, published a paper that incorporated Einstein's theory of special relativity into quantum mechanics. When you solve any equation with a square root, there are two solutions, one positive and one negative. For example, $\sqrt{4} = \pm 2$. Dirac found two solutions to a quantum equation for the electron. The negative solution corresponds to antimatter. Antimatter is matter that has all the same properties as "regular" matter except the electric charge is reversed, along with other "charge-like" properties.

Antimatter has the opposite charge

The *positron*, or antielectron, has the same mass as a normal electron but a positive charge. An *antiproton* has the mass of an ordinary proton, but a negative charge. As each new particle was discovered, each also turned out to have an antiparticle. For example, the antiparticle of the positive pion (π^+) is the negative pion (π^-). The pion with zero charge (π^0) is its own antiparticle because it has no electric charge.

Reactions between matter and antimatter

When ordinary matter and antimatter collide, both particles are turned into pure energy. The amount of energy is given by Einstein's formula $E = mc^2$, where m is the combined mass of the particles.

Charge-like properties

The discussion of antiparticles gets a little more complex because *particles have other "charge-like" properties besides electric charge*. An example of a property of particles that behaves like charge is the *baryon number*. A proton has a baryon number of +1 and an antiproton has a baryon number of −1. These other charge-like properties obey conservation laws, like electric charge. For example, the total baryon number before a reaction must equal the total baryon number after the reaction. Particles and antiparticles created from pure energy are usually created in pairs to conserve charge-like properties.

Symbols for antimatter

Sometimes an antiparticle is represented with a bar over the symbol. For example, an electron neutrino has the symbol v_e. The antielectron neutrino has the symbol $\bar{v}_e$. Similarly, a proton is represented by p and an antiproton by $\bar{p}$. The exception to this rule is when the charge of the particle is already included in the symbol, such as with the pions.

Energy from antimatter

Matter-antimatter reactions are the most energetic reactions in the known universe. For example, if you could take one barrel of water and add it to one barrel of antiwater, the energy released would be enough to power the entire United States for a year! That means every light bulb, every car, every factory, every furnace, and every type of energy used for any purpose.

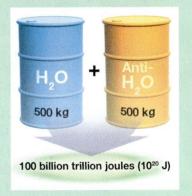

Of course, you could never contain the antimatter enough for it to react in a controlled fashion. Today, scientists work with antimatter in the most minute quantities, not much more than a few particles at a time. There is no permanent antimatter occurring in the natural universe, as far as we know. Any antimatter used in a laboratory must be created from pure energy.

The standard model of particles

Protons and neutrons are made of quarks

At first, people believed atoms were the smallest particles of matter. Then it was discovered that atoms have an internal *structure*, and were themselves made of protons, neutrons, and electrons. As physicists conducted more sophisticated experiments, they discovered that the *proton* and *neutron* also have an internal structure. Today, we believe only leptons, such as the electron, and bosons, such as the photon, are truly elementary particles. The heavier particles, baryons and mesons, are made of two or three simpler particles called **quarks**.

Six quarks and six antiquarks

There are six quarks (Figure 30.13) and each has an antiquark, making a total of 12 types of quark. The up, charm, and top quarks have a charge of +2/3. The down, strange, and bottom quarks have a charge of −1/3. A proton is made of two up quarks and a down quark, written "uud." The charge of the proton is +1, or +2/3 + 2/3 − 1/3. A neutron is made of one up quark and two down quarks and has a charge of 0, or +2/3 − 1/3 − 1/3. All of the baryon-type particles are made of three quarks. All of the mesons are made of quark-antiquark pairs (Figure 30.14). For example, the positive pion is made from an up quark and an anti-down quark. Like other antiparticles, antiquarks are written with a bar over their symbol.

The three families

The quarks and leptons are grouped into three families. The four lightest particles are in the first family, including the electron, electron neutrino, up quark, and down quark. Almost all of the matter we see is made from the first family because these are the lightest particles and therefore have the lowest energy. We only see the second and third families in very high-energy situations. Such situations include particle-accelerator experiments, cosmic rays, the environment of black holes, and exotic forms of matter found in stellar remnants like neutron stars.

We do not see isolated quarks

Quarks are never found by themselves. The force that binds quarks to each other is extremely strong and it increases with distance instead of decreasing. If you try to pull the two quarks in a meson apart, you have to do work and add energy. It takes so much energy to pull the quarks apart that there is enough energy to create *two new quarks*. Remember, according to Einstein's formula, you can create matter from energy according to the formula $m = E \div c^2$. The new quarks pair up with the ones you are trying to pull apart, creating a new meson out of the energy. As a result, you still do not have a quark by itself.

Leptons		Quarks	
Third family			
ν_τ	τ	t	b
Tau neutrino	Tau	Top quark	Bottom quark
Second family			
ν_μ	μ	c	s
Muon neutrino	Muon	Charm quark	Strange quark
First family			
ν_e	e	u	d
Electron neutrino	Electron	Up quark	Down quark
Electric charge			
0	−1	$+\frac{2}{3}$	$-\frac{1}{3}$

Figure 30.13: *The standard model for particles of matter.*

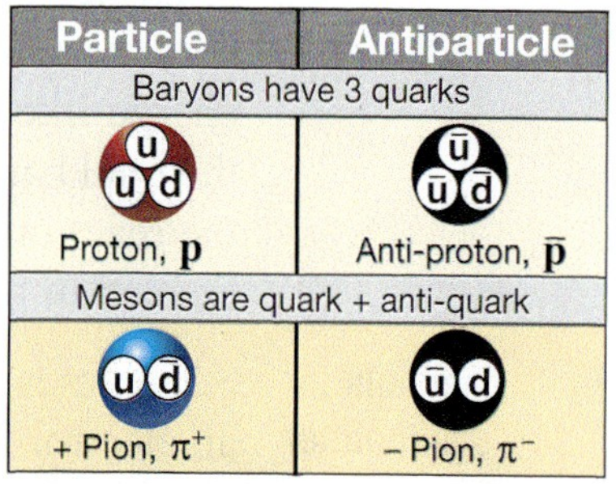

Particle	Antiparticle
Baryons have 3 quarks	
Proton, p (uud)	Anti-proton, $\bar{p}$ ($\bar{u}\bar{u}\bar{d}$)
Mesons are quark + anti-quark	
+ Pion, π^+ (u$\bar{d}$)	− Pion, π^- ($\bar{u}$d)

Figure 30.14: *The quark structure for baryons and mesons.*

The standard model of forces

The four forces in nature

We see four forces in nature. The interactions of matter and energy occur through the four forces. In order from strongest to weakest, the four forces are as follows.

1. The strong nuclear force, reference strength = 10.
2. The electromagnetic force, relative strength = 0.0073.
3. The weak nuclear force, relative strength = 10^{-7}.
4. The gravitational force, relative strength = 10^{-45}.

In Chapter 21, we learned about fields and forces. A field is an area of space that contains energy that can exert forces on particles. The quantum theory of fields requires that energy be quantized. You can think of a quantum of energy as a *particle*. We have already discussed the first field quantum, the photon.

Every field has an associated particle

All of the four forces in nature are transmitted through fields. Each of the fields has a particle associated with it. The particle that transmits the electric and magnetic field is the photon.

How a particle transmits force

To see how particles transmit force, consider two people on skateboards (Figure 30.15). One person throws a bowling ball to the other (A). The ball carries momentum away from the thrower (B), and gives the momentum to the receiver (C). The thrower moves to the left and the receiver moves to the right. If the ball was invisible, you might think a force of repulsion existed between the two people (D). Particles carry forces between other particles in a similar way.

The strong and weak nuclear forces

The strong nuclear force is transmitted by *gluons*, which also carry the force between quarks. The weak nuclear force is transmitted by the W and Z bosons. Figure 30.16 shows the particles that carry the four forces in the standard model.

Gravity is an unsolved mystery

The standard model includes a *graviton* to carry the gravitational force. However, no one has ever observed a graviton and the incomplete understanding of how gravitons interact with other particles is one of the biggest unresolved problems with the standard model. We know the standard model is an incomplete theory partly because it does not include gravity.

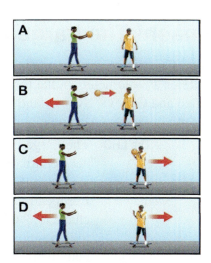

Figure 30.15: *Two people on skateboards throwing a ball back and forth. The motions of the people make it appear as if they are exerting a repulsive force on each other.*

Force carrying particles	
Gluons	Strong nuclear force Short range
γ Photon	Electromagnetic force Long range
W⁺ W⁻ Z⁰ Bosons	Weak nuclear force Short range
? ? Graviton	Gravitational force Long range

Figure 30.16: *The particles that carry force and their ranges.*

Chapter 30 Connection

Smash! The Large Hadron Collider

Albert Einstein said, "Knowledge is limited. Imagination encircles the world." It certainly encircles 27 kilometers in the form of a tunnel 100 meters underground near the border of France and Geneva, Switzerland (Figure 30.17). If you traveled the area in a car, you'd drive over the biggest man-made machine ever, and maybe not even know it!

The LHC The Large Hadron Collider (LHC) is the world's largest particle accelerator. It was built by CERN, the European Organization for Nuclear Research. CERN's mission is to "provide for collaboration among European States in nuclear research of a pure scientific and fundamental character." CERN's scientists are using the LHC to help them investigate the origin of the universe—the big bang. They also hope the LHC will help them understand some more elusive aspects of the universe, including antimatter, dark matter, and dark energy.

How it works The inside of the giant circular tunnel is a vacuum. It is even more empty than outer space. Scientists have made it that way to avoid unintended collisions with stray particles, which could be problematic for the LHC's experiments and the equipment.

A beam of particles called *hadrons* is generated to travel in one direction through the vacuum in the tunnel, while a second beam of hadrons is generated to travel in the other direction. The particles travel at nearly the speed of light in a strong magnetic field created by superconducting electromagnets (Figure 30.18). To keep the particles from losing energy as they travel, the magnets are cooled to a temperature colder than outer space!

Other magnets play a part as well. More than 1,200 magnets, each of which is 15 meters long, bend the beams around the circular tunnel, and 392 magnets, each of which is 5–7 meters long, focus the beams. When particles collide, the temperature caused by the collisions is more than 100,000 times hotter than the temperature at the core of the Sun!

Once a collision has occurred, a particle detector records the speed, mass, and electric charge of any particles created by the collision. It may also record the path and measure the energy of the particles. Physicists use the information to determine the identity of the created particles.

Figure 30.17: *Particles moving at nearly the speed of light make about 4,000 circles through this accelerator chamber in the time it takes to blink an eye!*

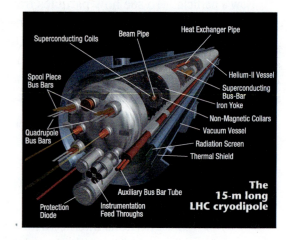

Figure 30.18: *The 15-meter-long magnets that bend the beams are quite complex.* Cryodipole *means "supercooled and having two poles of equal but opposite polarity."*

Chapter 30 Connection

What the LHC studies

The LHC is currently involved in several experiments, each designed to answer specific questions:

LHCb When the universe formed, the big bang should have produced equal amounts of matter and antimatter. But when matter and antimatter interact, they wipe each other out, producing energy in the process. That means that there should be neither matter nor antimatter left in the universe. But, of course, the universe contains a great deal of matter. The LHCb experiment is investigating why.

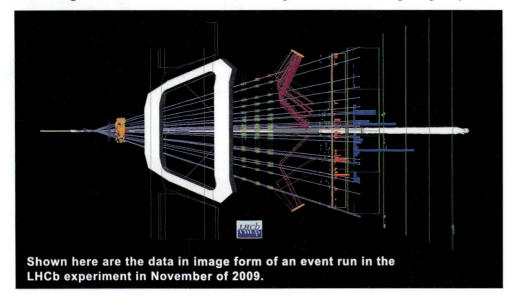

Shown here are the data in image form of an event run in the LHCb experiment in November of 2009.

ALICE Matter is made of atoms, which consist of electrons and a nucleus containing protons and neutrons. Protons and neutrons are made of other particles called quarks, bound together by particles called gluons. But, gluons could not have held quarks together in the hot, energetic conditions existing in the microseconds immediately after the big bang. Scientists believe that during this time, a hot, dense mixture called quark-gluon plasma was present. ALICE simulates conditions present just after the big bang in an effort to create and analyze this plasma (Figure 30.19).

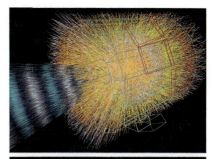

Figure 30.19: *These images of a simulated lead-ion collision were created as part of the ALICE experiment.*

30.3 THE STANDARD MODEL

Chapter 30 Connection

ATLAS and CMS In 2012, both ATLAS and CMS were successful in detecting and measuring the *Higgs boson*, a particle theorized by scientists as vital to the Standard Model of particles and forces. Discovery of the Higgs boson particle confirmed the theories of François Englert and Peter Higgs for how particles attain mass, for which they were awarded the 2013 Nobel Prize in Physics.

TOTEM TOTEM is designed to measure the size of protons. It also measures the LHC's luminosity, or the precision at which the LHC produces collisions.

LHCf Charged particles from outer space constantly bombard Earth's atmosphere. These cosmic rays collide with nuclei in Earth's upper atmosphere, creating a shower of particles on Earth. LHCf studies these particle showers to help scientists design larger-scale experiments with cosmic rays.

Analyzing the data When running at full capacity, the LHC experiments can gather 15 million gigabytes of data every year. That's enough to fill 3.4 million DVDs! To handle that much information, scientists use a grid computing system. Each of thousands of computers handles a chunk of information, then sends its findings on to a central computer and is given another chunk of information to analyze. Most of the computers are located at universities or scientific institutions. Each site can process information and has the opportunity to conduct specific research or collect specific data.

LHC's future Beset by problems early on, the LHC has only recently started to generate much data. Already it has accelerated particles to energy levels never before reached by any other machine. When the LHC gets going at maximum power, particles will travel around the ring more than 11,000 times per second, generating about 600 million collisions each second. Who knows what mysteries it will solve once it gets up and running at full capacity. Maybe even Einstein couldn't have imagined what we will learn from the LHC!

Photos and images courtesy of CERN.

Figure 30.20: *Workers are installing a particle tracker in the heart of the CMS detector.*

A

aberration – a distortion in an optical image when compared to the original object.

absolute zero – the theoretical temperature at which there is zero thermal energy and atoms completely stop moving; 0 K on the Kelvin temperature scale.

absorption – occurs when the amplitude of a wave decreases and/or disappears as it passes through a material due to the material taking up some or all of the wave's energy.

acceleration – the rate of change of velocity over time; acceleration is a vector quantity with a magnitude and a direction.

acceleration due to gravity (g) – the acceleration of an object due to Earth's gravitational field strength; equal to 9.8 m/sec^2 on Earth.

accuracy – the quality of being exact and free from error.

acoustics – the science and technology of how sound behaves.

action – one of the two equal and opposite forces in an action-reaction pair, according to Newton's third law of motion.

additive color process (RGB) – the process through which new colors are formed by the addition of multiple colors; the additive primary colors of light are red, green, and blue.

air resistance – the resisting effect on a moving object due to friction of air.

airfoil – a body with a shape that generates a large force normal to the direction of the surrounding fluid's motion.

alloy – a metal that is a mixture of more than one element.

alpha decay – a radioactive transformation in which a nucleus emits an alpha particle, decreasing the atomic number by two; an alpha particle consisting of two protons and two neutrons and signified by the Greek letter alpha (α).

alternating current (AC) – electric current that reverses its direction at regularly recurring intervals, usually many times per second.

ammeter – an instrument for measuring electric current flow in amperes.

ampere (amp or A) – the unit of measure for electric current.

amplify – to increase the magnitude of a signal by increasing its amplitude.

amplitude – the maximum value of a quantity that varies periodically from its base or equilibrium value to either extreme.

analog signal – a continuous electrical signal that varies in magnitude.

analysis – the detailed examination of experimental results to determine whether they support the hypothesis.

AND – a type of logic gate whose output is "1" only when both of its inputs are "1."

angle – the geometric figure formed by two lines extending from the same point.

angle of refraction – the angle formed between a refracted ray and the normal to the surface at the point of refraction.

angular acceleration – the change of angular velocity per unit of time.

angular displacement (θ) – the displacement of an object spinning on its axis, measured in radians or degrees.

angular momentum – the momentum due to an object's rotation or spin; it is the product of the object's moment of inertia multiplied by its angular velocity.

antimatter – material made of antiparticles of matter (positrons, antiprotons and antineutrons); when antimatter and matter meet, annihilation occurs, and both the matter and antimatter are converted to pure energy.

antinode – a point in a standing wave where the amplitude of the wave is at its maximum.

astronomical refracting telescope – an optical telescope arrangement consisting of a converging objective and eyepiece that produces a real inverted image.

at rest – a condition in which the speed of an object is zero.

atom – the smallest particle of an element that can exist alone or in combination with other atoms.

atomic mass – the average mass of all the known isotopes of an element, weighted for their relative natural abundance; usually given in atomic mass units (amu).

atomic mass unit (amu) – a unit of measure for expressing the mass of atoms, molecules, or nuclear particles equal to $1.661 \infty 10^{-27}$ kg; the atomic mass unit is defined as the mass of 1/12 of a carbon-12 atom.

atomic number – the number of protons that an atom contains.

average speed – the distance an object travels divided by the time elapsed regardless of variations in speed during the trip.

axis – (1) the line about which an object rotates; (2) one of the reference lines of a coordinate system.

B

background radiation – the radiation in the natural environment including cosmic rays and the radiation from naturally-occurring radioactive elements.

baryon – a group of heavy elementary particles that includes protons and neutrons.

base – (2) in scientific notation, the number between 1 and 10 that is multiplied by a power of 10.

base – (3) one of the three terminals of a transistor; the bias voltage between the base and emitter controls the current in the emitter and the collector.

battery – a device that uses chemical energy to generate electric energy.

beat – the pulsations of sound produced when two similar frequencies interfere with each other.

Bernoulli's equation – relates fluid pressure to its speed, density, and potential energy; the law of conservation of energy as applied to a steady flow of fluids.

beta decay – a radioactive transformation in which a neutron splits into a proton and an electron; the electron is emitted as a beta particle and the proton stays in the nucleus, increasing the atomic number by one.

bias voltage – a voltage across a semiconductor junction that is used to control current flow in the junction.

big bang – a theory of the origin and evolution of the universe.

binary – a system of numbers with two as its base and only using the digits "0" and "1."

bit – a single state of a digital signal, signified by a "0" or a "1."

black – the condition created by the absorption of all visible wavelengths of light; black objects reflect little or no light.

black hole – a theoretical region of time-space with such strong gravity that its escape velocity is equal to the speed of light.

blackbody – an ideal surface that reflects nothing, completely absorbs all radiant energy falling on it, and emits pure thermal radiation.

blackbody spectrum – the characteristic form of the graphic distribution of power vs. wavelength for a perfect blackbody.

blue – one of the primary colors of the additive color process; one kind of the three types of the primary color photoreceptor cone cells in the eye; the color of light that has wavelengths of approximately 455 to 492 nanometers.

boiling point – the characteristic temperature of a material at which the phase changes from liquid to gas; water boils at 100°C (212°F).

boson – an elementary particle that obeys Bose-Einstein statistics.

boundary – the interface or common surface between two adjacent materials that a wave travels through.

boundary condition – a set of requirements to be met at a specific place (the boundary) that provides a solution to the system's equations.

Boyle's law – a rule stating that the pressure of a gas is inversely proportional to its volume at constant temperature.

British Thermal Unit (Btu) – a unit of energy equal to the quantity of heat required to increase the temperature of one pound of water at standard pressure by 1°F; the abbreviation is Btu; equal to 1.055×10^3 J.

brittleness – the property of a material that describes its ability to fracture without being deformed; a brittle material breaks under a low strain with little deformation.

Brownian motion – the random, irregular motion of small particles in a fluid due to collisions with atoms and molecules.

buoyancy – an upward force a fluid exerts on a submerged object; the buoyancy force is equal to the weight of liquid displaced by the object.

C

calorie (c) – a unit of heat energy often used in chemistry; equal to the quantity of heat required to raise one gram of water by 1°C; a food calorie, or Calorie, is equal to 1,000 calories, or 4,200 J; equal to 4.2 J.

capacitance (C) – a measure of the ability of conductors and insulators to store electric charge; the unit of capacitance is the farad (F).

capacitor – a device consisting of two parallel conductors separated by insulating material that is used for storing electric charge.

carnivore – an organism that eats the flesh of animals, including decomposers, herbivores, and/or other carnivores.

Cartesian coordinate – a set of numbers (x, y) used to locate a point, where x is the distance to the point along the x-axis and y is the distance to the point along the y-axis; also known as a rectangular coordinate.

CAT scan – an x-ray process that produces a three-dimensional image; multiple x-rays are taken in different cross sections at different angles and combined into the final image.

Celsius scale – a temperature scale on which zero equals the temperature that water freezes (0°C) and 100 is the temperature that water boils (100°C) at standard temperature and pressure.

center of gravity – a fixed point at which the force of gravity acts on an object.

center of mass – the point about which an object moves as if the entire object's mass is at that point and all external forces were applied at that point; the point at which the object's three axes of spin intersect.

center of rotation – the point or line around which an object rotates.

centimeter (cm) – a metric unit of length equal to 0.01 meter.

central processing unit (CPU) – the digital circuits in a computer that interpret and execute the instructions of a computer program.

centrifugal force – the effect of inertia on an object moving in a curve; centrifugal force is not a true force.

centripetal acceleration – the acceleration of an object moving in a circular path; the centripetal acceleration is in the same direction as the centripetal force, toward the center of the circle.

centripetal force – a force that causes an object to move in a circular path, rather than continuing in a straight line; a centripetal force is directed toward the center of the circle.

charge – (1) another term for electric charge; (2) to load a capacitor with electrical energy by putting current through the capacitor.

charge by friction – the mechanical transfer of electric charge between two different objects by rubbing them together.

charge polarization – the separation of the positive and negative charge in an object's atoms due to the effect of an electric field.

Charles's law – a rule stating that the volume of a gas is directly proportional to its temperature at constant pressure.

chemical energy – the energy stored in chemical bonds.

chemical reaction – a process in which a substance is changed into one or more new substances; atoms are rearranged into new molecules, but atoms are not changed into other types of elements.

chromatic aberration – the distortion of an image by a lens caused by variations in the angle of refraction of different wavelengths of light resulting in color fringes.

circuit – see *electric circuit*.

circuit analysis – the process of calculating currents and voltages in a circuit.

circuit breaker – a safety device that interrupts the flow of current in a circuit when the current exceeds a predetermined limit; circuit breakers can be reset.

circuit diagram – a drawing using standard symbols that illustrates the arrangement of an electric circuit's components; also known as a schematic circuit diagram.

circular waves – waves that originate from a single point of disturbance and travel out in all directions; waves with concentric circular crests.

circumference – the distance of the enclosing boundary around a circle; for a sphere, the length of any great circle on the sphere.

closed circuit – an electric circuit through which current flows.

closed system – a system that is isolated so it cannot exchange energy or matter with its surroundings.

CMYK color process – see *subtractive color process (CMYK)*.

cochlea – a tiny, fluid-filled bone structure in the inner ear that contains the essential organs of hearing.

coefficient of friction – the ratio of the frictional force between two objects in contact (parallel to the surface of contact) to the normal force with which the objects press together.

coefficient of static friction – the ratio of the frictional force preventing two objects in contact with each other from sliding or rolling against each other to the normal force with which the objects press against each other.

coefficient of thermal expansion – a factor by which a material expands under temperature changes; units are 1/°C.

coil – a current-carrying wire made into loops; the magnetic field through the center of the coil is intensified as more loops are included, and the field strength is proportional to the current times the number of loops.

collector – one of the three terminals of a transistor where the current flows into the transistor and the magnitude of the current depends on the base bias voltage.

collision – occurs when two or more objects hit each other, transforming or converting the kinetic energy while conserving momentum.

color – the visual appearance of light corresponding to wavelength.

commutation – the process of reversing the current in an electric motor's electromagnet.

components – two or more vectors with the same effect as the given vector.

composite material – an engineered material made up of two or more materials with different characteristics; the resulting material has improved characteristics over its component materials.

compressed – squeezed or shrunk.

compression – a reduction in the volume of a substance due to pressure.

condensation – the gradual transformation of matter from the gas to the liquid phase when the temperature is above the melting point.

conduction – (1) the transfer of thermal energy through materials by the direct contact of particles of matter; this energy transfer does not involve movement of the materials themselves; (2) the transfer of electricity by a material.

conductivity – see *electrical conductivity*.

conductor – see *thermal conductor and electrical conductor*.

cones – photoreceptor cells in the eye that respond to color.

consonance – a harmonious or agreeable combination of frequencies heard when sounds are played simultaneously.

constant – a quantity that remains at the same value when others quantities are changing.

constant acceleration – an acceleration that does not change; it is due to a velocity that changes at a constant rate.

constant speed – a maintained speed that does not vary.

constructive interference – occurs when two or more waves in the same phase are added together to create a wave of larger amplitude than the original waves.

continuous – connected or unbroken.

control variable – a variable in an experiment that is kept the same throughout the experiment.

controlled experiment – experiment in which when one variable is changed and all the others are controlled or stay the same throughout the experiment.

convection – the transfer of thermal energy by the flow of liquids or gases in currents.

converging – the coming together of light rays.

converging lens – a type of lens that bends light so that parallel rays entering the lens are bent toward the focal point as they exit the lens.

conversion factor – the number by which you multiply or divide a quantity in one unit to express the quantity in a different unit.

convex – having a curved form that bulges out in the middle.

cosine – to determine the *x*-component of a vector, the cosine of the included angle is the ratio of the *x*-component to the vector.

coulomb (C) – the unit for electric charge.

Coulomb's law – the attraction or repulsion between two electric charges is inversely proportional to the square of the distance between them.

crest – the top or highest point of a wave.

critical angle – the angle at which light is totally reflected back into a material.

cross section area – the area of an imaginary or real cut made through an object.

current – see *electric current*.

cyan – a greenish light blue that is created when red light is absorbed from white light and blue and green light are reflected or when blue and green light are projected onto a common surface.

cycle – (1) in natural systems, a process of energy flows that maintains a system in steady state, (2) in oscillating systems, a unit of motion that recurs regularly.

D

damping – the gradual decrease of amplitude of an oscillation or wave

dark matter – a type of matter that theoretically exists to explain the motion of galaxies.

decibel (db) – the dimensionless unit of a logarithmic scale for expressing the relative intensity of sounds.

decomposer – an organism that causes organic matter to decay; decomposers process organic matter back into the form of inorganic nutrients and energy.

deformation – for a spring, a deformation is the alteration in length due to extension or compression.

delta (Δ) – the symbol that indicates the difference or change between two quantities.

density – the characteristic ratio of a material's mass to its volume.

dependent variable – the variable in an experiment that changes in response to changes made to the independent variable; this variable is plotted on the y-axis of a graph.

depletion region – under reverse bias conditions, it is the region of a p-n junction where there are no available charge carriers.

destructive interference – when two or more out-of-phase waves are combined to create a wave of smaller amplitude than the original waves.

detector (radiation) – a device used to indicate the presence of radiation.

diamagnetic material – a material whose atoms have equal numbers of electrons spinning in each direction, giving the material a net magnetic field of zero; these materials are weakly repulsed by external magnetic fields.

diffraction – when waves change shape and direction as they pass through openings, around obstacles, or through boundaries.

diffraction grating – an optical device consisting of an assembly of parallel narrow slits or grooves that interfere with incident radiation to produce areas of maxima and minima that can result in spectra.

diffuse reflection – the random scattering of light rays at different angles when a beam of light is reflected off a rough surface.

digital signal – a "continuous" electrical signal that varies between two arbitrary states such as "0" and "1."

dimension – any of the three directions of movement.

diode – a semiconductor device with a p-n junction that allows current flow in one direction and blocks current in the other direction.

direct current (DC) – electric current that flows in one direction only.

discharge – to remove charge from a capacitor or other electrical energy storage device.

dispersion – the variation in the amount of refraction that occurs when different wavelengths of light cross a boundary from one transparent medium to another, resulting in the breakdown of constituent wavelengths seen as spectrum.

displacement – the difference between an object's initial and final positions; it is a vector quantity.

dissonance – an unpleasant combination of frequencies heard when sounds are played simultaneously.

distance – the measure of space between two separate points; distance is a scalar value and does not depend upon direction.

diverge – the spreading apart of light rays.

diverging lens – a type of lens that bends light so that parallel rays are bent away from the focal point as they exit the lens.

Doppler effect – the change in the observed frequency of a wave due to the relative motion of the source and observer.

dose – the total amount of radiation received by a person, measured in rems.

ductility – property of a material which describes its capability to be deformed without fracturing; a ductile material is not brittle.

dynamic – relating to motion of an object or system under the influence of forces.

dynamics – the process of calculating three-dimensional motion from forces and acceleration.

E

ecosystem – a functional system that includes the organisms of a natural community and their environment.

efficiency – the ratio of a machine's useful energy output to the energy input; the ratio of the work done by a system or machine to the energy input.

elastic collision – occurs when objects bounce off each other without breaking, changing shape, or losing energy due to heat or sound.

elasticity – the property of a material that describes its ability to regain its original size and shape after a deforming force has been removed.

electric charge – a fundamental property of matter; the unit of charge is the coulomb (C).

electric circuit – an arrangement of interconnected paths capable of carrying electric currents.

electric current – a flow of electric charge; current is measured in amperes (A); also known as current.

electric field – a region of electric force surrounded by electrically charged objects.

electric force – a fundamental force that charged materials or objects exert on each other.

electrical conductivity – the ability of a material to conduct or carry electric current; the inverse of resistance.

electrical conductor – a material with very low resistance that is used to efficiently carry electric current.

electrical energy – the type of energy resulting from the position of an electrical charge in an electrical field.

electrical insulator – a material that is a poor conductor of electric current.

electrical symbol – a simple symbol used to represent a component of a circuit in circuit diagrams.

electrically neutral – (1) an object that has a balanced amount of positive and negative charge, possessing no net charge; (2) an object (such as a neutron) with no electrical charge.

electricity – a physical phenomenon involving electrical charges and their effects when at rest and in motion.

electromagnet – a magnet made by inserting a magnetic core into a current-carrying wire coil; the core is magnetized while the current is switched on and demagnetized when the current is switched off.

electromagnetic induction – the process of producing current by the relative motion between a conductor and a magnetic field.

electromagnetic spectrum – the range of electromagnetic wave frequencies from very low frequencies like radio waves up through higher frequencies like infrared, visible light, x-rays, and gamma rays.

electromagnetic wave – a type of wave propagated by an oscillating or vibrating electric charge; consists of oscillating electric and magnetic fields that move at the speed of light.

electron – a negatively charged particle that comprises the outer layers of the atom; its mass is $9.11 \infty 10^{-3}$ kg; it carries a negative charge of $1.6 \infty 10^{-19}$ coulombs.

electroscope – an instrument used to detect charged objects.

elementary particles – an indivisible particle that is one of the fundamental constituents of matter.

ellipse – an oval shape, formed by the path of a point that moves so that the sum of the distances from a pair of fixed points, called foci, is a constant; an ellipse is the shape of the orbital path of planets in our solar system.

emitter – one of the three terminals of a transistor, where all the current from the collector and base exits the transistor.

energy – the ability to make things change; energy is required to make a force do work, change motion, raise temperature, create new matter, break chemical bonds, or push electric current through a wire.

energy barrier – the input energy required to start radioactive decay; it consists of the strong nuclear force and the electromagnetic force within the nucleus.

energy conversion – the process of changing energy from one form to another.

energy flow – the conversion and/or transmission of one form of energy to another.

energy levels – a set of quantum states, all at approximately the same value of energy; electrons must absorb or emit energy to change levels.

engineering – the application of science to solve technical problems.

engineering cycle – a process used to build devices that solve technical problems; the four steps this cycle are creating a design, building a prototype, testing the prototype, and evaluating test results.

English system – a standardized system of measurement that uses distance units of inches, yards, and miles and weight measurements of pounds and tons.

equilibrium – (1) in physics, occurs when the forces on an object are balanced; (2) in chemistry, the state in which the solute in a solution is both dissolving and coming out of solution at the same rate.

equivalence principle – in general relativity, the principle that to an observer, the local effects of a gravitational field are indistinguishable from the effects from acceleration of the observer's frame of reference.

evaporation – the gradual conversion of matter from the liquid to the gas phase when the temperature is below the boiling point.

event horizon – the sphere around a black hole at which the escape velocity equals the speed of light.

experimental technique – the methods used to do an experiment.

experimental variable – a variable in an experiment that is changed by the experimenter; the experimental variable is plotted as an independent variable on the *x*-axis of a graph.

exponent – the small numeral shown to the upper right of a quantity that tells you how many times to multiply the quantity by itself.

extend – to stretch or elongate.

eyepiece – the lens or combination of lenses nearest the eye in an optical instrument; used to produce a final virtual magnified image of the previous image in the system.

F

Fahrenheit scale – a temperature scale on which 32 is the temperature at which water freezes (32°F) and 212 is the temperature at which water boils (212°F).

fallout – the radioactive material that descends to Earth from a nuclear explosion.

farad (F) – the metric, or SI, unit of capacitance.

Faraday's law of induction – states that a current is induced through a coil when there is a change in the magnetic field around it; the current flow is proportional to the rate of the field's change.

ferromagnetic material – a material that exhibits a strong attraction to an external magnetic field due to the internal magnetic moments of the material's atoms spontaneously organizing into a common direction.

fiber optics – the use of thin, transparent fibers to transmit light; optical fibers are used in bundles to transmit information.

field – a mathematical description of how forces are distributed between particles in space.

field lines – vectors used to indicate the direction of force in a field.

fission – a nuclear reaction in which an atomic nucleus is split, resulting in the release of large amounts of energy.

fixed boundary – a boundary that is stationary and does not move in response to a wave.

fluid – a substance that can change shape and flow; both gases and liquids are fluids.

fluid mechanics – the study of the motion of fluids.

fluorescence – occurs when energy supplied by electromagnetic radiation causes atoms to excite and emit light energy.

focal length – the distance from the center of a lens to the focal point.

focal plane – a plane passing through the focal point that is perpendicular to the optical axis of a lens or mirror.

focal point – the point at which light rays either meet or diverge after passing through a lens parallel to the principal axis.

focus – (1) another term for the focal point; (2) to adjust the eyepiece or objective of a telescope so that the image can be clearly seen.

food chain – a hierarchy of feeding relationships between the organisms of a biological community; each level of organisms feeds on the level below.

food web – interrelated food chains within a biological community.

foot (ft) – the unit of length in the English system of measurement; equal to 0.3048 meters.

force – any action on a body that causes it to change motion; force is a vector and always has a magnitude and direction.

force field – a field that exerts a force on objects in its vicinity; examples include magnetic fields, gravitational fields, and electric fields.

forced convection – occurs when the flow of gas or liquid results from being circulated by fans or pumps.

forward bias – a voltage required by a diode (semiconductor junction) in the direction that produces current.

Fourier's theorem – states that nearly every wave can be expressed by superimposing single frequency waves.

free convection – occurs when the flow of gas or liquid results from differences in density and temperature.

free fall – movement that is due only to the force of gravity.

free-body diagram – a diagram showing all the force vectors that are acting on an object.

frequency – (1) in harmonics, the number of repetitions or cycles made in a unit of time; (2) in waves, the number of wavelengths that pass a given point in a specific unit of time.

frequency spectrum – a graphic representation showing the relative contribution to an overall sound made by each component frequency.

friction – the force that opposes the relative motion of bodies.

fulcrum – the point about which a lever rotates.

fundamental – the standing wave that has the lowest frequency and longest wavelength in a series of standing waves.

fusion – a nuclear reaction in which two atomic nuclei join together to form a larger nucleus.

G

g – the acceleration due to gravity equal to 9.8 m/sec^2.

Geiger counter – a device used for detecting and counting ionizing radiation particles.

general relativity – Einstein's theory that relates the theory of special relativity to noninertial frames of reference and incorporates gravity; and in which events take place in a curved space.

generator – a device that uses induction to convert mechanical energy into electrical energy.

geometric optics – the branch of optics that describes the behavior of light in terms of light rays; geometric optics is concerned with the use of lenses and mirrors and how light is reflected and refracted.

gram (g or gm) – a metric unit of mass equal to 0.001 kilogram.

graph – a diagram that represents the change of a variable in comparison with one or more other variables.

graphical model – a model that shows the relationship between two variables on a graph so that the relationship is easily seen and understood.

gravitational constant (G) – the constant in the equation of Newton's law of universal gravitation; equal to $6.67 \infty\ 10^{-11} \text{Nm}^2\text{kg}^{-2}$, and is the same throughout the universe.

gravitational field – a region of space in which one body attracts other bodies as a result of their mass.

gravitational mass – the mass of an object as measured by the force of attraction between masses; inertial and gravitational masses are equal in a uniform gravitational field.

green – one of the primary colors of the additive color process; one of the primary photoreceptor cone cells in the eye; the color of light that has wavelengths of approximately 492 to 577 nanometers.

gyroscope – a spinning object that tends to maintain a fixed orientation in space due to its angular momentum.

H

half-life – the average time required for one half of the atoms in a radioactive material to decay.

hard magnet – a material that is difficult to magnetize or demagnetize.

harmonic – a frequency that is a multiple of the fundamental note; a standing wave that has a frequency that is a multiple of the fundamental frequency; a multiple of the natural frequency.

harmonic motion – motion that repeats itself as in the case of a pendulum, rotating wheel, or other oscillator.

heat – the form of energy that results from the random motion of molecules; thermal energy that flows or is moving.

heat of fusion – the amount of energy it takes to change one kg of solid to one kg of liquid; the amount of energy it takes to change the phase of the material without changing the temperature.

heat of vaporization – the amount of energy it takes to change one kg of liquid to 1 kg of gas; the amount of energy it takes to change the phase without changing temperature.

heat transfer – the flow of thermal energy from one object to another due to a temperature difference.

heat transfer coefficient (h) – a constant used in calculating heat transfers in the convection equation; the heat transfer coefficient takes into account the fluids, flow speed, and surface conditions.

herbivore – an organism that eats only producers or vegetation.

hertz (Hz) – the unit of one cycle per second used to measure frequency.

Hooke's law – states that the force applied to a spring is directly proportional to the deformation of the spring.

horsepower (hp) – a unit of power equal to 746 watts.

I

ideal gas law – an equation expressing the relationship between pressure, volume, mass and temperature of a gas.

image – a picture of an object that is formed when light rays given off or reflected from the object meet.

image relay – a technique for an optical system made up of multiple lenses and/or mirrors; the image produced by the first lens becomes the object for the second lens, and so on through the system.

impulse – the single application of an outside force that causes a change in momentum; it equals the product of the force and the time applied.

incandescence – occurs when visible light is produced by an object's high temperature.

inch – a unit of length commonly used in the United States; equal to 1/12 foot or 2.54 centimeters.

incident ray – a light ray from an object that strikes a surface.

incident wave – a wave coming into contact with a surface or boundary.

inclined plane – a flat, smooth surface at an angle to a force.

independent variable – the variable in an experiment that is manipulated by the experimenter and that causes changes in the dependent variable in the experiment; this variable is plotted on the x-axis of a graph.

index of refraction (n) – a ratio that expresses how much a ray of light bends when it passes from one kind of material to another.

induced current – the flow of electric current due to a changing magnetic field around the conductor.

induction (electrostatic) – the process of electrically charging an object by bringing it physically near another charged object.

inelastic collision – a type of collision where the objects stick together or change shape and lose kinetic energy; in any type of collision, momentum is always conserved.

inertia – the resistance of a body to a change in motion.

infrared – the section of the electromagnetic spectrum below the red end of visible light and above microwaves; infrared waves are invisible to the human eye and are usually in the form of radiant heat.

initial speed – the speed an object has at the beginning of an experiment.

input – the work or energy put into a machine; the resources used by a machine.

input arm – the distance on a lever between the fulcrum and the point where the input force is applied.

input force – the force applied to a machine.

inquiry – to explore and investigate through observation.

instantaneous speed – an object's speed measured at a precise moment in time.

insulator – see *thermal insulator and electrical insulator*.

integrated circuit – a circuit that incorporates many components into one functional unit.

intensity – a measure of the brightness of light that is related to the total number of photons per second.

interference pattern – the pattern of pressure, brightness, and darkness, or other wave characteristic resulting from the superposition of waves of the same kind and frequency.

inverse square law – any law in which a physical quantity varies with the distance from a source inversely as the square of the distance.

ionization – a process by which a neutral atom or molecule loses or gains electrons, acquiring a net charge and becoming an ion.

ionize – a process by which electrons are added to or removed from electrically neutral atoms.

ionizing radiation – electromagnetic radiation with enough energy to remove an electron from, or ionize, an atom.

irreversible process – a process that cannot run forward and backward using the same series of steps; a process for which the efficiency is less than 100 percent.

isotope – A different form of the same element with different numbers of neutrons and different mass numbers, but the same number of protons and the same atomic number.

J

joule (J) – a unit for measuring energy and work; it is equal to one newton of force multiplied by one meter of distance.

K

Kelvin scale – temperature scale starting at absolute zero and measuring the actual energy of atoms; on this scale water freezes at 273 K and boils at 373 K.

kilogram (kg) – the unit of mass in the SI system; equal to the mass of the international prototype kilogram stored at Sevres, France.

kilowatt (kW) – a unit of power equal to 1,000 watts or 1,000 joules per second.

kilowatt-hour (kWh) – power companies' convenient measure of energy equal to the energy transferred by one kilowatt of power in one hour; equal to $3.6 \infty 10^6$ joules.

kinetic energy – the energy a body or system possesses due to motion.

Kirchhoff's current law – states that the sum of the currents flowing into any point in a circuit equals the current flowing out of that point in the circuit.

Kirchhoff's voltage law – states that the sum of the voltages of all the voltage-generating components equals the sum of the voltages of all the voltage-consuming components.

L

laminar flow – a streamlined flow where the adjacent layers of a fluid flow smoothly together.

law of conservation of energy – energy cannot be created or destroyed although it can be changed from one form to another.

law of conservation of momentum – the principle that, in the absence of outside forces, the total momentum of a system is constant although momentum may be transferred within the system.

law of inertia – another term for Newton's first law of motion.

law of reflection – states that when a light ray reflects off a surface, the angle of incidence is equal to the angle of reflection.

law of universal gravitation – the force of attraction between two objects is directly related to their masses and indirectly related to the square of the distance between them; also known as Newton's law of universal gravitation.

length (l) – a measured distance.

lens – a specially-shaped optical device made of transparent material like glass that is used to bend light rays.

lepton – a type of elementary particle having a mass smaller than the proton mass and interacting with electromagnetic and gravitational fields.

lever – a type of simple machine consisting of a rigid bar able to turn around a fixed point called a fulcrum; it is used to move or lift a load.

lever arm – the distance between the line of action of the force and the center of rotation at the fulcrum.

lift force – the net force on an airfoil acting perpendicular to the fluid flow.

light ray – a beam of light that travels in a straight line and has a very small cross section.

line of action – an imaginary line that follows the direction of a force and passes through its point of application.

linear momentum – the momentum due to an object's linear movement; it is the product of the object's mass multiplied by its velocity.

liquid – a phase of matter that keeps a constant volume but can flow and change its shape; a liquid's atoms or molecules move about, and intermolecular forces are strong enough to keep a constant volume.

locomotion – progressive movement as of an animal or a vehicle.

logic circuit – a circuit that processes digital signals; its output depends upon the states of its inputs and the logic functions of its gates.

logic gate – a type of simple logic circuit with a single logic function.

longitudinal wave – a wave whose oscillations are in the same direction as the wave travels; an example is a sound wave.

lubricant – a substance used to reduce friction between parts or objects moving against each other.

M

m/sec – a metric unit of measure for velocity.

machine – a type of mechanical system capable of performing work.

magenta – a pink-purple color created when green light is absorbed and red and blue light are reflected or when blue and red light are projected onto a common surface.

magnetic – a property of materials describing their ability to exert forces on magnets or other magnetic materials.

magnetic declination – the angle that indicates the difference between magnetic south from true north.

magnetic domain – a region of a material in which atoms with similar magnetic orientations align in the same direction, increasing their magnetic field strength rather than cancelling each other out as they do when randomly aligned.

magnetic field – a region of magnetic force surrounding magnetic objects.

magnetic field lines – the vector arrows used to indicate the direction of magnetic force within a magnetic field.

magnetic flux – a measure of a magnetic field's strength through an area based on the number of the magnetic force's lines passing through the area; its unit is the weber (Wb).

magnetic force – a force exerted on a particle or object in a magnetic field; the magnetic force can be either attractive or repulsive depending upon the object's alignment to the magnetic poles, and its material properties.

magnetic north pole – one of the two regions of a magnetic field where the field forces are the strongest; magnetic poles have only been found in opposite pairs and have not been singly isolated.

magnetic south pole – one of the two regions of a magnetic field where the field forces are the strongest; magnetic poles have only been found in opposite pairs and have not been singly isolated.

magnetize – to develop or strengthen a magnetic field in an object in response to an external magnetic field.

magnification – the amount that an optical system changes the apparent size of an object; the magnification of a telescope is the ratio of the focal lengths of the objective to the eyepiece.

magnifying glass – a simple converging lens that produces an enlarged right-side-up image of the object being viewed.

magnitude – a quantity's size or amount without regard to its direction.

mass – a measure an object's inertia; the amount of matter an object has.

mass number – the total number of protons and neutrons in the nucleus of an atom.

mechanical advantage – the ratio of output force to input force.

mechanical energy – the energy possessed by an object due to its motion or its position.

mechanical system – a series of interrelated, moving parts that work together to accomplish a specific task.

melting point – the characteristic temperature of a material at which its phase changes from solid to liquid; the kinetic energy of the atoms overcomes the attractive force between the atoms.

memory – the digital circuit in a computer in which digital information can be stored and then retrieved.

meson – a type of elementary particle with strong nuclear interactions and baryon number equal to zero.

meter (m) – the international standard unit of length; equal to the length of the path traveled by light in 1/299,792,458 of a second.

metric system – a standardized system of measuring based on the meter and the kilogram and using multiples of 10.

microfarad (μF) – a unit of capacitance equal to 10^{-6} farads.

microphone – a device that transforms sound waves into electrical signals for the purpose of transmitting or recording sound.

mile (mi) – a unit of length commonly used in the United States; equal to 5,280 feet or 1,609.344 meters.

millimeter (mm) – a metric unit of length equal to 0.001 meter.

mirror – an optical device that reflects light.

mixture – a substance that contains a combination of different compounds and/or elements and be separated by physical means.

mode – a pattern of wave motion or vibration.

modulus of elasticity – the ratio of the stress on a material to the amount of strain produced.

molecule – a neutral group of atoms that are chemically bonded together; it is the smallest particle of a compound that can exist by itself and retain the properties of the compound.

moment of inertia – a vector quantity that describes the distribution of mass in an object; the rotational equivalent of mass.

momentum – a vector quantity that is the product of an object's mass and its velocity.

multimeter – a test instrument used for measuring voltage, current, and resistance.

musical scale – a series of musical notes arranged from low to high in a special pattern.

N

NAND – a type of logic gate whose output is "0" only when each of its inputs is "1."

natural frequency – the frequency at which a system tends to oscillate when disturbed.

natural law – the set of rules that governs the fundamental workings of the universe.

negative charge – one of two types of electric charge; electrons carry a negative charge.

net force – the amount of force that overcomes an opposing force to cause motion; the net force can be zero if the opposing forces are equal.

network circuit – a complex circuit containing multiple paths and resistors that are connected both in series and in parallel.

neutrino – a neutral elementary particle having zero rest mass and spin of ½.

neutron – an uncharged particle found with protons in the nucleus of atoms; it is approximately the same size as a proton with a mass of $1.67 \infty 10^{-27}$ kg.

newton (N) – the SI, or metric, unit of force.

Newton's first law of motion – states that an object at rest remains at rest until acted on by an unbalanced force; an object in motion continues with constant speed and direction in a straight line unless acted on by an unbalanced force.

Newton's second law of motion – states that the acceleration of an object is directly proportional to the force acting on it and inversely proportional to its mass.

Newton's third law of motion – states that whenever one object exerts a force on another, the second object exerts an equal and opposite force on the first.

node – a stationary point with zero amplitude in a standing wave.

nonionizing radiation – electromagnetic radiation that is absorbed by atoms without ionizing them.

NOR – a type of logic gate whose output is "1" only if both inputs are "0."

normal – a line that is perpendicular to an object's surface.

note – a frequency in a musical scale.

note – a sound with a certain duration and pitch; it is based on a single fundamental frequency and may include harmonics and other related frequencies.

n-type semiconductor – a semiconductor with an added impurity where the conduction electron density exceeds the hole density; the opposite of a p-type.

nuclear chain reaction – occurs when the neutrons from a fission reaction cause the fission of more atoms, starting a succession of self-sustaining nuclear fissions, resulting in the continuous release of nuclear energy.

nuclear energy – the type of energy derived from nuclear reactions.

nuclear reaction – a process involving changes in an atom's nucleus; atoms of one element can be changed into atoms of different elements.

nuclear waste – the unwanted radioactive by-products from the nuclear industry and from materials used in research, industry, and medicine.

nucleus – the central, positively-charged, dense core of an atom that contains protons and neutrons.

O

object – the source of light rays, either given off or reflected, in an optical system.

objective – the lens or combination of lenses nearest the object in an optical instrument.

octave – a frequency difference that is double or half the starting frequency.

ohm (Ω) – the unit of measurement for electrical resistance.

Ohm's law – the mathematical relationship in which current in an electric circuit is directly proportional to the voltage applied and inversely proportional to the resistance.

open boundary – a boundary that is free to move in response to a wave.

open circuit – an electric circuit with a broken pathway and no current.

optical axis – the line joining the centers of curvature of lenses and/or mirrors in an optical system.

optics – the study of the behavior of light.

OR – a type of logic gate whose output is "1" when any of its inputs are "1."

orbit – a regular, repeating, curved path that an object in space follows around another object due to the effects of gravity between the objects.

orbital – the region of space occupied by an electron in an atom; a term used by chemists for quantum state.

origin – the point where something begins; a fixed reference point; the point of a coordinate system where all the coordinate axes meet.

oscillation – a motion that varies periodically back and forth between two values.

oscillator – a system that periodically varies between two values, positions, or states; a system that has harmonic motion.

output – the work or energy produced by a machine; the product of a system or machine.

output arm = the distance on a lever between the fulcrum and where the output force is exerted.

output force – the force produced by a machine.

P

parabola – the distinctive, arched shape of a projectile's trajectory.

parallel circuit – a circuit in which components are connected so the current can take more than one path.

parallel plate capacitor – a capacitor consisting of two parallel metal plates with an insulator filling the space between them.

paramagnetic material – a material that can be magnetized when placed in an external magnetic field; its properties are due to the unpaired electron spins of the material's atoms.

pascal (Pa) – the metric, or SI, unit of pressure; one pascal is equal to one newton of force acting on one square meter of surface.

Pauli exclusion principle – the rule, according to quantum theory, stating that no two electrons in an atom can simultaneously occupy the same quantum state.

period – the amount of time it takes for one repetition of a cycle.

periodic force – a force that oscillates in strength or direction.

periodic motion – cycles of motion that repeat for every period of time; an example is harmonic motion.

periodic table of elements – a table that visually organizes all known elements; elements are grouped in the table in order of atomic number, and columns are grouped by the elements' properties.

permanent magnet – a magnetic object that retains its magnetism when an external magnetic field is removed.

phase – the point where an oscillator is in its cycle in relation to its starting point.

phase difference – the difference between two cycles with the same frequency; it is measured either as an angle or a time.

photoelectric effect – effect observed when light incident to certain metal surfaces causes electrons to be emitted.

photoluminescence – the emission of light from a substance that has absorbed light energy.

photon – according to quantum theory and the particle concept of light, a photon is the smallest discrete packet of energy that makes up light; the quantum of electromagnetic radiation.

photoreceptors – highly specialized, light-sensitive cells in the retina of the eye.

piezoelectric effect – the effect shown by certain types of crystals that produce voltage across their surfaces when they are compressed or distorted.

pigment – a solid that reflects color of a certain wavelength and absorbs colors of other wavelengths, allowing your eyes to perceive the reflected color.

pitch – the property of a sound determined by the frequency of the waves producing it; the highness or lowness of a sound.

pixel – the smallest part of an electronic picture image.

Planck's constant (h) – a fundamental constant that is the ratio of a photon's quantum energy to its frequency; equal to 6.63×10^{-34} J-sec.

plane wave – waves that originate from a straight line disturbance and move in a straight line direction.

plasma – an ionized gas phase of matter; the atoms or molecules of plasma are broken down into a mixture of free electrons and ions; examples of plasma include stars, lightning, and neon-type lights.

p-n junction – the interface between a p-type semiconductor and an n-type semiconductor; it allows current to flow in only one direction.

point charges – charges whose dimensions are small compared with their distance apart from one another.

polar coordinate – a group of numbers, or coordinate, $(r, l,)$ used to locate a point, where r is the distance from the origin and l is the angle between the positive x-axis and a ray from the origin to the point.

polarity – the property of a system that has two directions or signs with opposite characteristics; examples include electrical polarity with opposite charges or potentials or magnetic polarity with opposite magnetic poles.

polarization – limiting the orientation of a transverse wave, especially of light.

polarizer – a device or material that polarizes light.

position – an object's location in space compared to where it started.

positive charge – one of two types of electric charge; protons carry a positive charge.

potential energy – the work a body or system can do because of its position or state.

potentiometer – a mechanical, continuously-adjustable resistor.

power – the amount of work done per unit of time; work done divided by the time it takes to do the work; the rate of energy transfer or work done; the unit of power is the watt (W).

power factor – the loss of power in an alternating current circuit due to the phase difference between the voltage and the current.

power of 10 – the result when 10 is multiplied by itself; 1,000 is the third power of 10 ($10 \times 10 \times 10$ or 10^3); powers of 10 are positive for values greater than 1 and negative for values less than 1.

precision – the degree of mutual agreement among a series of individual measurements, values, or results.

pressure – the force exerted per unit area; fluid pressure acts in all directions.

pressure energy – the energy stored in a pressure differential.

prism – an optical device used to separate white light into its component colors by dispersion.

probability – mathematical rules governing the relative possibility of an event occurring.

procedure – all the experimental techniques used to run an experiment; procedures must be repeatable to ensure the results are accurate and true.

process – (1) a set of actions or steps performed to achieve a given purpose; (2) any activity that changes things and can be described in terms of input and output.

producer – an organism that manufactures organic nutrients directly from simple inorganic raw materials.

program – the set of instructions that tells a computer how to process data.

projectile – an object that is launched by an applied external force and whose motion ideally is only affected by gravity.

propagation – to spread out or travel through a medium; wave motion.

proton – a positively charged particle found with neutrons in an atom's nucleus; its mass is 1.67×10^{-27} kg; it carries a positive charge of 1.60×10^{-19} coulombs.

prototype – a working model of a design that can be tested to see if it works.

p-type semiconductor – a semiconductor with an added impurity where the hole density exceeds the conduction electron density; the opposite of an n-type.

Pythagorean theorem – in a right triangle, the square of the length of the hypotenuse equals the sum of the squares of the lengths of the other two sides.

Q

quantum – the smallest discrete quantity of energy released or absorbed in a process; the photon is the quantum of an electromagnetic wave.

quantum number – one of the quantities that specifies the value of a quantum state; the value is an integer or half integer.

quantum physics – the branch of science that explains the behavior of matter and energy on the atomic and subatomic scale.

quantum state – the set of characteristics completely describing the state of electrons in an atom.

quark – a hypothetical basic particle that is theoretically a constituent of elementary particles.

R

radian (rad) – a type of angle measurement; one radian is defined as the central angle of a circle where the two radii and the arc joining them are all equal in length; one radian equals 57.3 degrees.

radiant energy – another term for electromagnetic energy.

radiation – (1) the process of emitting radiant energy; (2) the particles and energy that are emitted from radioactive substances.

radio wave – a form of electromagnetic waves with a wavelength greater than a few millimeters (bordering with microwaves); its main application is carrying information for radios and televisions.

radioactive – an unstable atomic state in which the nucleus emits radiation in the form of particles and energy until it becomes more stable or disintegrates.

radioactive decay – the spontaneous disintegration of a material with the release of radiation.

ramp – a type of simple machine consisting of a uniformly sloping surface.

random motion – motion that is scattered equally in all directions.

range – the distance a projectile travels horizontally.

rate – the amount of change of a quantity per unit of time.

ray diagram – a diagram that shows how light rays behave as they go through an optical system.

reaction – one of the two equal and opposite forces in an action-reaction pair according to Newton's third law of motion.

real image – the reproduction of an object produced by light rays that converge through the image; it forms on the side of the lens opposite the object; for example, a slide projector produces a real image on a screen.

red – one of the primary colors along with blue and green of the additive color process; one of the primary photoreceptor cone cells; the color of light that has wavelengths of approximately 622 to 770 nanometers.

red shift – a Doppler shift of light towards longer wavelengths.

reference frame – a coordinate system assigning positions and times to events.

reflected wave – a wave that is bounced off a boundary.

reflection – the process by which waves or light return or bounce off a surface or a boundary between two materials.

refracted wave – the part of a wave that travels into and through a boundary; also known as a transmitted wave.

refraction – the process by which waves change direction while traveling into and through a surface or a boundary between two materials.

relative humidity – the ratio of the amount of water vapor in the air to the equilibrium amount at a given temperature; shows how close the air is to maximum saturation.

rem – a unit for measuring ionizing radiation which takes into account the damage to humans.

resistance – the opposition that a device or material offers to the flow of electric current.

resistor – an electrical device that regulates current in circuits.

resolution – (1) the ability of a lens system to reproduce the details of an object as details in the image; (2) the procedure of separating a vector into its components.

resonance – a large oscillation created when the frequency of a driving force matches the system's natural frequency.

rest energy – the energy equivalent to the rest mass of a particle or body; the quantity of mc^2 where c is the speed of light.

restoring force – a force that always acts to pull an oscillating system back toward equilibrium.

resultant – the single vector that represents the sum of a number of vectors and connects the starting position with the final position.

reverberation – multiple reflections of sound building up and blending together.

reverse bias – a voltage across a diode with polarity such that little or no current is produced; the opposite of forward bias.

reversible process – an ideal process that can be returned to its original starting condition through the same series of steps; a process that can run forward or backward.

revolve – to move around, or orbit, an external axis.

RGB color process – see *additive color process (RGB)*.

rhythm – the organization of a sound into regular time patterns.

right triangle – a triangle, one of whose angles is $90°$.

right-hand-rule – the rule used to determine a magnetic field's direction in the presence of a current; when the thumb of your right hand points in the direction of conventional current, your fingers wrap in the magnetic field's direction.

rods – photoreceptor cells in the eye that respond to differences in brightness.

rope and pulleys – a type of simple machine; a pulley is a wheel with a rim that rotates on a shaft and carries a rope, transmitting motion and force.

rotate – to spin around an internal axis.

rotational equilibrium – occurs when the torques on an object are balanced.

rotational inertia – see *moment of inertia*.

R-value – a measure of resistance to the flow of heat by conduction; a higher number indicates better insulating property.

S

safety factor – the ratio between the breaking load on something and the safe permissible load on it.

satellite – an object in orbit that is bound by gravity to another object.

saturation – when a solution contains as much of a dissolved solid, liquid, or gas as will dissolve into the solution at a given temperature and pressure; when the processes of evaporation and condensation are in equilibrium.

scalar – a quantity having magnitude only, and no direction; examples are mass, distance, and time.

scale – a ratio or constant factor used to change the size or magnitude of a quantity in a uniform way.

scientific notation – a mathematical abbreviation for writing very large or very small numbers; numbers are expressed as products consisting of a base between 1 and 10 multiplied by an appropriate power of 10.

screw – a type of simple machine consisting of a threaded body and grooved head that turns rotational motion into linear motion; it is used to fasten things together when its body is twisted into a material.

semiconductor – a solid crystalline material whose electrical resistance is temperature dependent and can be controlled or changed.

series circuit – a circuit where the components are connected one after the other so the same current moves through each component with only one possible path.

shielding – (1) radiation, reducing or blocking ionizing radiation by using a shield or other device; (2) electrical, a conductive metal casing used to shield electrical devices from stray electric fields.

shock wave – the compressed wave fronts that form in front of a supersonic object; the wavefronts form a boundary between sound and silence.

short circuit – a electric circuit where the resistance is very low, causing currents much higher than the circuit was designed to handle.

simple machine – an unpowered mechanical device that uses only one motion to change the size or direction of a force.

sine – to determine the y-component of a vector, the sine of the opposite angle is the ratio of the vector's y-component to itself.

slope – a line's vertical change divided by its horizontal change.

Snell's Law – a mathematical relationship that can be used to calculate the angle at which a light ray will bend as it moves from one material into another.

soft magnet – a magnetic material that is relatively easily magnetized or demagnetized; an example is iron.

solenoid – a device consisting of a current-carrying coil of wire with a movable central magnetic core; when the wire is energized, a magnetic field is produced within the coil, pulling the core into position.

solid – a phase of matter that has a definite volume and shape and does not flow; the atoms or molecules of a solid occupy fixed positions, and intermolecular forces are strong.

sonogram – a graphic representation of sound showing frequency, time, and intensity.

speaker – a device that converts electrical signals into sound; used to reproduce sound accurately.

specific heat – a characteristic property of a substance equal to the amount of heat energy, measured in calories, required to raise the temperature of one gram of the substance $1°C$.

spectral line – each specific wavelength of light emitted or absorbed by an element as it appears in a spectrometer.

spectrometer – an optical instrument used for producing, examining, and measuring the different wavelengths of the electromagnetic spectrum; also known as a spectroscope.

spectrum – the characteristic wavelengths of light emitted or absorbed by elements; the pattern of characteristic wavelengths of electromagnetic radiation emitted from a source such as sunlight dispersed by a prism.

specular reflection – reflection that occurs off smooth surfaces; the light ray is not scattered, and the reflected image is undistorted.

speed – the measure of distance traveled in a given amount of time without regard to direction; the magnitude of the velocity vector; distance divided by time.

speed of light (c) – the speed of light is a constant equal to 299,792,458 m/s in a vacuum.

spherical aberration – the distortion of an image from a spherical mirror or lens caused by the rays from a point object failing to converge to a point image.

spherical pattern – the distribution of light rays emitted from a source of light that travel in straight lines going in every direction from the source.

spring – an elastic device that can be compressed or extended, and that when released, returns to its former shape.

spring constant (k) – a constant that represents how much a specific spring deforms under a force.

stable equilibrium – a state of equilibrium in which a system tends to return to its original state when disturbed.

standing wave – a wave trapped in one spot that is formed when two identical waves travel in opposite directions between two boundaries.

static – without motion or change.

static electricity – a buildup of either positive or negative charge on an insulated object's surface.

steady state – occurs in a system when the total energy of that system remains the same over time.

stereo – a method of sound reproduction involving multiple speakers to approximate the spatial distribution of the original sound.

strain – the ratio of the change of an object's length to its original size.

streamline – a line indicating the direction of fluid flow in a laminar flow.

stress (σ) – force divided by the area on which it acts; its unit of measure is the pascal (Pa)

strong nuclear force – the strongest fundamental force in the atom that attracts neutrons and protons to each other, holding the atomic nucleus together.

subscript – a letter, symbol, or number written below and to the right of another symbol; used to indicate the number of atoms of a kind in a molecule for chemical reactions.

subsonic – motion that is slower than the speed of sound in air.

subtractive color process (CMYK) – the process through which colors are created by the removal of colors by absorption, allowing the reflection of the desired color; the subtractive primary colors of pigment are magenta, cyan, and yellow.

superconductivity – a condition of zero electrical resistance that exists for certain materials at very low temperatures.

superposition principle – states that when two or more waves overlap, the amplitude of the resulting wave is the sum of the amplitudes of the individual waves.

supersonic – motion that is faster than the speed of sound in air.

surface area – the measurement of the extent of an object's surface or area without including its thickness.

switch – a device for opening, closing, or changing the connections in an electric circuit.

T

tangent – the ratio of an angle's sine to its cosine.

temperature – the measurement used to quantify the sensations of hot and cold; a measurement of the average kinetic energy of molecules in a substance.

tensile strength – the measure of how much pulling, or tension, a material can withstand before breaking or deforming permanently.

tension – a force that stretches or pulls a material.

term – a component of an equation.

terminal speed – the maximum speed reached by an object in free fall; the speed at which forces of gravity and air resistance are equal.

terrestrial refracting telescope – a refracting telescope arrangement using an inverting lens between the objective lens and the eyepiece to form a right-side-up image.

tesla – a metric, or SI, unit of magnetic flux density equal to one weber per square meter.

test charge – a charge used for measuring electric fields.

theory – an explanation or hypothesis that has been well tested but is not yet proven.

theory of special relativity – a theory by Albert Einstein that natural laws are the same in all frames of reference and that the speed of light in a vacuum is constant for all observers, regardless of the motion of the source or observer.

thermal conductivity – a measure of a material's ability to conduct thermal energy.

thermal conductor – a material through which heat can easily flow.

thermal energy – the total kinetic energy contained in a material's atoms and molecules.

thermal equilibrium – occurs when two bodies have the same temperature; no heat flows because the temperatures are the same.

thermal insulator – a material that conducts heat poorly.

thermal radiation – the energy emitted by objects in the form of electromagnetic waves as a result of the thermal motion of their molecules.

thermal stress – mechanical stress induced by a change in temperature when part of a body is not free to contract or expand.

thermistor – an electrical component that changes its electrical resistance as the temperature changes.

thermocouple – an electrical sensor that measures temperature.

thermometer – an instrument that measures temperature.

thin lens formula – states that the sum of the inverse of the object and image distances equals the inverse of the focal length of the lens.

time (t) – a measurement of duration between events; all or part of the past, present, and future.

time dilation – according to Einstein's theory of special relativity, a clock appears to run slower to an observer moving relative to the clock than to an observer who is at rest with respect to the clock.

time interval – the time separating two events.

time of flight – the elapsed time an object or projectile spends in the air.

torque (τ) – a measure of how much a force acting on an object causes the object to rotate.

total internal reflection – occurs when light traveling from a more optically dense to a less optically dense medium approaches the boundary greater than the critical angle and reflects back.

trajectory – the curved path a projectile follows; its curve is due to a constant horizontal velocity and an accelerating vertical velocity.

transistor – a semiconductor device with three terminals used to control current.

translation – linear motion involving change of position without rotation.

transmission axis – the orientation of a polarizer filter.

transverse wave – a wave whose oscillations move perpendicular to the direction the wave travels.

trial – each time an experiment is run.

trough – the lowest or bottom point of a wave.

turbulent flow – an irregular and random motion of fluid flow.

U

uncertainty principle – according to quantum theory, the laws of physics can only control the probability of certain events; the act of observing at the quantum level changes the outcome of the event.

uniform acceleration – see *constant acceleration*.

unstable equilibrium – a state of equilibrium in which a disturbance or force acts to pull the system away from equilibrium.

V

variable – a symbol used to represent an undetermined component in an experiment or a function.

vector – a quantity that has both magnitude and direction; examples are weight, velocity, and magnetic field strength.

velocity vector – the speed and direction of motion at a point along a trajectory; the velocity vector changes direction and magnitude throughout the path of the projectile's trajectory.

virtual image – an optical image formed when rays of light appear to be coming from a place other than where the actual object exists; a virtual image cannot be projected on a screen.

visible light – a form of electromagnetic waves with wavelengths capable of being detected by the human eye, ranging in wavelength from approximately 400 nm (bordering on ultraviolet) to 700 nm (bordering on infrared).

volt (V) – the unit of measurement for electrical potential energy.

voltage – the amount of potential energy that each unit of electric charge has.

voltage drop – the loss of voltage potential across a component in a series circuit.

volume – a measure of the space occupied by a object.

W

watt (W) – the SI, or metric, unit of power defined as one joule per second; equals the voltage multiplied by the current of the circuit.

wave – an oscillation that propagates through a medium, transferring energy from one point to another.

wave front – another term for the crests of a wave.

wave function – a complex-valued function containing all the information that can be known about a quantum of matter or energy.

wave pulse – a single occurrence of a wave; not a continuous wave.

weak force – one of the fundamental forces in the atom that governs certain processes of radioactive decay.

weight – a force created by gravity; the gravitational force with which bodies attract each other.

weightless – having no net force from gravity; a body can be weightless because it is in free fall or because it is away from any source of gravity.

white light – the appearance of light that is the combination of all the colors of light.

wind chill factor – the loss of body heat due to a given combination of wind speed and temperature.

wire – the conductor for electric current in a circuit; it is usually made of metal and may have an insulating sheath surrounding it.

work – the quantity of force multiplied by distance; the energy used to move something.

X

x-axis – the horizontal axis of a graph.

x-component – a vector component in the east-west, or horizontal, direction.

Y

y-axis – the vertical axis of a graph.

y-component – a vector component in the north-south, or vertical, direction.

A

aberration 367
ABS system 95
absolute zero 529
absorption 342, 346
    microwave oven 302
    reflected wave 294
    wave 292, 294
acceleration 80, 81, 82, 103, 128, 170
    angular 191, 193
    average car acceleration 84
    calculating 105, 106, 158, 159, 172
    compared to speed 81
    constant acceleration 83, 215
    direction of 106
    due to gravity 90
    effects of friction on 125
    formula 81
    formula for experiments 86
    formula for finding position of accelerating
        object 88
    free fall 90
    from changing direction 83
    kinetic energy 215
    momentum 246, 250
    motion formulas 88
    natural frequency 274
    slope of speed vs. time graph 85
    uniform acceleration 83
    units of 80, 82
    vector 159
accident reconstruction 249
accuracy 46
acoustics 311
action force 109
actuator valve 95
additive color process 340, 343
additive primary colors 340
air bag 102, 250
air resistance 93
airfoil 576
airplanes 272
Aldrin, Buzz 177
alpha decay 614, 615
alternating current (AC) 433, 492
aluminum 45
Alvin 582
ammeter 407, 489, 490
ampere 404
Ampere, Andre-Marie 404
amplitude 267
    absorption 294
    calculating 268
    constructive interference 295
    damping 267
    destructive interference 295
    determining from a graph 268
    diagram 286
    energy 267, 278, 294, 299
    equilibrium 267
    force 294
    friction 278
    harmonic motion 267
    natural frequency 276, 278
    of a wave 286, 299
    pendulum 276
    period 268
    periodic force 276, 277
    resonance 277
    standing wave 297
    steady state 278
amplitude vs. time graph 295
amusement park ride 171
analysis 16
angle of a ramp 48
angle of incidence 355, 357
angle of reflection 355
angle of refraction 357
angular acceleration 191, 193
angular displacement 169
angular momentum 253, 254
    angular velocity 254
    calculating 255
    center of rotation 253
    formula 255
    friction 254
    gyroscope 256
    jet engine 257
    mass 254
    shape 254
    speed 254
    torque 254, 256
    units of 255
angular speed 166, 168
angular velocity 254, 255
antilock brake 94, 95
antimatter 605, 630, 647
antinode 300, 301
antiparticle 647
antiproton 630, 647
antiquark 648
apogee 178
application
    antilock brakes 94
    bicycle 194
    biomechanics 113
    color printing 347
    deep water submarine Alvin 582
    energy from ocean tides 238
    energy-efficient buildings 556
    hybrid car 414
    hydroelectric power 218
    jet engine 257
    maglev train 494
    magnetic resonance imaging 473
    microwave ovens 302
    nuclear power 631
    quartz crystals 279
    radioactivity 618
    refrigerator 538
    slow-motion photography 73
    telescope 371
    wiring 27, 435, 456
applied force 278
arc length 168
area 44
    associated with a speed vs. time graph . . 67
    rectangle 67
area of support 189
armature 488
Armstrong, Neil 177
astronomical refracting telescope 371
astronomy 26
atom 41, 522, 640
    and light 346
    diameter of 41

electron 441, 443, 444, 466
   energy of 344
   magnetic field 473
   mass of 39, 42
   neutron 441
   nucleus of 588
   proof of 521
   proton 441
   quantum state 598
   structure of 588–595
atomic clock 265
atomic mass 593
atomic mass unit (amu) 588, 593
atomic number 524, 591
average speed 64
axis 166

B

background radiation 621
bacteria 39
ball bearings 126
Balmer, Johann 597
Balmer's formula 597, 598, 599
banked turns 173
baryon 629, 646, 647, 648
baseline metabolic rate (BMR) 227
battery 402, 403, 405, 406, 422
Bay of Fundy 239
beats in music 323
Bequerel, Henri 615
Bernoulli's equation 575
beta decay 614, 615, 630
bias voltage 500
bicycle 194, 200, 256
big bang 639
big crunch 641
biological systems
   efficiency 227
   energy flow 237
black hole 645
blackbody 554
blackbody spectrum 554, 602
block and tackle 204
Bohr model 599
Bohr, Niels 596, 598, 599

boiling point 532
boson 646
bottom quark 648
boundaries 297
Boyle's law 579
brake pads 127
braking system 94, 214
branch current 423
breakdown voltage 412
bridge structure 134
British thermal units (Btu) 535
brittle 566
Brownian motion 521
brush 488
buoyancy 569, 578

C

calorie 227, 535
camera 343, 360, 366, 370
capacitance 455
capacitor 452
   amount of charge 454
   charging 452
   current and voltage graphs 453
   discharging 452
   milk carton 454
   parallel plate 454
carbon dating 618
carbon dioxide, as greenhouse gas 226
carbon-14 . . . . 614, 616, 617, 618
career
   electrical engineer 426, 429, 434
   engineering 494
carnivore 237
Cartesian coordinates 142
CAT scan 623
cause and effect relationship 52
cellular phone 277
Celsius scale 526
center of gravity 188, 189
center of mass 187, 272
center of rotation 182, 184
   angular momentum 253
   mass 254
centimeter 33

centrifugal force 173
centripetal acceleration 172
centripetal force 170, 171, 177
cesium 265
chain reaction 628
charge 450
   Coulomb's law 446, 447
   electric 440
   electric field 449
   force 447
   in a capacitor 452, 454
   inverse square law 446
   motion 443
   polarization 445
Charles' law 580
chemical energy 212, 227, 406
chemical reaction 595
chlorophyll 342
chromatic aberration 359
circuit
   battery 403, 422
   capacitor 453
   closed 403
   diagram 420, 423
   Kirchhoff's current law 423
   Kirchhoff's voltage law 422
   natural examples 401
   network 429
   open 403
   parallel 423, 428, 436
   resistor 413
   series 420, 428
   short 403, 424
   switch 403
circuit breaker 407, 424, 435, 436
circuit diagram 402, 420, 423, 425, 427
circular motion 166, 269
circular wave 290, 291, 293, 294
circumference 167
cleats 127
clock 273
   atomic 265
   pendulum 265
closed circuit 403
closed system 217
closed universe 641

CMYK *341, 343, 347, 348*
coefficient of friction *123, 124*
coefficient of static friction *124*
coefficient of thermal expansion *567*
coil *480–491, 493*
collision
    elastic *247*
    impulse *252*
    inelastic *247, 248*
    kinetic energy *249*
    momentum *247, 248, 249*
    one dimensional *247*
    two-dimensional *249*
    velocity *247*
color *338, 339, 341, 344, 345, 348*
color printing *347*
color separations *347*
commutator *487*
compass *142, 469, 470*
    biological *470*
    orienteering *472*
composite material *568*
compound *525*
compound microscope *369*
compression *564*
compression wave *289*
    transverse *289*
conceptual design *135*
condensation *533*
conduction *544–547*
conductivity *412, 502*
conductor *412, 445, 451, 502*
cone cell *339*
conservation of energy *211, 219, 422, 629*
    calculating *217*
    efficiency *225*
    friction *217*
    systems *217*
conservation of momentum *629*
constant acceleration *83, 215*
constant slope *65*
constant speed *61, 65, 67*
constructive interference *295*
control rod *631*
control variable *49*
controlled experiment *49*

convection
    equation *551*
    flow speed and surface area *549*
    forced convection *548*
    free convection *548*
    in the environment *550*
converging lens *353, 362, 365*
conversion factor *34*
converting units *34*
coordinate system *142*
Copernicus, Nicolaus *26*
copper *403, 443*
cosine *144*
cosmic background radiation *639*
coulomb *441*
Coulomb, Charles-Augustin de *441*
Coulomb's law *446, 447*
Crab Nebula *232*
critical angle *358*
cross section area *563*
cubic centimeter *39*
cubic meter *44*
cup holder *102*
Curie temperature *468*
Curie, Jacques *279*
Curie, Marie *614*
Curie, Pierre *279*
current *400, 404, 408, 430, 478, 479*
    alternating *433, 492*
    average *434*
    branch *423*
    capacitor *452, 453*
    changing magnetic field *489*
    direct *433*
    direction of *443*
    electric charge *443*
    formula *443*
    insulator *444*
    Kirchhoff's law *423, 426*
    magnetic flux *490*
    peak *434*
    relationship to voltage *410*
    semiconductor *444*
current vs. voltage graph *411*
curved space *644*
cycle *266, 268, 269*

    in harmonic motion *264*
    pendulum *264, 271*
    periodic force *276*
    wave *286*
cylindrical magnet *481*

D

Dalton, John *521*
damping *267, 277*
dark matter *641*
data *50, 51*
day *36*
DC electricity *433*
DeBroglie formula *604*
decomposer *237*
deformation of a spring *132*
degrees *269*
demagnetization *468*
Democritus *521*
density *562, 569, 578*
    common substances *45*
dependent variable *50*
design cycle *135*
destructive interference *295*
detector *624*
diamagnetic material *466*
diffraction *367*
    action of sound wave *313*
    waves *292, 294*
diffuse reflection *354*
digital camera *370*
dimmer switch *420*
diode *500*
Dirac, Paul *647*
direct current (DC) *433*
dispersion *359*
displacement *141*
    angular *169*
    measuring *141*
distance *32, 58, 62*
    compared to position *58*
    determining from a speed vs. time graph *87*
    range *153*
    vector *208*
    work *207, 208*

diverging lens 353, 362, 365
domain 467
Doppler effect 638
dose 620
down quark 648
drift velocity 443
ductile 568
dynamics 159

E

ear 321
Earth
    absorbing solar energy 226
    angular momentum 253
    energy flow 236
    gravitational field 448
    harmonic motion in 265
    linear momentum of 253
    magnetic field and poles 471
    mass of 39
    momentum of 253
    natural cycles 265
    tides 238
earthquake 277
eddy current 494, 495
Edison, Thomas 333
efficiency 224–227
    energy flow 235, 277
    fluorescent light 333
    formulas 224
    fuel 415
    of turbojet 258
    resonance 277
Einstein, Albert . .338, 521, 602, 603, 642, 644
Einstein's formula 616, 625, 629, 647, 648
Einstein's theory
    of general relativity 642
    of special relativity 647
El Niño 550
elastic
    collision 247
    impulse 252
elasticity 566
electric car 414, 431
electric charge 440, 443

current 404
    lightning 441
electric circuit 401
    closed 403
    diagram 425
    dimmer switch 420
    natural examples 401
    network 429
    Ohm's law 426
    open 403
    parallel 423
    resistor 413
    series 420
    short 403, 424
    switch 403
    voltage divider 427
    voltage drop in series 422
electric current . . . 400, 404, 408, 430, 478, 484
    ammeter 407
    branch 423
    calculating 421
    changing magnetic field 489
    direction 443
    formula 443
    magnetic flux 490
    relationship to voltage 410
electric field 449–451
electric force 442, 447, 448
electric force vs. distance graph 464
electric motor 434, 486, 487, 488
electric power 431
electric shielding 451
electrical conductivity 412
electrical conductor 412, 445, 502
electrical energy 212, 218, 486, 492
electrical engineer 426, 429, 434
electrical insulator 412, 444, 502
electrical resistance 408, 409, 430
electrical semiconductor 412
electrical symbol 402
electricity 400, 403
    amps 404
    battery 403, 405, 422
    circuit diagram 402
    circuits 401
    conductivity 412

conductor 412
    electric current 400, 404, 408, 430
    electrical symbol 402
    fuses and circuit breakers 407
    generating 492
    insulator 412
    measuring 430
    Ohm's law 410, 421
    open circuit 403
    parallel circuit 426
    power 430
    resistance 408, 409, 411, 430
    resistor 402
    semiconductor 412
    series circuit 422
    static 440
electromagnet 481, 484, 486, 488, 494, 495
    magnetic field 485
    wire and a nail 485
electromagnetic force 489, 590, 626, 649
electromagnetic induction 489
electromagnetic spectrum 381
electromagnetic wave 552
electron 441, 443, 444, 466, 588, 646, 648
    quantum number 598
    quantum state 596
electron beam accelerator 451
electron cloud 589, 601
electron neutrino 648
electronics 500
electroscope 445
element 42
    and atoms 523
    early Greek theory 523
    isotope 592
    periodic table of 524, 591, 600
    ratio in the universe 639
elementary particle 646
ellipse 176
emission 346
energy 4, 211, 332
    absorption 294
    amplitude 267, 278, 294, 299
    and temperature 528
    and work 207, 211, 213, 229
    black hole 645

calorie 227
chemical 406
conservation of . . . 211, 422, 574, 575, 629
damping 277
efficiency 225, 228
electrical 486, 492
flow of 234
forms of 211, 212, 234
friction 278
harmonic motion 275
heat of fusion 531
heat of vaporization 532
in fluids 574
kilowatt-hour 432
kinetic 214, 234, 267, 275, 406
light 334, 338, 344, 345
mechanical 486, 492
nuclear 626, 632
of a closed system 217
of antimatter reactions 630
of applied force 278
of condensation 533
of evaporation 533
of quantum states 599
of radioactive decay 616
of standing wave 299
of sun 232
of tides 238
of wind 232
oscillator 267, 278
paying for it 430
pendulum 234
potential 234, 267, 275, 406, 645
rate of use 229
resonance 275, 277
solar 226
sound 299
systems 207, 217
transformations . . . 212, 216, 218, 234, 406
wave 285, 294
weather 232
energy barrier 616
energy distribution graph 344
energy efficient buildings 556
energy input 224
energy level 599
energy output 224
energy transformation 228, 235, 237, 299
energy vs. time graph 275
engine 271, 414, 415
engineer 426, 494, 565
engineering 134
    see also technology
engineering cycle 135
English system 33, 34
equator 178
equilibrium 108, 128, 155, 185, 271
    amplitude 267
    antinode 300
    finding forces 130
    harmonic motion 267, 270, 272
    inertia 274
    node 300
    restoring force 270, 272, 274
    rotational 185
    spring 274
    stable 272
    unstable 272
    wave 286, 299
equivalence principle 643, 644
escape velocity 645
evaporation 533
event horizon 645
expanding universe 638
experiment design 17, 49
exponent 40
eye 335, 339, 360

F

Fahrenheit scale 526
fallout 621
farad 455
Faraday's law of induction 490, 491, 495
ferromagnetic material 467, 468
fiber optics 335, 358
field 448, 649
    electric 450
    force 448
    gravitational 450
    magnetic 465, 470
field line 449
first law of motion 102, 173, 190
first law of thermodynamics 537
fission 628, 631
fixed resistor 413
fluid 569
    and pressure 570
    energy equation 574
    friction 577
    motion of 573
    pressure due to gravity 571
fluid mechanics 573
fluorescent light 333
flux vs. time graph 491
focal plane 363
focal point 362
focus 363
food chain 237
foot 34
foot-pound 186
force 100, 200, 258
    action-reaction 109–111, 124
    amplitude 294
    applied 277, 278, 298
    applied at an angle 155
    calculating 107, 121
    calculating from momentum 251
    causing acceleration 103
    centrifugal 173
    centripetal 170
    charge 447
    comparison of electric and gravitational 448
    electric 442, 447
    electromagnetic 590, 626, 649
    equilibrium 130, 155
    fields 448
    force pairs 109
    force vector 154
    free-body diagram 129
    friction . 122, 123, 125, 126, 156, 157, 158
    g force 120
    gravitational 590, 649
    gravity 100, 156, 442
    impulse 252
    input and output 200
    line of action 183
    magnetic 463, 464

671

momentum 244, 246, 250, 251
net force 105, 108, 125
newton 104
normal 124, 157
of springs 131, 274
oscillating 297, 300
periodic 278
restoring 131, 271–274, 298, 474
strong nuclear 590, 626, 646, 649
tension 121, 204
thrust 258
unit 104
vector 208
weak 590, 649
force of gravity 209
force restoring 133
force vs. time graph 297
formulas
   accelerated motion 88
   acceleration 81, 82
   acceleration in experiments 82, 86
   adding resistance in parallel 425
   adding resistances in series 421
   angular momentum 255
   area of rectangle 67
   Balmer's 597, 598
   Bernoulli's equation 575
   Boyle's law 579
   capacitance 455
   Celsius and Kelvin conversions 529
   centripetal acceleration 172
   centripetal force 171
   Charles' law 580
   circumference 167
   convection equation 551
   Coulomb's law 446
   current 443
   DeBroglie 604
   density 562
   efficiency 224
   Einstein's 616, 625, 629, 647, 648
   electric power 431, 434
   Fahrenheit and Celsius conversions 526
   free fall motion 91
   frequency 266
   friction 123

heat conduction equation 547
heat equation 536
Hooke's law 132, 566
ideal gas law 581
impulse 252
inverse square law 619
kinetic energy 214
magnetic field 483
mechanical advantage 201, 202, 206
momentum 245
Newton's second law 105, 191, 251
Ohm's law 410, 426
period and frequency 266
Planck's constant 603
position of accelerating object 88
potential energy 213
power 230
pressure and force 572
pressure and potential energy 574
pressure in a liquid 571
second law of motion 103, 105
Snell's law 357
speed 68
speed of accelerating object 86
speed of wave 288
Stefan-Boltzmann formula 555
strain 566
stress 563
surface area 44
thermal expansion 567
thin lens 368
torque 183, 203
universal gravitation 175
voltage divider circuit 427
volume 44
wave speed 288
weight 119
work 207, 208
forward bias 500
four-color printing process 348
frames 73
free fall 90, 176
   motion formulas 91
   solving problems 92
   weightlessness 120
free-body diagram 129, 134

frequency 266
   antinode 301
   calculating 266
   diagram 286
   energy 299
   fundamental 298
   harmonic motion 266
   harmonics 298
   hertz 286
   mode 301
   natural 273, 276
   of sound 309
   of waves 286
   oscillator 265, 286
   period of 266
   piezoelectric effect 279
   reflection 293
   resonance 278
   resonant 474
   speed of 288
   standing wave 299
   wave 286, 288, 300
   see also natural frequency
friction 48, 122–127, 156–158, 170, 204, 206, 494
   air 93
   brake pads 127
   calculating 123, 124
   causes of 122
   coefficient of 123
   coefficient of static 124
   conservation of energy 217
   damping 267
   formula 123
   lubricant 126
   mechanical advantage 206
   mechanical systems 127
   momentum 246
   nail 127
   net force 125
   rolling 126
   sliding 126
   static 125
   tire treads 127
   traction 94
   transfer of charge 445
   usefulness 127

values of coefficient of friction *124*
fuel efficiency *415*
fuel rod *631*
fulcrum *202, 203*
fundamental frequency *298*
fuse *407, 424*
fusion *595, 627, 632*

G

g-force *120*
galaxy *640*
Galileo *26, 48, 371*
gamma decay *614, 615*
gas *41, 530, 578–581*
 and pressure *579*
 and temperature *580*
gas constant *581*
gauss *471, 482*
gear *195, 205*
Geiger counter *624*
generator *218, 492, 493*
geometric optics *352*
geostationary orbit *177*
geostationary satellite *177, 178*
Global Positioning System (GPS) *160, 161*
glow-in-the-dark *345, 346*
gluon *649*
gram *33, 39*
Grand Coulee Dam *219*
graph *50*
 absorption vs. energy for plants *342*
 amplitude vs. time *295*
 blackbody spectrum *554*
 current vs. voltage *411*
 current vs. voltage for a diode *500*
 current vs. voltage for transistor *501*
 electric force vs. distance *464*
 energy distribution of atoms vs. temperature *344*
 energy vs. time *275*
 finding area from *67*
 finding slope *85*
 fission energy *628*
 flux vs. time *491*
 force vs. time *297*
 harmonic motion *268, 269*
 heat transfer vs. flow speed *549*
 height vs. time *37*
 magnetic force vs. distance *464*
 nuclear energy *626*
 position vs. time *64, 268, 269, 276*
 power vs. wavelength *553*
 speed vs. distance *50, 51*
 speed vs. time *67, 83, 84, 85, 87*
 stable isotopes *594*
 voltage vs. time *433*
graphical model *51*
gravitational constant *175*
gravitational field *450*
gravitational force *448, 590, 649*
gravitational mass *642*
gravitational potential energy *213*
graviton *649*
gravity *48, 90, 118, 119, 156, 442*
 acceleration of *119*
 and tides *238*
 center of *188, 189*
 force of *209*
 law of universal gravitation *174*
 potential energy *213*
 strength of *119*
 universal gravitation formula *175*
 weightlessness *120*
Greenwich, England *58*
Ground fault interrupt (GFI) outlet *27, 435, 456*
guitar *301*
gyroscope *161, 256*

H

half-life *617*
halftone screen *347*
hang time *153*
harmonic motion *264–278, 284*
 amplitude *267*
 circular motion *269*
 clock *266*
 compared to circular motion *269*
 cycle *264*
 damping *267, 277*
 energy vs. time graph *275*
 equilibrium *270, 272*
 friction *270*
 graph *268, 269*
 inertia *271*
 kinetic energy *275*
 machine *271*
 measuring *266*
 natural frequency *273, 274, 275, 276*
 oscillator *264, 284*
 pendulum *264*
 period *266, 268*
 periodic forces *276*
 phase *269*
 potential energy *275*
 resonance *275, 277*
 restoring force *272*
 wave *284, 285, 287*
harmonics *298*
harmony *323*
heat *534, 535*
heat conduction equation *547*
heat equation *536*
heat of fusion *531*
heat of vaporization *532*
heat transfer
 calculation *547*
 conduction *544–547*
 convection *548, 551*
 design of buildings *556*
 radiation *552–555*
heat transfer coefficient *551*
heat transfer vs. flow speed graph *549*
height *209, 213*
height vs. time graph *37*
helicopter *112*
helium, discovery of *597*
herbivore *237*
hertz *286*
highly elliptical orbit (HEO) *178*
Hooke's law *132, 133, 566*
Hoover Dam *218*
horizontal and vertical motion *150*
horsepower *230, 231, 233, 431*
hour *36*
Hubble Space Telescope *176, 372*
hybrid car *414*

hydroelectric dam 492
hydroelectric power 218, 239
hydrogen spectrum 597
hypothesis 17

I

ideal gas law 581
ideal machine 225
image 360
    real 363
    virtual 361, 363
image pipe 358
image relay 369
impulse 252
incandescence 333
inch 33, 34
incident material 357
incident ray 354
incident wave 293
inclined plane 156
independent variable .50
index of refraction 356
induction 445, 489, 492
inelastic collision 247, 248
inertia . . 38, 101, 109, 173, 270, 273, 474, 642
    Latin derivation .101
    mass 273
    moment of 192
    restoring force 271
    rotational 190, 191, 192
inertial mass 642
inertial navigation system (INS) 161
information 285, 335
infrared 342
initial speed 48, 86
input 200
input arm 202
input force 201, 204, 206
input gear 205
input voltage 427
input work 210
inquiry 20, 21
instantaneous speed 64
insulator 412, 444, 502
intensity 334, 339, 345, 619

interference 295
    reflection 296
    sound 318
    superposition principle 295
internal reflection 358
inverse 203
inverse relationship 52
inverse square law 464, 619
    force between charges 446
    light 334
ionization 530, 620
ionizing radiation 620
isotope 592, 614
    graph of stable isotopes 594

J

James Bay 219
jet engine 257
jet propulsion 112
joule 208, 535

K

kayak 112
Kelvin scale 529
kilogram 39
kilometer 33
kilowatt 431
kilowatt-hour 432
kinetic energy 212, 214–218, 267, 406
    acceleration 215
    and evaporation .533
    and photons 603
    and temperature 528
    brakes 214
    calculating 214, 215
    collision 249
    derivation of equation 215
    energy flow 234
    momentum 247
    of emitted electrons 602
    scalar 244
    unit 214
    work 215
Kirchhoff's current law 423, 426

Kirchhoff's voltage law 422, 426, 427

L

Lake Mead 218
laminar flow 577
latitude 58
Lavoisier, Antoine 25
law of conservation of angular momentum . 253, 254
law of conservation of energy . . . 216, 574, 575
law of conservation of momentum
    accident reconstruction 249
    angular 254
    collisions 247
    gyroscope 256
    jet engines 257
    Newton's third law 246
    rocket 257, 258
    turbofan 258
    two-dimensional 249
law of inertia 101, 102
law of reflection 355
law of universal gravitation 174
laws of motion 110
length 32
    accuracy 46
    arc 168
    measuring 32, 34
    units 34
lens 353, 362
    focal plane 363
    focal point 362
    focus 363
    optical axis 362
    ray diagram 363, 364
    telephoto 369
    thin lens formula 368
lepton 629, 646, 648
Leucippus 521
lever 201–203
lever arm 183
lift force 576
light 332, 334, 338
    and atoms 346
    calculating speed of light 336

electricity 333
electromagnetic spectrum 381
energy 332, 334, 338
fluorescent 333
incandescent 333
intensity 334
inverse square law 334
momentum 245
photon theory 344
speed of 336
visible 342
wave theory of 290
light ray 337
lightning 441
line of action 183
linear momentum 253
linear motion 166
liquid 41, 468, 530
buoyancy 569
density 569
pressure 571
liter 39
locomotion 112
lodestone 469
longitude 58
longitudinal wave 289
lubricant 126

M

machine 200, 201
bicycle 200
block and tackle 210
complex 200
energy flow 235
harmonic motion 271
input and output 200
perpetual motion 125, 210
simple 200
work 210
maglev train 126, 494
magnet 486, 489, 490, 491
compass needle 470
demagnetization 468
Earth's magnetic poles 471
electromagnet 481, 485, 487, 488
liquid 468
materials of 468
north and south poles 462, 470
permanent . . 462, 468, 481, 487, 488, 494
soft 468
source 465
test 465
magnetic
materials 481
poles 464
magnetic declination 472
magnetic domain 467
magnetic field 465, 470, 471, 478, 494
atom 473
calculating 483
changing 489
coil 481
cylindrical magnet 481
drawing 465
Earth 471
electromagnet 485
magnetic flux 490, 491, 492, 493
magnetic force 463, 464
applications 463
force vs. distance graph 464
magnetic induction 492
magnetic levitation 126, 494
magnetic resonance imaging (MRI) . . . 465, 473
magnetism 481
magnetite 469
magnifying glass 352
Magplane 494, 495
mass 38, 101, 118
angular momentum 254
center of 187
center of rotation 254
collision 247
compared to weight 118
definition of 101
inertia 273
measuring 38
moment of inertia 254
momentum 245, 246, 251
natural frequency 273
range of object masses 39
relationship to gravitational force 175
resisting acceleration 103
spring 274
unit 39
mass number 592
Massachusetts Institute of Technology 494
matter 4, 38, 211, 520
atom 41
charge 440
element 42
mass 38
mixture 42
measurement 32
accuracy 46
ammeter 407
angles of incidence and reflection 355
circumference 167
conversion factor 34
converting 34
electric current 407
English system 33, 34
length 32
mass 38
metric system 33, 34
multimeter 409, 410
multimeter device 405
precision 46
radian 168
resistance 409
surface area 44
time 35, 36, 46
timing equipment 36
voltage 405
volume 44
weight 38
mechanical advantage 201–206
mechanical energy 212, 486, 492
mechanical engineering 134
mechanical system 201
efficiency 225
friction 127
ideal machines 225
machines 201
mechanical systems 200
melting point 531
meson 646
meter 33

eters per second 62
ers per second per second 80
c system 33, 34
phone 427
mile 33, 34
miles per hour 62
milk carton capacitor 454
Milky Way Galaxy 638, 645
milliliter 33
millimeter 33
minute 36
mirror 337, 353, 361
mixture 42
mode 300, 301
model 51
modulus of elasticity 566
molecule 522
   chemical reaction 595
   structure of 601
moment of inertia 192
   angular momentum 254
   center of rotation 254
   mass 254
   shapes 255
momentum 244
   acceleration 246, 250
   angular 253
   bounce 252
   calculating 245, 251
   collision 247, 248, 249
   conservation of 629
   Earth 253
   force 244, 246, 250, 251, 252
   formula 245
   friction 246
   impulse 252
   jet engine 257
   kinetic energy 247
   law of conservation of 246, 254
   linear 253
   mass 245, 246, 251
   moon 253
   positive/negative 245, 246
   rocket 257
   speed 244
   thrust 258
   transfer 247
   turbojet 257
   vector 244, 245
   velocity 244, 245, 246, 251
Montgomery, D. Bruce 494
moon 253, 265, 448
   angular momentum 253
   calculating weight on 175
   momentum 253
   tides 238
moon's gravity 238
motion
   axis 166
   changes caused by force 100
   circular 166, 269
   harmonic 264, 284
   linear 166
   of tossed ball 91
   orbital 176
   periodic 268
   projectile 150
   rotation 166, 182
   solving motion problems 89
   translation 182
   wave 291, 296
   see also harmonic motion and laws of motion
motor 434
motors 231
MRI 465, 473
multimeter 405, 409, 410
muon 630
music 322–323

N

natural frequency 297, 474
   acceleration 274
   amplitude 276, 278
   boundaries 297
   earthquake 277
   harmonic motion 273, 274
   harmonics 298
   inertia 273, 274
   mass 273
   mode 300, 301
   Newton's second law 274
   oscillation 273
   pendulum 273, 276
   piezoelectric effect 279
   position vs. time graph 276
   quartz 279
   reflection 297
   resonance 276, 277, 278, 296, 297
   restoring force 273, 274
   spring 274
   standing wave 297
   wave 296, 297
Nebula 232
negative charge 440
negative image 370
net charge 440
net force 105, 125, 128
network circuit 429, 432
neutrino 630, 646
neutron 441, 588, 640, 646, 648
newton 104
Newton, Sir Isaac 48, 100, 174, 244, 372
   biographical information 100
   laws of motion 110
Newton's first law of motion 101, 102, 173, 190
Newton's law of universal gravitation 642
Newton's second law of motion . 103, 128, 158, 159, ... 191, 250, 251, 274
Newton's third law of motion 110, 111, 112, 246, 572
Newtonian telescope 372
newton-meter 184
node 300, 301
nonionizing radiation 620
normal force 124, 157
normal line 355
north pole 462
note 322
nuclear energy 212
nuclear energy graph 626
nuclear reaction
   and atoms 595
   chain reaction 628
   energy 626
   equations for 625
   fission 628, 631

fusion 595, 627
    rules for 629
nuclear waste 628, 632
nucleus 589

O

octave 322
odometer 169
ohm 409
Ohm's law 410, 421, 426
open circuit 403
open universe 641
optical axis 362
optical instrument 352
optical system 366
    compound microscope 369
    telescope 371
optics 352
    camera 360
    critical angle 358
    eye 360
    image 360
    image pipe 358
    lens 362
    mirror 361
    object 360
orbit 176, 265, 482, 644
    apogee 178
    geostationary 177
    highly elliptical orbit (HEO) 178
    perigee 178
orbital shapes 601
orienteering 472
origin 58
oscillation
    harmonic motion 270
    inertia 271
    longitudinal 289
    natural frequency 273
    phase 269
    superposition principle 295
    transverse 289
    wave 289, 295
    wave pulse 287
oscillator 474

electrical 279
energy 267
examples of 265
frequency 265, 286
harmonic motion 264, 284
kinetic energy 278
mechanical 279
pendulum 264
periodic force 276
phase of 269
potential energy 278
wave 284, 285
output 200
output arm 202
output force 201
output gear 205
output voltage 427
output work 210, 227

P

parabola 146, 643
parallel circuit 423, 424, 426, 436
    calculating current 424
    calculating resistance 425
    comparison to series 428
    diagram 423, 425
    Kirchhoff's current law 423
    resistance formula 425
parallel hybrid car 414
parallel plate capacitor 454
paramagnetic material 466
pascal (Pa) 564, 570
Pauli exclusion principle 600
Pauli, Wolfgang 600, 630
pendulum 234
    amplitude 276
    clock 265, 273
    cycle 264, 271
    damping 267
    gravity 270
    harmonic motion 264
    inertia 271
    kinetic energy 275
    natural frequency 273, 276, 278
    oscillator 264

periodic force 276, 278
phase 269
position vs. time graph 268, 269
potential energy 275
resonance 278
restoring force 271
perigee 178
period 266, 268, 297
periodic table of elements 524, 591, 600
periods of cycles 266
    calculating 266
    determining from a graph 268
    of harmonic motion 266
permanent magnet 462, 468, 481, 487, 488, 494, 495
perpetual motion 125, 210
phase 269
phases of matter 530
phlogiston 25
phosphorus 345, 346
photo luminescence 345
photoelectric effect 602, 603
photography 73
photon 344, 345, 346, 603, 646
photoreceptor 339
photosynthesis 227, 342
piezoelectric effect 279
pigment 341
pinhole camera 366
pion 630, 647
pitch in music 322
pixel 343, 370
Planck, Max 602, 603
Planck's constant 603
Planck's formula 603
plane wave 290, 291, 293, 294
plasma 41, 529, 530
plastics 568
polar coordinate system 142
polarity 462
polarization 445
pole, north and south 470
position 58, 141
    calculating from speed and acceleration . 88
    compared to distance 58
    from speed vs. time graph 67

677

triangulation 161
...ion vs. time graph 64, 268, 269, 276
...e charge 440
....630, 647
...ential energy ..212, 213–217, 267, 406, 645
    calculating 213
    unit 213
potentiometer 413
pound 104
pounds per square inch (psi) 570
power 229, 430
    biological power 233
    calculating 229, 231, 233, 431
    calorie 233
    electric power formula 431
    energy flow 235
    force 231
    formula 230
    horsepower 230, 233, 431
    kilowatt 431
    light 334
    motors 231
    natural systems 232
    power ratings 231
    stars 232
    technology 231, 239
    tidal 238, 239
    transmission 235
    turbine 230
    typical ratings 432
    watt 430, 547
    wind 232
    work 229, 230
power factor 434
power of 10 40
power vs. wavelength graph 553
precision 46
pressure 570–574
pressure energy 212
primary coil 493
prism 353, 359
probability 606
problem solving 24, 72
    circuit problems 428
    four-step technique 72
    motion problems 89

procedure 49
process 16, 224
projectile 146
    horizontally launched 151
    range 153
    upwardly launched 152
projectile motion 146, 150, 176
    horizontal and vertical motion 150
    trajectories 146
proton 441, 588, 640, 646, 647, 648
prototype 135
Ptolemy 26
Pythagorean theorem 145

Q

quantum 649
quantum number 598
quantum physics 596, 602
quantum state 596, 601
quantum theory of light 602–604
quark 646, 648
quartz 273, 279

R

radian 168
radiation 614
    background 621
    cosmic background 639
    intensity of 619
    ionizing 620
    measuring 620, 624
    nonionizing 620
    sources of 621
radiation detector 624
radio 265, 266
radioactive decay 614
radioactive isotope 592
radioactivity 614–618
radius 168
rainbow 338, 359
ramp 201, 206
    input force 206
random motion 528
range 146, 153

rate 61
ray 354
ray diagram 355, 363, 364
reaction
    chemical 595
    nuclear 595
reaction force 109, 156
reactor core 631
real image 363
red shift 638
reference frame 643
reflected ray 354
reflecting telescope 372
reflection 292, 337, 341, 346, 352
    angle of incidence 355
    angle of reflection 355
    boundaries 294
    constructive interference 296
    diffuse 354
    frequency 293
    law of 355
    mirror 361
    natural frequency 297
    normal line 355
    resonance 296, 297
    specular 354
    total internal reflection 358
    wave 293
    wavelength 293
refracting telescope 371
refraction 337, 352, 356
    angle of incidence 357
    as a wave action 293
    calculating angle of 357
    dispersion 359
    incident material 357
    index of 356
    lens 362
    refractive material 357
    Snell's law 357, 362
    wave 292, 293
refractive material 357
refrigerator 538
relative humidity 533
rem 620
repeatable results 49

resistance 408, 409, 430
    adding in series 421
    calculating in parallel circuit 425
    calculating in series circuit 428
    common values 411
    parallel circuit 425
resistor 402, 413
resolution 142
resonance 277, 296, 474
    applied force 278
    earthquake 277
    energy 275, 277
    frequency 278
    harmonics 298
    natural frequency .276, 277, 278, 296, 297
    period 297
    steady state 278
resonant frequency 474
restoring force 133, 270, 271, 474
    airplane 272
    calculating 133
    equilibrium 270, 272, 274
    harmonic motion 272
    inertia 271
    natural frequency 273, 274
    pendulum 271
    stable system 272
    unstable system 272
resultant 141
resultant vector 143
retina 339
reverberation 318
reverse bias 500
revolve 166
RGB 340, 343
rhythm in music 322
right triangle 144
right-hand rule 478, 479, 484
robot 160
rocket 251, 257, 258
rod cell 339
ropes and pulleys 204, 210, 228
rotating motion 253
rotation 182
rotational equilibrium 185
rotational inertia 190, 191, 192

rotor 486, 488
Royal Observatory 58
RPM 166
R-value 556

S

safety factor 565
satellite 160, 176
    geostationary 177, 178
    weather 177
saturated 533
scalar 140, 244
scientific notation 40
screw 206
seat belt 102, 250
second 35, 36, 37
second law of motion . . 103, 128, 158, 159, 191
second law of thermodynamics 537
secondary coil 493
semiconductor 412, 444, 500
    conductivity 502
    silicon 502
    types of 500–501
series circuit 420, 422
    calculating current 421
    comparison to parallel 428
    diagram 420
    energy loss diagram 422
    Kirchhoff's voltage law 422
    Ohm's law 421
    resistance 421, 428
    voltage drop 422
series hybrid car 414
shielding 620
short circuit 403, 424
sigma particle 646
silicon 502
simple machine 200
    force 200
    jack 200, 201
    lever 202
    mechanical advantage 201
    mechanical system 201
    output force 201
    types of 201

work 207
sine 144
slope 65, 85
    calculating 85
    constant 65
    position vs. time graph 65
    speed vs. time graph 85
slow-motion photography 73
smoke detector 615
Snell's law 357, 362
soft magnet 468
solar energy 226
solenoid 480
solid 41, 530
    breaking strength 564
    density 562
    elasticity 566
    properties of 562–568
    strength of 563
    thermal expansion 567
    types of 568
solving circuit problems 428
sonograms 320
sound
    acoustics 311
    energy 299
    frequency 309
    guitar 301
    loudness 310
    mode 301
    properties of 308, 310
    sonograms 320
    speed of 316
    vibration 301
    wave 289, 299
    wavelength of 314
sound wave 289, 294
source charge 450
source magnet 465
south pole 462
space shuttle 256
speaker 427
special relativity 638
specific heat 536, 537
spectral line 596
spectrometer 596

spectrum 338, 638
    absorption 596
    emission 596
    hydrogen 597
specular reflection 354
speed 61, 81, 168
    angular 166
    angular momentum 254
    average 64
    calculating 62, 68, 73, 86, 166, 167, 168, 169
    compared to acceleration 81
    constant 61, 65, 67, 83
    decreasing 83
    determining graphically 65
    English unit 62
    formula 68
    frequency 288
    increasing 83
    initial 48, 86
    instantaneous 64
    metric unit 62
    momentum 244
    of sound 316
    of waves 287
    rolling wheel 169
    slope of position vs. time graph 65
    terminal 93
    tossed ball 91
    variables that affect speed 48
    wave 287, 288
speed of light 336
speed vs. distance graph 50, 51
speed vs. time graph 67, 83, 84, 85, 87
speedometer 169
sport utility vehicles (SUVs) 189
spring
    compressed 131
    constant 132
    deformation of 132
    equilibrium 274
    examples of 131
    force 274
    frequency 274
    Hooke's law 132
    mass 274
    natural frequency 274

restoring force 131
Sputnik I 176, 177
square inch 44
square meter 44
squid 112
stable isotope 592
standing wave 297
    antinode 300
    applied force 298
    energy 299
    frequency 299
    harmonics 298
    mode 301
    node 300
star 232, 337, 640
static electricity 440
static friction 124, 125
steady state 236, 278
steam engine 230
Stefan-Boltzmann formula 555
stop-motion photography 74
strain 566
strange quark 648
streamline 575
strength 464
stress 563
strobe photography 74
strong nuclear force 590, 626, 646, 649
submarine, Alvin 582
subscript 119
subtractive color process ... 341, 343, 347, 348
sun 265, 627
superconductivity 444
supernova 232
superposition principle 295
surface area 44
switch 403, 420
symmetry 489

T

tablecloth trick 102
tangent 148
technology 134, 431
telephoto lens 369
telescope 371, 372

television 343
temperature 526, 528, 529, 534
    effect on atoms 528
    effect on pressure of a gas 580
    measuring 527
    thermal energy 212
tensile strength 564
tension 204, 564
term 86
terminal speed 93
terrestrial refracting telescope 371
tesla 482
test charge 450
test magnet 465
theory 17
thermal conductivity 546
thermal conductor 545, 546
thermal energy 212, 534, 538
thermal equilibrium 544
thermal expansion 567
thermal insulator 545, 546
thermal radiation 552, 555
thermal stress 567
thermistor 527
thermocouple 527
thermodynamics
    first law of 537
    second law of 537
thermometer 527
thin lens formula 368
third law of motion 110, 111, 112, 572
tidal energy 238
tidal power 238
tides 238
time 35
    arrow of 228
    converting units 36
    horizontal axis of graph 37
    measurements of 36
    measuring 35, 46
    mixed units 36
    time interval 37
time interval 35
time of flight 92
toaster 484
torque 182, 185, 190, 203, 464

angular momentum 254, 256
   calculating 203
   differs from force 182
   English unit 186
   formula 183, 203
   friction 254
   gyroscope 256
   negative 203
   unit 184
total internal reflection 358
traction 94
trajectory 146
transformation 237
transformer 493
transistor 501
translation 182
transverse wave 289
trial 49
triangulation 161
trigonometry 144
tuning fork 279
turbine 218, 230, 257, 492
turbofan 258
turbojet 257, 258
turbulent flow 577

U

ultraviolet 338, 342, 620, 624
uncertainty principle 605
uniform acceleration 83
unit 32, 471
   acceleration 80
   ampere 404
   angular momentum 255
   angular speed 166
   angular velocity 255
   atomic mass unit (amu) 588, 593
   British thermal units (Btu) 535
   calorie 535
   centimeter 33
   converting 34
   coulomb 441
   cubic centimeter 39
   cubic meter 44
   day 36

   degrees 269
   farad 455
   foot 34
   foot-pound 186
   gauss 471, 482
   gram 33, 39
   hertz 266, 286
   horsepower 431
   hour 36
   inch 33, 34
   joule 208, 535
   kilogram 39
   kilometer 33
   kilowatt 431
   kilowatt-hour 432
   kinetic energy 214
   length 32, 34
   liter 39
   meter 33
   meters per second 62
   meters per second per second 80, 82
   mile 33, 34
   miles per hour 62
   milliliter 33
   millimeter 33
   minute 36
   newton 104
   newton-meter 184
   ohm 409
   pascal (Pa) 564, 570
   potential energy 213
   pound 104
   pounds per square inch (psi) 570
   rem 620
   RPM 166
   second 35, 36, 37
   speed 62
   square inch 44
   square meter 44
   surface area 44
   tesla 482
   time relationships 36
   torque 184
   volt 405
   volume 44
   watt 430, 547

   work 208
   yard 33, 34
   year 36
universal gravitation 175
universe 638, 639, 640, 641
unstable isotope 592

V

variable 47
   control 49
   dependent 50
   experimental 49
   independent 50
   inverse relationship 52
variable resistor 413
vector 140, 208
   acceleration 159
   adding vectors 141
   arrow 141
   calculating components from a resultant 145
   calculating components of a velocity vector
      148, ... 149
   calculating from a resultant 143
   Cartesian coordinates 142
   component 142
   distance 208
   drawing 141
   field line 449
   finding components graphically 144
   finding its components 142
   force 154, 208
   magnitude 154
   momentum 244, 245
   polar coordinates 142
   resolution 142
   resultant 154
   robots 160
   scale 141
   velocity 146, 147, 148
   x-component 142
   y-component 142
velocity
   calculating components of a velocity vector
      148, ... 149
   collision 247

drawing velocity vector 147
escape 645
momentum 244, 245, 246, 251
thrust 258
vector 146, 147, 148
vertical and horizontal motion 150
vibration 300, 301
virtual image 361, 363
viscosity 577
visible light 342, 381
vision 339, 340
volt 405
voltage 405, 424, 430
    average 433
    breakdown 412
    calculating for capacitor 455
    capacitor 453
    hybrid car 415
    input 427
    Kirchhoff's law 422, 426, 427
    output 427
    parallel circuit 424
    peak 433
    relationship to current 410
    voltage divider 427
    voltage drop in series 422
voltage vs. time graph 433
volume 44

W

water 302
water cycle 236
watt 231, 430
    horsepower 230
    typical power ratings 432
Watt, James 230
wave 292, 379
    absorption 292, 294
    amplitude 286, 299
    boundaries 292, 293, 294
    calculating speed of 287, 288
    circular 290, 291, 293
    constructive interference 295
    crest 290
    cycle 286

destructive interference 295
diffraction 292
energy 284, 285, 294, 299
equilibrium 286, 299
examples of 284
force 294
frequency 286, 299
front 291
harmonic motion 284, 285
harmonics 298
hertz 286
incident 293
information 285
interference 295
kinetic energy 299
light 290, 332
longitudinal 289
mode 300
motion 291
natural frequency 296, 297
one-dimensional 290
oscillation 289, 295
oscillator 284, 285
plane 290, 291, 293
potential energy 299
propagation 291
properties of 286
reflection 292, 293
refraction 292, 293
resonance 296
sound 289, 299
standing 297
superposition principle 295
three-dimensional 290
transverse 289
trough 290
two-dimensional 290
types 289
vibration 300
wavelength 286, 300
wave front 290, 291
wave function 606
wave propagation 291
wave pulse 287, 289
    oscillation 297
    reflection 296

resonance 296
wave theory of light 603
wavelength 286, 288
    of light 379
    of sound waves 314
    reflection 293
weak nuclear force 590, 649
weather satellite 177
weight 38, 118, 119, 174, 450
    calculating 121, 175
    compared to mass 118
    force 209
    formula 119
    measuring 38
weightlessness 120
wheel and axle 201, 205
wheel-speed sensor 95
white dwarf 641
white light 338, 344
windchill factor 549
windmill 239, 492
wire . . . 401, 409, 435, 443, 444, 478, 479, 485
work 430
    calculating 207, 208, 209
    energy 207, 211, 213, 229
    friction 210
    power 229, 230
work input 225
work output 225

X

x-axis 249, 621, 622, 623

Y

yard 33, 34, 249
year 36

Z

zero gravity 120
zero net force 108